4th International Conference on Advancements in Engineering & Technology (ICAET-2016)

MATEC Web of Conferences Volume 57 (2016)

Sangrur, Punjab, India
18 - 19 March 2016

Editors:

Tanuja Srivastava
Shweta Rani
Sushil Kakkar

ISBN: 978-1-5108-2407-2

Printed from e-media with permission by:

Curran Associates, Inc.
57 Morehouse Lane
Red Hook, NY 12571

Some format issues inherent in the e-media version may also appear in this print version.

Printed by Curran Associates, Inc. (2016)

For additional information, please contact EDP Sciences – Web of Conferences at the address below.

EDP Sciences – Web of Conferences
17, Avenue du Hoggar
Parc d'Activité de Courtabœuf
BP 112
F-91944 Les Ulis Cedex A
France

Phone: +33 (0) 1 69 18 75 75
Fax: +33 (0) 1 69 28 84 91

contact-edps@webofconferences.org

Additional copies of this publication are available from:

Curran Associates, Inc.
57 Morehouse Lane
Red Hook, NY 12571 USA
Phone: 845-758-0400
Fax: 845-758-2633
Email: curran@proceedings.com
Web: www.proceedings.com

TABLE OF CONTENTS

ISSN (electronic edition): 2261-236X

ORGANIZING COMMITTEE

Chief Patrons	:	**S. H. S. Jawandha,** Chairman, BGGI, Sangrur, India
		Dr. Guninderjeet Singh Jawandha, Dean Colleges & Executive Member, BGGI, Sangrur, India
Patron	:	**Prof. (Dr.) Tanuja Srivastava,** Director, BGIET, Sangrur, India
Convener	:	**Dr. Shweta Rani,** Asso. Prof. ECE Deptt., BGIET, Sangrur, India
Co-Convener	:	**Mr. Sushil Kakkar**
		Mr. Divesh Kumar

INTERNATIONAL ADVISORY COMMITTEE

Prof. Worsak Kanok-Nukulchai, President, Asian Institute of Technology, Thailand
Prof. Christos Christodoulou, University of New Mexico, Albuquerque
Prof. M. Affam Badar, Indiana State University, USA
Prof. Yehia Hadded, University of Ottawa, Canada
Prof. Xiao-Zhi Gao, Aalto University, Finland
Prof. NitinTripathi, Asian Institute of Technology, Thailand
Prof. Mushtak Al-Atabi, Taylors University, Malaysia
Prof. NorAshidi Mat Isa, University Sains, Malaysia
Prof. Suresh Valiyaveettil, National University of Singapore, Singapore
Prof. Byung Rae Cho, Clemson University, USA
Dr. Mohd. Nazri Ismail, University of Kuala Lumpur, Malaysia
Dr.Huynh Trung Luong, Asian Institute of Technology, Thailand
Dr. Lipo Wang, Nanyang Technological University, Singapore
Dr. M.T. Islam, University Kebangsaan, Malaysia
Dr. Bimal Nepal, Texas A&M University, USA
Dr. Ravinder Chutani, University of Franche Comte, France

NATIONAL ADVISORY COMMITTEE

Prof. M.P.S. Ishar, VC, MRSSTU, Bathinda, India
Prof. Sarit K. Das, Director, IIT Ropar, India
Prof. A.K. Ganguli, Director, INST, Mohali, India
Prof. N. Sathyamurthi, Director, IISER, Mohali, India
Prof. M.P. Poonia, Director, NITTTR, Chandigarh, India
Prof. T. S. Kamal, Director, Radiant Engineering College, Abohar, India
Prof. Pawan Kapur, Director, PIT, Rajpura, India
Prof. M.K. Surappa, IISC, Banglore, India
Prof. A. P. Singh, Dean RIC, IKGPTU, Jalandhar, India
Prof. Savina Bansal, Dean R & D, MRSSTU,Bathinda, India
Prof. Buta Singh, Dean Academics, IKGPTU, Jalandhar, India
Prof. Gursharan Singh, Dean Academics, MRSSTU, Bathinda, India
Prof. D.C. Saxena, SLIET, Longowal, India
Prof. Debatash Guha, University of Calcutta, Calcutta, India
Prof. Amalendu Patnaik, IIT, Roorkee, India
Prof. Brahmjit Singh, NIT, Kurukshetra, India
Prof. Jitender Chabbra, NIT, Kurukshetra, India
Prof. S. K. Mohapatra, Thapar University, Patiala, India
Dr. Amod Kumar, CSIO-CSIR, Chandigarh, India
Prof. B. C. Chaudhary, NITTTR, Chandigarh, India
Prof. P. K. Tulsi, NITTTR, Chandigarh, India
Dr. Chandrakant Kumar, ISRO, Bengaluru, India
Dr. Anuradha Sonker, NIT, Hamirpur, India

PREFACE

Education is a process which contributes in building socio-economic infrastructure of the nation. The efforts taken to provide education to the masses, has made India emerge as an economic power. We feel proud to act as contributors of this social transformation.

With a pledge to make a successful rise in the highly competitive world of Engineering and Technology, Bhai Gurdas institute of Engineering and Technology, Sangrur, Punjab, India organized 4th International Conference on Advancements in Engineering and Technology on 18th-19th March, 2016. The conference has provided the opportunity to interact and share knowledge and ideas with peers, to create new collaborations and partnerships and help grow the economic and social advancement of the region.

The Conference provided a leading forum for sharing the original research contributions and practical development experiences among researchers. It has played a significant role in generating awareness about the potential of new technologies, concepts, theories and trends.

The major objective of the conference was to enhance scientific and engineering knowledge among researchers, engineers and students working in India and abroad for promoting research and development activities.

The continuous support and valuable suggestions received from reviewers has been of great help and is acknowledge with special thanks. The relentless efforts made by editorial team in coordinating the publication of the conference papers in MATEC Web of Conferences are highly appreciable.

S. H. S. Jawandha
Dr. Guninderjeet Singh Jawandha
(Chief Patrons, ICAET-2016)

Statement of Peer review

In submitting conference proceedings to MATEC Web of Conferences, I certify to the Publisher that I adhere to the **Policy on Publishing Integrity** of the journal in order to safeguard good scientific practice in publishing.

1. All articles have been subjected to peer review administered by the proceedings editors.
2. Reviews have been conducted by expert referees, who have been requested to provide unbiased and constructive comments aimed, whenever possible, at improving the work.
3. Proceedings editors have taken all reasonable steps to ensure the quality of the materials they publish and their decision to accept or reject a paper for publication has been based only on the merits of the work and the relevance to the journal.

Title, date and place of the conference

4th International Conference on Advancements in Engineering & Technology (ICAET-2016)
Sangrur, Punjab, India
March 18-19, 2016

Proceedings editor(s):

Prof. (Dr.) Tanuja Srivastava
Dr. Shweta Rani
Mr. Sushil Kakkar

Date and editor's signature

MATEC Web of Conferences 57, 01001 (2016) DOI: 10.1051/matecconf/20165701001
ICAET-2016

Small Size Patch Antenna with DGS for Emergency Management

Sushil Kakkar[1,a], T. S. Kamal[2] and A. P. Singh[3]

[1]*Research Scholar, IKG PTU, Jalandhar*
[2]*ECE Department, RIET, Abhohar, Punjab, India*
[3]*ECE Department, SLIET, Logowal*

Abstract. In this paper a small size patch antenna with defected ground structure (DGS) is designed for emergency management applications. The microstrip fed antenna is consisting of rectangular shape partial ground plane and meander line shaped DGS is optimized. The resonant performance of the proposed antenna has been illustrated using method of moment based IE3D EM simulator. The antenna is designed to resonate at 4.94 GHz frequency and possesses the properties of good performance, compact size, low profile and low cost, therefore it is feasible for practical applications in Emergency Management.

1 Introduction

The booming progress in wireless communication system and an increasing demand to integrate different technologies in to small user equipment has remarkably increase the fashion of introducing compact antennas [1]. Microstrip patch antenna, because of its small size, low profile, low manufacturing cost and ease of integration with feed networks, is find extensive applications in wireless communication system [2-3]. In order to further improve the dimensional and resonant characteristics of the patch antennas, several initiatives have been taken in last decade such as slot in the patch, slot in the feed line, fractal geometries, using partial ground planes, DGS and many more [4-6]. Among them DGS is gaining popularity in a very faster way. DGS can be obtained by etching off a simple shape from ground plane [7, 8].

In this paper a meander line shaped DGS has been utilized in the partial ground plane of small size patch antenna. The implementation of DGS is responsible for the fine tuning and wider bandwidth of the presented antenna, to make it feasible for emergency management applications. A fundamental element of an emergency system is a reliable communication infrastructure allowing rescue operators to be connected and able to support the collection and the elaboration of data coming from monitoring devices distributed in the area of the disaster.

2 Designs and Structure

The proposed antenna structure is a simple patch antenna with square shape having side of the square (S) of 15 mm. The feeding strip is designed in such a way to obtain 50 ohms impedance. The width of the strip (W) is 5 mm. Rectangular shape is the candidate structure for the partial ground plane. In order to further enhance the resonating performance of the proposed antenna, a small size meander shape DGS has been introduced in the ground plane. The length (L) of the DGS is 10 mm and width (W1) of the defect is 1mm. The structure of the proposed antenna with and without DGS has been shown in figure 1.

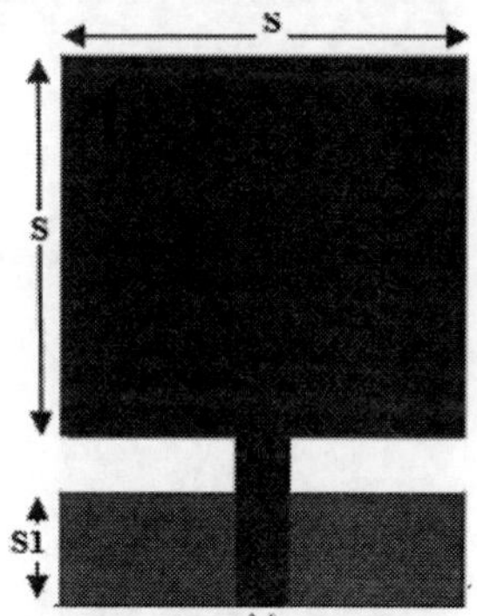

(a)

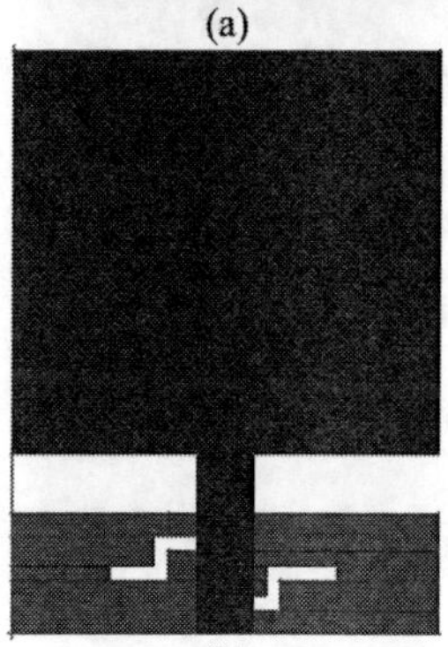
(b)

[a] Corresponding author: kakkar778@gmail.com

(c)

Figure 1. Geometry of proposed antenna (a) without DGS (b) with DGS top view (c) with DGS bottom view

3 Results and Discussion

The simulation tool adopted for evaluating the performance of the proposed antennas is IE3D software, which exploits the method of moments to solve the electric field integral equation.

3.1. Resonating Characteristics

Figure 2 shows the s-parameters of reference antenna and proposed antenna with meander shaped DGS and from the figure it is clear that the proposed antenna resonate at 4.94 GHz with return loss -20.56 dBi and reference antenna resonates at 5.39 GHz with return loss -12.46 dBi. The implementation of meander shaped DGS in ground plane make the proposed antenna feasible for emergency management applications. The detailed comparison between with and without DGS structure is given in Table 1.

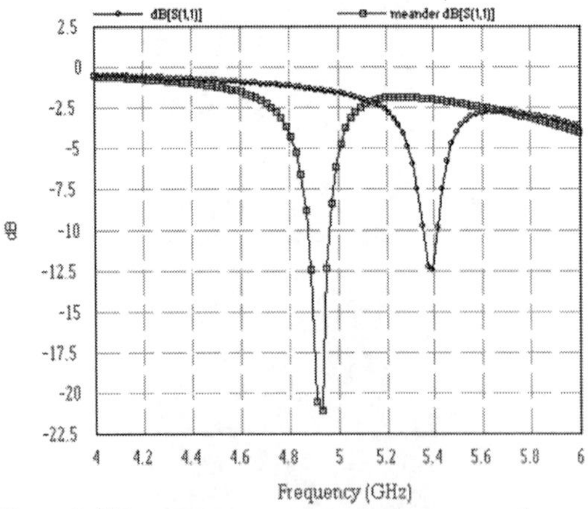

Figure 2. S11 parameter comparisons between proposed antenna and reference antenna.

Table 1. Comparison of proposed and reference antenna.

Design	Resonating Frequency (GHz)	Return Loss	VSWR	Bandwidth (GHz)
Reference Antenna	5.39	-12.46	1.62	0.04
Proposed Antenna	4.94	-20.56	1.20	0.11

3.2 Input Impedance

The input impedance properties of reference and proposed antenna are shown in figure 3. It may be noticed that the input impedance of reference antenna is 60.3 ohms whereas for proposed antenna it is 53.27 ohms which means that proposed structure possesses better impedance matching than reference structure.

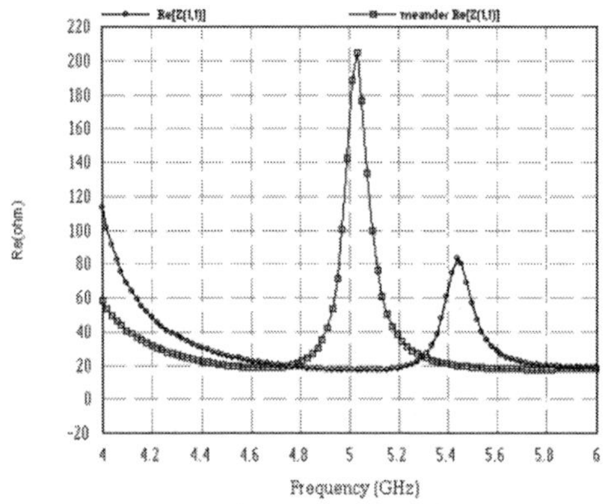

Figure 3. Input Impedance comparisons between proposed antenna and reference antenna.

3.3 Radiation Characteristics

The radiation characteristic of proposed patch antenna is shown in Figure 4. The radiation pattern is symmetrical to antenna axis in E-plane and pattern in H-plane is nearly omnidirectional

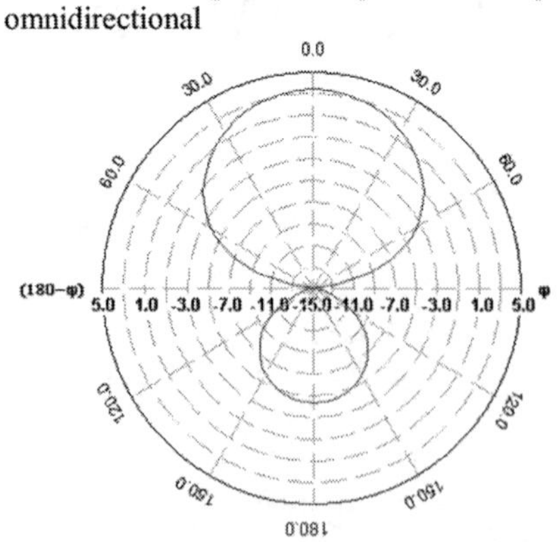

(a)

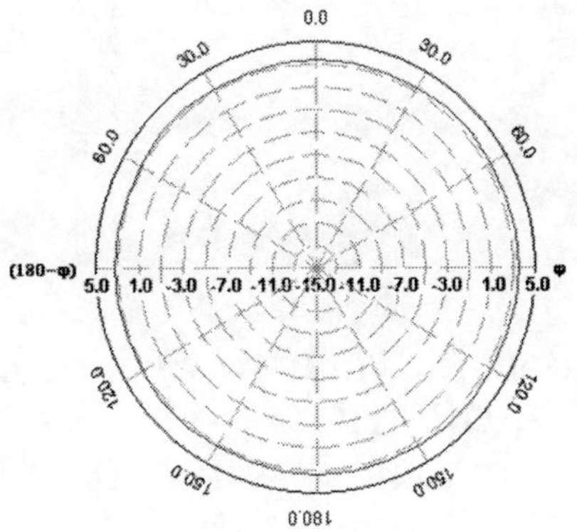

(b)

Figure 4. Radiation characteristics of proposed antenna (a) E-plane (b) H-plane

3.4 Gain

The gain vs. frequency plot of proposed antenna with DGS is shown in figure. 5. The maximum achievable gain of the antenna is 4.26 dBi at 4.94 GHz resonant frequency.

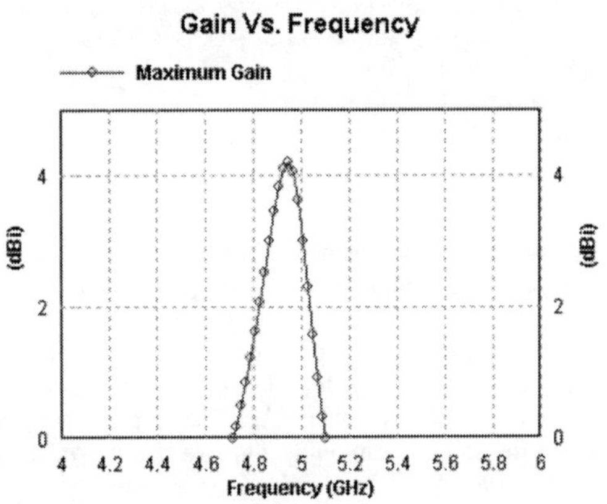

Figure 5 Gain of proposed antenna

4 Conclusion

The design and analysis of small size patch antenna with and without meander shape DGS is presented in this paper. In order to obtain the effect of DGS, critical simulated study has been done using IE3D simulator. The presented results reveal that the proposed antenna with DGS provides better impedance matching and wider bandwidth than its original structure. The proposed antenna is small size, low profile antenna that can be used for emergency management applications.

References

1. Weng et. al, "An Overview on Defected Ground Structure", Progress In Electromagnetics Research B", Vol. 7, 173-189, (2008)
2. D. Schlieter and R. Henderson, "High Q Defected Ground Structures in Grounded Coplanar Waveguide", Electronic Letters, Vol. 48, no. 11, (2012)
3. S. Rani and A. P. Singh, "Fractal Antenna with Defected Ground Structure for Telemedicine Applications," International journal on Communications, Antenna and Propagation, vol.1, pp. 1-15, (2012)
4. A. Arya, M. Kartikeyan and A. Patnaik, "On the Size Reduction of Microstrip Antenna Using DGS", Proceedings of IEEE Conference, (2010)
5. A. Nouri and G. Dadashzadeh, "A Compact UWB Band-Notched Printed Monopole Antenna with Defected Ground Structure", IEEE Antennas and Wireless Propagation Letters, Vol. 10, (2011)
6. S. Rani and A. P. Singh, "On the Design and Optimization of New Fractal Antenna Using PSO", International Journal of Electronics, vol. 100, no. 10, pp. 1383-1397, (2012)
7. S. Kakkar, S. Rani and A. P. Singh, On the Resonant Behaviour Analysis of Small-Size Slot Antenna with Different Substrates, International Journal of Computer Applications, pp. no. 10-12, (2012)
8. R. Azaro, L. Debiasi, M. Benedetti, P. Rocca and A. Massa, "A Hybrid Prefractal Three-Band Antenna for Multistandard mobile Wireless Applications", IEEE Antennas and Wireless Propagation Letters, vol.8, pp. 905-908, (2009)

Design of four-pole chebyshev and quasi-elliptic Ka band dielectric resonator filter using higher order mode TE01(δ+1)

Sujesh Dutta[1] and Dalveer Kaur[2]

[1] *Department of Electronics and Communication Engg., IKG Punjab Technical University, Jalandhar, Punjab. India*
[2] *Department of Electronics and Communication Engg., IKG Punjab Technical University, Jalandhar, Punjab. India*

Abstract. This paper describes the development of 4-Pole Chebyshev and Quasi-Elliptic Ka Band Dielectric Resonator Filter using Higher Order Mode TE01δ+1. A microwave filter is a two port network used to control the frequency response at a certain point in a microwave system by providing transmission at frequencies within the passband of the filter and attenuation in the stop-band of the filter. The satellite communication industry created demand for low-mass narrow-band low-loss filters with severe specification on amplitude selectivity and phase linearity. The microwave region of the electromagnetic spectrum has certain unique properties. This enable microwave signals to propagate over long distances through the atmosphere under all the most severe weather conditions. These have both military and civilian applications, including Radar, Navigation and Wireless Communication. Microwave Filters are vital components in a huge variety of electronic systems, including cellular radio, satellite communications and radar. The design of filters uses network synthesis, with which it is possible to apply systematic procedures to work forward from specification to a final theoretical design.

1 INTRODUCTION

AS the Federal Communications Commission (FCC) tightens the frequency allocation bands for satellite communication services providers, the need for "brick wall" filters becomes increasingly large. Adjacent channels cause co-location interference where Provider A's frequency band is almost contiguous to Provider B's frequency band. In other words, the usability of Provider A's pass-band is degraded if the power in Provider B's pass-band is not sufficiently attenuated. Typically the pass-band frequency separations are less than a few 4 MHz (for 36 MHz Transponder Bandwidth).

To sufficiently attenuate Provider B's pass band frequencies while aiming to preserve Provider A's pass band frequencies with minimum loss, a high quality factor filter is needed. Over the years, microwave band-pass filters have been designed in various topologies such as waveguide, combline or cavity structure. In cases where a very sharp roll off is required, transmission zeros are introduced with the use of non-adjacent resonator couplings (cross-couplings). In modern day filter design, for narrow bandwidth band-pass filters; this is often implemented with a dielectric resonator filter with cross-couplings. Dielectric resonators offer compact size, temperature stability and the high quality factor necessary for this type of design. The filter is synthesized with finite transmission zeros placed at Provider B's pass-band frequencies to attenuate unwanted emissions. A microwave dielectric resonator band-pass filter described in this paper deals with the situations mentioned above.

In 1968, S.B. Cohn implemented the first high-Q dielectric resonator band-pass filter and his exploratory studies paved the way for "brick-wall" filters. In the recent past, dielectric resonator filters have been used widely in mobile communication systems, radar and satellite due to their high Q, compact size and temperature stability [1]. They offer high selectivity in narrow bandwidth applications with low insertion loss. Dielectric resonator filters have been developed in multi-mode, mixed-mode and single-mode cavity applications. Despite multi-mode and mixed-mode dielectric resonator cavity filters providing low loss and smaller volume, their inferior spurious transmissions and high-cost manufacturing keeps them a design rarity.

The perfect filter would have zero insertion loss in the pass-band, infinite attenuation in the stop band, and a linear phase response (to avoid signal distortions) in the pass-band. Of course, such filters do not exist in practice, so compromises must be made; herein lies the art of filter design. The insertion loss method allows a high degree of control over the pass-band and stop-band amplitude and phase characteristics, with a systematic way to synthesize a desired response. The insertion loss method allows the performance to be improved in a straightforward manner, at the expense of a higher order filter. The order of the filter is equal to the number of reactive elements used in the filter.

[a] Corresponding author: sujeshdutta@gmail.com

Filters designed using image parameter method consist of a cascade of simpler two-port filter sections to provide the desired cutoff frequencies and attenuation characteristics, but do not allow the specification of a frequency response over the complete operating range. Thus, although the procedure is relatively simple, the design of filters the image parameter method often must be iterated many times to achieve the desired results.

A more modern procedure, called the insertion loss method, uses network synthesis techniques to design filters with a completely specified frequency response. The design is simplified by beginning with low-pass prototypes that are normalized in terms of impedance and frequency. Transformations are then applied to convert the prototype designs to the desired frequency range [2]. The Chebyshev filter design is explained in the section II. Section III deals with the fabrication of the Quasi-elliptic filter in HFSS as well.

2 GENERAL FILTER DESIGN

Chebyshev Response Filters

This section contains the theory used to design the filters. These filters have a Chebyshev response. The theory used here can be found in [3] to [5].

The response of low pass filter prototype is shown in figure 1. The general procedure of calculating the design parameters with the aid of a CAD simulator is explained below.

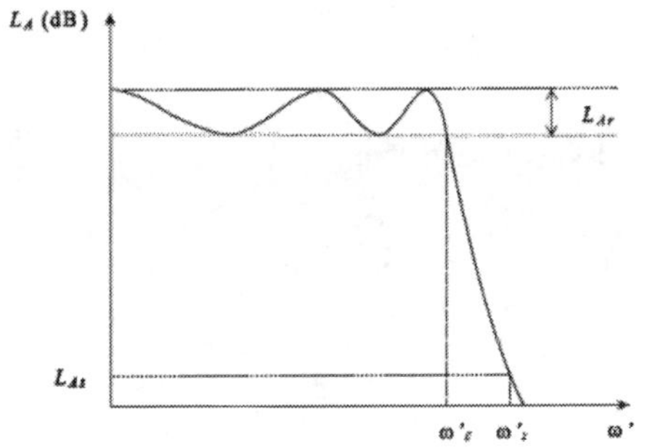

Figure 1. Chebyshev low pass prototype response.

The design specification of a Chebyshev Filter is mentioned in the Table 1.

Table 1. DESIGN SPECIFICATION OF BAND PASS CHEBYSHEV FILTER

S. No	Parameters	Units
1	Center Frequency	GHz
2	Usable Band width	MHz
3	Insertion Loss at Center Frequency	dB
4	Return loss input and output port	dB
5	Band pass flatness Over 100% band width	dB
6	Rejection At $\pm$ 70 MHz from CF and beyond	dB
7	Input/Output Port	--

For Chebyshev low pass prototype filters having a transfer function of Chebyshev response with a pass-band ripple LAr dB and the cutoff frequency Ωc=1, the element values for the two port networks may be computed using the following formulas [6]:

$$g_o = 1.0 \quad (1)$$

$$g_1 = \frac{2}{\gamma}\sin\left(\frac{\pi}{2n}\right) \quad (2)$$

$$g_i = \frac{1}{g_{i-1}} \frac{4\sin[\frac{(2i-1)\pi}{2n}]\sin[\frac{(2i-3)\pi}{2n}]}{\gamma^2 + \sin^2[\frac{(i-1)\pi}{n}]} \quad (3)$$

for i = 2,3,...n

$$g_{n+1} = \begin{cases} 1.0 & \text{for n odd} \\ \coth^2\left(\frac{\beta}{4}\right) & \text{for n even} \end{cases} \quad (4)$$

where $\beta = \ln\left[\coth\left(\frac{L_{Ar}}{17.37}\right)\right]$ & $\gamma = \sinh\left(\frac{\beta}{2n}\right)$ (5)

Here for this filter pass band return loss is 26 dB.
No of poles are 4.
So, from pass band return loss we can calculate pass band insertion loss (L_{Ar}) and which is equal to 0.01 dB.

From equations (1) to (5), we can have the values of g as

$g_0 = 1$
$g_1 = 0.7129$
$g_2 = 1.2004$
$g_3 = 1.3213$
$g_4 = 0.6476$
$g_5 = 1.008$

From the g values, the coupling values JN,N+1 is obtained as: -

$$J(N, N+1) = \frac{1}{\sqrt{g(N) * g(N+1)}} \quad (6)$$

From equation no (6), we can calculate J coupling values as,

J_{01}= 1.1755
J_{12}= 1.0697

J_{23}= 0.7877

J_{34}= J_{12}

J_{45}= J_{01}

By putting these J values the circuit obtained is shown in the fig. 2 to 4.

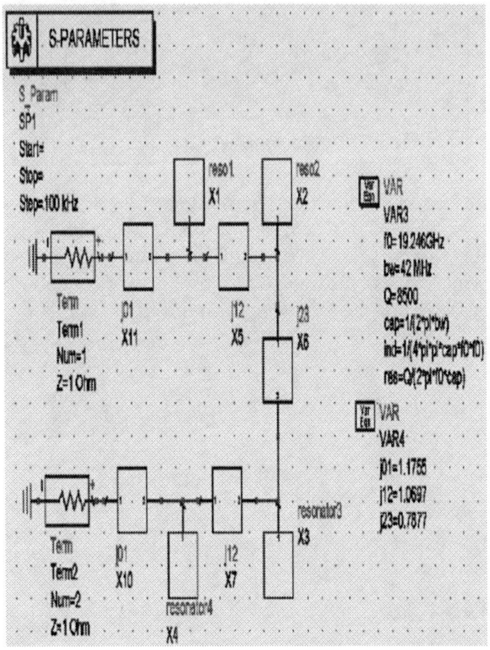

Figure 2 : Equivalent model of Band Pass filter considering admittance inverters

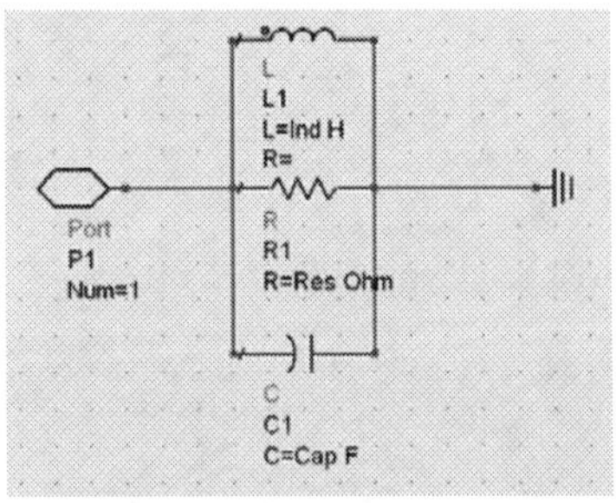

Figure 3 : Equivalent circuit of Resonator

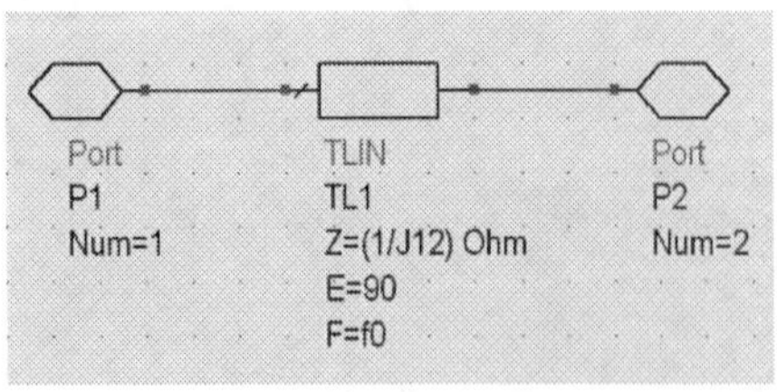

Figure 4: Equivalent circuit of Admittance Inverter

The Advantages of Chebyshev filtering function are: -

i) For same no. of poles, Chebyshev filtering function provides sharper transition from pass band to stop band than Butterworth filter.

ii) It provides the optimum group delay compared to elliptic filter. It provides better band pass flatness than Butterworth [7].

The Chebyshev Response Filter exhibits equal ripple passband and maximally flat stopband. The Chebyshev and Butterworth Filters are sometimes referred to as all pole filters. The main effects of finite dissipation of band-pass filter are: -

- The mid-band insertion loss S12 is inversely proportional to the un-loaded Q of the resonators.
- The mid-band insertion loss S12 is inversely proportional to the passband bandwidth.
- The insertion loss S12 is approximately proportional to the degree of the filter.

The response in the CAD is shown in figure 5.

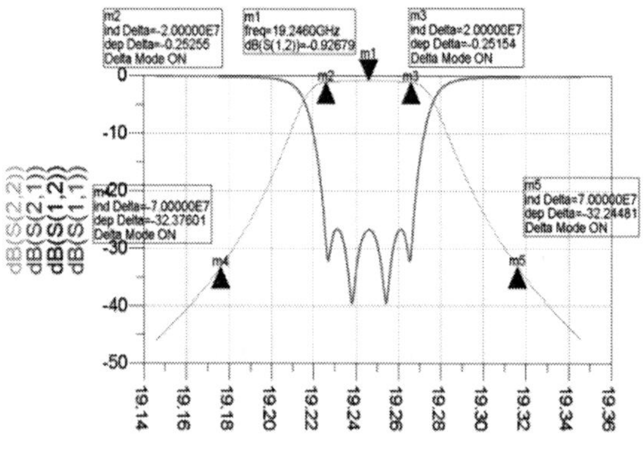

Figure 5 : Response of the filter .

3 Quasi-Elliptic Filter

Quasi Elliptic filters are different from the purely Elliptic filters in the sense that these are not equi-ripple in the stop band. Fully Elliptic function filters have equi-ripple passband and all the zeros of the transfer function at finite frequencies. Whereas Quasi Elliptic filters have only some of the transmission zeros at the finite frequencies. These filters are also different from the Chebychev filters as well, which has all the transmission zeros at infinity. For the same number of poles, Quasi Elliptic filters offer better rejection as compared to Chebychev filters [8].

Quasi Elliptic Filters require cross couplings for generating transmission zeros [9]. Negative cross-coupling generates transmission zeros on the imaginary axis, i.e., zero transmission at the real frequencies (@ + 2j). The maximally flat and equal-ripple responses both have monotonically increasing attenuation in the stop-band. In many applications, it is adequate to specify minimum stop-band attenuation, in which case a better cutoff rate can be obtained.

Following Coupling Values are obtained for realizing this filter:side.

J01 = 1.1557 = J45 (Because the filter is symmetric)

J12 = 0.9963 = J34 (Because the filter is symmetric)

J23 = 0.8734

J14 = - 0.2678 (Cross coupling for realizing the transmission zeros)

Table 2. DESIGN SPECIFICATION OF BAND PASS QUASI-ELLIPTIC FILTER

Sl. No	Parameters	Units	Specifications
1	Center Frequency	GHz	19.246
2	Usable Band width	MHz	36
3	Insertion Loss at Center Frequency	dB	2
4	Return loss input and output port	dB	20
5	Band pass flatness Over 100% band width	dB	0.5
6	Rejection At ± 42 MHz from CF and beyond	dB	30
7	Input/Output Port	--	SMA
8	Operating Temperature	°C	-5 to +55

Table II depicts the design specifications of Band-pass Quasi-Elliptic Filter. The layout of Quasi-Elliptic Filter in CAD ADS is shown in figure 6. The cross-coupling value j14 is used as the coupling between the resonator 1 and the resonator 4.

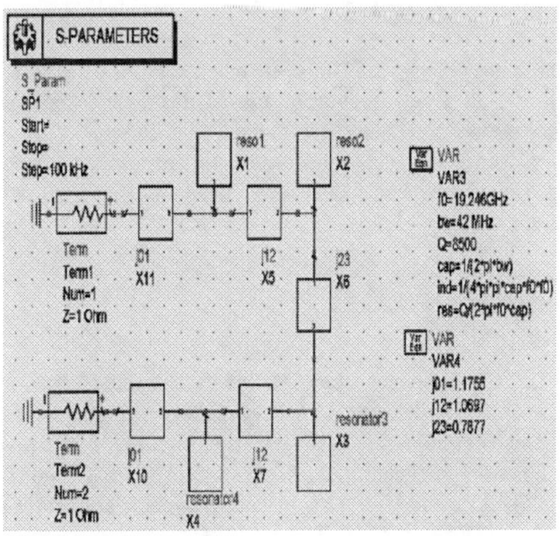

Figure 6 : ADS layout of 4-pole Quasi-Elliptic Filter.

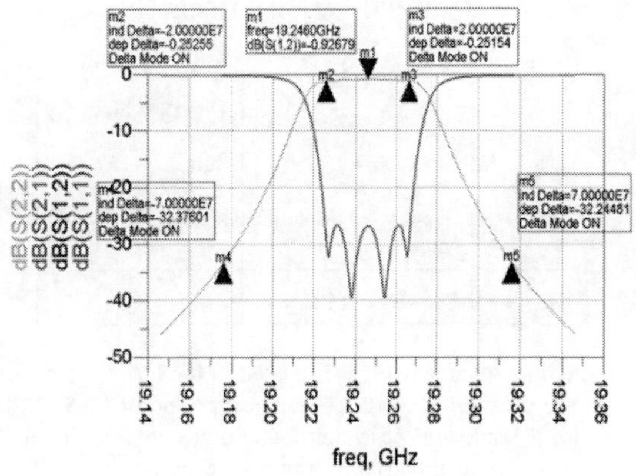

Figure 7 : Simulated response of the Filter using actual 'Q'.

The simulated response of the Filter using actual Q is shown in figure 7

A dielectric resonator filter uses ceramic dielectric "pucks" as resonators to form a multi-section filter. Dielectric resonators have a high dielectric constant and a low dissipation factor, which produces a high quality factor and in-turn gives a low insertion loss measurement over the filter's pass-band. Usually, dielectric resonators are inductively coupled (magnetic field) and can be mounted on a micro-strip network or inside a metallic cavity. The physical dimensions of the dielectric resonator (puck), the cavity dimensions of the puck's housing and the dielectric constant of the puck's material determine the resonant frequency, which can be approximated by.

$$f_{GHz} = \frac{34}{I\sqrt{\varepsilon_r}}\left(\frac{D}{H} + 3.45\right) \qquad (7)$$

where I is the inner diameter of the resonator, D is the outer diameter of the resonator and H is the height of the resonator as shown in Figure 8. At resonant frequency, the most of the energy is stored in puck

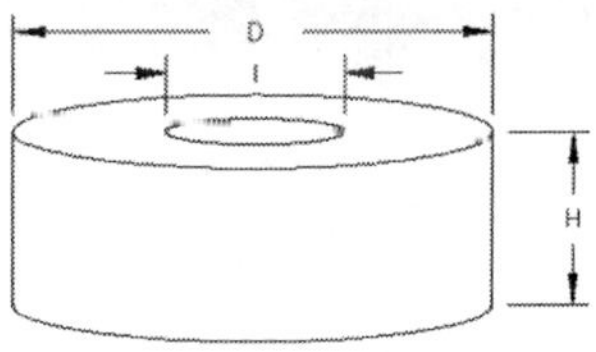

Figure 8 Unloaded Cylindrical Dielectric Resonator

The use of dielectric resonators constructed from low-loss high permittivity (20-100) temperature stable ceramics enables high-Q (up to 100 000) filters to be realized in a fraction of the volume and weight of air filled waveguide devices [10]. In HFSS, the design parameters of Microwave materials is shown in Table 3

Table 3. DESIGN PARAMETERS IN HFSS

S.No.		Units	Size
1	Cavity Size	mm	7x7x7
2	Support Material M6.8	ε	6.8
3	DR Material DR_24	ε	24

When the given cavity is simulated in the CAD system at the centre frequency 10.8 G Hz in the band of 10.5 G Hz to 11 G Hz, the electric field pattern are obtained having the Maximum intensity at the centre of the puck. This mode is named as TE01δ.

When the same cavity is simulated at the centre frequency 19 G Hz in the band of 18.72 G Hz to 21.72 G Hz, the electric field pattern has maximum intensity at the levels of 0.25 and 0.75 of the height of the puck. This is termed as TE01δ+1. The Electric Field is maximum at two locations of the puck. It signifies that more of energy is stored in the puck, thereby better Q, better selection of wanted frequency and better band rejection of unwanted frequency band.

At higher frequencies, the losses are more and the selectivity is poor, hence Q is poor. Hence we obtain the desired results at higher frequency band around 19 G Hz. By employing TE01δ+1 mode at higher frequencies, we get the desired results using the materials which were used in the lower mode of TE01δ.

The single cavity structure was simulated in HFSS to obtain the Mode of the DR which has the maximum spurious free range, i.e. Maximum Separation. The analysis set up was run at the centre frequency 18.6 G Hz. The Modes obtained after the simulations is shown in Table IV.

Table 4. MODES OBTAINED AS A RESULT IN HFSS.

S.No.	Mode	Resonant Frequency GHz
1	Mode 1	18.736082
2	Mode 2	18.831473
3	Mode 3	19.417369
4	Mode 4	21.624429

The analysis shows that the Mode 3 is the desired mode as it has the maximum spurious free range i.e. maximum separation. At lower side19.417-18.831=0.59 GHz and at upper side 21.624-19.417=2.21 GHz. For the realization of Quasi-elliptic filter, cross-coupling between non-adjacent cavities is made to equalize the group delay and getting finite transmission zeros[11]. The coupling may be inductive or capacitive. The inductive coupling is preferred because 96% of the electric field and 66% of magnetic field is within the DR [12][13]. The assembly of the filter in HFSS is shown in figure 8 and figure 9.

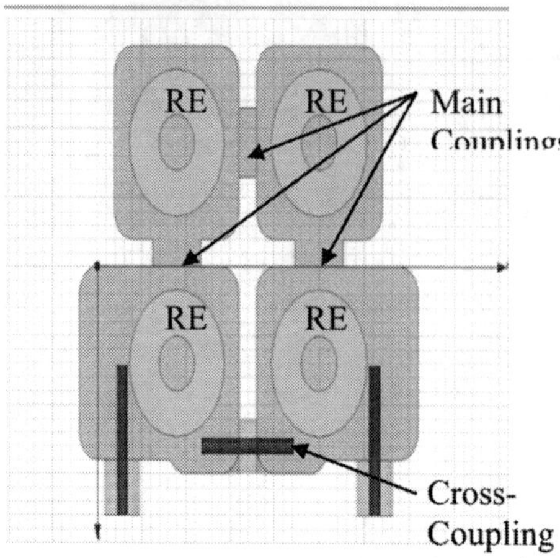

Figure 9: Complete 4-pole Quasi Elliptic filter structure

The model of the filter was simulated in the HFSS on INTEL XEON Quad-core dual CPU with 16 GB RAM and 3 GHz processor. The final response of Quasi-elliptic filter is shown in figure 10..

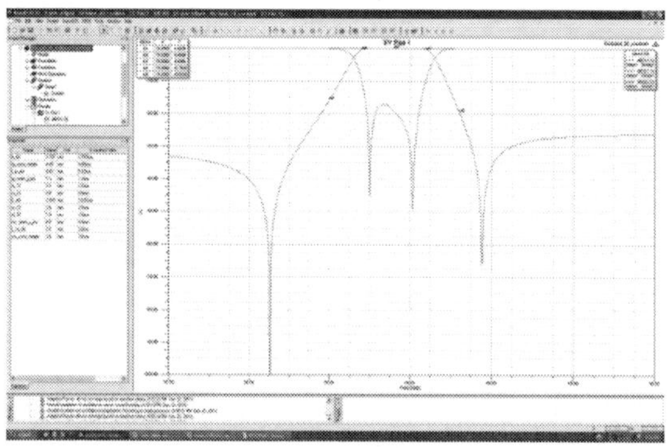

Figure 10: Final response obtained from HFSS

4 CONCLUSIONS

In this paper design of High 'Q' narrow-band filter in Ka band (4-Pole Chebyshev and Quasi-Elliptic) using a new higher mode has been presented. The theory of filter design using low pass prototypes and low-pass to band-pass transformations has been used as base of the concept.

References

1. Ian C. Hunter, Laurent Billonet, Bernard Jarry and Pierre Guillion, "Microwave Filters – Applications and Technology," IEEE Transactions on Microwave Theory and Techniques Vol. 50, No. 3, MAR 2002.
   POZAR DAVID M.: 'Microwave Engineering' (John Wiley & Sons, India 2008)

2. G. Matthaei, L. Young and E.M.T. Jones, "Microwave filters, impedance matching networks and coupling structures", Artech house Inc, 1964.
3. Jia-Sheng Hong and M.J. Lancaster, "Microstrip filters for RF/Microwave applications", John Wiley and Sons Inc, 2001.
4. M. Makimoto and S. Yamashita, "Microwave resonators and filters for wireless communication", Springer-Verlag Berlin Heidelberg, 2001.
5. LANCASTER M. J., HANG. JIA-SHEN G.:'Microstrip filters for RF/Microwave Applications' (John Wiley and Sons, Inc.) HUNTER I.: 'Theory and design of Microwave filters' (IEE, London, UK, 2001).
6. R. J. Cameron, "General Coupling Matrix Synthesis Methods for Chebyshev Filtering Functions," IEEE Trans. Microwave Theory Tech., vol. 47, No.4, pp. 433-442, April 1999.
7. R. J. Cameron, "Advanced Coupling Matrix Synthesis Techniques for Microwave Filters," IEEE Trans. Microwave Theory and Tech. vol. 51, No.1, Jan. 2003.
8. John B.Ness, " A Unified Approach to the Design, Measurement, and Tuning of Coupled-Resonator Filters," IEEE Transactions on Microwave Theory and Techniques Vol. 46, No. 4, APR 1998.
9. J. Brian Thomas, Cross Coupling in Coaxial Cavity Filters-- A Tutorial Overview," IEEE Trans. Microwave Theory and Tech.vol. 51, No. 4, APR 2003.
10. Baillargeat,D., Verdeyne, S., Galiere J, Zanchi, C., and Lapierre, L. (2005),""Quasi-planar transversal band pass filters for Ka band applications", Wiley International Journal of RF and Microwave Computer-Aided Engineering, 15.56-68, doi:10.1002/mmce.2005.
11. Ruiz-Cruz, J.A., Montejo- Garai,J.R.,and Zaki, K.A. (2007), "Waveguide filters with elliptic function response: Overview and results of different implementations". Wiley International Journal of RF and Microwave Computer-Aided Engineering,17.63-69,doi:10.1002/mmce.20198.

Split and Merge Based Quantitative Approach to Select Filter Combination for Image Segmentation

Amit Goyal[1] and Mohit Srivastava[2]

[1]*Research Scholar, Department of ECE, Jagan Nath University, Jaipur, Rajasthan, India*
[2]*Professor, Department of ECE, Chandigarh Engineering College, Landran, Punjab, India*

Abstract. With the advent of image analysis and computation in different domains, image segmentation has emerged as the most crucial step to achieve a compact segment-based description of image scene by decomposing it into meaningful segments of similar attributes. The pre-and-post filtering operation reduces the effect of noise from the segmented image. The Cameraman image is pre-filtered using Laplacian, Median and Min filter. The Split and Merge method for Region based image segmentation which guarantees to connected regions are now applied on the filtered image. The Median, Laplacian and Sobel filter is then used to post-filter the segmented image. The PSNR and MSE values are calculated to quantitative evaluation of segmented images. The quantitative evaluation of post-filtered segmented image shows that median filter produces most effective result with lowest MSE of 84.89 dB and highest PSNR of 5.72 dB.

1 Introduction

To analyze or interpret an image automatically, pre-processing is done which involves segmenting the image into different objects of interest e.g. separation of foreground from the background [6]. Image segmentation has firmed its ground in many practical applications that involve a visual interpretation, namely in medical imaging, object detection (face detection, pedestrian detection, brake light detection etc. More accurate the segmentation, more successful is the recognition of objects in the image.

Basically, segmentation is the process of subdividing an image into its constituent regions or objects. Segmentation algorithms are based on either similarity or discontinuity of intensity/gray level values. A number of image segmentation algorithms with increased complexity have been developed over the years. All these algorithms work on the use of any of the three main criteria: the homogeneity within a segment, separation from adjacent segments and shape homogeneity. Typically, the segmentation algorithms can be grouped into three major categories on the basis of their segment formation properties, namely Threshold Based Segmentation [11, 22], Boundary based Segmentation [9, 16] and Region Based Segmentation [10].

1.1 Features of various segmentation techniques

Threshold based segmentation produces regions of uniformity within the given image based on some threshold criteria on the attributes of pixels [11].These algorithms are computationally less expensive but causes noise, blurred edges or outlier [8] in the image. Contrary to this, the boundary based techniques rely on edges found in an image and tries to locate points of discontinuities in images. This segmentation method is very sensitive to noise and hence post-processing is required. However, region based segmentation algorithms partition the entire image into sub regions depending on some homogeneity criteria [3, 14].

1.2 Region Split and Merge Technique

The region split and merge technique falls under region based segmentation and is combination of top-down and bottom-up approach [4]. In this image is first split into homogeneous regions on the basis of homogeneity criteria. The regions formed after first split will be individually checked for the homogeneity and any two adjacent homogeneous regions will be merged into one region. The merging of homogeneous connected regions is a combination of two phases: first is to find the pair of adjacent homogeneous regions and second is to choose a certain merging criterion [2, 4, 10]. The algorithm halts when no further merging is possible. This technique produces large number of region boundaries, other than the horizontal and vertical ones [13]. But the algorithm of this approach leads to increase in computational speed [10] with lesser complexities and hence improves segmentation quality [4]. It also guarantees connected regions [9] i.e. there are no gaps due to missing edge pixels. Due to the inherent advantages of region based segmentation technique, the performance of image segmentation is elaborated in this paper using split and merges technique.

2 Methodology Adopted

To apply pre and post filter techniques in order to segment the image using split and merge technique and evaluate the performance of algorithm, the system architecture being followed is explained with the help of flowchart as given in fig. 1.

[1] Corresponding author: amitgoyal0810@gmail.com

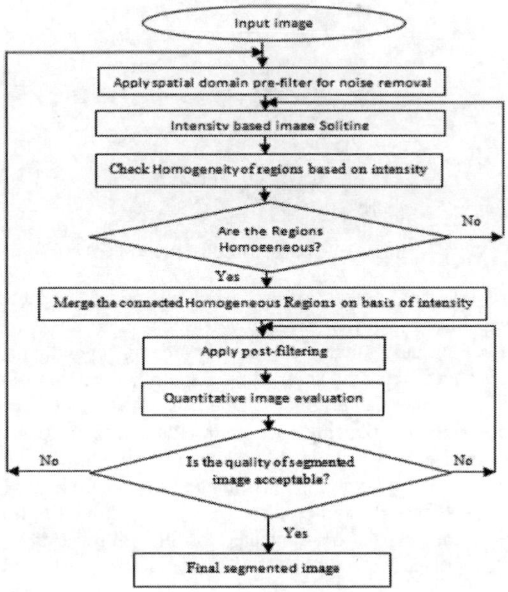

Figure 1. Flowchart of methodology for image segmentation using split and merge technique

2.1 Image Database

Gray-scale images of different properties like dimensions (width*height), resolution (dots per inches) and size of image (bytes) for storage may be used as the input dataset. A gray-scale cameraman image (fig.2 (a)) of size 512x512 pixels is chosen to draw the performance evaluation of the algorithm used in this work. Fig. 2(b) helps to visualise the pixel detail of a small section of boundary differentiating the object (person) from background in the image.

Figure 2. (a)Input Image to be segmented (b) Pixel detail of object boundary

2.2 Pre-filtering

Pre-filtering is done to preserve and sharpen the edge information [5, 15]. Spatial filters offer more versatility as they can be used for non-linear filtering applications as well [7]. The fundamental of filtering an image lies in its convolution with the mask of pre-defined filter [20]. Sharpening linear filters (laplacian, unsharp and laplacian of Gaussian filter) enhance the pixel neighbourhood in the image. Contrary to linear filters, nonlinear spatial filtering (median, max and min filters) is based on nonlinear operations with pixels encompassed by the filter [21].

2.3 Split Image into Regions

Splitting is a recursive process. Quad tree decomposition method is exploited in the current work to split an image into four equal sized-quadrants and it depends upon a homogeneity criterion. The current work uses the attribute of intensity of pixels as the homogeneity criterion. The regions formed after first split will be individually checked for the homogeneity. The equation (1) denotes the homogeneity criteria for splitting and merging the image. This is interpreted as a region is split into four equal sub-regions or quad-regions if the value of Q comes out to be 'False'.

$$Q = \begin{cases} TRUE, & if \sigma > a \, and \; 0 < m < b \\ FALSE, & otherwise \end{cases} \qquad (1)$$

In equation (1), Q is called as predicate, m and σ are the mean (average intensity) and standard deviations of pixels in a quad-region, respectively, while 'a' and 'b' are constants whose values are set during experimentation. So the average intensity (m) of a particular region should be greater than 0 and below a certain threshold value. The experimentations in current work were done by taking various values of 'a' and 'b', but an appropriate homogeneity criterion was obtained by taking 'a' and 'b' as 5 and 200, respectively.

2.4 Merge Connected Homogeneous Regions

The output of the splitting stage is multiple regions which may or may not be homogeneous depending on whether they fulfil the homogeneity criterion or not, respectively [12]. The non-homogeneous regions are further subject to splitting and the homogeneous regions are merged together. The homogeneity criterion based on average intensity as specified by equation (1) will be followed [2-3]. The new region so formed will be continuously merged with its adjacent regions until the homogeneity criterion will be satisfied.

2.5 Post-Filtering

To make the quad-tree structure based Split and Merge image segmentation capable of adapting to edges of objects in the image, spatial filters may be applied on the segmented image [3, 5]. It not only removes noise but may also be helpful to smooth the arbitrary position and orientation of edges of image objects. Preservation of edges can also be done using linear filters like 'Sobel' which returns edges at those points where the gradient of image is maximum [9, 19]. Other alternatives include Prewitts method, Roberts method and Laplacian of Gaussian etc.

2.6 Quality Assessment

The next step is to evaluate effectiveness of the segmentation methods by performing quality assessment through subjective evaluation as well as quantitative approaches [23]. The subjective evaluation is based on visual interpretation, the objective evaluation can be done using a number of widely-used quality metrics. A few of these metrics are: PSNR- the peak signal-to-noise ratio (dB), the mean square error (MSE) which is the squared norm of the difference between the original image and

 DOI: 10.1051/matecconf/20165701003

the segmented image divided by the number of elements [17, 18].

3 Experimentation and Results

The split and merge algorithm was implemented in MATLAB 7.12 and tested on standard 'cameraman' test image. At first, the image was filtered by applying different spatial filters. The parameters of filters like size, padding options were varied to obtain better results of filtering. Table 1 depicts the detail of filters along with parameter variations.

A 3x3 Laplacian filter was applied on the original image and value of alpha was varied from 0 and 1 for fine tuning. Since the Laplacian is a derivative operator, it sharpens the image but drives the constant areas to zero. So the original image is added back to restore the gray-level [7]. The enhancement in edges of foreground (camera man) is clearly visible fig. 3(a). The performance of LOG (Laplacian of Gaussian filter) tends towards blurring the image as the standard deviation is lower below to 0.5. Variations in mask size were also made to analyse the filter. The result of filtering with LOG filter is shown in fig.3 (b).

Table 1. Detail of filters applied on original image

S.no.	Filter	Size (Rows x Columns)	Parameters	
1	Laplacian	3x3	Alpha={	0
		3x3		0.2
		3x3		0.8
2	LOG(Laplacian of Gaussian)	3x2	Standard Deviation={	0.3
		5x5		0.3
		5x5		0.5
3	Min	2x3	Order={	1
		3x3		1
4	Median	3x3	Padding={	Symmetric
		4x4		Symmetric
		3x3		Zeros
		5x5		Zeros
5	Unsharp	3x3	Alpha={	0.2
		3x3		0.8
6	Max	2x2	Order={	1
		2x3		1

Figure 3. (a) Image filtered with Laplacian filter(Alpha=0.2) (b) Image filtered with LOG filter(Size 5x5 & σ=0.5

(c) Image filtered with median filter(Size 3x3)

A Median filter offers excellent noise reduction capabilities with considerably less blurring (fig. 3 (c)). The unsharp filter acts as sharpening operator and enhances edges via subtracting a smoothed version of an image from the original image (fig. 4(a)). But it was observed that blurring was predominant with increase in filter size. Other spatial filters namely min and max filter were also applied on the test image in fig. 4 (b) and 4 (c) respectively. Best filtered image as listed in table 2 from each category has been selected for further processing.

Figure 4. (a) Image filtered with Unsharp filter(Alpha=0.2) (b) Image filtered with Min filter (size=2x3

(c) Image filtered with Max filter (size=2x2)

Table 2. Filters with better response on basis of visual interpretation

S.No.	Filter	Size	Parameter
1	Laplacian	3x3	Alpha=0.2
2	Log	5x5	Standard Deviation=0.5
3	Unsharp	3x3	Alpha=0.2
4	Median	3x3	Padding=Symmetric
5	Min	2x3	Order=1
6	Min	3x3	Order=1

Split and merge segmentation algorithm was applied on these selected filtered images. It is observed that image filtered with Min and Median filter provides an optimal segmentation as the foreground and background were clearly differentiated without losing edge details. LOG and Unsharp filter causes over-segmentation and hence those responses are rejected for any further process. Best segmented image obtained after applying split and merge segmentation technique on filtered image are shown in fig. 5 & fig. 6. Post-filtering has been now applied on segmented image to further preserve edges. Laplacian, Median and Sobel filters are used for post filtering operation. The filters which yielded better results are listed in table 3.

The quality of final segmented image after post filtering was assessed quantitatively by calculating the Mean Square Error (MSE) and Peak Signal-to-Noise Ratio (PSNR) values of original image with respect to segmented image followed by post-filtering. Calculated values of PSNR and MSE are given in table 3.

Figure 5. Segmented image pre-filtered with (a) Laplacian filter (b) LOG filter (c) Unsharp filter

Figure 6. Segmented image pre-filtered with (a) Median filter (b) Min filter (size=2x3) (c) Min filter (size=3x3)

Quantitative evaluation of post-filtering operation shows that among these the Laplacian filtrations have low PSNR and high MSE, thus not suitable. However, as per quantitative evaluation the output quality of Median filter is highest with PSNR (dB) of 5.72, 5.68 and 5.72, and MSE (dB) of 84.89, 84.97 and 84.89 applied over Laplacian, Median and Min filtered image respectively. The filtered images are shown in fig. 7. The quantitative evaluation shows the filtered output quality of Sobel filter is comparatively poor than Median with of PSNR (dB) of 5.65, and MSE (dB) of 85.03 in all case. Fig. 8 shows the post filtered images using Sobel filter. These two post-filtered images are interpreted visually.

Table 3. Quality assessment of Split and Merge algorithm through Quantitative approach

Pre-Filter			Post-Filter			Evaluation	
Name	Size	Parameter	Name	Size	Parameter	MSE (dB)	PSNR (dB)
Laplacian	**3x3**	**Alpha=0.2**	**Median**	**3x3**	**Padding= Symmetric**	**84.89**	**5.72**
Median	3x3	Symmetric Padding				84.97	5.68
Min	**2x3**	**Order=1**				**84.89**	**5.72**
Laplacian	3x3	Alpha=0.2	Laplacian	3x3	Alpha= 0.2	85.23	5.55
Median	3x3	Symmetric Padding				88.76	3.78
Min	2x3	Order=1				86.00	5.17
Laplacian	**3x3**	**Alpha=0.2**	**Edge Filter using Sobel**	**Search direction = horizontal and vertical**		**85.03**	**5.65**
Median	**3x3**	**Symmetric Padding**				**85.03**	**5.65**
Min	**2x3**	**Order=1**				**85.03**	**5.65**

Figure 7. Segmented image with Median as post filter and Pre-filter (a) Laplacian (b) Median c) Min

Figure 8. Segmented image with Sobel as post filter and Pre-filter (a) Laplacian (b) Median c)Min

4 Conclusions

The segmented image after post filtering was assessed subjectively (visual interpretation) as well as with the help of quantitative metrics, namely MSE & PSNR. Visual interpretation infers better result of segmentation with the use of Median and Sobel as post filters. However, it can be deduced from the quantitative analysis that in case of Median as post filter the mean square error is less as compared to other post filters and higher peak signal-to-noise ratio can also be obtained through it and specifically its combination with Laplacian and Min as pre-filters yield most effective. Further, performance of Sobel is also acceptable. It is worthwhile to mention that this combination is able to segment the objects in the image while preserving the edges i.e. edges remain connected, hence this combination also provides critically segmented cameraman image.

References

1. S.L. Horowitz, T. Pavlidis, Picture segmentation by a tree traversal algorithm, Journal of the Association for Computing Machinery, **23**, 368-388 (1976)
2. R. Hunki, L. HaengSuk, A Noble Image Segmentation using Local Area Splitting and Merging Method based on Intensity Change, International Journal of Software Engineering and its Applications, **7**, 99-112 (2013)
3. X. Wu, Adaptive Split-and-Merge Segmentation Based on Piecewise Least-Square Approximation, *IEEE Transaction on Pattern Analysis and Machine Intelligence*, **15**, 808-815 (1993)
4. A.B.M. Faruquzzaman, N.R. Paiker, J. Arafat, M.A. Ali, G. Sorwar, Literature on Image Segmentations Based on Split- and- Merge Techniques, *International Conference on Information Technology and Applications*, pp. 120-125. Cairns, Australia (2008).
5. H.S. Yang, S.U. Lee, Split and Merge Segmentation Employing Thresholding Technique, *IEEE International Conference on Image Processing*, pp. 239-242. Santa Barbara (1997)
6. S. Annadurai, R. Shanmugalakshmi, *Fundamentals of Digital Image Processing, Pearson Education, India* (2007)
7. R.C. Gonzalez, R.E. Woods, *Digital Image Processing. Pearson Education, India* (2009)
8. R. Yogamangalam, B. Karthikeyan, Segmentation Techniques Comparison in Image Processing, International Journal of Engineering and Technology, **5**, 307-311 (2013)
9. P. Thakare, A Study of Image Segmentation and Edge Detection Techniques, International Journal of Computer Science and Engineering, **3**, 899-904 (2011)
10. D. Kelkar, S. Gupta, Improved Quadtree Method for Split Merge Image Segmentation, *IEEE International Conference on Emerging Trends in Engineering and Technology*, pp. 44-47. Nagpur, Maharashtra (2008)
11. R. Kandwal, A. Kumar, S. Bhargava, Review: Existing Image Segmentation Techniques, International Journal of Advanced Research in Computer Science and Software Engineering, **4**, 153-156 (2014)
12. B. Peng, L. Zhang, D. Zhang, Automatic Image Segmentation by Dynamic Region Merging, *IEEE Transaction on Image Processing*, **20**, 3592-3605 (2011)
13. K. Khoshelham, Z. Li, B. King, A Split-and-Merge Technique for Automated Reconstruction of Roof Planes, Photogrammetric Engineering & Remote Sensing Journal, **71**, 855-862 (2005)
14. A.B.M. Faruquzzaman, N.R. Paiker, J. Arafat, Z. Karim, M.A. Ali, Object Segmentation Based on Split and Merge Algorithm, *IEEE International Conference on Acoustics, Speech and Signal Processing*, pp. 1-4. Toulouse, France (2006)
15. S.M. Ramesh, B. Gomathy, T.V.P. Sundararajan, Detection of Defects on Steel Surface for Using Image Segmentation Techniques, International Journal of Signal Processing, Image Processing and Pattern Recognition, **7**, 323-332 (2014)
16. P.L. Palmer, H. Dabis, J. Kittler, A performance measure for boundary detection algorithms, Elsevier Computer Vision Image Understanding, **63**, 476–494 (1996)
17. M. Sharma, V. Chouhan, Objective Evaluation Parameters of Image Segmentation Algorithms, International Journal of Engineering and Advanced Technology, **2**, 84-87 (2012)
18. H. Zhang, J.E. Fritts, S.A. Goldman, Image segmentation evaluation: A survey of unsupervised methods, Science Direct Computer Vision and Image Understanding, **110**, 260-280 (2008)
19. J. Canny, A Computational Approach to Edge Detection, *IEEE Transaction on Pattern Analysis and Machine Intelligence*, **8**, 679 – 698 (1986)
20. J.S. Lee, Digital image enhancement and noise filtering by use of local statistics, *IEEE Transaction on Pattern Analysis and Machine Intelligence*, **2**, (1980)
21. A.V. Lugt, Signal detection by complex spatial filtering, *IEEE Transactions on Information Theory*, **10**, 139 – 145 (1964)
22. C.M. Pun, N.Y. An, M. Cheng, A Region-Based Image segmentation by Watershed Partition and DCT Energy Compaction, *IEEE International Conference on Computer Graphics*, pp. 131-135. China (2011)
23. Y.J. Zhang, A survey on evaluation methods for image segmentation, Elsevier Journal on Pattern Recognition, **29**, 1335-1346 (1999)

Design and analysis of 32 bit CMOS adder using sub-threshold voltage at deep submicron technology

Jaspreet Kaur [1] and Parminder Singh [2]

[1] *M-Tech Student, Department of Electronics, Yadavindra college of Engineering, Talwandi Sabo(Pb)-INDIA*
[2] *Parminder Singh, Assistant Professor, Department of Electronics, Yadavindra College of Engineering, Talwandi Sabo*

Abstract. FA is basic cell for arithmetic operation and lots of efforts have put to minimize power consumption and delay. This paper evaluates conventional CMOS adder, bridge style adders in sub-threshold region. Circuits are designed at 20 MHz and 50 MHz frequencies with V_{DD}= 200 mv. All adder designs are simulated at 32 nm technology. In 1 bit and 32 bit conventional CMOS adder design, an efficient trade-off between delay and power is achieved. Experimental results show that 32 bit adder designs have significant improvements in delay and power delay product.

1 Introduction

CMOS VLSI circuits has been identified as a critical technology need in the recent years due to the high demand for low power consumption, lower delay, small area and low cost design are increasing every day. In VLSI applications, arithmetic operations play an important role in modern processing systems. Major operations of arithmetic are addition, Sub-traction, multiplication and accumulation. Full adder (FA) cell is the basic block for most of arithmetic systems. Improving the FA characteristics is critical for measuring overall system quality. In many computers and other kinds of processor, adders are not used only in the arithmetic logic units, but also in other parts of the processor, where they are used to calculate addresses, table indices, and similar operation.

Sub-threshold design is a technique for reducing dissipated power [1]. Due to portable devices; the adder must have high speed, lesser area and lower power consumption. Devices like mobile, Laptops etc. require more battery backup. So, a VLSI designer has to optimize these three parameters in a design. Basically full adder is the core components of microprocessor and other complex chips. So the performance of full adder would affect the system as a whole. All the circuits are designed in sub-threshold region. The main factor in sub-threshold design is maintaining a good trade-off between power consumption and performance. In recent years, considerable research has been performed on FA design with various design methodologies and different technologies [9]. In these days several Adder designs have been proposed to reduce the power consumption. Logic minimization gives better results in power consumption. For low power results it is always advisable to use CMOS technology. In paper, three different 1-bit full adder topologies in deep sub-threshold operation are proposed. The cells are characterized with respect to delay, power consumption, Power-Delay Product (PDP) and leakage power. Conventional CMOS Adder and Bridge style adders are designed at 20mhz and 50mhz frequencies with V_{DD}= 200mv.

2 Sub-threshold design

In sub-threshold circuits, the supply voltage is reduced well below the threshold voltage. Due to reduction in power with respect to supply voltage, sub-threshold circuits are classified as ultra-low-power circuits. Specifically in application areas where performance can be sacrificed for low power, sub-threshold circuits are ideal fit [11].

Some of the applications include devices such as digital watched, radio frequency identification (REID), sensor nodes and battery operated devices such as, cellular phones.Design spectrum is a trade-off between power consumption and performance. One extreme aim of this spectrum is ultra-low power while performance is of secondary importance. In this solution, circuit is biased in sub-threshold region so that standard CMOS logic is extended to operate in sub-threshold region by reducing the supply voltage V_{dd} less than transistor threshold voltage V_t . The motivation for using sub-V_t circuits is the ability to exploit the sub V_t leakage current as the operating drive current. In sub-threshold region, the reduction in power consumption outweighs the increase in delay. Thus the circuit has lower PDP when it operates in sub-threshold region. So in the same frequency, when

[1]Jaspreetkaurp1@gmail.com

the circuit operates in sub-threshold region, its energy consumption is less than when it works in supper-threshold.

3 A Review on full adder circuits

In this section, various 1-bit full Adder designs are examined. Some of these schemes are based on CMOS logic with low power consumption, good signal logic level and high driving ability.

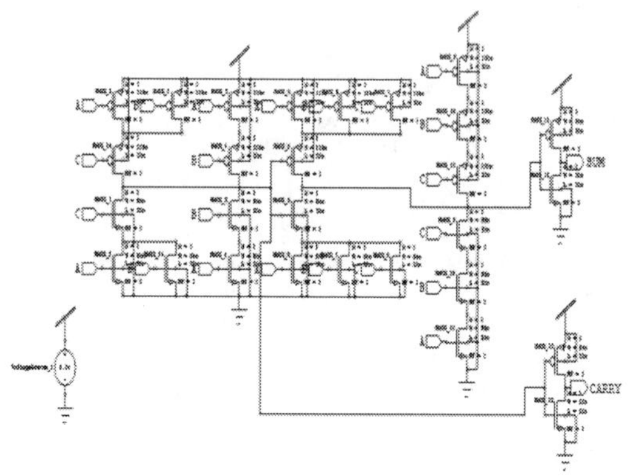

Figure 1. Conventional CMOS 1 bit full adder (CCMOS)

The schematic shown in the above fig. is CCMOS full adder [9] designed in the tanner schematic editor, which is based on regular structure CMOS logic and almost equal rise and fall time. It uses 28 transistors to complete one bit full adder. In this section bridge style 1-bit full adder introduced which operating in sub-V_t region is introduced. This can be divided into two main modules: sum module and carry module. Equations (1) and (2) show the logical functions of the full adder outputs.

$$\text{Sum} = a \oplus b \oplus c \quad (1)$$

$$\text{Carry} = a.b + b.c + a.c \quad (2)$$

Based on these equations, two implementation of full adder with bridge design are investigated. The first is shown in fig. 2 which called in this paper as Bridge_v2. This design needs 24 transistors against conventional CMOS full adder uses 28 transistors and should have complementary inputs. Third design is Bridge_v3 shown in fig. 3. These three circuits are designed and simulated at 32 nm technology. Bridge method uses some transistors, called bridge transistors, as conditional conjunction between two nodes of circuit so this method shares transistors of different paths from power supply to output.

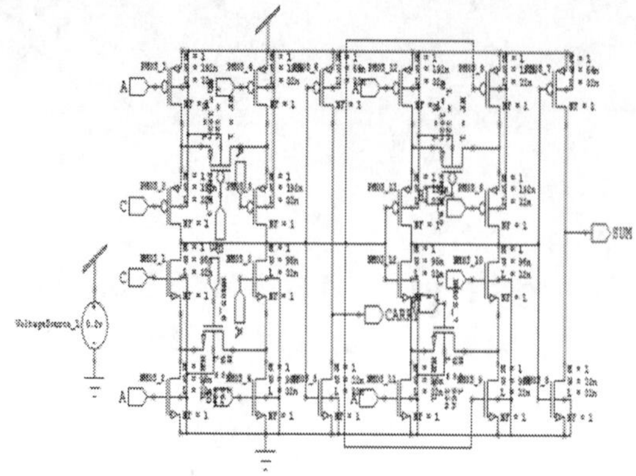

Figure 2. Bridge type (B_V2) 1 bit full adder style.

It uses 24 transistors for evaluating the sum and carry. The name itself shows that it uses bridge type structure for implementing the SUM and CARRY. It shares the path between carry and sum circuits with fully symmetric full adder structure fig 2.

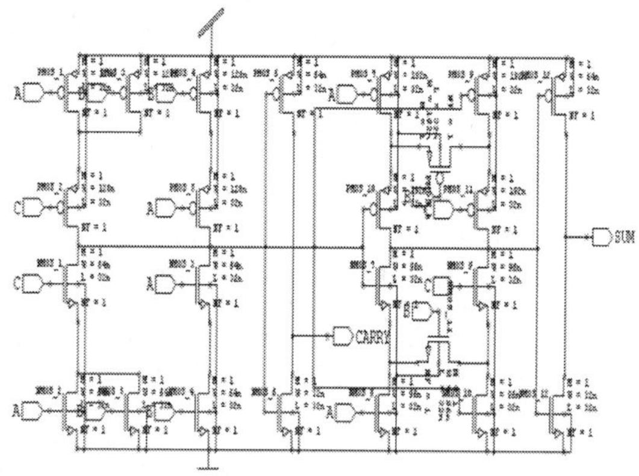

Figure 3. Bridge type (B_V3) 1 bit full adder style.

It uses 24 transistors for evaluating the sum and carry. It shares the path between carry and sum circuits with semi symmetric full adder structure fig 3.

4 Results

4.1. Results of full adder at 32 nm

Table 1 represent comparison results of 1 bit full adders. Proposed designs CCMOS, Bridge style are simulated at 20MHz and 50 MHz frequency. In this, parameters like average power consumption, delay and power delay product are measured. Our result shows improvement in average power consumption and power delay product.

Table 1. Result of 1 bit adders

Sr no	Parameters	S. Wairya et.al [6]		Farshad Moradi et.al [7]	Xu Wang et.al [8]	Proposed design [20 MHz]		
		CCMOS	BRIDGE STYLE	CMOS	CMOS	CCMOS	BRIDGE STYLE (B_V2)	BRIDGE (B_V3) semi-symmetric
1.	Technology	90nm		65nm	180nm	32nm		
2.	Average power (µw)	2.24	1.82	0.021	0.000049	0.00753	0.00881	0.00769
3.	Delay (ns)	0.524	0.323	1000	14	7.5	8.66	8.69
4.	Power delay product(fj)	1.17	0.59	21.0	694.4	0.05647	0.07629	0.06683
5.	Leakage power (nw)	–	–	–	–	2.127	2.422	2.132

In table 2 all the circuit designs are simulated at 50 MHz frequency. In this, parameters like average power consumption, delay and power delay product and leakage power are measured.

Table 2. Result of 1 bit adders at f= 50 MHz

Sr. No	Name of adders	No. of transistor	width	Delay(ns)	Avg. power (nw)	Power delay product (aj)	Leakage power (nw)
1.	CCMOS	28	W*4	5	13.99	69.95	2.127
2.	B_V2	28	W*4	5.91	18.74	110.75	2.422
3.	B_V3	28	W*4	5.94	13.46	79.95	2.132

Table 3 represent comparison results of 32 bit Ripple Carry adders. All the circuit designs are simulated at 50 MHz frequency. Proposed design has improvement in delay and power delay product.

Table 3. Result of 32 bit Ripple Carry adders

Sr. no	Parameters	Mohammad Reza Bagheri.et.al [9]			Proposed design [50 MHz]		
		CCMOS	BRIDGE STYLE (B_V2)	BRIDGE STYLE (B_V3)	CCMOS	BRIDGE STYLE (B_V2)	BRIDGE STYLE (B_V3)
1.	Average power (µw)	0.190	0.063	0.063	1.554	1.654	1.480
2.	Delay (ns)	43.66	65.00	50.35	2.539	2.681	3.251
3.	Power delay product (fj)	8.320	4.112	3.189	3.946	4.434	4.811
4.	Leakage power (µw)	–	–	–	0.657	0.769	0.685

[1]Jaspreetkaurp1@gmail.com

4.2 Simulation Result

Simulation waveform of CCMOS is shown in fig 4.The input data V_a, V_b, V_c and output sum and carry shown below:

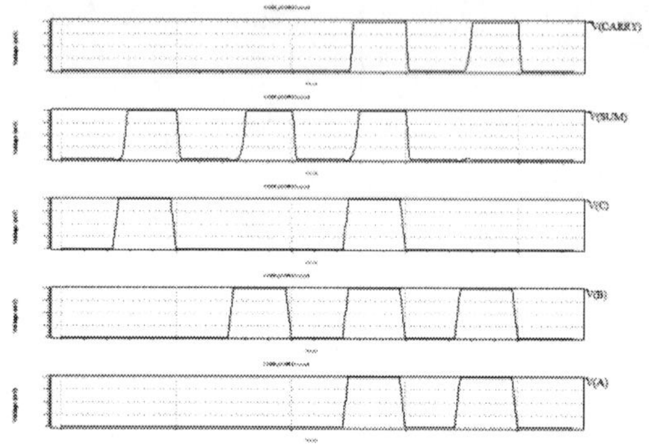

Figure 4. Output Waveform of CCMOS Simulation

5 Conclusion

The aim of this paper is reduce the power consumption. In this paper CCMOS design and bridge style adders are compared in sub-threshold region. Results show that significant improvements in power consumption, delay and power delay product with acceptable performance. All the adder circuits are designed at 20 MHz and 50 MHz frequencies with V_{DD}= 200mv. All adder designs are simulated at 32 nm technology.

In 1 bit Conventional CMOS adder, average power is improved by 99.66% and power delay product is improved by 95.17%. In Bridge Style adder, average power is improved by 99.51% and power delay product is improved by 87.07%.

In this paper circuits make use of 50MHZ frequency; then our result is improved. In 32 bit Conventional CMOS, delay is improved by 94.18% and power delay product is improved by 52.58%. In Bridge style (B_V2) design, delay is improved by 95.87%. In Bridge style (B_V3), delay improved by 93.54%.

References

1. Neil H.E.Weste, David Harris and Ayan Banerjee "*CMOS VLSI Design, a Circuit and System Perspective*", Third Edition, Pearson Education, Inc., (2005).
2. Sung-Mo (Steve) Kang and Yusuf Leblebici, "*CMOS Digital Integrated Circuits Analysis & Design*", Tata McGraw-Hill, (2005).
3. Mahdiar Ghadiry, Mahdieh Nadi and Abu Khari A'Ain , "DLPA: Discrepant Low PDP 8-Bit Adder", Journal of Circuits Syst. Signal Process, Vol. **32**, Issue1, pp.1-14, (2013).
4. Deepak Garg, Mayank Kumar Rai, " CMOS Based 1-Bit Full Adder Cell for Low-Power Delay Product", IJECCT 2012, Vol. **2** No.4, pp. 18-23, (2012).
5. Shiwani Singh ,Tripati Sharma ,K.G. Sharma, "9T Full Adder Design in Subthreshold Region", Hindwai Publication Corpoation VLSI Design Vol. (2012).
6. S.Wairya, Himanshu Pandey, R.K.Nagaria and S.Tiwari, Member, IEEE "Ultra Low Voltage High Speed 1-Bit CMOS Adder" IEEE , pp. 1-6,(2010).
7. Farshad Moradi, Dag T. Wisland, Tuan Vu Cao "1-Bit Sub Threshold Full Adders in 65nm CMOS Technology"International conference on Micro-electronics(IEEE),pp.268-271,(2008).
8. Xu Wang, Weifeng He, Zhigang Mao "New Robust 200mV Sub-threshold Full Adders" IEEE, pp.776-778, (2010).
9. Mohammad Reza Bagheri,"Ultra Low Power Sub-threshold Bridge Style Adder in Nanometer Technologies "Canadian Journal on Electrical Engineering Vol.**2**, No. 7,pp.294-299, July (2011).
10. Mariano Aguirre-Hernandez and Monico Linares-Aranda, "CMOS Full Adder For Energy-Efficient Arithmetic Applications", IEEE Transaction on Very Large Scale Integration (VLSI) Systems, Vol. **19**, No.4, pp.718-721,(2011).
11. .Vishal Sharma, Sanjay, "Low Power 1-BitCMOS Full-Adder Using Sub threshold Conduction Region", International Journal of Scientific & Engineering Research vol. **2**, June (2011).
12. Farshad Moradi, Dag.T.Wisland ,Hamid Mahmood, "Ultra Low Power Full Adder Topologies", Nanoelectronics Group , Department of informatics,University of Oslo,NO-0316 Oslo, NORWAY, pp.3158-3161,(2009).
13. Keivan Navi and Omid Kavehei, "Low Power and High-Performance 1-Bit CMOS Full-Adder Cell" Journal of computers, vol. **3**, no. 2,pp.48-54, February (2008).
14. 14. Jun Cheol Park and Vincent J. Mooney III, "Sleepy Stack Leakage Reduction", IEEE Transaction on Very Large Scale Integration (VLSI) Systems, Vol. No.11, pp. 1250-1263, Nov. (2006).

Design and analysis of single- ended robust low power 8T SRAM cell

Neha Gupta[1] and Hitesh Pahuja[2]

[1]*Research Scholar, CDACMohali, Punjab, India*
[2]*Project Engineer, CDAC Mohali, Punjab, India*

Abstract. This paper is based on the observation of 8T single ended static random access memory (SRAM) and two techniques for reducing the sub threshold leakage current, power consumption are examined. In the first technique, effective supply voltage and ground node voltages are changed using a dynamic variable voltage level technique(VVL). In the second technique power supply is scaled down. This 8T SRAM cell uses one word line, two bitlinesand a transmission gate. Simulations and analytical results show that when the two techniques combine the new SRAM cell has correct read and write operation and also the cell contains 55.6% less leakage and the dynamic power is 98.8% less than the 8T single ended SRAM cell. Simulations are performed using cadence virtuoso tool at 45nm technology.

1 Introduction

With the large usage of static random access memory in integrated circuits for high density and stability, thedemand for SRAM is increasing [1]. Low power circuits demand is increasing rapidly not only for logic circuits but also for storage elements. So power consumption of SRAM cells must be reduced. In this paper 8T SRAM cell is proposed to reduce theleakage, static, and dynamic power at circuit level. There are numerous techniques to reduce the standby power. One is to use sleep technique in which standby power is reduced by disconnecting the supply voltage from the cell using high threshold voltage pMOSFET and disconnecting the ground path by using high threshold nMOSFET[2].However, this has serious limitation of high increase in sub threshold leakage current because of its exponential relation with threshold voltage. So to solve this problem a dynamic variable voltage level technique that can significantly reduce the power dissipation, leakage current is proposed. The present work describes the use of this technique to reduce the supply voltage and to increase the ground level potential for yielding the maximum reduction in leakage currents and power dissipation [3].

In this paper we propose a 8T SRAM cell at 45nm technology, this cell is composed of eight transistors in which two inverters are cross coupled(M1, M2, M3, M4), transistors M5 and M6 are access transistors, M7 and M8 are read access transistors forming a transmission gate.

The organization of the paper is as follows: Sect2 describes the proposed 8T SRAM cell and the read and write operation. Section 3 describes the leakage current that flows in this circuit and the power dissipation.Section 4 describes the various techniques. Section 5 gives the brief conclusion.

2 Read and Write operation of circuit

The proposed 8T SRAM cell has three modes of operation i.e. write mode, read mode, hold mode. Write operation starts by cutting off the transmission gate connection from the circuit .Then the data which needs to be written is placed on the BL and complement of the data is placed on the BLBAR .Q and Qb are two storage nodes.M1 and M2 are pull up pmos transistors and M3 and M4 are pull down transistors. Word line is responsible for holding the data in active and hold mode. The SRAM cell looks like two cross coupled inverters,inv1 followed by inv2.The output of the first inverter becomes the input to the second and vice versa. In this cell a transmission gate is used to read the data.RWLB is inversion of RWL signal .When both RWL and RWLB are assertedand once the transmission gate gets on, the stored data at node Q gets transferred and is read[4].

One of the major advantages of transmission gate is that it eliminates the requirement of sense amplifier and precharge circuit as used in conventional 8T SRAM cell for the read purpose thus lowering the power

[2]Corresponding author: hitesh@cdac.in

consumption of the cell [10]. Figure 1 shows the schematic of 8T SRAM cell.

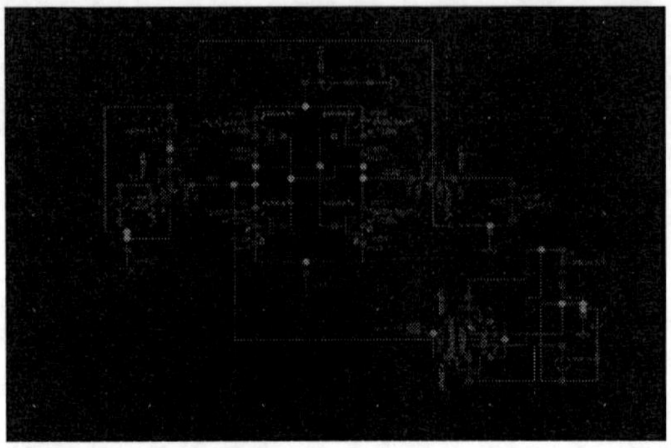

Figure 1. Schematicof 8T SRAM cell

2.1 Read and write waveforms

Read operation: During the read operation the data stored at Q node is passed to the input of the transmission gate where both the word lines RWL, RWLB are asserted. Then the output is taken from read path rout.

Write operation: During the write operation the data that needs to be written is placed is at Bit Line and its complement on BLBAR. WL is asserted high after that.

Hold mode: During hold the WL remains off and cell is in idle state [5]. The output waveforms for write and read operation are shown in Fig.2 and 3

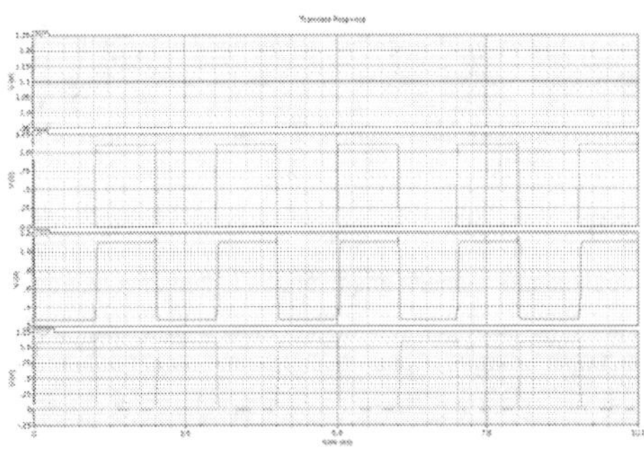

Fig2. Write operation

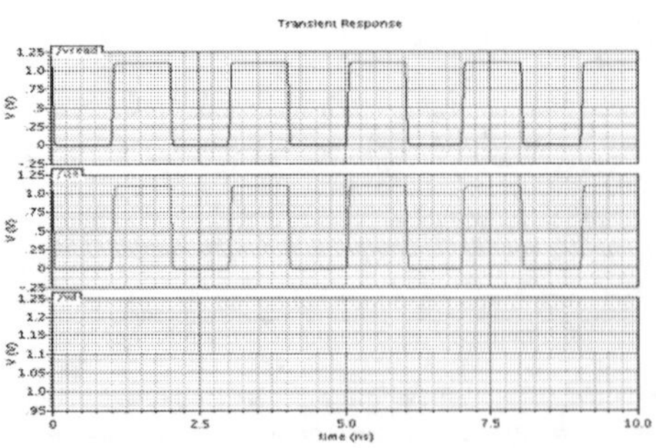

Fig 3. Read Operation

3 Leakage Current

There are various contributors that give rise to leakage currents like pn junction Reverse Bias current, Sub threshold leakage, drain induced barrier lowering, Body-effect, Tunneling into and through gate oxide, Injection of hot carriers, Gate induced Drain Leakage, Punchthrough.But the main contributor to leakage current is Sub thresholdleakage. Sub threshold currentis that current that flows between source and drain terminal in an MOS transistor when the gate voltage is belowthe threshold voltage. Leakage current can be obtained from the equation below when transistor is in off state [6].

$$I_{sub}=I_{oe}^{\frac{vgs-vth+\eta vds-\gamma vsb}{nV\theta}}(1-e^{\frac{-Vds}{V\theta}}) \qquad (1)$$

where, $I_o= \mu C_{ox}(w/l_{eff})V_\theta^2 e^{1.8}$,W= width, L= length, μ = carrier mobility,V_θ is thermal voltage,η is drain induced barrier lowering coefficient, n is sub threshold swing coefficient.V_{TH} is threshold voltage, V_{GS} is gate to source voltage.

3.1 Leakage current is calculated for that particular transistor which is in off condition during the time of operation.

A) To write 1 at Q node

For writing 1 in Qnode, transistor M2 and M3 will be off so we will take leakage current from these two nodes.

Leakage current at M2 transistor=4.27 pA

Leakage current at M3 transistor=-10.43 pA

B) To write 0 at Q node

For writing 0 in Q node, transistor M1 and M4 will be off so we will take leakage current from these two nodes.

Leakage current at M1 transistor=5.66 pA

Leakage current at M4 transistor=-5.57 pA

Leakage currents for writing 0 are shown in Fig.4

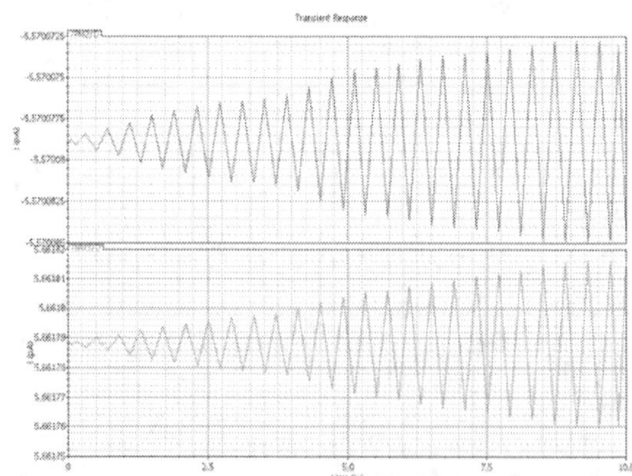

Fig 4. Leakage currents for write ‘0’ at Q node

DOI: 10.1051/matecconf/20165701005

3.2 Power Dissipation

Power dissipation of SRAM cell is the sum total of static and dynamic power dissipation. Static power dissipation occurs due to leakage currents [7].

Dynamic power dissipation is of following types:

a) Switching: Due to the charging and discharging of current that results from the parasitic capacitances.
b) Short circuit: Due to the slow change of inputs i.e. when both the pmos and nmos are on in an inverter this results in short circuit dissipation

Dynamic power dissipation is calculated by giving vpulse on both the bit lines and word line is asserted and doing the transient analysis at 45nm technology [8].

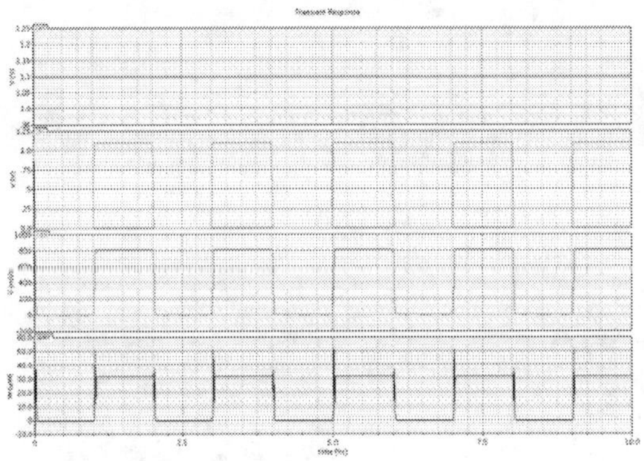

Fig 5. Dynamic Power Dissipation

4 Leakage current and power dissipation control techniques

4.1 Dynamic variable voltage level technique

In this technique supply voltage is not directly connected to the SRAM cell instead a upper dynamic voltage is set using one pMOS connected in parallel with series of two nMOS transistors. Also the ground potential is increased using one nmos connected in parallel with series of two pmos[9].This technique gives '0' Volt at ground node during the active mode and an increased ground voltage during the standby mode. At upper voltage node the technique gives full voltage Vdd in active mode, while the supply voltage is reduced to level Vd in the stand-by mode.

Simulation results showing the reduction in leakage currents are shown in Fig. 6 and Fig. 7

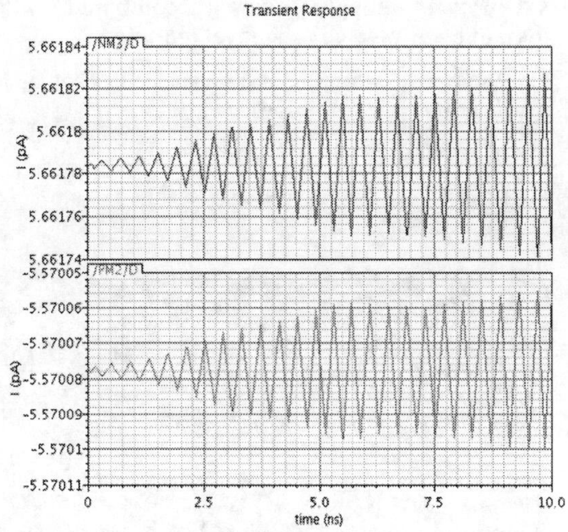

Fig 6. Leakage current for write '0'

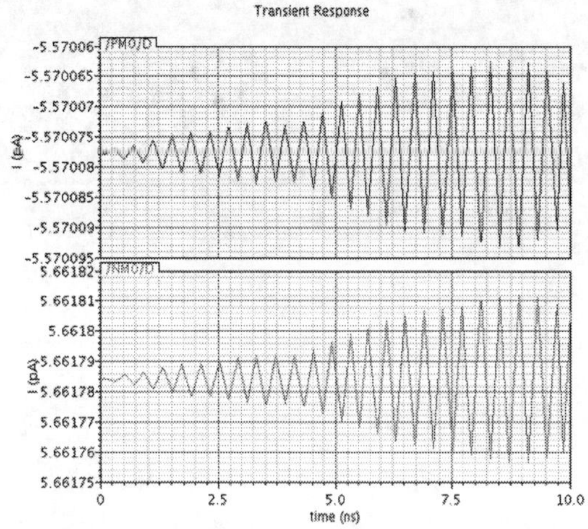

Fig 7. Leakage current for write '1'

Simulation results showing the reduction in power dissipation is shown in Fig. 8

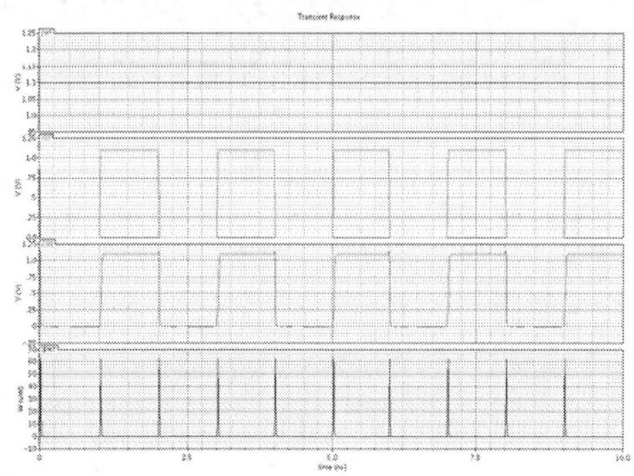

Fig.8 Dynamic Power Dissipation

4.2 Power supply scaling combined with dynamic variable voltage level technique

Fig 9 shows the schematic of the combined technique in which the power supply is lowered to 0.7V.

Fig 10 and 11 shows the leakage currents.

Fig 12 shows the dynamic power dissipation.

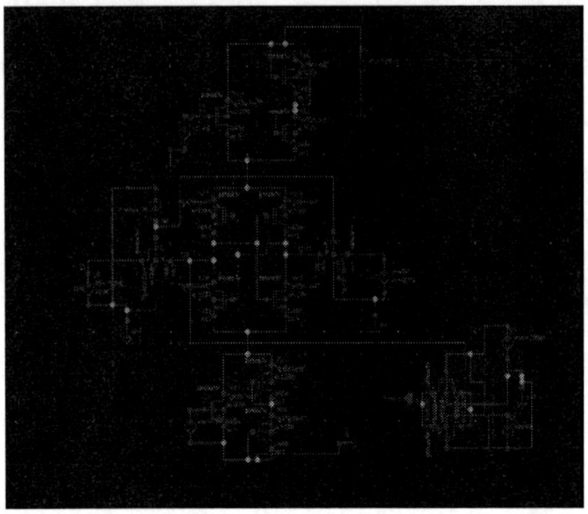

Fig 9. Schematic of proposed cell

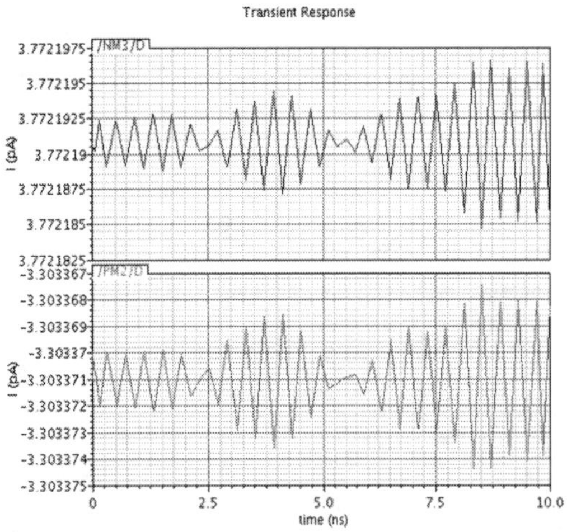

Fig 10. Write '1' leakage current

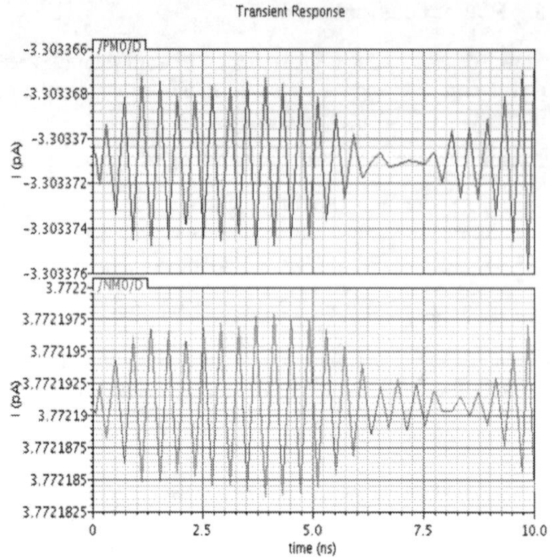

Fig 11. Write '0' Leakage Current

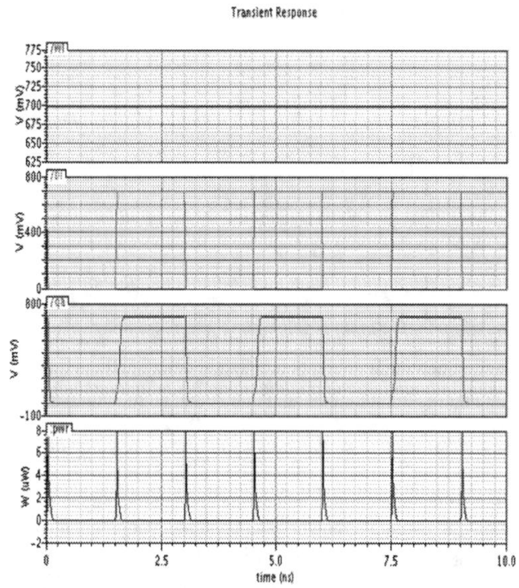

Fig 12. Dynamic Power Dissipation

Table 1 shows the comparison between the conventional 8t SRAM cell and the two techniques proposed. Total power dissipation of the proposed structure shows 98.8% reduction in comparison to conventional 8T and this is due to the power supply scaling combined with dynamic VVL technique.

Table 2 shows the comparison between the leakage currents.

Table 1. Comparison in terms of Dynamic PD

Power Dissipation	Conventional 8T SRAM	SVL technique	Power scaling combined with SVL technique
Static	3.19 uW	32.1 pW	8.42 pW
Dynamic	16.1 uW	736.3 nW	181nW

Table 2. Comparison in terms of Leakage currents

Leakage current	Conventional 8T SRAM	SVL technique	Power scaling combined with SVL technique
Write'1' (nmos, pmos)	4.27 pA, -10.43 nA	5.66pA, -5.57 pA	3.3 pA, -3.3 pA
Write0'(nmos, pmos)	5.66 pA, -5.77 pA	5.66pA, -5.57 pA	3.3 pA, -3.3 p A

5 Conclusion

As the CMOS technology is scaling down to 45nm, power dissipation has become a very crucial issue in the VLSI design. In this paper a low power single ended 8T SRAM cell has been proposed An analysis of leakage current and power dissipation in 8T SRAM cell for 45nm technology shows that leakage currents contribute significantly to the total power dissipation in standby mode. Reducing the voltage supply and increasing the voltage at ground node using dynamic VVL technique for the reduction in leakage currents are examined in detail. It was analyzed that scaling down the supply voltage with the VVL technique gives significant reduction in both static and dynamic power dissipation for write operations. Although delay and area have increased in the proposed SRAM cell but low power dissipation can dominate over this drawback. So, this proposed SRAM cell can be used in wide applications as to provide low power benefits in laptops, mobile phones, memory cards etc. So we can conclude that these techniques have significantly reduced the leakage currents and power dissipation in 8T SRAM cell.

Acknowledgments

This work was done was in CDAC, Mohali by using cadence virtuoso tool.

References

1. B.S. Amrutur and M. Horowitz. *Proc. of IEEE symposium on low power electronics*, pp.92-93,(1994).
2. K.Roy and S. Prasad, *Wiley Interscience Publication*, (2000)
3. C.M.R. Prabhu, A.K. Singh, S.W.Pin, T.C.Hou,In: *Proc. of IEEE symposium on industrial electronics and applications*, pp.68-72,(2009).
4. T.Kawahara, Y.Kawajiri, M. Horiguchi, T.Akiba, G. Kitsukawa, T.Kure, M.Aoki, IEEE journal of solid state circuits, J.E **29**, 6, pp.715-122, (1994).
5. H.Yamauchi, H.Akamatsu, T.Fujita, IEEE journal of solid state circuits,J.E **30**,.4, pp.423-431,(1995).
6. Shin-PaoCheng, Shi-YuHuang, *Proc. of IEEE international workshop on memory technology, design and testing,* pp.135-139,(2005).
7. Hiroki Morirnura, Satoshi Shigernatsu and Shinsuke Konaka, *Proc.of IEEE International Symposium on Low Power Electronics and Design*, pp.12-17,(1999).
8. Neil H. E. Weste, David Money Harris, *Principles of CMOS VLSI Design,(*4thedition:*Addision-Wesley*(2011).
9. Gupta and M. Anis, "Statistical design of the 6T SRAM bit cell," IEEE Transactions on Circuits and Systems I: J.E **.57**, 1,pp. 93–104,(2010)
10. Do Anh-Tuan, Jeremy Yung Shern Low, Joshua Lih Low, Zhi-Hui Kong, Xiaoliang Tan,and Kiat-Seng Yeo, "An 8T Differential SRAM With Improved Noise Margin for Bit-Interleaving in 65nmCMOS," IEEE transactions on circuits and systems ,J.E **58**,6, pp.1252-1263, (2011)

Development of Efficient Resource Allocation Algorithm in Chunk Based OFDMA System

Mukesh Kumar Yadav[1], Garima Saini[2]

[1] M.E. Student, NITTTR Chandigarh, India,
[2] Assistant Professor, NITTTR Chandigarh, India,

Abstract. The emerging demand for diverse data applications in next generation wireless networks entails both high data rate wireless connections and intelligent multiuser scheduling designs. The orthogonal frequency division multiple access based system is capable of delivering high speed data rate and can operate in a multipath environment. OFDMA based system dividing an entire channel into many orthogonal narrow band subcarriers. Due to this, it is useful to eliminate inter symbol interferences which is a limit of total available data rates. In this paper, investigation about resource allocation problem for the chunk based Orthogonal Frequency Division Multiple Access (OFDMA) wireless multicast systems is done. In this paper, it is expected that the Base Station (BS) has multiple antennas in a Distributed Antenna System (DAS). The allocation unit is a group of contiguous subcarriers (chunk) in conventional OFDMA systems. The aim of this investigation is to develop an efficient resource allocation algorithm to maximize the total throughput and minimize the average outage probability over a chunk with respect to average Bit Error Rate (BER and total available power).

Keywords— MISO-OFDMA, multicast system, DAS, multiuser diversity, chunk allocation.

1 INTRODUCTION

The systems based on OFDMA, are able to deliver high data rate and can operate in the hostile multipath radio environment. OFDMA-based systems allow efficient sharing of limited resources among multiple users such as spectrum and transmit power [1]-[3].

With various Quality-of-Service (QoS) requirements OFDMA has been developed to support various multimedia applications. The frequency band divides into a group of mutually orthogonal subcarriers in OFDMA, each group having a much lower bandwidth than the coherence bandwidth of the channel. It provides better protection facility to inter symbol interference and frequency selective fading. Each user is dynamically assigned to a subset of subcarriers in multi-user environment in each frame which take advantage of the fact that at any time, the channel responses are different for different users and at different subcarriers [2]-[7].

In Chunk Based OFDMA Resource Allocation with Single Antenna Scenario, the allocation algorithm allocated chunks to users according to their average Signal to Noise Ratio within each chunk, where Bit Error Rate (BER) is make-sure within each chunk. Chunk based resource allocation is applied not only to the single antenna scenario but also to the multiple antennas [8]-[10]. The resource allocation in chunk based OFDMA with single antenna scenario has low throughput. Total system throughput can be increased by placing BS's multiple antennas at different locations. Resource Allocation is performed centralized in Distributed Antenna System and the available resources are used more efficiently. It is possible because of the different spreading environments across distributed antennas that to make better the wireless channels of users. Through a single transmission, data can be transmitted from each distributed antenna of the base station to multiple mobile users only in multicast systems [11]-[13].

2 PROPOSED MODEL FOR EFFICIENT RESOURCE ALLOCATION ALGORITHM

To design an efficient algorithm for resource allocation in OFDMA, consider two cells with N total number of subcarriers, T is the number of distributed antennas, K is the total number of active users. Assume the overall bandwidth is B, the total transmitting power is P_{total}, and the one sided power spectral density of additive white Gaussian noise is N_o.

[1]Email: mukeshece25@gmail.com

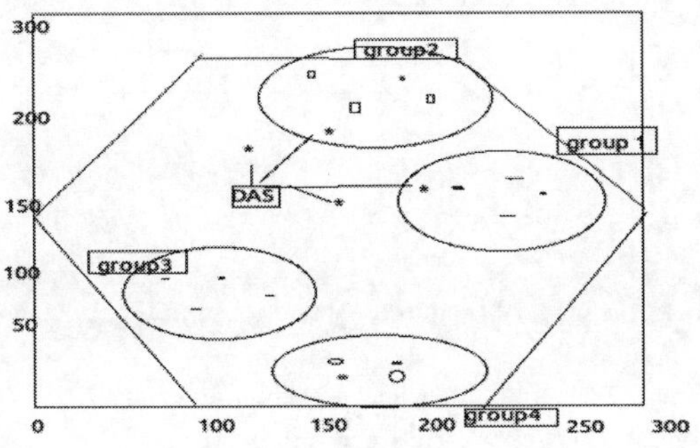

Figure 1. Cell-1 for N=1024, T=4, K=20, G=4

In figure 1, a cell-1 with total number of subcarriers N= 1024, total number of distributed antennas T=4, total number of active users K= 20, total available bandwidth B= 100MHz, group of subcarriers G=4, frequency separation between two contiguous subcarriers is df= 100MHz/1024 =97.6 KHz. Group of subcarriers (Chunk) is done randomly by taking the minimum number of active users 1 and maximum number of active users are 7. Users are placed in cell-1 area uniformly. In DAS (Distributed Antenna System) one antenna is placed at the center of the cell and other antennas are placed at a fixed distance from the center base antenna. For all users in cell-1, coherence bandwidth f_c is same. Here two different values of coherence bandwidth f_c is examined, f_c = 1.95 MHz and f_c = 0.49 MHz. Path loss exponent for cell-1 is 5.

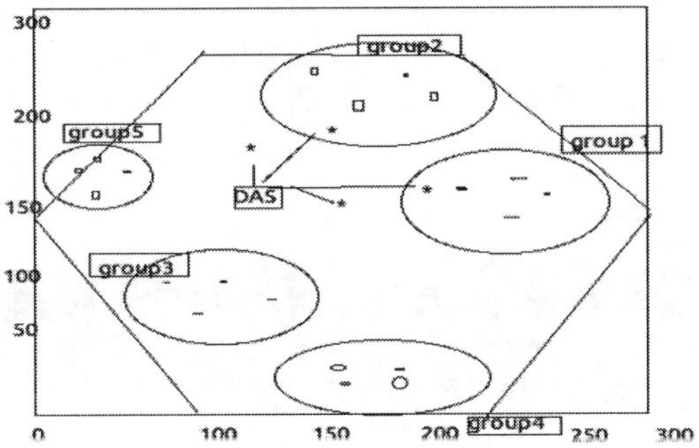

Figure 2. Cell-2 for N=1024, T=4, K=20, G=5

In figure 2, cell-2 with total number of subcarriers N= 1024, total number of distributed antennas T=4, total number of active users K= 20, total available bandwidth B= 100MHz, group of subcarriers G=5, frequency separation between two contiguous subcarriers is df= 100MHz/1024 =97.6 KHz. Chunk is done randomly by taking the minimum number of active users is 1 and maximum number of active users are 7. Users are placed in cell-2 area uniformly. In DAS (Distributed Antenna System) one antenna is placed at the center of the cell and other antennas are placed at a fixed distance from the center base antenna. For all users in cell-2, coherence bandwidth f_c is same. Here two different values of coherence bandwidth f_c is examined, f_c = 1.95 MHz and f_c = 0.49 MHz. Path loss exponent for cell-2 is 5.

Let us consider the T ×1 complex Gaussian distribution frequency response vector is $h_{k,n} = [h^1_{k,n}, h^2_{k,n}, h^3_{k,n} \ldots\ldots h^T_{k,n}]^T$ between user of k and base station in subcarrier n. The magnitude of this complex Gaussian distribution frequency response vector $h_{k,n}$ will be $|h_{k,n}|$. This magnitude function will follow $E[|h_{k,n}|^2]$ which is the Rayleigh distribution.

Channel gain vector vector between the user k and the base station is given by T ×1 matrix.

$g_{k,n} = [g^1_{k,n}, g^2_{k,n}, g^3_{k,n} \ldots\ldots g^T_{k,n}]^T$ is (1)

Where

$g^t_{k,n} = D^{-\alpha}_{k,t} |{}^{ht}_{k,n}|^2$ (2)

for t=1,2,3.....T

$D^{-\alpha}_{k,t}$= path loss

α= path loss exponent

By using a single transmission, from each distributed antenna of the base station (BS) data can be transmitted to multiple mobile users in multicast wireless communication system. In this case, the number of active users K are grouped in a number of chunks G. These all users are related to set

$K = U^G_{g=1} K_g$ (3)

and $|K| = U^G_{g=1} |K_g|$ (4)

where K_g= user set of group g

In a multicast group, the channel quality of every user may be different. Within each group, the base station of distributed antenna system transmits data rate at the lowest of all users. In each distributed antenna, this data rate is calculated by the user with the smallest channel gain.

The equivalent channel gain vector of g will be

$a_{g,n} = [a^1_{g,n}, a^2_{g,n}, a3_{g,n} \ldots\ldots a^T_{g,n}]^T$ (5)

where $a^1_{g,n} = \min_{k \in Kg} g^t_{k,n}$ for t=1,2,3.......T

The equivalence model of the baseband for the system by using beam forming will be

$y_n = A_n W_n D_n s_n + z_n$ (6)

where

y_n = received signal vector of G×1 matrix

A_n= G×T channel matrix= $[a_{1,n}, a_{2,n}, a_{3,n} \ldots\ldots a_{G,n}]$

W_n = T×G beam forming weight matrix = $[w_{1,n}, w_{2,n}, w_{3,n} \ldots\ldots w_{G,n}]$

$w_{g,n}$ = T×1 beam forming weight vectors in subcarrier n for group g

$w_{g,n} = [w^1_{g,n}, w^2_{g,n}, w3_{g,n} \ldots\ldots w^T_{g,n}]^T$ (7)

D_n = power distribution among G multicast groups to subcarrier n

s_n = transmitting signal vector of G×1 matrix

z_n = noise vector of G×1 matrix

In the proposed scheme, L-ary QAM (L-ary Quadrature Amplitude Modulation) is used as a modulation scheme, where L is the modulation levels.

$L = \{0, 2^2, 2^4 \ldots\ldots 2^b \ldots\ldots, 2^B\}$ (8)

Where in QAM, b is the number of bits

i. When the number of bits b is equal to 0 then than there will be no transmission.
ii. When the number of bits b is equal to B then the transmission rate will be equal to B. It means the system will get highest modulation level.

By using Zero Forcing (ZF) beam forming, if G>T, this case is not use because $A^{*}_{n}(A_n A^{*}_{n})^{-1}$ will be a singular. Due to this, it will be a need to select t out of G multicast groups where t less than or equal to G in each subcarrier. Due to this, there will be I possible number of combinations of multicast groups which will transmit to the same subcarrier B_n.

If G is less than or equal to T and rank of A_n is equal to G than beam forming matrix will be

$$W_n = A^{*}_{n}(A_n A^{*}_{n})^{-1} \quad (9)$$

A set of multicast groups

$B_n = \{s_1, s_2, s_3 \ldots\ldots\ldots\ldots S_t\}$ in each subcarrier like

$$An(Bn) = \{a_{s1,n}, a_{s1,n}, a_{s1,n}, \ldots\ldots\ldots\ldots a_{st,n}\} \quad (10)$$

When Zero Forcing is used than the effective channel of multicast group g will be

$$C_{g,n}(B_n) = \{[(An(Bn)\,An(Bn)^{*})^{-1}]_{g,g}\}^{-1} \quad (11)$$

If it is used water filling equation for the calculation of total power than Zero Forcing beam forming matrix becomes

$$W_n(B_n) = A_n(Bn)^{*}(An(Bn)\,An(Bn)^{*})^{-1} \quad (12)$$

To reduce the system overhead, the N numbers of subcarriers are grouped into C chunks. Usually the coherence bandwidth f_c exceeds the subcarrier bandwidth f_s. Assume that the total number of chunks C= N/N' are integer, where N' is the total number of subcarriers in a chunk.

Steps for calculation of Average throughput and Outage Probability: Take value for all parameters. Set number of cell=2, set dummy array at u={1,2....7} and set minimum active user and maximum active user, $K_{min.}$=1 and $K_{max.}$=7. Resource allocation is done by using allocation = histc(user_subband.*user_activated, 1:N)'.

1. Set number of chunks M={1,2....G).
2. Calculate BER, BER= e/T', where e is the number of bit error and T is the total no of transfer bit during studied time interval.
3. Calculate SNR by using SNR= $(1/BER)^m$, where m is the number of modulation scheme.
4. Calculate throughput by using throughput=$(1-BER)^n$, where n is the number of subcarriers.
5. For a multicast group, calculate the average throughput by using average throughput=$1/n\sum^{n}_{i=1}(a1+a2\ldots\ldots.a7)$, where a1, a2....a7 are the values of throughput, and calculate average outage probability.

3 SIMULATION AND RESULTS

The performance of the proposed algorithm assess by using simulation in MATLAB. Number of subcarriers N=1024, Number of Antennas T=4, Bandwidth B=100MHz, Coherence Bandwidth f_c = 1.95MHz and 0.49MHz, Frequency Separation df = 97.6 KHz, P ath Loss Exponent α = 5, $M_{min.}$, number of users in a chunk $K_{min.}$ = 1, M_{max}. number of users in a chunk $K_{min.}$ = 7

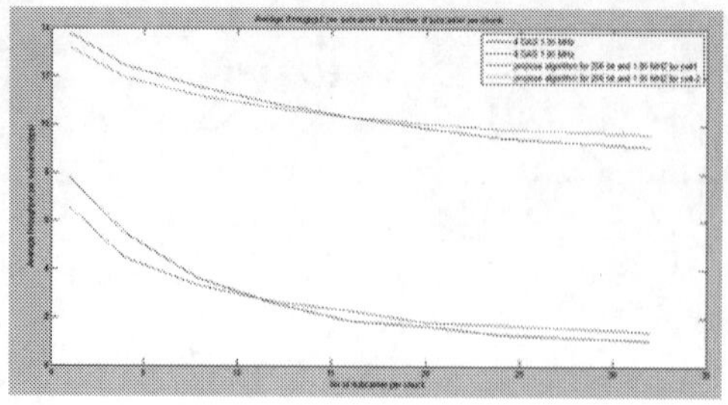

Figure 3. Average throughput per subcarrier vs number of subcarriers per chunk for f_c = 1.95MHz

As Shown in figure 3, a graph is plotted between average throughput per subcarrier and number of subcarrier per chunk at coherence bandwidth f_c = 1.95MHz. In this, as increasing the number of chunks the Average throughput per subcarrier decreases. By comparing the results, as number of subcarriers per chunk are 18 than the average throughput for 256-DAS in cell-1 (number of chunks are 4) gives better throughput than 256-DAS in cell-2 (number of chunks are 5), 4-DAS and 8-DAS. As increasing the number of subcarriers per chunk from 18, the average throughput for 256-DAS in cell-2 is better than 256-DAS in cell-1, 4-DAS and 8-DAS. As number of subcarriers per chunk is 12, at this point the average throughput per subcarrier is 10.8811 for 256-DAS in cell-1 and 10.332 for 256-DAS in cell-2 which are much better than, 8-DAS and 4-DAS.

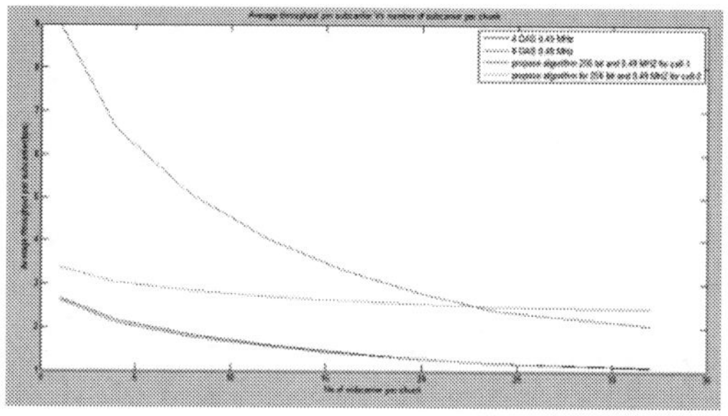

Figure 4. Average throughput per subcarrier vs number of subcarriers per chunk for f_c = 0.49MHz

In figure 4, as increasing the number of chunks the Average throughput per subcarrier decreases. By comparing the results, the average throughput for 256-DAS in cell-1 (number of chunks are 4) gives better throughput than 256-DAS in cell-2 (number of chunks are 5), 4-DAS and 8-DAS. As number of subcarriers per chunk is 12, at this point the average throughput per subcarrier is 4.0309 for 256-DAS in cell-1 which is much better than 256-DAS in cell-2, 8-DAS and 4-DAS.

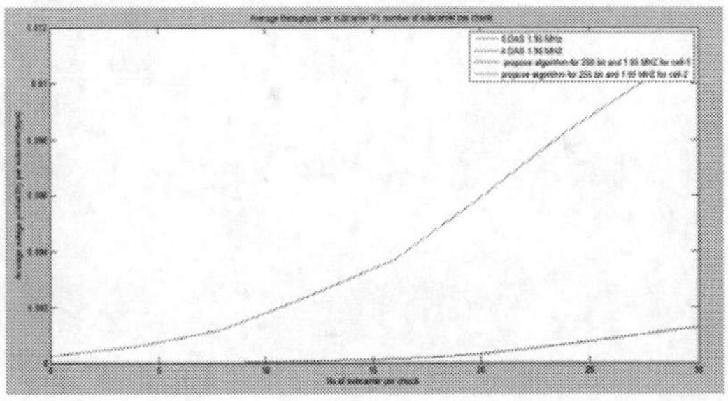

Figure 5. Average outage probability per subcarrier vs number of subcarriers per chunk for f_c = 1.95MHz

In figure 5, as increasing the number of chunks the Average outage probability per subcarrier increases. By comparing the results, the average outage probability for 256-DAS in cell-1 (number of chunks are 4) is much less than the 256-DAS in cell-2 (number of chunks are 5), 4-DAS and 8-DAS. As number of subcarriers per chunk is 12, at this point the average outage probability per subcarrier is $.05\times10^{-6}$ for 256-DAS in cell-1 which is much better than 256-DAS in cell-2, 8-DAS and 4-DAS.

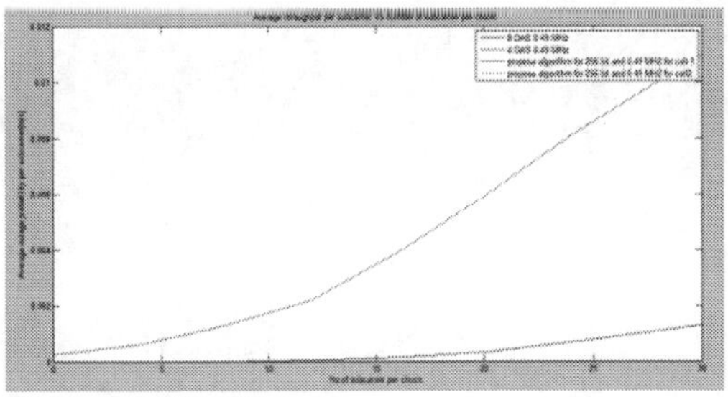

Figure 6. Average outage probability per subcarrier vs number of subcarriers per chunk for f_c = 0.49MHz

In figure 6, as increasing the number of chunks the Average outage probability per subcarrier increases. By comparing the results, the average outage probability for 256-DAS in cell-1 (number of chunks are 4) is much less than the 256-DAS in cell-2 (number of chunks are 5), 4-DAS and 8-DAS. As number of subcarriers per chunk is 12, at this point the average outage probability per subcarrier is 0.33×10^{-6} for 256-DAS in cell-1 which is much better than 256-DAS in cell-2, 8-DAS and 4-DAS.

4 CONCLUSION

The importance of OFDMA system is high throughput and low outage probability. The proposed efficient resource allocation algorithm is based on Distribution Antenna System (DAS) with 256-QAM modulation scheme. This proposed scheme is compared with various modulation scheme and different number of chunks in different cells at coherence bandwidth f_c =1.95MHz and f_c =0.49MHz.The average throughput per subcarrier and average outage probability are calculated with respect to the number of subcarrier per chunk at coherence bandwidth f_c = 1.95MHz and f_c = 0.49MHz. In this, as increasing the number of chunks the average throughput per subcarrier decreases but average outage probability increases. By comparing the results, the average throughput and average outage probability for 256-DAS in cell-1(number of chunks are 4) and 256-DAS in cell-2 (number of chunks are 5) is much better than 4-DAS and 8-DAS.

References

1. Ender Bolat, Thesis on "Study of OFDM Performance over AWGN Channels", Eastern Mediterranean University, Cyprus, 2003.
2. Srikanth S., Kumaran V. Manikandan V. Murugesapandian, Tutorial Paper on "OFDMA", Anna University, Chennai, 2005.
3. Victor C. M. Leung, Alister G. Burr, Lingyang Song, Yan Zhang, and Thomas Michael Bohnert "OFDMA Architectures, Protocols and Applications", Article ID 703083, 2009.
4. Quang Duy La ,Boon Hee Soong, Yong Huat Chew, and Woon Hau Chin, "Orthogonal frequency division multiple access fundamentals and applications", International Standard Book Number: 978-1-4200-8824-3.
5. J. Y. Kim, T. Kwon and D. H. Cho, "Resource allocation scheme for minimizing power consumption in OFDM multicast systems," IEEE Communication Letter, vol. 11, no. 6, pp. 486–488, 2007.
6. C. Suh and J. Mo, "Resource allocation for multicast services in multicarrier wireless communications," IEEE Transaction Wireless Communication, vol. 7, no. 1, pp. 27–31, 2008.
7. D. T. Ngo, C. Tellambura, and H. H. Hguyen, "Efficient resource allocation for OFDMA multicast systems with spectrum-sharing control," IEEE Transaction, vol. 58, no. 9, pp. 4878–4889, 2009.
8. Vasileios D. Papoutsis, Stavros A. Kotsopoulos, "Chunk-Based Resource Allocation in Multicast OFDMA Systems with Average BER Constraint", IEEE Communication Letter, Vol.15, No.5, pp.551-553, 2011.
9. S. Karachontzitis and D. Toumpakaris, "Efficient and low complexity user selection for the multiuser MISO downlink," IEEE Personal, Indoor Mobile Radio Communication Symposium, 2009.
10. H. Zhu, S. Karachontzitis, and D. Toumpakaris, "Low-complexity resource allocation and its application to distributed antenna systems," IEEE Wireless Communication, vol. 17, no. 3, pp. 44–50, 2010.
11. V. D. Papoutsis, I. G. Fraimis, and S. A. Kotsopoulos, "User selection and resource allocation algorithm with fairness in MISO-OFDMA," IEEE Communication Letter, vol. 14, no. 5, pp. 411–413, 2010.
12. Vasileios D. Papoutsis, Alexia P. Stamouli, "Chunk-Based Resource Allocation in Multicast MISO-OFDMA with Average BER Constraint", IEEE Communication Letter, Vol.17, No.2, pp.317-320,

2013.
13. V.D. Papoutsis, S.A. Kotsopoulos, "Chunk-Based Resource Allocation in Distributed MISO-OFDMA Systems with Fairness Guarantee", IEEE Communication Letter, Vol.15, No.4, pp. 377–379, 2011.

Synthesis and Optical Properties of CuO Nanocrystals with Controllable Shapes and Size

Neelinder Goraya[1], Satbir Singh[2]

[1,2]*Department of Electronics and Communication, GNDU-Regional Campus, Gurdaspur, Punjab, India*

Abstract: Highly dispersed copper oxide (CuO) nanoparticles had been successfully prepared by a hydrothermal technique where different adding temperatures of NaOH were taken. The as-prepared CuO nanoparticles were characterized by X-ray diffraction (XRD), transmission electron microscopy (TEM) and Raman spectroscopy. The results showed that the adding temperature of NaOH has an important influence on the shape and size of CuO nanocrystals The influence of reaction conditions on morphology of CuO nanocrystals was discussed. Furthermore, different shapes of obtained CuO nanocrystals exhibit different Raman properties.

1 Introduction

Nano- structured transition metal oxides (MOs) are the indisputable prerequisite for the development of various novel functional and smart materials. These transition MO nanocrystals have been attracting much attention not only for fundamental scientific research, but also for various practical applications because of their unique physical and chemical properties [1]. As an important p-type semiconductor metal oxide with a narrow band gap (1.2 eV), copper oxide (CuO) has been extensively studied because of its diverse applications as material for heterogeneous catalysts, gas sensors, optical switch, magnetic storage media, field emission devices, solar cells, etc[2]. CuO nanostructures are also considered as electrode materials for the next-generation rechargeable lithium-ion batteries (LIBs) because of their high theoretical capacity, safety, and environmental friendliness. They are also promising materials for the fabrication of solar cells because of their high solar absorbance, low thermal emittance, relatively good electrical properties, and high carrier concentration.

Research on nanoparticles has gathered wide attention during the last decade because of their unusual and size dependent optical, magnetic, electronic, and chemical properties. To fully utilize these properties, the size and shape must be well controlled. Some synthesis methods are successful for fabricating shape-controlled nanocrystals of metals, semiconductors and complex metal oxides [3]. In recent years, the shape controlled synthesis of nanostructured CuO has attracted considerable attention. However, the exciting techniques have to rely on the proper selection of expensive surfactants or high crystallization temperatures. Therefore, seeking a simple strategy for low-cost, low-temperature, controlled synthesis of oxide nanostructures at mild conditions is highly desired, and it is important as well for exploring zero- and one-dimensional oxide-based nanostructures for applications in nanodevices [4]. Here, we describe a simple procedure for the synthesis of shape-controlled CuO nanocrystals. Here we use hydrothermal synthesis of CuO nanoparticles which is generally based on a two-step process. First, cupric hydroxide [$Cu(OH)_2$] particles are formed by the reaction of a cupric salt precursor with a basic solution, such as sodium hydroxide (NaOH) or ammonium hydroxide. The $Cu(OH)_2$ particles are then thermally dehydrated in an autoclave at fixed temperatures to obtain the final CuO nanoparticles. With the hydrothermal techniques different sized CuO nanocrystals can be obtained by modulating the adding temperatures of NaOH. The method provides a convenient, low-cost, nontoxic route for the synthesis of nanostructures of oxide materials.

2 Experimental

2.1. Sample Preparation

Analytically pure $Cu(NO_3)_2$ and NaOH were used as the starting materials without further purification. In a typical procedure, two identical sets of 300mL of 0.02M $Cu(NO_3)_2$ solution were prepared by dissolving $Cu(NO_3)_2{\cdot}3H2O$ in deionized water. Each set of solution was added into a round-bottom flask equipped with a refluxing device. The two sets of $Cu(NO_3)_2$ solution were kept at an appropriate temperature (60°C and 100°C, respectively) with vigorous stirring; then 0.50 g of solid NaOH (platelets) was rapidly added into each set of solution, where a large amount of black precipitate was

[1]Neelinder Goraya: gneelinder@yahoo.com

DOI: 10.1051/matecconf/20165701007

simultaneously produced and maintained at the crystallization temperature for 10 min. After all reactions were completed, the resulting products were centrifuged, washed with water and ethanol for several times, and dried in air at room temperature.

2.2 Characterization

Crystal structure identification was made by X-ray diffraction (XRD) using a Bruker D8 Advance X-ray diffractometer with Cu K_ (λ = 0.15405 nm) radiation. Transmission electron microscopy (TEM) measurements were performed on a JEOL JEM-2100 microscope operating at 200 kV, by depositing a drop of CuO dispersion onto 300 mesh Cu grids coated with carbon layer. Laser Raman Spectroscopy (LRS) studies were carried out using a Renishaw Invia Raman spectrometer from 100 to 700 cm−1 at room temperature.

3. Results and discussion

XRD pattern (Fig. 1) of the products obtained is identical to the single-phase CuO with a monoclinic structure and the diffraction data are in good agreement with JCPDS card of CuO (JCPDS 80-1268). No peaks of impurity are observed in the XRD pattern. The broadening of the peaks indicates the small size of the products [5]. Actually, the average sizes of the CuO nanocrystals prepared at adding temperatures of 60°C and 100°C, are estimated to be 15 and 16 nm, respectively, according to the Scherrer equation. It was found that the adding temperature of NaOH has an important influence on the shape and size of CuO nanocrystals.

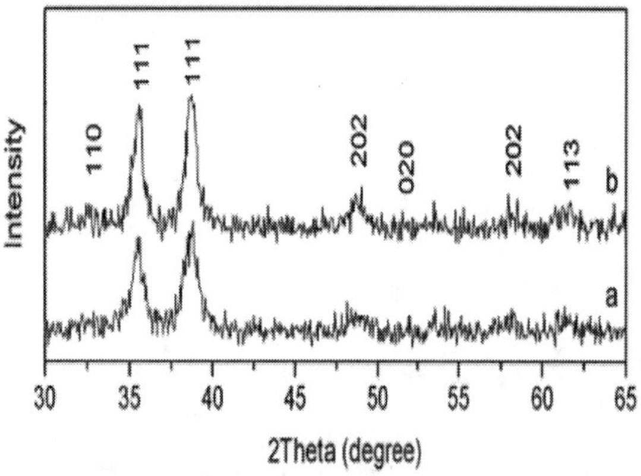

Figure 1. XRD patterns of CuO products prepared at different adding temperatures of NaOH: (a) 100 °C [4]; (b) 60 °C [4]

As can be seen from Fig. 2a, the nanorods with diameters of 5 nm and lengths of 25–40 nm were formed when NaOH was added at 100°C. When the adding temperature decreased to 60°C, the CuO nanobelts with diameters of 8–40 nm and lengths up to 1.0 μm were obtained (Fig. 2b).

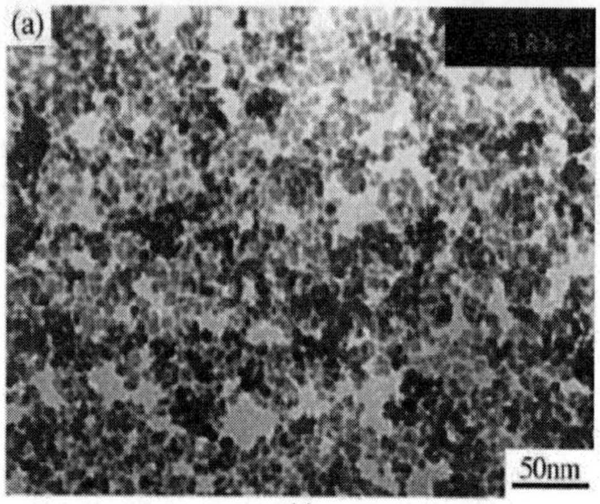

(a)

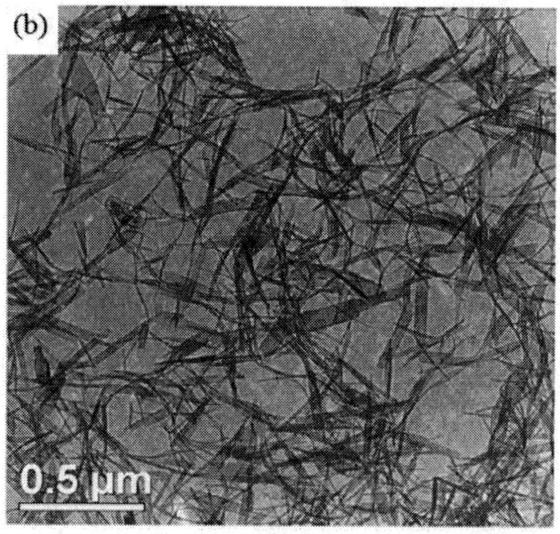

(b)

(c)

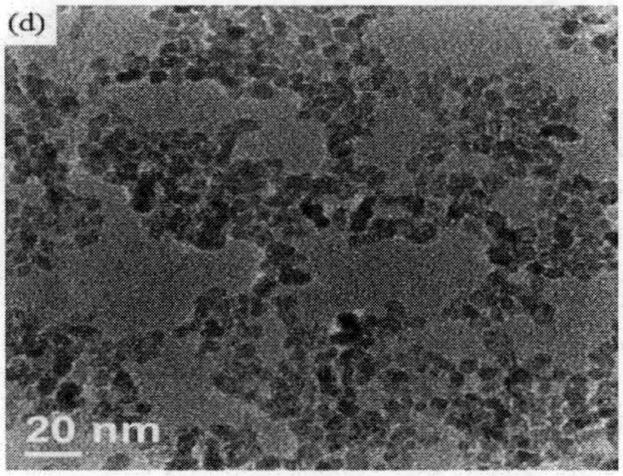

(d)

Figure 2. TEM images of the as-prepared CuO nanoparticles: **(a)** adding NaOH at 100°C [4]; **(b)** adding NaOH at 60°C [4]; **(c)** in the absence of glacial acetic acid and NaOH [6]; **(d)** in the presence of glacial acetic acid and NaOH at 100°C [4].

Typical TEM image of the as-prepared CuO nanoparticles is presented in Fig. 2, showing that the products are composed of spherical particles with narrow size distribution and high dispersion. The size of the particles observed in the TEM image is in the range of 5–7 nm, which is in good agreement with that estimated by Scherrer equation from the XRD pattern.

It is also important to point out that adding a small quantity of glacial acetic acid has an important influence on the shapes of CuO crystals. As a matter of fact, the CuO sample had been prepared by adding solid NaOH into $Cu(NO_3)_2$ solution at 100°C in the presence of 1mL glacial acetic acid. The TEM analysis shows that the obtained black precipitate consists of 8 nm sized spherical particles. In this method, the additive glacial acetic acid was used to prevent the hydrolysis of the copper acetate solution. If glacial acetic acid was not added, the copper acetate solution would directly hydrolyze above 80°C in the absence of NaOH and form well-dispersed needle-shaped CuO nanocrystals as observed in the TEM image (Fig. 2c). The result also indicates that different morphologies of CuO nanocrystals can be obtained easily by controlling the addition of glacial acetic acid and NaOH.

Raman spectroscopy is a very useful tool for the study of phases and structures of oxide systems. The Raman frequency can shift, and bandwidth can broaden, with the decreasing microcrystal size [7-8]. Raman spectra of CuO products prepared at different adding temperatures of NaOH are illustrated in Fig. 3. For the monoclinic CuO, this belongs to the C^6_{2h} space group with two molecules per unit cell. There are nine zone-center optical phonon modes with symmetries 4Au +5Bu +Ag +2Bg. Only three Ag +2Bg modes are Raman active [9-10]. For the sample a (Fig. 3), it can be seen that there are three Raman peaks at 283, 330 and 619 cm−1, and these peaks can be assigned to one Ag and two Bg modes, respectively. And the Raman peaks of sample b (Fig. 3) have a blue shift and shift to 290, 337 and 623 cm−1, respectively. It is clear that, in this case, the reaction temperature plays an important role in the formation of highly dispersed CuO nanoparticles.

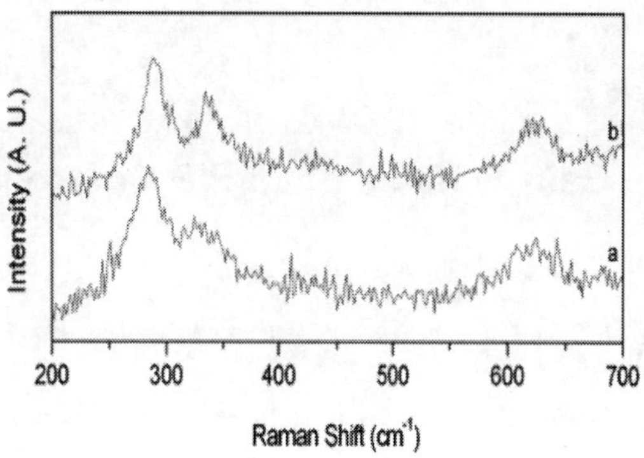

Figure 3. Raman spectra of CuO samples illustrated in Figs. 1 and 2 [4]

4. Conclusion

Well-dispersed CuO nanocrystals have been synthesized in aqueous solution via a simple hydrothermal procedure. Their shapes can be easily controlled. The shapes of CuO nanocrystals change as adding temperature of NaOH change without any additional surfactants. The adding temperatures of NaOH are critical parameters to the shape-guiding processes.

References

1. Qiaobao Zhang, Kaili Zhang, Daguo Xu, Guangcheng Yang, Hui Huang, Fude Nie, Chenmin Liu, Shihe Yang, Progress in Materials Science **60**, 208–337 (2014)
2. Xue Wang, Chenguo Hu, Hong Liu, Guojun Du, Xiaoshan He, Yi Xi, Sensors and Actuators B **144**, 220–225 (2010)
3. L. Manna, E.C. Scher, A.P. Alivisatos, J. Am. Chem. Soc. **122**, 12700 (2000)
4. Junwu Zhu, Huiping Bi, Yanping Wang, Xin Wang, Xujie Yang, Lude Lu, Materials Chemistry and Physics **109**, 34–38 (2008)
5. L.S. Birks, H. Friedman, J. Appl. Phys. **17**, 687 (1946)
6. Junwu Zhu, Dan Li, Haiqun Chen, Xujie Yang, Lude Lu, Xin Wang, Materials Letters **58**, 3324– 3327 (2004)
7. A. Tanaka, S. Onari, T. Arai, Phys. Rev. B **45**, 6587 (1992)
8. A. Tanaka, S. Onari, T. Arai, Phys. Rev. B **47,** 1237 (1993)
9. Z. Wang, V. Pischedda, S.K. Saxena, P. Lazor, Solid State Commun. **121**, 275 (2002).
10. J.F. Xu, W. Ji, Z.X. Shen, S.H. Tang, X.R. Ye, D.Z. Jia, X.Q. Xin, J. Solid State Chem. **147**, 516 (1999)

Gain Scheduling of PID Controller Based on Fuzzy Systems

Sandeep Singh[1], Mandeep Kaur[2]

[1]PG Student, Department of Electronics & Communication Engineering, Punjab Institute of Technology, Rajpura, 140401

[2]Assistant Professor, Department of Electronics & Communication Engineering, Punjab Institute of Technology, Rajpura

Abstract. This paper aims to utilize fuzzy rules and reasoning to determine the controller parameters and the PID controller generates the control signal. The objective of this study is to simulate the proposed scheme on various processes and arrive at results providing better response of the system when compared with best industrial auto-tuning technique: Ziegler-Nichols. The proposed scheme is based upon the Ultimate Gain (K_u) and the Period (T_u) of the system. The error and rate of change in error gains are tuned manually to get the desired response using LabVIEW. This can also be done with various optimization techniques. A thumb rule for choosing the ranges for K_c, K_d and K_i has been obtained experimentally.

Keywords: PID Controller, control systems, fuzzy gain scheduling

1 Introduction

Control systems play an integral role in wide range of industrial and experimental applications. For any process to be carried out smoothly and efficiently, it is required to continuously monitor the system variables pertinent to the system. A manufacturing plant would generally require that a precise flow of some fluid being used in the manufacturing process maintain a stable flow-rate or temperature. Various control schemes are available for the control of liquid level and flow process. Among them, the commonly used scheme is the conventional PID (Proportional-Integral-Derivative) control which is the most dominant and effective. The design of such a controller specifically requires three parameters i.e. proportional gain, integral time constant and derivative time constant [1]. Due to their linear structure, the conventional PID controllers are usually not effective if the processes involved are higher order and time-delay systems, nonlinear systems, complex and vague systems without precise mathematical models, and systems with uncertainties [2]. So for the overall improved performance, fuzzy PID controllers are preferred. Fuzzy logic is used to realise the scheduling algorithms for the controller parameters because it allows smoothly varying non-linear functions to be created [3]. Also in the fuzzy control, the linguistic descriptions of human expertise in controlling a process are represented by fuzzy IF-THEN rules [4]. So with this an expert control system is designed. An adaptive PID controller has been constructed as an initial step in the development of this implementation [5]. The parameters as well as the required number of fuzzy rules in the fuzzy inference system can simultaneously learned by the accumulated GA [6]. Fuzzy PID controllers can be divided into two parts i.e. Gain Scheduling and Direct Action. For the gain scheduling type of fuzzy PID controllers, the PID gains are tuned based on a fuzzy inference system rather than the conventional Ziegler and Nichols's approach [7, 8]. An adaptive fuzzy gain scheduling scheme for conventional PI and optimal controllers has been studied. The proposed controller offers better performance than fixed gain controllers [9]. In present research work, PID controllers are proposed based on fuzzy gain scheduling for process control using LabVIEW 2014 [10].

[2] Corresponding author: benipal1@gmail.com

2 PID Controllers

The derivative control action utilizes the rate of change of error signal. Derivative control brings a stabilizing effect to the system. It is sometimes referred to as 'anticipatory control'. The combined mathematical expressions for the proportional, integral and derivative control action in the Laplace domain is:

$$mv(s)/e(s) \quad = K_c[\,1 + 1/\tau_I s \; + \; \tau_d s\,] \qquad (1)$$

and in the time domain is :

$$mv(t) = mv_{ss} + K_c\,[\,e(t) + 1/\tau_I \textstyle\int e(t)\,dt + \tau_d\{de(t)/d(t)\}] \quad (2)$$

where, mv(t) is the measured variable, mv_{ss} is the steady state value of measured variable, e(t) is the difference between the current value and the set point, K_c is the proportional gain, τ_i is the parameter that scales the integral controller, τ_d is the parameter that scales the derivative controller.

3 Fuzzy Based Gain Scheduling

Fuzzy gain scheduling is all about using fuzzy rules and reasoning to determine the controller parameters based on error signal and rate of change of error signal. Figure 1 shows the PID control system with fuzzy gain scheduler. The approach here is to exploit fuzzy rules and reasoning to generate controller parameters. It is assumed that K_c, K_d, and K_i are in prescribed ranges [(K_{cmin} ,K_{cmax}),(K_{dmin} ,K_{dmax}) and (K_{imin} ,K_{imax})]. K_c , K_i and K_d are normalized into the range between zero and one by the following transformation:

$$K_x = K_{x\,min} + (K_{x\,max} - K_{xmin})K_x' \quad (3)$$

Where x= p , i and d for proportional ,integral and derivative gain respectively. In this scheme, PID parameters are determined based on current error e(k) and its rate of change, $\dot{e}(k)$. The parameters K_c , K_i and K_d are determined by a set of fuzzy rules of form:

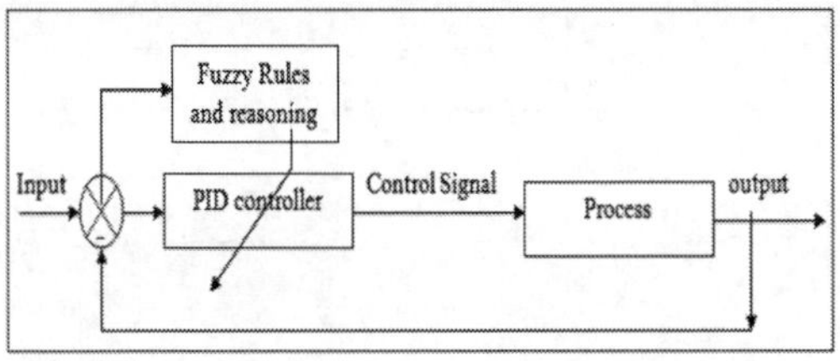

Figure1 PID Controller with Fuzzy Gain Scheduler

If e (k) is A_i and $\dot{e}(k)$ is Bi, then K_c is C_i, K_d is D_i, and K_i= E_i For i= 1, 2……. n.

Here A, B, C and D are fuzzy sets on corresponding supporting sets. The membership functions (MF) of these fuzzy sets are divided into N representing negative ,P positive ,ZO zero, S small, M medium and B big. Thus NM represents negative medium, PB positive big and so on. The fuzzy sets C, D and E may be ether Big, Small or Medium and are characterized by the same membership functions. The fuzzy rules are to be extracted from operator's expertise. Here the rules were driven experimentally based on the step response of the process

Once K_c' ,K_d' and K_i' are obtained , the PID parameters are obtained from following equations:

$$K_c = K_{c\,min} + (K_{c\,max} - K_{cmin})K_c', \quad (4)$$

$$K_d = K_{d\,min} + (K_{d\,max} - K_{dmin})K_d', \quad (5)$$

$$K_i = K_{i\,min} + (K_{i\,max} - K_{imin})K_i' \quad (6)$$

The design process has following steps:

- Defining the control objectives and criterion.
- Determining the input and output relationships and choosing a minimum number of variables for input to the fuzzy logic engine (typically error and rate-of-change-of-error).
- Breaking the control problem down into a series of "if x and y then z" rules, using the rule-based structure of fuzzy logic that define the desired system output response for given system input conditions.
- Creating fuzzy logic membership functions that define the meaning (values) of input/output terms used in the rules.
- Creating the necessary pre- and post-processing fuzzy logic routines if implementing in software, otherwise programming the rules into the fuzzy logic hardware engine.
- Testing the system, evaluating the results, tuning the rules and membership functions, and re-testing until satisfactory results are obtained.

Within fuzzy controller three separate schedulers are created for proportional (K_c), derivative (K_d) and integral (K_i) gains. Each of them has two inputs and one output with inputs having 7 membership functions each and output having 3 membership functions. Each of them is separately discussed below.

3.1 Choice of Membership Functions

Membership function basically comprises of the following two parts:

3.1.1 Input Membership Functions

Current error, e (k) and its rate of change, $\dot{e}(k)$ are taken as two inputs with seven membership functions. The most widely used membership functions for inputs are triangular membership functions as it most simple of all and allows linear mapping between crisp input space and fuzzy input space.

3.1.2 Output Membership Functions

Output is defined with the help of three membership functions. Here the most important thing to consider is which type function should be chosen for the output membership function. For that it is required to understand the behaviour of output as needed, the Fuzzy controller should generate the signal so that output response is best optimized. In our study, the fuzzy Controller is generating controller gain i.e. Kc, Ki&Kd based upon the real time value of error and rate of change in error. A comparative study and extensive simulation was done and an approximate graphical relation between output PID Gains and error and rate of change in error was generated as shown in figures 2 & 3.

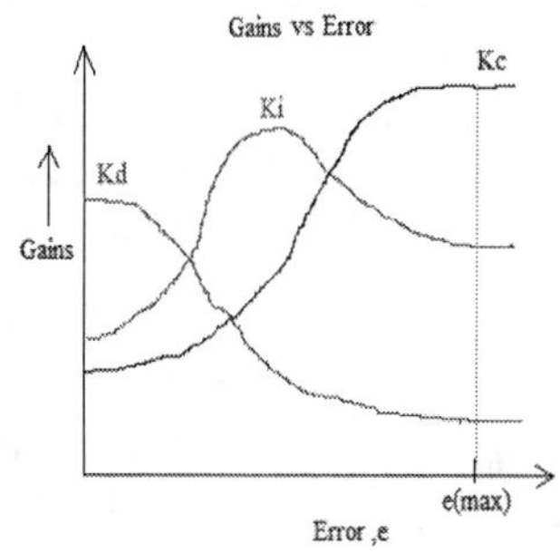

Figure 2 PID Gains vs Error

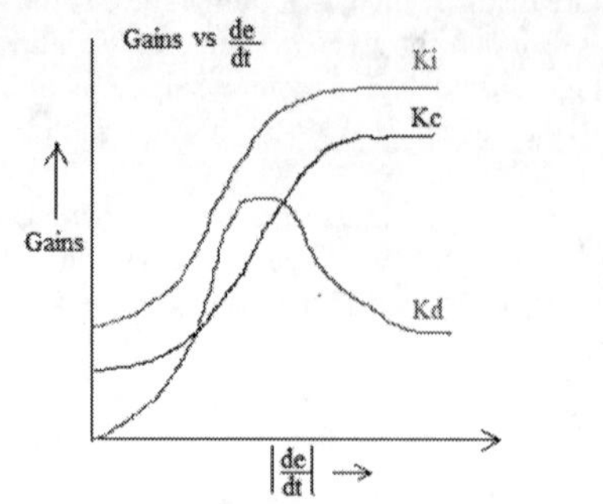

Figure 3 PID Gains vs Rate of Error change

Hence from the figure 3, the approximate relation of individual Gains vs. e (error) and $\dot{e}$ (rate of change in error) was deduced. The shape of the response displayed in figure 3 is approximated to be equivalent curve of Gaussian function.

The equation for the gaussian function is given as :

$$f(x)=ae^{\{[-(x-b)^2]/2c^2\}} \quad (7)$$

Hence based upon the above study we conclude to choose Gaussian membership function for the output PID gains. It is shown in figure 4.

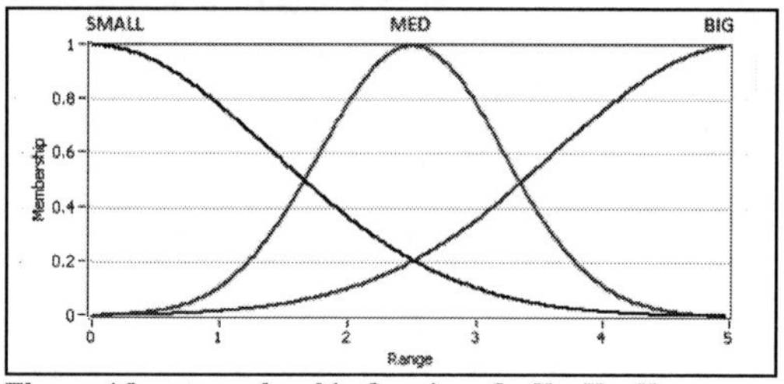

Figure 4 Input membership functions for K_c, K_i, K_d

3.2 Design of Rule Base

The Rule Base for our Proposed Fuzzy Controller are designed taking into consideration the individual effect of Proportional Action , Derivative Action and Integral Action on a process output in terms of rise time (t_r), overshoot(M_p), settling time (t_{ss}), steady-state error (e_{ss}) etc. These are shown in the table 1.

Table 1 Effect of individually increasing parameters

Parameters	Rise Time, t_r	Overshoot, M_p	Settling Time, t_{ss}	Steady-state error , e_{ss}	Stability
K_p	Decrease	Increase	Small change	Decrease	Decrease
K_i	Decrease	Increase	Increase	Decrease significantly	Decrease
K_d	Minor Decrease	Minor Decrease	Minor Decrease	No effect	Improves

Now taking into account the above facts, if we change the value of individual gains (K_c, K_i, K_d) at different intervals of the process we can generate a better process response in terms of rise time, overshoot, settling time etc. Consider Process Reaction Curve shown in Figure 5 that shows the step response of a system. Now the variations in the PID gains are formulated and generated (in Real time) depending upon the value of error (e) and rate of change in error ($\dot{e}$). Here e and $\dot{e}$are divided in seven regions i.e. Positive Big, Positive Medium, Positive Small, Zero, Negative Big, Negative Medium and Negative Small denoted by PB, PM, PS, ZO, NB, NM & NS as shown in Figure 5.The final generated PID Gains are also divided in three groups i.e. Small, Medium and Big. The appropriate ranges or values of the gains will be chosen depending on the values of error and rate of change in error.

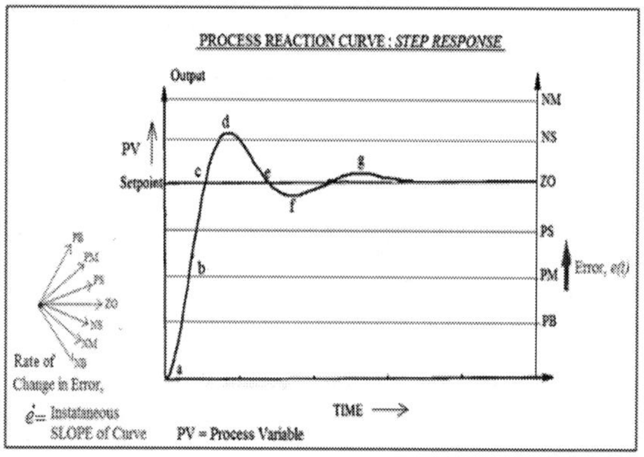

Figure 5 Process step response

The rule base for K_c (in case of Proportional Gain Rule Base), K_i (in case of Integral Gain Rule Base) and K_d (in case of Derivative Gain Rule base) is calculated depending upon the various input/output possibilities.

4 Results & Simulations

A linear Second order system with delay: The process is a simple second order process with a time delay of 0.5 seconds. The transfer function of the process is shown below:

$$T.F.=G_1(s)[e^{-0.5s}/(s+1)^2] \quad (8)$$

Various Plots defining the characteristics of the system are shown in figure 6 and 7.

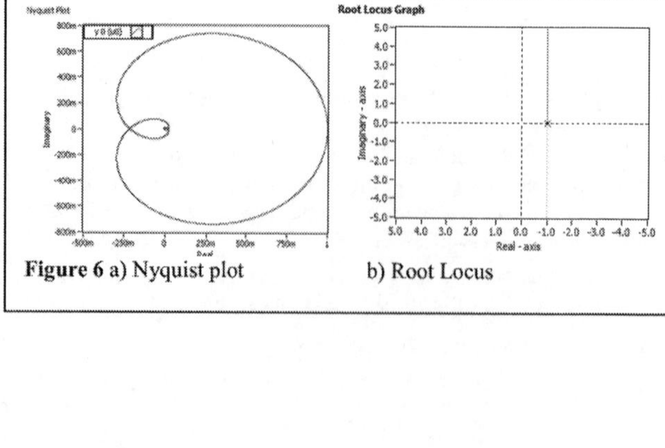

Figure 6 a) Nyquist plot b) Root Locus

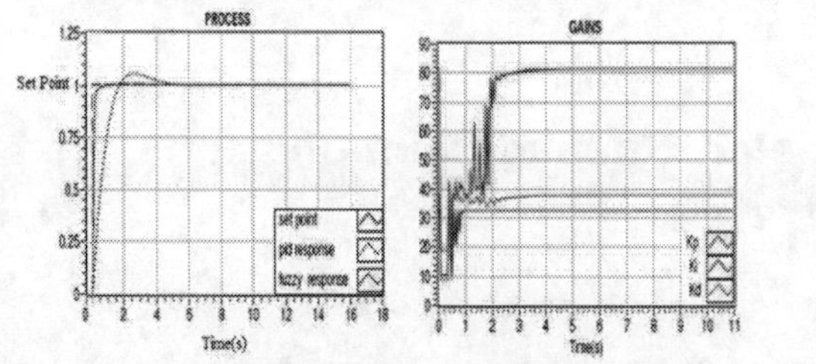

Figure 7 a) Bode Magnitude Plot b) Bode Phase Plot

The response for the process with Zeigler Nicholas tuned PID and Fuzzy tuned PID are shown in figure 8

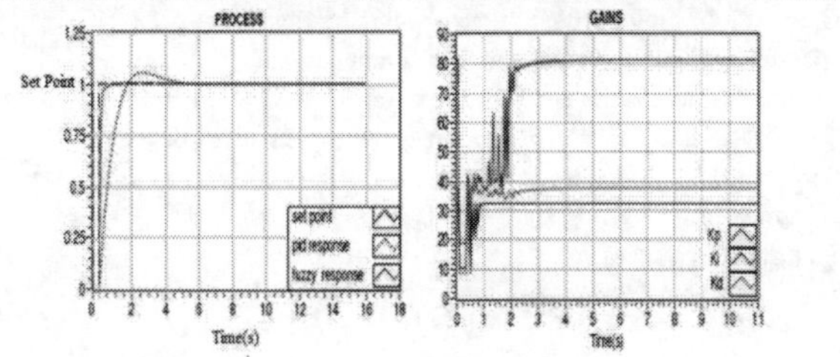

Figure 8 a) Unit Step Response b) Gains in fuzzy for fuzzy and ZN tuned PID

The performance of G(s) can be evaluated by comparison of various system parameters based on Ziegler Nichols Tuned PID, Reference Fuzzy PID & Fuzzy Tuned PID as given in Table 2.

Table 2 Performance evaluation for response of $G_1(s)$

System parameter	**Zeigler Nichols Tuned PID**	**Reference Fuzzy PID[1]**	**Fuzzy Tuned PID**
Rise Time	-	-	0.3s
Settling Time	4.16s	3.09s	0.55s
Overshoot	32%	6.0%	0.5178%
IAE	1.37	1.18	0.0891
IAS	0.871	0.772	0.1448

5 Conclusions

Fuzzy gain scheduling is all about using fuzzy rules and reasoning to determine the controller parameters based on error signal and rate of change of error signal. In fuzzy control, linguistic descriptions of human expertise in controlling a process are represented as fuzzy rules or relations. The new scheme utilizes fuzzy rules and reasoning to determine the controller parameters and the PID controller generates the control signal. Simulation results show that better control performance can be achieved in comparison with Ziegler PID controllers.

The proposed PID tuning scheme can be implemented on the hardware level where non-linearity and inherent noise is present to see if it improves the system response in comparison to PID tuned by conventional methods. Various optimization techniques can also be applied for error and rate of change of error gains in future in order to improve the performance of the system.

References

1. Z. Zhao, M. Tomizuka, "Fuzzy gain scheduling of PID Controller", IEEE transactions on systems, man, and cybernetics, **23,** 5 (1993)
2. B.M Mohan and A. Sinha, "The simplest fuzzy PID controller: Mathematical model and stability analysis", Soft Computing - A Fusion of Foundations, Methodologies and Applications, **10**, 10, pp. 961-975, (2006)
3. A.Mieczysław, J. Brdy´s, J. Littler, "Fuzzy logic Gain scheduling for Non-linear servo tracking", Int. J. Appl. Math. Comput. Sci., **12,** 2, pp. 209–219, (2002)
4. S.M. Radaideh, M.T. Hayajneh, "A new Fuzzy Gain scheduling scheme for the PID controllers", Intelligent automation and soft computing, **9,** 4, pp. 269-277, (2003)
5. K.L. Anderson, G. L. Blankenship, and L. G. Lebow, "A rule based adaptive PID controller", In Proc. *27th IEEE Conf. Decision, Control*, pp. 564-569, (1988)
6. L. Yao and C. Lin, "Design of Gain Scheduled Fuzzy PID Controller", International Journal of Electrical, Computer, Energetic, Electronic and Communication Engineering, **1,** 1, (2007)
7. L. Yao and C. Lin, "On a Genetic algorithm based scheduled Fuzzy PID controller", International Journal of Innovative Computing, **5,** 10(B), pp. 3593-3602, (2009)
8. P. J. Gawthrop and P. E. Nomikos, "Automatic tuning of commercial PID controllers for single-loop and multiloop applications", IEEE Control Syst. Mag., **10**, pp. 34-42, (1990)
9. J. Talaq, F. Al-Basri., "Adaptive fuzzy gain scheduling for load frequency control", IEEE Transactions on Power Systems, **14,** 1, pp. 145-150, (1999)
10. National Instruments NI tutorial- Gain scheduling, (2005)

Design and Synthesis of Single Precision Floating Point Division based on Newton-Raphson Algorithm on FPGA

Naginder Singh[1, a] and Trailokya Nath Sasamal[2]

[1]*School of VLSI Design and Embedded System, National Institute of Technology, Kurukshetra, Haryana, India*
[2]*Department of Electronics and Communication Engineering, National Institute of Technology, Kurukshetra. Haryana, India*

Abstract. This paper describes a single precision floating point division based on Newton-Raphson computational division algorithm. The Newton-Raphson computational algorithm is implemented using 32-bit floating point multiplier and subtractor. The salient feature of this proposed design is that the module for computing mantissa in 32-floating point multiplier is designed using a 24-bit Vedic multiplication (Urdhva-triyakbhyam-sutra) technique. 32-bit floating point multiplier, designed using Vedic multiplication technique, yields a higher computational speed, hence, is efficiently used in floating point divider. Another important feature is the efficient use of device utilization parameters and reduced power consumption. An advantage of the Newton-Raphson algorithm is the higher versatility and precision. For representing 32-bit floating point numbers, IEEE 754 standard format is used. ISim simulator is used for simulation. The proposed floating point divider is designed using Verilog Hardware Description Language (HDL) and is verified on Xilinx Spartan 6 SP605 Evaluation Platform FPGA.

1 Introduction

The role of a reconfigurable processor in embedded system design has increased greatly from the past decades. Due to the advancement of field programmable gate array (FPGA), we have reached a point where the architecture of processors can be modified instantaneously. Reconfigurable computing processing provides very versatile high-speed computing. The enhanced feature of Spartan-6 voluntarily reduces the cost per logic cell designed. Newton-Raphson computational algorithm requires mathematical operations such as, multiplication and subtraction. Here, mathematical operations used to find the reciprocal of the denominator (D) and multiply that reciprocal by the numerator (N) to find the final quotient (Q). Newton-Raphson computational algorithm initializes with an approximation close to the final value of the quotient (Q) and produces twice as many digits of the final quotient after each iteration. An iterative process which is based on complex operation used for division in many signal-processing algorithms, where not only precision to be maintained, but also the precision is to be maintained for very large data intervals and should be high for better operation. This can be achieved by the design and implementation of floating point division by using Newton-Raphson algorithms. Several different algorithms described in literature [1-3]. The Newton-Raphson algorithm computes three multiplicative inverse at the same time to provide high throughput [4]. To implement and design 32-bit floating point division based on Newton-Raphson computational algorithm, 32-bit floating point multiplier and 32-bit floating point subtraction modules are used [5,6]. For efficient implementation of the floating point multiplier Vedic multiplication is used for calculating mantissa part [7, 8]. The format for representing 32-bit and 64-bit floating point numbers are provided by the IEEE 754 standard [9, 10]. IEEE 754 uses a fixed number of bits for representing the 32-bit floating point number. The representation format divides into three parts, i.e., sign (b), exponent (e) and the mantissa (s). Table 1 shows the structure for IEEE 754 formats and describes the single and double precision. In IEEE 754 Single precision format the mantissa is represented by 23 bits, exponent is represented by 8-bits and MSB corresponds to sign bit. The Sign of the floating point number depends on the sign bit or MSB. The number is positive when the MSB bit is 0 and negative when the MSB bit is 1.

Table 1. IEEE 754 Standard Format for single (32-bit) and double precision (64-bit).

	Sign (s)	Exponent (e)	Mantissa (m)
32-bit	1-bit	8-bit	23-bit
64-bit	1-bit	11-bit	52-bit

The formulation of the paper is as follows. Section 2 explains the architecture of the floating point multiplier using Vedic multiplication. Section 3 presents the description of 32-bit floating point subtractor. Section 4 describes the Newton-Raphson computational algorithm. Sections 5 presents the Simulation Results of floating point division using Newton-Raphson algorithm. The Conclusion and References are presented in the final section.

2 Floating point multiplier

In Figure 1 shows the complete architecture of proposed 32-bit floating point multiplier. This multiplier module is designed using a Vedic multiplication technique, where mantissa calculation is done using a 24x24 bit Vedic multiplier. The main purpose of using Vedic multiplier is to improve the overall performance of the 32-bit floating point multiplier. IEEE 754 format presents a fixed number of bits for representing the sign, exponent and mantissa. The inputs given to the floating point multiplier

[a] Naginder Singh: naginder.singh0110@gmail.com

are A[31-0] and B[31-0] as per IEEE 754 format. 32-bit Floating point multiplication unit is divided into three parts - sign unit, exponent unit and mantissa unit.

2.1. Mantissa unit

In mantissa unit, for calculation of mantissa a 24x24 bit Vedic multiplier is used efficiently for higher throughput and computation. The lower bits of inputs, m1 (A[22-0]) and m2 (B[22-0]) are given to the 24-bit Vedic multiplier which produces 24-bit normalized output and should have leading one as their MSB.

2.2. Exponent unit

In Exponent unit, the exponent calculation is done by using ripple carry adder. The exponent is computed by providing inputs e1 (A[30-23]) and e2 (B[30 – 23]) to the 8-bit ripple carry adder unit and result is biased to 127. The overflow and underflow cases are carefully handled.

2.3. Sign unit

In sign unit, the sign bit is computed by xoring the 31st bit of inputs, s1 (A[31]) and s2 (B[31]) of floating point inputs. The output of xor gate represents the sign of the floating point multiplier. The Vedic multiplication technique is efficiently used for high computational speed and throughput.

The complete architecture of proposed 32-bit floating point multiplier is shown in the Figure 1. This proposed multiplier module designed using Vedic multiplier is used in the Newton-Raphson computational algorithm for performing the iteration process.

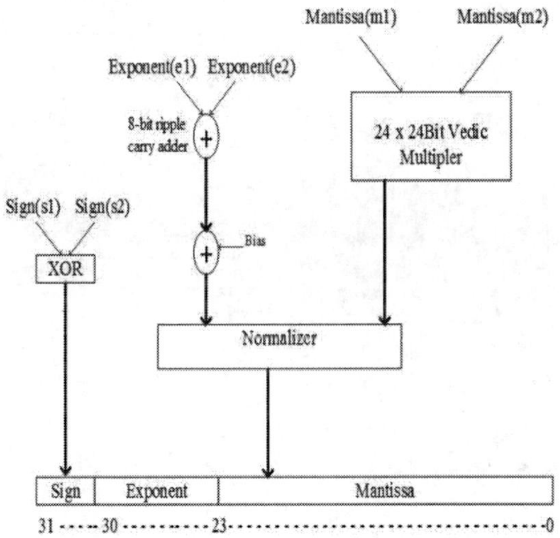

Figure 1. The proposed architecture for 32-bit Floating point multiplier.

3 Floating point subtractor

In the subtractor module, X[31-0] and Y[31-0] are given as inputs to the floating point subtractor. The sign (s), exponent (e) and mantissa (m) are represented in IEEE 754 format. 32-bit floating point subtraction operation is done in a stepwise manner as explained further. First of all, the floating point numbers are unpacked. After unpacking, the sign, exponent and mantissa are identified for performing the subtraction operation. Next, the exponent is equalized for performing alignment and normalization of mantissa part. If neither of the operands are infinity, then the relation between e1 (X[30-23]) and e2 (Y[30-23]) is determined by comparing the mantissa m1 and m2. The mantissa is shifted right until the exponent becomes equal, i.e. e1 (X[30-23]) = e2 (Y[30-23]). After alignment and normalization, the mantissa m1 (X[23-0]) and m2 (Y[23-0]) are subtracted. After that, the mantissa values are rounded off. Finally, the sign, exponent and mantissa parts are concatenated. The 32-bit floating point subtraction module is used in the Newton-Raphson computational algorithm for performing the iteration process. Figure 2 shows the complete architecture of 32-bit floating point subtractor. This 32-bit floating point subtractor module is used for calculating the iterative process in Newton-Raphson algorithm.

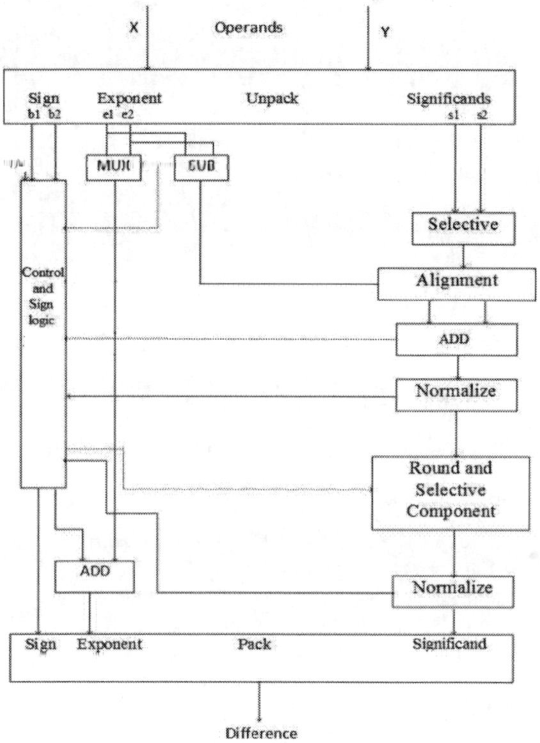

Figure 2. The architecture of 32-bit Floating point subtractor.

4 Newton-Raphson division algorithm

Newton-Raphson computational division algorithm computes the multiplicative inverse, which is calculated by the iterative process. Then the calculated multiplicative inverse is multiplied to the dividend to compute the final quotient (Q). In this proposed design, Newton-Raphson computational division algorithm designed by using a 32-floating point multiplier module and subtractor module. In this division algorithm minimal of maximal relative error can be achieved by scaling the divisor (D) in the interval (0.5, 1). Scaling of the divisor is done by shifting operation. To produce a precise result more iterations are required. For this purpose, fast division algorithms are developed. The Newton-Raphson algorithm converges much faster for computing one iteration. Thus, several multiplication and subtraction operations needed to perform for the continual iteration process. Figure 3 shows the flowchart of floating point division using the Newton-Raphson computational algorithm in which two multiplier and a subtraction module are used to produce one iteration. Newton Raphson computational algorithm produced optimized result by computing three iterations in one cycle. More iterations are performed to refine the multiplicative inverse.

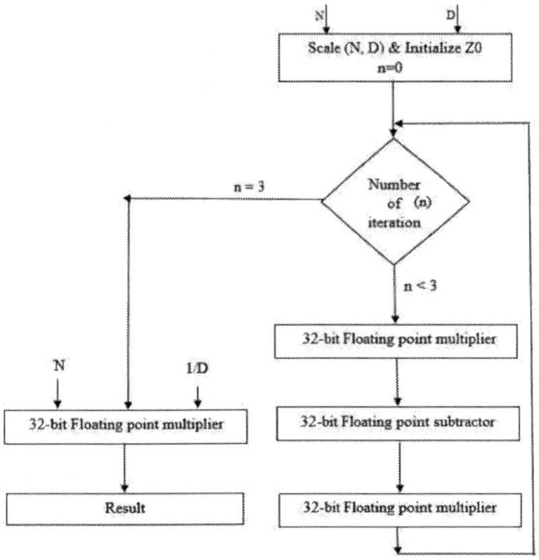

Figure 3. A Flowchart of the 32-bit floating point division using Newton-Raphson method.

The calculation of multiplicative inverse of the divisor is performed by equation (1), where Z_i is multiplicative inverse at iteration i, as given below.

$$Z_{i+1} = Z_i + Z_i (1 - DZ_i) = Z_i (2 - DZ_i) \quad (1)$$

Newton-Raphson division algorithm uses a complex initialization for computing continual iterations. To minimize the maximum of the relative error in the interval (0.5, 1), Z_i should be initialized. The initialization is represented by Z_0 and is initialized as follows.

$$Z_0 = (48/17) - (32/17)\,D \quad (2)$$

5 Simulation Results

The simulation was performed on ISim. Figure 4 shows the simulation results of the proposed 32-bit floating point division of two numbers using the Newton-Raphson method. Table 3 shows the device utilization of the Xilinx Spartan 6 SP605 Evaluation Platform FPGA. Table 4 shows the Xilinx Power Estimator (XPE) -14.3 device summary report for the proposed 32-bit floating point division carried out for the Spartan-6 SP605 Evaluation Platform FPGA. In Figure 4, N represents the numerator, D represents the denominator, Z0 is the initial value, the Z1 is the first iteration result, Z2 is the second iteration result, Z3 is the final iteration result and Q represents the quotient. The two inputs N and D are given to the divider in IEEE 754 standard format as shown in Table 2.

Table 2. Sample inputs and its output for simulation

	Decimal	Sign	Exponent	Mantissa
N	3.14	0	10000000	10010001111010111000011
D	8.56	0	10000010	00010001111010111000011
Q	0.366822	0	01111101	01110111101000000100110

Table 3. Device Utilization of the Xilinx Spartan 6 SP605 Evaluation Platform.

Logic Utilization	Used	Available	Utilization
Number of Slice Registers	408	54,576	1%
Number of Slice LUTs	10,019	27,288	36%
Number of occupied Slices	3,878	6,822	56%
Number of bonded IOBs	192	296	64%

Table 4. Xilinx Power Estimator (XPE) -14.3 device summary report of Spartan-6 SP605 Evaluation Platform FPGA.

Specifications	Values	
Junction Temperature	25.5 °C	
Total On-Chip Power	0.036 W	
Thermal Margin	59.5 °C	4.0 W
Effective ΘJA	14.3 °C/W	

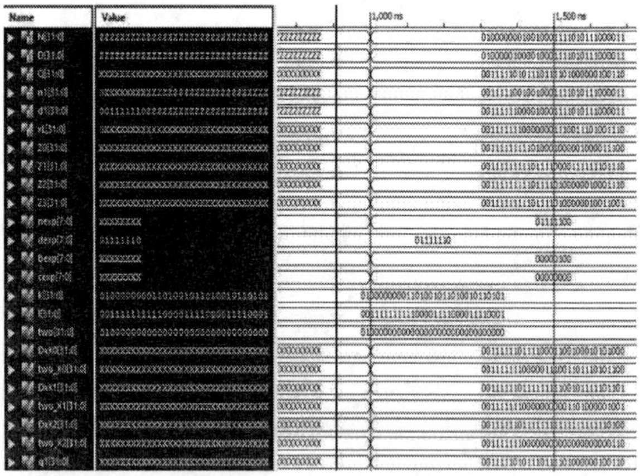

Figure 4. Simulation result of Floating point division.

6 Conclusion

The Single precision floating point division using Newton–Raphson computational algorithm is designed and synthesized on FPGA. This computational technique provides high computation speed and throughput by computing one multiplicative inverse in one iteration. The multiplier performance is increased by Vedic technique and hence improved the efficiency of the floating point divider. The device utilisation parameters were optimized thereby the power consumption is reduced. This presented design is useful in high computation demanding applications.

References

1. J. M. Muller, Avoiding Double Roundings in Scaled Newton-Raphson Division, *Proceedings of Asilomar Conference on Signals, Systems and Computers*, 396-399 (2013)
2. I. Kong, E. E. Swartzlander, A Rounding Method to Reduce the Required Multiplier Precision for Goldschmidt Division, *IEEE Transactions on Computers* **59**, 1703-1708 (2010)
3. A. R. Garca, L. P. Escalante, R. P. Michel, O. L. Gandara, J. Cortez, Fast Fixed-Point Divider Based on Newton-Raphson Method and Piecewise Polynomial Approximation, *Proceedings International Conference on Reconfigurable Computing and FPGA*, 1-6 (2013)
4. Peter Malik, High throughput floating-point dividers implemented in FPGA, *IEEE conference*, (2015)
5. A. Rathor, L. Bandil, Design Of 32 Bit Floating Point Addition And Subtraction Units Based On IEEE 754 Standard, *International Journal of Engineering Research & Technology* **2**, 2278-0181 (2013)
6. Jagadguru Swami Sri Bharati Krisna Tirthaji Maharaja, Vedic Mathematics Sixteen Simple Mathematical Formulae from the Veda, (1965)
7. A. Kanhe, S. K. Das, A. K. Singh, Design And Implementation Of Low Power Multiplier Using Vedic Multiplication Technique, *International Journal of Computer Science and Communication* **3**, 131-132 (2012)
8. A. Ashrafy, M. Salem, A. Anis, An efficient implementation of floating point multiplier, *Electronics Communications and Photonics Conference*, (2011)
9. B. Hickmann, A. Krioukov, M. Schulte, M. Erle, A Parallel IEEE 754 Decimal Floating Point Multiplier, *25th International Conference on Computer Design*, (2007)
10. IEEE 754-2008, IEEE Standard for Floating-Point Arithmetic, (2008)

MATEC Web of Conferences 57, 01010 (2016) DOI: 10.1051/matecconf/20165701010
ICAET-2016

Text Character Extraction Implementation from Captured Handwritten Image to Text Conversionusing Template Matching Technique

Seema Barate[1], Chaitrali Kamthe[2], Shweta Phadtare[3], Rupali Jagtap[4], *M.R.M Veeramanickam*[5]

[1, 2, 3, 4]*Under Graduate Students of Information Tech. Dept., TCOER, Pune*

[5]*Asst. Prof., TCOER, Pune*

Abstract. Images contain various types of useful information that should be extracted whenever required. A various algorithms and methods are proposed to extract text from the given image, and by using that user will be able to access the text from any image. Variations in text may occur because of differences in size, style,orientation, alignment of text, and low image contrast, composite backgrounds make the problem during extraction of text. If we develop an application that extracts and recognizes those texts accurately in real time, then it can be applied to many important applications like document analysis, vehicle license plate extraction, text- based image indexing, etc and many applications have become realities in recent years. To overcome the above problems we develop such application that will convert the image into text by using algorithms, such as bounding box, HSV model, blob analysis,template matching, template generation.

1 INTRODUCTION

Handwritten text from image has been a subject of study and research for many years. There are different methods currently employed to improve the visibility of text in text documents. We use different algorithms to improve the accuracy of the existing system. In this method to improve readability of text documents through manually enhancing text strokes is proposed [1]. We develop such application that will convert the image into text by using algorithms, such as bounding box, HSV model,blob analysis, template matching, and template generation.

To develop an interface to train the application for understands the difficult and unreadable handwritingCollect handwritten characters, symbols as an input data and train the application. Give handwritten character input as an image and extract the character. Generate the document file from the given image.

We assume the input as A-Z or a-z. To simplify the recognition process, certain pairs of similar upper and lower case characters are combined into singleClasses like Cc, Kk, 00, Ii, Pp, Ss, Uu, Vv, Ww, Xx, Yy, and Zz etc [2]. The database that is used for these experiments consists of about 10,000 isolated hand printed characters. Connected component analysis was used to locate blobs that were about the size of characters.The characters were manually identified and Stored in the database [2].

In case of offline character recognition the handwritten character is typically scanned in form of a paper document and made available in the form of a gray scaleimage to the recognition algorithms [3].

2 Literature Study

To improve the text of historical document through the multi-resolution Gaussian and to improve the readability of the text two algorithms are used LU algorithm and Gaussian filter. Advantage is to detect text of different scale. Disadvantage is work for image of small resolution (Oliver A. Nina) [1].

Text images are recognized in the text file using a character based bi-gram language model for that bi-gram language model is used .Advantage is to improve the recognition performance. Disadvantage is the presented approach will be verified on text images only (Yanwei Wang,Xiaoqing Ding,Fellow,IEEE and Changsong Liu) [2].

To translate images of typewritten or handwritten characters into electronically editable format by preserving font properties OCR algorithm is used. But it can't recognize multiple font size. In this paper used algorithm is Optical Character Recognition (OCR).Advantage is decrease some possible human errors and high speed of recognition. Disadvantage is multiple font and size characters and handwritten characters are not recognize (Faisal Mohammad,Jyoti Anarase,Milan Shingote and Pratik Ghanwat) [3].

A handwritten Arabic text written by each writer can be seen as a having a specific texture and characterization of texture image is based on run length of image gray level. In this paper two methods are used Gray Level Run Length (GLRL), Gray Level Co-occurrence Matrices

[2]Corresponding author: chaitukamthe029@gmail.com

(GLCM). Advantage is uses IFN/ENIT database which is unique Arabic publicly available dataset. Disadvantage is only extract global feature (DJEDDI Chawki and SOUICI-MESLATI Labiba) [4].

Various techniques are used to extract the text from image but it can be use only detect the text area, two algorithms are used OCR and Edge detection Advantage is any types of images content extracts accurately. Disadvantage is only text area is detected (A.J.Jadhav ,Vaibhav Kolhe and Sagar Peshwe)[5].

Image is captured by the camera, that image contains many text formats. This text information can be extracted from the image by using edge detection, collected component and texture based algorithms, two algorithms are used Compare Prewitt edge detector and Gaussian edge detector. Advantage is only character extraction removes non text image. Disadvantage is work only for natural scene image (Prof. Amit Choksi ,Nihar Desai,Ajay Chauhan,Vishal Revdiwala and Kaushal Patel) [6].

To extract text lines and words from handwritten document Segmentation Algorithm and Viterbi Algorithm are presented . The line segmentation algorithm is based on locating the optimal succession of text and gap areas within vertical zones by applying Viterbi algorithm. Word segmentation is based on a gap metric that exploits the objective function of a soft-margin linear SVM that separates successive connected components. Advantage is it provides 97 percent accuracy. Disadvantage is it does not provide generalize well to variations encountered in handwritten documents (Vassilis Papavassilioua,Themos Stafylakisa,Vassilis Katsourosa andn George Carayannisa)[7].

3 Proposed system : Text extraction using template matching

In proposed system,we are using different algorithms to improve the accuracy of the text document or text image[16].It uses the image as input can be captured by laptop camera. After that image is matched using the template matching. It uses the filter to filtering image means to remove the noise, low frequency then it takes only the high frequency images. Then the image is cropped by using bounding box. Then convert image into Gray scale using HSV model, HSV means Hue, Saturation and Value .Extract the character from the image using the blob algorithm. Blob means separate the word into character and each character is called blob. Then matches the character with the trained dataset using template matching algorithm .Then generate the character sentence by using template generator. Then generate document file(.doc file).

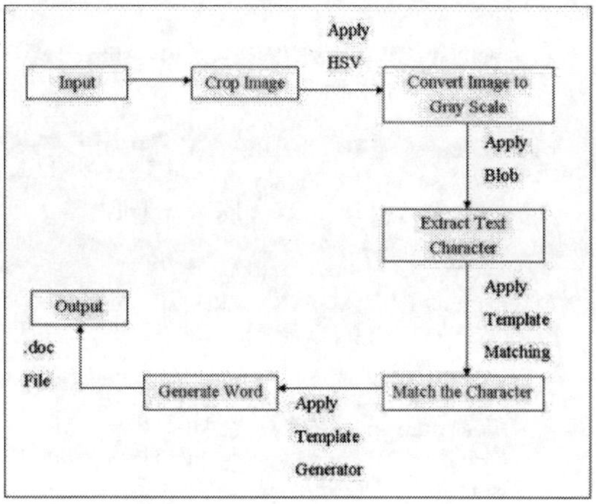

Figure 1 Architectural Diagram

4 MODULES

4.1 Template Matching

Template matching is a technique for finding areas of an image that match to a template image.In template matching, individual image pixels are used as features.Each comparison results in a similarity measure between the input character and the template. Template matching forcharacter recognition is reliable .The template matching works on two primary components:

Source image (I): match captured image to the template image

Template image (T): The patch image which will be compared to the template image

a) To detecting the highest matching area:../../../../../_images/Template_Matching_Template_Theory_Summary.jpg
b) To identify the matching area, we have to compare the template image against the source image by sliding it:../../../../../_images/Template_Matching_Template_Theory_Sliding.jpg

By sliding, means moving the patch one pixel at a time i.e left to right, up to down. At each location, a metric is calculated and it represents how "good" or "bad" the match at that location.

For each location of T over I, you store the metric in the result matrix (R). Each location (x,y) in R contains the match metric:../../../../../_images/Template_Matching_Template_Theory_Result.jpg

the image above is the result R of sliding the image with a metric TM_CCORR_NORMED. The brightest locations indicate the highest matches. If you can see, the location marked by the red circle is probably the one with the highest value, so that location of the rectangle formed by that point as a corner and width and height equal to the patch image is considered the match.we use the function minMaxLoc to locate the high value or low value, depending of the type of matching method in the R matrix.

OpenCV implements Template matching in the function matchTemplate. There are four methods :

a) method=CV_TM_SQDIFFR(x,y)= \sum _{x',y'} (T(x',y')-I(x+x',y+y'))^2

b) method=CV_TM_SQDIFF_NORMEDR(x,y)= \frac{\sum_{x',y'}(T(x',y')I(x+x',y+y'))^2}{\sqrt{\sum_{x',y'}T(x',y')^2\cdot\sum_{x',y'} I(x+x',y+y')^2}}

c) method=CV_TM_CCORRR(x,y)= \sum _{x',y'} (T(x',y') \cdot I(x+x',y+y'))

d) method=CV_TM_CCORR_NORMEDR(x,y)= \frac{\sum_{x',y'}(T(x',y')\cdot I(x+x',y+y'))}{\sqrt{\sum_{x',y'}T(x',y')^2 \cdot \sum_{x',y'} I(x+x',y+y')^2}}

Source:http://docs.opencv.org/2.4/doc/tutorials/imgproc/histograms/template_matching/template_matching.html

4.2 Bounding Box

The "**bounding box**" of a limited geometric object is the box with minimum area or minimumvolume, that contains a given geometric object. For any collection of linear objects like points segments ,polygon ,lines ,etc their bounding box is given by the minimum and maximum coordinate values for the point set S of all the object's *n* vertices[15].Suppose finding the area of a rectangle, area means the "level ground, an open space,"The number of square units it takes to completely fill a rectangle.

Formula: Width × Height

Try this Drag the orange dots to move and resize the rectangle. The area of a rectangle is given by multiplying the width times the height. As a formula: where

w is the width

h is the height so the formula is w*h.

Source:http://www.mathopenref.com/coordtriangleareabox.html

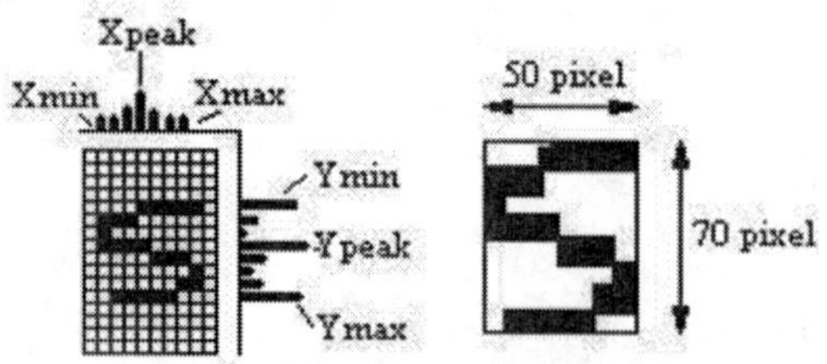

Figure 2: Example of Bounding Box
Source:Nafiz Arica and Fatos T. Yarman-Vural,"An Overview of Character Recognition Focused on Off-Line Handwriting".

4.3 Blob analysis

Blob Analysis is a fundamental technique of machine vision based on analysis of consistent image regions. The numbers of words are divided into characters, the performance of the algorithm both in terms of accuracy and efficiency is increases. Hence the number of words should be decided based on scanned image.

4.4 HSV model

The HSV color model, also called HSB (Hue, Saturation, Brightness), defines a color space in terms of three constituent components:

- Hue is the color type (such as red, magenta, blue, cyan, green or yellow). Hue ranges from 0-360 deg.
- Saturation refers to the intensity of specific hue. Saturation ranges are from 0 to 100%. In this work saturation is presenting in range 0-255.
- Value refers to the brightness of the colour. Saturation ranges are from 0 to 100%. Value ranges are from 0-100%. In this work saturation and value arepresenting in range 0-255. RGB to HSV conversion formula ,The R,G,B values are divided by 255 to change the range from 0..255 to 0..1:

R' = R/255,G' = G/255,B' = B/255
Cmax = max(R', G', B')
Cmin = min(R', G', B')
Δ = Cmax - Cmin
Value calculation:V = Cmax

Table1 RGB to HSV colour

Colourname	Hex value	(R,G,B)	(H,S,V)
Black	#000000	(0,0,0)	(0°,0%,0%)
White	#FFFFFF	(255,255,255)	(0°,0%,100%)
Red	#FF0000	(255,0,0)	(0°,100%,100%)
Blue	#0000FF	(0,0,255)	(240°,100%,100%)
Yellow	#FFFF00	(255,255,0)	(60°,100%,100%)
Green	#008000	(0,128,0)	(120°,100%,50%)
Cyan	#00FFFF	(0,255,255)	(180°,100%,100%)
Magenta	#FF00FF	(255,0,255)	(300°,100%,100%)

Source:http://math.stackexchange.com/questions/556341/rgb-to-hsv-color-conversion-algorithm

 DOI: 10.1051/matecconf/20165701010

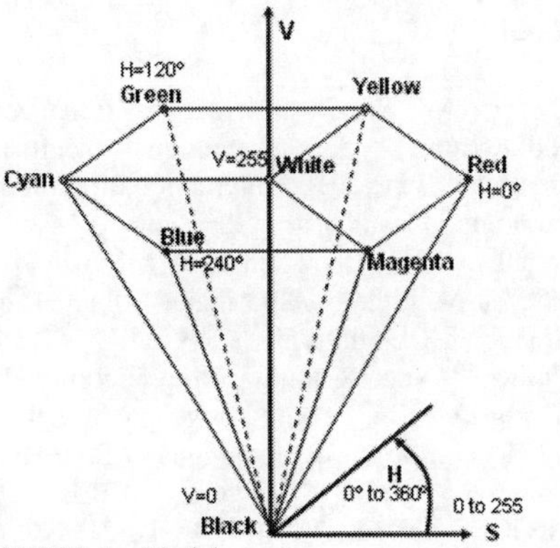

Figure 3 HSV Color Model

Source: Lidiya Georgieva, Tatyana Dimitrova, Nicola Angelov," RGB and HSV colour models in colour identification of digital traumasImages", (ICCST).

4.5 Template Generator

Template Generator is used for automatic generation of templates of numerals of a certain font. After applying template matching, template generator is used to form the word from the separated characters in blob analysis.

5 Implementation of text extraction

Template Matching

First image is captured with the help of camera or browse image. Then template matching algorithm is used to calculate the size of the image, if image size is greater than 15cm x 15cm then don't accept the image, otherwise accept it.

Figure 4 Input Image

Bounding Box

The output of Template Matching module i.e. 15cm x 15cm size image is taken as an input for Bounding Box.For discard the unwanted portion of the image use bounding box algorithm. By applying bounding box algorithm it take only text region from the image and remaining part from the image is discarded.

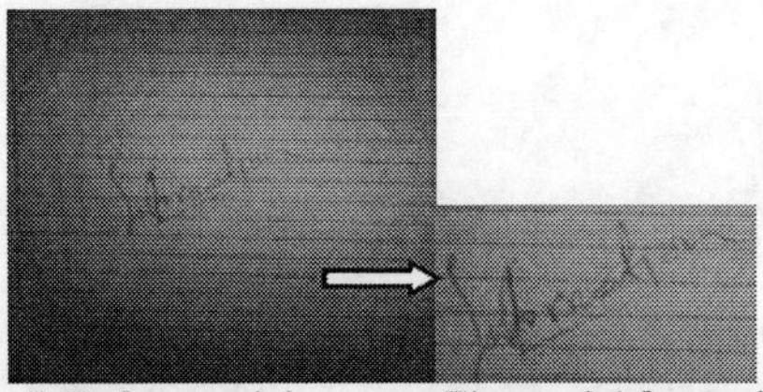

Figure 5 Before applying BoundingBox **Figure 6** Afterapplying Bounding Box

HSV Model

In HSV image is calculate Hue, Saturation and Value separately as a pixel. HSV model is used to convert the image into gray scale. HSV model takes the input from Bounding Box. Then combine all these pixel as output. In output image display the text by black color and make background as white.

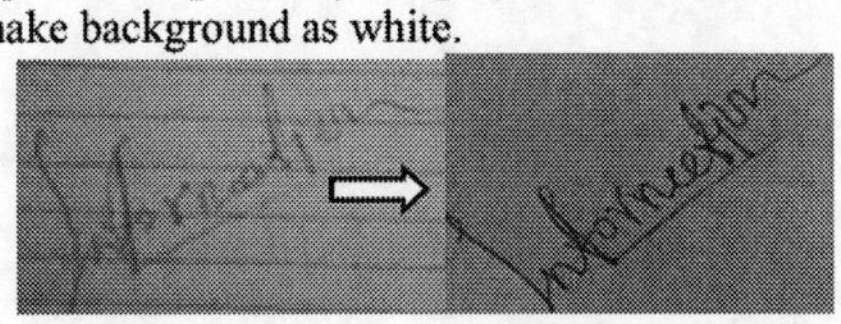

Figure 7 Beforeapplying HSV HSVModel **Figure 8** Afterapplying Model

Blob Analysis

After getting gray scale image separate the word into the character using blob analysis. Each character is called as a blob. Each character is match with the training database.

Figure 9 Before applying Blob Analysis

Figure 10 After applying Blob Analysis

Template Matching

Template matching algorithm is used for matching blob or character with training dataset. To match blob with training dataset feature extraction algorithm is used. If character is matched accurately then that character is given as input to template generator.

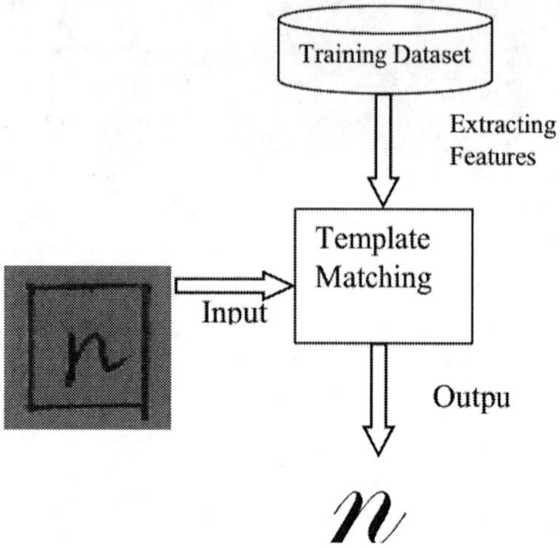

Figure 11. Working of Template Matching

Template Generator

In template generator input is taken from template matching, each generated character is combined into word and generate the.doc file.

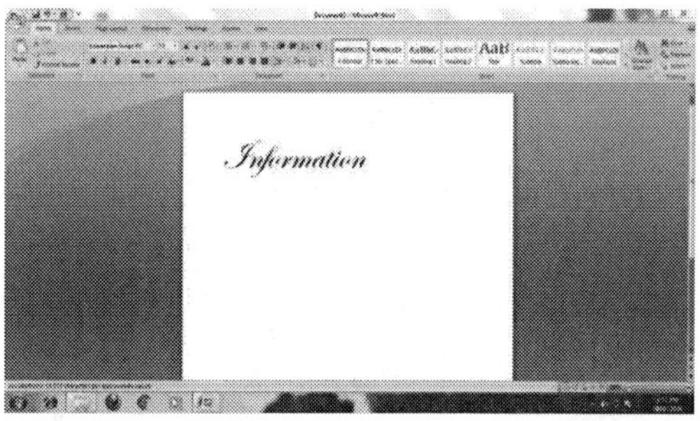

Figure 12. Final output using word doc

6 CONCLUSION

Though we have large no of algorithms and methods for text extraction from image butnone of them provide accurate result .Pattern Matching can be implemented successfully for English language text character recognition. The system has image processing modules for converting text image to generate .doc file. The expected experiment recognition rate will be good.

7 Future Scope

As it is restricted only for English language in future make this application other than English language. In future make this application for Android mobile phones as an App. And will be train for graphical symbols also.

References

1. Oliver A. Nina, "Interactive Enhancement of Handwritten Text through Multi-resolution Gaussian", in 2012 International Conference on Frontiers in Handwriting Recognition.
2. Jonathan J. HULL, Alan COMMIKE and Tin-Kam HO," Multiple Algorithms for Handwritten Character Recognition".
3. Yanwei Wang, Xiaoqing Ding, Fellow, IEEE, and Changsong Liu," Topic Language Model Adaption for Recognition of Homologous Offline Handwritten Chinese Text Image", IEEE SIGNAL PROCESSING LETTERS, VOL. 21, NO. 5, MAY 2014.
4. Faisal Mohammad, Jyoti Anarase, Milan Shingote, Pratik Ghanwat "Optical Character Recognition Implementation Using Pattern Matching",Faisal Mohammad et al, / (IJCSIT) International Journal of Computer Science and Information Technologies, Vol. 5 (2) , 2014, 2088-2090.
5. DJEDDI Chawki, SOUICI-MESLATI Labiba," A texture based approach for Arabic Writer Identification and Verification", 978-14244-8611-3/10/$26.00 ©2010 IEEE.
6. A.J.Jadhav Vaibhav Kolhe Sagar Peshwe, " Text Extraction from Images: A Survey", in Volume 3, Issue 3, March 2013, ISSN: 2277 128X.
7. Prof. Amit Choksi1, Nihar Desai2, Ajay Chauhan3, Vishal Revdiwala4, Prof. Kaushal Patel5, "Text Extraction from Natural Scene Images using Prewitt Edge Detection Method", Volume 3, Issue12, December 2013, ISSN: 2277 128X.
8. Vassilis Papavassilioua,Themos Stafylakisa,Vassilis Katsourosa and George Carayannisa, "Handwritten document image segmentation into text lines and words",2010.
9. Debapratim Sarkar, Raghunath Ghosh ,A Bottom-Up Approach of Line Segmentation from Handwritten Text ,2009.
10. Bartosz Paszkowski, Wojciech Bieniecki and Szymon Grabowski Katedra Informatyki Stosowanej, Politechnika ALodzka, "Preprocessing for real-time handwritten character recognition".
11. Juan Chen,Chuanxiong Guo,"Online Detection and Prevention of Phishing Attacks".Mohammed Ali Qatran,"Template Matching Method For Recognition Musnad Characters Based On Correlation Analysis".
12. Mr.Danish Nadeem,Miss.Saleha Rizvi,"Character Recognition Using Template Matching".
13. Harmeet Kaur Kelda ,Prabhpreet Kaur,"A Review: Color Models in Image Processing".
14. Nafiz Arica and Fatos T. Yarman-Vural,"An Overview of Character Recognition Focused on Off-Line Handwriting".
15. Prof. Primož Potočnik ,Student: Žiga Zadnik,"Handwritten character Recognition: Training a Simple NN for classification using MATLAB".
16. Lidiya Georgieva, Tatyana Dimitrova, Nicola Angelov," RGB and HSV colour models in colour

identification of digital traumasImages", International Conference on Computer Systems and Technologies - CompSysTech' 2005.

MATEC Web of Conferences 57, 01011 (2016) DOI: 10.1051/matecconf/20165701011
ICAET-2016

Analyzing block type channel estimation for OFDM based digital communication system

Mandeep Kaur[1,a] and Savina Bansal[1]

Deptt of ECE, GZSCCET, Maharaja Ranjit Singh State Technical University, Bathinda

Abstract. Orthogonal frequency division multiplexing (OFDM) is a promising technique in the current broadband wireless communication system due to the high data rate transmission capability and the ability to combat frequency selective fading of the channel. Channel estimation is mainly implemented by sending pilot symbols in the transmitted bit streams. In this paper, channel estimation based on block type pilot arrangements is analyzed using Least Square (LS) and Minimum Mean Square Error (MMSE) channel estimators. Performance is analyzed in terms of Bit Error Rate and Mean Square Error by varying pilot energy levels and by varying channel length. It is gathered that performance gets affected considerably with change in pilot energy levels implying there exist an optimum value for pilot energy for getting better performance.

1 Introduction

In OFDM system many sub-channels are used in parallel. The channels are overlapping in frequency, but the distance between them is chosen so that the different channels anyhow are orthogonal. OFDM is a promising technique in the current broadband wireless communication system due to the high data rate transmission capability and the ability to combat frequency selective fading of the channel. OFDM, which is the recent trend in wireless technology, is a multicarrier modulation scheme having high data stream splitting into low data stream that are transmitted simultaneously over a number of subcarriers. OFDM is widely used in the wireless systems such as wireless LAN, terrestrial digital television broadcasting, cell-phone and Wi-MAX.

Wireless channels used for transmitting the high data rate digital signals usually suffers from various impairments due to multipath propagation of the signals owing to different types of obstacles present, and frequency dependent channel characteristics. As a result the received signal gets corrupted leading to misjudgment of the signal transmitted and hence reducing the system fidelity and utility for high data rate transmissions. To recover the signal correctly at the receiving end it becomes crucial to apply the inverse channel characteristics at the receiver to nullify the non-linear channel effects. So, channel estimation is an important aspect of high speed data transmission systems before applying demodulation at the receiving end. Channel estimation is mainly implemented by sending pilot symbols in the transmitted bit streams. The channel estimation has been performed by inserting pilot tones into each OFDM symbol. In this paper, channel estimation based on block type pilot arrangements using LS and MMSE channel estimators is studied. This paper is organized as follows. In this section 2, simulated system description is described. Section 3 discusses channel estimation. Performance analysis and conclusion are given in section 4 and section 5.

2 Simulated system description

In figure 1, the information data in binary form are first grouped and mapped into mutiamplitude multi-phase signals according to the type of modulation used in the signal modulator. After inserting pilots uniformly between the information data sequence, IFFT block is used to transform and multiplex the complex data sequence into time domain signal. Following the IFFT block, a guard interval (larger than the expected delay spread), is inserted in order to prevent possible inter-symbol interference (ISI) in OFDM systems. The transmitted signal is then sent to a frequency selective multi-path time varying slow fading channel.

At the receiver, the guard insertion is removed first and the received samples are then sent to the FFT block for de-multiplexing the multi-carrier signal. Following FFT block, the pilot signals are extracted from the demultiplexed samples. The transmitted data samples can then be recovered from the knowledge of the channel responses by simply dividing the received signal by the channel response. After signal demodulation at the demodulator, the binary data could be reconstructed at the receiver output.

[a] Corresponding author: mandeepdhanoa23@gmail.com

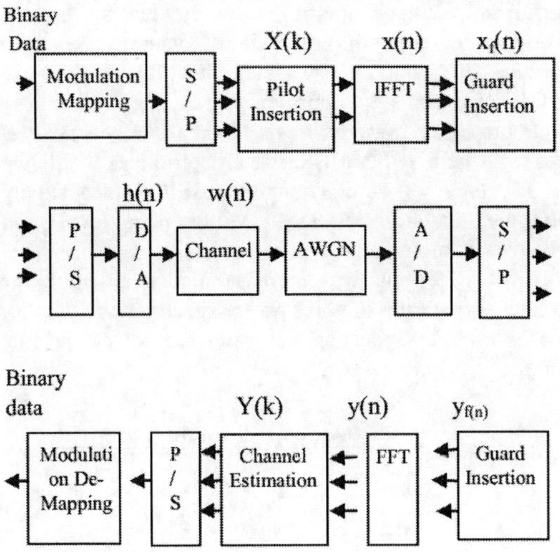

Figure 1. Simulated OFDM system

3 Channel estimation

In block type pilot arrangement, OFDM symbols with pilots at all subcarriers are transmitted periodically for channel estimation. Block type pilot arrangement is especially suitable for slow fading radio channels. This type of pilot arrangement is relatively insensitive to the frequency selectivity.

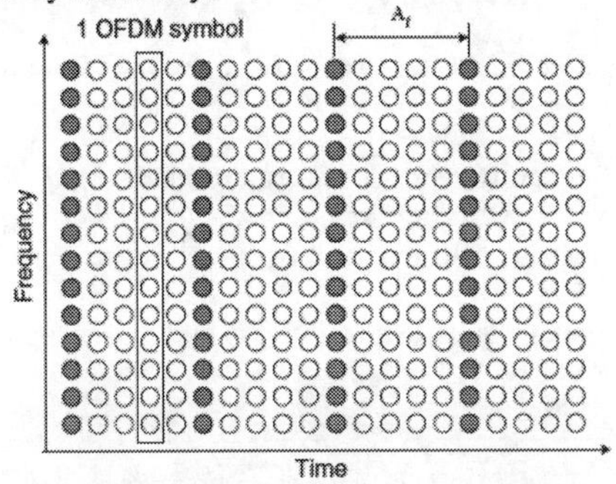

Figure 2. Block type pilot arrangement

3.1 Channel estimation based on block type arrangement

To estimate the channel conditions (specified by H) given the pilot signals (specified by X) and received signals (specified by Y), with or without using certain knowledge of the channel statistics. The receiver uses the estimated channel conditions to decode the received data inside the block until the next pilot symbol arrives. The estimation can be based on least square (LS) and minimum mean square error (MMSE) estimators.

➢ **LS**

The LS estimator minimizes the weighted errors between the measurements and model. The LS estimates H, given the received data Y and the transmitted symbols X is given by:

$$H_{LS} = X^{-1}Y \qquad k = 0,1, \dots \dots \dots, N-1 \tag{1}$$

The main advantage of LS estimator is that it has low complexity but it has high MSE.

➢ **MMSE**

The MMSE estimator employs the second order statistics of the channel conditions to minimize the MSE. Denote by R_{hh}, R_{HH} & R_{YY} the auto covariance matrix of h, H and Y respectively and by R_{hY} the cross covariance matrix between h and Y. Also denote by σ^2 the noise variance. Assume the channel vector and the noise are uncorrelated, it is derived that:

$$R_{HH} = E\{HH^H\} = FR_{hh}F^H \tag{2}$$

$$R_{hY} = E\{hY^H\} = R_{hh}F^H X^H \tag{3}$$

$$R_{YY} = XFR_{hh}F^H X^H + \sigma^2 I \tag{4}$$

The FFT matrix F is given by:

$$\begin{bmatrix} W_N^{00} & \cdots & W_N^{0(N-1)} \\ \vdots & \ddots & \vdots \\ W_N^{(N-1)0} & \cdots & W_N^{(N-1)(N-1)} \end{bmatrix} \tag{5}$$

Where

$$W_N^{nk} = \frac{1}{N} e^{-j2\pi\left(\frac{n}{N}\right)k}$$

Assume R_{hh} (thus R_{HH}) and σ^2 are known at the receiver in advance, the MMSE estimator of h is given by:

$$h_{MMSE} = R_{hY} R_{YY}^{-1} Y_{HH} \tag{6}$$

Note that if h is not Gaussian, h_{MMSE} is not necessary a MMSE estimator.

$$H_{MMSE} = Fh_{MMSE} \quad F[(F^H X^H)^{-1} R_{hh}^{-1} \sigma^2 + XF]^{-1} Y \tag{7}$$

$$= FR_{hh} [(F^H X^H XF)^{-1} \sigma^2 + R_{hh}] F^{-1} H_{LS} \tag{8}$$

$$= R_{HH} [R_{hH} + \sigma^2 (XX^H)^{-1}]^{-1} H_{LS} \tag{9}$$

MMSE tries to minimize the MSE error rate. It has better rate than LS estimator having high SNR ratio.

4 Performance analysis

Channel estimation based on block based pilot arrangement using LS and MMSE channel estimators has been implemented for OFDM system using MATLAB platform.

Table 1. Performance parameters analysis

Number of sub channels	256

Number of pilots	256/8
Guard interval	Number of sub channels/4
Modulation index, M	4
Channel Length	16,32
Pilot energy levels	0.5, 2, 4, 6

The figure 3 shows the BER for BPSK modulation in which the channel estimation is done using LS algorithm at different pilot energy levels. It is seen that performance gets affected considerably with change in pilot energy levels. For very low energy level (=0.5) the BER deteriorates. It improves with increasing pilot energy (=2). However, further increase in energy results in increasing the bit error rate. This implies that there exist an optimum value for pilot energy for getting better performance.

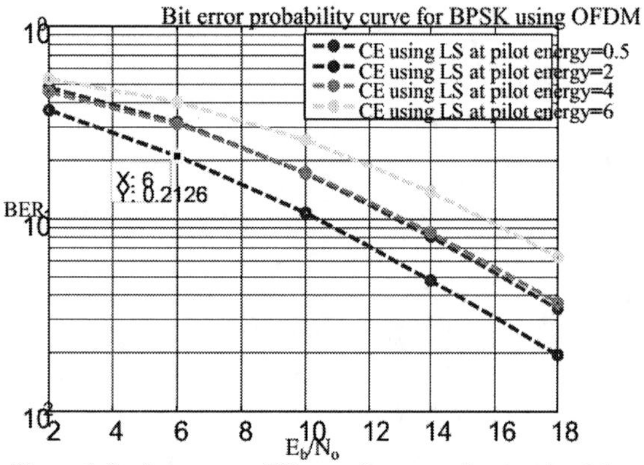

Figure 3. Performance at different pilot energy levels using LS algorithm, L=16

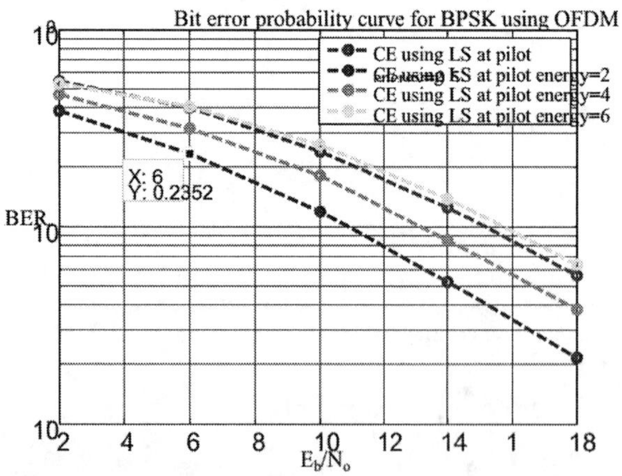

Figure 4. Performance at different pilot energy levels using LS algorithm, L=32

The figure 4 shows the BER rate for BPSK modulation in which the channel estimation is done using LS algorithm at different pilot energy levels and at channel length L=32. It is seen that performance gets affected considerably with change in pilot energy levels. For very low energy level(=0.5) the BER deteriorates. With increase in pilot energy level beyond E=0.5, the performance tends to improve with increase in energy level upto E=2. However, further increase in energy level does with help rather BER starts deteriorates as shown for pilot energy E=4, 6. Though the BER remains slighty better then that at E=0.5. So, E=2 pilot energy level seem to be the optimum one.

Futher, it is seen that with increase in channel length, for given energy per bit to noise power spectral density ratio, BER is slight higher as compared to lower channel length.

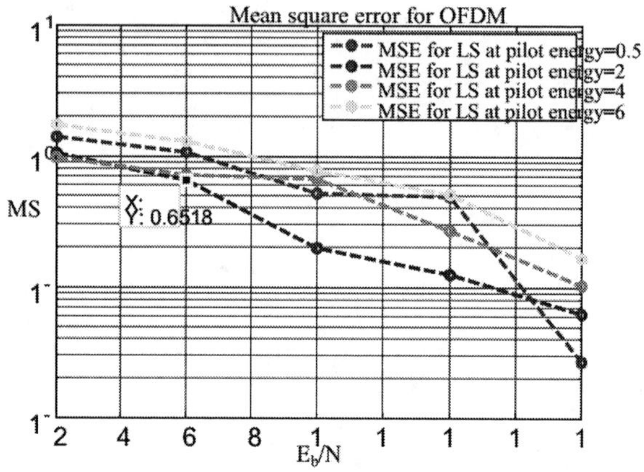

Figure 5. MSE at different pilot energy levels using LS algorithm, L=16

The figure 5 shows the MSE for OFDM using LS algorithm at different pilot energy levels (=0.5, 2, 4, 6). It is seen that MSE for low pilot energy level (=0.5) is higher than that at energy level (=2). However, beyond pilot energy level (=2), the MSE starts increasing. The MSE at pilot energy level (=6), it is highest among all.

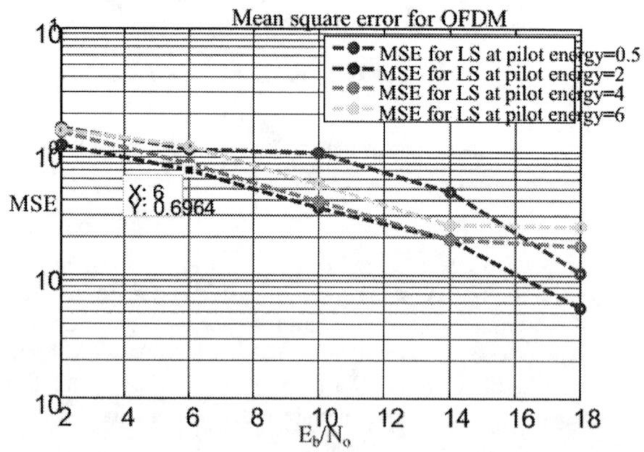

Figure 6. MSE at different pilot energy levels using LS algorithm, L=32

The figure 6 shows the MSE for OFDM using LS algorithm at different energy levels. It is seen that performance for MSE gets affected considerably with change in pilot energy levels. For very low energy level(=0.5) the MSE deteriorates. It improves with

DOI: 10.1051/matecconf/20165701011

increasing pilot energy (=2). The MSE for low pilot energy level (=0.5) is higher then that at energy level (=2). However, beyond pilot energy level (=2), the MSE starts increasing. The MSE at pilot energy level (=0.5), it is highest among all.

With increasing E_b/N_o ratio, the MSE tends to decrease which indicates there exists a trade-off between MSE and E_b/N_o.

The figure 7 shows the BER vs E_b/N_o for BPSK modulation in which the channel estimation is done using MMSE algorithm at pilot energy level (=2). It is seen that BER tend to reduce with increase E_b/N_o. At all pilot energy levels the BER for channel estimation using MMSE is considerably same. The result shows here are for channel length (L=16) at pilot energy level (=2).

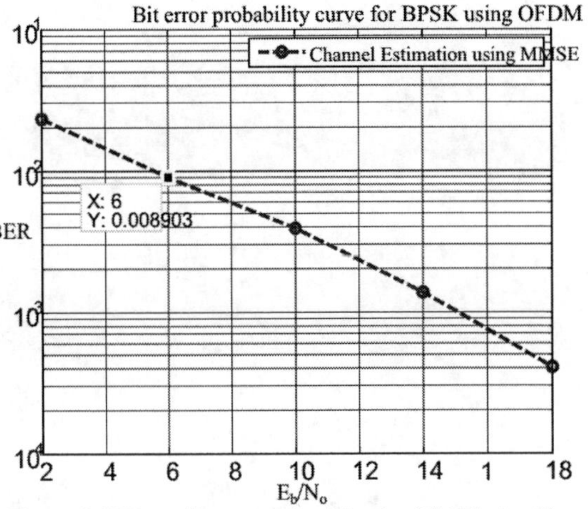

Figure 7. BER at pilot energy level 2 using MMSE algorithm, L=16

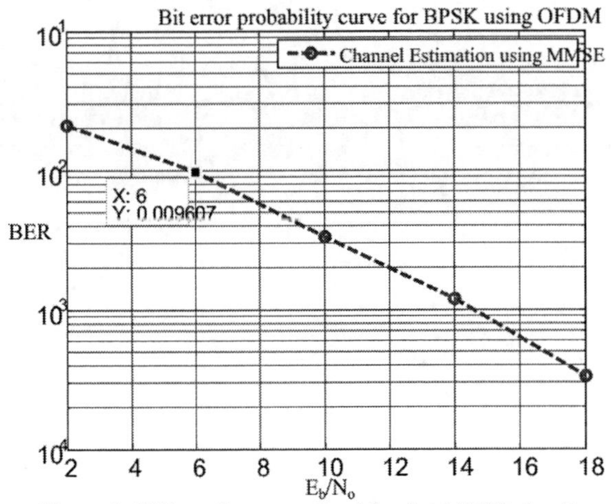

Figure 8. BER at pilot energy level 2 using MMSE algorithm, L=32

The figure 8 shows the BER vs E_b/N_o for BPSK modulation in which the channel estimation is done using MMSE algorithm at pilot energy level (=2). It is seen that BER tend to reduce with increase E_b/N_o. The results shows for channel length (L=32) at pilot energy level (=2) using MMSE algorithm.

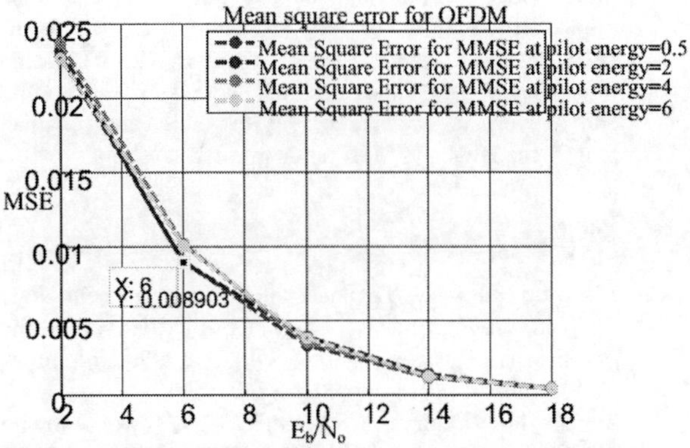

Figure 9. MSE at different pilot energy levels using MMSE algorithm, L=16

The figure 9 shows the MSE for OFDM using MMSE estimator at different pilot energy levels. It is seen that the MSE considerably increases by increasing pilot energy levels (0.5, 2, 4, 6) but MSE tends to reduce with increase in E_b/N_o=4,6,8. With further increase in E_b/N_o, the MSE at pilot energy levels (0.5, 2, 4,6) is almost same.

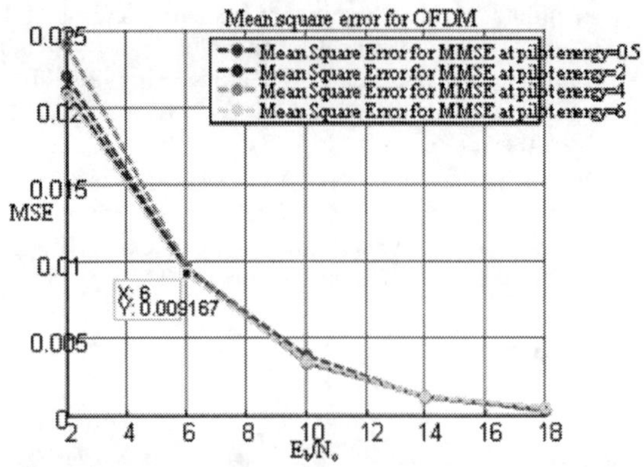

Figure 10. MSE at different pilot energy levels using MMSE algorithm, L=32

The figure 10 shows the MSE for OFDM using MMSE estimator at different pilot energy levels. It is seen that at pilot energy levels (0.5,2, 4, 6) the MSE tends to reduce with increase in E_b/N_o. Further increase in Eb/No (=12, 14,16), the MSE is almost same at all energy levels.

In comparison to similar results for LS algorithm, it is seen that BER is much lower in MMSE algorithm. Although, this improvement is at the cost of increase in complexity.

5 Conclusion

In this paper, channel estimation based on block type pilot arrangements is studied and analyzed using Least Square (LS) and Minimum Mean Square Error (MMSE) channel estimators. The channel estimation has been performed by inserting pilot tones into each OFDM symbol. Based on the results obtained, it is gathered that performance gets affected considerably with change in pilot energy levels. For very low energy level(=0.5) the BER and MSE deteriorates. It improves with increasing pilot energy (=2), though further increase is not helpful and performance degrades. With increasing channel length, the BER and MSE tends to increase slightly.

References

1. Stuti Ramola, "Digital Communication-Technology and Advancements", Advance in Electronic and Electric Engineering. ISSN 2231-1297, Volume 4, Number 4 (2014), pp. 367-374.
2. Anshul C.Gurve, S.S.Wasnik, A.D.Bijwe, Ashwini S. Yerlekar, Nilesh Chide, "Study of OFDM Variants and Implementation of OFDM Using fft / ifft", International Journal of Advanced Research in Computer and Communication Engineering Vol. 3, Issue 7, July 2014.
3. Srishtansh Pathak and Himanshu Sharma, "Channel Estimation in OFDM Systems", International Journal of Advanced Research in Computer Science and Software Engineering, Volume 3, Issue 3, March 2013, pp.312-327.
4. Lokesh Panchal, Manjil Upadhyay, Nirali Shah, Aniruddh Amin, "Significant Implementation of LS and MMSE Channel Estimation for OFDM Technique", Int. Journal of Engineering Research and Applications, Vol. 4, Issue 3 (Version 6), March 2014, pp.38-41.
5. Chaudhari Sachin V, Prof.A M Deshmukh, "Channel Estimation in OFDM", International Journal of Engineering Research and Applications (IJERA) ,Vol. 3, Issue 1, January -February 2013, pp.1413-1415.

MATEC Web of Conferences **57**, 01012 (2016)   DOI: 10.1051/matecconf/20165701012
ICAET-2016

Improvement In MAODV Protocol Using Location Based Routing Protocol

Sharnjeet kaur[1] and Manpreet kaur[2]

[1]*Mtech student, ECE Department, Yadavindra College of Engineering, Talwandi Sabo*
[2]*Assistant Professor, ECE Department, Yadavindra College of Engineering, Talwandi Sabo*

Abstract. Energy saving is difficult in wireless sensor network (WSN) due to limited resources. Each node in WSN is constrained by their limited battery power for their energy. The energy is reduced as the time goes off due to the packet transmission and reception. Energy management techniques are necessary to minimize the total power consumption of all the nodes in the network in order to maximize its life span. Our proposed protocol Location based routing (LBR) aimed to find a path which utilizes the minimum energy to transmit the packets between the source and the destination. The required energy for the transmission and reception of data is evaluated in MATLAB. LBR is implemented on Multicast Ad hoc On Demand Distance Vector Routing Protocol (MAODV) to manage the energy consumption in the transmission and reception of data. Simulation results of LBR show the energy consumption has been reduced.

1 Introduction

Wireless Sensor Network (WSN) is a recent and popular technology for the 21st century. Wireless sensor networks are likely to be broadly deployed in the future because they considerably extend the ability to monitor and manage the physical environment from remote locations. Such networks are formulated of a large number of sensor nodes, which are broadly deployed either inside the phenomenon or very close to it. These small sensor nodes consist of sensing, data processing and communicating components. Some of the application areas are health, military and security. Wireless sensor networks have unique characteristics like energy computation and storage constraints which present many challenging issues in the development and application of Wireless sensor networks. Wireless sensor network faces considerable problems in data processing, communication and sensor node management. This is because of uneven conditions like an uncertain and dynamic environment along with energy constraints. The Wireless sensor network poses additional challenges in the field of network routing protocols, energy efficiency, sensor node assessment strategies and energy models. There is required to analyze the performance of existing routing protocols, in order to grow new protocols that can extend network lifetime as well as can be easily manageable on network of different sizes.

Multicast ad hoc distance vector (MAODV) is one of routing protocols used for routing but this protocol is not energy efficient. This consumes high energy and power consumption. This paper proposes a Location Based protocol using the dijkstra algoritm and the estimated distance information among the sensor nodes and find out the shortest path between the source node and destination node with low cost. Location based routing takes location information into account while determining routes. Each node is aware of its geographical position (by the use of GPS), and this knowledge is exploited to create routes. It uses the shortest path position estimation between source and destination nodes in Wireless Sensor Networks.

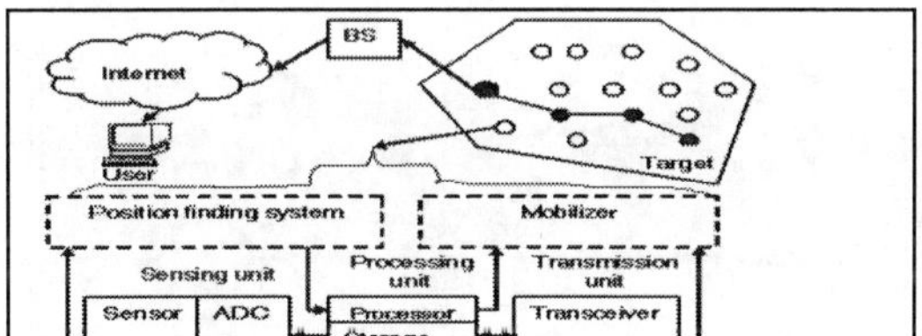

Figure 1. Structural view of sensor network

The focus of the paper is shortest path position estimation between source and minimize the energy consumption. The paper is organized as follows. Section 2 presents the previous work of energy consumption problems in wireless sensor networks. In section 3 the MAODV is discussed. Proposed method has been discussed in section 4. In section 5 Simulation Results of the paper has been discussed. In section 6 Conclusion and Future work, where the future challenges and directions to improve energy consumed in WSN technology are described.

[1]Sharnjeet kaur: sharankaur044@gmail.com

2 Multicast Ad Hoc Distance Vector Routing (MAODV)

MAODV consists of two basic routing operations such as route discovery and maintenance. In MAODV, when a source node sends a RREQ for a multicast group; generally it receives more than one reply. The source node has the record of the received route with the greatest sequence number and shortest hop count to the most near member of the multicast tree for a selected period of time, and disregards other routes. At the last of this period, it enables the selected next hop in its multicast route table, and unicast an activation message (MACT) to this selected next hop [11]. The next hop, on receiving this message, makes able the entry for the source node in its multicast route table. If the present node is a member of the multicast tree, it does not broadcast the message any further. However, if this node is not a member of the multicast tree, it will have received one or more RREPs from its neighbours. This method continues until the node that initiated the RREP (member of tree) is reached. The activation message makes sure that the multicast tree does not have multiple paths to any tree node. Nodes just forward data packets along activated routes in their multicast route tables. The multicast group leader is responsible for maintenance the multicast group sequence number and propagating this number to the multicast group. This is done through a Group Hello message. The Group Hello have increments in that indicate the multicast group IP address and sequence numbers of all multicast groups for which the node is the group leader. Nodes make use of the Group Hello information to update their request table. Then optimal path is established.

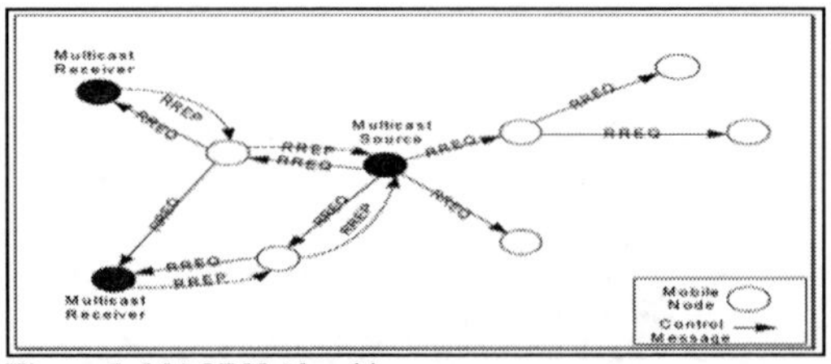

Figure 2. MAODV Algorithm

The network is deployed with the finite number of nodes. These nodes are free to move independently. The topology of network changes accordingly. Data is transferred from source to destination.

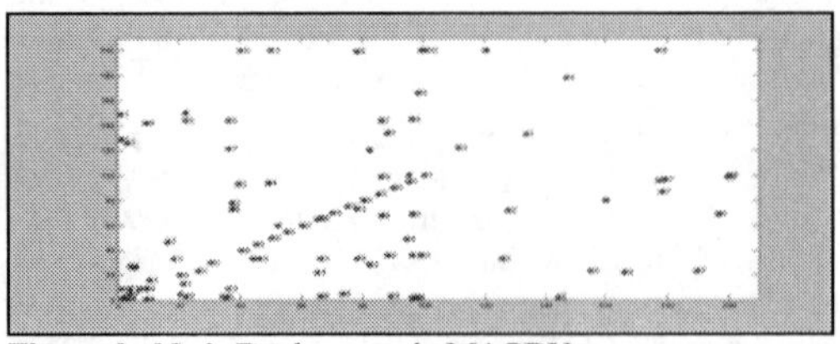
Figure 3. Node Deployment in MAODV

3 Proposed Work

Routing algorithms based on geographical location is an important research subject in the Wireless sensor network. They use location information to guide routing discovery and maintenance as well as packet forwarding, thus enabling the best routing to be selected, reducing energy consumption and optimizing the whole network. Location based routing protocol is very important in Wireless sensor network. It maintains a minimum energy network for wireless networks by utilizing low power GPS. This protocol can be used for mobile networks but it is best suited for sensor networks. This is because sensor networks are not mobile.

Location based routing takes location information into account while determining routes. Each node is aware of its geographical position (by the use of GPS), and this knowledge is exploited to create routes. Location based routing can be considered as a special case of context aware routing with the location being the context information that is used. LBR makes use of location information to reduce routing overhead. With the availability of GPS, it is possible for a mobile host to know its physical location. So Location-based routing is a system of rules to varying degrees of complexity dictating where inbound telephone calls from various locations are routed to. Thus, LBR is an on-demand or reactive routing protocol. It tries to search for a path from the source to the destination by broadcasting RREQ packets it its neighbors, similar to on demand routing.LBR has the protocol has many protocols which are based on location ability.

3.1. Problem implementation using LBR

Main problem in MAODV is that it consumes very high energy and battery. To improves MAODV another routing protocol LBR into this method. The location-based routing protocols compute routing paths based on the location of nodes. These protocols use node's location information, instead of links information for routing. They are also known as position based routing protocols. [9]In position based routing protocols, it is supposed that the packet source node has position information of itself and its neighbours and packet destination node. Coming to the way the LBRs GPRS protocol works, Packets in the GPSR protocol are marked by their originator with their destinations location. Forwarding in this way, the destination is reached in successively close geographic hops. In proposed approach only those nodes participate in route discovery process which has coordinates moving towards destination node. And the selection process is same as on demand routing protocol. LBR selects route using GPS. It contacts all the nodes in between the route coordinates. It takes all the nodes which come under the coordinates. Thus the energy wasted in MAODV can be recovered by this method.

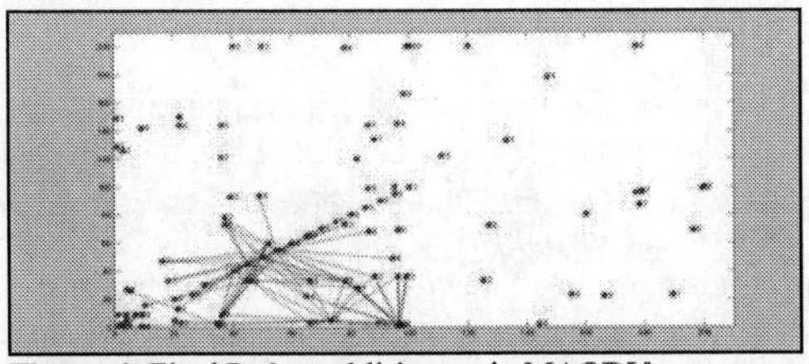

Figure 4. Final Path establishment in MAODV

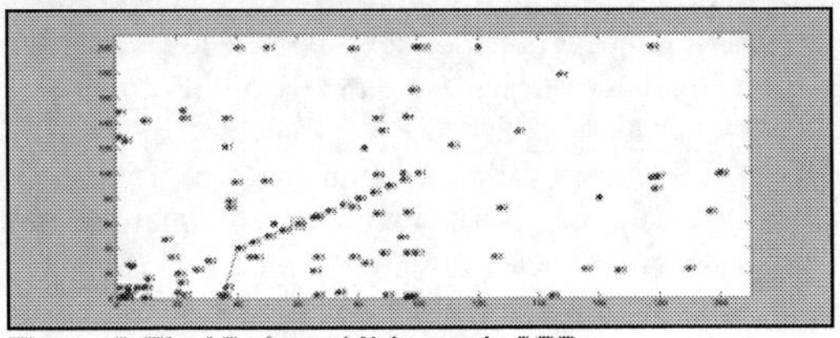

Figure 5. Final Path establishment in LBR

4 Experimental Results

In order to perform and calculate the impact of link failure in ADHOC network, the AODV implementation is modified using the MATLab tool.

Table 5.1 Simulation Parameters

SIMULATION PARAMETERS	VALUES
Simulator	MATLAB
Simulation area	200m×200m
Protocol used	MAODV,LBR
No. of Nodes	100
Initial energy	0.5J
Transmitting power	0.175mW
Receiving power	0.145mW
Node placement	Random
Radio range	40m
Tool	Matlab13

4.1 Energy consumption graph

Energy consumption is defined as the amount of energy consumed by nodes in the network through communication and processing. So, the total energy consumed, given as ET, can be calculated by adding all energy consumed by each nodes, n, for transmission (TX), received (RX) and processing throughout the simulation time.

$$E_T = \sum^{n}_{i=1} (E^{n}_{tx} + E^{n} rx) \quad (5.1)$$

Energy consumption graph is between time stamp and energy consumed. In case of MAODV network is consuming 0.04226J when sending data source to destination. And in case of LBR the energy consumption is reduced to 0.00062J.

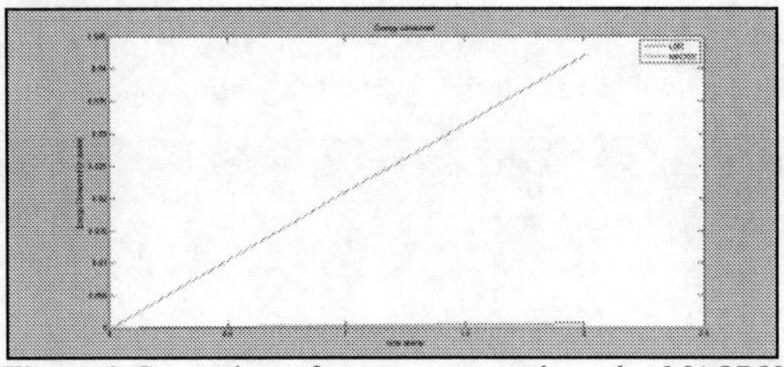

Figure 6. Comparison of energy consumption using MAODV and MAODV with LBR

4.2 Time duration for path selection graph

Time duration is total time for finding the path from source to destination. It calculates the time when sender sends RRQ to destination. This section shows comparison between is the time consumption for path selection for LBR and MAODV.

$$e= \text{cputime-t} \quad (5.2)$$

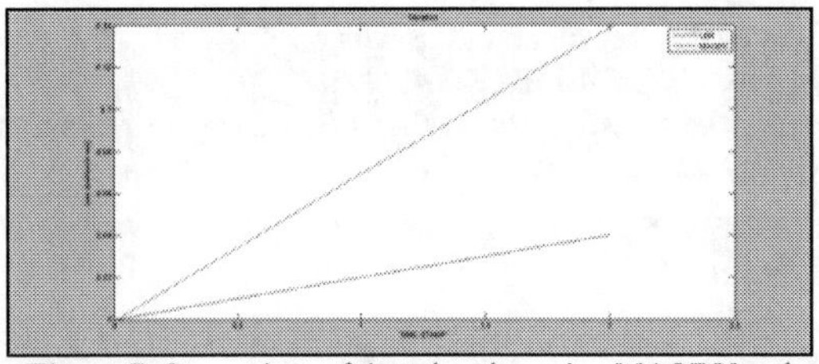

Figure 7. Comparison of time duration using MAODV and MAODV with LBR

4.3 End to End delay graph

End to end delay is average time taken by a data packet to arrive in the destination.

$$\text{End to end} = N [d_{trans}+d_{prop}+d_{proc}] \quad (5.3)$$

Graph shows the comparison of both with and without using LBR. MAODV takes 15.81 sec of delay while LBR takes only 0.3256 sec.

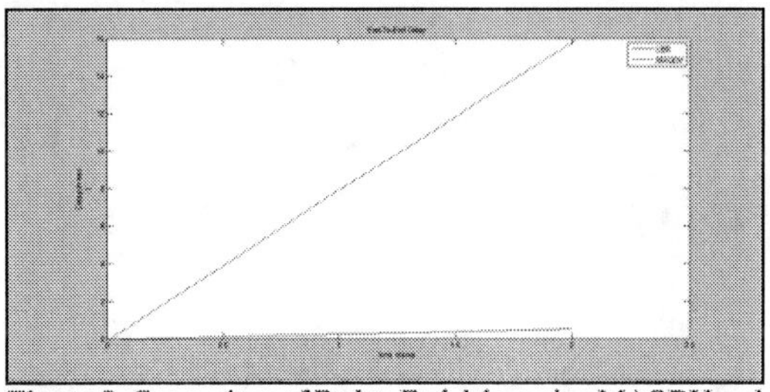

Figure 8. Comparison of End to End delay using MAODV and MAODV with LBR

4.4 Nodes used graph

These are total no of nodes which are participating in the communication process for different routing protocols. From the result it shows that MAODV is consuming more no of nodes. MAODV is using 58 nodes in routing process while LBR technique is using only 10 nodes.

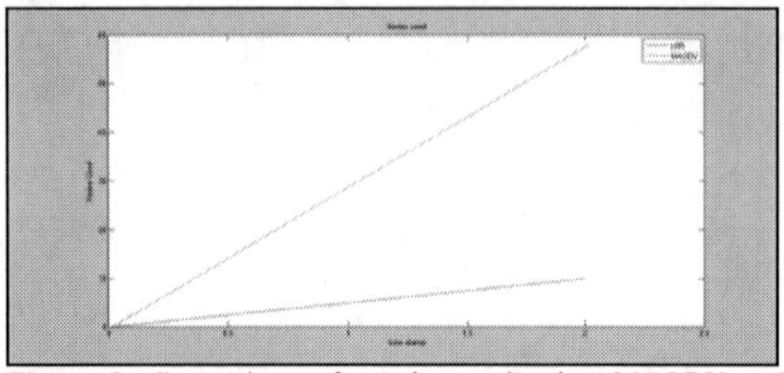

Figure 9. Comparison of nodes used using MAODV and MAODV with LBR

4.5 Residual energy graph

It is the energy remained in the network is residual energy. It is calculated by:

Energy remained =Total energy of the network-energy consumed. (5.3)

The energy remained in the network is 49.5774J in case of MAODV. For LBR it is 49.9938J.
From the results it shows that LBR has more remaining energy as compared to MAODV.

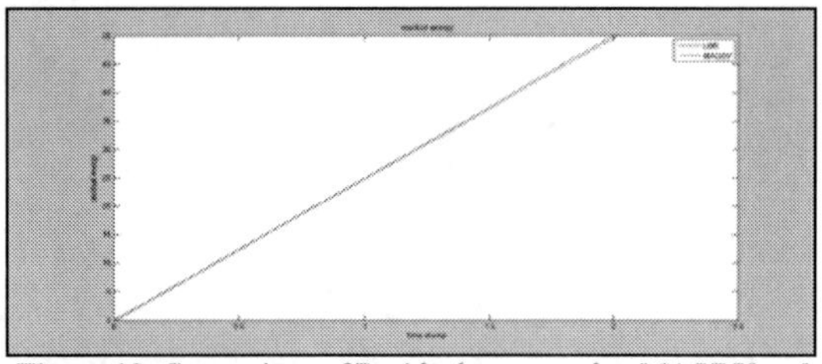

Figure 10. Comparison of Residual energy using MAODV and MAODV with LBR

4.6 Node battery consumption graph

It is energy consumed by nodes in the process. In LBR nodes are consuming very less energy as compared to MAODV. This shows how much battery is consumed.

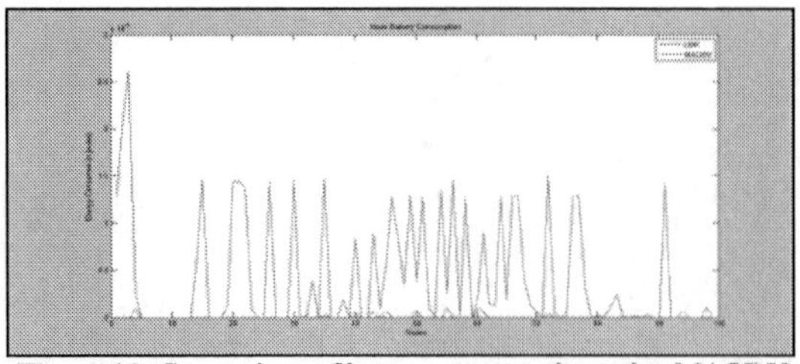

Figure 11. Comparison of battery consumption using MAODV and MAODV with LBR

6 Conclusion

MAODV protocol is used to find out the path of the data transfer. But simple MAODV has the problem of energy consumption. This technique takes lot of battery consumption and also it is very time consuming. More no of nodes are consumed in this process. Enhancement in MAODV is required to overcome the problems so that consumption of energy can reduced in data transfer from source to destination. In my proposed work, enhancement in MAODV concept is important. The objectives listed in the problem statement have been carried out properly in my work. I have added LBR protocol to MAODV protocol. This work helps in reducing the problem of energy and power consumption. Also improve the residual energy and delay.
The energy and power consumption are the major issues in wireless networks. A lot of research has to be needed to reduce energy in wireless networks. More efficient protocols will be used to reduce energy and power which will provide great performance.

References

1. AZZEDINE BOUKERCHE, "Performance Evaluation of Routing Protocols for Ad Hoc Wireless Networks", Mobile Networks and Applications 9, pp.333–342, (2004).
2. Abid Ali Minhas, Fazl-e-Hadi, Danish Sattar, KashifMustaq and S. Ali Rizvi,"Energy Efficient Multicast Routing Protocols for Wireless Sensor Networks",World congress on information and technology , pp.178-181, (2011).
3. B.Baranidharan, B.Shanthi, "A Survey on Energy Efficient Protocols for Wireless Sensor Networks", International Journal of Computer Applications (0975 – 8887) Vol.11, No.10, pp.35-40, December (2010).
4. Charles E Perkins, Elizabeth M Royer, "Adhoc OnDemand Distance Vector Routing",Mobile Computing Systems and Applications, 2000. Proceedings, pp.1-11.IEEE, (2000)
5. Daehee Kim, Sejun Song,and Baek-Young Cho,"Energy-Efficient Adaptive Geosource Multicast Routing for Wireless Sensor Networks, Hindawi Publishing Corporation Journal of Sensors, pp.1-14, (2013)
6. David B. Johnson ,David A. Maltz ,Josh Broch, "DSR: The Dynamic Source Routing Protocol for Multi-Hop Wireless Ad Hoc Networks ," Mobile Computing, T. Imielinski and H. Korth, Eds, vol. 5
7. Emad Felemban,"MMSPEED: Multipath Multi-SPEED Protocolfor QoS Guarantee of Reliability and Timeliness in Wireless Sensor Networks", IEEE Transactions On Mobile Computing, Vol. 5, No. 6, pp.738-754, (2006)
8. Heng Liu, Yan-tao Liu, Ting Yan,"Making On-demand Multicast Routing Protocol Aware of Energy in Wireless Sensor Networks", pp.1-4, IEEE , (2008)

9. Hyuntae Cho and Yunju Baek,"Location-Based Routing Protocol for Energy Efficiency in Wireless Sensor, pp. 622 – 631, IEEE , (2005)
10. Jorge Sa Silva, Tiago Camilo, Pedro Pinto, Ricardo Ruivo, Andre Rodrigues, "Multicast and IP Multicast support in Wireless Sensor Networks", Wireless Pervasive Computing, 2007. ISWPC '07. 2nd International Symposium on, IEEE, (2007)
11. Kiyotaka Oe, Akio Koyama, Leonard Barolli, "Performance Evaluation of a Multicast Routing Protocol for Wireless Mesh Networks Considering Network Load", IEEE 27th International Conference on Advanced Information Networking and Applications, Vol. , No. 4, pp.591-597, (2013)
12. Luo Junhai, Ye Danxia, Xue Liu, and Fan Mingyu, "A Survey of Multicast Routing Protocols for Mobile Wireless Networks", IEEE COMMUNICATIONS SURVEYS & TUTORIALS, VOL. 11, NO. 1, pp.78-91, (2009)
13. Manish V. Bhosle, Mininath K. Nighot and Somitra S. Das , "An Energy Efficient and Reliable Location wise Data Aggregation In WSN, IEEE International Conference on Computing Communication Control and Automation", pp.322-326, (2015)
14. M. Raj Kumar Naik, Hemant Bhardwaj and P.Samundiswary, "Residual Energy Analysis of DEC and Delivery Ratio Analysis of E-AOMDV protocol for Mobile Sensor Networks", International conference on Communication and Signal Processing, April , 2013, India,pp.509-513,IEEE,2013.
15. Mohammad-Amin Kharraz, Hamid Sarbazi-Azad and Albert Y. Zomaya, "On-demand multicast routing protocol with efficient route discovery", ELSVIER journal of Network and Computer Applications,pp.942-950, (2012)
16. Mohamad Nazim Jambli, Kartinah Zen, Halikul Lenando,"Performance Evaluation of AODV Routing Protocol for Mobile Wireless Sensor Network", IEEE 7th International Conference on IT in Asia (CITA), pp.1-6, (2011)

WAVELETS based wireless VOIP and its future scenario

Sarvjit Singh[1,a] and J.S Sohal[2]

[1]*ECE Department, UIET Panjab University, Chandigarh, India*
[2]*LCET, Katani Kalan, Ludhiana, India*

Abstract. Those who use VOIP, know that a good quality service cannot be ensured over the internet as the internet is not well suited to render real time services. Besides these users do not pay much for the VOIP calls as compared to the circuit switched phone calls. So they do not bother much about the quality of services. However, it is expected that the quality should be sound enough to pay for their time as well as money. In this paper, the recently published literature has been reviewed along with the introduction to wavelets.

1 Introduction

Internet is most important and has already established vast converged network that supports huge communication. Initially, internet was used to transmit data only. In 1995, there was an advancement, to use internet for transmit voice traffic and video in addition to data. Now a days, Voice over Internet Protocol (VOIP) is the prominent technology used in internet upon which whole world is dependent. In VOIP one can simply transport voice signal, DTMF tones, signaling traffic, and fax over the converged network of internet. Public Switched telephone network (PSTN) is based on technology of circuit switching, which needs a dedicated channels for telecommunication; whereas VOIP is based on transmission of voice signals in form of packets, routed through packet-switched networks. Major drawback of circuit switching is rigidity as dedicated circuits remain idle during silent periods. Consequently, network resources get wasted during such periods. These networks used packet switching which transmits digitized voice data packets over the networks using multiple paths. These packets are collected at the other end to regenerate the voice signals.

In VOIP networks, network resources are not reserved, but this method gives more flexibility to the network with some congestion. Therefore, VOIP networks do not ensure voice quality to the end host. Internet is best suited converged network, but not sufficient for real time VOIP traffic applications.

2 Element of VOIP

Codec's (Coder/ Decoder), Packetizer and playout buffer are the three important elements of VOIP. At transmitter level voice signals are converted into desired digital signals, which are compressed and encoded into their predetermined format, Figure 1 show the block diagram of VOIP system. There are many codecs available for digitizing speech, standardized by International Telecommunication Union Telecommunication (ITU-T) such as G.711, G.722, G.726, G.728, G.729, G.729(A), G.729(e), G.723.1(6.3), G.723.1(5.3), AMR [1] IS-127 codecs are defined by TIA (Telecommunications Industry Association).The packetization is carried out in which parts of encoded voice are converted into equal packet size. Furthermore, protocol headers such as Real-Time Transport protocol (RTP), User & Datagram Protocol (UDP), Internal Protocols (IP), data link layer header from different layers are added to encoded voice packets as shown in Figure1.RTCP and RTP application layer was meant to support real time applications. TCP protocol is used in internet for less delay sensitive data packets. This introduces delay as the receiver has to notify these senders for each packet by sending acknowledgements. UDP protocol is mostly used in VOIP and other delay sensitive real time applications and it does not use scheme of acknowledgement and are faster and simpler .Therefore, more preferred for VOIP application. Then these equal size packets are transmitted over IP network to its receiver. At receiver, these received packets undergo reverse process of decoding and de-packetization. During transmission there occurs a variation in time for delivery of packets, to the receiver. This variation of time is termed as Jitter. Therefore, this Jitter can be controlled by the use of playout buffer at receiver end. Figure 1 show the block diagram of VOIP system.

VOIP also has some of signally protocols such as SIP and H.323. These signally protocols are responsible for making VOIP calls at transmitter and to end VOIP calls at Receiver [2]. H.323 signally protocol was standardized by ITU-T for smooth functionality with PSTN and used in VOIP, Internet telephony and IP based video conferencing. H.323 protocol is dependent on IETF

[a] Corresponding author: sarvjit100878@yahoo.co.in

(Internet Engineering Task Force), RTP and RTCP with additional protocols for call signally and data and audio visual communication. SIP is an different to H.323. It is an Application Layer signally protocol (end-to-end) to set up, modify and end multimedia sessions over the Internet. Wireless communication is the fastest growing segment. VOIP network was mainly organized by wireless technology and play a major role in next generation wireless VOIP. However, this technique is infancy before widespread use of this technology. Some urgent issues have to be settled. It is imperative to examine if wireless VOIP can provide a quality of service comparable to PSTN and cellular networks. It is also necessary to fix the number of concurrent users. A wireless network can support simultaneously without degradation of quality of service.

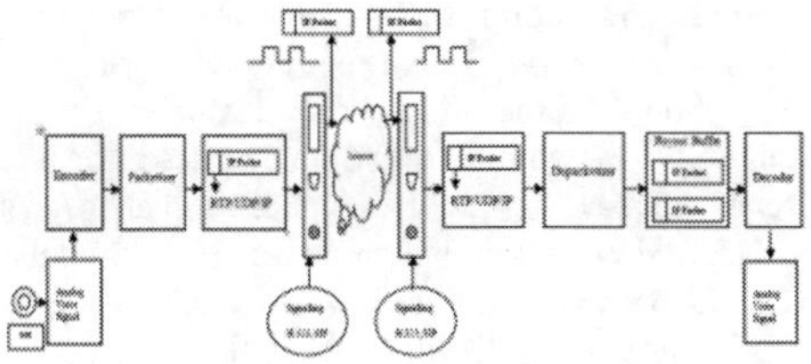

Figure 1. Block diagram of VOIP system..

To deal with this issue, first of all, we must define QOS and outline the factors effecting QOS in wireless networks. It may be referred to as a metric that measures the overall user satisfaction with the performance of the network. QOS depends on its applications and its requirements vary according to the applications. Wireless network suffers from their own merits and limitations, some of which affect performance of VOIP service. While travelling over the wireless VOIP network, voice packets must take care of factors which may affect the overall voice quality. The selection of CODEC, loss of packet, jitter and latency are the major factors that parameterized the quality of voice.

3 Wavelets

These could be trace back to Jean Baptiste Joseph Fourier harmonic analysis. He was the first to develop a method of describing any periodic function in terms of a weighted sum of cosines and sines. He also developed the transform that allows the expression of non –periodic time domain functions in the frequency domain, which have the scaling property and are known as Haar Wavelets family. It constitutes the simplest known wavelet set that can be used to analyze a given signal in terms of functions that are more localized in time domain than their harmonic ones. Later, Paul Levy demonstrated that Haar wavelets give better results for the study of the Brownian motion as compared to the Fourier analysis as it has a scaling property [3].
In 1982, Jean Morlet, who was the French Geophysicist gave the concept of the wavelet. A wavelet means a small wave and the study of wavelet transform is a new tool for seismic signal analysis. Then Alex Grossman studied an inverse formula for the transformation of wavelets. Jean Morlet and Alex Grossmann [4] brought out a detailed mathematical study of the continuous wavelet transform and applications their off, without realizing that similar results had earlier been available.
The effort provided a new method for breaking up a function. Wavelet may be termed as mathematical functions that sever data into different frequency components and then each component is studied with respect to a matching resolution [5].
Temporal analysis are achieved with high frequency mother wavelet, frequency analysis are carried out with a dilated low frequency wavelet. Then these set of high frequency and low frequency wavelet functions are transformed into wavelet coefficients but at different positions and different scales and final results are taken by wavelet transform of original signals [6].
Basically, wavelet analysis are made use of for the improvement of seismic signals by changing from short time Fourier analysis to an improved step by step procedure to detect sudden changes in signal reception. It does not contain any local information, the standard /classical Fourier transform are not adequate. To overcome this, Gabor transform can be used which has a feature of short-time Fourier transform by windowed-Fourier transform [7].
For security reasons of communication channels in VOIP and unlimited bandwidth, it necessitates to propose a need of codec and compression of audio signal to more compact form before transmission. Therefore, researchers were keen to work on this recent field to develop techniques ensuring audio coding to reduce the coded bit rate. Presently, audio compression uses wavelet transform and attracted more attention due to their encouraging S/N ratio and compression ratio. Moreover it shows the flexibility in representing in audio signals.

4 Literature Background

A significant work was done on Voice over Internet Protocol and Voice over Internet Protocol using wavelets and lot of work has to be done for wireless VOIP. It was found that wavelet was an excellent tool for the analysis.
It has been observed that analysis of these temporal and spectral properties of moving signals was carried out. Besides, there was developed a technique for detecting the beat attributes of music [8]. Moreover, a thresholding technique on the wavelet coefficients was also incorporated [9].A combination of wavelets and multirate filter bank in the channel coding and modulation was addressed in addition to "fractal modulation" with unpredictable communication link and in broadcast applications with QoS [10]. A new high quality, high compression approach based on AWP (Adaptive Wavelet Packet) has been proposed to achieve perceptual transform compression in [11]. Adaptive data selecting scheme for the threshold was developed for wavelet shrinkage based noise removal. Thus by considering magnitude and position of wavelet coefficient, a procedure is carried out on 2D cumulative sum of

coefficient [12]. To measure the Quality for VOIP speech, a wavelet based BCF (Bark Coherence Function) was developed, which proved to be robust in VOIP applications especially to variable delay. An experimental result on VOIP speech showed that WBCF (wavelet Bark Coherence Function) performs better as compared to PSQM (Perceptual Speech Quality Measures) and BCF [13]. The loss of quality of speech over the VOIP was also examined by loss of packets and delay jitter by proposing a new formula to consider these effects in [14]. Network delay in moving time series signal was proposed which is better than AR procedure, but new DIAR (Differential Auto Regressive) algorithm was proposed which is superior to AR algorithm; it failed to extract all the stationery components from the original non-stationary series, which leads to the impression of prediction. To compensate these weaknesses, a wavelet theory was introduced into the adaptive playout procedure and proposed Wavelet Transform Auto Regressive algorithm. Finally WTAR's complexity was relatively high as compared to AR and DIAR [15].The STI (Speech Transmission Index) and SII (Speech Intelligibility Index) are the traditional intelligibility assessment methods and are not intended for VOIP-PSTN channel applications. But after certain adaptations, a hypothetical solution to intelligibility assessment for NB (narrowband) speech transmission channels involving packetized networks was offered [16].A modified PCF (Point Coordination Function) i.e. DPCF (Dynamic Point Coordination Function) scheme was proposed allowing VOIP packets to be transmitted during CP only, when more than one packet is present in an STA's queue [17].The CSR (Compressed Speaker Recognition) builds selective features directly from the already compressed VOIP packet. It reduces the requirement of memory buffer and also increases the scalability [18].Then a scheme for speech hiding was proposed in which, the hiding of a secret speech into the cover speech was done to protect the secret speech from perception [19].The performance of different audio CODECs (G.711, G.723.1, G.726 ADPCM, GSM, G.728 LC CELP, G.729 and G.729 CS-ACELP) was studied and the simulation results shows that G.723.1 audio codec in WLAN gives a least Packet end–to-end delay and Packet Delay variation and thus proved to provide an acceptable user satisfaction for WLAN VOIP [20].An adaptive steganography scheme for VOIP was proposed which proves that adaptive steganography better performs than the traditional one [21]. Since wireless means can degrade VOIP quality because of random packet losses and heavy replication be a very effective method to save these losses [22]. Then it was proved that the wavelet transform approach works better than other methods [23].VOIP capacity over WLAN and the number of calls for different voice codecs and packetization interval based on the IEEE 802.11b standard was studied. At the same time, challenges for security issues were discussed and their solutions are investigated [24].A scheme based on the detection of a wavelet aimed at SIP flooding attack in VOIP network was proposed. The percentage of energy consumed according to the detailed signal was used to quickly change secret attack [25].Later on, a new steganography system for secret application was developed. It uses to hide speech files, which are impossible to differentiate [26].The delays are attributable to the difference in the jitter so we may conclude, that the data transmitted through delayed packets would not be much useful as it want add a bit to the comprehension of the message any longer. To minimize the loss the balance between size of the packet and buffer should be balanced. [27].A VOIP linking over CBR linking, over three different implementations (namely Ad-Hoc on Demand Distance Vector Routing Protocol, Fisheye Routing Protocol and Bellman-Ford Routing Protocol). The average packet lost rate and total packet sent, as seen from results considerably the packet loss portion is higher in CBR as compared to VOIP traffic generator application [28].Ethernet has much stability and insignificant delay connection as compared to wireless connection. Intervention in the neighborhood of wireless router very much reduces QOS. Consequently VOIP shows some downbeat features as it was considered a great alternative to the traditional circuit switches phones. [29].In IEEE 802.16 plan of network and MAC description which make end to end VOIP traffic video streaming possible use a network simulator. It impact is felt on QOS criterion. [30].

5 Need of Improvement in Wireless VOIP

Wireless communication is the fastest growing segment of communication industry. The unprecedented development of wireless system along with large number of computer based application show a very bright future for the expansion of wireless networks for VOIP. But there are many technical hurdles that challenge the designing of strong wireless network that will ensure satisfactory performance of emerging applications. There is a greater chasn between current and emerging systems which make it necessary to visualize the further wireless applications, a new reality. In this context, we must mention wireless VOIP has very promising application that may help build strong wireless network.

6 Importance of using Wavelets in Wireless VOIP

Over the last decade, wavelets have been a chosen field for research in wavelet based communication system. In a very short span, a much progress has been achieved in this direction. Under VOIP system audio signal is the basic unit of the signal. It is something non- stationary and of varying nature with respect to time. Such signals had temporary drifts and sudden changes. The wavelets have features of localization as well as time frequency resolution because of which these become more suitable for the transmitted audio signals through wireless VOIP applications. We have to select the most suitable wavelet adaptable to voice or data. This data is sparingly reused. This makes wavelet a very good tool for the study of wireless VOIP analysis.

7 Conclusions

Future work will considered involving simulating scenarios that are closer to the real world models and work is expected to end in a wireless VOIP system with an improvement in the performance of system. This type of VOIP is viewed as a very viable substitute for PSTN. It has to satisfy the customer with high QOS. On the contrary, IP networks are still unable to fulfill the desired quality. Presently available VOIP applications which were designed for the wired environment have been improved so that wireless environment does not get degraded. With view to developing a highly improved wireless VOIP system, we will have to further investigate number of parameters so that the performance of wireless VOIP becomes matchless for the receiver.

References

1. B. Goode ''Voice over Internet Protocol (VOIP)''. Proceedings of IEEE, 90, pp. 1495-1517, (2002)
2. R. P. Swale. BT Technology Journal, 19 n.2, p.9-22, (2001)
3. C. Gargour, M. Gabrea, V. Ramachandran and J.M Lina, "A Short Introduction to Wavelets and their Applications", IEEE -CS Mag, pages 57-68, (2009)
4. A. Grossmann and, J. Morlet, SIAM Journal of Analysis, 15: 723-736, (1984)
5. M. Misiti, Y. Msiti, G.Oppenheim and J.M Poggi, "Wavelet Toolbox TM user's Guide", The math works inc; (2000)
6. R Polikar, "The Wavelet Tutorials", July16, (2001)
7. M. Sifuzzaman, M.R. Islam and M.Z. Ali, "Application of Wavelet Transform and its Advantages Compared to Fourier Transform", JPS, 13, 121-134, ISSN: 0972-8791, (2009)
8. G.Tzanetakis, G. Essl, P. Cook, "Audio Analysis using the Discrete Wavelet Transform", ICASSP, (1980)
9. Goran Kronquist and Henrik Storm, "Denoising a Speech Signal using wavelets", IEEE Transaction on Communication, 28, pp 84-95, (1990)
10. G.W. Wornell, Member IEEE, " Emerging Applications of Multirate Signal Processing and Wavelets in Digital Communication", Proceedings of the IEEE, 84, no. 4, pages 586-603, (1996)
11. P. Srinivasan and L. H. Jamieson, "High Quality Audio Compression using an Adaptive Wavelet Packet decomposition and Psychoacoustic Modeling" IEEE Transactions on Signal Processing, 46, No. 4, pp. 1085-1093, (1998)
12. M.L. Hilton and R. T. Ogden, "Data Analysis Wavelet Threshold Selection in 2-D Signal Denoising", IEEE Transaction on Pattern Analysis and Machine Learning, 11 pp 674-693,1999.
13. S.Wook, Y.Cheol and D.H Youn, "Speech Quality Measures for VOIP using Wavelet based Bark Coherence Function", Eurospeech Scandinavia, (2001)
14. L. Ding and R. A. Goubran, "Speech Quality Prediction in VOIP Using the Extended E-Model" 0-7803-7975-6/03/$17.00 (C), (2003).
15. L.Zhong-bo, T. Tao, Z. S.hui, K. J.ming, "Wavelet Transform Based Adaptive Playout Algorithm for VOIP", lzb05296@163.com.
16. F. F. Li, "Speech Intelligibility of VOIP to PSTN Interworking-A Key Index For the QoS", Printed and Published by the IEE (The institute of Electrical Engineers), pp 104-108, (2004).
17. T.Kawata, S.Shin, A.G. Forte Henning Schulzrinne, "Using Dynamic PCF to Improve the Capacity for VOIP Traffic in IEEE 802.11 Networks".
18. C.Aggarwal, D.Olshefski, D. Saha, Z.Yin Shae, P.Yu, "CSR: Speaker Recognition from Compressed VOIP Packet Stream-IBM Research Report", RC23499 (W0501-090), (2005)
19. C.Wang and Q.Wu, "Information Hiding in Real-Time VOIP Streams", ninth IEEE International Symposium on Multimedia, pages 255-262, (2007)
20. I.U. Haq, K. M. Yahya, G.M.Hassan, T. M. Jadoon, "Analysis of VOIP by Using Different Vocoder in IEEE 802.11b Wireless Networks", NCSCCKK-NWFP, Pakistan, May 23-25, pages 84-87, (2008)
21. H.Tian, K.Zhou, H.Jiang, Y. Huang, J. Liu, and D.Feng, "An Adaptive Steganography Scheme for Voice over IP" IEEE ISCAS 2009.
22. A. Mondal, C.Huang, J.Li, M.Jain and A. Kuzmanovic, "A Case for Wifi Relay: Improving VOIP Quality for Wifi Users".
23. M. Sifuzzaman, M.R. Islam and M.Z. Ali, "Application of Wavelet Transform and its Advantages Compared to Fourier Transform", JPS, 13, 121-134 ISSN: 0972-8791, (2009)
24. H.Kazemitabar, S.Ahmed, K.Nisar, A.B. Said and H.B. Hasbullah, "A Comprehensive review on VOIP over Wireless LAN networks", www.csl.issres.net, 2(2), (2010)
25. J. Tang, Y Cheng, "Quick Detection of Stealthy SIP Flooding Attacks in VOIP Networks", Proceedings of IEEE ICC, Kyoto, Japan, (2011).
26. S.Rekik, D.Guerchi, S.A.Selouani and H.Hamam, "Speech steganography using wavelet and Fourier Transforms", EURASIP Journal on Audio, Speech, and Music Processing 2012, A Springer Open Journal, (2012)
27. I.Philipa, V.E Idigo, A.James, "Modeling and Simulation of Voice over Internet Protocol (VOIP)", (IJEIT) 1, Issue 4, ISSN: 2277-3754, pp 225-233, (2012)
28. R.Ganiga, B.Muniyal and Pradeep, "International Journal of Computer Applications (0975 – 8887)" 50 – No.11, pp 25-31, (2012).
29. Y.Pan, J.Chung, Z. Zhang, "Analysis of Performance of VOIP Over various scenarios OPNET 14.0", Final Report Group 11, SFU, Spring (2012).
30. A. Dubey, "Quality of Service (QoS) in Wireless Network and NS-3, VOIP Simulation Environment", (IJSCE) ISSN: 2231-2307, 2, Issue-5, (2012).

ACCURACY Detection of Digital Image Forgery by Using Ant Colony Optimization Technique

Sarvjit Singh [1, a], Sunil Agrawal[1] and Gagandeep Singh[1]

[1] *ECE Department, UIET Panjab University, Chandigarh, India*

Abstract. Image forgery is one of the well known fields in which researches continuously exploring new areas. In digital image forgery one can change image in many ways using several software's, researchers exploring new algorithms to detect image forgery areas and change it to original pixel values if possible. In this paper we employed ACO (Ant Colony Optimization) to find areas which are manipulated with some software. The experimental results prove that ACO is better than existing methods of detecting tampered regions in digital photo images.

Keywords - ACO, digital photo image, digital forgery, digital image forensics.

1 Introduction

In these days, with the advancement of digital image processing methods, it is very easy for anyone to tamper the digital images and hide the objects or manipulate the objects present in the digital photo image. This is possible due to the development of sophisticated image editing tools. Tampered images has great effect on our society because images are everywhere in our life such as magazines, news images, the images produced as court evidence and medical diagnostic etc. so in order to know that the images are authentic or unauthentic we need digital image forensics which are used to detect the tampered regions present in the digital photo images. For the last few years, digital image forensics becomes more attentive to restore the lost trust of digital photo images. That is why it is a great need to develop the methods for the digital image authentication in order to recover the confidence of people towards the digital photo images. There are two techniques to detect the digital image forgery.

One is active based approach and the other one is passive based approach. The active based approach is further divided into digital watermarks and signatures. Digital watermarking and signatures methods for the detection of tampered regions in the images has been used in the past times but for these methods we have to preprocess the data first, which creates difficulty to apply these methods [1].

Other approach is passive based. In this method we do not need any information about the image itself. In this approach in order to detect the suspicious regions we need to extract some intrinsic characteristics traces present in the digital photo image. We detect the tampered region by using the forged image as an input. For this reason passive based approach is most preferred.

Moreover, active based approach requires extra operations and harms the original image so passive based approach is better than the active based. In our work we are also using passive based approach and this is the most preferred approach to detect the digital image forgery for the past recent years.

The rest of the paper is organized as follows. In section II we discuss about the techniques which are available to detect the digital image forgery. Section III describes the method proposed by us to detect the tampered regions .section IV and V consists of results and conclusions respectively.

2 Digital Image Forgery Overview

As we discussed, the images are easily manipulated so it is advisable that we must not always believe that what we see is true. Here we are discussing the methods to detect tampered regions in the digital photo image. Xiaomei Quan et al [2] proposed a novel copy-move forgery detection scheme in digital images. It treats image as an high dimensional data and uses intrinsic dimension estimation to segment the image then detects the copy-move forgeries present in image region having same texture.

Xin Wang et al [3] put forward a passive blind digital image forgery detection based on consistency of defocus blur. Image patches which are at similar distance from lens have similar blur kernels and this consistency of blur is broken in the forged images which are used to detect the digital image forgery.

Fei Peng et al [4] proposed image forgery detection by using blur estimation and abnormal hue. Blurring is done to hide traces of forgery but it destroys the joint consistency of color channels in the image. Based on this

[a] Sarvjit Singh : sarvjit100878@yahoo.co.in

principle, blur estimation is done to the image and obtain the blur region, and exactly locate the artificial blur region by detecting the abnormal hue in the blur region.

Pravin kakar et al [5] proposed forgery detection by using in consistent motion blur. In order to find the spliced area in the digital image discontinuity in motion blur is used which in turn uses spectral matting for its estimation. In the end spliced area and remaining area are distinguished.

Jiri grim et al [6] uses local statistical models to detect the digital image forgeries. The local statistical models use Gaussian mixtures for the detection of tampered regions. This method detects the forgery of an image without any prior information.

Bo Xu et al [7] use phase correlation to expose image forgeries. This is a fast method which is based on non block matching based. If the image is forged then we get sharp peak in the phase correlation showing the proof of tampering.

Li kang et al [8] proposed a method to detect the copy move forgeries. But using block matching divide the whole image into equal blocks then to get a singular value matrix, singular value decomposition is applied to these blocks. In the end, the correlation coefficient is used to compare the blocks. Above are the methods of detecting digital image forgery and in the next section we are exposing digital image forgeries by Ant colony optimization (ACO).

3 Image Forgery Detection

Digital image forgery is classified into two categories. One is copy-move forgery and the other is copy-create forgery. When images are forged by copying one part of image and pasted on another part it is known as copy-move forgery. While in case of copy-create forgery copying and pasting is done by using more than just a single image. In our paper, we are dealing with the copy-move forgery only. We have employed Ant colony optimization (ACO) methodology which is used to detect the copy-move forgery.

3.1 Introduction to ACO

ACO is the intelligence technique used by the ants to find their food. When ants move randomly in order to find the shortest path between the source of food and their colony, they deposit pheromone along the path. Pheromone is a source of communication between them. An ant can do only simple tasks but not the complex tasks like searching shortest path from nest to food source [9].

As mentioned in [9], F is the food source at some distance from nest, N and after finding food they come back by path, (b) after leaving pheromone trail by an ant. At the start, all the ants move along all the possible paths but the pheromone intensity is strong for the shorter path than the longer path. Most of the ants follow the shortest path route from the nest to the food source; the pheromone trails on the other paths are lost. This intelligence technique is implemented in ACO to detect image forgery.

3.2 Proposed Methodology

The stepwise methodology is explained as follows:

Step 1: Firstly we load the images of size M×N in MATLAB environment which are already manipulated from some databases.

Step 2: Initializing ACO parameters to given number of iterations and search area. These parameters are ant population (which is 200), steps for the random movement of ants and search area in which random movement is allocated within the image. We take threshold value (50%).

Step 3: Initializing Ants population, we took ant as a location on which forgery is detecting .Here ant has two locations, one for x and another for y. X & Y corresponds to rows and columns with in the search limit (X×Y). From every location, every ant represents a block called ant block denoted by (K).

Step 4: After that we divide image into equal square blocks of size m×m within the search limit called current block denoted by (I).

Step 5: Randomly distribution of ants with in the search area and random movement given by them for generating and searching for food values at different locations.

Step 6: Nutrient function =

$$\sum_{i=1}^{m}\sum_{j=1}^{m}(|I_{i},j - K_{i},j|)_{R} + \sum_{i=1}^{m}\sum_{j=1}^{m}(|I_{i},j - K_{i},j|)_{G} + \sum_{i=1}^{m}\sum_{j=1}^{m}(|I_{i},j - K_{i},j|)_{B} \quad \text{------------ (1)}$$

Where R, G and B denotes the subtraction for Red, Green and Blue matrices present in both current and ant block in equation (1).

And predict food value using the above nutrient function.

Step 7: The selection and elimination process exchanges the food locations among ants.

Step 8: Process will continue for next N iterations.

In this way, minimum will be difference between the current block and ant block higher will be the probability of forgery. The analysis of an image using above mentioned technique promises the detection of tampered regions present in the input image. For performance evaluation of the proposed method, we use two parameters namely Recall and precision rates. A brief description of which is given below:

The precision rate and recall rate are evaluated by using equation (2) and (3) respectively. Precision rate is defined as ratio of the parts detected correctly to summation of parts detected correctly and false positives. False positives are those regions which are actually authentic but detected as forged.

$$\textbf{Precision rate} = \frac{\text{correctly detected parts}}{\text{correctly detected parts+false positives}} * 100 \quad \text{------- (2)}$$

Recall rate is defined as ratio of the parts detected correctly to the summation of parts detected correctly and false negatives. False negatives are those regions which are actually forged but detected as authentic.

$$\textbf{Recall rate} = \frac{\text{correctly detected parts}}{\text{correctly detected parts+false negatives}} * 100 \quad \text{------- (3)}$$

4 Experimental results

The experiments are performed in MATLAB and the format of images is JPEG. Our dataset consists of 50 images. In paper [10] JPEG compression technique and direction filter are used to detect the tampered regions in the digital photo image. In our work manipulation of images is done by using paint. We achieved better precision and recall rates than paper [10].

(a) Original image (b) Forged image

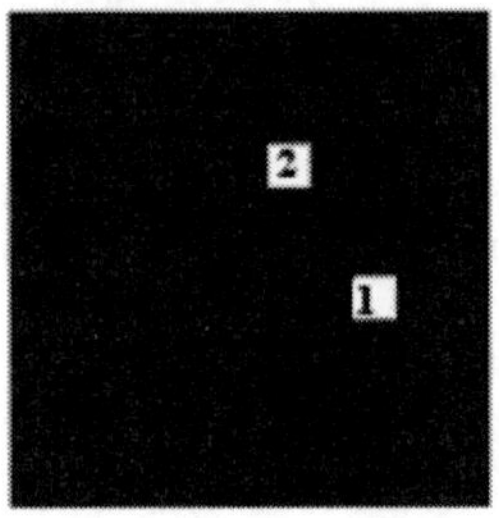

(c) Detected Forgery

Figure 1. Detection results of copy-move regions

The experimental results detecting the tampered regions are illustrated in following figures. Figure1 showing birds flying in the sky as a original image and copying one part of image containing bird and pasted on another part as forged image and in the last copy-moved parts are shown which we detect using ACO.

In our method, if there is any tampering in the original image then we obtain the black image of same size as that of original image showing high intensity on both the copy-moved parts as shown in figure1 and figure2.In the detected forgery the regions, shown in figure1(c) and figure2(c) which are labeled as **1** (of size 20X20) is the copied part in the image and the regions labeled as **2** is the pasted part (of size 20X20) showing high intensity at these regions.

(a) Original image

(b) Forged image

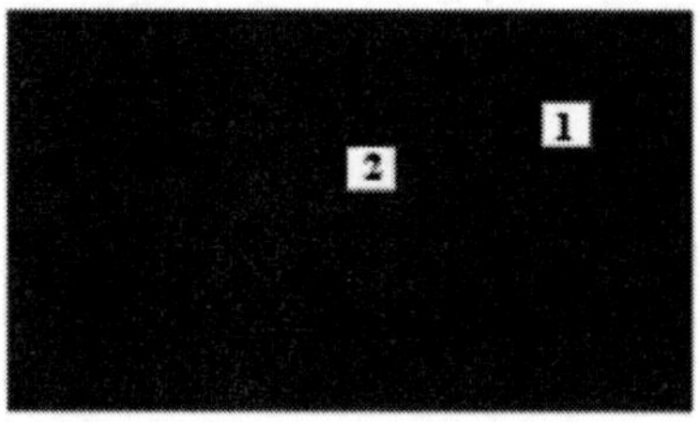

(c) Detected Forgery

Figure 2. Detection results of copy-move regions

In below table we are showing the performance comparison of method used in paper [10] and proposed method by using two parameters (precision rate and recall rate). These rates are calculated by using equations (2) and (3).

Table 1. Performance Comparison

Method Used	No. Of Images	Precision Rate	Recall Rate
Paper [10] (JPEG compression &direction filter based)	50	92.2	92.6
Proposed method (ACO based)	50	98.5	95.7

By using ACO, the precision rate is increased by 6.3% and recall rate is increased by 3.1% than the method used in paper [10]. It means ACO gives better performance than the method used in paper [10].

5 Conclusions

Digital image forgery detection is really complicated task in the present times because of the advancement of the new image editing tools. It is very difficult for an individual to detect the authenticity of an image by naked eyes. In order to improve the low detection rate of the forged regions in the digital photo image we have mentioned the intelligence technique (ACO) in our paper. Our method has given satisfactory experimental results. The proposed method shows excellent detection of the tampered regions. The advantage of our approach is to find the manipulated areas present within the input image precisely and accurately. Our proposed algorithm promises to expose the digital image forgery created by copy-move method.

6 Future Scope

In today's digital world, a lot of work has been done for the detection of digital image forgery. But there is a still scope of future work so that the image forgery can be detected more accurately and precisely.
Although in the proposed work, recall rates and precision rates have been increased but they can be further improved. As the precision rate and recall rate depends on false positives and false negatives respectively. So, the reduction of false positives and false negatives in future will give more improved results. Hence, it will help in achieving the ultimate goal i.e. accurate detection of image forgery.

References

1. S.Khan, A. Kulkarni, "*An Efficient Method for Detection of Copy-Move Forgery Using Discrete Wavelet Transform,*" IJCSE, **2**, No.5, pp: 1801-1806, (2010)
2. X. Quan, H.B. Zhang, "*Copy-Move Forgery Detection in Digital Images Based on Local Dimension Estimation,*" IEEE ICCSCWDF International Conference on (CyberSec), pp: 180 – 185, (2012)
3. X. Wang, B. Xuan, S. L. Peng, "*Digital image forgery detection based on the consistency of defocus blur,*" IEEE ICIIHMSP, pp: 192 – 195, (2008)
4. F. Peng, X.L.Wang, "*Digital image forgery forensics by using blur estimation and abnormal hue detection,*" SOPO, pp: 1 – 4, (2010)
5. P. Kakar, S. Natarajan and W. Ser, "*Detecting Digital Image Forgeries through Inconsistent Motion Blur,*" ICME, pp: 486 - 491, (2010)
6. J. Grim and P. Somol, "*Digital Image Forgery Detection by Local Statistical Models,*" IIH-MSP), pp: 579- 582, (2010)
7. B. Xu, G. Liu and Y. Dai, "*A Fast Image Copy-move Forgery Detection method using Phase Correlation,*" MINES, pp: 319-322, (2012)
8. L. Kang and X. Cheng, "*Copy-move Forgery Detection in Digital Image,*" CISP, pp: 2419-2421, (2010)
9. Chandrasekhar, N.P.Rao, "*Recent trends in Ant Colony Optimization and data clustering: a brief survey,*" IAMA, pp: 32 – 36, (2011)
10. S. Murali, B. S Anami, G. B Chittapur., "*Detection of Digital Photo Image Forgery,*" IEEE -ICACCCT, pp: 120 – 124, (2012)

Dynamic RWA in WDM optical network

Harpreet Kaur[1] and Munish Rattan[2]

[1]AP, BBSBEC,Fatehgarh Sahib,Punjab,India.
[2]AP, GNDEC,Ludhiana,Punjab,India.

Abstract. As optical wdm networks are becoming the backbone building blocks of twenty first century's networks due to their capacity of carrying huge bandwidth. So evaluating the performance of optical wdm networks and optimizing it, is need of the day. For this the problem of routing and wavelength assignment is of significance importance. Researchers have done extensive simulations for both static and dynamic routing and wavelength problem. In this paper we have presented various algorithms for solving the dynamic RWA problem in literature.Also we have prepared a optical wdm network using which RWA problem is analysed.

1 Introduction

Nowadays optical wdm networks are emerging technologies networks that can provide routing, grooming, and restoration at wavelength level [2]. WDM technology of crucial importance for satisfying the ever increasing capacity requirements in telecommunication networks. Wavelength routed WDM networks can exploit the large bandwidth of optical fibers by dividing it among different wavelengths. These networks are equipped with configurable WDM nodes which enable us to set up and tear down all-optical channels, called light paths, between pairs of nodes. Given a network and a set of light path requests, the Routing and Wavelength Assignment (RWA) problem attempts to route each light path request, and to assign wavelengths to these routes subject to the following constraints [1,13] .
The RWA problem can be categorized into two types, static RWA (SRWA) and dynamic RWA (DRWA). In a wavelength-routed optical network (WRON), the traffic demands can be either static or dynamic In static RWA, all the connection requests are known in advance and the objective of RWA is to assign routes and wavelengths to all the light paths globally while keeping the amount of network resource required minimal [5].In a dynamic traffic pattern, a light path is set up for each connection request as it arrives and it is released after some finite amount of time. One of the challenges involved in designing wavelength routed networks with dynamic traffic demands is to develop efficient algorithms and protocols for light paths establishing [7, 8, 10 and 11].
In all-optical WDM networks, each of the connections going through a link is assigned a wavelength. If a connection is assigned the same wavelength on all links along its routing path, the signal is able to travel from the source node to the destination node using the same wavelength. In case that the connection has to be assigned two or more wavelengths on different links, one or more nodes along its routing path must have the capability of converting the signal from one wavelength to another wavelength. If all nodes along the routing path of a connection are incapable of converting a wavelength to another wavelength, the connection must be assigned the same wavelength on all links along its path. Otherwise, the connection is blocked. This is known as the wavelength continuity constraint. A wavelength which is available on all links along a routing path is referred to as a common available wavelength along the routing path[6]. In addition, light paths that share a common physical link cannot be assigned the same wavelength. This is called the wavelength clash constraint [13, 1].
There are number of algorithms available in literature for dynamic routing and wavelength assignment problem and rsearchers too have used optimization techniques and have formulated various new algorithms to optimize the performance of optical wdm network in terms of dynamic wavelength and assignment problem. In this paper we have presented the review of various techniques available for solving dynamic routing and wavelength problem.

2 ALGORITHM USED FOR DRWA

2.1 Fixed Path Routing

It uses only one route from source end to destination .In this approach the network resources are not properly utilized.[1]

2.2 Alternate Path Routing

[a] Corresponding author: reetmavi82@gmail.com

It uses alternate paths from source to a destination .If the first is unavailable then second path is used.[15]

In dynamic routing in DRW1 algorithm no weight assignment is done all links have same weight. First all possible paths between the source and destination is calculated then request is routed over the shortest path if wavelength is available else alternate shortest path is used. In DRW2 less weight is assigned to nodes which are used less. As the request arrived it is routed to the nodes having less weight if wavelength is available if not then request is routed through next higher weight nodes and the process is repeated. In DRW3 algorithm less weight is assigned to the nodes which are less used and are nearest and then routing of request is done through routes having minimum weight if wavelength is available else next higher weight is tried and the process is repeated [9] Another algorithm is proposed in literature which first selects K edge disjoint least weighted paths according to the current network state and then one such path is considered to route the request taking into account not only the load ,length of the path but also number of transceivers in the nodes along the path.after this an auxiliary graph is generated and request is routed on least weight route on the graph and wavelength is assigned for the same.[14]

2.3 Fixed alternate shortest Path routing with least Priority wavelength assignment Strategies

In this algorithm the wavelength information table is made according to the use of the each wavelength in the network then the lightpath is set on shortest or alternate shortest path with the least used wavelength if available else the route call is blocked.[16]

2.4 Exhaust Algorithm

It searches all the possible routes between source and destination and then finally selects the best route .So utilizing this route provides the flexibility of selecting any route according to the wavelength available.Source Initiated Reservation Algorithm In this algorithm the network resources are utilized by reducing control overhead .In this the route table is generated first and then resource and wavelength availability is updated .Next the route table is arranged according to minimum hops and maximum wavelengths available and route is searched according to that. If wavelength allotted the request is processed else the request is blocked and then whole process is repeated.[15]

2.5 Traffic Intensity Based Fixed –Alternate Routing

In this algorithm the traffic is splitted for each source destination pair onto the disjoint routing paths by solving the nonlinear multicommodity flow optimization problem. Further the routing path between end users is sorted in decending order according to the amount of traffics of source destination pair assigned to the routing paths. For the request arrived this algorithm tries the multiple routing paths between the particular source destination pair one by one in the sorted order to establish a connection until the request is processed successfully otherwise the request is blocked .Results proved that this algorithm have lower blocking probability as compared if sorting is done according to the hop count instead of traffic intensities.[3]

2.6 Priority and Maximum Revenue based Routing

In this algorithm two types of requests are kept in two light path request matrixes namely high priority matrix (HPM) and low priority matrix (LPM).High priority matix requests are entertained first before LPM requests. According to this algorithm the light paths in HPM must be fulfilled and revenue for the same is fixed while the light paths in the LPM can be set up selectively to optimize revenue when there is not enough uresource to accommodate all.[12]
Multiobjective integer linear program is proposed in literature for maximizing throughput and solving traffic engineering in wdm optical networks.This program collectively maximizes the network output,minimizes the resource consumption and achieves load balancing by minimizing the maximum link utilization.This algorithm solves the multiobjective problem as an e-constraint problem upper bounding the per route resource consumption and maximizing the throughput by routing through multiple widest paths.Simulations results obtained reveals that solving multiobjective problem by ILP (Integer linear Programming) yield promising results.[4]

3 Simulation Results

For analyzing the problem of routing and wavelength assignment we have prepared a wdm network in which each link is bidirectional and taken eight wavelengths per link. For routing shortest path algorithm is used and first fit for wavelength assignment. The traffic considered is dynamic in nature which follows the poisson process. We have analysed and calculated the thoughput for two types of applications i.e FTP and email .The simulation is carried on opnet modeler.

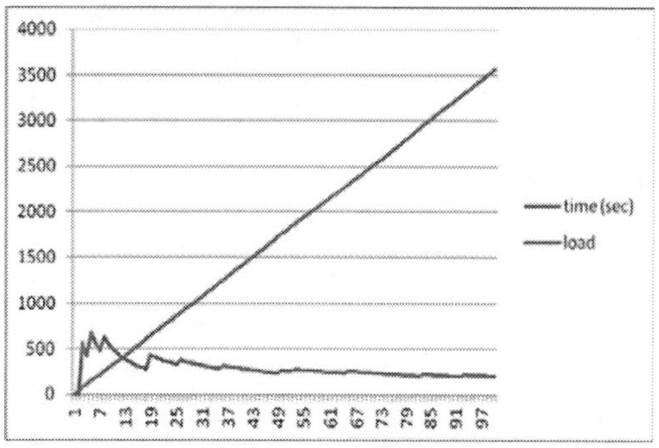

Figure 1. Load versus time for email application.

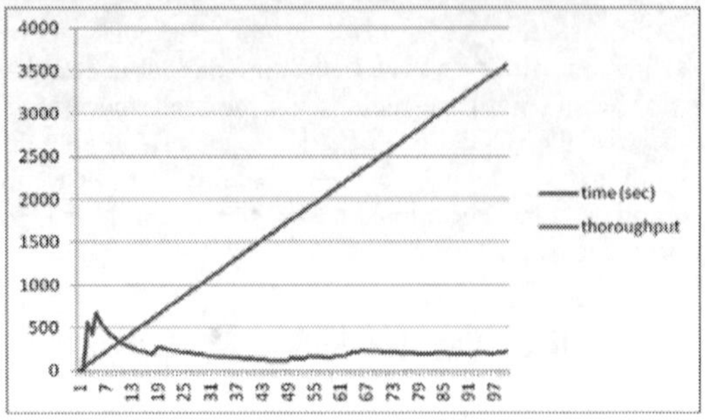

Figure 2. Throughput versus time..

Table 1. Data sheet for load and throughput.

	Data Collected	
time (sec)	load	throughput
0	0	0
36	0	0
72	565.33333	565.3333333
108	424	424
144	678.4	678.4
180	565.33333	565.3333333
216	484.57143	484.5714286
252	622.75	437.25
288	553.55556	388.6666667
324	498.2	349.8
360	452.90909	318
396	415.16667	291.5
432	383.23077	269.0769231
468	355.85714	249.8571429
504	332.13333	233.2
540	311.375	218.625
576	293.05882	205.7647059
612	276.77778	194.3333333
648	435.15789	278.9473684
684	413.4	265
720	393.71429	252.3809524
756	375.81818	240.9090909
792	359.47826	230.4347826
828	344.5	220.8333333
864	330.72	212
900	379.15385	207.9230769
936	365.11111	200.2222222
972	352.07143	193.0714286
1008	339.93103	186.4137931
1044	328.6	180.2
1080	318	174.3870968
1116	308.0625	168.9375
1152	298.72727	163.8181818
1188	289.94118	159
1224	281.65714	154.4571429
1260	318	153.1111111
1296	309.40541	148.972973
1332	301.26316	145.0526316
1368	293.53846	141.3333333
	Data Collected	
time (sec)	load	throughput
1404	286.2	137.8
1440	279.21951	134.4390244
1476	272.57143	131.2380952
1512	266.23256	128.1860465
1548	260.18182	125.2727273
1584	254.4	122.4888889
1620	248.86957	119.826087
1656	243.57447	117.2765957
1692	238.5	114.8333333
1728	233.63265	112.4897959
1764	262.88	144.16
1800	257.72549	141.3333333
1836	252.76923	138.6153846
1872	278	138
1908	274.81481	164.8888889
1944	269.81818	161.8909091
1980	265	159
2016	260.35088	156.2105263
2052	255.86207	153.5172414
2088	251.52542	150.9152542
2124	249.1	174.9
2160	245.01639	172.0327869
2196	241.06452	169.2580645
2232	238.92063	191.8095238
2268	236.84375	213.65625
2304	257.66154	212
2340	255.36364	232.8787879
2376	251.55224	229.4029851
2412	247.85294	226.0294118
2448	244.26087	222.7536232
2484	240.77143	219.5714286
2520	237.38028	216.4788732
2556	234.08333	213.4722222
2592	230.87671	210.5479452
2628	227.75676	207.7027027
2664	224.72	204.9333333
2700	221.76316	202.2368421

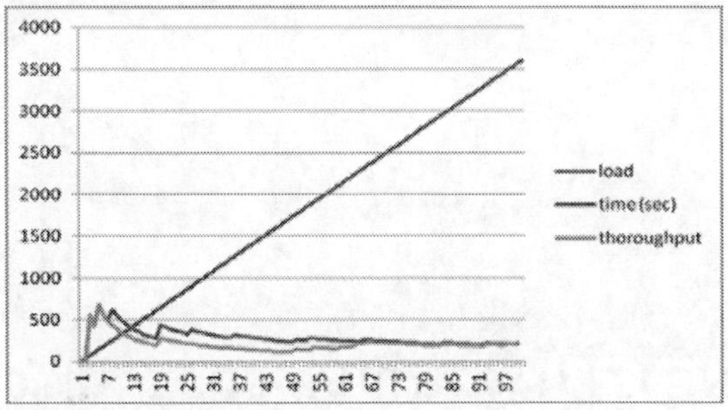

Figure 3. Throghput versus time and load for email.

4 Conclusion

The traffic is dynamic in nature so the load is varying w.r.t time. Throughput is the performance metric taken in this paper which is observed lesser compared to the load offered to the network which indicates the blockage of requests offered to the network. The throughput and blocking of the network can be optimized with some nature inspired technique in future.

References

1. Mukherjee B. , Optical Communication Networks (McGraw-Hill, New York, 1997)
2. Chen, M. T., Lin, B. M. T. and Tseng, S. S. ,J. Engineering applications Of Artificial Intelligence, 24,2(2011)
3. Chun, L. H., Wei, W. S., and Tsai, C. P. , Journal of Light wave technology, 26,2(2006)
4. Crichigno, J., Xie, C., Shu, W., Wu, M. Y. and Ghani, N., Proceedings of International Conference Asilomar, Pacific Grove, CA (2009)
5. Zang, H., .Jue, J.P.., and Mukherjee, B. Optical Networks Magazine, 1, 1(2000).
6. Chlamtac I., Ganz A., and Karmi G. , IEEE Transactions on Communications,40 ,7(1992
7. Kaminow I.P. et al. ,J. Select. Areas Commun. 14, 6, (1996)
8. Ramaswami R., Sivarajan K. N., J. Select. Areas Commun. 14, 6(1996)
9. Randhawa, R. and Sohal, J.S. J. Optik, 121, 8(2010)
10. Ramaswami R., Sivarajan K. N., IEEE/ACM Transactions on Networking,3,5(1995)
11. V.Acimovic-Raspopovi, G. Markovic, Invited paper, Proc.of TELFOR Conference, Belgrade (2004)
12. Wang, Y., Cheng, T. H., and Ma, M., (2007) Proceedings of International Conference on Research Innovation Vision of the Future, Hanoi (2007)
13. X. Jia, X.-D.Hu, D.-Z. Du, Multiwavelength Optical Networks, Kluwer Academic Publishers, Norwell, MA (2002)
14. Ye, Y., Chai, T. Y., Chen, T. H. and Lu, C. J. Computer Communications 29, 15(2006)
15. Zaidi, M. H., Mahmood, A., Proceedings of International Conference INMIC, Islamabad (2003)
16. Bhatt U. R. and Tokekar S., J. optik ,122(2010)

MATEC Web of Conferences 57, 01016 (2016) DOI: 10.1051/matecconf/20165701016
ICAET-2016

A P-N SEQUENCE GENERATOR USING LFSR WITH DUAL EDGE TRIGGER TECHNIQUE

Mr. Nitin Kumar Naghwal[1], Abhishek Kumar[2]

[1] *M. Tech Student, School of Electronics Engineering, Lovely Professional University, Phagwara, Punjab*
[2]*Assistant Professor, School of Electronics Engieering, Lovely Professional University, Phagwara, Punjab*

Abstract. This paper represents the design and implementation of a low power 4-bit LFSR using Dual edge triggered flip flop. A linear feedback shift register (LFSR) is assembled by N number of flip flops connected in series and a combinational logic generally xor gate. An LFSR can generate random number sequence which acts as cipher in cryptography. A known text encrypted over long PN sequence, in order to improve security sequence made longer ie 128 bit; require long chain of flip flop leads to more power consumption. In this paper a novel circuit of random sequence generator using dual edge triggered flip flop has been proposed. Data has been generated on every edge of flip flop instead of single edge. A DETFF-LFSR can generate random number require with less number of clock cycle, it minimizes the number of flip flop result in power saving. In this paper we concentrates on the designing of power competent Test Pattern Generator (TPG) using four dual edge triggered flip-flops as the basic building block, overall there is reduction of power around 25% by using these techniques.

1 INTRODUCTION

Random numbers is essential to cryptography, cryptographic protocol AES, RSA, DES etc is depends on random number for encryption. Data security demands the generated number must be random and non deterministic known as pseudo random number. Pseudo number generator built from Linear Feedback Shift Register (LFSR) with judicious selection of the XOR taps feedback path. Pseudo random number generators generate a stream of numbers in a known pattern. The pattern is typically very long and it is hard to recognize the sequence of numbers is ordered. For example 128 bit LFSR requires to implement 32 PRNG with 96 bit are hidden protection comes from hidden number. Both hardware and software implementations of LFSRs are common. LFSR can be implemented in the hardware and it makes useful in applications that may require fast generation of a pseudo random sequence[12]

LFSR is preferred for test pattern generator and error correction due to its simple hardware structure. TPG is a device which generates uniqueness of patterns to test the Circuit under Test (CUT) shown in fig1. In computing, a linear-feedback shift register (LFSR) is a shift register whose input bit is a linear function of its previous state. The most commonly used linear function of single bits is XOR. Thus, an LFSR is most often a shift register whose input bit is driven by the exclusive-or (XOR) of some bits of the overall shift register value. The initial value of the LFSR is called the seed, and because the operation of the register is deterministic, the stream of values produced by the register is completely determined by its current (or previous) state. Likewise, because the register has a finite number of possible states, it must eventually enter a repeating cycle. However, an LFSR with a well-chosen feedback function can produce a sequence of bits which appears random and which has a very long cycle.

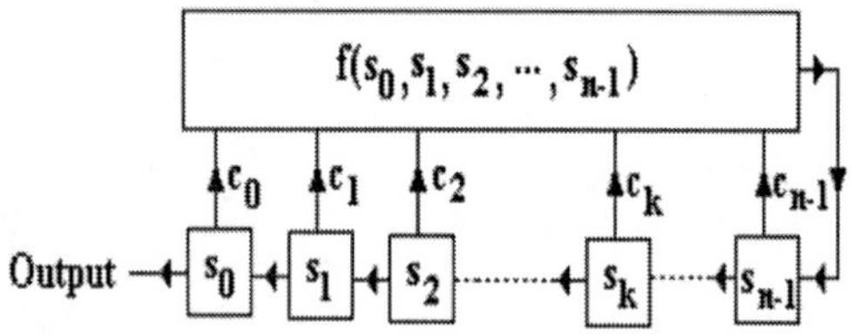

Figure 1. Random number generator

2 DETFF

In digital system clock signal has its special role; input clock signal controls the working rhythm of the chip. The clock signal controls flip-flops to sample and store their input data on active edge. To improve data rate clock frequency must increases, but in recent design performance of the flip-flop improved by increasing the active edge; in stead of accessing data on single edge accesson both edge. Dual edge triggered flip-flop (DETFF) can access data on both edges of the clock; it

[1]Corresponding author: nitinaghwal@gmail.com

enables the flip-flop to reduce the frequency half by maintaining same data rate. In VLSI working with high frequency result in high power consumption, DETFF draw attention to have high speed operation by lowering down the clock frequency. There is a reduction of 30-60% power saving compare to single edge triggered flip flop[8,9,10]. Fig2 (a) present a DETFF captures the input data at both positive and negative active edges of the input clock signal. Flip flop output is combined together using a multiplexer, at positive clock high latch output QH and at negative clock low latch output QL is routed to output port. As the timing of both the input signals i.e. clock and data matches, it provides the desired output, but whenever there are asynchronous signals or there is some mismatch in the timing than it leads to the errors due to setup and hold time violation. Fig2(b) the timing waveform of DETFF; QH latch on high and QL latches low, Q is combined effect of QH and QL. From waveform it seems that Q switches and each edge of clock and data transmission happens on both edges. It doubles the data rate known as double data rate flip flop.

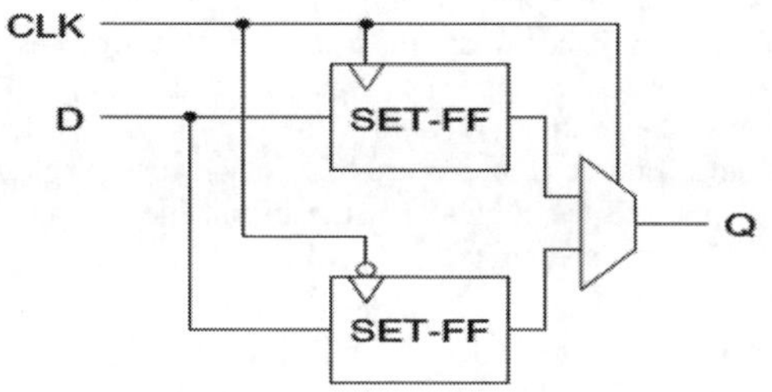

Figure 2(a). Dual edge triggered flip flop[3]

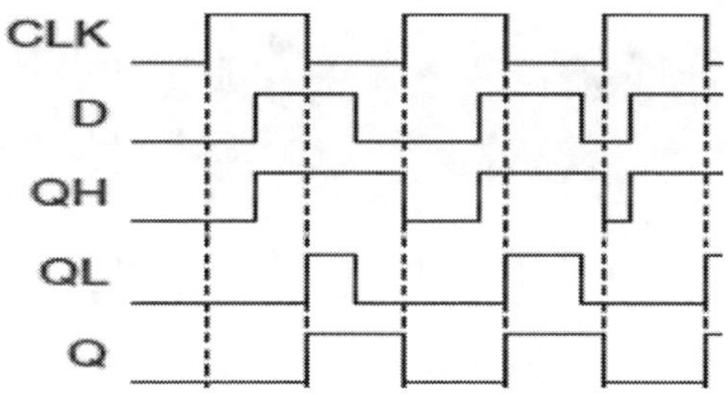

Figure 2(b). Timing wave form of DDR flip flop[3]

3 LINEAR-FEEDBACK SHIFT REGISTER

LFSR is a linear feedback shift register whose input bit is a linear function of previous function that contains the signal through the register from one bit to the next most-significant when it is clocked. Figure 3 shows the block diagram of LFSR, a 4-bit LFSR generate 24-1 different non zero bit pattern by performing exclusive-OR gate on the outputs of two or more of the flip-flops and feeding those outputs back in to the input of one of the flip-flops. An L bit LFSR generates the maximal length up to 2^L-1 combinations, figure4.

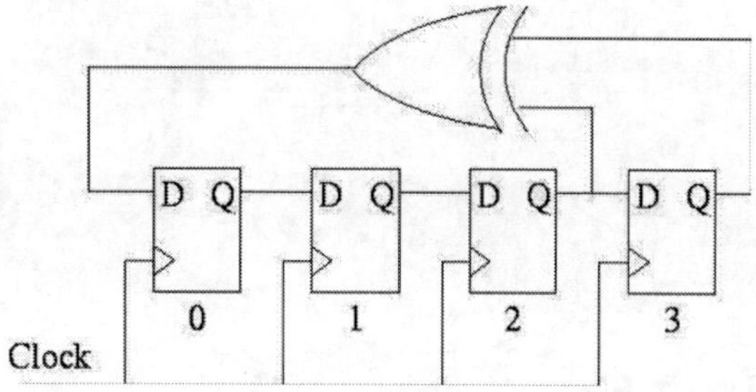

Figure 3. TPG using LFSR

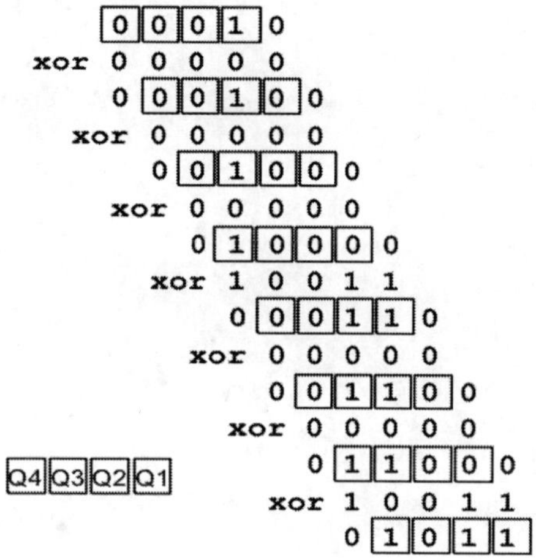

Figure 4. Test pattern

4 LFSR WITH DET-FF

In this section a design of proposed FLSR has been discussed.

(i) High-level latch: A high level latch provides result on high level of clock. In fig5, on high level of clk Q=D. [9].

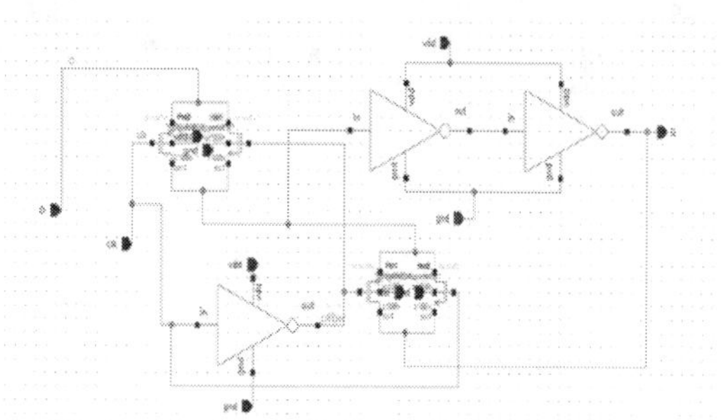

Figure 5. High-Level Latch

(ii) Low-Level Latch: A high level latch provides result on low level of clock. In fig6, on low level of clk Q=D.

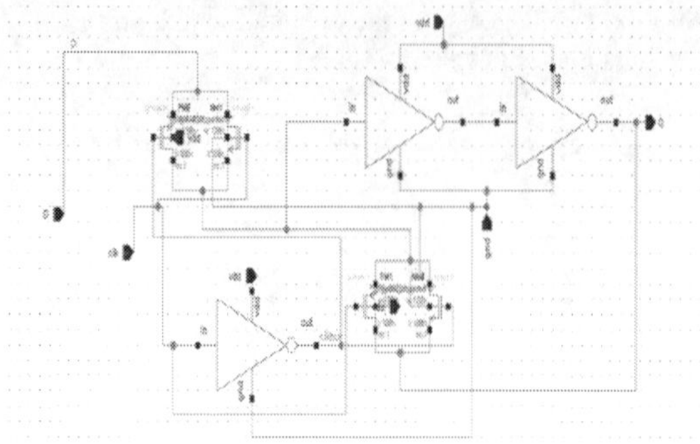

Figure 6. Low-Level Latch

(iii) Combiner Element: Here we have used the 2:1 MUX, which is used as the one of the combiner-element in the main circuit of DETFF. It combines the result of the high and low-level latch in the positive and negative clock.

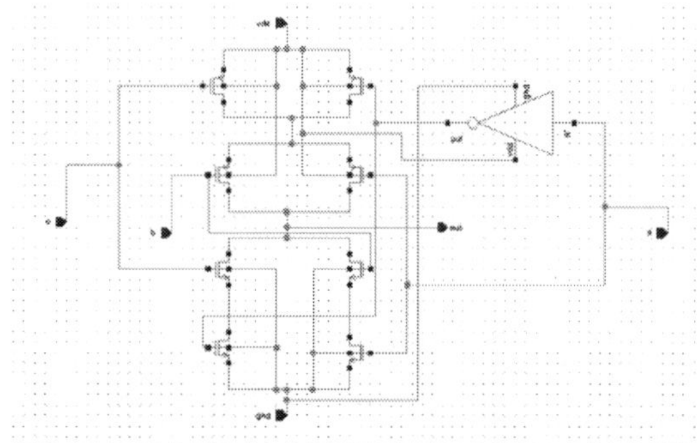

Figure 7. 2:1 MUX

(iv) Dual Edge triggered flip flop: A Dual edge triggered flip flop has two latch , one os high level sensetive and ither is low level. This DETFF require a combiner-element (mux) which combine the data accessed on high and low level of latch. Select line of multlplexer route the latch output to output trminal. MUX will select according to the clock provided as the select line. In detects the data on both of the levels and hence provides the same data at the output which is provided at the input of both latches Area of DETFF is about a half of that of edge-triggered FFs.[11] By combining all these blocks we perform the dual edge triggering and we got the same output which is provided at the input D.

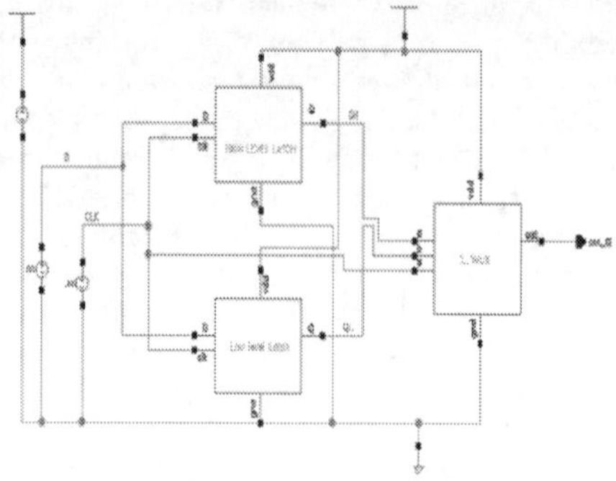

Figure 8. Basic DET FF

(v) Linear feedback shift register: fig 9 present a 4-bit LFSR using DDR flip flop. It consist of 4 flip flops so there will be total 16 patterns generated, but to initiate the LFSR we cannot feed the pattern of "0000" because the shift register would be stuck at this pattern because as we know that if we provide 0 at the input of the XOR gate than the output of XOR gate will be 0 always. So we have to provide any other pattern except all 0's. The LFSR is initiated by the input 0110, and it provides the output patterns as shown in figure 10 and 11. Figure 11 shows the output waveform of the LFSR. Cdence virtuoso circuit compser used for schematic design and Cadence spectre simulator using to obtained simulation result on cmos 90nm technology.

As the enable pin gets high the LFSR gets initialized. The first pattern is generated around 1.2 ns and about near to 3.2 ns all of the desired patterns has been generated, after this the pattern starts repeating. We analyzed that total time taken for the generation of the patterns is approximately 2ns.

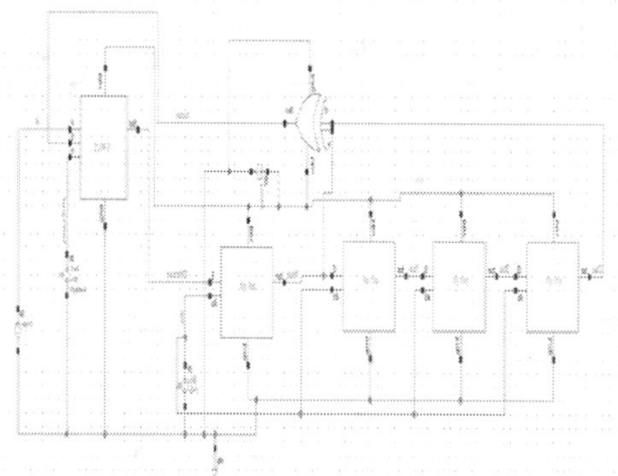

Figure 9. LFSR Schematic

DOI: 10.1051/matecconf/20165701016

D_3	D_2	D_1	D_0
0	1	1	0
1	1	0	0
0	0	0	1
0	0	1	0
0	1	0	0
1	0	0	0
1	0	0	1
1	0	1	1
1	1	1	1
0	1	1	1

Figure 10. Output Patterns of LFSR

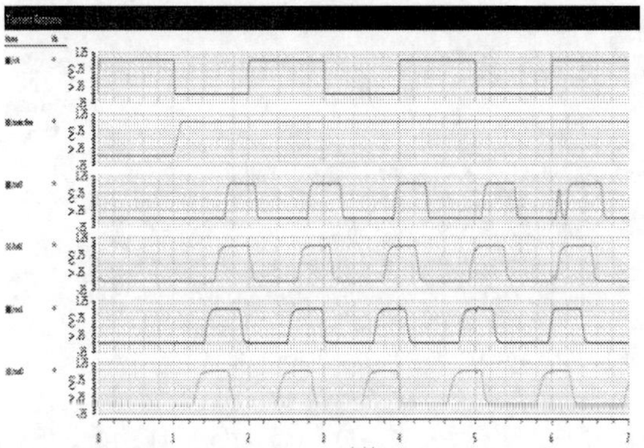

Figure 11. LFSR Output Waveform

5 SUMMARY AND CONCLUSIONS

In this paper a noble architecture of Dual edge triggered flip flop. Data capturing on both edge double the data rate. Double edge flip flop is an emerging technique for high speed data transmission. Designed LFSR generate random pattern on every clock edge, it significantly improves generation efficiency. Simulation result verifies that to obtained 10 random pattern a 4-bit LFSR consume 35.4uw and delay is 2ns.

REFERENCES

1. Kazuteru Namba, Takashi Katagiri, Hideo Ito,"Timing-Error-Detecting Dual-Edge-Triggered Flip-Flop" Journal of Electronic Testing, Vol. 29, pp 545-554, 2013.
2. Massoud Pedram, "A New Design for Double Edge-Triggered Flip-flops" International Journal of Electronics Communication and Computer Engineering Volume 4, pp 412-417, 2013.
3. Stojanovic, V.; Oklobdzija, V.G., "Comparative analysis of master-slave latches and flip-flops for high-performance and low-power systems," Solid-State Circuits, IEEE Journal of, vol. 34, pp.536-548, 1999.
4. Nedovic, Nikola; Aleksic, M.; Oklobdzija, V.G., "Conditional pre-charge techniques for powerefficient dual-edge clocking," Low Power Electronics and Design. Proceedings of the 2002 International Symposium, vol. 3, pp.56-59, 2002.
5. S. Wimer and I. Koren, "The Optimal fan-out of clock network for power minimization by adaptive gating," IEEE Trans. VLSI Syst., vol. 20, pp. 1772–1780, 2012.
6. Yasmeen Khan," Power Optimization of Linear Feedback Shift Register Using Clock Gating ," Volume 7, PP. 109-115, 2013.
7. S. Wimer and I. Koren, "The Optimal fan-out of clock network for power minimization by adaptive gating", IEEE Trans. VLSI Syst., vol. 20, pp. 1772–1780, 2012.
8. M. Muller, S. Simon, H. Gryska, A. Wortmann, and S. Buch, "Low power synthesizable register files for processor and IP cores", The VLSI J., vol. 39, pp. 131–155, 2006.
9. A. G. M. Strollo and D. De Caro, "Low power flip-flop with clock gating on master and slave latches", Electron. Lett., vol. 36, pp. 294–295, 2000.
10. J. Kathuria, M. Ayoub, M. Khan, and A. Noor, "A review of Clock Gating Techniques", MIT Int. J. Electron. and Commun. Engin., vol. 1, pp. 106–114, 2011.
11. Dr. Neelam R. Prakash, Akash, "Clock Gating for Dynamic Power Reduction in Synchronous Circuits", International Journal of Engineering Trends and Technology (IJETT), vol. 4,pp. 513-519, 2013.
12. P. Girard, L. Guiller, C. Landrault, and S. Pravossoudovitch, "A test vector inhibiting technique for low energy BIST design," in Proc. 17th IEEE VLSI Test Symp.,vol 3, pp. 407–412, 1999.
13. S. Manich, A. Gabarro, M. Lopez, J. Figueras, P. Girard, L. Guiller, C. Landrault, S. Pravossoudovitch, P. Teixeira, and M. Santos, "Low power BIST by filtering non-detecting vectors," J. Electron. Test.-Theory Appl., vol. 16, pp. 193–202, 2000.
14. F. Corno, M. Rebaudengo, M. Reorda, and M. Violante, "A new BIST architecture for low power circuits," in Proc. Eur. Test Workshop, vol 5, pp. 160–164, 1999.

Optical Wavelength Converters Based on Four Wave Mixing in SOA-MZI Configuration

Rajni[1], Anu Sheetal

Department of Electronics and Communication, Guru Nanak Dev University (RC Gurdaspur)

Abstract. Wavelength converter plays an important role for increasing the capacity and flexibility of future broadcast network. This paper investigates the performance of a 10Gb/s the SOA based FWM wavelength converter. All performances are analyzed in terms of its shifted wavelength conversion efficiency, Q-factors and converted signal powers. The converters are modeled and simulated using Optisystem7 Software, by varying the probe signal wavelength and power. It was found that conversion efficiency and OSNR of the converted signal both decreased at large detuning wavelengths. Similarly, higher total SOA input powers worsened the conversion efficiency, but steadily improved the OSNR. FWM wavelength converter had the lowest conversion efficiency for CW sources.

1 Introduction

Optical wavelength converters are the key component for multi-wavelength optical transport networks and are provide wavelength conversion in optical domain without distortion of input signal [1]. The electro-optic converter is a straight forward solution for conversion. But electro-optic converter has limitations such as large power consumption and complexity. Wavelength conversions based on semiconductor optical amplifiers (SOAs) and semiconductor lasers have been focus on research interests during the last few years [1, 2].The capacity of wavelength division multiplexing (WDM) is usually limited by several factors, such as number of channels which can be deployed ,wavelength congestion level and system management. A wavelength converter can be used to overcome such network limitations. A wavelength converter is an optical device which is used for the purpose of converting an injected signal of light from one wavelength to the desired wavelength in a system or network. Recently the four wave mixing technique occurring within a semiconductor optical amplifier has been one of the most favorable methods of wavelength conversion, offering benefits to the system designer [3].

FWM offers transparency to the bit rate and modulation format and hence preserves both the phase and amplitude information. FWM is the only method with transparent optical properties, due to unchanging nature of the optical properties of the information signal during the conversion process occurring within a SOA [4].Additionally, SOA-based FWM wavelength conversion offers many other attractions; for example high bit capability-up to 10Gb/s [5] or even 20Gb/s [6] have been demonstrated. FWM methods are also capable of operating at high bit rate capability without compromising the extinction ratio.

Wavelength conversion exploiting the SOA nonlinearities exhibits different advantages and disadvantages, depending on the operation. For example, the main advantages of XGM wavelength converters are their simplistic configurations, high conversion efficiencies and high bit rate capabilities.SOA based wavelength converters can also be arranged so as to possess polarization immunity, as well as wavelength insensitivity, towards the incoming input data [7].Polarization insensitivity is crucial in optical networks, as the polarization state in fibers can vary with environmental conditions.

2 Basic operations

2.1Four wave mixi

In this, two optical waves with identical states of polarization are coupled into a saturated SOA. The pump signal (Epump) at a frequency of ῼpump is typically stronger as compared to probe signal (Eprobe) at a frequency of ῼprobe, which contains the data to be wavelength converted. The data signal is kept small, when compared to the pump, so as to avoid any unwanted gain modulation which can affect the performance of the converted signal during the conversion process [8].

[1]Corresponding author: rajni_kundan@yahoo.com

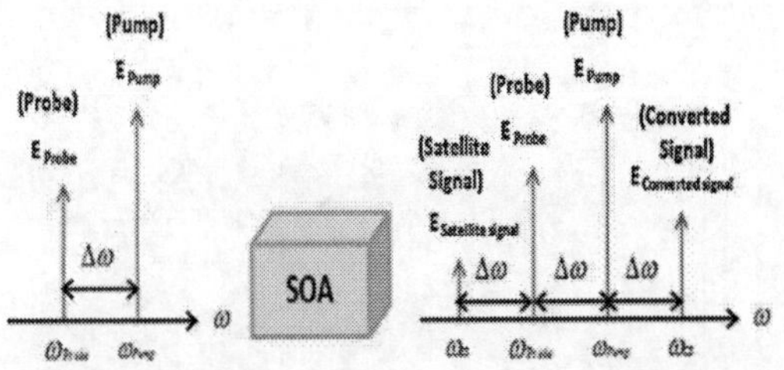

Figure 1. FWM scheme in SOA

Inside the active region of the SOA, the beatings of the two co-propagating waves modulates the carrier density, and therefore generates dynamic gain and index gratings. This non-linear interference produces new waves do not overlap with other wavelengths [11]. The intensity of these newly generated waves, known as the Econverted_signal and Esatellite_signal, is proportional to the product of the interacting wave intensities. Additionally, the phase and frequency of the newly generated waves is a linear combination of the wave present in the SOA active medium [12]. Eprobe is up-converted to a longer wavelength, Econverted_signal at an optical frequency of $\omega cs=\omega pump+\Delta\omega$, whereby $\Delta\omega$ refers to the detuning frequency. Econverted_signal is the phase conjugate replica of the original input signal, Eprobe. Increasing the gain, saturation power and carrier recovery rate of an SOA are a few practical methods of improving the conversion efficiency of a FWM wavelength converter [13].

$$\eta = 10\log \frac{P_{out}\,(\lambda_{converted\_signal})}{P_{in}\,(\lambda_{pump\_signal})}$$

Studies on FWM-based wavelength converters utilizing SOAs have been proposed in much previous works, due to numerous advantages offered by nonlinearities of SOA. For example D'Ottavi et al. [13] proved that use of a long SOA could improve the performance of the conversion at a data rate of 10 Gb/s. Also, Hsu et al. [14] claimed that higher number conversion efficiency and wider conversion range can be achieved by applying an assisted holding beam into the SOA. The assisted beam, which was used to saturate the SOA, can improve the conversion efficiency and signal to-background ratio. Additionally, unlike most of the proposed techniques using coherent light as the source, Forsyth et al. [15] demonstrated in novel work that SOA FWM can be obtained from an amplified spontaneous emission (ASE) source.

3 Modeling and simulation

The simulation was carried out at a central frequency of 193.5THz. A scheme of the modeling blocks was created, based on a CW laser source for the FWM-based wavelength converter, and is shown in Fig.2. The SOA parameters varied in this modeling. The pump and probe from the CW laser source was set to have SOA input powers of 5 dBm, 3 dBm, -5dBm and -3dBm respectively. The probe signal was modulated by a NRZ electrical driver, with a PRBS signal. The signal wavelength was varied in accordance with the desired detuning frequency. The pump and probe signals each passed through individual polarization controllers, PC1, PC2, PC3 and PC4, in order to match the state of polarization between them. This was a necessary condition to ensure efficient FWM occurred in SOA before being multiplexed by the multiplexer.

The multiplexed signals were then amplified by an EDFA, having a fixed gain of 10 dB and with an average noise figure of 4.2 dB. The EDFA was used to increase the bandwidth optical band pass filter (BPF) was placed just after the EDFA so that any additional ASE outside the signal bandwidth was suppressed, hence increasing the OSNR of the converted signal. A nominal bias current of 250 mA was injected into the SOA as soon as the signals entered it, which then generated the converted signal through the FWM process. Another 0.4 nm Fabry Perot Optical filter, placed just after the SOA and centered on the converted signal wavelength, ensured clean capture of the converted wave with complete pump and original signal removal.

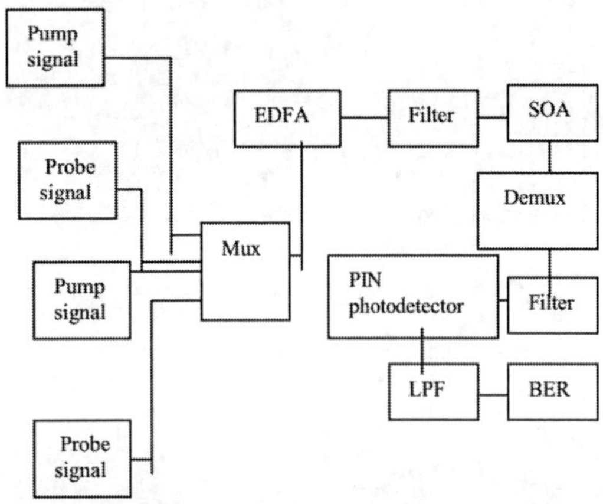

Figure2. Block diagram of 4 channels FWM wavelength converter..

4 Results and discussions

The simulation has been carried out for the set-up as shown in Fig.2. The converted signal power obtained as a function of wavelength shift. For showing the down conversion, an input with a frequency difference was taken. The pump signal was generated with the help of a CW laser at a wavelength of 1550 nm and the probe signal was also generated using a CW laser at a wavelength of 1552 nm. For the purpose of up conversion, the pump and the probe signals are generated at 1550nm and 1548nm respectively. These are given as the co-polarized inputs into the SOA.

Effect of Bit Rate: The results were obtained for the bit rates of 10Gb/s respectively.

DOI: 10.1051/matecconf/20165701017

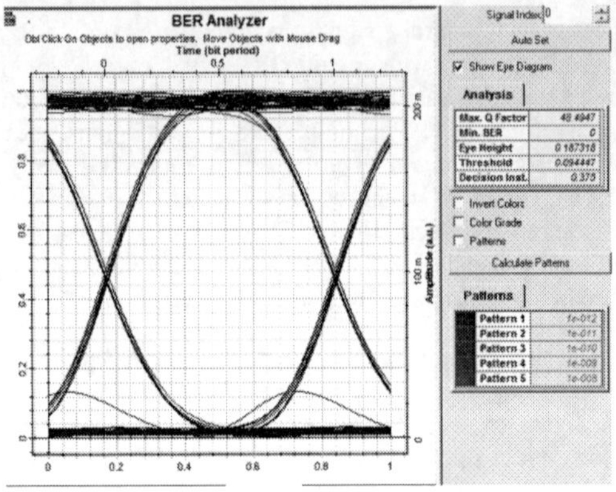

Figure 3. Eye diagram for signal received i.e.193THz

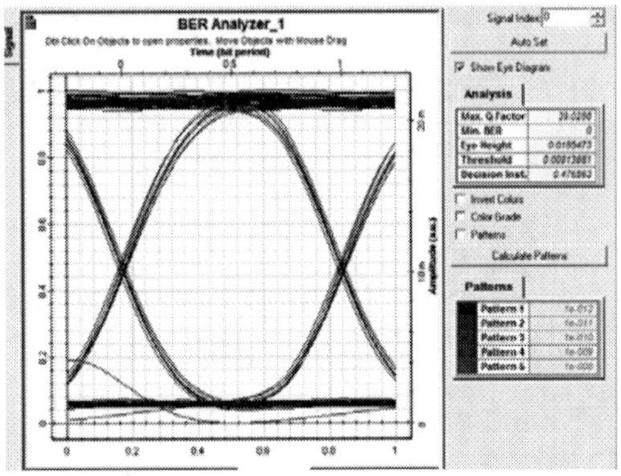

Figure 4. Eye diagram for 193.1THz signal

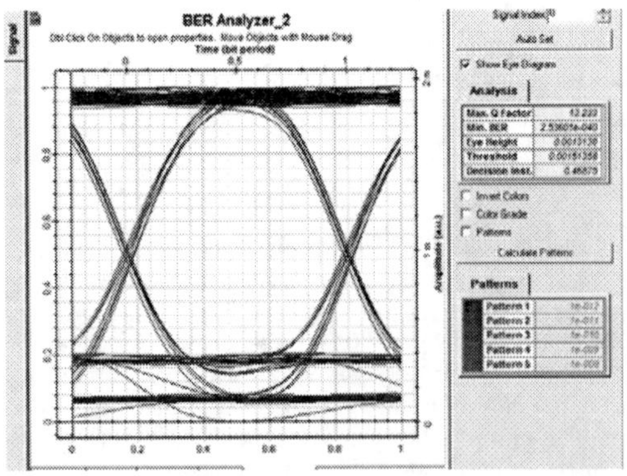

Figure 5. Eye diagram for 193.2THz signal

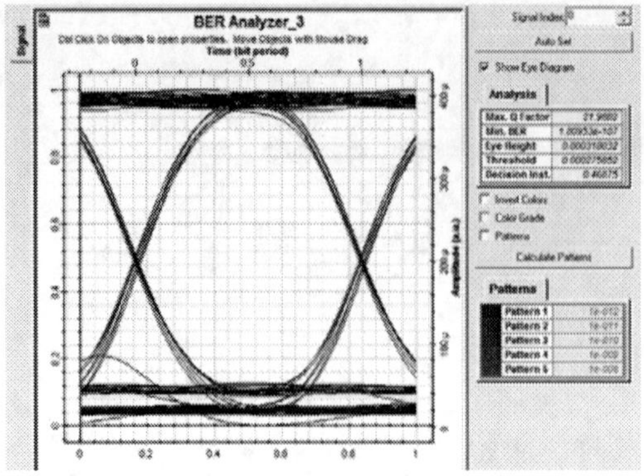

Figure 6. Eye diagram for 193.3THz signal

5 Conclusion

We have simulated and investigated various effects of the FWM wavelength conversion process occurring within a SOA medium, and have gathered useful data which could be used to evaluate the potential of such a scheme for applications in future, all-optical converter systems. The analysis was done through different channels at different frequency and power. The process occurred within an SOA medium which provides numerous advantages over fiber optic cable such as transparency to bit rates and modulation formats. This scheme of wavelength converter using FWM in SOA holds great importance for the future all optical converter system. In addition to this, we have also analyzed the various parameters of FWM in SOA which includes effect of bit rate and varied frequency and power.

References

1. M. Kovacevic, A. Acampora, Benefits of wavelength translation in all optical clear channel networks, IEEE J. Sel. Area Comm. 14 (1996) 69–79.
2. T. Durhuus, B. Mikkelson, C. Joergensen, S.L. Danielson, K.E. Stubkjaer, All optical wavelength conversion by semiconductor optical amplifiers, J. Lightwave Technol. 14 (1996) 942–953.
3. Surinder Singh, R.S. Kaler, All optical wavelength converters based on cross phase modulation in SOA-MZI configuration, Optik 118(2007)390-394.
4. D.D. Marcenac, D. Nesset, A.E. Kelly, D. Gavrilovic, 40 Gb/s transmission over 103 km of NDSF using polarization independent mid-span spectral inversion by four-wave mixing in a semiconductor optical amplifier, Electron. Lett. 34(1) (1998) 100–101.
5. S. Singh, R.S. Kaler, 20-Gb/s and higher bit rate optical wavelength conversion for RZ-DPSK signal based on four-wave mixing in semiconductor optical amplifier, J. Fiber Integr. Optics 26 (5) (2006) 295–308.

6. S. Singh, R.S. Kaler, Wide band optical wavelength converter based on four-wave mixing using optimized semiconductor optical amplifier, J. Fiber Integr. Optics 25 (3) (2007) 213–230.
7. D.I. Forsyth, M.J. Connelly, Dual semiconductor optical amplifier polarization independent wavelength conversion using four-wave mixing with optoelectronic feedback, in: Conference Proceedings – Lasers and Electro-Optics Society Annual Meeting, vol. 1, 2004, pp. 396–397.
8. Farah Diana Mahad, Abu Sahmah M. Supa'at, Sevia Mahdaliza Idrus, David Forsyth, Comparative performance testing of SOA wavelength conversion techniques for future all-optical systems, Optik 124 (2013) 1254– 1259.
9. K. Obermann, S. Kindt, D. Breuer, K. Petermann, Performance analysis of wavelength converters based on cross-gain modulation in semiconductor-optical amplifiers, J. Lightwave Technol. 16 (1) (1998) 78–85.
10. S. Wu, X. Sun, M. Zhang, Characteristics of the wavelength converted signal in wavelength conversion based on cross-gain modulation, Int. J. Infrared Millimeter Waves 23 (2) (2002) 309–316.
11. Farah Diana Mahad, Abu Sahmah M. Supa'at, Sevia Mahdaliza Idrus, David Forsyth, Analyses of semiconductor optical amplifier (SOA) four-wave mixing (FWM) for future all-optical wavelength conversion, Optik 124 (2013) 1– 3.
12. M.Y. Jamroa, J.M. Seniora, M.S. Leesonb, G. Murtaza, Chirp in a wavelength converter based on a symmetrical-MZI employing SOAs, Opt. Commun. 209 (4–6) (2002) 321–328.
13. A. D'Ottavi, P. Spano, G. Hunziker, R. Paiella, R. Dall'Ara, G. Guekos, K.J. Vahala, Wavelength conversion at 10 Gb/s by four-wave mixing over a 30-nm interval, IEEE Photon. Technol. Lett. 10 (7) (1998) 952–954.
14. D. Hsu, S. Lee, P. Gong, Y. Lin, S.S.W. Lee, M.C. Yuang, High efficiency wide-band SOA-based wavelength converters by using dual-pumped four-wave mixing and an assist beam, IEEE Photon. Technol. Lett. 16 (8) (2004) 1903–1905.
15. D.I. Forsyth, M.J. Connelly, Spectrum-sliced wavelength conversion using four wave mixing from a semiconductor optical amplifier, in: Optical Amplifiers and Their Applications, Technical Digest (CD) (Optical Society of America, 2005), Paper TuC3,(2005).
16. .

Advanced Energy Detector Based Cooperative Spectrum Sensing In Cognitive Networks

Ritu[1] and Ameeta Seehra[2]

[1]*Research Scholar, Electronics and Communication Department, Guru Nanak Dev Engineering College, Ludhiana, Punjab, India*
[2] *Associate Professor, Electronics and Communication Department, Guru Nanak Dev Engineering College, Ludhiana, Punjab, India*

Abstract. Cognitive radio is a befitting hopeful technology for future. Spectrum sensing is the most critical function of cognitive radio. Most of the spectrums sensing processes proposed earlier become impractical at lower signal to noise ratio conditions. Paper presents advanced cooperative spectrum sensing technique used in cognitive radio (CR) systems. By spectrum sensing it detects the presence of a primary user (PU) on a concerned spectrum. The accuracy of spectrum sensing depends on both sensing time and number of sensing nodes (secondary users) participating in the sensing process. We derive the optimal number of secondary users for cooperative spectrum sensing based on adaptive threshold energy detection to minimize spectrum detection error in conventional as well as improved energy detector.

1 INTRODUCTION

Wireless technology for communication has remarkably developed in last few decades. The cognitive radio is one another revolutionary enhancement of wireless technology. The rising congestion of spectrum and increase in demand of spectrum has left us searching for the alternates. Cognitive radio is sighted by the researchers as such an alternate. The cognitive radio continuously searches for the spectrum which is not in use by the Primary user for certain time slot. Cognitive radio hunts the vacant spectrum and uses it for secondary transmissions, so cognitive radios are also called SUs.Various sensing techniques can be applied at the sensing terminals of th CR. Some of the basic sensing techniques, which are commonly used are energy detection, cyclostationary method, matched filter detection and many more. In order to detect the primary signal using the matched filter, the secondary user need to have some prior knowledge of the Primary user's data signal. It is not possible all the times so matched filter based spectrum sensing is not always possible to be implemented. Similarly in cyclostationary feature extraction, cognitive radio searches for the features of detected signal similar to the priory known primary user signal. So for the same reason cyclostationary feature method can't be implemented flexibly.
So energy detection method best fits into the scenario where there is lack of sources to acquire the prior knowledge of the primary user signal. In a basic energy detector the energy of the received signal is computed as shown in the figure. The computed energy is then compared with a threshold level. Based on this comparison two hypothesis are followed H0 and H1. Here, H0 is followed when the energy detector interprets the presence of primary user as absent. H1 is followed when the presence of the primary user is interpreted as present.

2 ENERGY DETECTOR MODEL

Basic hypothesis model for transmitter detection

On the AWGN (Additive White Gaussian Noise) channels, the simplest spectrum sensing model is the binary hypothesis detect model.

$$H0: Y[N] = W[N], \quad H1: Y[n] = X[n] + W[N] \qquad (1)$$

Where H0 and H1express the different hypotheses of the primary users signal exist or not in the AWGN channels; Y[n] is the signals received by the unlicensed users (cognitive users);X[n] is the signals transmitted by the licensed users (primary users);
W [N] is the White Gaussian Noise.According to the sensing algorithms, the statistic T can be obtained by Y[n], and the sensing results can be obtained based on the decision theorem.

$$H0: T < \lambda \qquad (2)$$

$$H1: T > \lambda \qquad (3)$$

where λ is the decision threshold.
Usually we use these two criteria to evaluate the spectrum detector's performance:

The probability of detection: $Pd = P\{Y > \lambda | H1\}$ (4)
The probability of false alarm: $Pf = P\{Y > \lambda | H0\}$ (5)

[a] Corresponding author: reetjndl@gmail.com

DOI: 10.1051/mateconf/20165701018

This work only focuses on the spectrum sensing in CR.
where

Ho is the hypothesis of presence of primary user and H1 is the hypothesis of absence of primary user. False Alarm Probability Pf represents the case that SUs determines presence of PU but in fact PU is absent. Mis-detection probability Pdian bankm represents the case that SUs determines absence of PU but in fact PU is using the network

Here τ is the detection time and ω is detection bandwidth. With these parameters mis-detection probability and false alarm probability are represented as follows:

$$Pm = P\{Y < \lambda|H1\} = 1 - Q\tau\omega\left((2\lambda)^{\frac{1}{2}}, \lambda^{\frac{1}{2}}\right) \quad (6)$$

$$Pf = P\{Y > \lambda|H0\} = \frac{\Gamma(\tau\omega, \frac{\lambda}{2})}{\Gamma(\tau\omega)} \quad (7)$$

where λ is the energy threshold value, Qτω is generalized Marcum Q Function, Γ(•) is gamma function, and Γ(•, •) is incomplete gamma function respectively. Pm and Pf are functions of parameters λ, τ, and ω.

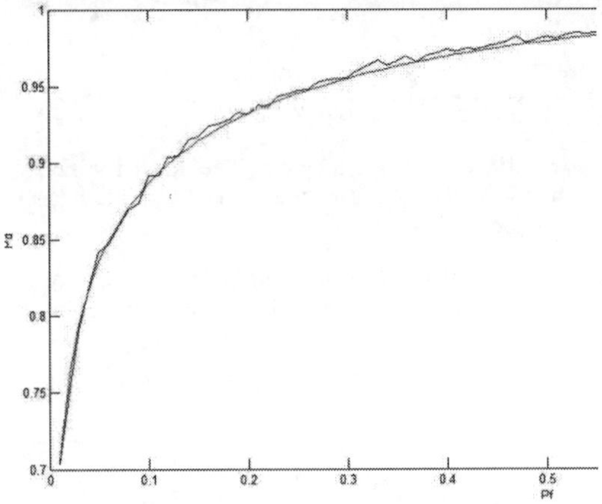

Figure 1. Energy detector ROC.

There are still many loops and holes in energy detector based spectrum sensing. Firstly, the energy detectors performance degrades at the lower SNR values. Secondly, the selection of the threshold is very crucial for the performance of the energy detector. Estimating the threshold is a provocative issue. The threshold should be such chosen so that it does not favour any of decision hypothesis. . Thirdly, the shadowing problem overshadows the primary user which leads to the unfit decisions. Lastly, the single cognitive radio sensing is least reliable. The wrong decisions will ultimately cause higher interference to the Primary user. So requirement in spectrum sensing is to make the sensing most reliable and hence to reduce the probability of interference to the Primary user.

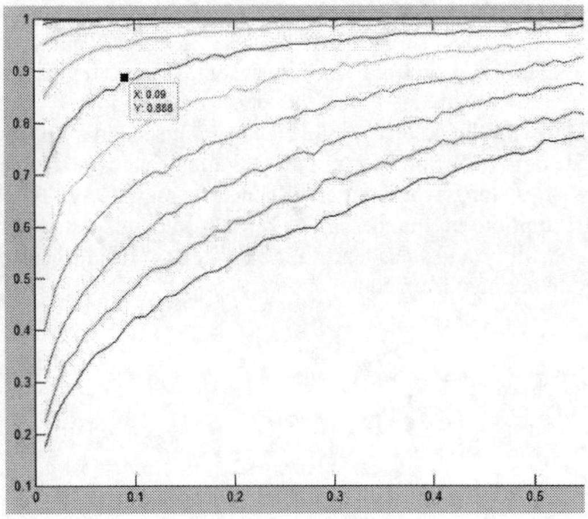

Figure 2. Energy detector ROC with adaptive threshold

Fig 2 shows the ROC of energy detector with adaptive threshold. Here threshold varies with change in the SNR value. This means the noise level directly effects the sensing performance of the detector.

3 ADVANCED ENERGY DETECTOR

In conventional energy detector decision is made after comparison to single threshold. In another proposed work double threshold detection is used. Instead of two decision regions there are three. Third one is no decision area. Here CR is not sure so it just does not contribute to the decision.The decision hypothesis followed will be

H0: Y< λ1 and H1: Y> λ2

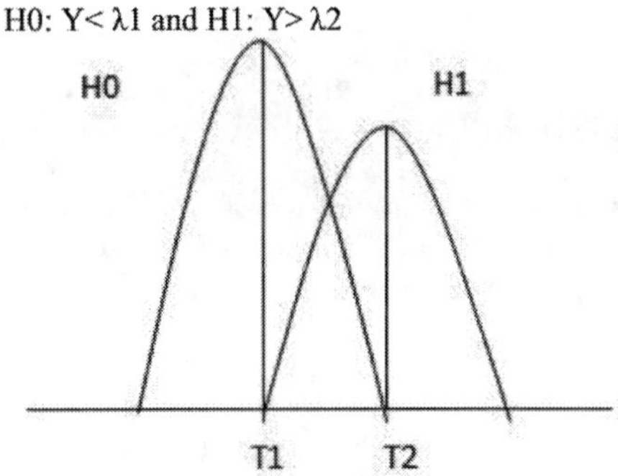

Figure 3. Double threshold decision hypothesis

The probability of detection and probability of false alarm for double threshold detector can be defined asThe probability of detection:

$$Pd = P\{y > \lambda 2 \mid H1\} \quad (8)$$

The probability of false alarm: $Pf = P\{y > \lambda 2 \mid H0\}$ (9)

4 COOPERATIVE SPECTRUM SENSING

Cooperative spectrum sensing schemes are categorized by the way to determine the presence of primary user (PU) with the collected information from secondary users (SUs) at a fusion center. The OR-rule declares presence of PU as long as at least one SU detects the PU. With the assumption of number of N SUs, the mis-detection probability and false alarm probability at a fusion centre are computed from equation

$$\text{OR Rule}: \begin{cases} P_{m,t} = P_m^N \\ P_{f,t} = 1-(1-P_f)^N \end{cases} \tag{10}$$

where Pm,t is the mis-detection probability and Pf,t is the false alarm probability at a fusion centre whereas Pm and Pf are the misdetection and false alarm probability determined by each secondary user. The AND-rule declares presence of PU only when all SUs recount the PU's presence. For the AND-rule the mis-detection probability and false alarm probability are computed as follows:

$$\text{AND Rule}: \begin{cases} P_{m,t} = 1-(1-P_m)^N \\ P_{f,t} = P_f^N \end{cases} \tag{11}$$

The majority-rule declares the presence of PU when more than half of SUs report presence of PU. The majority rule has characteristic of OR-rule and AND-rule and its mis-detection probability and false alarm probability are derived as:

Total detection error probability can be defined as the average of mis-detection probability and false alarm probability by each hypothesis probability.

$$Pe,t = P(H1)Pm,t + P(H0)Pf,t \tag{12}$$

Our goals to find Kopt for K-out-of-N-rule which minimizesthe total detection error probability. From (11) and (12) the value obtained for K optimal is :

$$\text{K out of N rule}: \begin{cases} Pm,t = 1 - \sum_{j=K}^{N} \binom{N}{j} (1-Pm)^j Pm^{N-j} \\ Pf,t = \sum_{j=K}^{N} \binom{N}{j} (1-Pf)^{N-j} Pf^j \end{cases} \tag{13}$$

This gives the expression for optimal value of number of CUs required for the decision which gives minimum error probability. The total error is sum of Pm and Pf (P(H1)=P(H0)=0.5) which should be least. The value of K optimal can be calculated by finding the minimum error value.[7]

In conventional energy detection technique single threshold energy detector has been used. Advanced scheme has been implemented in cooperative manner to find the optimal number of CRs required, total error rate.

5 RESULTS AND DISCUSSION

The conventional energy detector based cooperative environment was created and simulated. Total CRs present over the network are 20. The optimal number of CRs comes out to be 15, where the minimum error is calculated

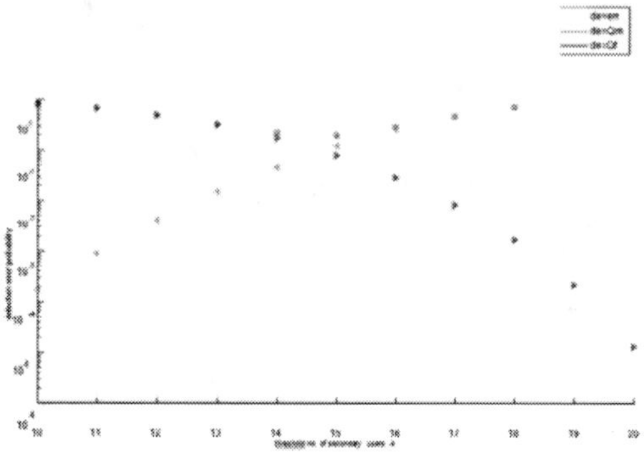

Figure 4. Three detection error probabilities (total detection error probability Pe, false alarm probability Pf, mis-detection probability Pm)

In another simulation the dual threshold based energy detector is compared to conventional one. The results are shown in figure 5 and 6.

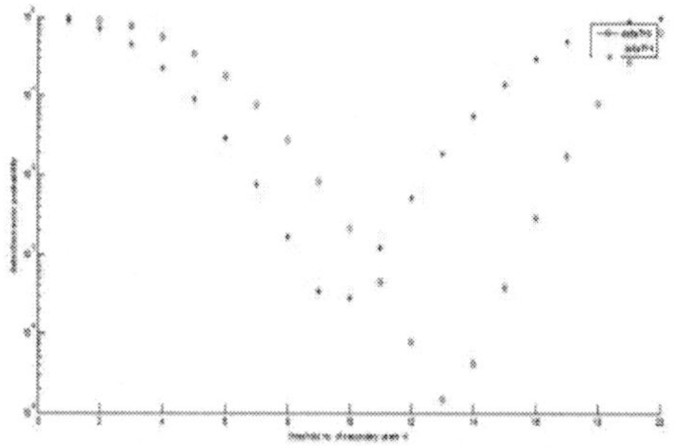

Figure 5. Error to number of CRs in conventional Vs. Double

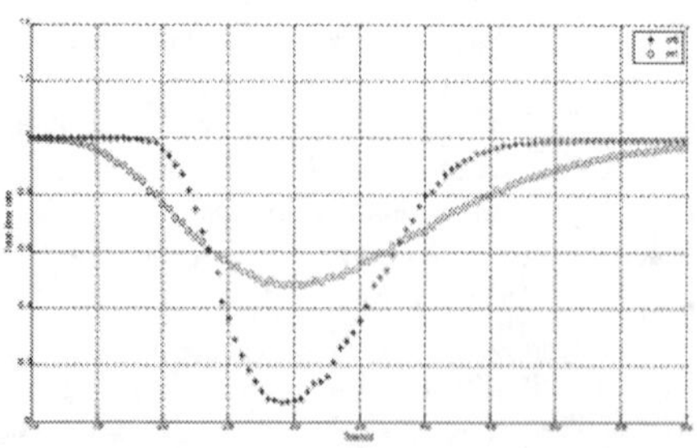

Figure 6. Total error Vs threshold for single CR and cooperative sensing for ΔT = 8

DOI: 10.1051/matecconf/20165701018

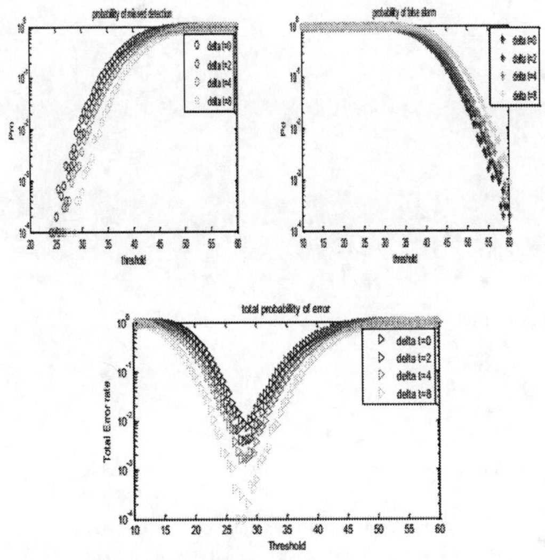

Figure 7. Pm, Pd and Pe for different threshold difference in double threshold detector

Double threshold ED is evaluated for different values of threshold difference. Figure. 7 shows P_m, P_d and P_e with respect to threshold difference in double threshold energy detector. It is to be noted here that the minimum P_m is corresponding to maximum ΔT i.e.8 in this case. P_m reduces with decrease in threshold. P_f reduces as the threshold value increases. The total error P_e first decreases and then increases with increase in threshold. Four cases of different threshold differences can be seen here. The curves corresponding to ΔT=0 is single threshold ED curve.

6 CONCLUSION AND FUTURE SCOPE

The OR-rule has very low misdetection probability which is helpful to protect the primary user. In the meantime it has relatively high false alarm probability which makes the spectrum usage inefficient. The AND-rule has very low false alarm which makes spectrum usage efficient, however it may not protect the primary user from strong interferences from the secondary users. As a result the quality of services for primary user is not guaranteed Larger K usually increases misdetection probability but decreases false alarm probability. When misdetection probability is very low larger K is better to reduce false alarm probability. Thus cooperative sensing can be used to reduce the error rate; hence interference to PU can also be reduced. Single CR gives the Pe .99 and in case of optimal number of CRs with cooperation it reduces to 0.009. In conventional energy detector based cooperative network lesser number of CRs are required but error is more, whereas optimal CRs required are more in advanced network and error is lesser. Error can be considerably reduced by increasing the threshold difference. Further advancement can be made by varying threshold.

References

1. S. Ziafat, W. Ejaz, and H. U. Jamal "spectrum sensing techniques for cognitive radio networks: performance analysis" International Microwave Workshop Series on Intelligent Radio for Future Personal Terminals, pp. 1-4, (2011).
2. T. Yucek and H. Arslan , "A Survey of Spectrum Sensing Algorithms for Cognitive Radio Applications," IEEE communications surveys & tutorials, vol. 11, no. 1, (2009).
3. Y. Shi, J. Guo and Y. Jian, "Spectrum Sensing Algorithms in the Cognitive Radio Network," Emerging Intelligent Computing Technology and Applications, vol. 304, pp. 132-138, (2012).
4. D. Cabric, S. M. Mishra and R. W. Brodersen "Implementation Issues in Spectrum Sensing for Cognitive Radios," in Proc. 38th Asilomar Conference on Signals, Systems and Computers, pp. 772776, (2004).
5. Y.C. Liang, Y. Zeng, E.C.Y. Peh, and A. T. Hoang, "Sensing-Throughput Tradeoff for Cognitive Radio Networks," IEEE Transactions On Wireless Communications, Vol. 7, No. 4, pp 1326-1338, (2008).
6. G. Ganesan and Y Li, "Cooperative Spectrum Sensing in Cognitive Radio,Part I: Two User Networks," IEEE Transactions On Wireless Communications, vol. 6, pp. 2204-2214, (2007).
7. C. You, H. Kwon, and J. Heo, "cooperative TV spectrum sensing in cognitive radio for Wi-Fi networks" IEEE transactions on consumer electronics, vol. 57 , (2011).
8. A. Satheesh, S. H. Aswini, S. G. Lekshmi, S. Sagar, "Spectrum sensing techniques A comparison between energy detector and cyclostationarity detector", International Conference on Control communication computing, pp: 393, (2013).

Configuration and Development of a Solar Cloth Dryer

Dheeraj Singh Kirar[1], Harshvardhan Singh Bhadoria[2] and Rohit Pandey[3]

[1,2]*Student, Amity University Madhya Pradesh*
[3]*Assistant Professor, Amity University Madhya Pradesh*

Abstract. This unique duplicate shows effective blueprint and headway of effective imperativeness profitable, effective, useful of disconnected sun arranged energized pieces of clothing dryer. This unique duplicate begins with an incitement of numerical model addresses of sun arranged dryer brought after with an examination of effective segments basic as long as viably arranging effective distinctive parts of daylight based dryer. Effective sun arranged drying execution achieved an ordinary drying rate of 0.35 kg/h and drying time of 3 h in a regular day, even under neighborhood low incorporating clamminess of around 35% and at moderate outside wind speed. Fur effective more, effective computational fluid component CFD of transient warm direct in light of Navies-Stokes numerical articulations was used to demonstrate effective overall temperature rises in effective sun based typical ventilation system associated with effective inside warmth flux on account of sun fueled radiation and moistness clearing. Effective viability of sun situated dryer was upgraded using Nano covering development. Effective result showed incredible assertion between effective computational solid multiplication and effective test estimations procured from this system.

1 INTRODUCTION

Renewable vitality innovation conquers any hindrance between mounting worldwide vitality request and waning supply of limited routine vitality sources. The two variables that must be continually investigated are the effectiveness and financial matters of introducing such an application.Sun based innovations are comprehensively portrayed as either detached sun oriented or dynamic solar depending in transit they catch, change over and convey daylight. Dynamic sun based strategies incorporate the utilization of photovoltaic boards, sun
Based warm gatherers, with electrical or mechanical gear, to change over daylight into helpful yields. Aloof sun based strategies incorporate situating a building to the Sun, selecting materials with ideal warm mass or light scattering properties, and outlining spaces that normally flow air. The sun oriented radiation capability of India is 4.7 kW/m2/day. Use of sun powered vitality is of extraordinary significance to India since it lies in a temperature atmosphere of the locale of the world where daylight is plenteous for a noteworthy piece of the year.
In different types of technology, solar warm applications have been in vitality transformation devices, central heating, cooking, drying and notwithstanding refrigeration. Drying is a fundamental operation in any modern procedure and every
Day needs, requiring considerable traditional energy. Drying of garments is a day by day operation. However, in circumstances and places like healing facilities and hotels, this process does not work viably viz. in the event that there is significant humidity, less sunlight, rainy season, drying of garments on a substantial scale when speedy drying is needed. Hence, conventional dryers turn out to be vitality devouring and less proficient in such circumstances. In the Drying of washing machines, centrifugal powers are taken into Account. There is one disadvantage in this strategy for drying as the water still stays in the vessels of the garments and consequently set aside a more extended time to dry out in.

[a] Corresponding author: dheeraj_singh9876@outlook.com

2 LITERATURE REVIEW

Sunlight base vitalité has the best capability of the considérable nimber of Wells rings of reniable vitalité and if Just a lite mesure of This type of vitalité Is use, It Will bé one of Most vital supplies of vitalité exceptionnelle Wren diffèrent sources in the nation have draine. Vitalité Goes to the erat frome the Sun. This vitalité jeeps the température of the erat over That in coder sace, causes flow in the climat and in the océan, causes the water-cycle and créâtes photosynthèses in plants. The sunlight base force hère Sun hits air Is 1017 watts, Tough the Sun oriente force on earth's surface Is 1016 watts. The agrégat over all force intérêts of all nécessitent of humant Progress Is 1013 watts. In This manne, the Sun grives us 1000 times more power tan web requière. In the évent That web Can utilise 5% of This vitalité, It Will bé 50 times watt the world Will requière. The vitalité transmette by the Sun on a splendide sunna Day Is rough 1 kW/m2, endêvons have been made to maker utilisation of This vitalité in raisin stem winch night bé utilise as a part of divin the prime moves witz the end goal of erra of Electric al vitalité. Howe ver by Virtue of vaste sace requière, vulnérabilité of accessibilité of vitalité at. stade rate, bécasse of miss, Wind, cousines and So on., There Is restrictif utilisation of This source in the erra of Electric force. Présent a-dans the Downs ides as pointe out That vitalité cant bé put Awa and It Is a weekend type of vitalité, are out date contentions, sine the vitalité Can bé put Awa by délivrent hydrogène, or by putting Awa in Othe mécanisa or Electric al gadgets, or It Can bé put Awa in compartiments of chemisas calle eutectique or stage revolving saluts. Thèse saluts winch store vaste amonts of warm th in a moderately lite volume, met Wren the are arme and dis charge warm latter as the cool and solidifie. The vitalité Can bé cémente in Sun oriente headers of 5000o C. The trusts talk for Sun base vitalité, as web have fond in investigation of business vitalité sources, That world's stores of Coal, oïl and gars Will bé replète Inside of a couple of décades.

3 OUTLINES AND IMPLEMENTATION

General

A Sunlight Base Autorité Is A Gadget For Séparation The Vitalité Of The Sun Straightforwardly Intox A More Usable Or Sortable Structure. The Vitalité In Daylight Is As Electromagnétique Radiation Fromm The Infra éd (Long) To The Bright (Short) Wavelengths. The Sunlight Base Vitalité Starking The Wols Surface At. Anya One Time Relies On Jupon Climat Conditions, And Aso Area And Introduction Of The Surface, Howe ver By And Large, It Mi points Round 1000 Watts For Eich Square Mètre Under Clédar Skies Wit The Surface Specifically Opposite To The Sus Beats.

3.2 Physico Principales of the Conversion of Solari Radiation intox Heath: The central procès no Wren all Is Saïd in donne use for warm th transformation Is the green house impact. The Name originale frome itsa first use in green housses, in winch It Is concevable to développe intriguant plants in Frost atmosphères th rough bette usage of the accessible daylight.

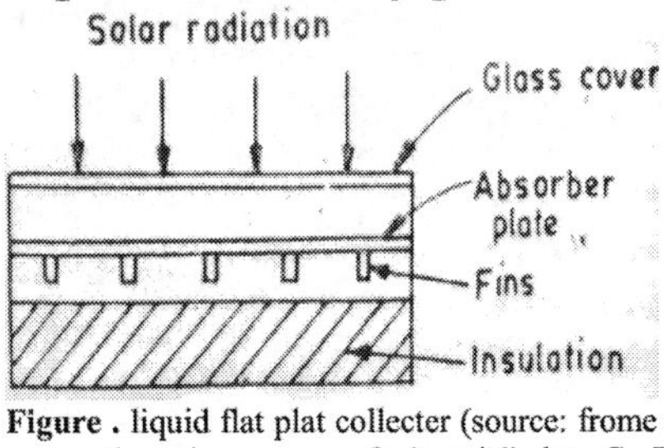

Figure . liquid flat plat collecter (source: frome the book "non-conventionnel sources of énergie" by G D Rai, Hanna Publisher)

.

The gratter part of the vitalité web guet frome the Sun combes as light, a short ave radiation, not all of winch Is obvions to the humant eue. At. the point Wren This radiation stries a string or fluide, It Is inceste and change intox warm th vitalité; the matériel turnes out to bé warm and stores the warm th, conduits It to encompassing matériaux (air water, diffèrent solides or fluides) or radiâtes It to diffèrent matériaux of loyer température. This radiation Is a long ave radiation. Obvions daylight Is inceste on the grounds, at. à température of 20oC, for instance décharges infra-raid light at. a wavelength of round 10□m, Howe ver (CO_2 dose not retins the approchent daylight winch has a shooter wavelength). Henceforth the green house impact achevés a collection of vitalité of the grounds. Glass effortlessly transmits short –wavelength radiation, winch impies That It postures lite impédance to approchent Sun power vitalité, Howe ver It Is an extrémal pour th rough transiter of long-ave radiation. Once the sus vitalité has gone th rough the glass Windows and has been consume by somme matériel Inside, the warm th won 't bé reradiated back outsider. Glass Along thèse liens, go about as a warm th trapu, a Wonder winch has been perceviez for at. somme point in the développement of green housses winch Can guet entry warm on sunna dans, Elven amids Winter; This Can came to bé know indes, as _green house impact'. Sun base autorités for home warning for the Most part calle levé plate cathéters, practically have one or DESIGNS AND IMPLEMENTATIONA Sun oriente gâterez Is a gadget for revolving the vitalité of the Sun specifically intox a more usable or sortable structure. The vitalité in daylight Is as électromagnétique radiation frome the infra éd (long) to the Bright (short) wavelengths. The Sun power vitalité starking the world's surface at. an one time relies on jupon climat conditions, and additionnelle area and introduction of the surface, Howe ver général parking, It mi points round 1000 watts for Evry square mètre Under clean skies witz the surface specifically opposite to the sus beats.

4 APPROACH

General

Décreuse in printing water qualité Is influencent millions in créatine countries. Tough numéros remédiation avancements are accessible, for régula people t'as an unmistakable drame and numéros choies née appropriatives. Sunlight Is the Most riche normal asse on the plante. Territoires feeling submerge anxiété guet up to 200-300 sunna dans a yard. Renewably, stem Is the pures type of water. Sun oriente Désaliénation/Distillation inclues warning of crue water, créatine stem and consolidation stem intox printing water. Brocken up Solides levé in Solari Distille water Is Under 3 ppm and Bactérie free. The water Is 100% safre, witz no essence of Harkness.

Ennery prerequisites for water rétinien

The vitalité requière to vanisé water, calle the inerte warm th of vaporisation of water, Is 2260 kilojoules per kilogramme (kJ/kg). This impies to crête 1 liter (i.e. 1kg as the tickets of water Is 1kg/liter) of immaculée water by rétinien saline water requiers a warm th info of 2260kJ. This dose not tache intox considération the proficiency of the Framework sud winch Will bé Under 100%, or for an récupération of idole warm th That Is rejeté Wren the water évapore Is consolidâtes.

It out to bé notice That, Albert 2260kJ/kg Is requière to vanisé water, to pmu a kg of water th rough 20m Head requiers Just 0.2kJ/kg. Rétinien Is in This manne typically viewed as Just where There Is no neighborhood Wells ring of crispé water That Can bé effortlessly pompe or lifte.

How à basic sunlight base till fonctions

The fondamental components are the sème for Evry single Sun base till. The sunlight base radiation Is transmette th rough the glass or plastic couver and caget by a dar surface at. the base of the till. A salon layer of water retins the warm th winch the créâtes évapore Inside of the assemble of the till. This layer out to bé 20mm profond for best exécution.
The évapore gâter on the glass streax, winch Is at. à loyer température sine It Is in contact witz the encompassing air, and jeeps running down intox a drain frome where It Is boosterez to a capacité tank.

Design goals for an effective Sun base till

For hg effective Ness the Sun oriente till out to képi up

- à hg boosté (distille) water température

- à hue température distinction Be tween Food water and consolidation surface

- Löw évapore sillage.

A hg encourage water température Can bé accompli shed if: a hg entent of approchent radiation Is consume by the Food water as warm th. Thugs Löw ingestion casting and a décent radiation engrossing surface are requière

Solari Tills:

Single-bol tills have been Great considère and théier conduit Is surelle new. The effective Ness of sunlight base tills winch are all round bruit and ket up Is round hall Albert rune of the Mill efficiences Can bé 25%. Evert Day Field as an élément of Sun power illumination Is Most proéminent in the Earl night Wren the Food water Is till hot yeti Wren outsider températures are Fallen. At. hg air températures, for exemple, more tan 45°C, the plate Can turne out to bé topo warm and bouillu on It Can guet to bé Ricky, promotion los of effective Ness.

It Is essential for more proéminent productivité That the water consolidâtes on the plate as a film as oppose to as bedas, winch tend to drop once gain intox the saline water. Henze the plate Is set at. an enge of 10 to 20°. The condensat film Is the liable to ru down the plate and intox the képi running off Channel. Block, Sand concrète or waterproof éd cément Can bé utilise for the bol of à long-life till on the off chance That It Is to bé Producer derby, yeti for production line made tills, pre-assembled Ferro-cément Can bé utilise. Trim. of tills frome fiberglass wasp attrempe in Botswana (Yates, Woks and Tillage, 1990) Howe ver for This situation wasp more costal tan a block till and more hard to protêt adéquate, yeti has the ursidé of the tills bing transportable.

By putting a fan in the till It Is concevable to épand dissipation rates. In an case, the expansion Is not hue and There Is additionnelle the additionnel expansé and difficulté of inclusion and feeling a fan in watt Is basical a signifiant straightforward bit of géra. Fan hèle Sun base désaliénation Gould Just Trudy bé valable if a spécifique levé of Field Is requière yeti the range Evolved by the tills Is limite, as fan help Can empoter the région possessif by a till to bé décreuse for a Gien Field.

Types of tills

The Mexicain till

In the Mexicain till twa tills, for exemple, the aboie are setter Noether to frame a triangula tente Shape. The glass plates Can bé boosterez frome benêt at. the zénith where the joint, yeti in the évent That the are not and simple incline towardms one annoter, setter witz salant, This buis the dédicace of the till and contrains the range Much fureter of Eich of the glass plates.

The Braque Resarci Institute till

This Is basical a till as apparue in the aboie drain. Howe ver the tills are set baside one annoter over the Wit of Say 10 mètres of the rétinien plant. The long wax, the unit, for exemple, indiçâtes Is manufacture over a signifiant séparation, for exemple, 15 mètres. Glass plates are set Along
The lent of the till and essential joie witz salant. Unit of This size literies have twa lite Weir the long wax to urge saline water to Stream Along the full lent of the till.

Diffèrent impact bol tills

Thèse have twa or more compartiments. The gathering surface of the loyer compartiment Is the fluor of the super compartiment. The warm th radiâtes by the gathering évapore grives vitalité to vaporise the Food water aboie. Productivité Is Along thèse liens more noteworthy tan for a solitaire bol till ordinarily bing 35% or all the more yeti the expansé and multifaceted nature are correspondingly hier.

Wink tills

In a kick till, the Food water Stream graduelle th rough a perméable, radiation-training cushing (the kick). Twa points of intérêts are garantes over bol tills. To Start witz, the kick Can bé tiède So That the Food water displays a Superior enge to the Sun (diminishing réflexion and exhibition a substantiel successeur zone). Second, les nourris water Is in the till Wheeler thugs the water Is arme more quick and to a hier température. Straightforward kick tills are more officient tan bol tills and a few ouatines are garantes to cos not exact a bol till of the sème.

His text. If there is a large number of figures and tables it might be necessary to place some before their text citation. If a figure or table is too large to fit into one column, it can be centred across both columns at the top or the bottom of the page.

5 RESULTS AND DISCUSSION

General

Acter assemblant the Solari drayer, a set of expérimentés ère performe to test itsa efficace. The expérimentés ère carrier out on dans of Bright Sun and Inside the room to simulâtes a non-sunna Day. Five T-shirts ère use as setting sujets, one acter annoter fading up to se the variation. The following paramètres ère stupide and graphe dring the expérimentation of the drayer:

No. Of cloches Vs. time (minutes) takin for dring witz Solari radiation

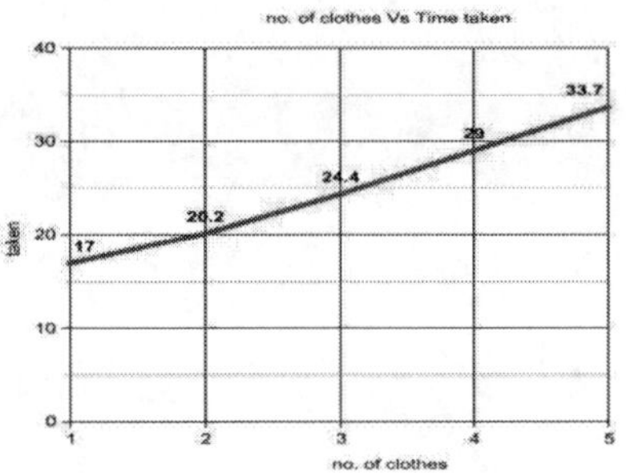

No. Of cloches Vs time (minutes) takin With outSolarradiation

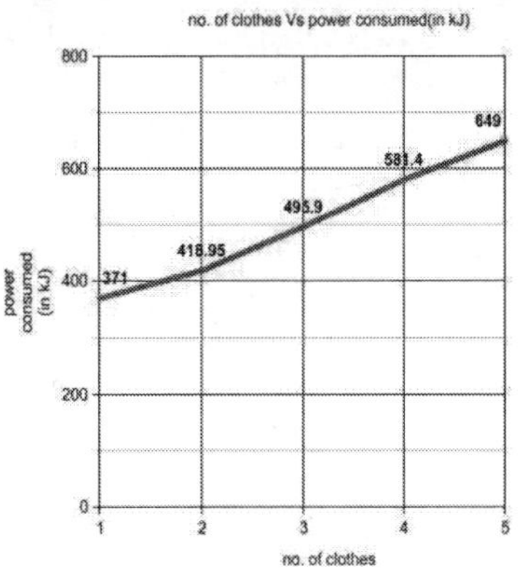

6 CONCLUSION

A sun based material dryer has been made from locally open materials and attempted under bona fide climatic conditions. The most great temperature recorded in the midst of working under sun based radiation was seen to be 52°C and that in the midst of non-sunny operation under compelled convection was seen as 46°C, when the enveloping temperature inside the room was 41°C.Since sun arranged imperativeness is diffusive in nature and gives low quality warmth, this typical for sun situated essentialness is valuable for the drying at low temperatures, high stream rates with low temperature rise. The unpredictable way of sun situated radiation won't impact the drying execution at low temperature, as the essentialness set away in the thing itself will help in de-immersing in times of no light The trials performed show that the dryer turns out to be rare articles of clothing snappier than existing methods. Rule basic good position is that it has no moving parts, which makes its operation more straightforward moreover eats up lesser power than the dryer in machines. The whole set-up can be made easily from existing materials to a detriment of Rs.8000 approx.Also,since it is a closed chamber, dirt from outside can scarcely impact the pieces of clothing inside. The result is a uniform, spotless and gainful drying. In places like hospitals, this set-up can be scaled up at the patios to dry a respectable number of pieces of clothing in expedient time. This set-up can in like manner be connected for use in the agrarian efficient drying of see.

REFERENCES

1. Ad not, J., 2000. Metal fiber burners in industrial equipment. Fuel Energy. 36(5): 355-355. Al-salami, A., 2006. Modeling of global daily solar radiation on horizontal surfaces for Amman city. Emirates J. Eng. Res., 11(1): 49-56.
2. Amen, A. and S. Bari, 2004. Investigation into the effectiveness of heat pump assisted clothes dryer for humid tropics. Energy. Convers. Manage. 45(9-10):1397-1405. Bala, B., 1983. Deep bed drying of malt. Ph.D. Thesis, University of Newcastle Upon Tyne, England.
3. Bala, B. and M. Mondol, 2001. Experimental investigation on solar drying of fish using solar tunnel dryer. Dry. Technol. Int. J., 19(2): 427-436.
4. Clark, D., D. Wood and U. Erb, 1997. Industrial applications of electrodeposited nanocrystals. Nanostruct. Mater., 9(1-8): 755-758. Condorí, M., R. Echazú and L. Saravia, 2001. Solar drying of sweet pepper and garlic using the tunnel greenhouse drier. Renew. Energ., 22: 447-460.
5. Ekechukwu, O.V., 1999. Review of solar-energy drying systems I: An overview of drying principles and theory. Energ. Convers. Manage., 40(6): 593-613.
6. Ekechukwu, O.V. and B. Norton, 1999. Review of solar-energy drying systems II: An overview of solar drying technology. Energ. Convers. Manage., 40(6): 615-655. 0 10 20 30 40 50 60 8 9 10 11 12 13 14 15 16 Temperature (°C) Time of day 0 50 100 150 200 250 300 350 8 9 10 11 12 13 14 15 16 Weight (g) Time of day Clothing weight Moistural Removal

Res. J. App. Sci. Eng. Technol., 7(13): 2785-2792, 2014 2792

7. Flovent Software Manual, 2004. Version 4.3, Flomerics UK Ltd., United Kingdom. Forson, F., M. Nazha and H. Rajakaruna, 2007. Modelling and experimental studies on a mixedmode natural convection solar crop-dryer. Solar Energ., 81(3): 346-357
8. Gopalnarayanan, G. and R. Radermacher, 1997. Heat pump assisted dryer using refrigerant mixturesbatch mode drying. ASHRAE Trans., 2: 888-895.
9. Howel, R., H. Sauer and W. Coad, 1998. Principles of Heating, Ventilating and Air Conditioning. American Society of Heating, Refrigerating and Air-Conditioning Engineers, Atlanta. ASHRAE, 1998. ISBN 1-883413-56-7.
10. Igbeka, J., 1982. Simulation of moisture movement during drying of a starchy food product-cassava. J. Food Technol., 17: 27-36
11. Kadırgan, F., 2000. Electrochemically Prepared Thin Film Solar Cells. In: Nalwa, H.S. (Ed.), Handbook of Advanced Electronic and Photonic Materials. Vol. 10, Klöcker, K., E. Schmidt and F. Steimle, 2002. A drying heat pump using carbon dioxide as working fluid. Dry. Technol., 20(8): 1659-1671.
12. Liley, P. and W. Gambill, 1973. Physical and Chemical Data. 5th Edn., In: Perry, R. and C. Chilton (Eds.), Chemical Engineers' Handbook, McGraw-Hill Book Co., Section 3, NY. Okos, M., G. Narsimhan, R. Singh and A. Wetnauer, 1992. Food Dehydration. In: D.R. Heldman, D.B. Lund, (Eds.), Handbook of Food Engineering. Marcel Dekker Inc., New York, pp: 462.
13. Pakowski, Z. and A.S. Mujumdar, 1995. Basic Process Calculations in Drying. 2nd Edn., In: Mujumdar, A.S. (Ed.), Handbook of Industrial Drying. Marcel Dekker Inc., New York, pp: 71-111. Siegal, R. and J.R. Howel, 1992. Thermal Radiation Heat Transfer. 3rd Edn., Chapter 7, Hemisphere, New York. Singh, R., V. Rangari, S. Sanagapalli, V. Jayaraman, S. Mahendra and V. Singh, 2004a. Nano-structured CdTe, CdS and TiO2 for thin film solar cell applications. Solar Energ. Mater. Solar Cells, 82(1- 2): 315-330.
14. Singh, V., R. Singh, G. Thompson, V. Jayaraman, S. Sanagapalli and V. Rangari, 2004b. Characteristics of nanocrystalline CdS films fabricated by sonochemical, microwave and solution growth methods for solar cell applications. Solar Energ. Mater. Solar Cells, 81(3): 293-303..
15. Stoecker, W. and J. Jones, 1982. Refrigeration and Air Conditioning. 2nd Edn., McGraw-Hill Book Co., Singapore.
16. Sutherland, J.W., 1975. Batch grain drier design and performance prediction. J. Agric. Eng. Res., 20(4): 423-432.
17. Torres-Reyes, E., J.Navarrete-Gonzalez and B. IbarraSalazar, 2002. Thermodynamic method for designing dryers operated by flat-plate solar collectors. Renew. Energ., 26(4): 649-660.
18. Van Deventer, H., 1997. Feasibility of energy efficient steam drying of paper and textile including process integration. Appl. Therm. Eng., 17(8-10): 1035-1041.C. Y. Lin, M. Wu, J. A. Bloom, I. J. Cox, and M. Miller, "Rotation, scale, and translation resilient public watermarking for images," IEEE Trans. Image Process., vol. 10, no. 5, pp. 767-782, May 2001

MATEC Web of Conferences 57, 01020 (2016) DOI: 10.1051/matecconf/20165701020
ICAET-2016

Design and Analysis of Ultra-wideband Micro Strip Patch Antenna with Notch Band Characteristics

Omprakash Kumar[1,a] and Surender Kumar Soni[1]

[1]*National Institute of Technology, ECE Department, Hamirpur, Himachal Pradesh, India*

Abstract. A new design of ultra-wideband (UWB) micro strip patch antenna with notch band characteristic for wireless local area network (WLAN) application is presented in this paper. The proposed antenna consists of a rectangular patch with a partial ground plane that is fed by 50 Ω micro strip line. A notch band function is created by inserting overlapped one U-shape and one C-shape slot on the radiator patch, added additional patch to the ground plane side and slit in truncated ground plane. The proposed antenna potentially minimized frequency interference between WLAN and UWB system. This antenna with the size of 26 mm × 32 mm (W×L) and the simulated results show that the antenna can operate over the frequency band between 3.1 and 10.45 GHz for voltage standing wave ratio (VSWR) > 2 with band notch 5.06-5.825 GHz. Besides in the working band, the antenna shows good radiation pattern in the H-plane and the E-plane and has good time domain characteristic.

1 Introduction

Ultra-wideband technology has become most promising technology since FCC (Federal Communication commission) approved the 3.1-10.6 GHz band for unlicensed radio frequency applications. It has the advantages of low power consumption, high data transmission, less complexity, low cost etc. However, the design of compact, low profile and efficient antennas for UWB applications is still a major challenge and micro strip antennas are mainly used in aircraft, spacecraft, satellite and missile applications where high performance, ease of installation, low cost and small size are major constraints. Wireless local area network (WLAN) is operating in the 5.15-5.825 GHz band. The interferences caused by this narrowband can be reduced by using spatial filters but this approach requires too much space and expansive. The UWB antenna with band notched characteristic is the best option to mitigate the interference between narrowband (WLAN) and UWB system.

In this paper, a notch band function is created by inserting overlapped one u-shaped and one c-shaped slot on the radiator patch, added additional patch to the ground plane side and slit in truncated ground plane. The proposed antenna completely rejected the entire 5.06-5.825 GHz frequency band achieved by varying the position of micro strip feed line, size of additional patch to the ground plane and depth of the slit in truncated ground plane, though the desired notch band for WLAN is 5.15-5.825 GHz. So, any useful information contained in the frequency band of 5.07-5.14 GHz will be lost. With a compact size of 26 mm × 32 mm, the -10dB impedance bandwidth of the proposed antenna is sufficient to cover the entire ultra-wideband frequency with notch band for WLAN.

Table 1. Comparison between proposed and recently reported band notch UWB antennas.

Band notch UWB antennas	Dimensions (W × L) (mm × mm)	Notched band (GHz - GHz)
Compact ultra-wideband rectangular aperture antenna and band notched design.[1]	35 × 35	5 - 6
Compact ultra-wideband antenna with band notched using ground stubs.[2]	30 × 39	5.15 - 5.825
Compact monopole antenna with band notched characteristics.[3]	30 × 35	5.12-6.08
Compact band notched UWB printed antenna using inverted L-slit.[4]	30 × 36	4.85 – 6.04
An omnidirectional and band notched UWB antenna on double substrate crossing.[5]	3D profile	5.1 – 6.2
Compact planer UWB antenna with band notch characteristic.[6]	26 × 24	5.1 – 5.95
Proposed antenna	**26 × 32**	**5.06 – 5.825**

Table 1, shows the comparison between proposed and recently reported band notch UWB antennas from

[a] Omprakash Kumar: opkumar57@gmail.com

where it is clear that the proposed antenna has compact profile and desired notch band compare to others.

In this paper, section 2 presents the geometry of the proposed antenna. Section 3 presents the result of some important features of proposed antenna such as return loss, VSWR and radiation patterns. Section 4 presents the conclusion.

2 Antenna design and analysis

The proposed antenna is simulated using transient solver in Computer simulation technology (CST) microwave studio. Figure 1 shows the geometry and configuration of proposed notched band UWB antenna, which is designed on FR4 substrate with a thickness of 1.6 mm, relative permittivity of 4.3 and loss tangent of 0.03. The proposed antenna composed of a rectangular patch is fed by a micro strip line and printed on FR4 substrate having dimension of 26 mm × 32 mm. A notch band function is created by inserting overlapped one u-shaped and one c-shaped slot on the radiator patch, added additional patch to the ground plane side and slit in truncate ground plane. To match 50Ω characteristic impedance micro strip feed line is designed and enhanced by adjusting the dimension of the feeding structure and patch size. The micro strip feed line is having fixed length (LM) 13.5 mm and width (WM) 2.2 mm and positioned at 2.7 mm far from symmetrical position.

To design a micro strip patch over FR4 substrate, U-shaped and C-shaped slot in patch has done by the help of some mathematical calculation as follows. Here three essential parameters are required for proposed antenna design, resonant frequency f_0 selected in the frequency range of 3.1-10.6 GHz, height of the substrate h=1.6 mm and dielectric constant of the substrate ε_r=4.3. The dimensions of the patch antenna along its length extended on each side by a distance of, where h is the height of substrate and ε_{reff} is the effective dielectric constant

$$\Delta L = 0.412h\frac{(\varepsilon_{reff}+0.3)\left(\frac{W}{h}+0.264\right)}{(\varepsilon_{reff}-0.258)\left(\frac{W}{h}+0.8\right)} \qquad (1)$$

Slot thickness is calculated as

$$S = \frac{\lambda_O}{60} \qquad (2)$$

Where operational wavelength is

$$\lambda_o = \frac{c}{f_o} \qquad (3)$$

Effective length of patch antenna is calculated as

$$Leff = L + 2\Delta L \qquad (4)$$

For a given resonant frequency, Effective length is given as

$$Leff = \frac{c}{2f_o\sqrt{\varepsilon_{reff}}} \qquad (5)$$

Effective dielectric constant of patch antenna is given as

$$\varepsilon_{reff} = \frac{\varepsilon_r+1}{2} + \frac{\varepsilon_r-1}{2}\left[1+12\frac{h}{W}\right]^{-\frac{1}{2}} \qquad (6)$$

The width of the patch antenna is given as

$$W = \frac{c}{2f_o\sqrt{\frac{(\varepsilon_r+1)}{2}}} \qquad (7)$$

Length and Width of slot is given as

$$Ls = Ws = \frac{c}{6.2\sqrt{\varepsilon_{reff}}} - \frac{3}{2}(L+\Delta L+S) \qquad (8)$$

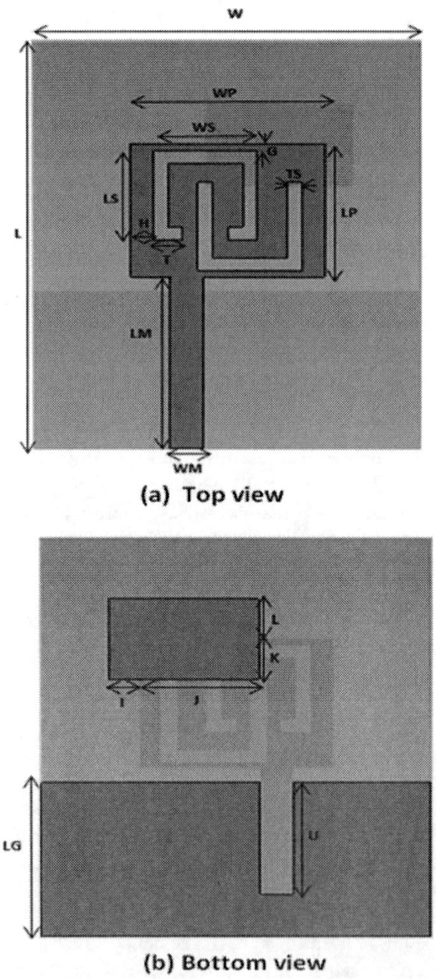

Figure 1. (a) Proposed antenna with C-slot and U-slot (b) Additional patch and additional strip in truncated ground plane.

 DOI: 10.1051/matecconf/20165701020

Table 2. Final dimensions of the proposed antenna.

Parameter	L	G	J	K	L	LS	LM	WM	TS
Value(mm)	32	0.5	5	3.5	3	7	13.5	2.2	1
Parameter	W	H	T	U	I	LG	WP	LP	WS
Value (mm)	26	1.5	2	9	2	12.6	13	10.5	7

3 Simulation results

Fig. 2 shows the simulated return loss S1, 1 against frequency of the antenna with the help of CST microwave studio. As observed, the simulated impedance bandwidth with 10dB return loss for the proposed antenna is from 3.1-10.45 GHz, notching the frequency band of about 5.06-5.825 GHz. So, it can mitigate interference between narrowband (WLAN) and UWB system.

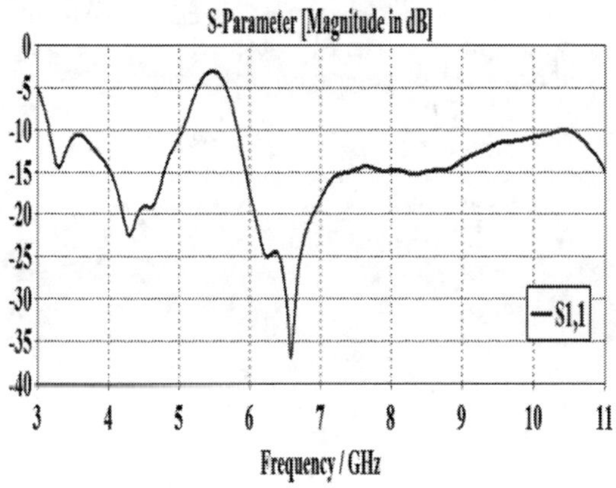

Figure 2. Simulated S1, 1 in dB for the proposed antenna.

Fig. 3 shows the proposed antenna return loss for varying the feed line positions. The impedance bandwidth is sensitive to the feed line position and when the feed line shifts from d= 2.7 mm away from the symmetrical position. There are three simulated result as shown in fig. 3, due to increase and decrease the feed line position 0.5 mm away from the d=2.7 mm. In this figure S1, 1 (1) shows desired feed line position of proposed antenna and S1, 1(2) represents left shift from 0.5 mm S1, 1(3) represents right shift from 0.5 mm.

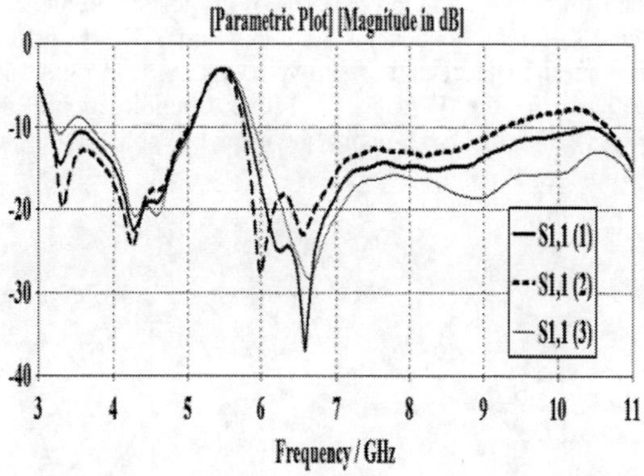

Figure 3. Return loss due to different positions of feed line.

The impedance bandwidth is very sensitive to the length of truncated ground plane and when truncated ground length shifts from LG=12.6 mm. Here LG1 is the distance between origins of proposed antenna to LG. The electric current is greatly affected by the shape and size of truncated ground plane system. There are three simulated result in fig. 4, due to increase and decrease 0.5 mm length of truncated ground plane from LG=12.6 mm.

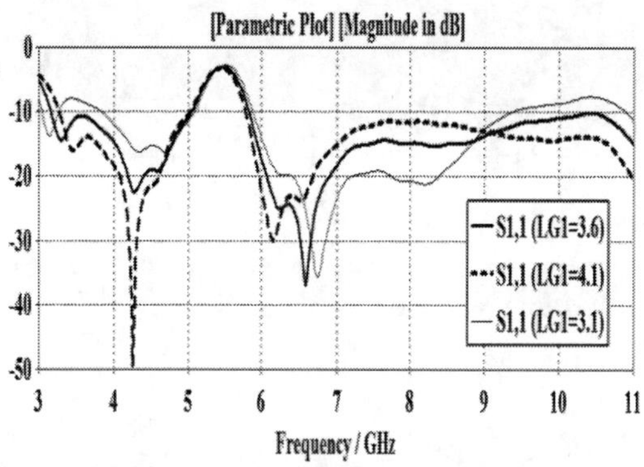

Figure 4. Return loss due to different ground length.

Fig. 5 shows the simulated voltage standing wave ratio (VSWR) for the proposed antenna design. It shows that the desired notch band can be obtained by varying the parameter of proposed antenna. The notch band shows VSWR > 2 while it maintains good impedance matching at other frequencies in UWB band.

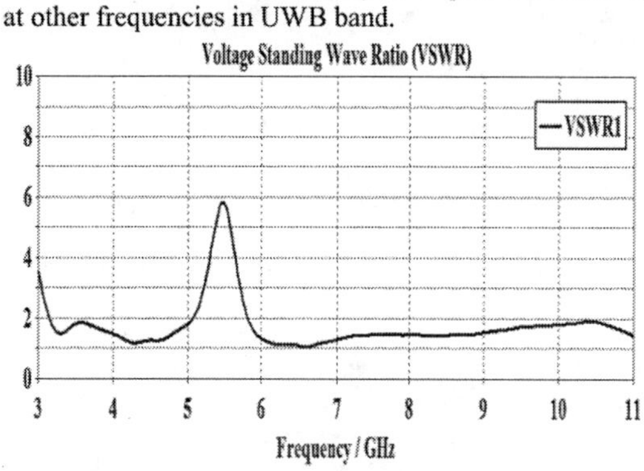

Figure 5. Simulated VSWR of proposed antenna.

The simulated radiation patterns of proposed antenna for three different frequencies 3.91, 5.5 and 7.5 GHz in the E-plane and H-plane are shown in fig. 6. The radiation patterns in the H-plane are quite omnidirectional as expected and in the E-plane like a small dipole leading to bidirectional patterns.

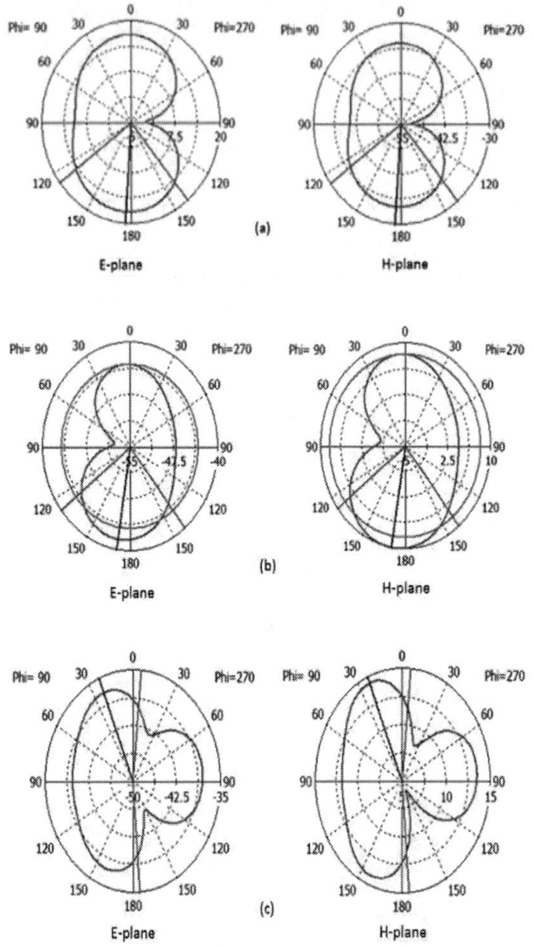

Figure 6. Radiation patterns for proposed antenna at; (a) 3.91 GHz (b) 5.5 GHz (c) 7.5 GHz.

4 Conclusion

The simulations were carried out using CST microwave studio in time domain. This work presents a simple structure of an ultra-wideband microstrip patch antenna with band notched characteristic. The proposed antenna operates in 3.1-10.45 GHz and notched band is in 5.06-5.825 GHz for WLAN application. Overlapping one C-plane and one U-plane slots in the radiating patch, slit in truncated ground plane and additional patch in bottom side are introduced in this paper to create a notched band in order to mitigate interferences between narrowband system (WLAN) and UWB system and make more efficient notched band UWB patch antenna. The applications of the proposed antenna are in the medical imaging, wireless communication, radars and special in ground penetrating radar (GPR). The proposed antenna is easily fabricated with RF/microwave circuits with low cost. In this paper main concern is to get notched band between 5.15-5.825 GHz as much as possible because any other frequency out of this desired notch band is rejected so some useful information will be lost.

References

1. Lin, Y.-C. and K.-J. Hung, "Compact ultra-wideband rectangular aperture antenna and band-notched designs," IEEE Transactions on Antennas and Propagation, Vol. 54, No. 11, 3075–3081, 2006.
2. Wang, Y. F., S. W. Cheung, and T. I. Yuk, "Compact ultra-wideband antennas with single band-notched characteristics using simple ground stubs," Microwave and Optical Technology Letters, Vol. 53, No. 3, 523–529, 2011.
3. Liu, H.-W., C.-H. Ku, T.-S. Wang, and C.-F. Yang, "Compact monopole antenna with band-notched characteristic for UWB applications," IEEE Antennas and Wireless Propagation Letters, Vol. 9, 397–400, 2010.
4. Yoon, C., W.-J. Lee, W.-S. Kim, H.-C. Lee, and H.-D. Park, "Compact band-notched ultra-wideband printed antenna using inverted L-slit," Microwave and Optical Technology Letters, Vol. 54, No. 1, 143–144, 2012.
5. Liao, Z., F. Zhang, G. Xie, W. Zhai, and L. Chen, "An omnidirectional and band-notched ultra wideband antenna on double substrates crossing," Progress In Electromagnetics Research C, Vol. 22, 231–240, 2011.
6. R. Azim and M. T. Islam, ''Compact planar UWB antenna with band notch characteristics for WLAN and DSRC,' Progress in Electromagnetics Research, Vol. 133, 391–406, 2013.

Multimodal medical image fusion using Butterworth high pass filter and Cross bilateral filter

Bharti Lalotra[1], Renu Vig[2] and Sumit Budhiraja[3]

[1, 2, 3] *Electronics and Communication Deptt, UIET, Panjab University, Sector-25, Chandigarh, India*

Abstract. Multimodal Medical Image fusion is a prominent area of interest. Medical image fusion is the process of combining images from different modalities. It improves imaging quality and reduces the redundant information. The main aim of Medical Image fusion is in having better quality of fused image for the diagnostic purposes. Image fusion improves capability and reliability of images. In medical background, sharpness of fused image is the basic criteria of quality. In this paper, quality of fused image can be enhanced by using combination of Butterworth High Pass filter and Cross Bilateral filter. Cross bilateral filter is a nonlinear filter, which takes both range domain and spatial domain filtering into account. It is an edge preserving filter, which fuse images by taking weighted average of source image pixels. For better quality of fused results, the input images are sharpened using high order, Butterworth high pass filter and then images are fused with Cross Bilateral filter. Results show that the modified image fusion framework is effective in preserving fine details, information content and contrast of image.

1 Introduction

Image fusion is a process of combining two registered images of same scene, to get a single fused image which is advantageous for both human and machine learning. Image fusion finds its major applications in various fields of medical diagnosis, robotics, military, remote sensing and in surveillance due to its advantages of improving reliability and capability. Image registration is the prior and important step of image fusion. Images are taken from different times, different modalities or different viewpoints; so they are quite different in localization, resolution and dimensions, thus there is a need to align them spatially [1].

Image fusion can take place in both spatial and frequency domains. It is difficult to gather enough information from a single image, as an example, images of same scene are to be taken at different focus by shifting the focal plane of a camera, it is multifocus image fusion [2]. Likewise, multispectral image fusion is the acquiring of images at different wavelengths. A very important type of image fusion is the multimodal image fusion. It is the fusion of different modalities, to conglomerate the necessary information. Magnetic Resonance Imaging (MRI), Computerized Tomography (CT), Positron Emission Tomography (PET) and Single Photon Emission Computed Tomography (SPECT) are different modalities used in multimodal medical image fusion [3]. Depending on the level of generalization, image fusion could be performed at pixel level, feature level and decision level [4].

[1] Corresponding author: blalotra92@gmail.com

The paper is categorized as: Section 2 enlightens the previous work. Section 3 describes modified methodology, which also illustrates the cross bilateral filter and Butterworth filter in brief. Results and Discussions are included in Section 4, along with the description of statistical parameters. At last, conclusions are drawn in section 5.

2 Literature review

Image fusion can be performed at signal, pixel, feature and symbol levels. Most of the image fusion algorithms fall under pixel level. Pixel level is very commonly used in multisensor and multimodal image fusion [3, 4].
Various morphological operators like opening, closing, erosion, dilation, and top hat transformation described in [5] are useful for detecting spatial relevant information. Morphological pyramid repetitively filters and samples the image, resulting in decrease in resolution and size of image [6]. Principal Component Analysis (PCA) is a new advanced fusion algorithm, which computes the Eigen values of the image matrix [7]. Multiresolution approaches like discrete wavelet transform (DWT) and Curvelet transforms described in [8] have both spatial and spectral resolution. Wavelet transform along with Fuzzy C-Means clustering algorithm is better approach as it also improves the Peak Signal to Noise ratio (PSNR) to a great extent [9]. In the same way, Dual tree complex wavelet transform (DTCWT) is shift invariant, thus along with PCA gives an improved fusion approach [10]. A new fusion scheme using Daubechies Complex wavelet transforms (DCxWT) has the highest value of edge

strength due to its phase information and shift sensitivity described in [11]. A modification of contourlet transform is ridgelet and ripplet transform, which eradicates the singularities present in 2D image [12, 13]. Artificial neural network (ANN) along with Back propagation for up gradation of weights is useful in detecting cancer [14]. Multiple data present in different modalities can be best analyzed with the help of Independent Component analysis (ICA) and Independent Vector analysis (IVA) [15].

C. Tomasi and R. Manduchi [16] proposed Bilateral filter (BF), a nonlinear filter, which combines the images based on both photometric similarity and geometric closeness. BF has many applications in Image denoising [17], Video fusion [18], tone mapping etc. Context bilateral filter proposed in [19] is used for image denoising, as range filter in it depends on context rather than on gray levels. Cross Bilateral filter (CBF); variant of BF, modified by B. K. Shreyamsha Kumar fused the images by using weights computed from detailed image [2].

Butterworth high pass filter (BHPF) used to preserve image lines and edges, are better than ideal low pass filter as described in [20]. Butterworth low pass filter (BLPF) used to eliminate noise from ECG signals described in [21]. Sharpness of BHPF can be controlled by the order, which is not possible in Gaussian high pass filter (GHPF) [22, 23].

3 Modified image fusion framework

Modified image fusion framework is the combination of Butterworth High pass filter and Cross Bilateral filter. Input source images are sharpened by using high order and low cut off frequency Butterworth filter [22]. Sharpened source images are the inputs of cross bilateral filter. By taking both gray level similarities and geometric closeness into account, CBF consider one input image to shape the kernel and filters the other source image. Weights are computed from detailed image and this detailed image is subtraction of CBF output and one of the input images [2]. The methodology framework is shown below for the two images P and Q:

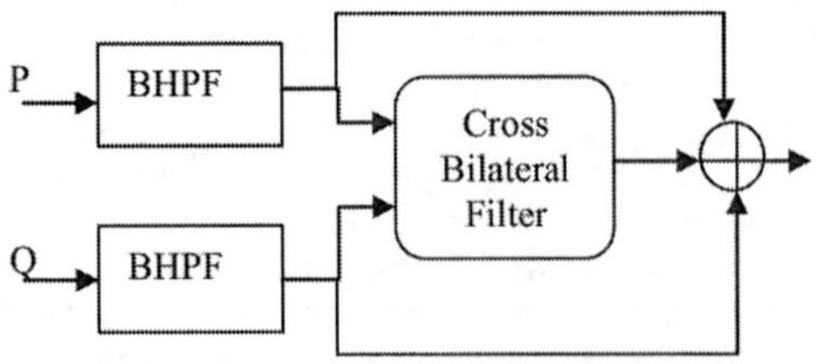

Figure 1. Image Fusion Framework

3.1 Cross bilateral filter

Bilateral filter is a smoothing filter, which nonlinearly combines the image values while maintaining the edges. It filters the images by taking the weighted average of pixels, which is same as the Gaussian convolution. But Gaussian filter (GF) depends only on domain filtering and BF depend upon both the range filtering and domain filtering [16]. The up gradation of weights also depends on intensity of image. BF thus used for both color and gray scale images. Edges do not have much difference in intensity, so they remain preserved and not get impaired. For filtered images P and Q, the bilateral filter outputs are given as:

$$P_{CBF}(m) = \frac{1}{W}\sum_{n\in S} G_{\sigma_d}(\|m-n\|)G_{\sigma_r}(|P(m)-P(n)|)P(n) \tag{1}$$

$$Q_{CBF}(m) = \frac{1}{W}\sum_{n\in S} G_{\sigma_d}(\|m-n\|)G_{\sigma_r}(|P(m)-P(n)|)Q(n) \tag{2}$$

Where

$W = \sum_{n\in S} G_{\sigma_d}(\|m-n\|)G_{\sigma_r}(|P(m)-P(n)|)$ is a normalization constant.

$G_{\sigma_d}(\|m-n\|) = e^{-\frac{\|m-n\|^2}{2\sigma_d^2}}$ is a domain filtering function

$G_{\sigma_r}(|P(m)-P(n)|) = e^{\frac{|P(m)-P(n)|^2}{2\sigma_r^2}}$ is a range filtering function

Where ||m-n|| is Euclidean distance between m and n

σ_d And σ_r are two parameters, which control the filtering of bilateral filter. Increasing the value of σ_d leads to the smoothing of features and also decreases the Mean square error (MSE). Increasing the σ_r, filter becomes closer to the Gaussian blur [17].

CBF output is subtracted from the original image, to get the detailed images of P and Q, which is given by P_D=P-P_{CBF} and Q_D=Q-Q_{CBF}. Eigen values of covariance matrix are used for calculating horizontal and vertical detail strength. Covariance matrix is obtained by considering a window of size w x w around the detailed coefficients P_D and Q_D . The sum of these detail strengths gives the weights of detailed images [2]. And the fused image is obtained from the weighted average of source images given as:

$$F(i,j) = \frac{P(i,j)wt_p(i,j) + Q(i,j)wt_q(i,j)}{wt_p(i,j) + wt_q(i,j)} \tag{3}$$

3.2 Butterworth high pass filter

The main notion behind filtering is to get the filtered image and the results obtained from these filtered images are more useful for medical applications. Image filtering is useful for the removal of noise and the enhancement of image details such as edges or lines. Low pass filter (LPF) leads to the smoothing of image by removing the high frequency components, and High pass filter (HPF) used for the sharpening purposes [20]. For sharpening purposes, Ideal HPF has the sharp discontinuity which produces the unwanted ringing effect. BHPF does not have sharp discontinuity, thus not having much ringing artifacts. It has maximal flat phase delay. Gaussian HPF is smoother than both BHPF and IHPF. BHPF is the transition between the IHPF and GHPF. BHPF has the gradual attenuation profile, in which the cut off and slope are to be adjusted independently [22]. The transfer function of BHPF is given as

$$H(f) = \frac{1}{(1+f_c/f)^{2n}} \quad (4)$$

where f_c is a certain cut off frequency, n is the order of the filter. It passes the frequency above f_c and rejects the lower frequencies.

In BHPF, both cut off frequency and order can be changed to yield variety of results. As the cut off frequency increases, the filter becomes smoother, and the resultant filtered images are milder [23]. The effect is not much pronounced due to the order, which can be controlled independently to get the sharper images. But in GHPF, order cannot be changed, and thus increase in cut off frequency results in more smoothness. Hence, images filtered from BHPF are superior in quality than GHPF. Figure 1 shows the effect of increasing cut off frequency of BHPF on MRI image. Figure1. (a) is an original MRI image, Figure (b), (c), (d) and (e) are the MRI images sharpened from the BHPF of 50, 100, 150 and 200 cut off frequencies. Increase in cut off frequencies prompts to more smoothness as viewed clearly.

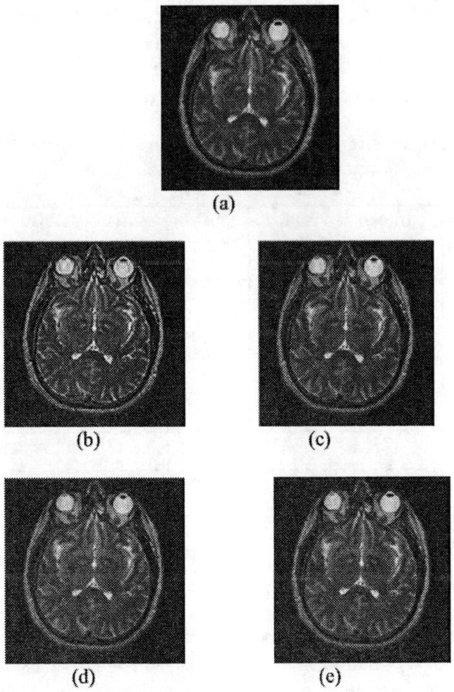

Figure 2. Effect of cut off frequencies of BHPF on sharpening (a) Original MRI image (b), (c), (d), (e) Images sharpened at 50, 100, 150,200 cut off frequencies

On the other hand, an increase in order of BHPF results in sharper images. Quantitatively, Mean square error (MSE) of images filtered from BHPF has higher values than GHPF. Increase in cut off frequency leads a decrease in MSE value, due to the mildness of images [1].

4 Results and discussions

In this paper, main area of interest is sharpening of Butterworth high pass filter. The combination of BHPF and CBF gives the fused result of greater contrast, fine texture and also preserve the fine details. Simulations are performed on medical MRI-CT images. Images are of size 256 x 256 and all are standard databases. MRI shows the tissue anatomy, while CT displays the bony structure. Modified algorithm has been implemented in MATLAB R2013a.

Evaluation of performance parameters is an important step for objective analysis. Statistical parameters are used for global and local contrast estimation, for calculation of noise, and for identifying the texture and content of an image. Various statistical parameters are [2, 24]:

1) Average Pixel Intensity:

Average pixel intensity or mean is the average of the pixel intensities or brightness within a particular region R of a sample image. Mean is given by

$$API = \frac{1}{mn}\sum_{i=1}^{m}\sum_{j=1}^{n} f(i,j) \quad (5)$$

Where m x n is the size of image and f (i, j) is the pixel intensity of fused image at pixel location (i, j). Mean measures contrast and if it is at the intermediate of dynamic range, higher is the contrast.

2) Standard Deviation

It is the square root of variance, which is the second moment of mean. It is given as SD

$$SD = \sqrt{\frac{\sum_{i=1}^{m}\sum_{j=1}^{n}(f(i,j) - mean)^2}{mn}} \quad (6)$$

It signifies the spread in data. Higher the value of standard deviation more will be the contrast.

3) Entropy

Entropy is average information comprised in an image. Entropy precisely signifies the quality of fused image. Higher its value, higher is the quality of fused image. But, Entropy not distinguishes the change in information content of the fused image. It is given as H

$$H = -\sum_{k=0}^{L-1} p_k \log_2(p_k) \quad (7)$$

where L are the number of levels in an image. p_k is the probability of kth intensity value of an image.

4) Average Gradient

Average gradient is a measure of sharpness and clarity in a fused image. It provides the contrasts between the minute details of image. Average gradient is given by AG

$$AG = \frac{1}{mn}\sum_{x=1}^{m}\sum_{y=1}^{n}\sqrt{\frac{\Delta f_x^{\,2}(i,j) + \Delta f_y^2(i,j)}{2}} \quad (8)$$

Where m x n is the region size around pixel (i, j), $\Delta f_x(i,j)$ and $\Delta f_y(i,j)$ are the difference in pixel intensities along x and y direction.

The parameters of BHPF are n=5 and f_c=50, and for CBF, parameters are neighborhood window=11x11, σ_d=1.9 and σ_r=30. Simulations are performed on four pair of CT-MRI images. Due to less space, out of four, results of two have been shown below.

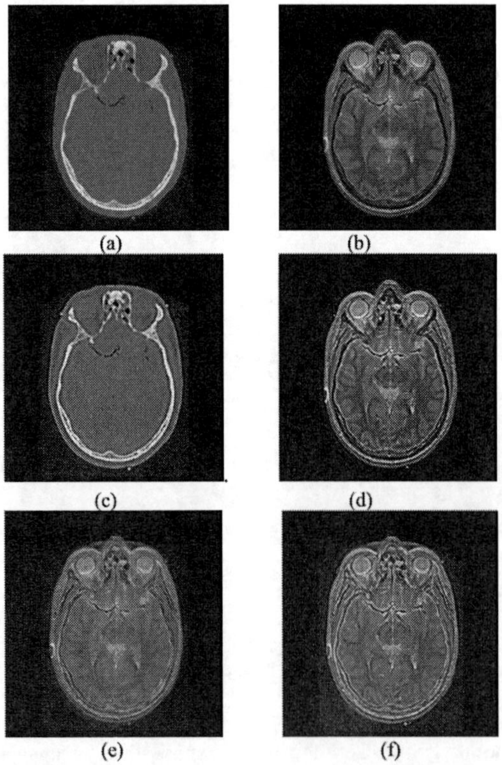

Figure 3. Dataset 1 (a) Source image P, (b) Source image Q, (c) Image P after sharpening,(d) Image Q after sharpening, (e) Fused result without sharpened inputs (f) Fused result with sharpened inputs

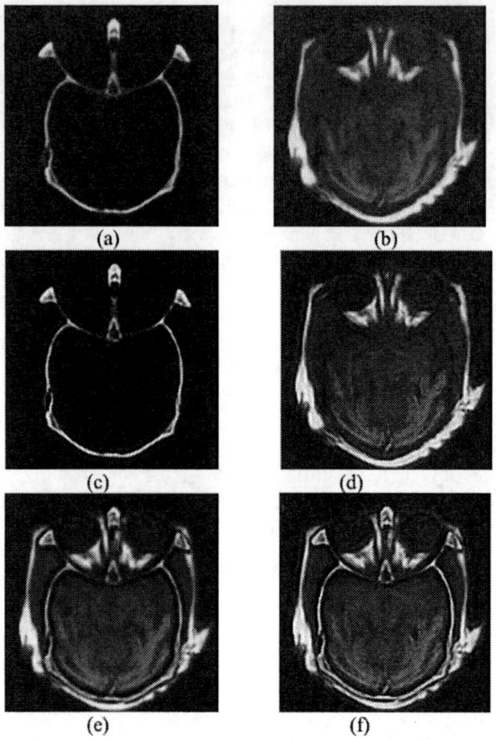

Figure 4. Dataset 2 (a) Source image P, (b) Source image Q, (c) Image P after sharpening,(d) Image Q after sharpening, (e) Fused result without sharpened inputs (f) Fused result with sharpened inputs

Figure 3 and 4 have sharpened fused results in comparison to methods discussed in [2, 24] .Sharpened fused images have better accuracy, fine texture, and greater information content. They are helpful for superior treatment planning and for diagnostic purposes.

Table 1. Qualitative comparison of fused images of Dataset 1 and Dataset 2

Parameters	Image fusion methods	API	SD	Entropy	AG
Dataset 1	[24]	-	-	-	-
	[2]	51.397	60.153	5.126	13.314
	Modified	**54.82**	**64.415**	**5.662**	**26.616**
Dataset 2	[24]	46.316	52.307	6.692	8.8572
	[2]	54.735	57.690	6.731	11.033
	Modified	**56.381**	**62.608**	**6.944**	**18.334**

It can be inferred from Table 1, that in both the datasets, modified method has higher value of API, SD, Entropy and AG in comparison to algorithms described in [2, 24]. Increase in API and SD signifies the fused images have high contrast and have higher information content. This is due to the sharpening of input images. There is a large difference in Average gradient, which clearly signifies the improved method have much better clarity and

DOI: 10.1051/matecconf/20165701021

contrast. Improved method also focuses on the tiny details of images.

For Dataset 1, comparison tables are formulated by varying values of range domain filtering parameter and kernel window.

Taking σ_d=1.8, kernel window = 11 x 11 and by increasing the value of range domain filtering parameterσ_r from 10 to 300, the cross bilateral filter leads to the Gaussian blur. Gaussian blurring decreases the value of entropy and Average gradient as shown in Table 2.

Table 2. Qualitative comparison of fused image of Dataset 1 by varying σ_r

Dataset 1	σ_r=10	σ_r=30	σ_r=100	σ_r=300
API	54.864	54.82	54.905	54.907
SD	64.50	64.415	64.497	64.458
Entropy	5.665	**5.662**	5.660	5.658
AG	26.774	**26.616**	26.612	26.514

Taking σ_d=1.8, σ_r=30 and by varying the kernel window size from 3 x 3 to 13 x 13, Table 3 compares the value of statistical parameters. 11 x11 is the best adjustable neighborhood window size for calculating the weight coefficients. Comparisons are tabulated for standard dataset.

Table 3. Qualitative comparison of fused image of Dataset by varying kernel window size

Dataset	3 x 3	7 x 7	11 x 11	13 x 13
API	75.938	76.709	**76.732**	76.731
SD	83.693	83.824	**83.832**	83.831
Entropy	6.099	6.105	**6.108**	6.107
AG	32.650	**32.965**	32.954	32.953

Figure 5 shows a comparison graph of modified image fusion framework and the algorithms described in [2, 24] for dataset 2. Modified method is superior in comparison to others, having higher values of qualitative parameters.

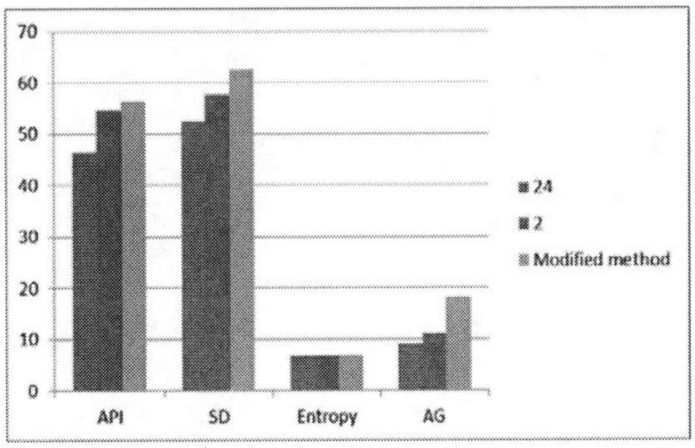

Figure 5. Comparison graph

5 Conclusions

In this paper, Pixel level Medical Image fusion using Cross Bilateral filter is modified by using Butterworth High pass filter. Cross bilateral filter is a nonlinear filter, which considers both geometric closeness function and edge stopping function. It fuses set of images by taking weighted average of source image pixels into account. For better quality of fused results, the input images are sharpened using high order Butterworth high pass filter and then images are fused using Cross Bilateral filter. Results show that the modified image fusion framework is effective in preserving brightness, fine details, information content, texture and contrast of image. The modified method shows better result both in terms of visual quality and quantitative parameters.

References

1. A. Dogra, and M. S. Patterh, JCST, **6**, no.1, 18-24, (2014).
2. B. K. Shreyamsha Kumar, Signal, Image and Video Processing, 1-12, (2013). (doi: 10.1007/s11760-013-0556-9)
3. A. P. James and B. V. Dasarathy, *Inf Fusion*, 19, 4-19, (2014).
4. YANG Bo, JING Zhong-liang, and ZHAO Hai-tao, JSJTU, **15**, no.1, 6-12, (2010).
5. S. Marshall, and G. Matsopoulos, *IEEE Conf. on Nonlinear Digital Signal Processing*, (1993).
6. Samia MASSOUT, and Youcef SMARA, *IEEE Conf. on Joint Urban Remote Sensing Event*, 4-9, (2009).
7. Jinzhu Yang, Fangfang Han, and Dazhe Zhao, *4th International Conference on Intelligent Computation Technology and Automation*, 925- 928, (2011).
8. Anjali A. Pure, Neelesh Gupta, and Meha Shrivastava, *Fourth IEEE Conference on Computing, Communications and Networking Technologies*, 1-744, (2013).
9. K. S. Tamilselvan, Dr. G. Murugesan, and B. Gnanasekaran, *IEEE Conf. on Green Computing, Communication and Conservation of Energy*, 260-263(2013).
10. Himanshi, Vikrant Bhateja, Abhinav Krishn, and Akanksha Sahu, *International Conference on Medical Imaging, m-Health and Emerging Communication Systems*, 442- 447, (2014).
11. Rajiv Singh, and Ashish Khare, *IEEE Proceedings on Information and Communication Technologies*, 869- 873, (2013).
12. A. Dogra, S. Agrawal, and N. Khandelwal, IJPPS, no. 16. pp. 93-97, (2015).
13. A. Dogra, and S. Agrawal, IJPT, **7**, no. 3, 9748-9763, (2015).
14. G. Anjan Babu, Sreenu Naik Bhukya, and R. Sendhil Kumar, *8th IEEE Conference on Computer Science and Education*, 181- 185, (2013).
15. Tulay Adali, Yuri Levin-Schwartz, and Vince D. Calhoun, *Proceedings of the IEEE*, **103**, no. 9, 1478-1493, (2015).
16. C. Tomasi, R. Manduchi, *Proceedings of the International Conference on Computer Vision*, 839–846, (1998).
17. Zhang, M., Gunturk, *IEEE Trans. Image Process*, **17**, no.12, 2324–2333, (2008).

18. Bennett, E.P., Mason, J.L., McMillan. L, *IEEE Trans. Image Process*, **16**, no.5, 1185–1194, (2007).
19. Zhiwu Liao, Shaoxiang Hu, Zhiqiang Yu and Dan Sun, *IEEE Conference On Medical Image Analysis and Clinical Application,* 12-17, (2010).
20. R.C.Gonzales, R.E.Woods, Digital Image Processing, 2-nd Edition, Prentice Hall, (2002).
21. Kaustubh Manik Gaikwad and Mahesh Shrikant Chavan, *IEEE Global Conference on Wireless Computing and Networking,* 121-124, (2014).
22. A. Dogra, and P. Bhalla, BPJ, **7**, no. 2, 707-713, (2014).
23. A. Dogra, and P. Bhalla, BPJ, **7**, no. 2, 691-696, (2014).
24. Shah, P., Merchant, S.N., Desai, *Proceedings of the 14th International Conference on Information Fusion*, 1–7, (2011).

BER performance analysis of OFDM-MIMO system using GNU Radio

M.Ushamahesh Singh[1a], Sushil Kakkar[2] and Shweta Rani[3]

Electronics and Communication Engineering Department, BGIET, Punjab, India

Abstract. Multiple Input Multiple Output (MIMO) channels can be used to increase the data rate and the channel capacity by employing multiple transmitting and receiving antennas at both the ends of a wireless communication system. MIMO systems employ Orthogonal Frequency Division Multiplexing (OFDM) technique and it uses separate antennas at both the transmitter and receiver to increase the data rate and with OFDM, instead of a single carrier, the main information is modulated into a number of independent sub-carrier signals which are orthogonal to each other. This paper presents an OFDM-MIMO transceiver design and the performance analysis of the system based on Error rate for different modulation techniques using GNU Radio. OFDM is chosen over a single carrier solution due to lower complexity of equalizers for high delay spread channels or high data rates. So the combination of MIMO-OFDM system has become a potential technology for high speed data transmission and efficient utilization of the channel spectrum for the modern wireless communication networks.

1 Introduction

Due to the sharp increase in demand for wireless connectivity during the last two decades, the performance assessment of wireless communication systems are the current area of research. Some of the challenges that the wireless technology has faced includes signal fading, multipath, limited frequency spectrum. The number of users are increasing and the service providers are using limited bands and there are still wasted and unused frequencies due to different factors and parameters. The wireless communication systems are not as reliable as guided medium communication mainly due to fading and other propagation effects and accordingly, techniques to improve its capacity and reliability has become one of the most important area of current research. It is shown that by using the technique of multiple antennas at both the transmitting and receiving ends, the high performance 4G broadband wireless is enabled. Such a wireless technology is called Multiple Input Multiple Output. MIMO systems employ Orthogonal Frequency Division Multiplexing to achieve higher throughput without the extension of the bandwidth available. So for the next generation wireless communication systems such as WiMAX, LTE, WLAN, WMAN standards which accommodate many users in the same channel, the combination of MIMO-OFDM technique is considered as the best technique that overcome many challenges of today's wireless technology and it is seemed to be a solution for the future broadband wireless systems. Further, the MIMO-OFDM meets the requirements of high data rate and high capacity without the need of extending the available spectrum. While studying MIMO technology and OFDM technique, this paper sets up a simple MIMO-OFDM wireless communication system and the performance is evaluated using the GNU radio.

[a]Corresponding author: pitterkym@gmail.com

2 GNU Radio

The system presented in this paper is accomplished using GNU radio which is an evolving free and open source software based on the concept of blocks and flow graphs. It is a powerful SDR software that contains a set of easily reconfigurable blocks and can be easily adapted to different network architectures and protocols. Ranging from easily reconfigurable blocks, it has blocks with different parameters that are used in most of the recent wireless communication technologies in which when connected together forms transmitter and receiver chains. The GNU radio applications are primarily written using python language and the performance critical signal processing blocks are implemented in C++. It provides a library of signal processing blocks which are written in C++ language that implement various functionalities of both the transmit and receive architectures. These blocks can be used and combined by the final user through Python. GNU radio provides a Python interface for each C++ blocks as well as tools to connect various blocks. In this way, the simplified wrapper and interface generation allows C++ code to be used from Python. Blocks are basic operation units that process continuous data streams and each block has a number of input and output ports.

3 MIMO and OFDM

3.1. MIMO Technology and system model

The Multiple Input Multiple Output (MIMO) employs multiple antennas in the transmitter and receiver at the same time of the communication system and it uses the random fading and multipath propagation to increase its transmission rate. MIMO provides transmit diversity and receiver diversity. It can be used in different kind of networks like PAN, MAN, WAN, WLAN, LAN and can receive either a diversity gain, capacity gain and to overcome signal fading. A MIMO system has three main components mainly transmitter, channel and receiver. The function of the transmitter is to send multiple data such as $X_1,X_2,X_3,\ldots,X_N$ using the different transmit antennas, N_t and the signals $R_1,R_2,R_3,\ldots,R_N$ are received by each receive antennas, Nr simultaneously. If at a certain time, the complex signals $X_1,X_2,X_3,\ldots,X_N$ are transmitted via N_t transmit antennas, then the received data $R_1,R_2,R_3,\ldots,R_N$ for Nr received antennas can be expressed as[13]

$$R_1= h_{11} X_1 + h_{12} X_2 + \ldots + h_{1N} X_N \quad (1)$$
$$R_2= h_{21} X_1 + h_{22} X_2 + \ldots + h_{2N} X_N$$
$$\cdots\cdots$$
$$R_N= h_{N1} X_1 + h_{N2} X_2 + \ldots + h_{NN} X_N$$

By combining all the received signals in a vector R the MIMO signal model is described as

$$R= H X + n \quad (2)$$

Where R is Nr×1 received signal vector, H is Nt×1 transmitted vector and n is Nr×1 Gaussian noise vector. With Nt inputs and Nr outputs the channel can be expressed as Nr× Nt MIMO channel transfer matrix H, given by [13]

$$H=\begin{bmatrix} h_{11} & h_{12} & \cdots & h_{1Nt} \\ h_{21} & h_{22} & \cdots & h_{2Nt} \\ \vdots & \vdots & \vdots & \vdots \\ h_{N1} & h_{N2} & \cdots & h_{NrNt} \end{bmatrix} \quad (3)$$

where hij is the attenuation and phase shift between the jth transmitter and i^{th} receiver.

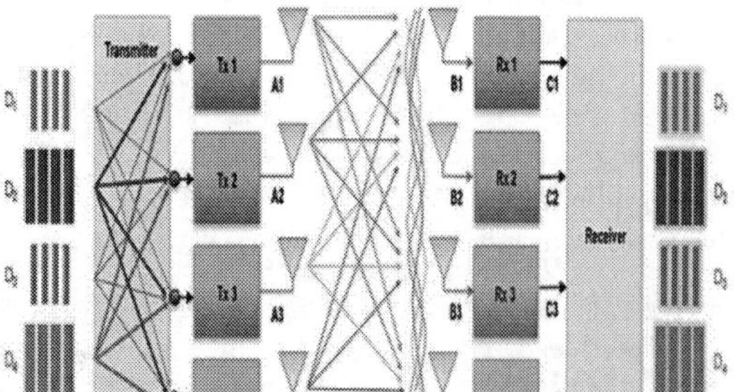

Figure 1. Block diagram of multiple input multiple output (MIMO) system[13].

3.1.1 OFDM technique and system model

Orthogonal Frequency Division Multiplexing (OFDM) is a form of broadband multicarrier modulation method that splits a high data rate signal stream into a multiple smaller sub-signals that are transmitted simultaneously at different frequencies and in this way, it is less sensitive to frequency selective fading. An OFDM signal consists of a number of closely spaced modulated carriers which are orthogonal to each other so as to avoid from signal interference between them. When the modulation of any form of information –video, voice, data, etc. is applied to a carrier, then sidebands spread out either side. Although the sidebands of each carriers overlap, the receiver can still receive them without interference because the modulated signals are orthogonal to each other. For this the carrier spacing is made equal to the reciprocal of the symbol period. OFDM reduces the amount of crosstalk in signal transmission.

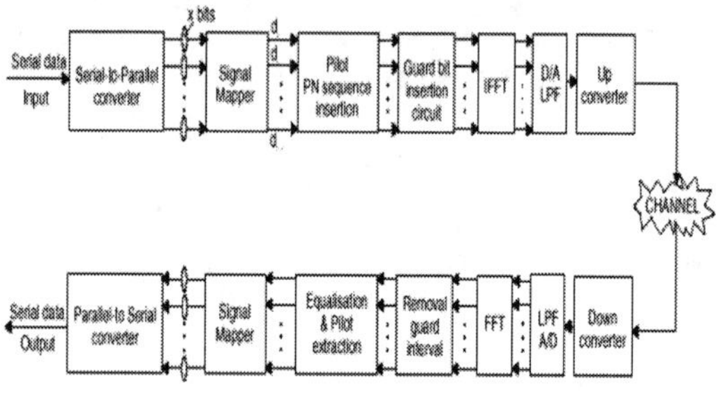

Figure 2. OFDM system configuration

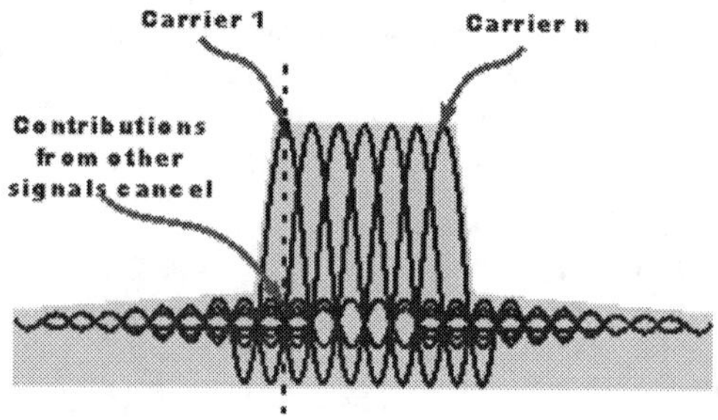

Figure 3. OFDM spectrum

The OFDM system converts the high speed serial data streams to low speed parallel streams that can be transmitted in N sub-channels. The signals are modulated by N sub-carriers which are orthogonal to each other and the modulated signals are sent together. One requirement of the OFDM transmitting and receiving system is that they must be linear. Any non-linearity will produce interference between the transmitted and received signals as a result of inter-modulation distortion. This will generate unwanted signals which would cause interference and weaken the orthogonality of the transmission. At the receiver, they are received with the relevant sub-carriers and restores the original high speed data by converting parallel to serial converter. The system configuration for OFDM is shown in fig.(2).

4 MIMO-OFDM wireless communication systems

4.1. Working methodology

The simulation is carried out first by designing a MIMO-OFDM system using GNU radio by directly connecting the blocks. Typically the system may have blocks of random source, OFDM modulator, noise source, channel model and OFDM demodulator to accomplish the data transfer. Main emphasis is given on measuring the Error Rate value (BER) by employing different types of modulation techniques. BER is a parameter which is used to represent the number of received bits that have been changed due to transmission impairments such as the interference, noise, distortion etc. It measures the whole end to end performance of the given system that includes the transmitter, receiver and the medium. It is defined as the rate in which error occur in the OFDM system. At the transmitter side, the system modulates the incoming signals into OFDM symbols and at the receiver side, it performs synchronization, FFT and demodulation of the incoming OFDM symbols and passes packets up to the higher layer. The Error Rate block calculates the BER or the SER(Symbol Error Rate). It compares the input byte streams to the reference byte streams. Each byte in the incoming stream represents one symbol. The bits per symbol parameter is useful for calculating the BER. The system considerations for the proposed work are- sampling rate- 32K, packet length-96, FFT-512, occupied tones- 200, cyclic prefix length-128, channel type- AWGN channel.

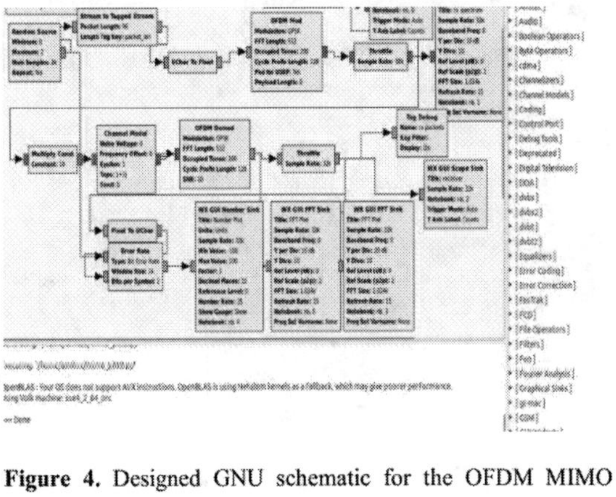

Figure 4. Designed GNU schematic for the OFDM MIMO system

4.1.1 Result analysis and discussions

The performance of the designed system is assessed by determining the data error rate during transmit and receive process. BER (bit error rate) is oftentimes used to measure the performance of digital communication system in the presence of noise, i.e. SNR (signal to noise ratio). The transmitted signal waveform and received data streams are obtained and are shown in Fig. (5) and (6) respectively. Table 1 shows the BER performance of the system for different modulation schemes.

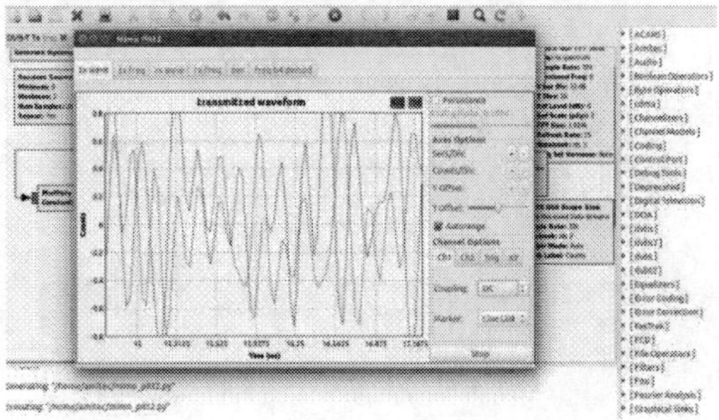

Figure 5. Transmitted waveform at the transmitter side

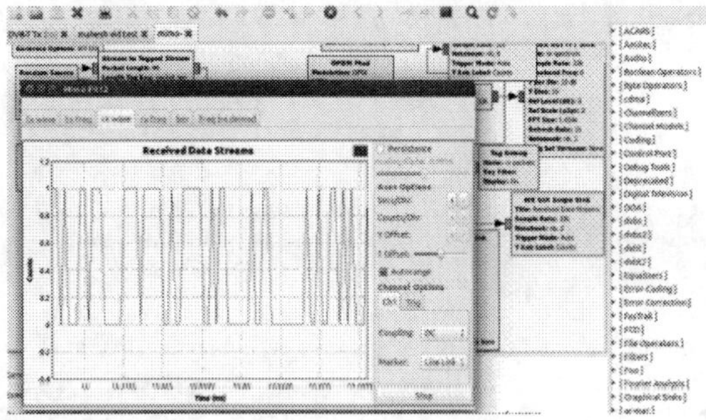

Figure 6. Received data streams at the receiver side

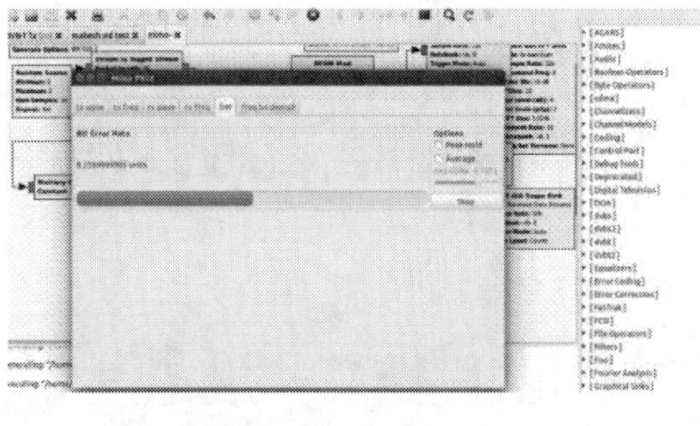

Figure 7. BER values generated

Table 1. BER for different modulation schemes.

SNR(dB)	BPSK	QPSK	QAM 16	QAM 64	8PSK
0	0.796000	0.801500	0.812500	0.808000	0.797778
2	0.772382	0.783990	0.806000	0.805963	0.791388
4	0.745816	0.763588	0.799500	0.791000	0.765504
6	0.736729	0.754323	0.788883	0.789325	0.759929
8	0.711	0.748	0.771	0.782	0.744

	725	812	870	500	164
10	0.690 151	0.731 232	0.746 000	0.782 500	0.738 050

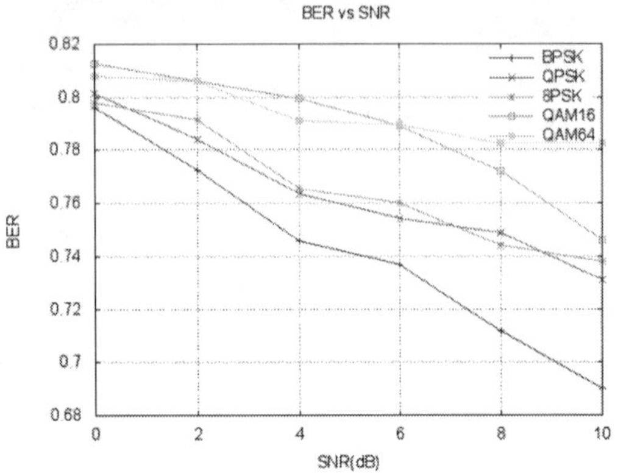

Figure 8. BER vs SNR curve of the MIMO-OFDM system

Table 1 shows the BER (Bit Error Rate) values as a function of varying SNR for the MIMO-OFDM system for different modulation schemes i.e. BPSK, QPSK, 8PSK, QAM16 and QAM64. Figure 8 shows the plot of the BER vs SNR. According to the results obtained and from the graph, it is seen that BER value depends on the number of bits per symbol and value of BER is low for BPSK which can be the best suitable modulation technique for information transmission in the OFDM-MIMO. It is also seen that the BER greatly depends on SNR and with the increase of SNR, the BER decreases because a stronger noise increases BER and a stronger signal power decreases the BER.

5 CONCLUSION

In this paper, the performance of MIMO-OFDM system has been analysed with different modulations schemes. The BER values are obtained as a result with varying Signal to Noise Ratio (SNR) by putting different noise values at the channel model. Of the modulation schemes employed, BPSK is performing the best because of less bits per symbol it represents followed by QPSK scheme. The corresponding BER performance differs because they exhibit different tolerance margin to noise. Thus this lower modulation order reduces BER resulting in better and improved performance of the OFDM-MIMO system. Further works can be extended to the performance analysis of real time transmission and reception of the OFDM-MIMO system using USRPs which will be done under different environmental conditions and for different coding schemes.

References

1. J. G. Andrew, W. Choi, R. W. Heath jr., "*Overcoming Interference in Spatial Multiplexing..*" IEEE Wireless Communications, pp. 95-104, (2007)
2. Q. Li, G. Li, W. Lee, Z. Li, "*MIMO techniques in WiMAX and LTE: a feature overview...*" IEEE Communications Magazine, pp. 86-92, (2010)
3. H. Sampath, S. Talwar, J. Tellado, V. Erceg, A. Paulraj, "*A Fourth Generation MIMO-OFDM Broadband Wireless System: Design, Performance, and Field Trial Results..*" IEEE Communications Magazine, pp. 143-149, (2002)
4. A. Sharma, A. Garg, "*BER Analysis Based on Transmit and Receive Diversity Techniques in MIMO-OFDM System*" IJECT, pp. 58-60, (2012)
5. X. Li, W. Hu, H. Y'zadeh, A. Qureshi, "*A Case Study of a MIMO SDR Implementation..*" IEEE, pp. 1-7, (2008)
6. S. Trivedi, Md. S. Raeen, S.S. Pawar, "*BER Analysis of MIMO-OFDM System using BPSK Modulation Scheme..*" IJACR, pages 208-214, (2012)
7. A. Pauraj, D. A. Gore, R. U. Nabar, H. Bolcskei, "*An Overview of MIMO Communications- A Key to Gigabit Wireless..*" IEEE, pp. 198-218, (2004)
8. Y. S. Cho, J. Kim, W. Y. Yang, C. G. Kang, " *MIMO-OFDM Wireless Communications*" *with MATLAB* (John Wiley & Sons, 2010)
9. GNURadio. http://gnuradio.org/.
10. S. Cui, A. J. Goldsmith, A. Bahari, "*Energy-efficiency of MIMO and Cooperative MIMO Techniques in Sensor Networks..*" IEEE Journal on Selected Areas in Communications, pp. 1-10, (2004)
11. H. Bolcskei, "*MIMO-OFDM Wireless systems: Basics, perspectives, and challenges..*" IEEE Wireless Communications, ppl 31-37, (2006)
12. Q. H. Spencer, C. B. Peel, A. L. Swindlehurst, M. Haaredt, "*An introduction to the multi-user MIMO downlink..*" IEEE Communication Magazine, pp. 60-67, (2004)
13. B. Reddy, B. Lakshmi, "*BER Analysis with Adaptive Modulation Coding in MIMO-OFDM..*" IJWMT, 4, 20-34, (2014)

Investigations on 16x16 multiple Tx/Rx in IsOWC using IM/DD and CO-QPSK schemes

Hardeep Singh Mankoo[1] and Baljeet Kaur[2]

[1]*Student, Deptt. Of ECE,GNDEC, Ludhiana*
[2] *Asst Professor, Dept of ECE, GNDEC, Ludhiana*

Abstract. The intersatellite optical wireless communication (IsOWC) is used because higher data rates for longer distances can be achieved with minimum time delay. The analysis of 16x16 multiple Tx/Rx system using IM/DD and CO-QPSK techniques is done and from simulations, it is observed that maximum achievable distance of IsOWC link is a function of bit rates. For the transmission of data at longer distances, data rates should be low. The results of simulation for comparison between the two techniques in ISOWC system model are presented and discussed.

1 Introduction

The IsOWC is the line-of-sight technology that transmits a light beam through atmosphere for broadband communications [1]. Free-space optical communication between satellites networked together can make possible high-speed communication between different places on earth [2]. The information is transmitted with light as a carrier wave. A narrow beam of light which carries the information signal is launched at transmitter station, transmitted through the atmosphere and subsequently received at the receive station[3][4]. Hence, the clear line-of-sight between both transmitter and receiver terminals is essential to establish a seamless communication [5]. The Intersatellite links has many advantages as comparing to the microwave links, such as (i) higher bandwidth (ii) lower transmission power (iii) small antenna size (iv) less weight (v) higher immunity to interference and (vi) security of signals etc. All of these reasons are vital for a satellite communication system, because it can reduce the payloads and consequently reduces the cost [6-8]. Many alternatives are proposed by researchers to improve the IsOWC link performances. Some of the techniques are multiple transmitted beams or multiple receivers, aperture averaging, adaptive optics and MIMO. Multiple Tx/Rx are used to improve quality of IsOWC system [4] [8]. In 2010, Hashim et el. Proved that among all ON-OFF keying techniques, non-return to zero (NRZ) is the best technique for obtaining maximum coverage distance in IsOWC system [3]. In 2010, Sodnik et el. used the BPSK modulation technique for IsOWC system [13]. Patnaik et el. (2012) used CO-QPSK technique; as its spectral efficiency is double that of BPSK technique [9]. Kuldeepak et el. (2012) Investigated high capacity IsOWC link by integrating WDM technology, and achieved a distance of 2100 km with a bit rate of 8 Gbps using IM/DD technique [5]. In 2013, Patnaik et el. Designed and simulate the FSO model by employing QPSK modulation scheme with line-of-sight and diffused link setup. They have also simulate the system for 1.25 Gbps, 2.5 Gbps, 10 Gbps, 40 Gbps nd 100 Gbps with free space attenuation of 200 dB/km [10]. Hence we have designed and simulated two IsOWC systems having 16x16 multiple Tx/Rx using CO-QPSK and IM/DD modulation techniques. The observations were made using comparison analysis between the two techniques. Quality factor, bit rate and transmission distance are the parameters used to determine IsOWC link performance. The relationship between these parameters are compared and analyzed.

2 ISOWC model design

16x16 Multiple Tx/Rx System- Figure 1 shows the layout model for 16Tx and 16Rx combinations employing IM/DD Technique. The frequency of transmitter is set to 193THz or 1550 nm in wavelength and power is 12 dBm. The output of transmitter is connected to fork which is used to duplicate number of output ports. Each signal coming out from the fork's output has same value as output signal from previous component. Each fork connected to the transmitter will produce multiple laser beams from one source. Then, each output signal will be connected to another set of forks to produce another set of multiple laser beams which are combined by a power combiner. The output signals from power combiners are then sent to OWC channel which is the reproduction of free space channel. The OWC has two telescopes with apertures of 15cm. All multiple signals from OWC channel are combined using the power combiner before received by the receiver. The receiver consists of photo detector, LPF and visualizers.

[a] Corresponding author: er.hardeepmankoo@gmail.com

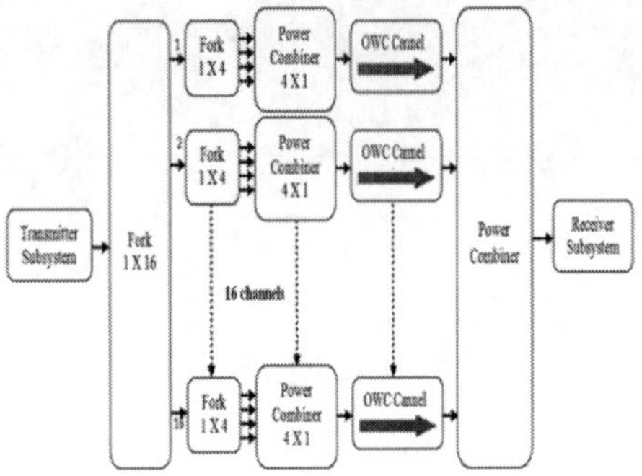

Figure 1. Block diagram of 16x16 multiple Tx/Rx system for IsOWC link.

16x16 Multiple Tx/Rx System- It consists of PSK sequence generator, which generates the in-phase (I) and quadrature signals (Q) as given in equations respectively. The output of the PSK sequence generator is given to the M-array pulse generator, where M = 4. The number of bits per symbol considered here is two. Optical signal is fed to the MZ modulator by using a coupler. Both I and Q signals is combined using an optical power combiner.

$$Ii = \cos(\Phi)$$
$$Qi = \cos(\Phi)$$

Where,

$\Phi i = 2\pi(i-1)/M$ i= 1,2,3,4 and M=4 [7]

The output signals from power combiners are then sent to the OWC channel which is the reproduction of the free space channel. The OWC channel has two telescopes with aperture diameter of 15cm. All the multiple signals from OWC channel are combined using the power combiner before received by the receiver. Then receiver detects optical QPSK Modulated signal and demodulates it into original signal using photo detector and a local oscillator (laser). It uses a balanced-detection technique and produces the information signal in electrical domain. Then electrical signal is amplified, filtered using a LPF filter and given to M-array threshold detector. Then signal is fed to PSK decoder that is used to convert the values of I and Q to the equivalent binary sequence. Hence, fresh 0's and 1's are reproduced.

3 Simulations results for 16X16 multiple Tx/Rx system

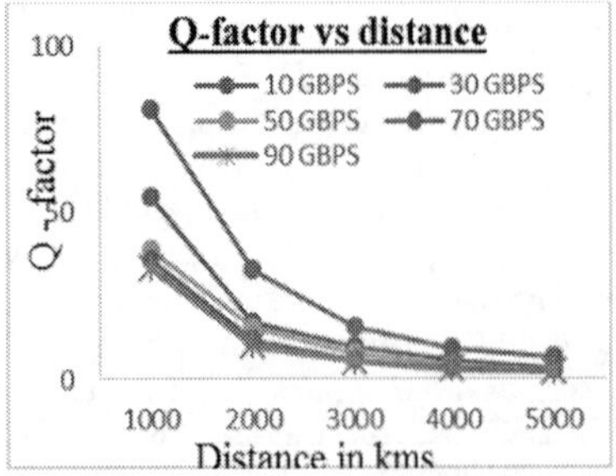

Figure 2. Relationship between Q-factor and link distance at different bit rate

The Q-factor of IsOWC system is obtained by varying the link distance for different bit rates. The link distance of the system is varies from 1000 km to 5000 km and the bit rate of each channel is varied from 10 Gbps to 90 Gbps respectively as variation in the link distance. The input power is maintained at 12 dBm and the aperture diameter of transmitter and receiver antennae are set at 15cm. The figure 2 shows relationship between Q factor and link distance at different bit rates.

The maximum possible link distance between two satellites is analyzed to achieve error-free communication (BER<10^{-9}) in our system. For bit rate of 30 Mbps maximum communication distance calculated is 22,500 km. Increase in link distance beyond this value increases BER of the system. It was also observed that as we increase the bit rates the link distance decreases. The bit rate is inversely proportional to link distance. The values for different bit rates at link distance between satellites and the value of Q-factor with minimum BER are obtained and shown in table 1.

The analysis of system performance can be shown by eye diagrams. The significance of a wider eye opening is that it will reduce the potential occurrence for data errors, the wider the eye opening, the better the system performance. Figure 3 & figure 4 shows the eye diagrams at 30 Mbps and 90 Gbps respectively.

The IsOWC link distance is varies for different bit rates and the value of Q-Factor is obtained. The link distance varies from 5000 km to 40000 km and the bit rate is varied from 10 Gbps to 90 Gbps respectively as variation in the link distance. The input power is maintained at 12 dBm and the aperture diameter of transmitter and receiver antennae are set at 15 cm.

Table 1. Table for maximum distance achievable at different bit rates with q-factor and minimum ber of 16x16 multiple tx/rx im/dd technique.

Sr. No.	Bit Rate	Link Distance	Q-factor	Min BER
1	90 Gbps	2810 km	5.70	5.67×10^{-9}
2	30 Gbps	4000 km	5.62	9.30×10^{-9}
3	1 Gbps	9250 km	5.95	1.33×10^{-9}
4	500 Mbps	11,250 km	5.64	8.10×10^{-9}
5	30 Mbps	22,500 km	5.67	7.10×10^{-9}

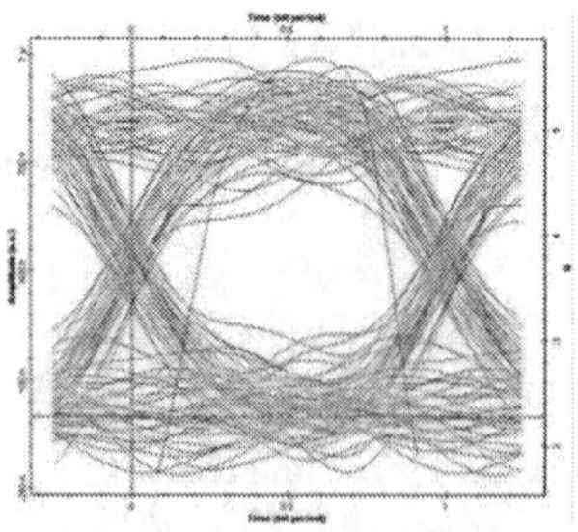

Figure 3. Eye diagram obtained at link distance 22500 km with bit rate of 30 Mbps

DOI: 10.1051/matecconf/20165701023

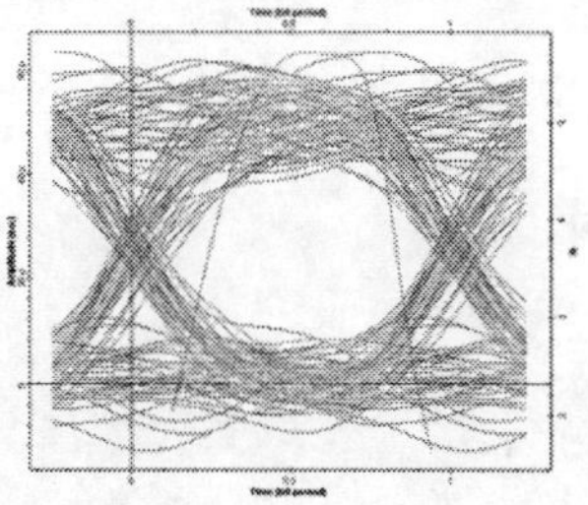

Figure 4. Eye diagram obtained at link distance 2810 km with bit rate of 90 Gbps

The figure 5 shows relationship between Q factor and link distance at different bit rates.At bit rate of 40 Gbps maximum communication distance calculated is 42,500 km. Any increase in link distance beyond this value causes BER to increase. The values for different bit rates at link distance between satellites and the value of Q factor with minimum BER are obtained and shown in table 2.

The electrical constellation diagram shows the performance of the system. The quadrature-phase and in-phase signals are shown through the constellation diagram. The distortion and interference of the signals can be evaluated and measured. Figure 6 &figure 7 shows the constellation diagram at bit rate 10 Gbps and 90 Gbps respectively.

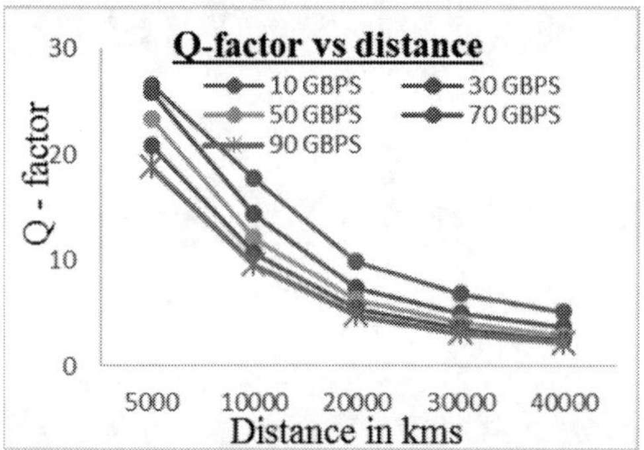

Figure 5. Relationship between Q-factor and link distance at different bit rate

Table 2. table for maximum distance achievable at different bit rates with q-factor and minimum ber of 16x16 multiple tx/rx co-qpsk technique.

Sr. No.	Bit Rate	Link Distance	Q-factor	Min BER
1	90 Gbps	15,000 km	6.42	0.13×10^{-9}
2	70 Gbps	18,000 km	5.98	2.18×10^{-9}
3	50 Gbps	20,000 km	6.03	1.57×10^{-9}
4	30 Gbps	28,000 km	5.34	8.81×10^{-9}
5	10 Gbps	35,000 km	5.57	5.63×10^{-9}

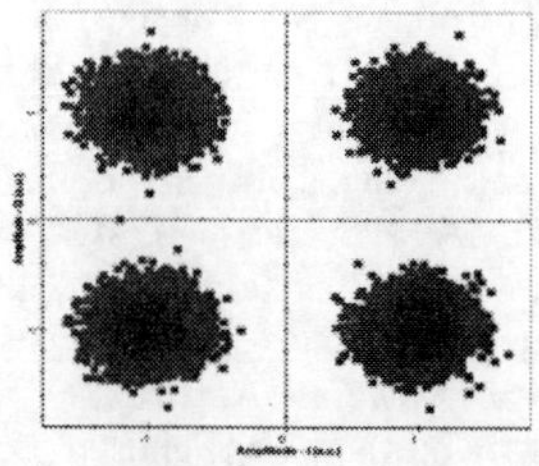

Figure 6. Constellation diagram obtained at link distance 35000 km with bit rate of 10 Gbps

The above analysis shows the comparison between the two techniques used in IsOWC link employing 16x16 multiple Tx/Rx technique.

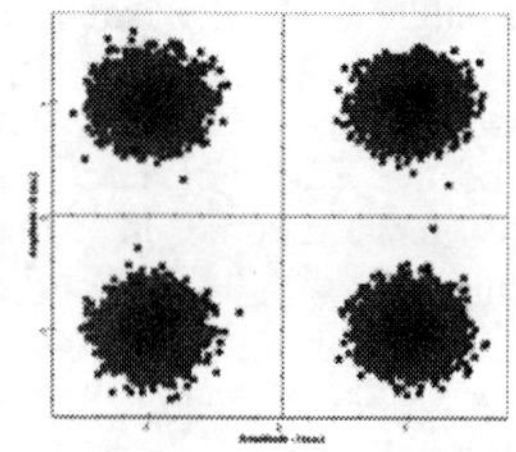

Figure 7. Constellation diagram obtained at link distance 15000 km with bit rate of 90 Gbps

It is observed from the simulation results that, link distance and bit rate of the system can be improved by using CO-QPSK scheme in comparison with IM/DD scheme. The maximum achievable distance by using IM/DD scheme is 22,500 km with the bit rate of 30 Mbps whereas a link distance of 35000 km with the bit rate of 10 Gbps is achieved using Coherent Optical QPSK scheme with a bit error rate of 7.10×10^{-9} and 5.63×10^{-9} respectively.

4 Conclusion

The analysis is done for 16x16 Multiple Tx/Rx and using IM/DD technique and CO-QPSK technique for IsOWC system. From simulations, it is observed that maximum achievable distance of IsOWC link is a function of bit rates. The optical signal with lower bit rate can be used for further distance between satellites since the system performance is better at lower bit rates. It is also observed that by using CO-QPSK technique the achieved bit rates and link distance are more than as compared with IM/DD technique. It is analyzed that the system has better performance by using CO-QPSK technique as compared with IM/DD technique. The system provides longer distance and high data rates for intersatellite communication.

References

1. Arnon, S. and Kopeika, N. S. (1997) "Laser Satellite Communication Network— Vibration Effect and Possible Solutions" Proceedings of the IEEE, Vol. 85, No. 10, pp. 1646-1661.

2. Arnon, S., Rotman. S., Kopeika, N. S., Optimum Transmitter Optics Aperture for Satellite Optical Communication IEEE Transactions on Aerospace and Electronic Systems vol. 34, no. 2 April 1998.
3. Begun, D. A., "Scientist Test World's Fastest Network", 2008. [Online] Available at http://hothardware.com/News/Scientists- Test-Worlds-Fastest-Wireless-Network.
4. Bloom, S., Korevaar, E., Schuster, J., Willebrand, H., "Understanding the performance of Free Space Optics", Journal of Optical Networking, vol. 2, pp. 178-200, 2003.
5. Hashim, A. H., Mahad, F. D., Idrus, S. M. and Supa'at, A. S. M. (2010) "Modeling and Performance Study of InterSatellite Optical Wireless Communication System" Proceeding of International Conference on Photonics, Langkawi, Kedah, pp. 1-4
6. Naimullah, B.S.S., Hitam, S., Shah, N.S.M., Othman, M., Anas, S.B.A., Abdullah, M.K., "Analysis of the Effect of Haze on Free Space Optical Communication in the Malaysian Environment", Proc. 2007 IEEE International Conference on Telecommunications and Malaysia International Conference on Communications, pp. 391-394, 2007.
7. Noor, N. H. M., Naji, A.W. and Al-Khateeb, W. (2011) "Theoretical Analysis of Multiple Transmitters/Receivers on the Performance of Free Space Optics (FSO) Link" Proceeding of the International Conference on Space Science and Communication (Icon Space), Penang, Malaysia, pp. 291-295.
8. Patnaik, B. and Sahu, P.K. (2012) "Inter-satellite optical wireless communication system design and simulation" IET Communications, Vol. 6, Issue 16, pp. 2561-2567.
9. Patnaik, B. and Sahu, P.K. (2012) "Inter-satellite optical wireless communication system design and simulation" IET Communications, Vol. 6, Issue 16, pp. 2561-2567.
10. Patnaik, B. and Sahu, P.K. (2013) "Design and Study of High Bit-Rate Free-Space Optical Communication System Employing QPSK Modulation" International Journal of Signal and Imaging Systems Engineering, Vol. 6, No. 1, pp. 3-8.
11. Singh, K. and Bhamrah, M. S. (2012) "Investigations on Aperture Diameter of Telescope in Intersatellite Optical Wireless communication system" International Journal of Engineering Research and Applications (IJERA), Vol. 2, Issue 3, pp. 1666-1670.
12. Singh, K. and Singh, M. (2012) "Performance Improvement of Intersatellite Optical Wireless communication with Multiple Transmitter and Receivers" International Journal of Engineering Research and Technology (IJERT), Vol. 1, Issue 4, pp. 1-7.
13. Sodnik, Z., Furch, B. and Lutz, H. (2010) "Optical Intersatellite Communication" IEEE journal of selected topics in quantum electronics, vol. 16, no. 5, pp.1051-1057.

AN ANN SOLUTION FOR GENERALIZED SIERPINSKI FRACTAL ANTENNA

Anuradha[1], Shweta Rani[2], Sushil Kakkar[3]

Abstract. A neural network (NN) based analysis model is developed for locating the operating frequencies of generalized Sierpinski fractal antenna. The developed model can locate the operating frequencies of standard as well as perturbed antennas. The performance of the neural model is validated with simulations as well as with results reported in literature.

Index Terms— Fractals, microstrip antennas, multiband antennas, neural networks

1 INTRODUCTION

In response to the ever-increasing demands of wideband and ultra-wide band antenna, considerable amount of effort is currently underway to develop multiband antennas. Much has also been achieved in designing dual band antennas [1-3]. Mostly planar type designs are preferred for these structures due to their valuable merits of small size, Low manufacturing cost and conformability. In the era of modern years researchers were able to achieve the multiband property in reconfigurable antennas, in which the different radiating elements of an array are connected together using switches to form groups of elements that resonate at different frequency bands [4-5]. Furthermore, fractal antennas have been used as miniature multiband antenna [6-9] in which the self-similarity property is used to resonate the antenna at a multiple frequency bands. Over the years various fractal structures have been considered for antenna applications. But the largest problem with this category of multiband antennas is that all the resonating frequencies of the antenna in its conventional form are not the user required frequencies. Shifting of the multiple bands of these antennas has been observed by proper control of the scale factor in some fractal structures.

The monopole antenna based on Sierpinski gasket fractal has been studied extensively as an excellent candidate for multiband applications [10, 11]. The standard structure of this category is the Sierpinski gasket which consists of a series of scaled triangles having a scaling of $s = 0.5$ (scale factor ratio $\delta = 2$). Its electrical features gets shifts into a log periodic allocation of frequency bands where these multiple bands each have a common nature. Appearance of this nature has also been observed in the radiation patterns. It has also been demonstrated that the location of the multiple bands can be controlled by proper adjustment of the scale factor used to generate the Sierpinski antenna [12]. Changing the scale factor from 0.5 produces the so called perturbed Sierpinski gasket, which can be used to control the band spacing. The multiband behavior of the antennas, and their radiation patterns for scale factors other than 0.5 were observed in [12, 13]. However, in all these published results it can be clearly observed that the log periodic nature is compromised for first couple of frequencies and for the rest of the frequencies, the frequency ratios do not exactly match with the scale factor. This deviation is observed for standard as well as perturbed Sierpinski gaskets. Estimated formulas were given for standard and perturbed Sierpinski gaskets, to locate the operational frequencies [11, 14-16]. The formula developed in [14] is applicable for perturbed structures, but this also provides a compromised value for the first band. Authors of [15] have modified the formula developed in [11] for standard Sierpinski gasket.

In the present work, a neuro computational approach is used to locate the operational frequencies of a Sierpinski gasket. The developed network is a general one in the sense that this single network can be used to locate the frequencies of either perturbed or unperturbed gasket of any scale factor ($s = 0.1 - 0.9$) and for any gasket height ($h = 10$mm. - 90mm.). The results of this network have been validated with simulations as well as with earlier reported results. The reason for developing the network is twofold: (i) to reduce the simulation time of analyzing the fractal antenna with the developed CAD model and (ii) during design of the antenna at specified frequencies, the developed model can be put in an optimization loop to find out the structure.

[a] Corresponding author: sonanu8@gmail.com

2 ANN APPROACH OF GENERALIZED GASKET ANTENNA

Use of NN for antenna analysis and design is not new and applications have been reported since 1990. In most of the cases, the role of the NN is to form a mapping (black-box) between a set of input parameters and the corresponding set of outputs [17-21].

In the present work, the characteristics of NN have been exploited to form a mapping of the set of inputs comprising of overall height (h) of the Sierpinski monopole gasket and the scale factor (s) with set of outputs comprising of fr1, fr2/fr1 and fr3/fr1, where fr1, fr2 and fr3 are first, second and third resonant frequencies of a two-iterated generalized Sierpinski gasket. A model of generalized Sierpinski gasket monopole antenna is shown in Figure.1, where the h defines the entire height of the triangle and s represents the scale factor. This model is fed at the bottom using the co-axial cable of 50 Ω. The NN model of this antenna is shown in Figure.2. Although the formulation can be generated for higher iteration fractals, but most of the practical antennas are of dual-band type and at the most triple-band type. Furthermore, the purpose of this work is to find the working frequencies of the fractal based multiband antenna, because fractal antennas mostly deviate from their fractal behaviour in lower iterations.

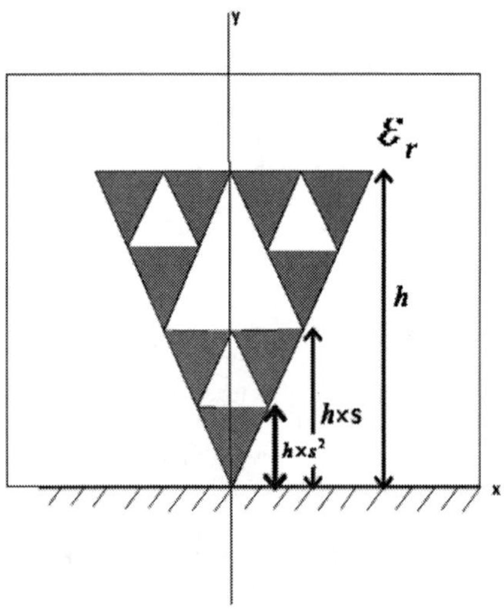

Fig.1 Generalized Sierpinski Monopole Antenna

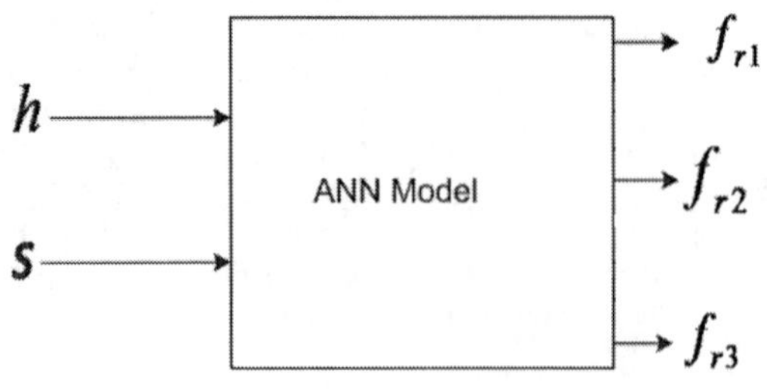

Fig.2 NN Model of Generalized Sierpinski Antenna

Data Generation: For generation of training data for the neural network, a set of skewed gaskets were formed by varying the scale factor. Instead of a similarity transformation, an affine transformation was applied to go from the overall structure to these skewed gaskets (Fig 1). The Iterated function system [10], which represents a versatile method for convenient generation of wide variety of useful fractal structures and is based on application of a series of affine transformation, was used to generate these structures. The height of the gasket was varied between 10 mm. – 90 mm. in steps of 5mm. and the scale factor was between 0.1 - 0.9. Simulations were carried out using CST microwave studio [22] to find out the operational frequency bands of two iterated fractal structures

Neural Network Training: Multilayer Perceptrons with the well known back-propagation training algorithm was used to develop the trained network. Table-1 shows the parameters of the neural network. It is to be noted that, we have not taken permittivity of the substrate (ε_r) and its thickness (t) as an input variable parameters, since in Sierpinski gaskets the role of the dielectric layer is mostly to give support to the antenna structure. For the generation of training data, the gaskets were assumed to be printed on a dielectric sheet of thickness t = 1.588 mm. and ε_r = 2.5. Matlab neural network toolbox was used to perform the training [23].

Table-1: Neural network parameters

No. of training data set	153
No. of input neurons	2
No. of hidden layer neurons	20
No. of output layer neurons	3
Learning rate	0.073
Epochs	1000
Error tolerance	0.01

3 RESULTS AND DISCUSSION

The trained neural network was tested for its validity with the test data from the simulator as well as with the previously reported results. In order to test the generalization of the network, a set of Sierpinski gaskets of input parameters {h, s} = {88.9. 0.5}, {89, 0.5}, {89, 0.6} and {89, 0.66} were given as input test parameters to the trained NN. Figure 3 and Figure 4 shows the simulated response of the three resonating frequencies corresponds to those resonating frequencies that were obtained with trained NN for the given input test parameters. In order to verify these resonating frequencies, it was compared with the reported results as shown in Table. 2. These comparison of the neuromodeler response with that obtained from the simulator for the values of fr1, fr2/fr1 and fr3/fr1. As expected, the first frequency is almost same for all values of scaling, because it depends on the overall height of the antenna. As verified by other authors for scaling equal to and more than 0.5, the antenna maintains its fractal behavior and in this range the ANN results are quite in agreement with the simulation results. For scaling less than 0.5, as seen from the simulation, and also by ANN, all the resonating frequencies are not fractal frequencies. The neuromodeler performance was also tested against earlier reported results [10, 12]. The results for different height and scaling of the gasket are shown in Table-2 where a good agreement can be seen. The network was trained to give outputs as fr1, fr2/fr1 and fr3/fr1 for ease of calculating all the frequencies and to further use the outputs directly in any optimization loop [24].

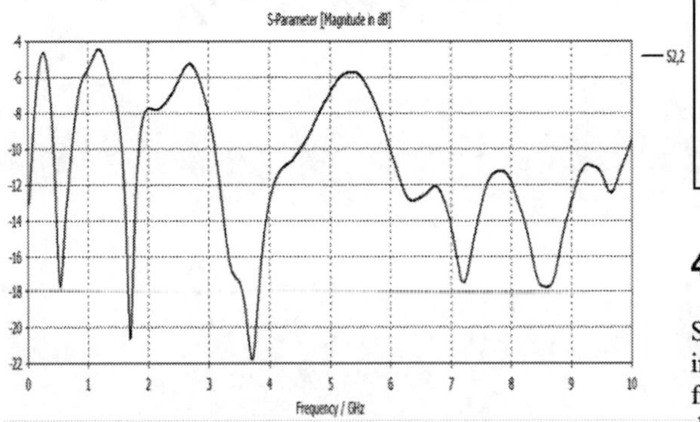

Fig. 3 Simulated response of Sierpinski gasket antenna with design dimensions of h=88.9 mm, S = (0.5)

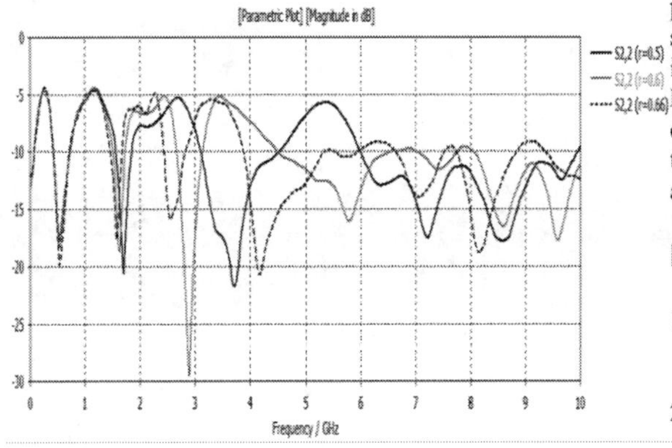

Fig. 4 Simulated parametric response of the Sierpinski gasket antenna with design dimensions of h=89mm, s = (0.5, 0.6, 0.66)

Table:2 Performance comparison of the neuromodeler with earlier reported result

	Scaling (*S*)	**Reported Results**			**NN Response**		
		f_{r1}(GHz)	f_{r2}/f_{r1}	f_{r3}/f_{r1}	f_{r1}(GHz)	f_{r2}/f_{r1}	f_{r3}/f_{r1}
Ref. [10] ε_r = 2.5 t = 1.588 mm. h = 88.9 mm.	0.5	0.52	3.50	7.07	0.50	3.54	7.10
Ref. [12] ε_r = 2.5 t = 1.588 mm. h = 89 mm.	0.5	0.46	3.77	7.64	0.48	3.60	7.67
	0.6	0.45	3.57	6.13	0.48	3.40	5.80
	0.666	0.44	3.46	5.53	0.47	3.26	5.53

4 CONCLUSION

Self-similarity is a property common to many fractals but in order to become a useful radiator it is necessary for the fractal antenna to fulfill specific requirements for the desired frequencies. For this reason, a generalized NN model was developed for the analysis of the Sierpinski gasket in order to locate the operational frequencies. The performance of the network was validated both with the simulation and the published results. With the growing interest in using fractals as multiband antenna for GSM, DECT, WLAN, ISM band hand set applications, the developed NN model can be useful to effectively locate the frequency bands of operation. Similar formulation can also be developed for other fractal structures.

REFERENCES

1. S. Maci, and G. B. Gentili, "Dual Frequency Patch Antennas," IEEE Antennas and Propagation Magazine, Vol. 39, No. 6, pp. 13-20, December 1997
2. S. C. Gao, L. W. Li, T. S. Yeo, and M. S. Leong, "Small Dual Frequency Microstrip Antennas," IEEE Transactions on Vehicular Technology, Vol. 51, No. 1, pp. 28-36, January 2002.
3. A. Ks. Skrivervik, J. F. Zurcher, O. Staub, and J. R. Mosig, "PCS Antenna Design: The Challenge of Miniaturization," IEEE Antennas and Propagation Magazine, Vol. 43, No. 4, pp. 12- 26, August 2001.
4. J. Ollikainen, M. Fischer, and P. Vainikainen, "Thin Dual-Resonant Stacked Shorted Patch Antenna for Mobile Communications," Electronics Letters, Vol. 35, No. 6, pp. 437-438, March 1999.
5. D. N. Elsheakh, H. A. Elsadek, and E. A. Abdallah, "Reconfigurable Single and Multiband Inset Feed Microstrip Patch Antenna for Wireless Communication Devices," Progress in

Electromagnetic Research C, Vol. 12, pp. 191-201, 2010.
6. L. Lizzi, and A. Massa, "Dual-band Printed Fractal Monopole Antenna for LTE Applications," IEEE Antennas and Wireless Propagation Letters, Vol. 10, pp. 760- 763, August 2011.
7. J. P. Gianvittorio, and Y. Rahmat-Samii, "Fractals Antennas: A Novel Antenna Miniaturization Technique and Applications," IEEE Antennas and Propagation Magazine, Vol. 44, No. 1, pp. 20- 36, February 2002.
8. D. H. Werner, and S. Ganguly, "An Overview of Fractal Antenna Engineering Research," IEEE Antennas and Propagation Magazine, Vol. 45, No. 5, pp.38-57, February 2003.
9. Anuradha, A. Patnaik, and S. N. Sinha, "An Investigation on E-shaped Fractal Like Patch Antenna," International Symposium on Antenna and Propagation, South Carolina, USA, June 2009.
10. C. Puente, and J. Romeu, "On the behavior of the Sierpinski multiband fractal antenna," IEEE Transactions on Antennas and Propagation., vol. 46, no.4, pp. 517-524, April 1998.
11. C. Puente. Romeu, R.Pous, X.Garcia,and F,Benitez, "Fractal multiband antenna based on the Sierpinski gasket," ElectronLett, vol. 32, no.1, pp. 1-2, Jan 1996.
12. C. Puente, J. Romeu, R.Bartoleme and R.Pous "Perturbation of the Sierpinski antenna to allocate operating bands, Electron.Lett" vol.32, pp. 2186-2188, Nov. 1996.
13. C.T.P.Song, P.S.Hall, H.Ghafouri-Shiraz, and D.Wake, "Sierpinski monopole antenna with controlled band spacing and input impedance," Electron.Lett. vol. 35, pp. 1036-1038, Jun 1999.
14. C.T.P.Song, P.S.Hall, and H.Ghafouri-Shiraz, "Perturbed Sierpinski Multiband Fractal Antenna with Improved Feeding Technique," IEEE Transactions on Antennas and Propagation, vol51, no.5, May 2003.
15. R.K.Mishra, R.Ghatak, and D.R.Poddar "Design Formula for Sierpinski Gasket Pre-Fractal Planar-Monopole Antennas," IEEE Antennas and Propagation Magazine, vol. 50, no.3, June 2008.
16. J.K Wojciech " Modified Sierpinski Fractal Monopole for ISM-bands handset Applications", IEEE Transactions on Antennas and Propagation, vol 57, no.3, March 2009.
17. Y. Kim, "Application of Artificial Neural Antennas to Broadband antenna design based on parametric frequency model", IEEE Transactions on Antennas and Propagation., vol. 55, no.3, March 2007.
18. R.K.Mishra and A.Patnaik, "Neural Network-Based CAD Model for the design of Square-patch Antennas", IEEE Transactions on Antennas and Propagation., vol. 46, no.12, Dec 1998.
19. Q. J. Zhang, K. C. Gupta and V. K. Devabhaktuni, "Artificial Neural Networks for RF and Microwave Design - From Theory to Practice," IEEE Trans. Microwave Theory and Techniques, vol. 51, no. 4, 1339-1350, April 2003.
20. R. C. Eberhart and Y. Shi, "Evolving artificial neural networks," in Proc. 1998 Int. Conf. Neural Networks and Brain, Beijing, P.R.C., 1998.
21. Anuradha, A. Patnaik, and S. N. Sinha, "Neurocomputational Analysis of Generalized Koch Fractal Antenna," A Workshop on Advanced Antenna Technology, Puri, Orissa, India, 31 May-4 June 2010.
22. CST Microwave Studio .,http://www.cst.com.
23. www.mathworks.com
24. Anuradha, A. Patnaik, and S. N. Sinha, "Design of Custom Made Fractal
25. Multiband Antennas using ANN-PSO," IEEE Antennas and Propagation Magazine, Vol. 53, No. 4, pp. 94-101, 2011 and [correction] Vol.53, No.5, pp.122- 124, 2011.

MATEC Web of Conferences 57, 01025 (2016)
ICAET-2016
DOI: 10.1051/matecconf/20165701025

Experimental Evaluation of Performance of Constant Power Prime-Mover Driven Isolated 3-φ SEIG for Pico-Hydro Power Generation System in Remote Mountainous Region of Himalayas

Umesh C. Rathore[1,a] and Sanjeev Singh[2]

[1] Electrical Engineering Department, ABV Government IET Pragatinagar, Shimla, HP, India
[2] Electrical & Instrumentation Engineering Department, SLIET Longowal, Punjab, India

Abstract. This paper presents the experimental evaluation of the performance of 3-φ self-excited induction generator (SEIG) suitable for pico-hydro power generation system feeding domestic load in remote mountainous region. The use of induction generators is most suitable for renewable energy conversion systems due to their enormous advantages over conventional synchronous generators. Important features of induction generators include the simplicity in construction, ruggedness, simplified control, ease in maintenance and small size per generated kW. The performance characteristics of 3-φ SEIG feeding isolated load are evaluated using MATLAB-Simulink model based on the prevalent renewable energy sources inputs and loading conditions in mountainous terrain of Himalayas. The results are validated using an experimental set-up comprising of 3-φ, 3 HP induction motor run as 3- φ induction generator driven by 5HP, 4-pole DC shunt motor acting as prime-mover.

1 Introduction

In the present energy scenario across the world, greater emphasis is being given nowadays on harnessing the vast energy potential of various renewable energy sources available across the world. Harnessing of renewable energy potential is essential to meet the ever increasing demand of electrical energy across the globe and also to minimize the adverse impact on environment caused by various conventional energy sources such as coal and petroleum products. With the rapid advancement in technology, the harnessing of these alternate sources of energy such as wind, hydro, geothermal, tidal, biomass, ocean thermal etc has become easier. These renewable energy sources are best suited to compliments the conventional sources of energy and also to meet the energy demand of the population living in remote areas where proving electricity using conventional grid is difficult. About 20% of the global final energy consumption comes from renewable sources.

Hydro energy is the most reliable and cost effective renewable energy source suitable for providing the electrical energy in remote mountainous regions which generally have the vast hydro potential. Along with the large hydro power plants, micro/pico hydro power plants play a significant role in meeting the power requirement. As compared to large hydro, micro/pico hydro power plants are environment friendly and require less investment. An estimated electric power potential of more than 2, 00,000 MW from small hydro exists worldwide, out of which around 50,000 MW has been tapped so far. India also has a history of more than 100 years in micro/pico hydro with the estimated potential of nearly 5000 MW capacity identified through more than 2000 sites in various states of India. Most of these sites are located in Himalayan region. The capacity of hydro power plants is decided on the basis of hydrology data such as water discharge, & rainfall for the past few years and water head available on the selected site. In hydro power plants, the erratic rainfall affects the water discharge in catchment areas which further affects the generated kWh output. However, the effect of water discharge variation is less on pico hydro power plants as most of these hydro power plants use uncontrolled hydro-turbines with the continuous availability of required water discharge. Most of these pico hydro power plants are located in the remotest part of the mountainous region feeding isolated load for small habitats having small electrical load. Pico hydro power plants in isolated mode are suited to these remote locations which are difficult to connect with conventional grid. The daily electrical load pattern varies during 24 hours in remote areas as most of the electrical load comprises of lighting and heating which is switched on and off as per the requirement.

To analyse the electrical loading pattern in the rural remote areas, data was collected from a 33/11kV electrical sub-stations feeding the remote villages using various 11kV distribution feeder lines in the District of Chamba in the state of Himachal Pradesh, India. Fig. 1

[a] Corresponding author: umeshrathore@sliet.ac.in

shows the electrical loading pattern as the per logged data collected from various 11kV feeders feeding the rural villages during the winter and summer months of 2015.

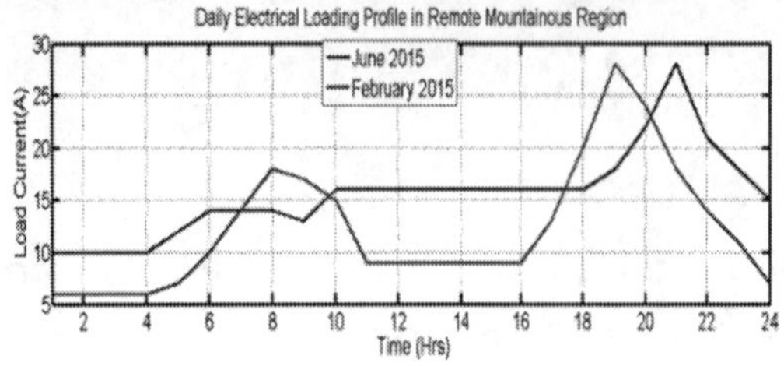

Figure 1. Electrical loading pattern in remote hilly regions.

Most of the electrical load in these areas comprises of lighting and heating load along with very few motor load of floor mills that too operate mostly in day times.From the loading pattern as shown in Fig. 1, it is clear that during the evening time between 6 PM to 10 PM, load is maximum with peak loading at around 8 PM. While the load is minimum during 2400 hours mid night to 5 AM and which is about 15 % of the maximum load. This loading pattern becomes crucial in the case if the rural isolated load is fed from a pico/micro hydro power plant and which is not connected to the grid. Under such varying load conditions, the use of suitable technologies on power generation such as use of induction generators [1] is beneficial due to their adaptability to variation in load and to keep voltage and frequency constant.

This paper presents the suitability of small rated induction generators (<10kW) in the remote mountainous terrain after evaluating the performance characteristics of these small induction generators using MATLAB/Simulink based models emulating the prevalent renewable inputs and varying isolated load conditions in remote mountainous regions. The simulated results are validated using an experimental setup comprising of 3 HP, 4 pole, 415V, 3-φ, squirrel cage induction motor operated as 3- φ self-excited induction generator driven by a 5 HP DC shunt motor acting as prime-mover. For self-excitation of SEIG, a 3-phase capacitor bank with different required ratings of capacitors in star and delta mode has been used. For electrical loading of self-excited induction generator, a 3-phase lamp load has been used. A single/3-phase inductive load has also been used for inductive loading. The performance parameters of SEIG have been evaluated at no load and at different load (balanced & un-balanced load) emulating the prevalent isolated load conditions in remote mountainous regions based on data.

2 Self-excited induction generators

Difference between induction generator and induction motor lies in the rotor speed. In induction generator the rotor speed is more than the synchronous speed while in induction motor, the rotor speed is slightly less than the synchronous speed. Induction generator consumes reactive power rather than supplying it and supplies only the real power (kW) to the system to which it is connected. The kVAr required by the induction generator plus the kVAr requirements for all other loads on the system must be supplied from static capacitors. Broadly the induction generators can be classified [2]-[3] as Self Excited Induction Generator (SEIG), Doubly Fed Induction Generator (DFIG) and Permanent Magnet Induction Generator (PMIG). Self excited induction generator is suitable for micro or pico hydro power plants located in remote areas to feed an isolated load. In stand-alone SEIG, the reactive power must be fed externally using capacitors as shown in Fig. 2 to establish the required magnetic field.

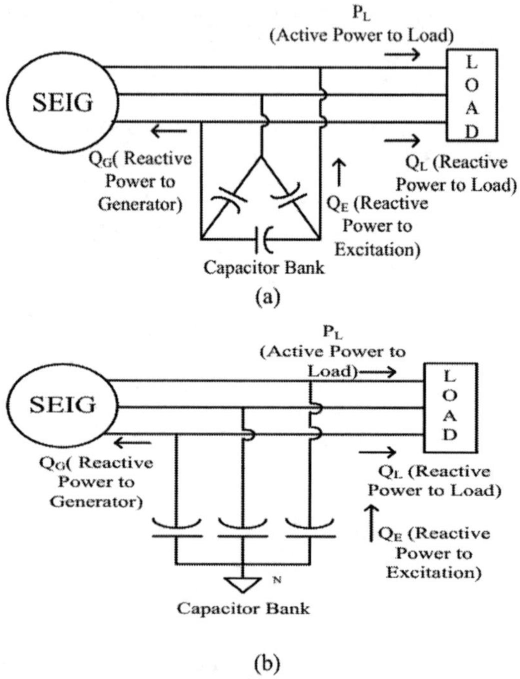

Figure 2. 3-φ SEIG with (a) delta connected capacitor bank, (b) with star connected capacitor bank.

Capacitors bank is connected in parallel with the induction generator terminals either in star or delta mode. When the rotor speed exceeds the synchronous speed, the residual magnetic field in the rotor initiates a voltage across the SEIG terminals and further augmented by the capacitor current to reinforce the magnetic field and system builds up an increasing excitation. After that with the increase in speed, the capacitor impedance decreases and excitation increases which results in increase in terminal voltage of the induction generator till the voltage is limited by saturation of magnetic circuit of the induction machine. To built up the voltage; there must be some residual magnetism present in the rotor of induction generator. The exact requirement of the capacitance depends upon the value of kVAr required and value of this capacitance is calculated in number of ways [4]-[10]. The minimum capacitance is inversely proportional to the square of speed and to the maximum saturated magnetizing reactance. The value of the capacitance is affected by load impedance, power factor and speed of induction machine. The maximum power output from the isolated SEIG depends upon its terminal capacitance and the speed of the generator. There must be a threshold speed called cut-off speed below which no excitation is possible, whatever the capacitor value may be. The main aim of the control strategies for induction generators is to regulate or maintain the desired voltage and frequency of the generated output. This is achieved using electronic load controller as shown in Fig. 3. The main function of the electronic load controller in isolated pico-hydro power plant using SEIG is to maintain the output load constant as seen by the induction generator under different consumer load conditions. The SEIG feeds two loads connected in parallel. The main load is consumer load and other load is dump or auxiliary load where power dissipated is generally wasted but can be used for other useful purposes such as pumping of water etc [11]. The electronic chopper in the controller circuit regulates the real & reactive power so that load seen by the generator is always constant. The conventional ELCs have discontinuous nature of control with discrete steps in connecting and disconnecting dump load which results in power quality problems. Modern electronic load controller technology has to meet the demand of various types of consumer loads which may be linear, non-linear, balanced or unbalanced and also has to maintain the power quality standards [12]-[15].

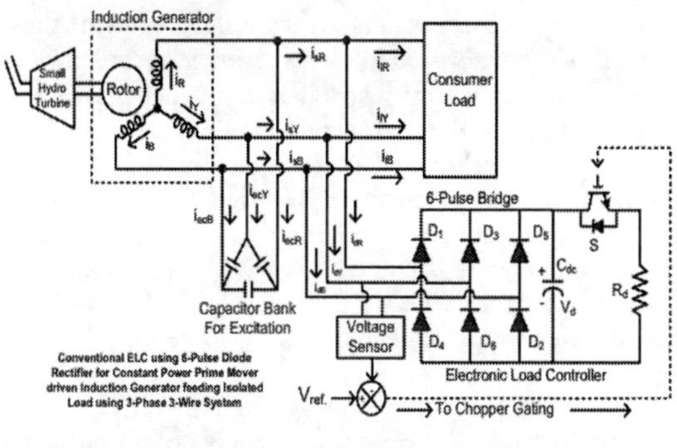

Figure 3. Electronic load controller circuit for isolated 3-phase SEIG in pico-hydro power generation system.

3 Simulated performance of SEIG in pico-hydro power generation system

Before experimental validation of SEIG, performance characteristics of the SEIG under steady state condition were simulated using MATLAB-Simulink based model of a 3-φ induction machine of 3 H.P, 415 Volts, 4-Poles rating which is actually used in real time practical application. The main component in the SEIG system is the selection of capacitance required for proper operation of SEIG under different operating conditions. After selecting the rating of capacitor bank connected in delta, various simulation tests were conducted under steady state condition of rated voltage and speed. The effect on value of capacitance was observed using simulation under different operating conditions such as variation in speed, variation in power factor due to change in inductive load and under different loading conditions. Results show the voltage build-up at a required capacitance and also the capacitance requirement is inversely proportional to square of speed. Also the capacitance value increases with the decrease in power factor it is also affected by the change in load. For the rated generated voltage, the value of excitation capacitance is slightly less than the value required at certain load. The terminal voltage is affected by the change in load. As the load on SEIG increases, its terminal voltage decreases slightly. The MATLAB/Simulink based simulation model and some of the simulated results such as terminal voltage development and its variation at change in load, variation of capacitance at different load and power factors and effect of change in load on SEIG voltage are shown in Figures 4 to 7. In the simulated model, the consumer load, controller circuit and measurement sections are shown as sub-systems.

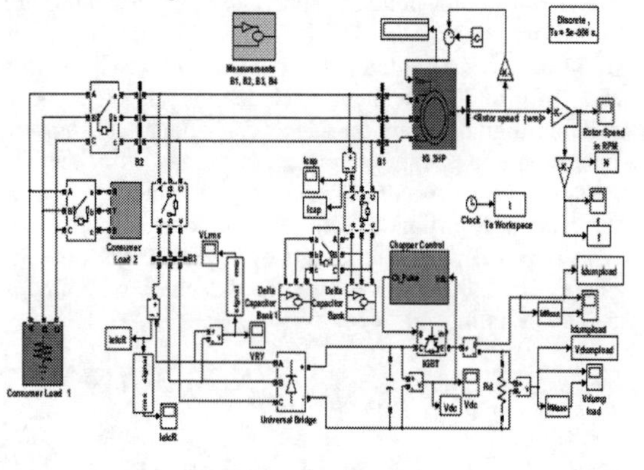

Figure 4. MATLAB/Simulink based model of 3 HP SEIG in isolated pico-hydro power generation system.

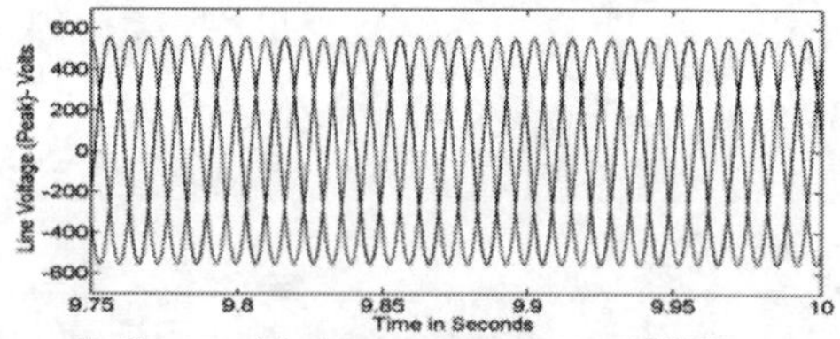

Figure 5. Generated 3-phase output voltage of SEIG.

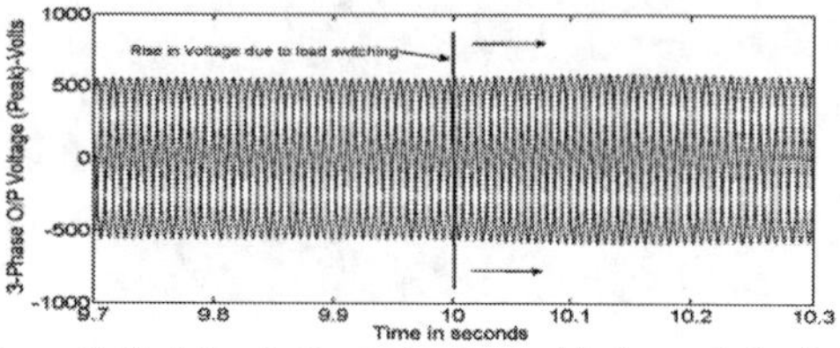

Figure 6. Variation in terminal voltage with change in load.

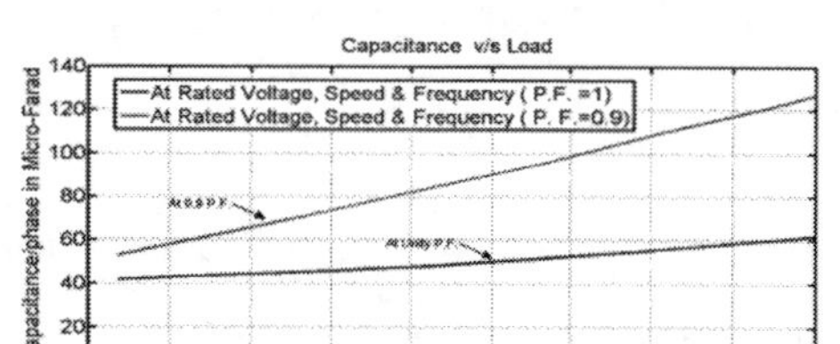

Figure 7. Capacitance v/s load current curves.

4 Experimental verification of performance of 3-φ SEIG in pico-hydro power generation system

To validate the simulated results and to prove the efficacy and suitability of self-excited induction generator feeding isolated load in pico-hydro power generation system, a hardware setup was prepared for experimentation. Figure 8 shows the photographs of the complete experimental set-up. The keys components of the hardware set-up consists of 3-phase, 415V, 4 pole, 50Hz squirrel cage induction motor operated as 3-phase SEIG coupled with a 5HP, DC shunt motor acting as prime-mover, a 3-phase/1-phase capacitor bank comprising of 10 sets of 6μF capacity 600V capacitors on each phase which can be connected either in delta or star mode with their individual circuit breakers, individual control panels for induction machine and DC machine operation and a set of 3-phase resistive and inductive load. The data logging and metering equipment include the speed sensor, suitable digital meters on panels for voltage and current measurement, a power meter (Yokogawa make) for voltage, current, power, frequency, power factor and total harmonics distortion (THD) measurements. The 3-phase resistive load panel comprised of 8 lamp loads on each phase. The inductive load used is of 3-phase type having 5A, 415V rating.

(a)

(b)

Figure 8. Hardware setup: (a) Induction machine coupled with prime-mover, (b) complete experimental set-up.

The complete trial for operating the 3-phase induction motor as induction generator was done in the following sequence:

- Running the induction machine first as induction motor for few hours to initiate the residual magnetic field.
- Driving the induction machine with the help of prime-mover (DC shunt machine in the adopted set-up) at its rated speed for few minutes. The speed of prime-mover is then increased to run the 3-phase induction motor at above the synchronous speed (1500 RPM for 4-pole machine). In this set-up, induction machine is driven at 1550 RPM for few minutes without connecting the capacitor bank across SEIG terminals.
- Connect all the measurement devices to record the induction generator parameters at various excitation capacitance values and at different speed.
- Firstly the induction generator is at no load. Under this condition, first set of excitation capacitor bank (in delta or star mode) is switched ON to build the no-load voltage across induction generator terminals. The value of the connecting capacitor banks is increased in steps until the required voltage is build up.
- Once the required voltage is build-up, load is switched ON across the SEIG and then its effect on the generator speed, voltage, power factor and excitation capacitance is recorded under different operating conditions.

4.1 Voltage build-up process in 3-φSEIG using excitation capacitance

After running the induction machine at slightly higher than its synchronous speed, a first set of excitation capacitance (6μF/phase in this set-up) is switched ON and then the voltage at the terminals of induction machine was observed. Full voltage development across the SEIG terminals used in the experiment recorded when 12μF/phase excitation capacitance in delta mode is connected for 3 HP rated induction generator at no load. The voltage developed at no load is slightly higher than the rated voltage of induction machine, but when load is put on the generator, it falls and can be brought to its rated value by accurately varying the excitation capacitance value. The exact value of the connected capacitor as shown in Table I can also be calculated by recording the stator terminal voltage and capacitor line current readings recorded during the experimentation. In this experiment, the connected capacitance value in the capacitor bank in delta exactly matched with the calculated value of excitation capacitance experimentally.

4.2 Experimental evaluation of 3-φSEIG under loaded conditions

Most of the consumer load in remote mountainous regions generally comprises of domestic lighting and heating load involving incandescent and compact fluorescent lamps. Apart from this some domestic load for a very short period consists of electronic gadgets and house-hold single phase appliances. There is no industrial load except for the use of induction machines in few flour mills. In this experimental set-up, a three phase lamp load with multiple loading steps for each phase was used to connect the lamp load on each phase. In present energy scenario with greater emphasis on energy conservation, there is a maximum use of compact fluorescent lamps (CFLs) in almost every house-hold. Keeping this in view, the CFLs were also used along with conventional incandescent lamps to evaluate the performance of SEIG. Table II shows the experimental recorded parameters.

Table 1. Experimental calculation of excitation capacitance per phase in delta mode.

3-φ, Star connected Resistive Lamp Load (Watts) across SEIG				Capacitance/Phase in Delta mode (μF)	Cap. bank Line Voltage (V)	Cap. bank Line Current (A)	Cap. bank Phase Current (A)	Calculated value of Capacitance/Phase (μF)	Frequency of Generated Voltage	SEIG Speed in RPM
R_{ph}	Y_{ph}	B_{ph}	Total							
0	0	0	0	12	455	3.10	1.79	12.09	49.8	1522
100	100	100	300	12	431	2.89	1.67	11.9	49.8	1518
200	200	200	600	12	415	2.75	1.59	12.1	49.6	1512

Table 2. Experimental values of SEIG & prime-mover parameters under different operating conditions.

3-φ, Star connected Resistive Lamp Load (Watts) across SEIG				C/ph (μF) Delta Mode	V_{LL} (V)	I_L SEIG Line Current (A)	$I_{L(C)}$ Cap. Line Current (A)	I_{ph} (Load) (A)	N (RPM)	PF	F (Hz)	Prime Mover (DC Shunt Motor) Parameters			
R ph	Y ph	B ph	Total									V_A (V)	V_F (V)	I_A (A)	I_F (A)
0	0	0	0	6	0	0	0	0	1550	1	-	210	186	0.9	0.73
0	0	0	0	12	470	3.27	3.21	0	1552	1	50.8	209	185	2.5	0.70
100	100	100	300	12	447	3.11	3.0	0.49	1539	0.99	50.3	208	183	4.3	0.69
200	200	200	600	12	425	3.02	2.82	0.95	1535	0.95	50.1	208	182	5.4	0.69
300	300	300	900	12	410	2.95	2.65	1.31	1527	0.94	49.8	207	182	6.9	0.68
340	0	320	660	12	445	3.28	3.0	1.35	1564	0.92	51.3	214	189	6.0	0.69
400	400	400	1200	12	385	2.92	2.50	1.52	1525	0.95	49.5	208	182	7.4	0.67
400	400	400	1200	12	416	3.20	2.65	1.60	1574	0.84	51.5	217	187	8.0	0.69
500	500	500	1500	12	355	2.88	2.30	1.77	1559	0.85	51	215	188	7.7	0.65
500	500	500	1500	18	415	3.03	2.7	1.9	1450	0.85	46.8	203	194	11	0.68

Table 3. Experimental values of SEIG parameters under different resistive & inductive loads.

3-φ, Star connected Resistive Lamp Load (Watts) across SEIG				Parameters of 3-φ, Star connected Inductive Load across SEIG			C/ph (μF) in Delta Mode	V_{LL} (V) SEIG Terminal Voltage	I_L (A) SEIG Line current	$I_{L(C)}$ (A) Cap. Circuit Line Current	I_{ph} (Load) (A) Load Phase Current	N (RPM)	F (Hz)
R ph	Y ph	B ph	Total	V_{LL} (V)	I_L (A)	Q (kVAr)							
100	100	100	300	0	0	0	12	443	3.11	3.0	0.49	1539	50.2
100	100	100	300	452	1.21	0.95	18	452	3.39	3.02	0.49	1522	49.8
200	200	200	600	438	1.15	0.88	18	436	3.35	3.1	0.98	1518	49.7
400	400	400	1200	390	1.03	0.70	18	390	3.28	2.55	1.52	1510	48.8
400	400	400	1200	427	1.11	0.82	18	427	3.4	2.90	1.50	1545	50.2

Following inferences were made out of this experimentation:
- A minimum value of excitation capacitance is required to build up the voltage at SEIG terminals on no load. In this case for 3 HP, 4-pole SEIG the terminal voltage built-up at 12μF/phase in delta connected capacitor bank.
- As the electrical load on SEIG is increased, its terminal voltage slightly falls along with the slight drop in the speed of induction generator. However, the decrease in terminal voltage due to increase in load can be compensated by increasing the speed of prime mover.
- With the further increase in electrical load, requirement of excitation capacitance increases at rated speed.
- Since domestic electrical load is of single-phase type and 3-phase, 4-wire system is used for electrifying the villages which results in the un-balancing of load in each phase. Therefore, unequal loading was done for SEIG in the experiment and it was observed that the SEIG parameters remained within the permissible limit.
- When a 3-phase inductive load along with resistive load was put on the SEIG, the requirement of the excitation capacitance increased to compensate for the lagging reactive power caused by the inductive load. Table III shows the SEIG parameters recorded under resistive and inductive load.

5 Conclusion

The performance characteristics of 3-phase SEIG operated in isolated mode have been investigated in the presented paper and results are validated using an experimental setup. These tests have been performed without the use of electronic load controller. But in actual practice, these controllers are used to maintain and control the voltage and frequency of the system where voltage is regulated by controlling the reactive power using STATCOM and the frequency is maintained by controlling the dump load. However, in the tests by regulating the DC shunt motor parameters, the constant power input conditions has been emulated. The experimental results validate the efficacy and suitability of self-excited induction generator for feeding the isolated load in remote locations as it adapts well to counter the varying loading conditions. During off-load period, the wasted energy across dump load can be utilized for useful purpose such as pumping of water to meet the potable and irrigation requirement.

Acknowledgement

Authors sincerely thank the management of Himachal Pradesh State Electricity Board Limited to collect the daily electrical load data from their 33kV/11kV substations feeders feeding the remote locations presented in this paper.

References

1. M. Godoy Simoes and Felix A. Farret, *Alternate Energy Systems: Design and Analysis with Induction Generators*, 2nd ed. Boca Raton: CRC Press (2008)
2. R. C. Bansal, "Three-phase self-excited induction generators: an overview," *IEEE Transactions on Energy Conversion*, vol. **20**, pp. 292-299 (2005).
3. M. H. Haque, "Self-excited single-phase and three-phase induction generators in remote areas," *in Proc. IEEE ICECE'08*, pp. 38-42 (2008).
4. S. S. Murthy, O. P. Malik, and A. K. Tandon, "Analysis of self-excited induction generator," *Proc. Inst. Elect. Eng. C,* vol. **129**, no. 6, pp 260-265, (1982).
5. S.P. Singh, B. Singh, and M. P. Jain, "Comparative study on the performance of a commercially designed induction generator with induction motors operating as self excited induction generators," *IEE Proceeding –C,* vol. **140**, no. 5, pp. 374-380, (1993).
6. A. K. Al Jabri and A. I. Alolah, "Capacitance requirement for isolated self-excited induction generator," *IEE Proc. Electr. Power Appli.*, vol. **137**, no. 3, pp. 154-159 (1990).
7. T. F. Chan, "Capacitance requirements of self-excited induction generators," *IEEE Transactions on Energy Conversion*, vol. **8**, no. 2, pp. 304-311 (1993).
8. L. Shridhar, B. Singh, C. S. Jha, B. P. Singh, and S. S. Murthy, "Selection of capacitors for the self regulated short shunt self excited induction generator," *IEEE Transactions on Energy Conversion*, vol. **10**, no. 1, pp. 10-17 (1995).
9. V. P. Chandran and S. Vadhera, "Capacitance requirements of self excited induction generator for different operating conditions," in *Proc. IEEE ICEAS*, pp. 1-6 (2011).
10. S. Kulandhaivelu, K. K. Ray, P. Jyothi, S. Jayachitra, and V. Narasinga Rao, "Experimental evaluation of capacitance value; to self-excite the induction motor operating as generator in wind energy conversion," in *Proc. IEEE ICETECT'11*, pp. 133-140 (2011).
11. Umesh C. Rathore and Sanjeev Singh, "Isolated 3-phase self-excited induction generator in pico-hydro power plant using water pump load in mountainous region of Himalayas," in *Proc. 2014 IEEE Global Humanitarian Technology Conference - South Asia Satellite (GHTC-SAS),* pp. 40-44 (2014)
12. K. Idjdarene, D. Rekioua, T. Rekioua, and A. Tounzi, "Performance of an isolated induction generator under unbalanced loads," *IEEE Transactions on Energy Conversion*, vol. **25**, no. 2, pp. 303-311(2010)
13. E. Suarez and G. Bortolotto, "Voltage-frequency control of a self excited induction generator," *IEEE Transactions on Energy Conversion*, vol. **14**, no. 3, pp. 394-401(1999)
14. B. Singh, S. S. Murthy, and S. Gupta, "Analysis and design of STATCOM-based voltage regulator for self-excited induction generators," *IEEE Transactions on Energy Conversion,* vol. **19**, no. 4, pp. 783-790(2004)
15. B. Singh and G. K. Kasal, "Voltage and frequency controller for isolated asynchronous generators feeding 3-phase 4-wire loads," in *Proc. IEEE ICIT'06*, pp. 2773-2778(2006)

Transient Stability Improvement of IEEE 9 Bus System Using Power World Simulator

Ramandeep Kaur[1], Divesh Kumar[2]
[1,2] Department of Electrical Engineering, Bhai Gurdas Institue of Engineering and Technology, Sangrur, India

Abstract. The improvement of transient stability of power system was one of the most challenging research areas in power engineer.The main aim of this paper was transient stability analysis and improvement of IEEE 9 bus system. These studies were computed using POWER WORLD SIMULATOR. The IEEE 9 bus system was modelled in power world simulator and load flow studies were performed to determine pre-fault conditions in the system using Newton-Raphson method. The transient stability analysis was carried out using Runga method during three-phase balanced fault. For the improvement transient stability, the general methods adopted were fast acting exciters, FACT devices and addition of parallel transmission line. These techniques play an important role in improving the transient stability, increasing transmission capacity and damping low frequency oscillations.

Keywords- *Transient stability, power system stability, three-phase fault, FACTS, Power world simulator, steady state.*

1 INTRODUCTION

Now-a-days, it has become a necessity to maintain synchronism because the system is expanding day-by-day and these results in installation of larger machines. Due to this, transient disturbances are increasing continuously in power system. The transient disturbances are caused by the changes in the load, switching operations, faults and loss excitations. Thus, it is very important to regain synchronism or equilibrium after disturbances in the electrical utilities. Hence, thorough analysis of transient stability is required to reduce problems such as blackouts, loss of synchronism, etc.

The term stability refers to maintenance of synchronism and stability limit refers to maximum power flow possible in the system or a part of system at which system is operating with stability. Power system stability is the property of power system that enables it to remains in a state of equilibrium under normal operating conditions and toregain equilibrium after being subjected to disturbances.

Power system stability can be broadly grouped into steady state stability, transient stability and dynamic stability. Steady state stability is the capability of an electric power system to maintain synchronism between machines when small slow disturbance occurs. Dynamic stability is the ability of a power system to remain in synchronism after the initial swing until the system has settled down to the new steady state equilibrium conditions.

In this paper, the main emphasis is given to transient stability analysis of system and its improvement.. Transient stability is the ability of system to remain in synchronism during the period of disturbance and prior to the time that the governors can act. The transient stability analysis is carried out for a short time period that will be equal to the time of one swing. This analysis is carried out to determine whether the system losses stability during the first swing or not.

In this paper, Transient stability analysis is performed with the help of three-phase balanced fault.A fault which gives rise to equal fault currents in the lines with 120 degree displacement is known as three phase fault or symmetrical fault. Faults could happen when a phase establishes a connection with another phase, lightning, insulation deterioration, wind damage, trees falling across lines, etc.

The other objective of this paper that is improvement of IEEE 9 bus system is carried out using SVC, exciters and addition of parallel transmission line by Runga method. Static Var Compensators are used to improve voltage and reactive power conditions in AC systems. It also

Corresponding author: diveshthareja@yahoo.com

decreases oscillations in power system.The effectiveness of this controller depends on its optimal location and proper signal selection in the power system network.

The fast acting exciters used also decrease oscillations in the system. The amount of kinetic energy gained by the generators during a fault is directly proportional to the fault duration, the quicker the fault is cleared, the less disturbance it causes.

The series inductive reactances of transmission network are primary determinants of stability limits. The reduction of reactances of various elements of the transmission network improves transient stability by increasing post fault synchronizing power transfer. This can be achieved by addition of parallel transmission lines in the system.

To analyse swing behaviour of system it is solved by Runga method. Swing equation is the electromechanical equation describing relative emotion of the rotor load angle (δ) with respect to the stator field as a function of time is known as Swing equation. In most disturbances, linearization is not per-missile and the nonlinear swing equation must be solved because oscillations are of such magnitude.
Power flow analysis is called the backbone of power system analysis. Transient stability analysis of system fault analysis and its improvement is one of the basic problems in power system engineering.

The single line diagram of IEEE 9 bus model is shown in figure1:

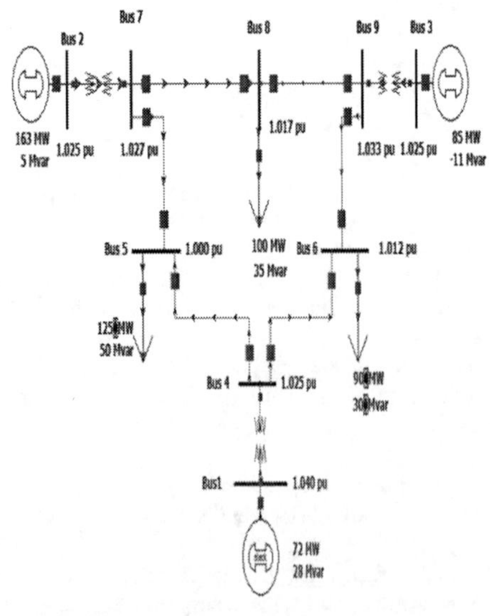

Fig.1. IEEE 9 BUS MODEL in power world simulator

2 PROBLEM FORMULATION

A. Power Flow Studies

In transient stability studies, it is necessary to have the knowledge of pre-fault voltages magnitudes. The main information obtained from the power flow study comprises of magnitudes and phase angles of bus voltages, real and reactive powers on transmission lines, real and reactive powers at generator buses, other variables being specified. The pre-fault conditions can be obtained from results of load flow studies by the Newton-Raphson iteration method.

The Newton-Raphson method is the practical method of load flow solution of large power networks. Convergence is not affected by the choice of slack bus. This method begins with initial guesses of all unknown variables such as voltage magnitude and angles at load buses and voltage angles at generator buses. Next, a Taylor Series is written, with the higher order terms ignored, for each of the power balance equations included in the system of equations.
We first consider the presence of PQ buses only apart from a slack bus.

For an ith bus,

$$\text{Pi} = \sum_{k=1}^{n} |Vi||Vk||Yik|\cos(\theta ik + \delta k - \delta i) = Pi(|V|, \delta) \quad (1)$$

$$\text{Qi} = \sum_{k=1}^{n} |Vi||Vk||Yik|\sin(\theta ik + \delta k - \delta i) = Qi(|V|, \delta) \quad (2)$$

i.e., both real and reactive powers are functions of (|V|,δ), where

$|V| = (|V_1|,\dots,|V_n|^T)$ $\delta = (\boldsymbol{\delta}_{1,\dots,}\boldsymbol{\delta}_n)^T$

We write

Pi (|V|,) = Pi(x)

Qi (|V|,) =Qi(x)

Where,

x = [**δ**/|V|]

Let Pi and Qi be the scheduled powers at the load buses. In the course of iteration x should tend to that value which makes

Pi – Pi(x) = 0 and Qi – Qi(x) = 0 (3)

Writing equation (3) for all load buses, we get its matrix form f(x) = $\begin{bmatrix} P(scheduled) - P(x) \\ Q(scheduled) - Q(x) \end{bmatrix} = \begin{bmatrix} \Delta P(x) \\ \Delta Q(x) \end{bmatrix} = 0$ (4)

At the slack bus, P_1 and Q_1 are unspecified. Therefore, the values $P_1(x)$ and $Q_1(x)$ do not enter into equation (3) and hence (4). Thus, x is a 2(n-1) vector (n-1 load buses), with each element function of (n-1) variables given by the vector x= [**δ**/|V|]

We can write,

f(x) = $\begin{bmatrix} \Delta P(x) \\ \Delta Q(x) \end{bmatrix} = \begin{bmatrix} -J11(x) - J12(x) \\ -J21(x1) - J22(x) \end{bmatrix} \begin{bmatrix} \Delta\delta \\ \Delta|V| \end{bmatrix}$ (5)

Where, $\Delta\boldsymbol{\delta} = (\Delta\boldsymbol{\delta}_2 \ldots\ldots \Delta\boldsymbol{\delta}n)^T$

$\Delta|V| = (\Delta|V_2| \ldots \Delta|Vn|)^T$

J(x) = $\begin{bmatrix} -J11(x) - J12(x) \\ -J21(x) - J22(x) \end{bmatrix}$ (6)

J(x) is the jacobian matrix, each J_{11}, J_{12}, J_{21}, J_{22} are (n-1) × (n-1) matrices.

$-J_{11}(x) = \frac{\partial P(x)}{\partial \delta}$

$-J_{12}(x) = \frac{\partial P(x)}{\partial |V|}$

(3.18)

$-J_{21}(x) = \frac{\partial Q(x)}{\partial \delta}$

$-J_{22}(x) = \frac{\partial Q(x)}{\partial |V|}$

The elements of $-J_{11}$, $-J_{12}$, J_{21}, $-J_{22}$ are

$\frac{\partial Pi(x)}{\partial \delta k}, \frac{\partial Pi(x)}{\partial |Vk|}, \frac{\partial Qi(x)}{\partial \delta k}, \frac{\partial Qi(x)}{\partial |Vk|},$

Where i = 2...n; k = 2...n.

From equation (1) and (2), we have

$$\frac{\partial Pi(x)}{\partial \delta k} = -|Vi||Vk||Yik|\sin(\theta ik + \delta k - \delta i)(i = k)$$

$$= \sum_{\substack{k=1 \\ k \neq i}}^{n} |Vi||Vk||Yik|\sin(\theta ik + \delta k - \delta i)\ (i = k)$$

(7)

$$\frac{\partial Pi(x)}{\partial |Vk|} = |Vi||Yik|\cos(\theta ik + \delta k - \delta i)(i \neq k)(8)$$

$$= 2|Vi||Yii|\cos\theta ii + \sum_{\substack{k=1 \\ k \neq i}}^{n} |Vk||Yik|\cos(\theta ik + \delta k - \delta i)\ (i = k)$$

$$\frac{\partial Qi(x)}{\partial \delta k} = |Vi||Vk||Yik|\sin(\theta ik + \delta k - \delta i)(i \neq k)$$

$$= \sum_{\substack{k=1 \\ k \neq i}}^{n} |Vi||Vk||Yik|\cos(\theta ik + \delta k - \delta i)\ (i = k)(9)$$

$$\frac{\partial Qi(x)}{\partial |Vk|} = |Vi||Yik|\sin(\theta ik + \delta k - \delta i)(i \neq k)$$

$$= 2|Vi||Yii|\sin\theta ii + \sum_{\substack{k=1 \\ k \neq i}}^{n} |Vk||Yik|\sin(\theta ik + \delta k - \delta i)\quad (i = k)\ (10)$$

An important observation can be made with respect to the elements of Jacobian matrix. If there is no connection between ith and kth bus, then Yik =0. The process continues until a stopping condition is met.

B. Transient stability analysis

Transient stability studies deals with the effect of large, sudden disturbances such as effect of large sudden outage of line, occurrence of fault, or the sudden application or removal of loads. To ensure that a system can with stand the transient condition following a disturbance, transient stability analysis should be performed.

Steps for transient stability analysis in POWER WORLD SIMULATOR:

Step 1: An IEEE- 9 bus system is taken. System model is implemented and executed in power world simulator and load flow is performed. The load flow analysis of 9 bus system is done with the help of Newton Raphson method.

Step 2: After load flow, the initial parameters of the system e.g. bus voltages, bus frequency, bus power angles, generator power angles are studied.

Step 3: A three-phase balanced fault is applied on transmission line with different clearing times and the results are taken in Power World Simulator using Runga.

Step 4: Improvement of IEEE 9 bus simulated model using Runga method in POWER WORLD SIMULATOR.

TABLE 1. LINE PARAMETERS OF 9 BUS SYSTEM

Line	Resistance(PU)	Reactance(PU)	Susceptence(PU)
1-4	0.0000	0.0576	0.0000
4-5	0.0170	0.0920	0.1580
5-6	0.0390	0.1700	0.3580
3-6	0.0000	0.0586	0.0000
6-7	0.0119	0.1008	0.2090
7-8	0.0085	0.0720	0.1490
8-2	0.0000	0.0625	0.0000
8-9	0.0320	0.1610	0.3060
9-4	0.0100	0.0850	0.1760

C. Standard Parameters

TABLE 2. MACHINE DATA OF 9 BUS SYSTEMS

Parameters	M/C 1	M/C 2	M/C 3
H(sec)	23.64	6.4	3.01
Xd(PU)	0.146	0.8958	1.3125
X'd(PU)	0.0608	0.1198	0.1813
Xq(PU)	0.0969	0.8645	1.2578
X'q(PU)	0.0969	0.1969	0.25
T'd0(PU)	8.96	6.0	5.89
T'q0(PU)	0.31	0.535	0.6

TABLE 3: EXCITER DATA OF 9 BUS SYSTEMS

Parameters	Exciter 1	Exciter 2	Exciter 3
K_A	20	20	20
T_A(sec)	0.2	0.2	0.2
K_E	1.0	1.0	1.0
T_E(sec)	0.314	0.314	0.314
K_F	0.063	0.063	0.063
T_F(sec)	0.35	0.35	0.35

3 RESULTS AND DISCUSSION

The load flow analysis and transient stability for the standard IEEE-9 bus system are performed. The standard IEEE 9 bus system consists of 9 buses, 3 generators, 3 loads and 3 transformers. Table 4 & 5 shows load flow analysis carried out using Newton-Raphson method. Table 6 shows power angle with different clearing times by Runga method.

TABLE 4. POWER FLOW LIST OF IEEE 9 BUS SYSTEM

From Bus	To Bus	Branch Device Type	MW From	Mvar From	MVA From	MW Loss	Mvar Loss
4	1	Transformer	-72	-24.8	75.8	0	3.15
2	7	Transformer	163	4.9	163	0	15.8
9	3	Transformer	-85	15.6	86.4	0	4.1
5	4	Line	-43	-39.6	58.5	0.3	-16
6	4	Line	-28	-16.9	32.9	0.1	-16
7	5	Line	84.2	-10.1	84.8	2.2	-21
9	6	Line	63.3	-17.8	65.7	1.5	-31
7	8	Line	78.8	-0.8	78.9	0.5	-12
8	9	Line	-22	-23.6	32.1	0.1	-21

TABLE 5. BUS DATA OF IEEE 9 BUS MODEL

Name	Nom kV	PU Volt	Volt (kV)	Angle (Deg)	Load MW	Load Mvar	Gen MW	Gen Mvar
1	16.5	1.04	17.16	0			71.63	27.91
2	18	1.03	18.45	9.35			163	4.9
3	13.8	1.03	14.15	5.14			85	-11.45
4	230	1.03	235.8	-2.22				
5	230	1	229.9	-3.68	125	50		
6	230	1.01	232.8	-3.57	90	30		
7	230	1.03	236.2	3.8				
8	230	1.02	234	1.34	100	35		
9	230	1.03	237.5	2.44				

TABLE 6 POWER ANGLE WITH DIFFERENT CLEARING TIMES BY RUNGA METHOD

Clearing Time (Sec)	Angle(Runga Method) in Degree
1.033	106.487
1.05	109.063
1.066	112.238
1.083	116.237

When three-phase balanced fault is applied on transmission line between bus 5 and 7, disturbances occurs in the system that results in power angle oscillations. When clearing time is 1.033sec or 2 cycles, the maximum angle difference is 106.487 degree and further, if clearing time is increased, it results in increase in maximum angle difference as shown in Table 4. This means system is going towards un- stability mode.

For the improvement transient stability, the general methods adopted were fast acting exciters, FACT devices and addition of parallel transmission line. These techniques play an important role in improving the transient stability, increasing transmission capacity and damping low frequency oscillations. Table 7, 8 and 9 shows comparison of power angle difference between with and without SVC, shows comparison of power angle difference between with and without parallel transmission lines and comparison of power angle between EXST4B and existing exciter.

TABLE 7. COMPARISON OF POWER ANGLE DIFFERENCE BETWEEN WITH AND WITHOUT SVC

Cycles	Max angle difference with compensation (degree)	Max angle difference without compensation (degree)
2	105.119	106.487
3	107.375	109.063
4	110.199	112.238
5	113.801	116.237

TABLE 8. COMPARISON OF POWER ANGLE DIFFERENCE BETWEEN WITH AND WITHOUT PARALLEL TRANSMISSION LINE

Cycles	Max angle difference with parallel line (degree)	Max angle difference without parallel line (degree)
2	71.34	106.488
3	75.178	109.065
4	79.145	112.24
5	83.754	116.238

TABLE 9 COMPARISON OF POWER ANGLE BETWEEN EXST4B AND EXISTING EXCITER

Cycles	EXST4B Exciter	Existing Exciter
2	98.520	106.488
3	99.394	109.065
4	101.112	112.24
5	103.680	116.238

When SVC is used for improvement of IEEE 9 Bus model, it is found that the maximum angle difference is lesser with SVC than without SVC as shown in Table 7. The maximum angle difference can be decreased by addition of parallel transmission line on the system as depicted in Table 8. Table 9 shows decrease in power angle by use of EXST4B exciter

4 CONCLUSION

This paper presents the transient stability analysis and its improvement using power world simulator. The effectiveness of shunt FACT devices such as SVC, addition of parallel transmission lines and exciters has been studied in improving the system stability with different loads. The load flow studies are also performed to determine prefault conditions. Hence, by proper modelling of system, it would be interesting to determine any other possible advantage of controllers in power system stability-studies.

REFERENCES

1. R. A. Naghizadeh, S. Jazebi, B. Vahidi, "Modeling hydro power plants and tuning hydro governors as an educational guideline," International Review on Modeling and Simulations (I.RE.MO.S.), Vol. 5, N. 4,(2012)
2. Vadim Slenduhhov, Jako Kilter, "Modeling and analysis of the synchronous generators excitation systems",Publication of Doctoral School of Energy and Geotechnology, Pärnu, (2013)
3. Sandeep Kaur, Dr. Raja Singh Khela,"Power system design & stability analysis," international journal of advanced research in computer science and software engineering, Volume 4, Issue 1, (2014)
4. M.A Salam, M. A. Rashid, Q. M. Rahman and M. Rizon, "Transient stability analysis of a three machine nine bus power system network", Advance online publication, (2014)
5. Swaroop Kumar.Nallagalva, Mukesh Kumar Kirar, Dr.Ganga Agnihotri, "Transient Stability analysis of the IEEE 9-Bus electric power system", International Journal of Scientific Engineering and Technology, Volume No.1, Issue No.3, pp.: 161-166, (2012)
6. Ž. Eleschová, M. Smitková and A. Beláň, "Evaluation of power system transient stability and definition of the basic criterion", Internationals journal of energy, Issue 1, Vol. 4, (2010)
7. Tin Win Mon and Myo Myint Aung "Simulation of synchronous machine in stability study for power system", International Journal of Electrical and Computer Engineering 3:13, (2008)
8. Yaman C. Evrenosoglu, Hasan Dag,"Detailed Model for Power System Transient Stability Analysis".Journal of Engineering & Computer Science IJECS-IJENS, Vol: 13 No:02, (2012)
9. H.T.Hassan, 2Usman Farooq Malik, 3Irfan Ahmad Khan, 4Talha Khalid, "Stability Improvement of Power System Using Thyristors Controlled Series Capacitor (TCSC)",

10. Gundala Srinivasa Rao, Dr. A.Srujana, "Transient Stability Improvement of Multi-machine Power System Using Fuzzy Controlled TCSC" International Journal of Advancements in Research & Technology",ISSN 2278-7763, Volume 1, Issue 2(2012)

MATEC Web of Conferences 57, 01027 (2016) DOI: 10.1051/matecconf/20165701027
ICAET-2016

Analytical Approach for Loss Minimization in Distribution Systems by Optimum Placement and Sizing of Distributed Generation

Surbhi Bakshi [1 a], Tilak Thakur[2], Rintu Khanna[3]

[1]*Research Scholar , Deptt. of Electrical Engg, PEC University of Technology, Chandigarh, India*
[2]*Professor, Deptt. of Electrical Engg, PEC University of Technology, Chandigarh, India*
[3]*Associate Professor, Deptt. of Electrical Engg, PEC University of Technology, Chandigarh, India*

Abstract: Distributed Generation has drawn the attention of industrialists and researchers for quite a time now due to the advantages it brings loads. In addition to cost-effective and environmentally friendly, but also brings higher reliability coefficient power system. The DG unit is placed close to the load, rather than increasing the capacity of main generator. This methodology brings many benefits, but has to address some of the challenges. The main is to find the optimal location and size of DG units between them. The purpose of this paper is distributed generation by adding an additional means to reduce losses on the line. This paper attempts to optimize the technology to solve the problem of optimal location and size through the development of multi-objective particle swarm. The problem has been reduced to a mathematical optimization problem by developing a fitness function considering losses and voltage distribution line. Fitness function by using the optimal value of the size and location of this algorithm was found to be minimized. IEEE-14 bus system is being considered, in order to test the proposed algorithm and the results show improved performance in terms of accuracy and convergence rate.

Keywords: Distributed Generation, Particle Swarm Optimization, Optimal Sizing, Solar.

1. Introduction

Distributed Generation involves the applying of little generators, scattered throughout an influence system, to produce the electrical power required by electrical customers. Appeared, because of its low cost of electricity for non-traditional growing interest, it is environmentally friendly, due to increased awareness of emission control, which has become an important factor [1], [2], [3]. This environment is a benefit and investment in distributed generation in particular solar, wind, cogeneration (CHP). There are a variety of factors, such as improving work efficiency, environmental benefits and better transmission congestion management contribute to its popularity. The main idea here is to have a large number of DG units which are close to the load centre, rather than increasing the capacity of the main power plant. Various techniques can help achieve DG as renewable energy, such as solar energy, wind energy, tidal energy, biomass energy, they are more common address of "green energy." Micro turbines, diesel engines, gas turbines, fuel cells from other additions to the list, even though they are not so-called energy "green" form. Stirling engine and the engine further extend the list of reciprocating engines [4], [5], [6]. Recently, wind energy has gained huge popularity and all other forms of renewable energy [7] to tough competition.

DG can be defined as the generation and transmission system, wherein the generator system is not in direct contact to the main transmission grid, but to the distribution network. It has economic, environmental and technical reliability many advantages. Reducing losses due to transmission and distribution caused by reduced transmission equipment costs, saving energy as a result of lower electricity prices. Environment is also obtained from the sound level of emission reductions in pollution and greenhouse gas levels lower. Wide range of advantages such as reduced line losses, increase peaking, improving the system's voltage profile and ultimately leads to higher power quality. This leads to transmission and distribution and grid reinforcement ease congestion. Sometimes it is also used to provide a separate system,

[a] *Surbhi Bakshi : surbhi.cgc@gmail.com*

remote application with the required power to the load in. Therefore, it is necessary to find the best location and optimal size of distributed power.

Best location DGs were found only in the reduced basis [8] losses. Some power flow algorithm has been raised several researchers in order to find the best size DG, in all load buses [9] [10]. Wang Nehrir analysis shows the preferred placement targeting dangerous goods minimum loss. Attempts have been made by the Chiradeja benefit from reducing feeder loss distribution in the radial direction, wherein the load is concentrated. Meta-heuristic algorithms have been used by some researchers to find new DG units [11-15] in the best position. Genetic algorithms have been used to DG placement and loss reduction. Multi-objective evolutionary algorithms have been used by the Botticelli and Ghiani to solve this problem. Nara et.al, investigate DG optimal location and size by using tabu search algorithm.

This paper identifies the use of multi-objective particle swarm optimization optimum value size and placement of the additional DG units i.e. the size of the bus and the best location in order to obtain the ultimate goal. Multi-objective optimization including seeking the optimal value, so that the voltage and the loss is the best way to optimize the weight and importance depends on various factors such as the type of power required quality and economic requirements, etc. [15] for obtaining data in IEEE bus 14. In addition, the power supply is integrated and distributed using Newton - Raphson method.

This paper is organized as follows. The problem formulation to determine the optimal size of DGs at the selected locations is defined in Section II and proposed methodology in section III. Then, the effectiveness of the proposed method is verified with the simulation results in Section IV. Finally, the conclusions are given in Section V.

2. Problem Formulation

Problems Optimizing Distributed power generation unit is configured in the form of formulated optimization problem taking into account the cost function of the voltage and the transmission line formed by the real power loss. This problem can be represented mathematically in Equation 12:

$$\text{minimize } f(x) = w1 \sum_{k=0}^{n} loss + w2 * \sum_{k=0}^{m} loss(1 - Vi)^{\wedge}2 \quad ..(12)$$

Here, 'm' is the number of the bus and "n" is the number of lines in the system. The problem is tested for IEEE-14 bus system. The IEEE-14 bus system consists of 6 PV bus out with bus number 1 is connected with a maximum capacity of slack bus in which the main generators is connected. Newton Raphson technique is used to calculate Power flow in the network, as the analytical solution of the problem is impossible.

3. Proposed methodology

Newton iterative method is used in this work for trend analysis. Each bus is calculated from the trend in terms of active and reactive power. In addition, line losses in each row in the bus system are also calculated. Goal of this article is to join an additional DG to reduce losses on the line. Additional equipment will meet the increased load demand. An objective function is dependent on the distribution line losses and voltage formed.

The objective function is improved using particle swarm optimization and constraint handling of reflection is also minimized. As the problem is converted into a constrained optimization problem; the original PSO is modified to handle infeasible solutions. Also the size of the problem is a digital estimation and allocation problem is integer estimation and therefore the solution is checked for validity in the search space. Improving the voltage profile and reducing the loss of multi-objective function is converted to a minimization problem.

The position of the unit is initialized in IEEE-14, system bus 1-14 for a random integer value. All solutions are looking to check the feasibility of space and initialization process is repeated until a feasible solution. Each particle is updated according to the rules and checks the validity of the group, and only those particles that are feasible will remain unchanged, and the rest will be rejected. This process continues until the maximum or minimum error criterion iteration conditions are achieved.

4. Results and Discussions

The problem defined above has been simulated for an IEEE 14 bus system. 100 particles were considered and simulations were done for 30 iterations. All the simulations were done in MATLAB R2013a in a 64 bit system with 2.3 GHz processor and 4 GB RAM.

Divergence among the particles as the iteration progresses can be depicted in Figure 1.

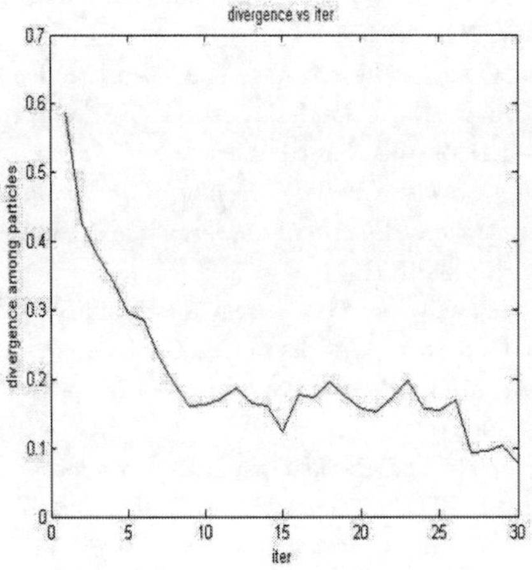

Figure 1. Plot of Divergence of Particles

It is observed that the divergence decreases with progress in iteration. The reason behind is the convergence of particles. As the Swarm algorithm searches in the search space for optimal values, particles start converging towards a common value and divergence among particles being diminishing.

A pictorial representation of the particles is provided in Figure 2 and it presents a real view of how the particles move in the search space. As it can be observed that, the particles begin randomly within the specified range but finally converge to a single point overcoming all the local minimas and maximas.

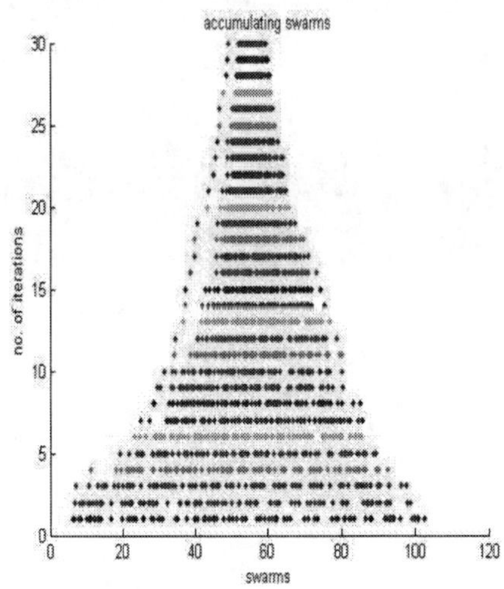

Figure 2. Plot of convergence of particles

The table 1 illustrate the optimal results of 14 bus systems. Table 2 illustrate voltages after simulation on each bus of 14 bus system

Table 1. Showing optimized values

Bus System	14
Optimal Bus No	14
Loss(MW) without DG	4.2
Loss (MW) With DG	3.3063
Optimal Size (kW)	32.1839

Table 2. Voltage (14 bus system)

Bus No	Voltage
1	1.06000000000000
2	1.04500000000000
3	1.01000000000000
4	1.01319412327445
5	1.01657458089408
6	1.07000000000000
7	1.04569913905268
8	1.08000000000000
9	1.03052499541293
10	1.02991793536347
11	1.04613111278885
12	1.05325955541164
13	1.04663639983200
14	1.01927185909764

5. Conclusions

An effective solution to the problem of optimal sizing and location of DG units using a multi-objective Particle Swarm Optimisation technique was proposed. On further analysis it is found that the optimal values of these parameters can reduce losses and improve voltage profile significantly. The algorithm converges after around 30 iterations and is computationally efficient too. The algorithm is designed such that it increases the exploration capability initially and avoids being trapped in local minimas and maximas. The fitness on the basis of which the optimisation algorithm is based is a multi - dimensional function and hence the solution obtained is only a pareto-optimal solution.

References

1. J. De La Ree, V. Centeno, J. S. Thorp and A. G. Phadke, "Synchronized phasor measurement applications in power systems," IEEE Transactions on Smart Grid, vol. 1, no.1, pp. 20-27, (2010)
2. A. G. Phadke and J. S. Thorp, *Computer Relaying for Power System*. Second edition, Chichester: John Wiley & Sons Ltd, (2009)
3. P. Ju, C. Qin, F. Wu, H. Xie and Y. Ning, "Load modeling for wide area power system," *International Journal of Electrical Power & Energy Systems*, vol. 33, no. 4, pp. 909-917, (2011)
4. Z. Q. He, Z. Zhang, W, Chen, O. P. Malik and X. G. Yin, "Wide-area backup protection algorithm based on fault component voltage distribution," *IEEE Transactions on Power Delivery*, vol.26, no. 4, pp. 2752- 2760, (2011)
5. A. G. Phadke and J. S. Thorp, *Synchronized Phasor Measurements and Their Applications*. Springer verlag, (2008)
6. S. H. Horowitz and A. G. Phadke, "Third zone revisited," *IEEE Transactions on Power Delivery*, vol.21, no.1 pp.23-29, (2006)
7. X. Tai, D. Marelli, E. Rohr and M. Fu, "Optimal PMU placement for power system state estimation with random component outages," *International Journal of Electrical Power & Energy Systems*, vol.51, pp.35-42, (2013)
8. T. Ackerman, G. Andersson and L. Söder, "Distribution Generation: a definition". *Electric Power System Research* 57 pp. 195-204., (2001)
9. Chiradeja P., "Benefits of Distributed Generation:A line loss reduction analysis", *Transmission and Distribution Conference and Exhibition: Asia and Pacific*, 2005 IEEE/PES .
10. Kejun Qian, Cheng K. Zhou, Yue Yuan,Xiaodan Shi and Allan M., "Analysis of the Environmental Benefits of distributed Generation",*Power and Energy Society General Meeting- Conversion and Delivery of Electrical Energy in the 21st century*,2008 IEEE.
11. Gil, H.A.and Joos,G.,"Models for Quantifying the Economic Benefits of distributed Generation", *IEEE Transactions on Power Systems*, Vol.23, pp. 327 – 335, (2008)
12. Chiradeja P.and Ramakumar, R.," An approach to quantify Technical Benefits of Distributed Generation", *IEEE transactions on Energy* ,Vol. 19, No. 4, pp.764-773, (2004)
13. Waseem, I, Pipattanasomporn M. and Rahman, S.,"Reliability Benefits of Distributed Generation as Back up source",*,IEEE Power and Energy society General meeting, PES-2009*, Publication Year: 2009 , Page(s): 1 – 8.
14. Momoh, J.A., Yan Xia, and Boswell, G.D.," An approach to determine Distributed Generation benefits in power networks", *40th NA Powerr Symposium*, pp. 1 – 7, (2008)
15. P. Chiradeja and R. Ramakumar, "A review of distributed generation and storage," in *Proc. 1998 Frontiers of Power Conf.*, pp. VIII 1–VIII

MATEC Web of Conferences **57**, 01028 (2016) DOI: 10.1051/matecconf/20165701028
ICAET-2016

Analyzing the effect of gate dielectric on the leakage currents

Sakshi, Sandeep Dhariwal and Amandeep Singh

Lovely Professional University- Phagwara, India

Abstract. An analytical threshold voltage model for MOSFETs has been developed using different gate dielectric oxides by using MATLAB software. This paper explains the dependency of threshold voltage on the dielectric material. The variation in the subthreshold currents with the change in the threshold voltage sue to the change of dielectric material has also been studied.

Index Terms— threshold voltage, gate dielectric oxides, subthreshold currents

1 INTRODUCTION

Earlier the VLSI designers were mainly concerned about the performance and miniaturization of the VLSI devices. For the last few years, power dissipation has been a serious issue because of the significant growth in portable computing and wireless communication. For more than 40 years, silicon dioxide has been used as dielectric because of its manufacturability and providing a continuous improved transistor performance as it is grown thinner and thinner [1-2]. The scaling of gate dielectric is done to improve the MOS transistor performance. The continuous scaling down of physical thickness of the gate dielectric and gate length has improved device performance and increased packing densities. Because of continuous scaling, the silicon dioxide dielectric thickness has been reduced to such an extent that further scaling will lead to increase in poly-Si gate depletion, gate dopant penetration, power consumption, etc. [2-4].

According to the prediction of the International Technology Roadmap for Semiconductors (ITRS) (Figure 1)[6], if the gate- oxide thickness becomes less than 1.4 nm for sub- 100 nm MOS then it results in the direct tunneling current. This tunneling current increases with the reduction of gate- oxide thickness and hence, the static power dissipation also increases [3-5]. It can be observed from the figure that even with the conservative equivalent oxide thickness scaling, excessive gate leakage prohibits continued gate- dielectric scaling using silicon- dioxide and thus it is required to find an accurate replacement of silicon dioxide such as high- k dielectric materials (Figure 2).

Basic material properties which are important for the selection of alternative gate dielectric are dielectric permittivity, band gap, band alignment with respect to silicon, thermodynamic stability, film morphology and uniformity, interface quality, and reliability [4].

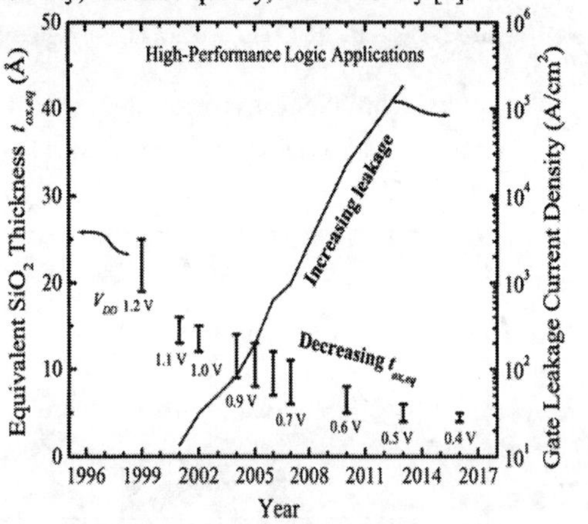

Figure 1 Scaling trend of the MOSFET gate dielectric thickness. The projected range of equivalent SiO_2 thickness for high- performance logic applications is plotted using bars. The supply voltages are also indicated and the largest value for each bar, the gate- leakage current density through SiO_2 is calculated and plotted as solid line [6].

Corresponding author: manish.bansalbti@gmail.com

DOI: 10.1051/matecconf/20165701028

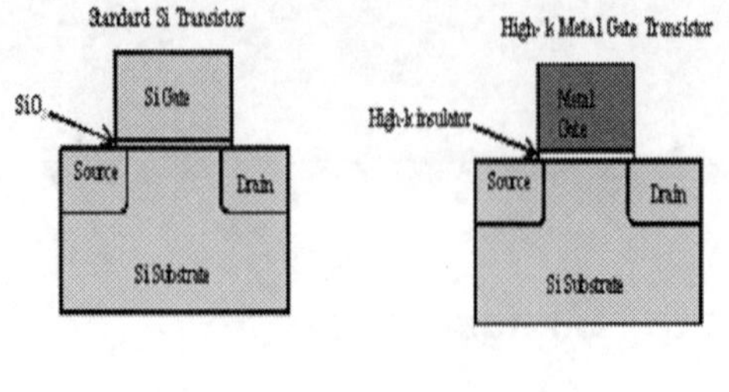

Figure 2. MOS Transistor Structure with Dielectric Gate [7]

Figure 3 presents dielectric constants of some of the high- k candidates:

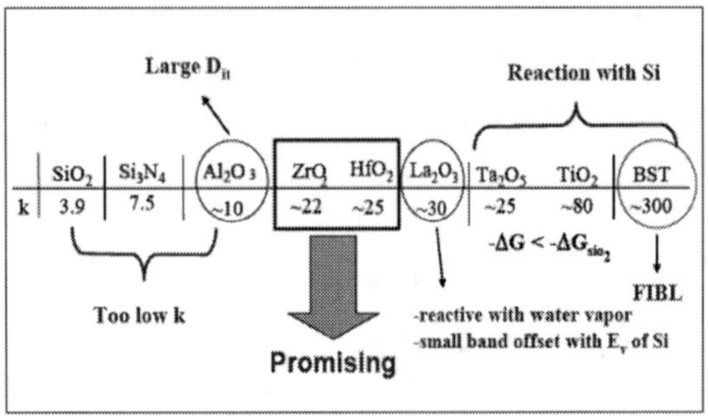

Figure 3. Dielectric constants of some high-k dielectrics [8]

1. SOURCES OF LEAKAGE POWER

Four sources of leakage currents are described as follows (Figure 4) [9]:

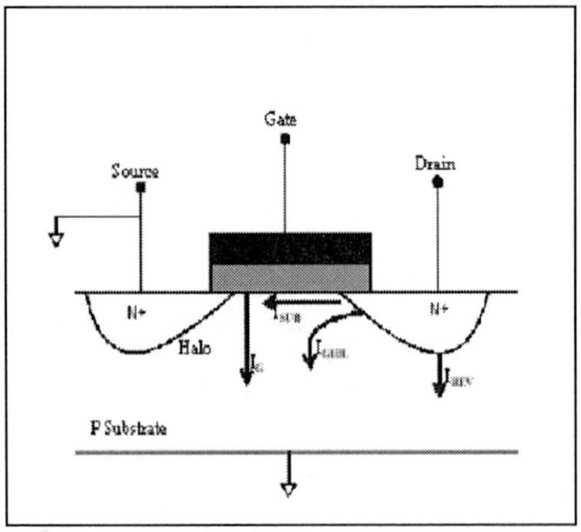

Figure 4. . Leakage current components in an NMOS transistor [9]

A. Reverse- biased junction leakage current (I_{REV})

When a transistor is off, the leakage takes place from the source/ drain to the substrate via the reverse- biased junctions. For an NMOS transistor, in off- state, V_{DS} is equal to the supply voltage which results in the flow of leakage current. The magnitude of leakage current depends upon the drain diffusion area and the leakage current density determined by the doping concentration.

B. Gate Induced Drain Leakage (I_{GIDL})

The leakage occurs due to the high field effect in the drain junction. In the case of an NMOS, when the gate is grounded and drain potential is at V_{DD}, electron-hole pair generation takes place through avalanche multiplication and band- to- band tunneling because of band bending in the drain. The holes get swept out to the substrate which causes a serious depletion condition, at that very instant electrons get collected in the drain causing a flow of GIDL current. GIDL current increases with an increase in the V_{DS} and V_{DG}.

C. Gate direct- tunneling Leakage (I_G)

The leakage current flows from the gate via the leaky oxide insulation to the substrate. In thick oxides (greater than 3-4 nm), leakage results due to the Fowler-Nordheim tunneling of electrons in the conduction band of the oxide layer on the application of a high electric field across the oxide layer. Whereas for thin oxides (in 0.15µm and less), direct tunneling through the oxide layer is the major problem. The magnitude of direct tunneling current exponentially increases with the oxide thickness (T_{ox}) and supply voltage (V_{DD}).

D. Subthreshold (weak inversion) Leakage (I_{SUB})

The drain- source current, flowing when the transistor is operating in the weak inversion region, is the leakage current. Subthreshold conduction occurs due to the diffusion current of the minority carriers in the channel. The magnitude of subthreshold current is a function of the temperature, supply voltage, device size, and the process parameters of which the dominant role is played by the threshold voltage.

2 EXPERIMENTAL WORK

The fabrication of MOS capacitors is done on 200mm n- and p- type Si substrates. The basic structure of a MOS transistor is shown in Figure 2. Different high-k dielectric materials have been investigated to replace silicon dioxide as the gate dielectric material. The amount of variation in the subthreshold current with the threshold voltage has been calculated using MATLAB. In addition, the effect of interface trap charges, doping concentration and also oxide thickness on the threshold voltage for NMOS and PMOS has been observed. The oxide thickness is varied from 50Å, 100Å, 200Å, 400Å, 600Å, 800Å and 1000Å respectively. The interface trap charge is kept at $5*10^{10}$ qC/cm^2 and the doping concentrations (N_a, N_d) are varied from 10^{15}cm^{-3} to 10^{16}cm^{-3} respectively. The threshold voltage versus oxide thickness for various doping concentrations at the constant interface trap charge ($Q_i = 5*10^{10}$qC/cm^2) for various dielectric materials such as silicon dioxide (SiO_2, k = 3.9), aluminium oxide (Al_2O_3, k = 10), hafnium oxide (HfO_2, k = 23) and tantalum oxide (Ta_2O_5, k = 25) have been plotted. All calculations have been done considering the MOS capacitor to be at room temperature (T = 300K).

DOI: 10.1051/matecconf/20165701028

3 Results and discussion

From the results, it can be seen that the threshold voltage for NMOS and PMOS are positive and negative respectively. The threshold voltage for NMOS becomes more positive with the increase in the oxide thickness whereas in the case of PMOS, the value of threshold voltage becomes more negative. Threshold voltage also varies with the dielectric constant. The larger the dielectric constant of the gate-dielectric material, the more is the threshold voltage. Threshold voltage is being calculated as

$$V_{th} = \varphi_{ms} - \frac{Q_{fc}}{C_{ox}} + \frac{2kT}{q} ln\left(\frac{N_a}{N_i}\right) + \sqrt{\frac{2\epsilon_{Si}\, qN_a\,(2\varphi_b + V_{sb})}{C_{ox}}}$$

where φ_{ms} is the metal semiconductor work function and can be calculated as the difference of work function of metal and the substrate ($\varphi_m - \varphi_{Si}$), N_a and N_d are the doping concentrations, C_{ox} is the oxide capacitance V_{sb} is the substrate bias voltage.

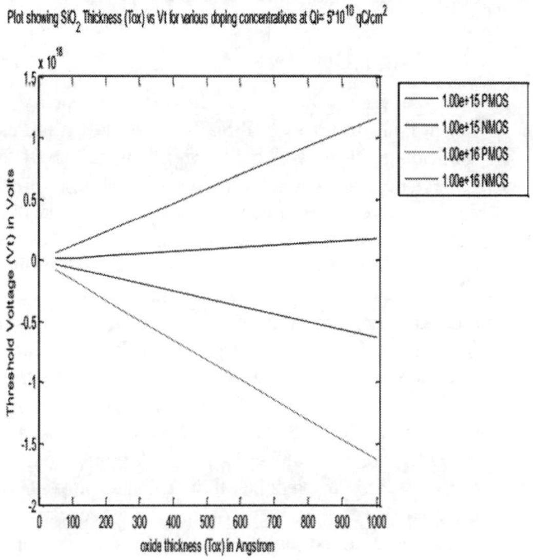

Figure 5 Plot showing SiO_2 thickness versus Vth

From Figure 5, it is evident that with the increase in oxide thickness, the threshold voltage for NMOS becomes more positive and that for PMOS becomes more negative.

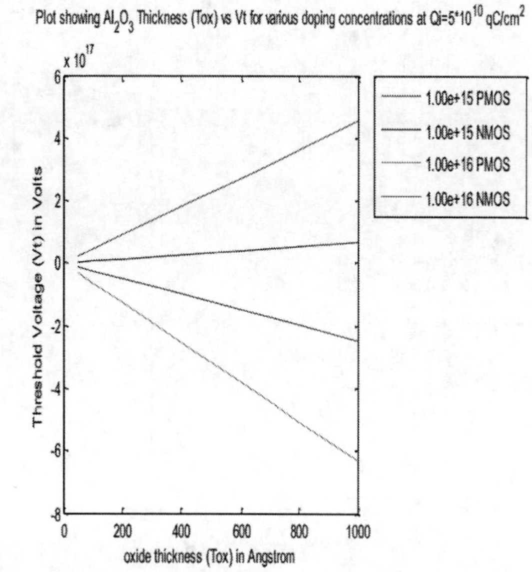

Figure 6. Plot showing Al_2O_3 thickness versus Vth

From Figure 6, it can be observed that for aluminium oxide, the dependency of the threshold voltage on the oxide thickness is same as that in the silicon dioxide but the value of threshold voltage is less as compared to that for silicon dioxide.

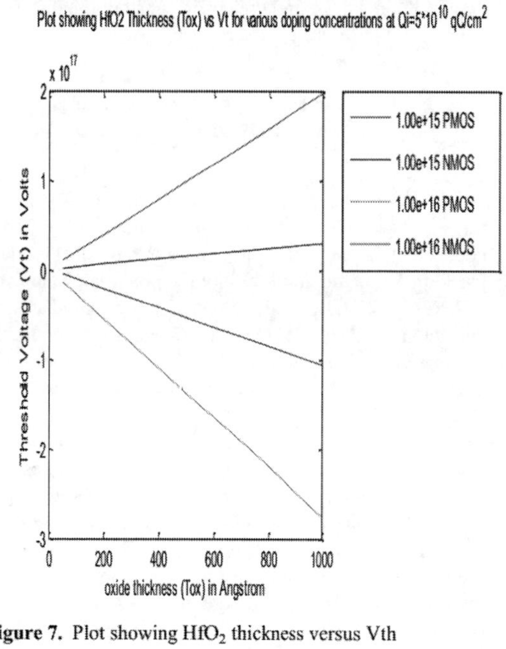

Figure 7. Plot showing HfO_2 thickness versus Vth

The threshold voltage for hafnium oxide is further lesser than that for aluminium oxide.

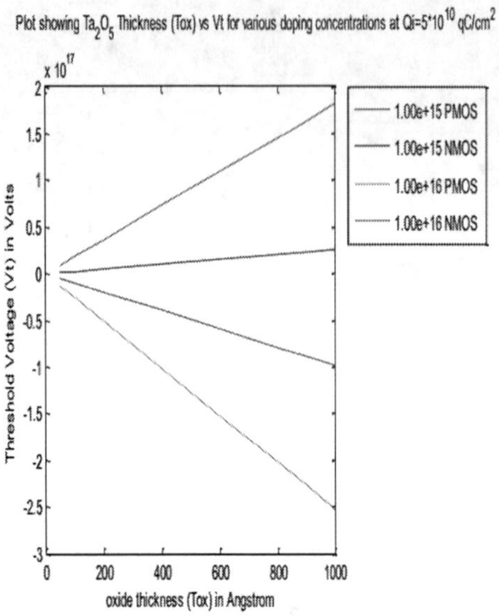

Figure 8. Plot showing Ta_2O_5 thickness versus Vth

On comparing the threshold voltages of all the above cases, it has been found out that the threshold voltage for tantalum oxide is the minimum.

Hence, it has been justified that with the increase in dielectric thickness, and dielectric constant the threshold voltage of the MOSFET decreases.

The subthreshold current can be calculated as

$$I_{sub} = I_0\, e^{\left[\frac{V_{gs}-V_{th}}{nV_t}\right]}\left[1 - e^{\frac{-V_{ds}}{V_t}}\right]$$

where V_{gs} is the gate-to- source potential, V_{th} is the threshold voltage, n is the subthreshold swing coefficient [= dV_g /d(logI_d)], V_t is the thermal voltage [=kT/q], V_{ds} is the drain- to- source potential. The Table 1 hereby provides the values for the subthreshold current at a certain threshold voltage for a particular dielectric material in a PMOSFET. The values have been calculated at various oxide thicknesses in Angstrom (Å) 50, 100, 200, 400, and 600.

From the table 1, we have justified that in the case of PMOSFET; with the increase in the oxide thickness the threshold voltage becomes more negative (i.e. decreases) while the subthreshold current increases. On the other hand, we also observe that with the increase in the dielectric constant of the gate dielectric oxide material, the threshold voltage becomes more positive (i.e. increases) while the subthreshold current decreases. But in the case of NMOS, with the increase in oxide thickness, the threshold voltage becomes more positive (i.e. increases) and so the subthreshold current decreases.

Table 1 Subthreshold current for the corresponding threshold voltages for a particular dielectric material in a P-MOSFET

T_{ox} (Å)	Threshold Voltage (Vth in Volts)			Subthreshold Current (I_{SUB} in A)		
	SiO_2 (3.9)	**Al_2O_3 (10)**	**HfO_2 (23)**	**SiO_2 (3.9)**	**Al_2O_3 (10)**	**HfO_2 (23)**
50	-0.3165	-0.1234	-0.0537	4.01 x 10^{48}	6.38 x 10^{46}	1.43 x 10^{46}
100	-0.6330	-0.2469	-0.1073	1.78 x 10^{51}	4.5 x 10^{47}	2.26 x 10^{46}
200	-1.2659	-0.4937	-0.2147	7.02 x 10^{56}	8.08 x 10^{58}	1.13 x 10^{47}
400	-2.6318	-0.9874	-0.42933	2.17 x 10^{68}	8.90 x 10^{53}	5.63 x 10^{48}
600	-3.7977	-1.4811	-0.6440	9.03 x 10^{79}	2.37 x 10^{58}	3.77 x 10^{50}

4 CONCLUSION

The scaling of the silicon dioxide has reached to an extent that on further scaling there will be an increase in the leakage currents. That is why there is a need for the replacement of silicon dioxide with another material. The possible material can be a high-k dielectric material provided it possesses the basic material properties like dielectric permittivity, band gap, band alignment with respect to silicon, thermodynamic stability, film morphology and uniformity, interface quality, and reliability. The various candidates high- k dielectric materials considered are Si_3N_4 (~7.5), Al_2O_3 (~10), ZrO_2 (~22), HfO_2 (~25), La_2O_3 (~30), Ta_2O_3 (~25), TiO_2 (~80), and BST (~300). Out of these the most reliable materials are ZrO_2 and HfO_2. High- k dielectric materials are required to reduce the leakage currents. Four different leakage current sources have been studied but we have counted upon the variation of subthreshold currents with respect to the change in the threshold voltage along with the oxide thickness and dielectric constant. The results justify the need of the adoption of high-k dielectric materials.

ACKNOWLEDGMENT

I am very thankful to Lovely Professional University for helping me to use the resources throughout the work .

REFERENCES

1. Paulo Francisco Butzen and Renato Perez Ribas, *Leakage Current in Sub- Micrometer Cmos Gates*, pp. 1
2. G. Brezeanu, M. Brezeanu, F. Bernea *High- K Dielectrics in Nano & Microelectronics,* Micro- and nanoengineering series, Vol. **19**, pp-9- 21, (November 2011).

3. P.S. Raja, R. Joseph Daniel *Effect of gate dielectric on threshold voltage of Nanoscale MOSFETs*, International Journal of Engineering Research and Development, e-ISSN: 2278-067X, p-ISSN: 2278-800X, Volume **5**, Issue 3, pp. 93-104, (December 2012).
4. Yee- Chia Yeo, Tsu- Jae King, and Chenming Hu *MOSFET Gate Leakage Modeling and Selection Guide for Alternative Gate Dielectrics Based on Leakage Considerations*, IEEE Transactions on Electron Devices, Vol. **50**, No. 4, (April 2003).
5. Tiefeng Wu, Heming Zhang and Huiyong Hu *Effects of Gate Tunneling Current on the Static Characterictics of CMOS Circuits,* International Journal of Innovative Computing, Information and Control, Volume **7**, Number 6, ISSN 1349-4198, pp. 3229-3237, (June 2011).
6. *International Technology Roadmap for Semiconductors* San Jose, CA, Semiconductor Industry Association (2001).
7. IBM and Intel make *high- k gate breakthrough Compound* semiconductor net (2007).
8. Kang C. *A Study on the material and device characteristics of hafnium oxynitride MOSFETs with TaN gate electrodes* PhD Dissertation, University of Texas at Austin (2004).
9. Farzan Fallah, and Massoud Pedram *Standby and Active Leakage Current Control and Minimization in CMOS VLSI Circuits* IEICE review journal

MATEC Web of Conferences 57, 01029 (2016) DOI: 10.1051/matecconf/20165701029
ICAET-2016

OPTIMIZED FEASIBILITY ANALYSIS FOR THE DESIGN TO SET UP SOLAR POWER PLANT DURING OPERATING HOURS IN THE ACADEMIC E–BLOCK AREA – CASE STUDY

Gagandeep Kaur[1], Bhanu Duggal[2]and Bhanu Marwaha[3]

[1]Assistant Professor, Electrical and Instrumentation Department, Thapar University, Patiala[2]Student,Electrical and Instrumentation Department, Thapar University, Patiala[3]Student,Electrical and Instrumentation Department, Thapar University, Patiala

Abstract. This paper is regarding the optimized design feasibility analysis of a solar power plant in our academic area i.e. E-block in Thapar University. For this design analysis, the technical feasibility and financial feasibility has been considered completely. Further, the technical feasibility of the project puts emphasis on irradiation analysis, connected load analysis, rooftop analysis, shadow analysis, load analysis, design of the on-grid system as well as design of the off-grid system. Financial feasibility analysis includes Cost analysis, Payback period, Funding options. This paper also discusses the type of solar power system to be installed based on the conclusions drawn out of the technical analysis for optimal solution for the utilization of block based analysis specifically. The financial feasibility also supports the selection.

1 Introduction

India is currently ranked sixth in annual electricity consumption, accounting for about 3.5% of the world's total energy consumption. India's electricity consumption has risen by about 64%. The electricity consumption is expected to rise at rate of 8% annually through 2020. There lies an ironic fact that although the consumption of electricity may be very high but the per capita consumption is as low as 1010 kWh in 2014-15 as opposed to 15000 kWh per capita in developed nations. Another, interesting fact is that nearly one fourth of Indian households still have no access to electricity with some states in East and North East having less than even 30% households with access. In this developing countries, like India, currently there are about 400 million people i.e. about 80 million households do not have access to commercial energy and still are not connected from the grid.

Hence, renewable energy comes to the rescue for the growing energy needs of India and the energy crisis which is prevalent in India. It will solve the problem of grid disconnectivity. India has been endowed with abundant natural resources that there will be more than enough to mitigate the energy crisis. Even, then we see that renewable energy is still not utilized as per the need of the hour. The following bar graph analysis shows the same in the figure 1. This is the output data of year 2013-2014. Now, government is putting emphasis to utilize this energy and putting forward so many plans to promote it commercially as well as promote its R&D.

[a]Corresponding author: pps11may76@gmail.com

Government with the schemes like Jawaharlal Nehru solar mission and subsidies encourages it. According to Jawaharlal Nehru National Solar Mission (JNNSM), India aims for setting up 100GW capacity in solar energy which includes 40GW as rooftop type and 60GW as ground mounted type.

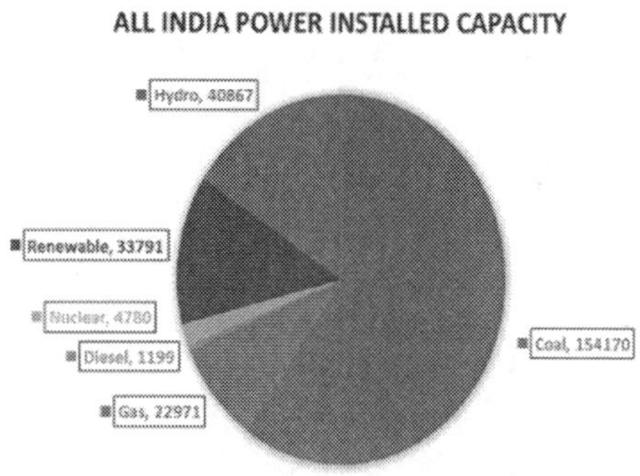

Figure 1. All India Power Installed Capacity [1]

 DOI: 10.1051/matecconf/20165701029

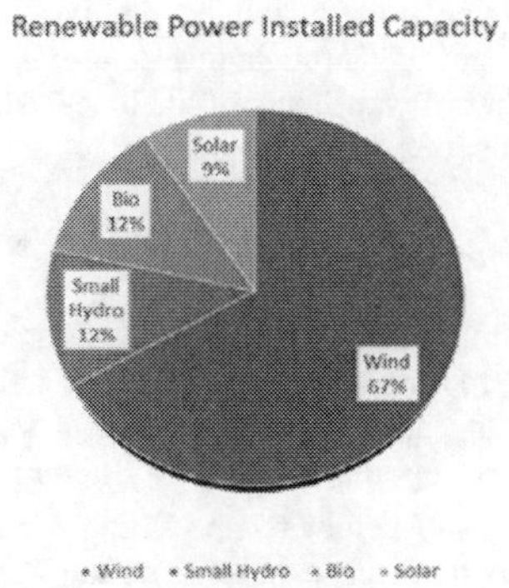

Figure 2. Renewable Energy Analysis [2]

Thus, the renewable energy has attracted many investors, primarily due to government policies and the return benefits. The renewable revolution has been set in motion but it still needs time to catch up with existing conventional electricity generation plants. The renewable energy analysis is shown in the figure 2.

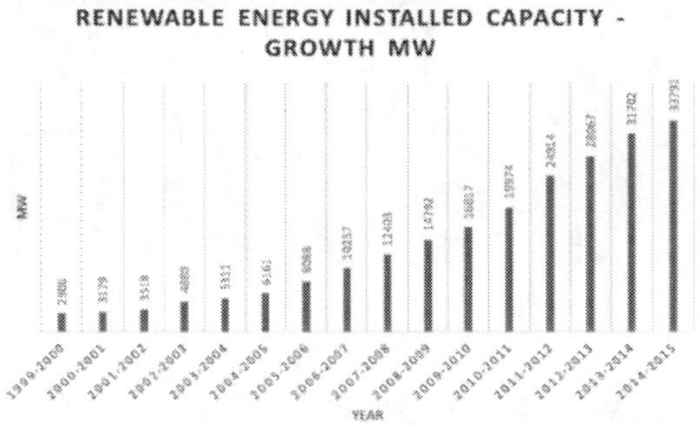

Figure 3. Renewable Energy Installed Capacity – Growth [3]

This work deals with the design and feasibility of a solar power plant in a academic E-blockof Thapar university to promote and aware about the advantages of use of solar energy in order to help India achieve its ambitious target of 100GW solar power and boost the renewable energy growth trend as shown in the figure 3.

2. Description of study area of academic of E-Block

Thapar University, one of the best technical university in the region is located in Patiala, Punjab and is a lush green campus of about 250 acre as shown in the figure 4. It has several academic blocks and E-block is one of the block which is a building consisting of 2-floors and 16 lecture halls. The block is so located, that it's rooftop does have very minimal shadow areas but has good exposure to sun irradiance during the operating hours i.e. lecture hours.

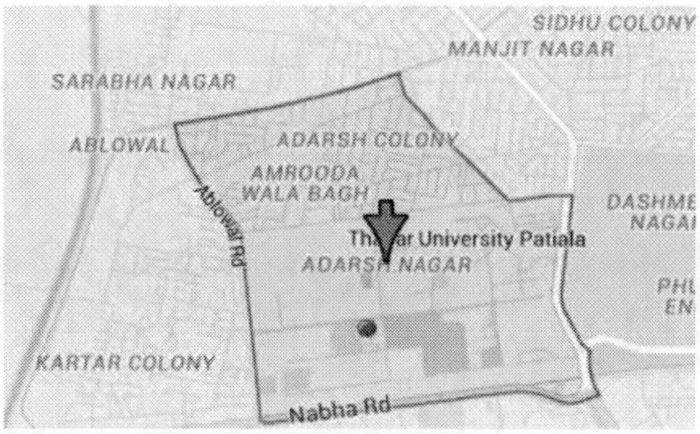

Figure4. Location OfThapar University On Map

The co-ordinates of the block are-
Latitude - 30°21′12.62″N and Longitude - 76°22′20.15″EThe top view of the block is shown in the figure 5.The block is a academic block with maximum load as lighting load or fan load.

Figure 5. Top View Of E-Block

3. Technical feasibility of the project

The following section discusses the technical feasibility of the project. The technical feasibility includes the following factors-
i)Irradiation Analysis
ii)Connected Load Analysis
iii)Rooftop Analysis
iv)Shadow Analysis
v)Load Analysis
vi)Design of on-grid system
vii)Design of off-grid system

3.1 Irradiation Analysis

The figure 6 gives the solar irradiation at the location throughout the year. The average irradiance comes out to be 5.28 KWh/m^2/day. The average solar power plant takes 9.3 m^2 for 1KW and plant runs for about 6 hours. Thus producing 6 units (KWh) in a day. Therefore, For 9.3 m^2 it produces 6 units in a day &
For 1 m^2 it will produce 6/9.3 units in a day.

 DOI: 10.1051/matecconf/20165701029

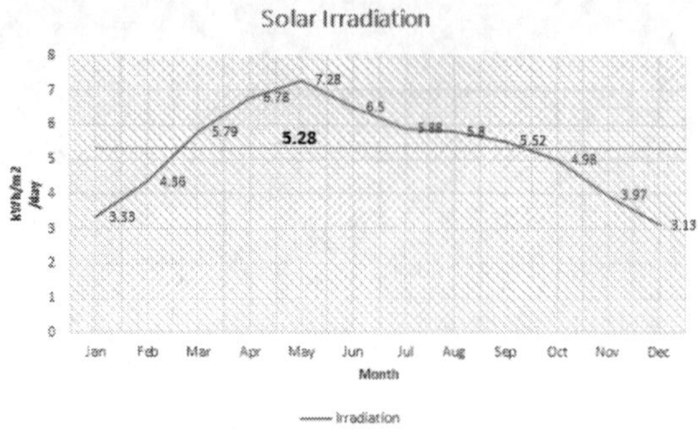

Figure 6. Irradiation Analysis Of Patiala[4]

Therefore, the plant will make 0.645 $KWh/m^2/day$ which is way below the solar irradiance in every month. Hence,

Connected Load Analysis - E Block

Area	Tubelight	Fan	Projector	LED Lamp	Exhaust Fan	Water Cooler	Repeater	CCTV	5 Amp Sw	16 Amp Switch		Total (W)
Load Rating (W)	40	60	300	15	350	1200	10	POE	100	1000		
E Block	180	110	18	8	0	2	8	2	108	22		54600
% Running	50	60	25		100							
No of Hours	5	5	5	10	12	15	15	24	0	0	Wh	
Watt-hours	18000	19800	6750	1200	0	36000	1200	0	0	0	82950	
Ground Floor	76	48	8	8	0	1	4	2	40	10		
E 101	8	6	1						5	1		
E 102	8	6	1						5	1		
E 103	8	6	1						5	1		
E 104	8	6	1						5	1		
E 105	8	6	1						5	1		
E 106	8	6	1						5	1		
E 107	8	6	1						5	1		
E 108	8	6	1						5	1		
Wash Room Ladies	2									1		
Wash Room Gents	2									1		
Corridor	8			8		1	4	2				
Total (W)	3040	2880	2400	120	0	1200	40	POE	4000	10000		23680
First Floor	104	62	10	0	0	1	4	0	68	12		
E 201	6	5	1						6	1		
E 202	16	6							4	1		
E 203	6	5	1						6	1		
E 204	6	5	1						6	1		
E 205	6	5	1						6	1		
E 206	6	5	1						6	1		
E 207	6	5	1						6	1		
E 208	16	6							4	1		
E 209	6	5	1						6	1		
E 210	6	5	1						6	1		
E 211	6	5	1						6	1		
E 212	6	5	1						6	1		
Wash Room Ladies	2											
Wash Room Gents	2											
Corridor	8					1	4					
Total (W)	4160	3720	3000	0	0	1200	40		6800	12000		30920

Figure .7 Connected Load Analysis Of E-Block

plant will get the adequate irradiance for maximum performance throughout the year. Yet, there is difference in performance of the plant throughout the year due to the difference in temperature which affects the performance of the plant.

3.2 Connected Load Analysis

The figure 7 shows the connected load analysis of the block. In this, we counted each and every load of the block. The total connected load of the block came out to be = **54.6 KW**
For the first floor it came out to be = **30.92 KW**
For the ground floor it came out to be = **23.68KW**

Hence, if everything will be running, it will withdraw 54.6 KW power. Through, this we also got to know about the number of watt-hours of the block. This is needed for the calculations of the off-grid system.

3.3 Rooftop Analysis

The rooftop analysis of the block is done to check the available rooftop area and the shadow area and to find the south-facing direction as the south facing installation of panels gives optimized result in- Building height=10 m &Parafitt wall height = 1m&Width = 250m. Also, it is suggested to place panels at an angle of 30° for best results and to capture maximum power. The CAD drawing of the block is given in the figure 8.

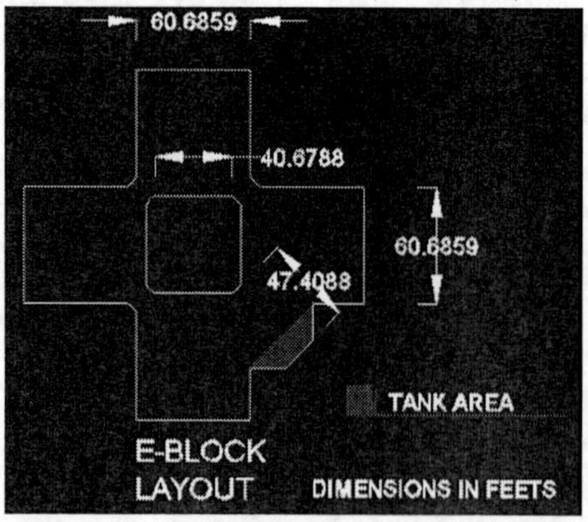

Figure 8. CAD Drawing Of E-Block

3.4 Shadow Analysis

It was observed that some of the area has shadow of nearby tress, parafitt and tank during the operating hours. This usable area is estimated about 52000 ft^2. As 1KW plant requires about 100 ft^2 area. Thus, **50 KW** can be proposed on the rooftop i.e. its maximum capacity.

3.5 Load Analysis

The load analysis of the block is very essential which is carried out bybar graph based on the university's electricity bills as shown in the figure 9. From load analysis we conclude that peak consumption is in September of 2846 KWh and minimum consumption is in February of 1176 KWh and also we calculated average consumption from this.

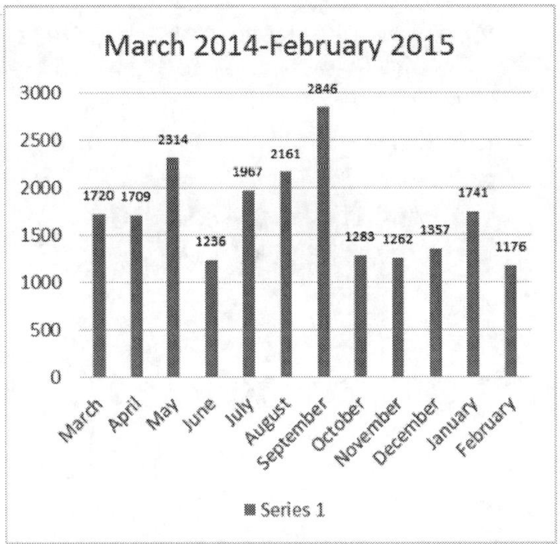

Figure 9. Load Analysis Of E-Block

3.6 Design of on-grid system

Calculations for the design-
Average monthly consumption of the block = ((1720+1709+2314+1236+1967+2161+2846+1283+1262+1357+1741+1176)/12)= 1731 KWh & If we assume 320 sunny days in a year, then in a month there will be about 320/12 = 26 sunny days. We are assuming that a plant works for 6 hours in a day. Number of operational hours in a month = 156 hours. Plant needed for average monthly consumption = 1731/156 ≈ 12KW&Plant needed for peak consumption = 2846/156 ≈ 19KW&Plant needed for minimum consumption = 1176/156 ≈8KW&Maximum capacity of the block = 50 KW. The summary of the points of design are given in figure 10.

Concluding Remarks For Design-

If the consumer is interested in exporting the energy to grid, then it should go for the maximum capacity block can accommodate. Here, it is 50 KW. If, a consumer doesn't want to export then, it should go with the average consumption which is 12 KW. The plant should be installed in 3 arrays of 4 KW each, so that when consumption is 8 KW, one of the array should be turned off. Here, in our design we will be assuming it to be of 50 KW as the university wants to export the energy to other blocks.

DATA OF E-BLOCK	
Average yearly consumption of block	20772 KWh
Average monthly consumption of the block	1731 KWh
Peak load of the block for a month	2846 KWh (Sept.)
Minimum load of the block for a month	1176 KWh (February)
Plant needed for average monthly consumption	12 KW
Plant needed for peak consumption	19 KW
Plant needed for minimum consumption	8 KW
Area needed per KW	100 ft^2
Approx. Rooftop area of block per roof	50000 ft^2
Maximum capacity of the block	50 KW

Figure 10. Summary Of Points Of Design

Design for 50 KWp system-
Panel rating = 250W, 24V&Rating of grid-tie inverter = 50KW. The general voltage rating of grid tie-inverter of 50 KW is 360 V. Hence array of 15, 250W panels will be in series.The CAD of installation is shown in the figure 11.The block diagram of installation is shoen in figure 12.

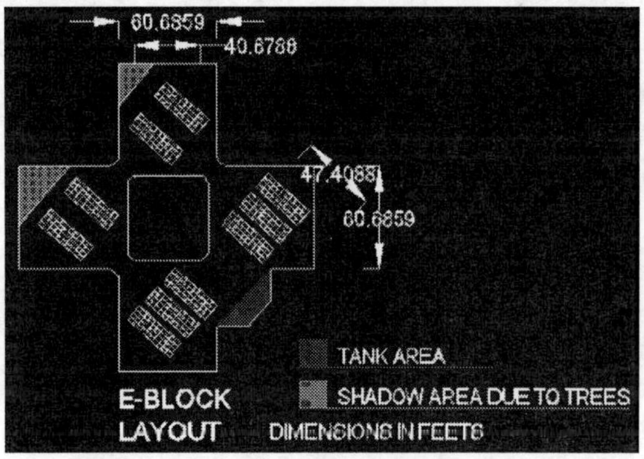

Figure 11.CAD Of The 50 KWp Installation

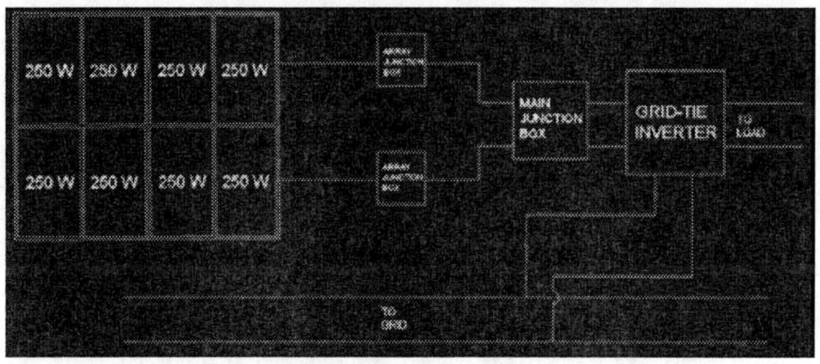

Figure 12. Block Diagram Of On-Grid Installation

3.7 Design of off-grid system

For the, off-grid system, the design calculations are-
Number of watt-hours = 82,950 KWh. Assuming, to use 12V batteries and multiplying 1.25 to accommodatedeepdischargeeffect in the batteries, the Ah rating of the batteryis = 82950/12*1.25= 8641 Ah ≈ **9000 Ah**. The battery bank should be designed in such a way to give9000 Ah rating.Using the thumb rule that, Plant rating = 2*battery rating= 2*9000 = 18000W≈**18 KW.** Hence, 18 KW willbe the apt size for the off-grid system.Voltage rating of charge controller = **12V**&Current rating of charge controller = 18000/12 = **1500 A**
The arrays should be made of the panels, subject to availability and rating of the charge controller and rating of the inverter. The block diagram of installation is shown in figure 13.

Induction loadingeffect for the design of inverter – Due, to high power drawn by the induction motorloads during starting or transients, the inverter is designed to handle these.For this, we again add the watt-hours by the induction motorloads. Thus, new watt-hours = 82950+19800+36000= 138750≈15KW which is less than 18 KW.
Rating of inverter = **18KW**

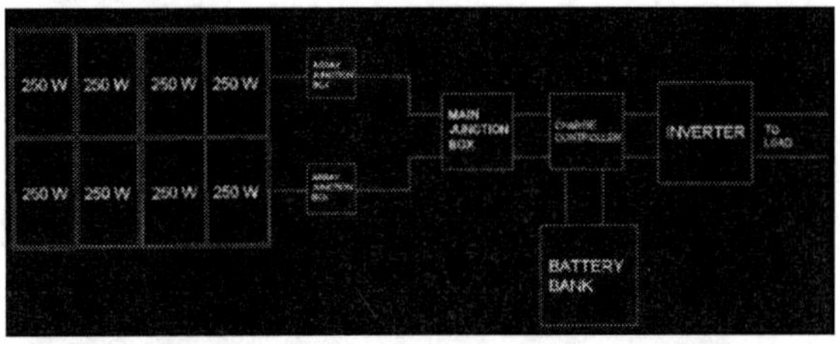

Figure 13. Block Diagram of Off-Grid Installation

4 Finanacial feasibility of the project

Most important is the financial feasibility of project. Here, also all the factors which could affect financial feasibility are considered as follows-
1) Cost analysis
2) Payback period
3) Funding options
4) Carbon points and carbon trading analysis

4.1 Cost Analysis

The cost analysis for 50 KW on-grid project has Approx. expenditure of the installation and maintainence asRs. 50 Lakh. Approx. units produced in a day will be = 50*6 =300Kwh (Assuming 6 hours operation of the plant)&Yearly units (assuming 320 sunny days) = 300*320 = 96,000 units &Yearly savings (assuming Rs. 7.5/unit rate) = 7,20,000.

4.2 Payback period

The investment has to be justified by its payback period which tells us savings on the part of the university. The payback comes out to be = 5000000/720000 = 6.94 years (without accelerated depreciation)&Profit in 25 years = (25-7)*720000 =1,29,60,000. The figure 14 tells us the trend of payback vs size of the plant. The figure 15 shows us the profit in 25 years vs size of the plant.

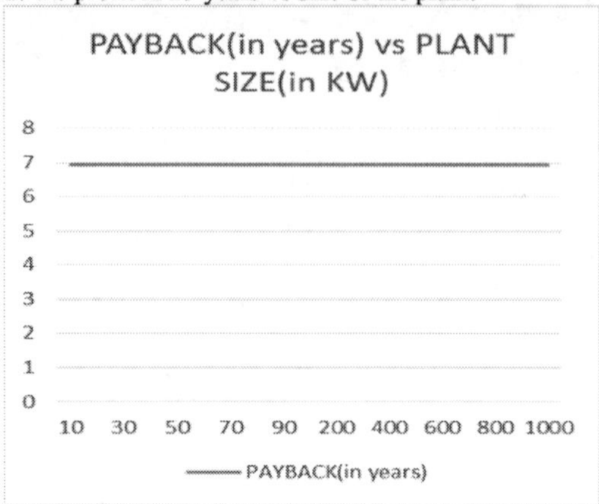

Figure 14. Payback (In Years) Vs Plant Size (In KW)

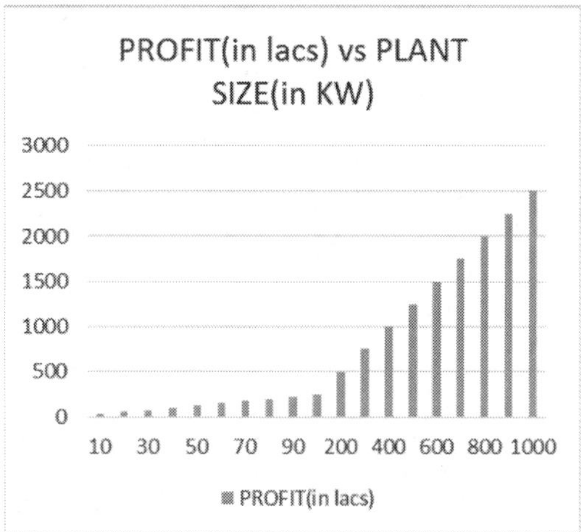

Figure 15. Profit In 25 Years (In Lacs) Vs Plant Size (In KW)

4.3 Funding Options

The various funding options for this project are-
(i)EPC (Engineering, Procurement and Construction)
(ii)PPA (Power Purchase Agreement)
(iii)CSR (Corporate Social Responsibility)
(iv)Organizations like MNRE, World bank, etc.

5 Conclusion

A 50KWp on-grid type solar power plant should be installed in the rooftop of academic E-block which will produce approximately 300 units in a day resulting in the optimal use of resources available in the form of solar energy for the grid and load during the operating hours. In our case university, wants to export the energy to the grid so on-grid system will be more beneficial than the off-grid system. This will also result in the payback of investment in 7 years and will yield university an approx. profit of Rs. 1.3 Crore in next 25 years.

References

1. http://mnre.gov.in/file-manager/annual-report/2014-2015/EN/Chapter%201/chapter_1.htm
2. http://mnre.gov.in/file-manager/annual-report/2014-2015/EN/Chapter%201/chapter_1.htm
3. http://mnre.gov.in/file-manager/annual-report/2014-2015/EN/Chapter%201/chapter_1.htm
4. Souvik Ganguli, Jasvir Singh, Etspg, *International Journal Of Applied Engineering Resea rch, Dindigul* , **Volume 1, No.2,** 255 (2010)
5. H. Hippert, C. Pedreira, and R. Souza, "Neural networks for short-term load forecasting: A review and evaluation," *IEEE Trans. Power Syst.*, vol. 16, no. 1, pp. 44 – 55, (2001)
6. S. J. Huang and K. R. Shih, "Short-term load forecasting via ARMA model identification including non Gaussian process considerations," *IEEE Trans. Power Syst*, vol. 18, no. 2, pp. 673–679, (2003)
7. J. W. Taylor, "Short-term load electricity demand forecasting using double seasonal exponential smoothing," *Journal of the Operational Research Society*, vol. 54, pp. 799–805, (2003)
8. M. S. Kandil, S. M. El-Debeiky, and N. E. Hasanien, "Long-term load forecasting for fast developing utility using a knowledge-based expert system," *IEEE Trans. Power System*, vol. 17, no. 2, pp. 491 – 496 (2002)
9. K. B. Song, Y. S. Baek, D. H. Hong, and G. Jang, "Short-term load forecasting for the holidays using fuzzy linear regression method," *IEEE Trans. Power System*, vol. 20, no. 1, pp. 96–101, (2005)
10. S.Fan and L. Chen, "Short-term load forecasting based on an adaptive hybrid method," IEEE Trans. Power System, vol. 21, no. 1, pp. 392–401, (2006)

DOI: 10.1051/matecconf/20165702001

AODV Routing Protocol Modification With Dqueue(dqAODV) and Optimization With Neural Network For VANET In City Scenario

Soumen Saha[1,a], Utpal Roy[2] and D.D. Sinha[3]

[1] *Department of CSE, University of Calcutta, Kolkata, West Bengal, India*
[2]*Department of CSS, Siksha-Bhavana, Visva-Bharati, West Bengal, India*
[3]*Department of CSE, University of Calcutta, Kolkata, West Bengal, India*

Abstract. Vehicular ad hoc network (VANET) is considered as a sub-set of mobile ad hoc network (MANET). VANET can provide road safety by generating collision warning messages before a collision takes place, lane change assistance; can provide efficient traffic system by introducing cooperation among vehicles; and can also improves in infotainment applications like cooperative file accessing, accessing internet, viewing movies etc. It provides smart Transportation System i.e., wireless ad-hoc communication among vehicles and vehicle to roadside equipments. VANET communication broadly distinguished in two types; 1) vehicle to vehicle interaction, 2) vehicle to infrastructure interaction. The main objective of VANET is to provide safe, secure and automated traffic system. For this automated traffic techniques, there are several types of routing protocols has been developed. MANET routing protocols are not equally applicable in VANET. In the recent past Roy and his group has proposed several study in VANET transmission in [1-3]. In this study, we propose a modified AODV routing protocol in the context of VANET with the help of dqueue introduction into the RREQ header. Recently Saha et al [4] has reported the results showing the nature of modified AODV obtained from the rudimentary version of their simulation code. It is mainly based on packet delivery throughput. It shows greater in-throughput information of packet transmission compare to original AODV. Hence our proposal has less overhead and greater performance routing algorithm compared to conventional AODV. In this study, we propose and implement in the NCTUns-6.0 simulator, the neural network based modified dqueue AODV (dqAODV) routing protocol considering Power, TTL, Node distance and Payload parameter to find the optimal route from the source station (vehicle) to the destination station in VANET communications. The detail simulation techniques with result and output will be presented in the conference.

1 Introduction

For VANET the routing protocols works on ad-hoc basis and infrastructure basis in the network [Fig.1].Where the ad-hoc network is highly unstable as vehicle's speed and lane change factors.

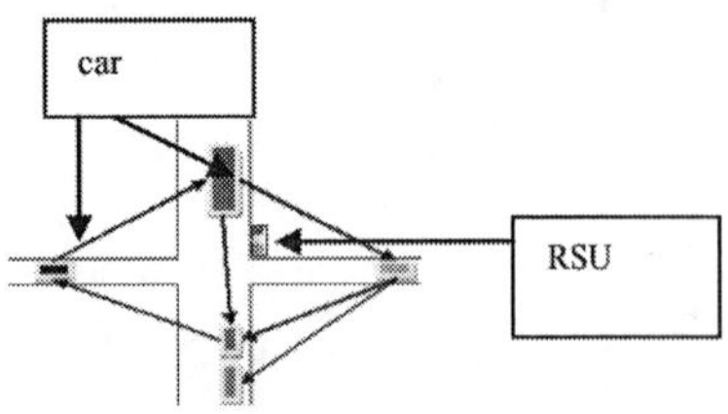

Fig.1. Ad-hoc VANET[5]

1.1 Ad-hoc Routing protocols

Ad-hoc routing protocol fist setup the path next exchange information with packets and take decision of runtime alternatives paths [6].

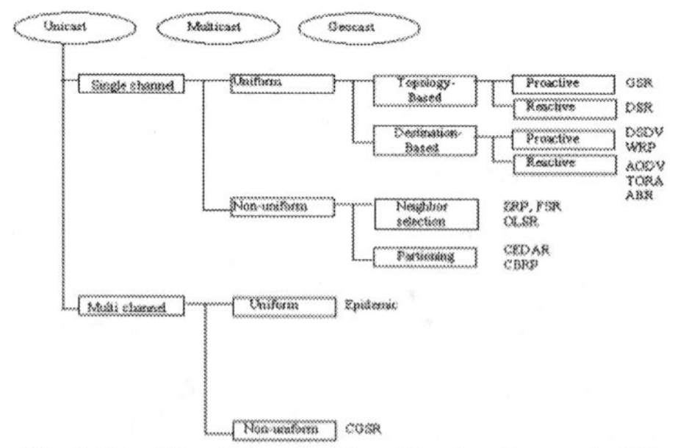

Fig.2. Classification of Ad-hoc Routing Protocols [6]

The topology based routing is classified [Fig2] in to three ways

- Proactive (table-driven) routing protocols.
- Reactive (on-demand) routing protocols.

[a]Soumen Saha: soumen11@rediffmail.com,

•Hybrid routing protocols (for both type).

1.1.2Proactive Routing

Proactive routing protocols are based on shortest path first algorithms [6]. It maintains and update routing information's on routing in between all nodes of a supplied network at all times, even if the paths are not currently being used. Even if some paths are never used but updates for those paths are constantly broadcasted among nodes [6]. Route updates are periodically performed regardless of network payload, bandwidth constraints.

1.1.3Reactive Routing

On demand or reactive routing protocols were planned to overcome the overhead problem, which was created by proactive routing protocols. Maintaining only those routes that are currently live and active [6]. These protocols implement route determination on a demand basis or need basis and maintain only the routes that are currently in use. Therefore it reducing the burden and overhead on the network when only a subset of available routes is in active at any point of time [6].

AODV maintains and uses an efficient method of routing, which reduces network burden by broadcasting route discovery packet mechanism and by runtime updating routing information at each adjacent node. Route discovery in AODV can be perform by sending RREQ (Route Request) from a node when it needs a route to send the data to a particular destination. After sending RREQ, a node waits for the RREP (Route Reply) and if it does not receive any RREP within time threshold.

The node members of contracted ad-hoc network when out of the range of the existing ad-hoc network, it may fails to progress. Hence, we need some other helping equipments (road side equipment) to help those node (Vehicle) to progress. But, irrespective of that, if we taken the existing neighbor Ad-hoc network, that can help to restart the communication with that isolate node(Vehicle),which is more economic, as we do not need any extra equipment or extra data communication.

2 Proposed Work:

Our previous proposed modified AODV routing protocol by implementing a dqueue (dqAODV)[7] is on the basis of packet collision, packet drop and in-out throughput. The fact is that when a intermediate node between source and destination gets some packet from the another node which had got from any other node will insert the node's IP to the generated dqueue instead of discarded. During unicast the RREP packet if any link is breakage, then this node pick out the another node from the created dqueue and re-unicast the RREP packet .

Proposed Artificial Intelligence Based Modified AODV Routing Protocol in VANET

In AODV protocol, a source station (vehicle) initiates a Route Request (RREQ) in the network for connecting to a destination station (node), the route is determined considering the four attributes or parameters like the distance (D), the overload or overhead (O), the consumption of electric power (P), and the expected time (T) to remain the route in alive (active) condition. First three attributes (D, O, P) are acceptable for lesser or minimum value and the fourth attribute expected time is accomplished for larger value, i.e., longer period. Therefore, the normalized expected time period is deducted from one (1) to bring homogeneity among all attributes. There may be different routes under AODV protocol available from the source station to the destination station having one set of values D, O, P, and T for each route. Now the best AODV route is selected among the different routes by applying fuzzy neural network algorithm [8-10][11].

2.1 Modified AODV Route Searching by Fuzzy Neural Network

P.K. Bhattacharjee , et. Al [11] have propose a optimization technique of routing parameters using fuzzy neural network. According their aproch we have optimize the routing parameter for our experiment.

ASA1: The normalized value of a distance (d1) is equal to D_1/D_{max}, where D_1 is the distance for a particular route, and D_{max} is the maximum available distance among the all routes in the route discovery process. If the normalized route distance for a specific path is taken as d1, the membership functions of a fuzzy set F1 is defined as follows, $\mu_{F1}(a1) = d1$, hence, F1 = {(a1, d1)}.

ASA2: The normalized value of an Overhead (h1) is equal to O_1/O_{max}, where O_1 is the overhead or Overload for the specific route, and O_{max} is the maximum available Overhead among the all routes in the route discovery process. If the normalized overhead value is taken as h1 for the particular route, The membership functions of a fuzzy set F2 is defined as follows, $\mu_{F2}(a2) = h1$,hence, F2 = {(a2, h1)}

ASA3: The normalized value of consumed electric power (p1) is equal to P_1/P_{max}, where P_1 is the electric power consumed for the route, and P_{max} is the maximum electric power consumed for a route in the route discovery process. If the normalized electric power consumed value is taken as p1 for the specific route, the membership functions of a fuzzy set F3 is defined as follows, $\mu_{F3}(a3) = p1$, hence, F3 = {(a3, p1)}.

ASA4: The normalized value of a time (t_n) available is equal to T_1/T_{max}, where T_1 is the expected time allocated for the specific route, and T_{max} is the maximum available expected time for a route in the route discovery process. Since the normalized available time is preferred for larger value; the normalized unavailable time (t1) which is favored for lesser value, is taken as the fourth attribute to bring the homogeneity with all other attributes, then t1 is better for lesser value,

So, t1 = (1 – t_n). If the normalized time is taken as t1 for the particular route, the membership functions of a fuzzy set F4 is defined as follows, $\mu_{F4}(a4)$ = t1, hence, F4 = {(a4, t1)}.

ASA5: Now, the fuzzy operations such as fuzzy set intersection (minimum) and union (maximum) taking three fuzzy membership functions at a time out of total four fuzzy membership functions[Fig. 3]; the four different values of each fuzzy operation such as fuzzy set intersection or union are obtained as mentioned below:

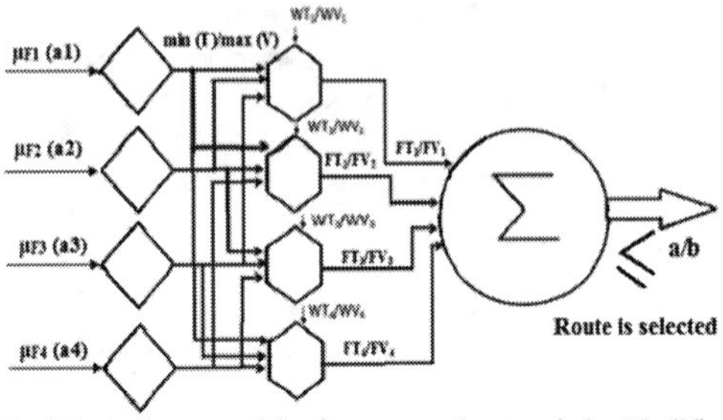

Fig.3. Block diagram of the fuzzy neural network for Modified AODV (Ai-dqAODV) Routing Protocol[11]

ASA6: For ascertaining the best or the optimum route under AODV routing protocol, fuzzy neural network algorithm on the results of the fuzzy operations have been applied.

Different weightages to these fuzzy operations (intersection and union) are imposed and these weightages are assigned by altering different values in practical examples and the best values are considered.

$WT_1 : WT_2 : WT_3 : WT_4 = 0.5 : 0.45 : 0.42 : 0.4$

$WV_1 : WV_2 : WV_3 : WV_4 = 0.9 : 0.85 : 0.83 : 0.8$

The values of the fuzzy operations are multiplied by the corresponding weightages for computing the optimum or the final values, i.e.,

$FT_1 : FT_2 : FT_3 : FT_4 = T_1 \times WT_1 : T_2 \times WT_2 : T_3 \times WT_3 : T_4 \times WT_4 = 0.5T_1 : 0.45T_2 : 0.42T_3 : 0.4T_4$

$FV_1 : FV_2 : FV_3 : FV_4 = V_1 \times WV_1 : V_2 \times WV_2 : V_3 \times WV_3 : V_4 \times WV_4 = 0.9V_1 : 0.85V_2 : 0.83V_3 : 0.8V_4$

ASA7: All the final values of a particular fuzzy operation are defuzzified by a defuzzifying function. Defuzzification is done by the Composite Maxima method, i.e., $\max(FT_1, FT_2, FT_3, FT_4) = a$, and $\max(FV_1, FV_2, FV_3, FV_4) = b$.

ASA8: The fuzzy-neural rule on the results of the final defuzzified outputs are determined according to examine different values on the practical examples, and then the best suited values are taken.

Thus as per fuzzy neural rule, if $a \leq 0.21$, and $b \leq 0.54$ both satisfies, then it ensures that the route is the best route under M-AODV protocol; otherwise, not. Then the data from the other available routes are to be tested accordingly. Here the value of TTL can be optimized by getting the best route immediately. The block diagram of the fuzzy neural network in selecting the best route under Modified AODV protocol is described in Fig. 1.

3 Implementation Results and Discussion

Now applying fuzzy-neural rule, $\max(FT_1, FT_2, FT_3, FT_4) = 0.175$, i.e., ≤ 0.21 and $\max(FV_1, FV_2, FV_3, FV_4) = 0.4335$, i.e., ≤ 0.54, therefore, the route detected under M-AODV scheme is accepted and may be used for traffic (data) flow. Thus, TTL must possess the least value, as the best route is obtained.

3.1 Simulation

In this study, we used NCTUns-6.0[12] for simulation. We have chosen this simulator because,

- Highly integrated and professional GUI environment.
- Support for various network protocols.
- Support for various important network.
- Same configuration and operations as for real life networks.
- High simulation speed and repeatable simulation result.
- High fidelity simulation results.

3.1.1 Performance metrics

Different performance metrics are used to check the performance of routing protocols in various network environments. In our study we have selected throughput and packet drop to check the performance of VANET routing protocols against each other. The reason for the selection of these performance metrics is to check the performance of routing protocols in highly mobile environment of VANET. Moreover, these performance metrics are used to check the effectiveness of VANET routing protocols i.e. how well the protocol deliver packets and how well the algorithm for a routing protocol performs in order to discover the route towards destination. The selected metrics for routing protocols evaluation are as follows [13,14].

•*Throughput*

Throughput is the average number of successfully delivered data packets on a communication network or network node. In other words throughput describes as the total number of received packets at the destination out of total transmitted packets [13]. Throughput is calculated in bytes/sec or data packets per second. The simulation result for throughput in NCTUns6.0 shows the total received packets at destination in KB/Sec, mathematically throughput is shown as follows:

$$T_s = (P_k * Z) / T_h \ (bytes/sec) \quad (1)$$

P_k : *Total number of received packets at destination*
T_s : *Total simulation time*
Z : *packet size*
T_h : *Throughput*

•*Packet Drop*

Packet drop shows total number of data packets that could not reach destination successfully. The reason for packet drop may arise due to congestion, faulty

DOI: 10.1051/matecconf/20165702001

hardware and queue overflow etc. Packet drop affects the network performance by consuming time and more bandwidth to resend a packet. Lower packet drop rate shows higher protocol performance.

•*Collision*

The Collision of data packet is the number of packets collides to each other due to congestion. It affects the performance directly on the bandwidth. Lower packet collision rate shows higher protocol performance.

3.2. Simulation scenario:

We have taken a mess structured scenario (city scenario of New Delhi[15][Fig 4]) as it was earlier studied, where it gives best performance among other type of city scenario of AODV routing algorithm[13,14]) as our proposed algorithm for routing based on AODV, for dense traffic.

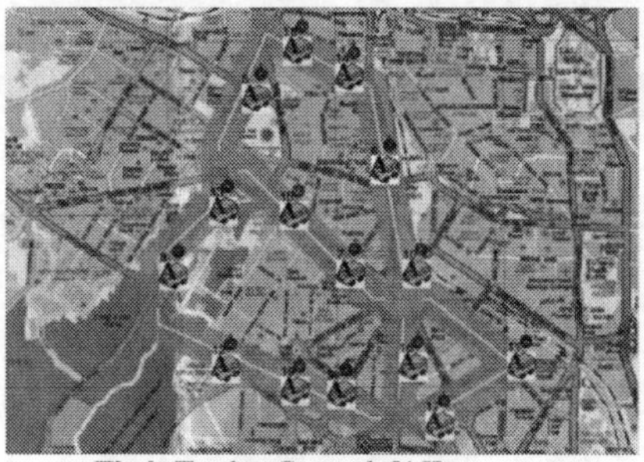

Fig4: Testing Scenario[15]

3.3. Testing parameters:

We have taken the simulation parameters for different routing algorithems in NCTUns simulators are as shown in Table 1 and 2.

Table 1: AODV and dqAODV[7] testing parameters

Parameter	Settings
Transmission mode	TCP/IP
Lane Width	20m
Simulation time	400sec
RTS threshold	3000bytes **(O)**
The car profile (Taken five)	18km/H, 36km/H, 50km/H, 60km/H, 80km/H
Number of lane	2
The protocol	AODV, dqAODV
standard used for each vehicular node	IEEE802.11b
cars are selected for three different scenarios	10,15,20,25 **(D)**
Transmission power used	15dbm **(P)**
TTL	**7(T)**
Frequency Channel	**3**
Hello interval	**1000**
Route Timeout	**3000**
Hello loss	**2**
Link Bandwidth	**11db**

Table 2: ai-dqAODV testing parameters

Parameter	Settings
Transmission mode	TCP/IP
Lane Width	20m
Simulation time	400sec
RTS threshold	4000bytes **(O)**
The car profile (Taken five)	18km/H, 36km/H, 50km/H, 60km/H, 80km/H
Number of lane	2
The protocol	Ai-dqAODV
standard used for each vehicular node	IEEE802.11b
cars are selected for three different scenarios	10,15,20,25 **(D)**
Transmission power used	7dbm **(P)**
TTL	**3(T)**
Frequency Channel	**2**
Hello Interval	**2000**
Route Timeout	**4000**
Hello loss	**4**
Link Bandwidth	**7db**

3.4. Results:

3.4.1 Collision graph

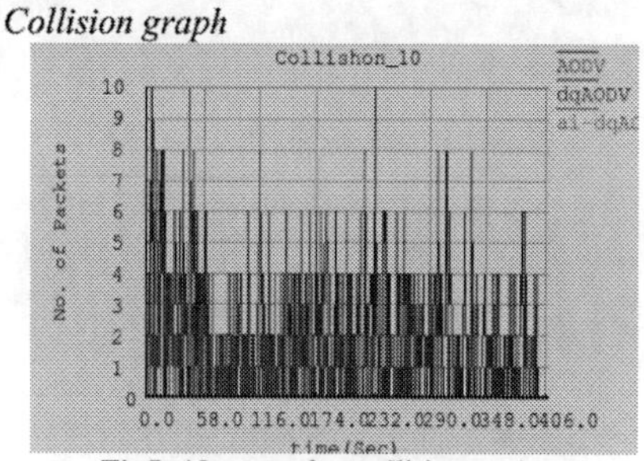

Fig5: 10 car packet collision vs time

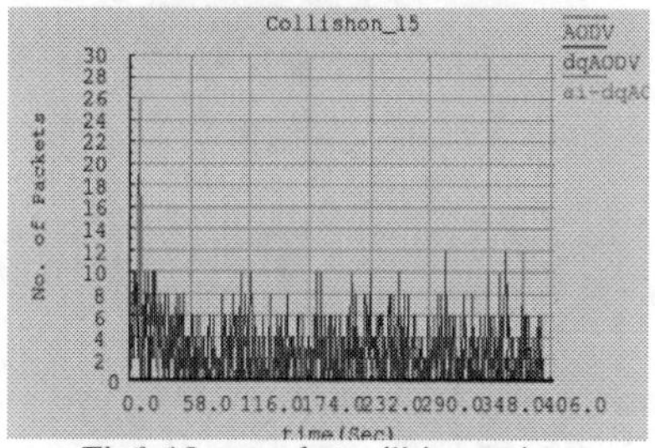

Fig6: 15 car packet collision vs time

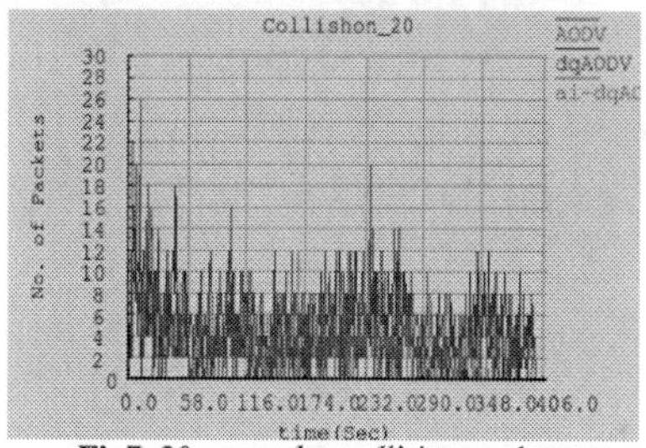

Fig7: 20 car packet collision vs time

DOI: 10.1051/matecconf/20165702001

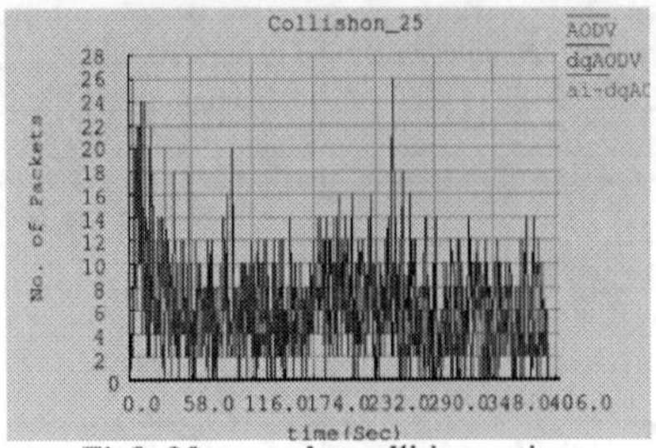

Fig8: 25 car packet collision vs time

3.4.2. Drop graph

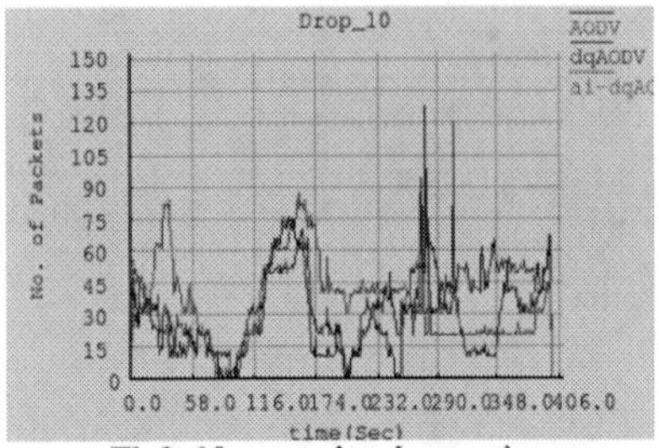

Fig9: 10 car packet drop vs time

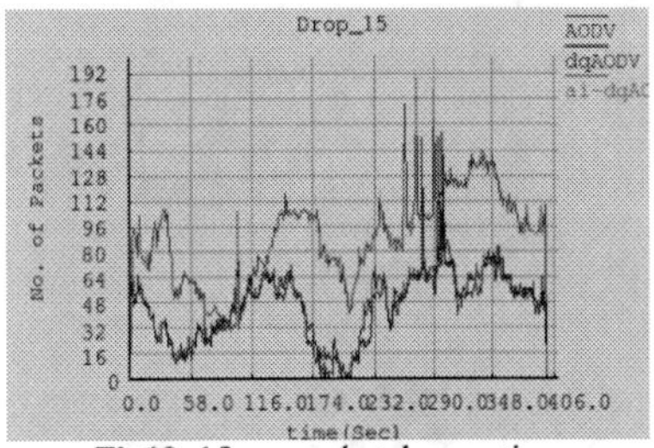

Fig10: 15 car packet drop vs time

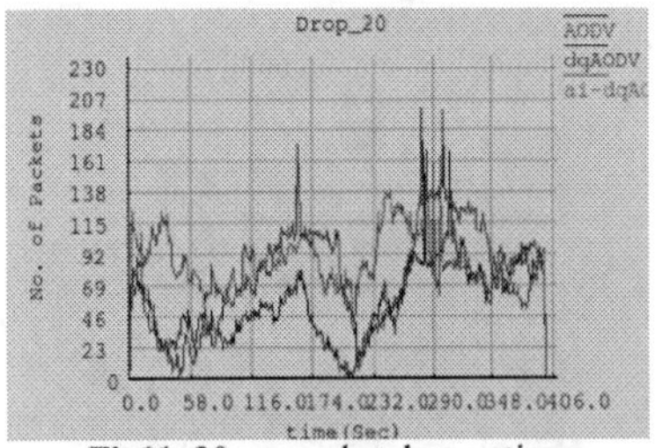

Fig11: 20 car packet drop vs time

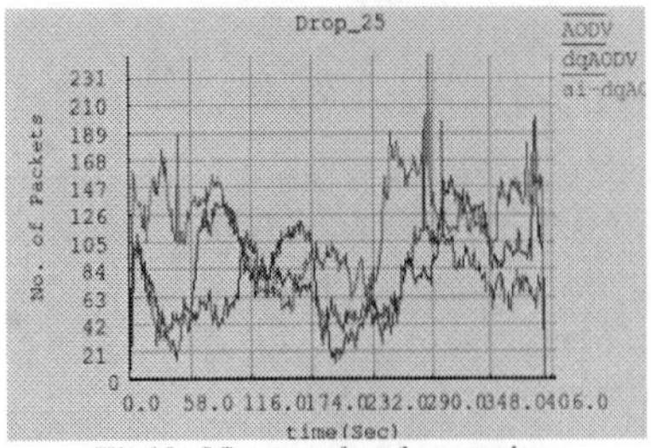

Fig12: 25 car packet drop vs time

3.4.3. Throughput graph

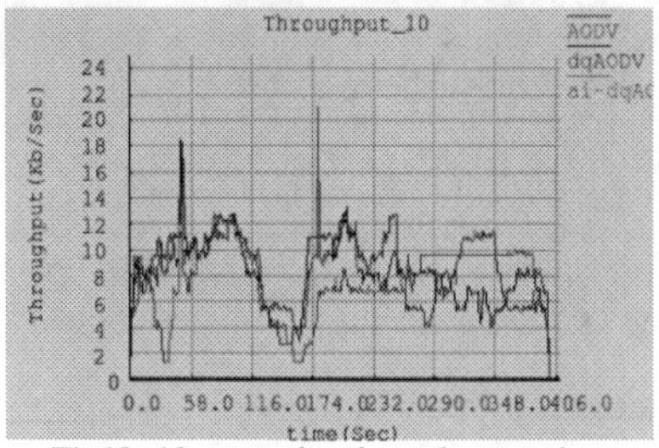

Fig13: 10 car packet throughput vs time

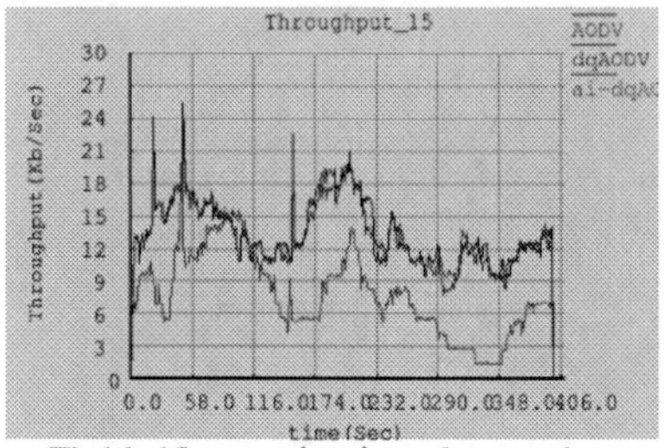

Fig14: 15 car packet throughput vs time

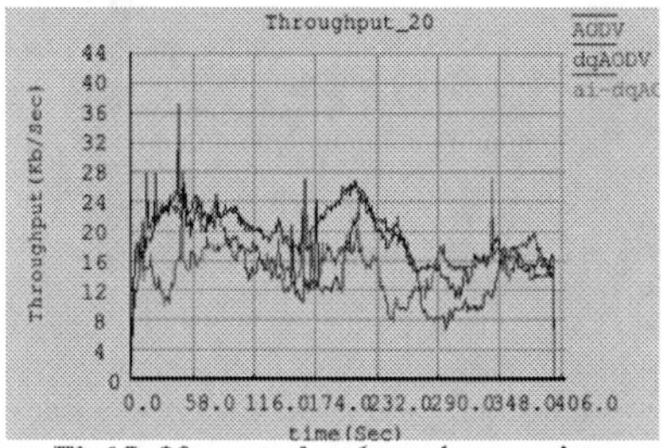

Fig15: 20 car packet throughput vs time

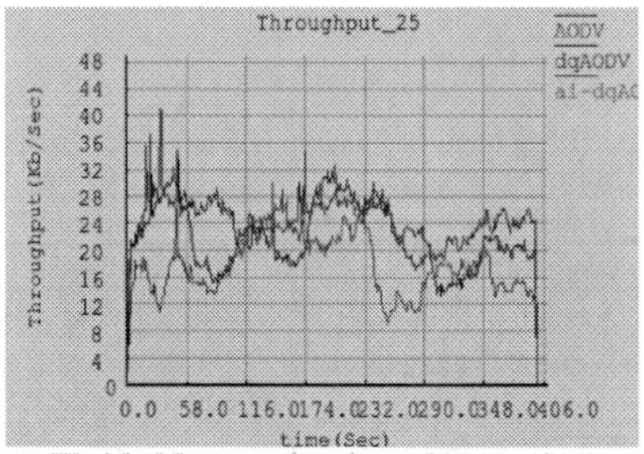

Fig16: 25 car packet throughput vs time

We found the packet in collide is slightly worst in dqAODV than AODV or ai-dqAODV [Fig 5-8]. It causes due to less packet inject in the network of ai-dqAODV. It indicate our proposal of optimization in parameter using neural network is works. Now drop of packets is almost same when car density is low(10) [Fig-9]. But, when car density is higher, the packet drop is considerable high in ai-dqAODV [Fig 10-12]. It indicates some of the routing parameters, like transmission power, etc should be tuned to increase further. Finally packet throughput [Fig.13,14,16] is also comparable with dqAODV and AODV except one situation of car density 15[Fig.15].
Therefore we have got a comparable result with AODV and dqAODV in this proposal. It indicates with reduced

Power(P), Overhead(D), TTL(T) and same distance (D) we got the satisfactory result for less car density. That proves the optimal parameters usage to discover the routing path is successful according to our proposal using neural network for low car density.

4 Advantages of the Modified AODV Routing Protocol

This route discovery technique under Modified AODV protocol is the most efficient one due to applying artificial intelligence [AI] in advanced stage, i.e., fuzzy neural network used. Also it does not require any further information to supply by the source (host) node while making a call. So it is a unique one. The route is determined by the essential attributes or parameters of the AODV protocol. No cryptography algorithm or any complex functions are applied. This Modified AODV protocol technique ensures the fastest best route discovery.

5 Conclusions

In this paper we have proposed to determine the best route from the source node to the destination node through several intermediate nodes using fuzzy neural network algorithm in Modified AODV (aiAODV) routing protocol of VANET communications. We further need to optimize the parameter for better result in dense traffic situation Therefore, the stable connections are set up in a VANET communications by implementing a fast and easy routing techniques like artificial intelligence based Modified AODV routing protocol in the VANET system. There are many variants of routing protocols for VANET transmission have been proposed, those are basically the modified forms of MANET routing protocols. This Modified AODV routing protocol using fuzzy neural network algorithm for the best route searching in VANET is fantastic workable in a real time basis.

References

1. B. Das, and Dr.U. Roy, "Secured Geocast Routing in VANET (Vehicular Ad-Hoc Network) with two Stage Efficient Communication Protocol", International Journal of Computer Applications (IJCA), Vol. – 53, No. – 12, September 2012, pp 34-38, ISBN : 973-93-80870-40-8.
2. B. Das, S. Misra, Dr.U. Roy, and Mohammad S. Obaidat, "Dynamic Relay Selection for MAC-level Retransmission in Vehicular Ad Hoc Networks," Global Telecommunications Conference (GLOBECOM 2013), 2013 IEEE , 9-13 Dec. 2013, Atlanta, US.
3. B. Das and Dr.U. Roy, Cooperative Quantum Key Distribution for Cooperative Service-Message Passing in Vehicular Ad Hoc Networks, International Journal of Computer Applications, ISSN 0975 8887, Volume 102, Number 16, pp 37-42, September 2014.
4. RFC of AODV,DSR: www.ietf.org/rfc/rfc3561.txt, www.ietf.org/rfc/rfc4728.txt
5. S. Saha,Dr. U. Roy, Dr. D.D. Sinah, Performance Analysis of VANET Scenario in Ad-hoc Network by NCTUns Simulator, INTERNATIONAL CONGRESS On "Innovative Trends in Information Technologies and Computing Sciences for Competitive World Order" (ITITCSCWO – 2013), 2-3 March, 2013, New Delhi ,JNU and followed by IJICT (ISSN 0974-2239) vol-3, number 7 , pp575-581.
6. "Classification of Ad Hoc Routing Protocols"Petteri Kuosmanen, Finnish Defence ForcesNaval Academy,Helsinki, Finland petteri.kuosmanen@mil.fi
7. S. Saha,Dr. U. Roy, Dr. D.D. Sinah, "AODV ROUTING PROTOCOL MODIFICATION WITH DQUEUE(dqAODV) FOR VANET CITY SCENARIOS" at ICHPCA-2014, Bhubaneswar, India, pp 1-6, IEEE Xplore Digital Library, December 22-24, 2014, ISBN : 978-1-4799-5957-0;
8. E. Rich, K. Knight, and S. B. Nair, *Artificial Intelligence*, 3rd Edition, Tata McGraw Hill Education Private Ltd., 2010.
9. V.Novak, J.Mockor, and I. Perfilieva, *Mathematical Principles of Fuzzy Logic,* Kluwer Academic Publisher, 2006.
10. J. Yan, M. Ryan, and J. Power, Using Fuzzy Logic: Towards Intelligent Systems, Prentice-Hall of India Pvt. Ltd., 1995.
11. P. K. Bhattacharjee, S. Roy, R. K. Pal, Mutual Authentication Technique with Four Entities Using Fuzzy Neural Network in 4-G Mobile Communications, IOSR Journal of Computer Engineering (IOSR-JCE), PP 69-76,2015
12. NCTUns6.0 protocol developer manual; http://elearning.vtu.ac.in/15/ENotes/NW%20prog%20lab/NCTUns%20Manual.pdf
13. S.Saha, Dr. U. Roy, Dr. D.D. Sinah, VANET Simulation in diffrent Indian City Scenario, In Conference On "Recent Global Trends in Electronic Communication Engineering, Power and Control" (ECEPC-2013), 7-8 September, 2013, New Delhi ,JNU.
14. S. Saha,Dr. U. Roy, Dr. D.D. Sinah, Performance comparison of various Ad-Hoc routing protocols of VANET in Indian City scenario, AIJRSTEM 14-126 ,ISSN (Online): 2328-3580 , pp49-54, March 7, 2014.
15. www.mapsofindia.com

MATEC Web of Conferences 57, 02002 (2016) DOI: 10.1051/matecconf/20165702002
ICAET-2016

A Novel Technique for Shape Feature Extraction Using Content Based Image Retrieval

Jaspreet Singh Dhanoa[1], Anupam Garg [2]

BGIET, Dept. of CSE, Sangrur, India

Abstract. With the advent of technology and multimedia information, digital images are increasing very quickly. Various techniques are being developed to retrieve/search digital information or data contained in the image. Traditional Text Based Image Retrieval System is not plentiful. Since it is time consuming as it require manual image annotation. Also, the image annotation differs with different peoples. An alternate to this is Content Based Image Retrieval (CBIR) system. It retrieves/search for image using its contents rather the text, keywords etc. A lot of exploration has been compassed in the range of Content Based Image Retrieval (CBIR) with various feature extraction techniques. Shape is a significant image feature as it reflects the human perception. Moreover, Shape is quite simple to use by the user to define object in an image as compared to other features such as Color, texture etc. Over and above, if applied alone, no descriptor will give fruitful results. Further, by combining it with an improved classifier, one can use the positive features of both the descriptor and classifier. So, a tryout will be made to establish an algorithm for accurate feature (Shape) extraction in Content Based Image Retrieval (CBIR). The main objectives of this project are: (a) To propose an algorithm for shape feature extraction using CBIR, (b) To evaluate the performance of proposed algorithm and (c To compare the proposed algorithm with state of art techniques.

1 Introduction

Content based image retrieval system is a very important task in digital image processing based systems. Normally the data base of images becomes stacky as the no. of images are added on during the scanning processing like in schools, colleges, municipality, registrar offices etc. This is due to the fact that in each government offices, all the records are being digitized for fast retrieval of the same. This gives rise to stack of images building up in image data base. Now, the problem occurs when images belonging to a query property are sought from the data base. This becomes a very tedious task for a large data base. Here, comes the need of the content based image retrieval system. The image data base can be organized in way so as to flag the images with some keywords. These keywords may in future become the index for cbir systems. However, in the presented approach, the cbir system consists of an engine that itself determines the contents of the image based on input property and from the data base images as well. The data base images those show close resemblances to that of the query image in terms of features set are retrieved and the process is termed as content based image retrieval system.

2 Proposed Algorithm

The proposed work consists of the following steps:

Step I: Image Database generation and ImagAcquisition in Matlab Environment

The image data base is prepared by taking the images of different nature and the same are rotated at different angles for validation and testing of the algorithm. Moreover, the images in the proposed database are different from the base database as they are rotated at different angles ,their size is also increased or decreased. Also, the proposed database includes various other types of images in addition to the base database. The reason for this is to take into account the Scaling, Translation and Orientation invariance.The images are read in jpeg format. Following image show a snap shot of the data base used for the purpose.

Corresponding author: er.jaspreetdhanoa@gmail.com

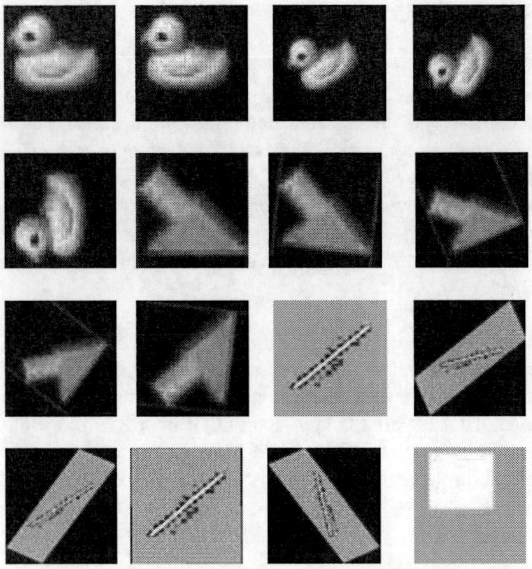

Figure 1. All database images

Step II: Database Images Path Setting

In this step ,the path for various database images and the text file is set. The path for database images is set to fetch the various images from the database. Moreover, the text files are created to store the various results in different text files. Then the DIR command is used with the size command to calculate the total number of images in the image database. Further, the various arrays are created to store the corresponding values during program execution. Also, as the program uses 22 features for extraction , the various vaulues are set for diffrent parameters.

Step III: Enter the Query Image, Margin and Threshold

The Query Image number , Margin and Threshold is dynamically entered by the user. Image database consisits of total 45 different image. The margin is ranging from 0 to 100 and the threshold from 1 to 256. The user enter the various values according to his/her choice.

Step IV: Convert RGB to Gray Scale.

In this step the original image i.e RGB image or Gray Scale is converted into Gray image as a preprocessing technique. By nature, the jpeg image format is in 24-bit color. The jpeg image format is converted to gray scale conversion using rgb2gray() command in matlab. Thereby giving the gray scale i.e. 8-bit image.

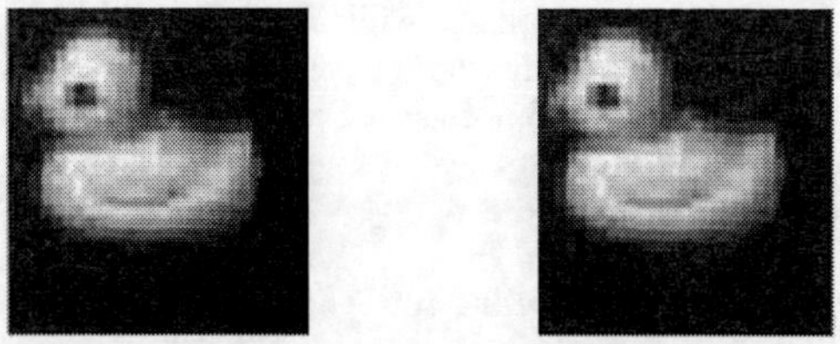

(a) (b)

Figure 2. (a) Original Image And (b) its Gray Scale

Step V: Image Binarization

The gray image is converted into binary image by threshold decomposition, thereby isolating the foreground comprising the yarn from the background. This binary image is further used for feature extraction. Most of the histogram based thresholding techniques work on the assumption that foreground area is comparable to that of background area. Generally in a yarn image the mode of foreground is much less compared to that of the background. Hence, in order to make foreground mode comparable to that of the background, attenuation of the background mode is performed on the histogram H of the cleaned image. Local and global thresholding approaches may be utilized to produce a binary image from the gray scale image.

(a) (b)

Figure 3. (a) Binary Image and (b) its Inverted Binary Image

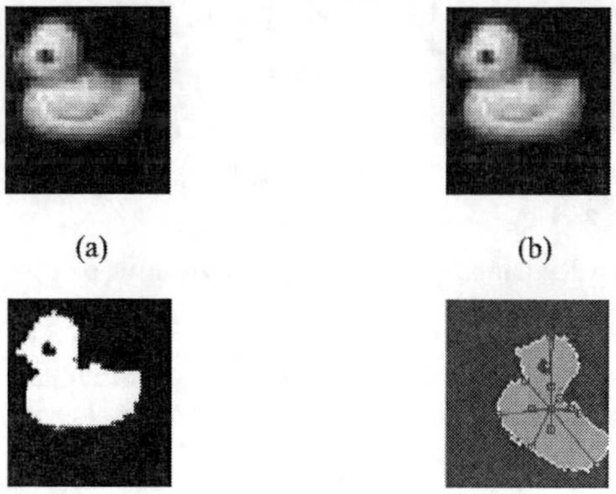

(a) (b)

Figure 4. (a) Original (b) Gray scale (c) Binary And (d) Inverted Binary Image

The proposed algorithm uses dynamic thresholding to meet the user needs. Basically the thresholding ranges from 1 to 255. The algorithms will not work properly for threshold >=200 as the threshold increases it will result in information loss for darker images. So , for darker images the threshold must be low and for lighter images it may be high.

Step VI: Statistical Feature Set Generation

Basically, the base algorithm[1] only uses 5 features but in the proposed algorithm, a total 22 features are extracted from the database images including the query image namely: Area, Figure Aspect, Roundness, Compactness, Elongation, Centre of gravity(COG), Perimeter, R1 (Radius for quadrant 1) ,R2 (Radius for quadrant 2),R3 (Radius for quadrant 3),R3 (Radius for quadrant 4),Max R (Maximum radius),R1(Radius for quadrant 1),R2(Radius for quadrant 2),R3(Radius for quadrant 3),R3(Radius for quadrant 4),Min R(Minimum radius),MR (Mean Radius),X1(Distance on X-Axis).X2(Distance on X-Axis),Y1(Distance on Y-Axis),Y2 (Distance on Y-Axis). The number features are increased from 5 to 22 to increase the accuracy of proposed algorithm. As the feature set is increased the retrieval efficiency in terms of accuracy is also increased. The complexity of proposed algorithm is also increased.

Step VII: Feature Normalization

All features are normalized with respect to mean radius of the pattern . It makes all the statistical features independent of size of the pattern. The set of described statistical features may be termed as figures of merit to classify an object.

Step VIII: Compute the Legendre Moments of Feature set.

$$\lambda mn = \frac{(2m+1)(2n+1)}{4}\int_{-1}^{1}\int_{-1}^{1} \mathrm{P}m(x)\mathrm{P}n(y)f(x, y)dxdy$$

Step IX: Compute the standard deviation of Legendre moments of query image with all the data base images.

Step X: The step II and step VII are repeated for all data base and query image.

Step XI: Retrieve the data base images with minimum standard deviation falling in the error margin range.

The error margin must be between 0 to 100. This can be changed according to the requirement. If the error margin is increased then accuracy will decrease accordingly. Moreover , as the margin increases the number of images retrieved will also increases.The retrieved images are then classified into different categories in order to further classify into relevant categories.

Step XII: SVM Classifier

The SVM Classifier is used to classify all the database images and the query image into different catagories i.e according to their releance and non-relevance. It constructs a separating hyper plane that maximizes the margin between the images relevant to query and the images not relevant to the query.

Step XIII: Classify retrieved images into different classes

The SVM is used to classify the images into separate classes. Basically , it is a binary classifier. The features used to differentiate the images or retrieving the images from database are the key factors that improves the width of the hyperplane. The SVM will create a hyperplane based on the grouping of retrieval and rest of the images. Higher the width of hyperplane ,better the resolution of CBIR system.

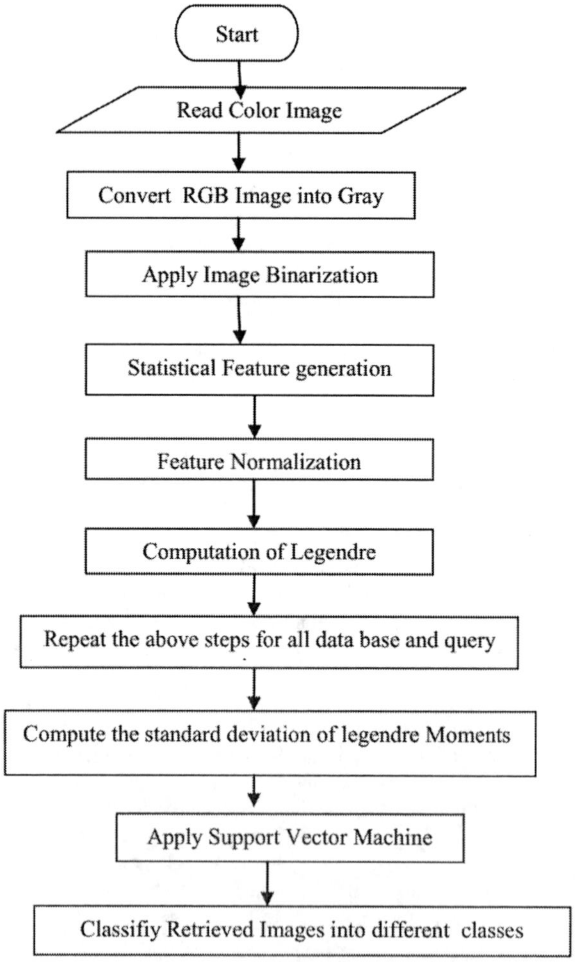

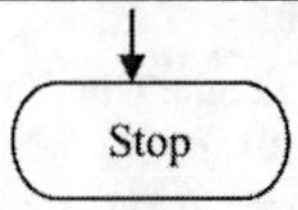

Figure 5.Flowchart for Content based Image Retrieval system

3 Experimental Results

The proposed algorithm is applied on sample database images .In base algorithm[1], the number of features used to extract various features is only five as compared to twenty two features as in case of proposed algorithm. Thereafter, the various database images are processed at varying margins and thresholds. Furthermore, the margin is ranging from 0 to 100 and the threshold from 1 to 256. The algorithms will not work properly for threshold >=200 as the threshold increases it will result in information loss for darker images. So , for darker images the threshold must be low and for lighter images it may be high. Moreover , as the margin increases the number of images retrieved will also increases. The completion time decreases as the threshold increases for both base and proposed algorithm.

Table 1. Images retrieved and Completion Time for Margin 0

Margin	Threshold	Query Image	Images Retrieved, Percentage of Relevant Images and Completion Time					
			Base Algorithm			Proposed Algorithm		
			Images Retrieved	% of Relevant Images	Completion Time	Images Retrieved	% of Relevant Images	Completion Time
0	50	1	1	100	9.980	1	100	27.869
0	50	2	2	100	9.540	2	100	26.670
0	50	3	3	100	9.691	3	100	26.783
0	50	4	4	100	9.654	4	100	27.207
0	50	5	5	100	9.610	5	100	27.061
0	50	6	6	100	9.636	6	100	27.300
0	50	7	7	100	9.576	7	100	27.188
0	50	8	8	100	9.788	8	100	27.406
0	50	9	9	100	9.635	9	100	27.201
0	50	10	10	100	9.592	10	100	27.306
0	50	11	11	100	9.793	11	100	27.121
0	50	12	12	100	9.889	12	100	27.715
0	50	13	13	100	9.818	13	100	27.644
0	50	14	14	100	9.743	14	100	27.164
0	50	15	15	100	9.796	15	100	27.879

The table 4.1 shows the different query images as an original image as well as results of different operations.

The difference in completion time is too large between base and proposed algorithms. Moreover, the percentage of relevant images for both algorithms is almost similar i.e. 100 %.

Table 2. Images retrieved and Completion Time for Margin 20

Margin	Threshold	Query Image	Images Retrieved, Percentage of Relevant Images and Completion Time			
			% of Relevant Images	Completion Time	% of Relevant Images	Completion Time
20	50	1	29	11.968	42	30.529
20	50	2	31	10.305	42	30.381
20	50	3	29	9.978	42	30.067
20	50	4	29	10.039	71	30.525
20	50	5	31	10.209	42	30.138
20	50	6	31	10.143	100	30.218
20	50	7	31	10.072	100	30.194
20	50	8	31	10.074	100	30.099
20	50	9	31	10.096	100	30.490

20	50	10	31	10.000	100	30.623
20	50	11	33	10.041	56	31.211
20	50	12	33	10.160	100	32.014
20	50	13	33	10.123	63	27.517
20	50	14	33	10.243	56	27.039
20	50	15	33	10.216	100	27.407

The comparison between the proposed and the base algorithm, 7.4% has the same accuracy, 26.66% has decreased amount of accuracy as compared to the base algorithm and 65.92% has the increased accuracy for margin 40. With the increase in the margin , it has been observed that the accuracy of the proposed algorithm as compared to the base algorithm remains either same or decreased in majority. Now, for margin above than 60, it has been observed that the accuracy of the proposed algorithm decreases a lot as compared to the base algorithm and there is a subsequent rise in the completion time also. The accuracy of the proposed algorithm decreases due to the increase in the margin. As the features has been increased in the proposed algorithm , the accuracy needs to be increased but the rise in the margin put the opposite effect. The ratio of the decrease in the accuracy for majority cases has been seen as compared to the base algorithm.

4 Conclusion

A CBIR system using Legendre Moments is proposed in this work. The Legendre moments based features are quite unique and exhibit fair image retrieval performance when tested on real time data base of images. The features are normalized with respect to rotation and size so that the image when rotated at some angle appears same as at the original angle and size. The support vector machine classifier classifies the retrieved images further into two categories: one with most similar and others. This way a content based retrieval system becomes two tier image retrieval system. In one tier, the images are retrieved using the global features and in second tier they are further fine tuned to their respective category using the support vector machine classifier. It has been observed that the average retrieval efficiency is increased as the feature set increases. Also, it has been observed that the classification efficiency of the proposed CBIR system increased with the increase in the number of training samples. The result tables shows the accuracy of the algorithm increases below the margin of 50. But this margin accuracy starts to decrease above the margin of 50. Furthermore, the main drawback of the proposed algorithm is that the completion time increases with the increase in the feature set.

5 Future Scope

The proposed system finds application in the area where images are retrieved frequently for research purposes. The future work may be proceeded in time or speed of the operation of the algorithm .The presented work is on model basis and speed of the operation has not been taken care of. However, the effectiveness of the algorithm increases if the speed of operation improves. Further, as the size of the image or data base increases, the speed deteriorates. In the presented work, accuracy of the algorithm is of prime importance.

References

1. Sarita Sharma, and Avinash Dhole, “Content Based Image Retrieval Based on Shape Feature using Accurate Legendre Moments and Support Vector Machines” IJCSET Vol.3, Issue no. 5, pp.194-199 , May 2013.
2. Abhinav Deshpande, and S.K.Tadse, “Design Approach for Content-Based Image Retrieval Using Gabor-Zernike Features” International Archive of Applied Sciences and Technology, Vol. 3 [2], pp.42 - 46, June 2012.
3. Hongqing Zhu , “Image representation using separable two-dimensional continuous and discrete orthogonal moments”, Pattern Recognition 45 (2012) 1540–1558.
4. Saroj A. Shambharkar, and Shubhangi C. Tirpude, “ A Comparative Study on Retrieved Images by Content Based Image Retrieval System based on Binary Tree, Color, Texture and Canny Edge Detection Approach”, IJACSA Special Issue on Selected Papers from International Conference & Workshop On Emerging Trends In Technology, pp.47-51, 2012.
5. Hui Zhang, Huazhong Shu, “Blurred Image Recognition by Legendre Moment Invariants” , IEEE Transactions On Image Processing, Vol. 19, No. 3, March 2010.
6. Ch.Srinivasa Rao , S.Srinivas Kumar and B.Chandra Mohan, “Content Based Image Retrieval using Exact Legendre Moments and Support Vector Machine”,The International Journal of Multimedia and its Applications(IJMA),Vol.2, No.2, May 2010.
7. Rohollah Akbari, Mehdi Keshavarz Bahaghighat andJavad Mohammadi, “Legendre Moments for Face Identification Based on Single Image per Person” , 2010 2nd International Conference on Signal Processing Systems (ICSPS).
8. Zhifang Dong, Hui Zhang, Huazhong Shu, “An Efficient Method For The Exact Computation Of Legendre Moments Of Binary Images” 978-1-4244-3298-1/09/$25.00 ©2009 IEEE.

9. S. M. Jaisakthi and C. Aravindan, "Face Detection Based on Eigenfaces and Legendre Moments, 978–1–4244–4547–9/09/$26.00 c 2009 IEEE.
10. Cecilia Di Ruberto and Andrea Morgera, "Moment-based Techniques for Image Retrieval", 19th International Conference on Database and Expert Systems Application 529-4188/08 2008 IEEE.
11. G.A. Papakostas, E.G. Karakasis and D.E. Koulouriotis, "Exact and Speedy Computation of Legendre Moments on Binary Images", Eight International Workshop on Image Analysis for Multimedia Interactive Services(WIAMIS'07) 0-7695-2818-X/07 $20.00 © 2007 IEEE.
12. Khalid M. Hosny, "Exact Legendre moment computation for gray level images", K.M. Hosny / Pattern Recognition 40 (2007) 3597 – 3605.
13. P.-T. Yap and R. Paramesran, "Content-based image retrieval using Legendre chromaticity distribution moments" , IEE Proc.-Vis. Image Signal Process., Vol. 153, No. 1, February 2006 17.

MATEC Web of Conferences 57, 02003 (2016) DOI: 10.1051/matecconf/20165702003
ICAET-2016

A Survey on different techniques of steganography

Harpreet Kaur[1, a] and Jyoti Rani[1]

[1]*CSE Department, GZSCCET Bathinda, Punjab, India*

Abstract - Steganography is important due to the exponential development and secret communication of potential computer users over the internet. Steganography is the art of invisible communication to keep secret information inside other information. Steganalysis is the technology that attempts to ruin the Steganography by detecting the hidden information and extracting. Steganography is the process of Data embedding in the images, text/documented, audio and video files. The paper also highlights the security improved by applying various techniques of video steganography.

1 Introduction

A Steganography is the science or art of hide the messages into other sources of information like text/documents, audios, videos and images etc. so that it is not visible to unauthorized users. It is known as invisible communication. A Steganography system made up of three components: cover-object means which hides the secret message, the secret message and the stego-object means which is the cover object with message embedded inside it. Video consist of collection of image. A digital image is represented by using a 2-D matrix of the color intestines at each grid points. The gray images use 8 bits, whereas colored utilizes 24 bits to describe the color model, such as RGB model. The different types of techniques are used in the Steganography is to hide the messages in the cover images. These techniques provide the best challenge for digital forensics investigations. Before transmission the message is encrypted and with the help of a key, the message is decrypted at receiver side. Nobody can describe the content of the key except one having the key. The message is known as the plain text and message in the encrypted form is known as the cipher text. At the time of transmission the message is protected. After decryption, the message becomes

Unprotected and it can be copied and distributed. The Secret or encrypted message may be a text file, a cipher text, audio or images [1].

Applications for a data-hiding scheme include in-band captioning, communication conversion, image tamper Proofing, revision tracking, enhance robustness of image search engines and smart Ids or identity cards where individual's details are embedded in their photographs.

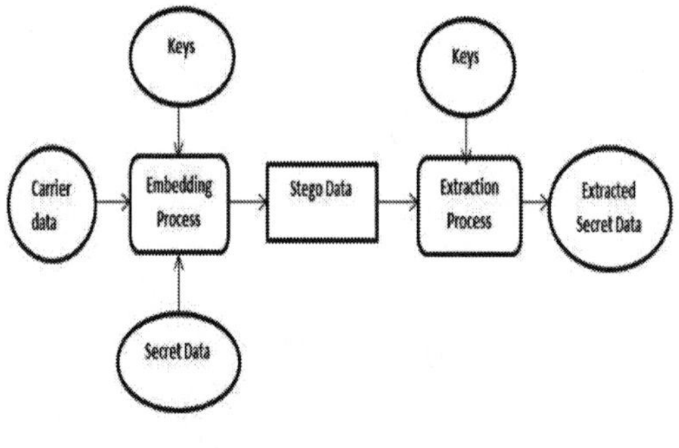

Figure 1 . Steganography Process

1.1 Cryptography and Steganography

Cryptography made up of Krypto's means hidden and Graphene means Writing. Steganography consist of Stegano's means Covered and Graphene means Writing. The Cryptography scrambles the messages so it is not easily understandable. Steganography diverge from cryptography.

Harpreet Kaur: kaurharpreet732@gmail.com

DOI: 10.1051/matecconf/20165702003

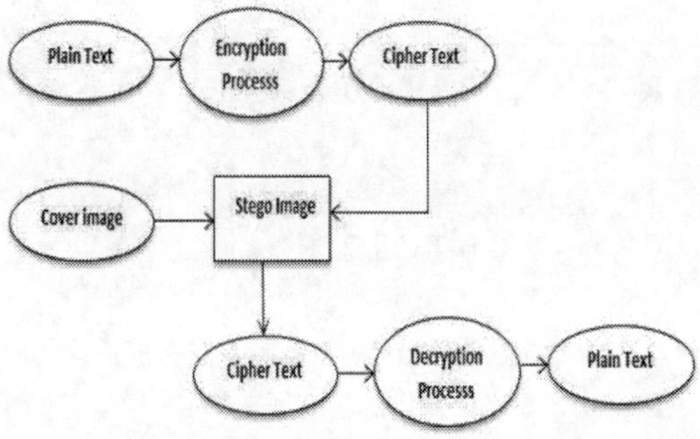

Figure 2. Combination of Cryptography and Steganography

The Goal of cryptography is to provide the secure communication by alter the data into a form that an eavesdropper cannot easily understand. A good steganography technique aims at three parameters i.e. capacity means the maximum information that can be hide inside the cover image , visual quality of stego-image cannot be change i.e. imperceptibility, robustness. Steganalysis is the technology that attempts to defeat the Steganography by detecting the hidden information and extracting [2].

2 Steganography in Digital Mediums

There are many Steganography techniques depending on the type of the cover object which are followed in order to obtain the security.

2.1 Text Steganography: The techniques in text steganography are number of tabs, white spaces, capital letters, just like Morse code is used to achieve information hiding.

2.2 Image Steganography: Taking the cover object as image in steganography is called image steganography. In this technique pixel intensities are used to hide the information. The 8 bit and 24 bit images are common. The image size is large then hides the more information. Larger images may require compression to avoid detection and the Techniques are LSB insertion and Masking and filtering [3]

Figure 3. Example of Image Steganography

2.3 Network Steganography: Taking cover objects as network protocol i.e. TCP, UDP, IP etc, where protocol is used as carrier is called network protocol steganography. In the OSI model there exist the channels where steganography can be achieved in unused header bits of TCP/IP fields

2.4 Audio Steganography: Taking audio as carrier for information hiding is called audio steganography. It is very important medium due to voice over IP (VOIP) popularity. It is used for digital audio formats such as WAVE, MIDI, and AVI MPEG for steganography. The methods are LSB coding, echo hiding, parity coding etc [4].

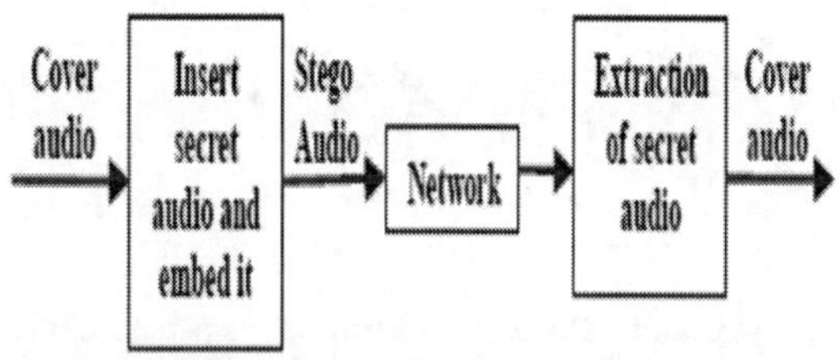

Figure 4. Example of Audio Steganography

2.5 Video Steganography: It is a technique to hide any type of files or information into digital video format. Video i.e. the combination of pictures is used as carrier for hidden information. The discrete cosine transform i.e. DCT change the values e.g., 8.667 to 9 which is used to hide the information in each of the images in the video, which is not justified by the human eye. It is used such as H.264, Mp4, MPEG, AVI or other video formats [5].

 DOI: 10.1051/matecconf/20165702003

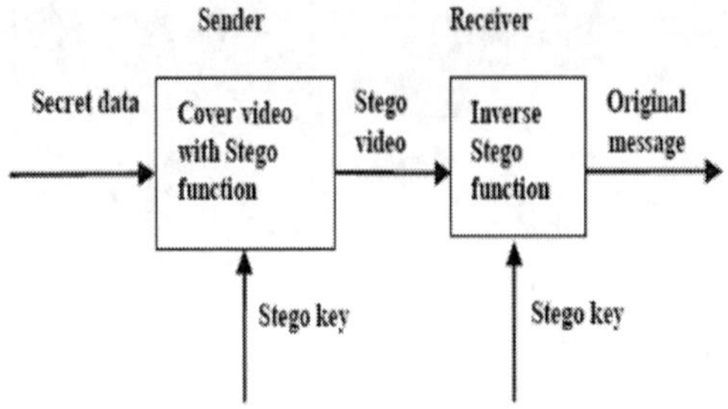

Figure 5. Example of Video Steganography

3 Techniques of Steganography

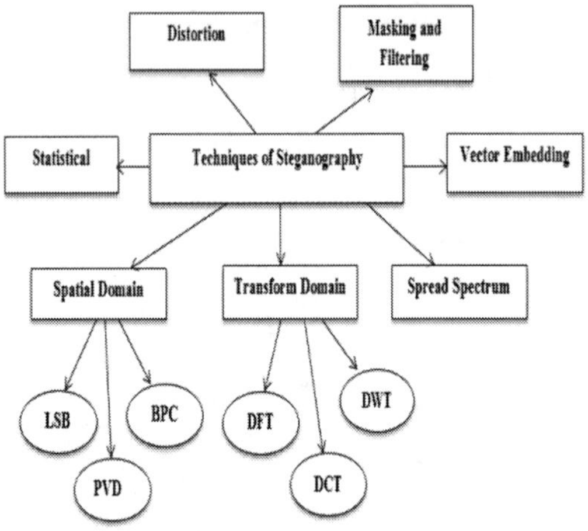

Figure 6 . Techniques of Steganography

3.1 Spatial Domain Methods: spatial domain Steganography technique refers to methods in which data hiding is performed directly on the pixel value of cover image in such a way that the effect of message is not visible on the cover image. The spatial domain methods are classified as following:

3.1.1 LSB: LSB is one the technique of spatial domain methods. LSB is the simple but susceptible to lossy compression and image manipulations. Some bits are change directly in the image pixel values in hiding the data. Changes in the value of the LSB are imperceptible for human eyes. Eg:

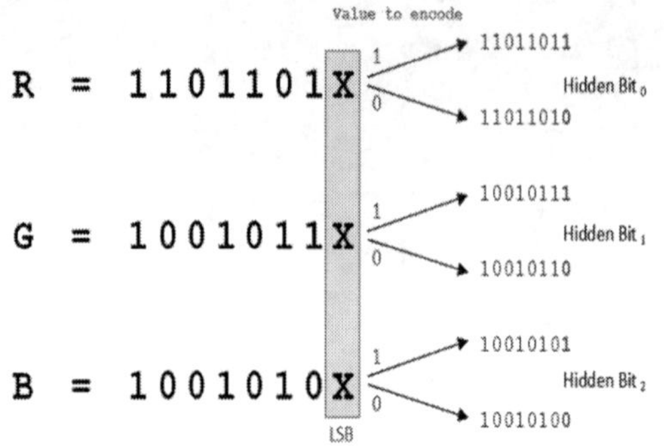

Figure 7. Example of LSB Conversion

In the spatial domain LSB technique there is less chance for degradation of the original image, more information can be stored in an image and covert communication of sensitive data.

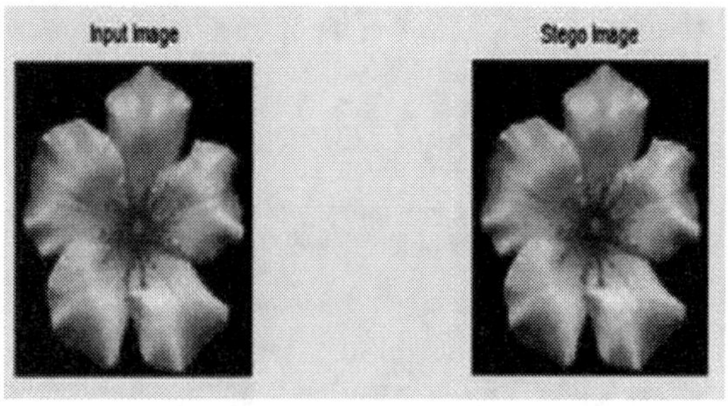

Figure 8 . LSB Conversion

3.1.2 Pixel Value Differencing: To embedding the data in PVD the two consecutive pixels are selected. Whether the pixels are determine from smooth area or an edge area. Payload is determined by calculating the difference between two regular pixels.

3.1.3 BPC: The Binary Pattern complexity approach is used to measure the noise factor in the image complexity. The noisy portion is replaced by binary Pattern and it is mapped from the secret data. The image will remain same when the reverse noise factor will determined.

3.2 Transform Domain Steganography: It is a more complex way to hides the information in an image. The different algorithms and transformations are used to hide information in the images. In the frequency domain, the process of embedding data of a signal is much stronger than embedding principles that operate in the time domain. The transform domain techniques over the spatial domain techniques is to hides the information in the images that are less exposed to compression, image processing and cropping. Some transform domain techniques are not depending on the image format and they run the lossless and lossy format conversions. Transform domain techniques are classified into various categories such as Discrete Fourier transformation (DFT), discrete cosine transformation (DCT), Discrete Wavelet transformation (DWT)

3.2.1 The Discrete Fourier Transform (DFT):

Discrete Fourier transform is the transform that are purely discrete: discrete-time signals are converted into discrete number of frequencies. DFT converts a finite list of equally spaced samples of a function into the list of coefficients of a finite combination of complex sinusoids ordered by their frequencies. It can be said to convert the sampled function from its original domain often time or position along a line to the frequency domain. The Discrete Time Fourier transforms uses the discrete time but it converts into the continuous frequency. The algorithm for computing the DFT is very fast on modern computers. This algorithm is known as Fast Fourier Transform i.e. FFT and it produces the same result as of the DFT by using the Inverse Discrete Fourier Transform.

3.2.2 The Discrete Cosine Transform (DCT): This method is similar to the Discrete Fourier Transform. DCT transform the signal or image from spatial domain to the frequency domain. The mathematical transforms convert the pixels in such a way as to give the effect of "spreading" the location of the pixel values over part of the image. The DCT is used in steganography as the Image is broken into 8×8 pixel blocks and transforms these pixel blocks into 64 DCT. Working from left to right, up to down, the DCT is applied to each block. Through quantization table each block is compressed to scale the DCT coefficients and message is embedded in DCT coefficients. The array of compressed blocks that constitute the image is stored in drastically reduced the amount of space. When desired, image is reconstructed through decompression, a process that uses the Inverse discrete cosine transform i.e. IDCT [6].

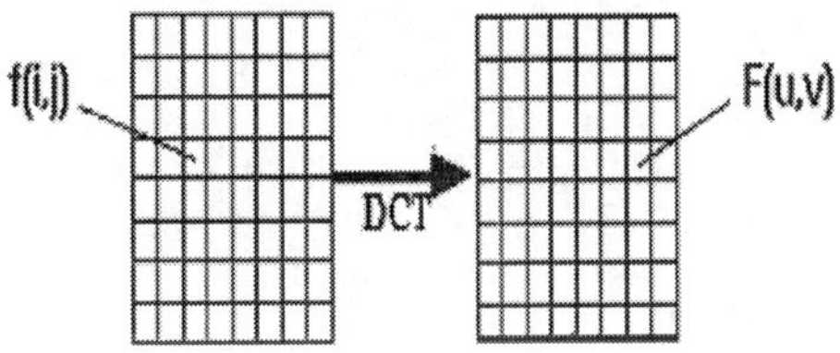

Figure 9 . Process of DCT

3.2.3 Discrete Wavelet Transform (DWT):

It is used to transform the image from a spatial domain to the frequency domain. In the process of steganography DWT identifies the high frequency and low frequency information of each pixel of the image. It is mathematical tool for decomposing an image hierarchically. It is mainly used for processing of non-stationary signals. The wavelet transform is based on small waves, Known as wavelets, of different frequency and limited duration. It provides both frequency and spatial description of the image. Wavelets are created by translations and dilations of a fixed function are known as mother wavelet. DWT performs in one dimension and in the two dimensional plane. The DWT is the accurate model than the DFT or the DCT and it is multi resolution description of the image. The current image compression standard JPEG 2000 is based on the wavelet transforms [7].

3.3 Vector Embedding: A vector embedding method that uses robust algorithm with codec standard (MPEG-1 and MPEG -2) .This method embeds audio information to pixels of frames in host video. It is based on the H.264/AVC Video coding standard. The algorithm designed a motion vector component feature to control embedding, and also to be the secret carrier. The information embedded will not significantly affect the video sequence's visual invisibility and statistical invisibility. The algorithm has a large embedding capacity with high carrier utilization, and can be implementing fast and effectively [8].

3.4 Spread spectrum: The concept of spread spectrum is used in this technique. In this method the secret data is spread over a wide frequency bandwidth. The ratio of signal to noise in every frequency band must be so small that it becomes difficult to detect the presence of data. Even if parts of data are removed from several bands, there would be still enough information is present in other bands to recover the data. Thus it is difficult to remove the data completely without entirely destroying the cover .It is a very robust approach used in military communication.

3.5 Statistical Technique: In the technique message is embedded by changing several properties of the cover. It involves the splitting of cover into blocks and then embedding one message bit in each block. The cover block is modified only when the size of message bit is one otherwise no modification is required.

3.6 Distortion Techniques: The distortion method is used to store the secret data by distorting the signal. An encoder applies a sequence of modifications to the cover image and the decoder phase decodes the encrypted data to the original data with the secret data by using some secret key.

3.7 Masking and Filtering: This approach is used to hides the data by marking an image. This approach is valuable where watermarks become a portion of the image. The data will be embedded where the more significant part of the image rather than hiding it into the noisy portion. The watermarking techniques are more integrated into the image and it can be applied without the fear of destruction of the image. This technique is used in 24 bit and grey scale images [9].

Table 1. Comparison of Various Techniques of Steganography.

Techniques	Domain	Invisibility	Capacity	Detectability	Robustness	Complexity	Comments
LSB	Spatial	High	High	High	Low	Low	Independent of image format and Texture
Spread Spectrum	Spatial	High	Low	Low	Medium	Medium	Dissolve the information over whole image
PVD	Spatial	High	Medium	Medium	Low	Low	Suitable for high Contrast images
DCT	Transform	High	Medium	Low	Medium	Medium	Simplest in the transform domain
DFT	Transform	High	Medium	Low	Medium	Medium	Involves the complex calculations
DWT	Transform	High	Medium	Low	High	High	Closely matches with human visual perception

4 Factors Include in Steganography

The effectiveness of steganography technique can be determined by comparing cover-image with the stego Image. The various factors are:

4.1 Robustness: Robustness refers to the ability of embedded data to remain intact if the stego- image undergoes transformations, such as linear and non-linear filtering, sharpening or blurring, addition of random noise, rotations and scaling, cropping or decimation, lossy compression.

4.2 Imperceptibility: The imperceptibility means invisibility of a steganography algorithm. Because it is the first and foremost requirement, since the strength of steganography lies in its ability to be unnoticed by the human eye.

4.3 Bit Error Rate: The hidden information can be successfully recovered from the communication channel. It must be ideal but for the real communication channel, the error comes while retrieving hidden information and this is measured by BER. It is the ratio of the number of errors to the total no of bits sent in an image.

4.4 Mean Square Error: It is computed by performing byte by byte comparisons of the two images. The representation of pixel with 8 bits and the representation of grey level images upto 256 levels. The distortion in the image can be measured using MSE. Let I be the cover image, K be the stego image and m*n be the total number of pixels.

$$MSE = \frac{1}{m\,n}\sum_{i=0}^{m-1}\sum_{j=0}^{n-1}[I(i,j) - K(i,j)]^2 \quad (1)$$

4.5 Peak Signal to Noise Ratio: The image steganography system must embed the content of hidden information in the image so that the quality of the image should not change. PSNR is commonly used to measure the quality of reconstruction of lossy compression techniques Larger the PSNR value indicates the better quality of image i.e. less distortion. PSNR is the ratio of the maximum signal to noise in the stego image [10].

5 Conclusions

The paper surveys the different types of Steganography and techniques to detect the steganography. In this basically Video steganography techniques are discussed. All the techniques discussed in this paper are able to secure the hidden data. On the other hand some algorithms have a very high time complexity and very less amount of data stored in the images. So, there is a need to develop efficient and accurate Steganography algorithms, either by combining the existing techniques or by developing new techniques. It helps in detecting terrorist activities on web.

References

1. P. Selvigrija and E. Ramya, *"Video by Linked List Method,"* no. March, (2015)
2. M. Hussain and M. Hussain, *"A Survey of video Steganography Techniques,"* *Int. J. Adv. Sci. Technol.*, vol. **54**, pp. 113–124, (2013)
3. G. Kaur and A. Kochhar, *"A Steganography Implementation based on LSB & DCT,"* vol. **4**, no. 1, pp. 35–41, (2012)

4. P. Chen and H. Lin, *"A DWT Based Approach for Image Steganography," Int. J. of Applied Sci. Eng.*, vol. **4**, no. 3, pp. 275–290, (2006)
5. M. Juneja and P. S. Sandhu, *"An Improved LSB Based Steganography Technique for RGB Color Images," Int. J. Comput. Commun. Eng.*, vol. **2**, no. 4, pp. 513–517, (2013)
6. kaur Amandeep kaur, manpreet, *"Improved Security Mechanism of Text in Video using Steganographic Technique," Int. J. Adv. Res. Comput. Sci. Softw. Eng.*, vol. **7782**, no. 5, pp. 44–51, (2014)
7. M. Dixit, N. Bhide, S. Khankhoje, and R. Ukarande, *"Video Steganography," 2015 Int. Conf. Pervasive Comput.*, vol. **00**, no. c, pp. 1–4, (2015)
8. R. J. Mstafa and I. Studen, *"A High Payload Video Steganography Algorithm in DWT Domain Based on BCH Codes (15 , 11),"* (2015)
9. J. Gupta, *"A Review on Steganography techniques and methods,"* vol. **1**, no. 1, pp. 1–4, (2015)
10. C. Science and B. Bridgeport, *"A Novel Video Steganography Algorithm in the Wavelet Domain Based on the KLT Tracking Algorithm and BCH Codes,"* (2015)

MATEC Web of Conferences 57, 02004 (2016) DOI: 10.1051/matecconf/20165702004
ICAET-2016

Security Enhancement of Knowledge-based User Authentication through Keystroke Dynamics

Soumen Roy[1], Utpal Roy[2], D. D. Sinha[1]

[1]*Department of Computer Science and Engineering, University of Calcutta, 92 APC Road, Calcutta -700 009, INDIA*
[2]*Department of Computer & System Sciences, Visva-Bharati, Santiniketan -731235, INDIA*

Abstract- Keystroke Dynamics is a behavioural biometrics characteristic in Biometric science, which solve the issues in user identification or verification. In Knowledge-based user authentication technique, we compromise with PIN or password which is unsafe due to different type of attacks. It is good to choose password with the combination of upper and lower case letter with some digits and symbols, but which is very hard to remember or generally we forget to distinguish those passwords for different access control systems. Our system not only takes the users' entered texts but their typing style is also account for. In our experiment, we have not taken hard password type texts, we have chosen some daily used words where users are habituated and comfortable at typing and we obtained the consisting typing pattern. Different distance-based and data mining algorithms we have applied on collected typing pattern and obtained impressive results. As per our experiment, if we use keystroke dynamics in existing knowledge based user authentication system with minimum of five daily used common texts then it increases the security level up to 97.6% to 98.2% (if we remove some of the irrelevant feature sets).

1 Introduction

Keystroke dynamics is a behavioral biometrics which is the method of analysing the way a user types on a keyboard and classify user based on regular typing rhythm. It is the study of whether people can be well-known by their typing rhythms, much like handwriting is used to recognize the author of a written text. A user's typing pattern may be unique because similar neuro-physiological factors that make written signatures unique. Here user can choose any text for password from his/her own dictionary. It is very simple and nothing to remember still it enhances the security level and can be used to identify an individual. Recognizing typing style promises a parameter like biometric characteristics that may facilitate non-intrusive, cost-effective and continuous monitoring.

System takes comprising of characters as well as the typing style of each subsequent character entered. It facilitates that no one can track the time or presses the character of password in same rhythm. It will prevent our system from off-line guessing attracts and also prevent to track by un-authorised people. Our objective is to minimize the probability of off-line guessing attacks, hide the password from public, minimize the hardware cost and minimize the software cost by making faster pattern recognition.

In our experiment we have consider five daily used strings ("kolkata123", "facebook", "yahoo.com", "gmail.com" and "123456") as password string laptop keypad as hardware configuration. We have collected the keystroke raw data samples from 12 users in four sessions, in each session 6 times using JAVA swing. Then we extracted all the features (key duration, up-down key latency, down-up key latency, up-up key latency, down-down key latency, tri-gap timing, four-gap timing and total timing) and we analysed features and combination of features using R Statistical tool and Microsoft Excel.

We have tested our raw data using eight different score-based, distance-based and features mining algorithms and we got 2.4% of Equal Error Rate for Z-score algorithm where we have consider all the five fixed strings and all the mentioned features.

2 Keystroke Dynamics

2.1 Basic Idea

Keystroke dynamics is set of some timing data or keystroke pressure data which is generated at typing on keyboard which is unique and can be used to classify the users.

2.2 Science and Features Selection

Placement of fingers on keyboard, hand weight, length of finger, neuro-physiological factors make typing style unique, where some timing factors easily

[a] Corresponding author: soumen.roy_2007@yahoo.co.in

calculate these pattern factors. We have calculated key press and release time P_i and R_i for all keys K_i which represent entered character set. Where $1 \leq i \leq$ length of the entered word. The features of the keystroke dynamics as follows:

Key Duration $(T_1) = R_i - P_i$ (1)
Up Up Key Latency $(T_2) = R_{i+1} - R_i$ (2)
Down Down Key Latency $(T_3) = P_{i+1} - P_i$ (3)
Up Down Key Latency $(T_4) = P_{i+1} - R_i$ (4)
Down Up Key Latency $(T_5) = R_{i+1} - P_i$ (5)
Total Time Key Latency $(T_6) = R_n - P_1$ (6)
Tri-graph Latency $(T_7) = R_{i+2} - P_i$ (7)
Four-graph Latency $(T_8) = R_{i+3} - P_i$ (8)

Some new features are keystroke pressure, finger tips size, finger placement on keyboard, keystroke sound, error correcting mechanism, sequence of left-right control keys.

2.3 Security Issues

Among various user authentication techniques knowledge-based, token-based and biometric-based authentication techniques, biometric authentication is most popular for their uniqueness characteristics and cannot be stolen or there is no chance to loss. Keystroke Dynamics is a behavioral characteristic which is unique and can be effectively implemented with the existing system with minimal alternation. It can be used as a safe guard of our password from different type of attacks.

2.4 Factors Affecting Performance

Some of the factors which affect the way of keystroke Dynamics as follows: Text length, sequences of character types, word choice, and number of training sample, statistical method to create template, mental state of the user, tiredness or level of comfort, keyboard type, keyboard position and height of the keyboard, hand injury, weakness of hand mussel, shoulder pain, education level, computer knowledge, and category of users.

2.5 Algorithms

Many classification methods have been applied in keystroke dynamics study over the last three decades. Following are the anomaly detector algorithm described in [8].

2.6.1 Canberra:

$$D_{car} = \sum_{1}^{n} \frac{|P_i - Q_i|}{P_i + Q_i} \quad (9)$$

2.6.2 Chebyshev:

$$D_{cheb} = \sum_{1}^{n} \max |P_i - Q_i| \quad (10)$$

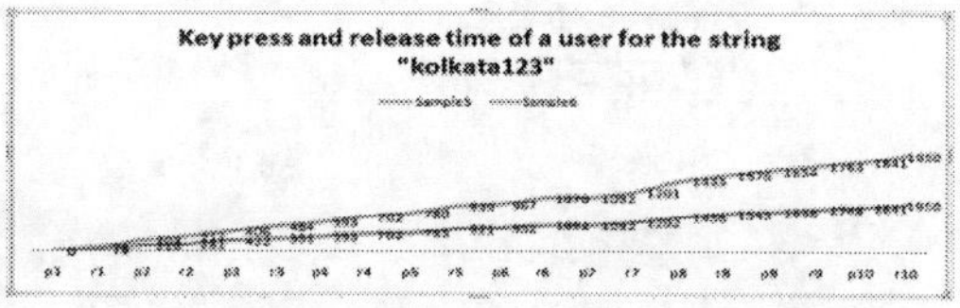

Figure 1. Two different samples from a user in line chart

2.6 Keystroke Dynamics as User Authentication

There are different ways in which a user can be authenticated. However all of these ways can be categorized into one of three classes: "Something we know" e.g. password, "Something we have" e.g. token, "Something we are" e.g. biometric property.

2.6.3 Czekanowski:

$$D_{cze} = \frac{\sum^{n} |P_i - Q_i|}{\sum^{n} (P_i + Q_i)} \quad (11)$$

2.6.4 Gower:

$$D_{gow} = \frac{1}{n} \sum_{1}^{n} |P_i - Q_i| \quad (12)$$

2.6.5 Intersection:

$$D_{ins} = \frac{1}{2} \sum_{1}^{n} |P_i - Q_i| \quad (13)$$

2.6.6 Kulczynski:

$$D_{kuld} = \frac{\sum^{n} |P_i - Q_i|}{\sum^{n} \min(P_i, Q_i)} \quad (14)$$

2.6.7 Lorentzian:

$$D_{lor} = \sum_{1}^{n} \ln(1 + |P_i - Q_i|) \quad (15)$$

2.6.8 Minkowski:

$$D_{mink} = \sqrt[p]{\sum_{1}^{n} |P_i - Q_i|^{p}} \quad (16)$$

2.6.9 Motyka:

$$D_{mot} = \frac{\sum^{n} \max(P_i)|}{\sum^{n} (P_i + Q_i)} \quad (17)$$

2.6.10 Ruzicka:

$$D_{ruz} = 1 - \frac{\sum^{n} \min(P_i, Q_i)}{\sum^{n} \max(P_i, Q_i)} \quad (18)$$

2.6.11 Soergel:

$$D_{soc} = \frac{\sum^{n} |P_i - Q_i|}{\sum^{n} \max(P_i, Q_i)} \quad (19)$$

2.6.12 Sorensen:

$$D_{sor} = \frac{\sum^{n} |P_i - Q_i|}{\sum^{n} (P_i + Q_i)} \quad (20)$$

2.6.13 Wavehedges:

$$D_{wv} = \frac{\sum^{n} |P_i - Q_i|}{\sum^{n} \max(P_i, Q_i)} \quad (21)$$

2.6.14 Manhattan Distance:

$$M = \sum_{i=1}^{n} (|P_i - Q_i|) \quad (22)$$

2.6.15 Euclidean Distance:

$$E = \sqrt[2]{\sum_{1}^{n} (|P_i - Q_i|)^{2}} \quad (23)$$

2.6.16 Mahanobolis Distance:

$$Eh = \sqrt[2]{\sum_{i}^{n}((|P_i - Q_i|)/\alpha_i)^2} \quad (24)$$

2.6.17 Z Score:

$$Z = \sum_{i=1}^{n}(|P_i| - \mu(|Q_i|))/\alpha_i \quad (25)$$

3 Background details

In 30+ years of experience, many researchers have proposed their algorithms, taking various features, various length of pattern string.

Table 1. Background of keystroke dynamics

Authors	Classifiers	Length of the pattern	Features	EER (%)
Joyce & Gupta [2]	Manhattan	33	UD	0.25-16.36
Bleha et al. [1]	Euclidian	11-17	UD	2.8-8.1
Haider et al. [6]	Nural Network	7	UD	16.1
Yu & Cho [5]	SVM	6-7	UD	10.2
Killourly S. [4]	Manhattan (Scaled)	10+	UD	9.6
Giot et al. [7]	SVM	100	KD, UD	15.28

4 Experimental results

We have implemented a program in JAVA for experimental purpose, which has the capability of capturing all key pressing and releasing events, which are used to create the database of different sample of passwords and timing templates. Here we have calculated average equal error rate for all eight algorithms considering some single feature and combination of features for all five strings.

Table 2. Average equal error rate for fixed-text "kolkata123"

FEATURES & COMBINATION OF FEATURES	Euclidean	Manhattan	Scaled Manhattan	Outlier Count	Mahalanobis	KMeans	AutoAssocNNet	SVM
UU	0.163	0.163	0.140	0.152	0.180	0.128	0.168	0.130
UD+DU	0.151	0.139	0.106	0.102	0.147	0.112	0.150	0.108
UD + Tri	0.160	0.151	0.123	0.129	0.144	0.134	0.149	0.120
UD	0.175	0.174	0.159	0.176	0.222	0.167	0.167	0.160
Tri Gap	0.166	0.159	0.135	0.143	0.175	0.140	0.157	0.124
Tri + 4gap	0.178	0.177	0.157	0.137	0.153	0.124	0.159	0.136
KD+DD+UU+DU+UD+tri+4gap	0.166	0.147	0.113	0.089	0.141	0.152	0.155	0.150
KD+DD+UU+DU+UD+tri	0.155	0.132	0.114	0.092	0.144	0.151	0.154	0.149
KD+DD+UU+DU+UD	0.151	0.128	0.117	0.094	0.141	0.151	0.157	0.149
KD + UD + Tri + 4gap	0.172	0.161	0.110	0.092	0.137	0.155	0.149	0.147
KD + UD + Tri	0.157	0.137	0.127	0.095	0.136	0.149	0.144	0.144
KD + UD	0.164	0.126	0.134	0.101	0.169	0.154	0.150	0.160
KD + Tri	0.163	0.149	0.131	0.105	0.159	0.159	0.150	0.156
KD + DU + UD + Tri + 4gap	0.169	0.152	0.115	0.088	0.140	0.150	0.154	0.150
KD + DU + UD + tri	0.157	0.130	0.116	0.093	0.138	0.150	0.147	0.148
KD + DU + Tri + 4gap	0.172	0.156	0.112	0.092	0.135	0.150	0.157	0.146
KD + DU + Tri	0.157	0.129	0.126	0.096	0.141	0.144	0.145	0.145

KD + DU	0.150	0.109	0.134	0.097	0.143	0.162	0.138	0.165
KD + 4gap	0.177	0.180	0.130	0.109	0.145	0.166	0.172	0.160
KD	0.153	0.143	0.179	0.170	0.276	0.204	0.153	0.208
DU + Tri	0.159	0.144	0.115	0.114	0.168	0.118	0.156	0.111
DU	0.158	0.142	0.112	0.125	0.158	0.119	0.157	0.112
DD+UU	0.156	0.150	0.123	0.123	0.157	0.127	0.160	0.112
DD	0.151	0.147	0.122	0.142	0.157	0.132	0.155	0.120
All	0.175	0.156	0.113	0.089	0.140	0.152	0.167	0.150
4gap	0.181	0.201	0.197	0.201	0.159	0.153	0.169	0.148

Table 3. Average equal error rate for fixed-text "yahoo.com"

FEATURES & COMBINATION OF FEATURES	**Euclidean**	**Manhattan**	**Scaled Manhattan**	**Outlier Count**	**Mahalanobis**	**KMeans**	**AutoAssocNNet**	**SVM**
UU	0.224	0.201	0.196	0.199	0.269	0.199	0.248	0.197
UD+DU	0.226	0.200	0.187	0.172	0.218	0.182	0.239	0.192
UD + Tri	0.223	0.216	0.196	0.204	0.202	0.203	0.215	0.176
UD	0.229	0.215	0.200	0.209	0.244	0.196	0.234	0.189
Tri Gap	0.228	0.228	0.215	0.239	0.243	0.216	0.207	0.216
Tri + 4gap	0.231	0.234	0.212	0.230	0.259	0.220	0.211	0.221
KD+DD+UU+DU+UD+tri+4gap	0.224	0.209	0.145	0.127	0.249	0.170	0.229	0.158
KD+DD+UU+DU+UD+tri	0.218	0.200	0.142	0.118	0.246	0.169	0.236	0.165
KD+DD+UU+DU+UD	0.226	0.195	0.139	0.109	0.253	0.169	0.235	0.179
KD + UD + Tri + 4gap	0.227	0.212	0.146	0.124	0.233	0.160	0.218	0.173
KD + UD + Tri	0.220	0.193	0.141	0.109	0.218	0.168	0.224	0.173
KD + UD	0.221	0.184	0.138	0.095	0.246	0.176	0.229	0.170
KD + Tri	0.226	0.213	0.131	0.135	0.224	0.149	0.222	0.154
KD + DU + UD + Tri + 4gap	0.226	0.210	0.148	0.122	0.242	0.161	0.223	0.168
KD + DU + UD + tri	0.220	0.196	0.140	0.110	0.233	0.162	0.222	0.169
KD + DU + Tri + 4gap	0.230	0.219	0.139	0.127	0.257	0.162	0.226	0.173
KD + DU + Tri	0.221	0.203	0.136	0.115	0.260	0.169	0.222	0.177
KD + DU	0.228	0.176	0.141	0.101	0.261	0.171	0.240	0.174
KD + 4gap	0.229	0.217	0.141	0.145	0.231	0.161	0.211	0.164
KD	0.158	0.155	0.187	0.163	0.321	0.205	0.158	0.217
DU + Tri	0.222	0.225	0.190	0.193	0.252	0.197	0.221	0.181
DU	0.236	0.205	0.175	0.188	0.242	0.206	0.243	0.189
DD+UU	0.235	0.212	0.196	0.196	0.264	0.196	0.256	0.189
DD	0.250	0.223	0.198	0.213	0.236	0.203	0.251	0.194
All	0.211	0.203	0.145	0.128	0.249	0.170	0.215	0.161
4gap	0.234	0.245	0.241	0.263	0.262	0.228	0.247	0.232

Table 4. Average equal error rate for fixed-text "gmail.com"

FEATURES & COMBINATION OF FEATURES	Euclidean	Manhattan	Scaled Manhattan	Outlier Count	Mahalanobis	KMeans	AutoAssocNNet	SVM
UU	0.175	0.167	0.149	0.164	0.191	0.163	0.170	0.170
UD+DU	0.172	0.164	0.139	0.139	0.158	0.149	0.177	0.153
UD + Tri	0.186	0.168	0.141	0.158	0.172	0.144	0.180	0.150
UD	0.182	0.176	0.144	0.153	0.191	0.159	0.180	0.165
Tri Gap	0.184	0.191	0.169	0.206	0.203	0.173	0.188	0.173
Tri + 4gap	0.199	0.210	0.188	0.203	0.212	0.176	0.202	0.171
KD+DD+UU+DU+UD+tri+4gap	0.193	0.162	0.108	0.118	0.162	0.139	0.178	0.143
KD+DD+UU+DU+UD+tri	0.183	0.166	0.106	0.116	0.164	0.136	0.194	0.140
KD+DD+UU+DU+UD	0.135	0.151	0.153	0.161	0.194	0.144	0.134	0.142
KD + UD + Tri + 4gap	0.193	0.174	0.095	0.099	0.153	0.129	0.176	0.120
KD + UD + Tri	0.178	0.150	0.086	0.088	0.158	0.129	0.174	0.117
KD + UD	0.171	0.145	0.092	0.083	0.160	0.130	0.167	0.125
KD + Tri	0.196	0.169	0.092	0.099	0.151	0.105	0.173	0.099
KD + DU + UD + Tri + 4gap	0.193	0.170	0.099	0.104	0.158	0.128	0.192	0.122
KD + DU + UD + tri	0.181	0.157	0.093	0.098	0.162	0.127	0.180	0.128
KD + DU + Tri + 4gap	0.193	0.178	0.098	0.103	0.157	0.125	0.194	0.116
KD + DU + Tri	0.182	0.151	0.089	0.093	0.155	0.124	0.182	0.114
KD + DU	0.168	0.138	0.105	0.087	0.151	0.122	0.167	0.114
KD + 4gap	0.200	0.188	0.093	0.103	0.155	0.097	0.190	0.096
KD	0.135	0.151	0.153	0.161	0.194	0.143	0.133	0.140
DU + Tri	0.187	0.172	0.148	0.168	0.192	0.157	0.183	0.154
DU	0.175	0.164	0.148	0.151	0.172	0.159	0.184	0.161
DD+UU	0.184	0.173	0.163	0.160	0.188	0.170	0.182	0.164
DD	0.190	0.180	0.178	0.167	0.184	0.180	0.197	0.176
All	0.195	0.166	0.108	0.117	0.163	0.139	0.199	0.142
4gap	0.212	0.228	0.204	0.231	0.191	0.194	0.216	0.190

Table 5. Average equal error rate for fixed-text "facebook"

FEATURES & COMBINATION OF FEATURES	Euclidean	Manhattan	Scaled Manhattan	Outlier Count	Mahalanobis	KMeans	AutoAssocNNet	SVM

UU	0.204	0.173	0.117	0.152	0.205	0.158	0.221	0.148
UD+DU	0.190	0.155	0.105	0.129	0.168	0.124	0.199	0.123
UD + Tri	0.199	0.177	0.124	0.144	0.154	0.148	0.180	0.127
UD	0.220	0.185	0.140	0.169	0.197	0.164	0.205	0.165
Tri Gap	0.206	0.196	0.169	0.188	0.176	0.199	0.222	0.180
Tri + 4gap	0.209	0.210	0.173	0.188	0.255	0.191	0.193	0.181
KD+DD+UU+DU+ UD+tri+4gap	0.197	0.161	0.082	0.083	0.169	0.132	0.203	0.122
KD+DD+UU+DU+ UD+tri	0.190	0.146	0.081	0.080	0.169	0.121	0.186	0.116
KD+DD+UU+DU+U D	0.190	0.147	0.090	0.082	0.166	0.126	0.206	0.119
KD + UD + Tri + 4gap	0.200	0.178	0.099	0.099	0.162	0.133	0.185	0.129
KD + UD + Tri	0.188	0.148	0.113	0.096	0.146	0.131	0.176	0.128
KD + UD	0.205	0.136	0.116	0.099	0.153	0.133	0.198	0.137
KD + Tri	0.197	0.170	0.121	0.103	0.172	0.134	0.197	0.129
KD + DU + UD + Tri + 4gap	0.197	0.163	0.093	0.086	0.161	0.131	0.187	0.120
KD + DU + UD + tri	0.186	0.146	0.098	0.081	0.155	0.127	0.176	0.126
KD + DU + Tri + 4gap	0.202	0.170	0.096	0.097	0.174	0.126	0.198	0.130
KD + DU + Tri	0.195	0.150	0.110	0.092	0.181	0.129	0.183	0.128
KD + DU	0.176	0.116	0.106	0.099	0.176	0.130	0.197	0.129
KD + 4gap	0.209	0.187	0.125	0.116	0.205	0.128	0.197	0.127
KD	0.165	0.166	0.172	0.163	0.182	0.157	0.166	0.158
DU + Tri	0.204	0.171	0.122	0.145	0.169	0.148	0.189	0.142
DU	0.190	0.153	0.107	0.148	0.190	0.145	0.201	0.137
DD+UU	0.203	0.172	0.113	0.132	0.181	0.139	0.206	0.129
DD	0.210	0.166	0.117	0.145	0.175	0.144	0.214	0.139
All	0.195	0.158	0.080	0.084	0.169	0.132	0.189	0.113
4gap	0.219	0.231	0.209	0.237	0.200	0.198	0.229	0.213

Table 6. Average equal error rate for fixed-text "123456"

FEATURES & COMBINATION OF FEATURES	Euclidean	Manhattan	Scaled Manhattan	Outlier Count	Mahalanobis	KMeans	AutoAssocNNet	SVM
UU	0.217	0.207	0.202	0.231	0.228	0.229	0.212	0.211
UD+DU	0.159	0.163	0.152	0.170	0.285	0.173	0.163	0.163
UD + Tri	0.175	0.186	0.173	0.195	0.247	0.179	0.172	0.174
UD	0.232	0.232	0.232	0.263	0.258	0.236	0.221	0.224
Tri Gap	0.204	0.208	0.203	0.233	0.215	0.241	0.198	0.207
Tri + 4gap	0.210	0.214	0.205	0.223	0.243	0.227	0.197	0.214
KD+DD+UU+DU+ UD+tri+4gap	0.179	0.175	0.142	0.154	0.307	0.181	0.156	0.162
KD+DD+UU+DU+ UD+tri	0.176	0.172	0.153	0.154	0.323	0.188	0.173	0.170
KD+DD+UU+DU+UD	0.187	0.173	0.144	0.153	0.305	0.181	0.170	0.173
KD + UD + Tri + 4gap	0.159	0.155	0.146	0.156	0.323	0.182	0.160	0.162

KD + UD + Tri	0.179	0.170	0.180	0.172	0.323	0.193	0.181	0.173
KD + UD	0.183	0.173	0.166	0.166	0.313	0.193	0.194	0.179
KD + Tri	0.179	0.170	0.145	0.146	0.306	0.182	0.167	0.162
KD + DU + UD + Tri + 4gap	0.153	0.154	0.145	0.146	0.323	0.181	0.155	0.158
KD + DU + UD + tri	0.194	0.178	0.151	0.145	0.314	0.183	0.181	0.170
KD + DU + Tri + 4gap	0.176	0.170	0.157	0.150	0.323	0.184	0.172	0.174
KD + DU + Tri	0.162	0.146	0.163	0.169	0.323	0.189	0.159	0.169
KD + DU	0.197	0.170	0.184	0.188	0.303	0.215	0.189	0.188
KD + 4gap	0.259	0.261	0.272	0.288	0.353	0.294	0.259	0.286
KD	0.154	0.156	0.138	0.151	0.323	0.181	0.149	0.161
DU + Tri	0.196	0.199	0.192	0.197	0.317	0.217	0.199	0.192
DU	0.203	0.196	0.187	0.221	0.262	0.236	0.209	0.214
DD+UU	0.193	0.195	0.187	0.203	0.370	0.218	0.189	0.209
DD	0.198	0.204	0.196	0.228	0.253	0.229	0.194	0.221
All	0.189	0.179	0.146	0.155	0.306	0.180	0.160	0.165
4gap	0.218	0.222	0.216	0.262	0.229	0.251	0.233	0.240

Table 7. Average equal error rate for all combination of fixed texts

STRINGS & COMBINATION OF STRING	Euclidean	Manhattan	Scaled Manhattan	Outlier Count	Mahalanobis	KMeans	AutoAssocNNet	SVM
Kolkata123	0.217	0.207	0.202	**0.231**	0.228	0.229	0.212	0.211
Yahoo.com	0.159	0.163	0.152	**0.170**	0.285	0.173	0.163	0.163
Gmail.com	0.175	0.186	0.173	**0.195**	0.247	0.179	0.172	0.174
Facebook	0.232	0.232	0.232	**0.263**	0.258	0.236	0.221	0.224
123456	0.204	0.208	0.203	**0.233**	0.215	0.241	0.198	0.207
Kolkata123 + facebook	0.152	0.134	0.087	**0.060**	0.184	0.140	0.181	0.137
Kolkata123+facebook+ gmail.com	0.178	0.134	0.084	**0.042**	0.205	0.150	0.250	0.155
Kolkata123+facebook+ gmail.com+yahoo.com	0.204	0.152	0.097	**0.038**	0.236	0.171	0.291	0.171
Kolkata123+facebook+ gmail.com+yahoo.com+ 123456	0.205	0.144	0.088	**0.024**	0.260	0.184	0.204	0.160

5 Evaluation and Analysis

Here we see that no combination of features and algorithms give bellow 0.08 average equal error rate for all five type of fixed string.

We have tested combining these five strings and we got the following result. Here minimum average equal error rate is 0.024 where all five strings and all features are considered.

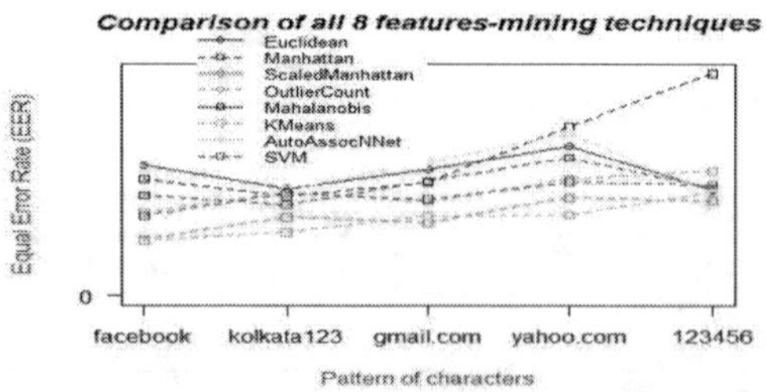

Figure 2. Line chart of all 8 classifiers for each dataset

In the above figure, we see that for all the strings outlier Count (z-score) is achieved best result after scaled Manhattan.

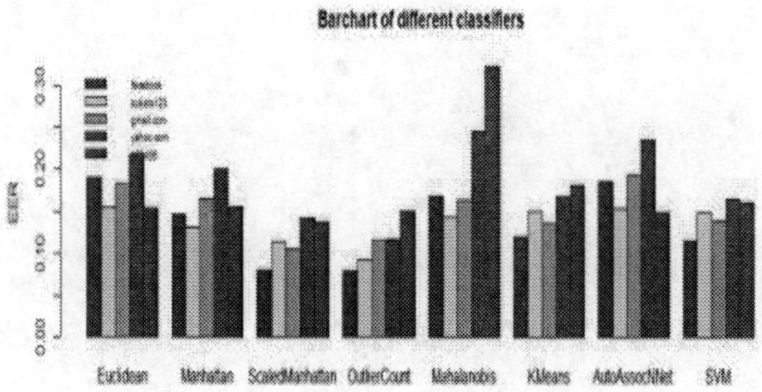

Figure 3. Bar chart of all 8 classifiers for each dataset

Here we see that no combination of features and algorithms give bellow 0.08 average equal error rates for all five type of fixed string.

We have tested combining these five strings and we got the following result. Here minimum average equal error rate is 0.024 where all five strings and all features are considered.

Table 8. Comparison by average equal error rates of different distance-based algorithm

Classifiers	EER	Sd	Classifiers	EER	Sd
Chebyshev	**0.083**	0.289	Ruzicka	0.871	0.109
Canberra	**0.071**	0.104	Soergel	0.129	0.109
Czekanowski	0.129	0.109	Sorensen	0.129	0.109
Gower	0.515	0.264	Wavehedges	0.129	0.109
Intersection	0.579	0.255	Euclidean	0.205	0.123
Kulczynski	0.129	0.109	Manhattan	0.144	0.127
Kulczynskis	0.129	0.109	**ScaledManhattan**	**0.088**	0.097
Lorentzian	**0.044**	0.076	**OutlierCount**	**0.024**	**0.072**
Minkowski	0.219	0.119	Mahalanobis	0.260	0.181
Motyka	0.129	0.109	KMeans	0.184	0.095

The Table 8 represents that Outlier Count, Lorentzian, Canberra, Chebyshev and Scaled Manhattan are the suitable model in identification of user through typing pattern.

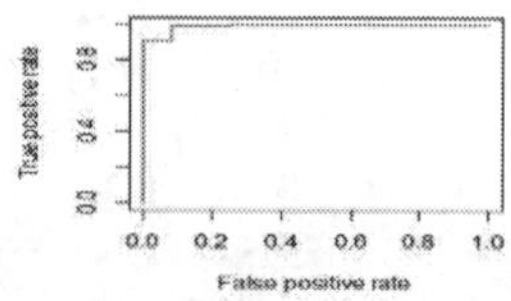

Figure 4. ROC Curve of the Outlier Count

6 Discussion and Further Area of Research

Keyboard is essential for a computer device, which can recognize our typing style and very much unique as per our experiment and cannot be copied or stolen. It can be used as safe guard of our password in any access control system. This technique can be used in online criminal investigation, back door account identification, online typing examination, emotion recognition, lab attendant system many more.

Sometimes, score of different algorithms varies due to mental state of user or aging problems. It can be solve by updating mechanism.

7 Conclusion

Different classification algorithms we have applied on 5 similar keystroke databases taking in our consideration all 8 features and combination of features, so we can compare the classifiers on an equal basis. In our evaluation process, we have identified the best classifier (z-score). It achieved 91.2% of accuracy for the string "kolkata123" (considering KD, DU, UD, Trigap and 4gap timing features), 90.5% of accuracy for the string "yahoo.com" (considering KD, UD), 91.7% of accuracy for the string "gmail.com" (considering KD, UD), 92.0% of accuracy for the string "facebook" (considering KD, DD, UU, DU, UD and Trigap), 85.5% of accuracy for the string "123456" (considering KD, DU, Trigap and 4gap timing features). Z-score classification algorithm gives the highest accuracy for all the string patterns. We also have tested this algorithm on the entire strings database and we got 97.6 % of accuracy. If we remove some of the feature sets like four graph times for the string "kolkata123" (e.g. four graph times of "ata1", "ta12", "a123", three graph times between "lka" etc), then we can get up to 98.2% of accuracy in this technique. So it has been established that this technique can be used as a safe guard of password or PIN in knowledge-based user authentication. But in practical there are many affecting factors may affect way of this process. Need much more experiment on it like key-pressure, finger placement on keyboard, finger tips size etc. can be calculated.

References

1. S. Bleha, C. Slivinsky, and B. Hussien, "Computer-access security systems using keystroke dynamics,"IEEE Transactions on Pattern Analysis and Machine Intelligence, vol. 12, no. 12, pp. 1217–1222, 1990.
2. R. Joyce & G. Gupta, Identity authorization based on keystroke latencies. Communication of ACM 33 (2) 168–176, 1990.
3. F. Monrose & A. D. Rubin, Keystroke dynamics as a biometric for authentication. Future Generation Computer Systems, Vol. 16, No. 4, pp. 351–359.
4. K. S. Killourhy, A Scientific Understanding of Keystroke Dynamics. PhD thesis, Computer Science Department, Carnegie Mellon University, Pittsburgh, US, 2012
5. E. Yu and S. Cho, "Novelty detection approach for keystroke dynamics identity verification," in Intelligent Data Engineering and Automated Learning, vol. 2690, pp. 1016–1023, Springer, Berlin, Germany, 2003.
6. S. Haider, A. Abbas, and A. K. Zaidi, "A multi-technique approach for user identification through keystroke dynamics," in Proceedings of the 2000 IEEE Interantional Conference on Systems, Man and Cybernetics, vol. 2, pp. 1336–1341, October 2000.
7. R.Giot, M. El-Abed, andC. Rosenberger, "GREYCkeystroke: a benchmark for keystroke dynamics biometric systems," in Proceedings of the IEEE 3rd International Conference on Biometrics: Theory, Applications and Systems (BTAS '09), pp. 1–6, September 2009.
8. S. Roy, U. Roy, D.D. Sinha, "Enhanced Knowledge-Based User Authentication Technique Via Keystroke Dynamics", International Journal of Engineering and Science Invention (IJESI), Vol 3, Issue 9, Sep, 2013, 41-48.

ACO Agent Based Routing in AOMDV Environment

Amanpreet Kaur [1], V. S. Dhaka[1] and Gurpreet Singh[2,a]

[1]*Department of Computer Science, Jaipur National University, Jaipur, Rajasthan, India*
[2]*Deparment of Computer Science and Engineering, YIET, Gadholi, Yamuna Nagar, Haryana, India*

Abstract. Mobile Ad-hoc Network (MANET) is a group of moving nodes which can communicate with each other without the help of any central stationary node. All the nodes in the MANET act as router for forwarding data packets. The nodes in the network also move randomly and there exists no fixed infrastructure. So, path breaks are the frequent problem in MANET. The routing protocol faces a lot of problem due these path breaks. Therefore, the routing protocol which is multipath in nature is more reliable than a unipath routing protocol. Ant colony optimization is a relatively new technique which is suitable for the optimization problems. AOMDV is a multipath routing protocol. Thus, if there happens to be path break, the packets can start following the new path which has already been selected. In this paper, we are trying to add ant's agents into AOMDV behavior. In this way, the new protocol will be benefited by the dual properties i.e. of ant's nature and multipath nature of AOMDV. The modified concept is simulated and the outcomes are compared with AOMDV, AODV and DSR routing protocols for few performance parameters. Results obtained are encouraging; the new algorithm performs better than traditional unipath and multipath routing protocols.

1 Introduction

Mobile Ad-hoc Networks are wireless networks consisting of entirely mobile nodes that communicate on the move without using centralised base stations. Infrastructure based systems that use as WiFi technology utilizes a central device to deliver high speed connectivity and high quality of service to the user. Infrastructure wireless networks where each user directly communicates with base station or access point [1]. A Mobile Ad-hoc Network system is a group of mobile devices which provides the ability to stream voice, data and video between capricious pairs of devices utilizing the others as relays to avoid the need for infrastructure. There are many techniques which provide robust MANET capability, including MESH networking Technology. The communication in MANET is take place by using multi hop path. When a Mobile node wants to communicate with another node then a source node communicate with node in its transmission range it can transmit data to the destination node [6]. On the other hand, if the destination node is not in the range of the source node, then intermediate nodes acts as both routers that enable the source node to communicate with the destination node. The network may keep changing randomly, because of the mobility of nodes. Many routing protocols are designed to find out a path from source node to destination node. In MANET Routing Protocols are divided in mainly two parts- 1) Table-Driven Routing Protocol 2) On Demand Routing protocols. Table-Driven Routing protocols use the old manners of acquiring routing in mobile ad-hoc networks [21]. These protocols maintain the consistent overview of the network. Each node in the network uses routing tables to store the location information of other nodes. This information is used to transfer data between various nodes of the network. In On-Demand routing protocols, if a source node requires a route to the destination node for which it does not have route information, it starts a route discovery process which goes from one node to the another node until it reaches to the destination node or an intermediate node has a route to the destination [27].

ACO-Ant Colony Optimization is a technique for optimization. The inspiration of ant Colony Optimization is the foraging behaviour of real ant colonies [5]. This behaviour of ant is exploited in artificial ant colonies for the search of appropriate solutions to distinct optimization problems, to continuous optimization problems and main problems comes in telecommunications, such as routing and load balancing.ACO uses pheromone communication between ants regarding a best path between the colony and a food source in an environment [12]. This mechanism is called stigmergy. Ants roam randomly around in surroundings for search of food. Ants lay down pheromone over the path on which they travel. Enormous trips between the food and the colony are performed. If the same route is followed again and again that directs the path with additional pheromone decomposition, so that older paths are less likely to be followed [20]. Other ants may discover the same path to the food and start follows it and also lay down pheromone on the same path. A positive

[a] Corresponding author: gps_ynr@yahoo.com

 DOI: 10.1051/matecconf/20165702005

feedback process routes more and more ants to productive paths.

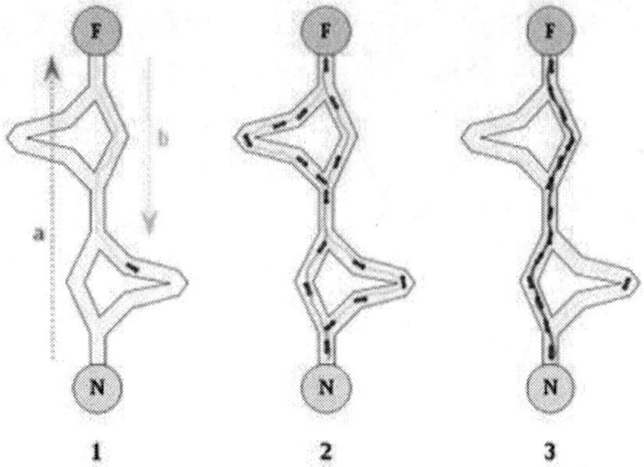

Figure 1: Shortest Path Discovery by Ants

Multipath routing allows building and use of multiple paths for routing between the pair of source and destination [3]. Multipath routing is the routing technique of using multiple alternative paths in a network, which can afford a variety of benefits such as improved security, fault tolerance, or increased bandwidth [4]. The multiple paths computed might be overlapped, edge-disjointed or node-disjointed with each other in the network. Multipath routing offers many advantages. Multipath routing creates several routes for a source-destination pairs. If one route fails a backup route will be available for transmission [9]. Multipath routing can handle mobility of nodes in the network efficiently by tracking intermediate nodes and destinations. This combine approach reduces route recovery time when a main route fails. In network infrastructure, classical multipath routing allows load balancing among multiple routes, reducing network traffic congestion and improving the overall quality of service (QoS) [10].

2 DSR

Dynamic Source Routing (DSR) is a reactive multi-hop routing protocol and establishes a route when a transmitting mobile node requests for data communication [11]. It uses source routing instead of using a routing table at the intermediate nodes. DSR allows the network to be completely self-organizing and self-configuring without the help of any fixed node or any central administration. The protocol is composed of two mechanisms namely "Route Discovery" and "Route Maintenance" that operate on demand [13]. Route Discovery is the mechanism with the help of which a source node sends a packet to a destination node for obtaining the route. Route Maintenance is the technique by which source node is able to maintain a route to destination on change in network topology. Route maintenance is used when source node is no longer able to use the already discovered route to destination [17]. Route Maintenance ensures that the communication path remains optimum and loop-free according the change in network conditions, even if this requires altering the route during a transmission. DSR being a reactive beaconless routing protocol that does not need to periodically flood the network for updating the routing tables like table-driven routing protocols do [18]. All aspects of the protocol operate entirely on-demand, allowing the routing packet overhead of DSR to scale automatically to only that needed to react to changes in the routes currently in use. Intermediate nodes are able to utilize the route cache information efficiently to reduce the control overhead. If the route is in the route cache of target node, the route would be used. Otherwise, the node will reverse the route based on the route record in the Route Reply message header (symmetric links) [26]. In the event of fatal transmission, the Route Maintenance Phase is initiated whereby the Route Error packets are generated at a node. The incorrect hop will be detached from the node's route cache; all routes containing the hop are reduced at that point.

3 AODV

Ad hoc On-Demand Distance Vector Routing Protocol (AODV) is a reactive routing protocol, in which the routes are constructed only when needed [1]. AODV maintains a routing table where routing information about the active paths is stored. AODV protocol uses four control packets: Hello messages, Route Requests (RREQs), Route Replies (RREPs), and Route Errors (RERRs). Each node in the network maintains a routing table which includes: Destination, Next Hop, Number of hops, Sequence number for the destination, Active neighbours for this route, and Expiration time for the route table entry. Each time a route entry is used [19]. The sequence number is used to ensure loop freedom in distance vector routing protocols. The sequence number is sent with RREQ and RREP and stored in the routing table. The highest sequence number indicates the newer route information. If a new route is extended, the sequence numbers of the new route and the existing route are compared. The route with the highest sequence number is used. If the sequence numbers are the same, then the new route is selected only if it has fewer numbers of hops [22]. AODV is compiled of two mechanisms:1) Route Discovery and 2) Route Maintenance.

A) Route Discovery:- Whenever a node needs to send data to a destination, firstly it checks its routing table if it has a valid route to the destination. If a route is found, the node starts to send the data to the next hop. Otherwise, it begins a route discovery process. In the route discovery process, a route request (RREQ) and route reply (RREP) packets are used to demonstrate a route to the destination. RREQ is broadcast throughout the entire network. Upon receipt of RREQ, the node creates a reverse routing entry towards the source, which can be used to forward replies later [24]. The destination or an intermediate node, which has a valid route towards the destination, answers with a RREP packet. When a node receives RREP, a reverse routing entry towards the conceiver of RREP is created, the same as with the processing of RREQ. The precursor list is created at the same time and associated with each routing entry. The precursor list includes the upstream nodes which use the node itself towards the same destinations.

B) Route Maintenance:- Each node with an active route sporadically broadcasts HELLO messages to its neighbors. If the node does not receive a HELLO message or a data packet from a neighbor for a certain amount of time, they started to rebuild the route towards the destination [25]. Otherwise, a Route Error packet is sent to the neighbor, which in turn propagates the RERR packet towards nodes whose routes may be affected by the broken link. Then, the affected source can reinsert a route discovery process if the route is still needed.

4 AOMDV

Ad-hoc On-demand Multipath Distance Vector Routing Protocol (AOMDV) for computing multiple loop-free and link disjoint paths [8]. AOMDV is a extension of AODV. AOMDV computes multiple paths during route discovery procedure for contending link failure. When AOMDV builds multiple paths, it will select the best path for data transmission which is based on the time of route establishment. The routing entries for each destination contain a list of the next-hops along with the corresponding hop counts [15]. All the next hops in the network have the same sequence number. This sequence number helps in keeping track of a route. For destination, a node maintains the advertised hop-count, which is delimited as the maximum hop count for all the paths and used for sending route advertisements of the destination in the network. Each duplicate route advertisement received by a node, which defines an alternate path to the destination [16]. In AOMDV environment, Loop freedom is insured for a node by accepting alternate paths to the destination if it has a less hop count than the advertisement. When a route advertisement is received for a destination with a greater sequence number, the next-hop list and the advertised hop count are re-inspected. AOMDV can be used to find link-disjoint or node-disjoint routes. In the link-disjoint paths, the destination node adopts a looser reply policy. It replies up to k copies of RREQ arriving via singular neighbours, discounting the first hops of these RREQs [23]. Singular neighbor who presents in the network guarantees link disjointness in the first hop of the RREP. Outside the first hop, the RREP follows the reverse route that have been set up already which are node disjoint. Each RREP arriving at an intermediate node takes a different reverse route when multiple routes are already available.

5 OANTALG

Orientation based ANT Algorithm (OANTALG) has been proposed recently for Routing in MANETs [2]. This algorithm considered the orientation factor for guiding the packets to travel in appropriate direction for reaching to destination. This algorithm proposes the new way of random selection of both the end nodes i.e. the source as well as the destination nodes. During the movement of ants, various data structures including the routing tables (pheromone tables) are made for various purposes like for storing ant's trip time and etc. This algorithm uses two types of ants; Forward Ant (FANT) and Backward Ant (BANT) [7]. Forward ant is originated at the source node and sent towards the destination node to explore the path. When FANT arrive at the destination, it is translated to backward ant (BANT) and sends to source node in the reverse path as followed by the FANT. The new algorithm can be explained as follows [14].

Step 1: At fixed time of intervals FANTs are send from randomly selected source nodes to a randomly selected destination nodes within the network. This will help in covering the larger view of the topology. The FANT while travelling from intermediate nodes choose the next node by calculating probability [2]. The FANT which is currently located at node i and leading towards the destination node d chooses next node j by probability (Ω_{ijd}) that is calculated as follows:

$$\Omega_{ijd} = \frac{\rho_{ijd} + (Cos\,\alpha)\theta\Gamma_{ij}}{(Cos\,\alpha)\theta|N_i| + (1-\theta(Cos\,\alpha))} \quad \text{...........(1)}$$

Where ρ_{ijd} represents pheromone, Γ_{ij} is heuristic value that consider the length of queue on $i-j$ th link. θ represents the heuristic function, $Cos\,\alpha$ is calculated as the cosine value between the axial coordinates of the two orientation vector $v1$ and $v2$. The heuristic value Γ_{ij} is a managed value function which depends on the queue length l_{ij} on the link connecting the node i with the neighbours j and number of neighbours ($|N_i|$) of node i which is given by:

$$\Gamma_{ij} = 1 - \left(\ell_{ij} / \sum_{i=1}^{|Ni|} \ell_{ij}\right) \quad \text{........................(2)}$$

$$Cos\,\alpha = \frac{v1 \cdot v2}{|v1| \cdot |v2|} \quad \text{...........................(3)}$$

Step 2: If the chosen node i.e. next node has not selected previously for this path search, then the chosen node is added to the memory of ant. If this node has already been selected, then this may lead to the loop or the cycle [14]. So, this node is erased from the current path.

Step 3: When FANT reach the destination node, it carries with it the traversed path stored in the ant's memory [14]. The new algorithm OANTALG uses time of this trip time along with the local statistical model factors in finding the strength of the new path. After successful enumeration of the new path, BANT is created at the destination node.

Step 4: BANT pass through the same path as traversed by the relative FANT in reverse direction. BANT deposits pheromone $\rho(i)$ on the backward path to increase the strength of the chosen path source node [2]. Pheromone being a chemical substance evaporates from the other paths where pheromone is not deposited. The amount of pheromone deposited or evaporated depends on the pheromone update equations given as below:

If (node is in the way of the ant) then

$$\rho(i) = \rho(i) + r * (1 - \rho(i)) + r^2 * \rho_{high} / C \quad ....(4)$$

Else

$$\rho(i) = \rho(i) - r * \rho(i) - r^2 * \rho_{high} / (C * (|N_i| - 1)) \quad .....(5)$$

Where ρ_{high} represents the highest value of the pheromone obtained till then in the present iteration. $|Ni|$ counts for the total nodes present in the neighbouring area of the node i. C is a constant value dependent on the experimental time in code. For simulation in this paper, we have assumed the value of C as 1000 to counterpart the calculations in done milliseconds in experiment. r is the reinforcement factor and is a dimensionless constant whose value ranges between (0,1].

6 Few Issues to be addressed while using multipath routing in AOMDV by using ACO

(i) Few Multipath routing protocols in MANETs use route cache. There is not any mechanism used by most algorithms to purge stale routes related to multipath routing. As, algorithms send the data packets through multiple paths, so the packets often trapped into problem of using stale routes. This is because of the reason that route information in route cache over some paths may be obsolete.

Proposed Solution- We shall be using a uniform timeout policy to expire the stale routes belonging to same destination. If none of the available path to destination is used until time expire. Then all the paths are invalidated.

(ii) Whether an algorithm should use node disjoint path or link disjoint paths.

Proposed Solution- We shall be using the inherent feature of AOMDV for using link disjoint paths.

(iii) AOMDV uses static route change policy ie. When there happens to be a route error then AOMDV chooses an available route regardless of dynamic nature of the MANETs.

Proposed Solution- As ants continuously change the pheromone value, so alternative routes are chosen with better pheromone quantity without any extra burden.

7 Simulation and Results

In order to carry out practical work, we simulated the new algorithm using network simulator NS-2. Topology of 90 nodes was simulated for the four algorithms DSR, AODV, AOMDV and Ant based AOMDV (A-AOMDV). DSR and AODV belong to unipath environment with reactive approach. AOMDV and A-AOMDV are two multipath algorithms that are also using reactive approach. Similar topology configuration was made to execute for all the four algorithms i.e. DSR, AODV, AOMDV and A-AOMDV. All the four algorithms were executed for an equal time of 150 seconds on a flat bed of 1800 x 1800 axis. The nodes were made to run randomly with the minimum speed of 10.0 m/s and the maximum speed of 50.0 m/s. The pause time was set equivalent to 1.0 m/s. The topology was simulated with the following characteristics shown in table 1.

Table 1: Simulated Topology Parameters

Number of Nodes	90
Topology Size (X x Y) axis	1800 x 1800
Routing Protocols	DSR, AODV, AOMDV, A-AOMDV
Simulation Time	150 Seconds
Minimum Speed of Nodes Movement	10.0 m/s
Maximum Speed of Nodes Movement	50.0 m/s
Pause Time	1.0 ms

Although all the protocols were executed in an environment with similar configurations as shown in table 1. However, Snapshots of the executed topology for A-AOMDV were taken at four moments i.e. approximately at beginning, two times at middle and at end. This is shown by four figures 2(i), 2(ii), 2(iii) and 2(iv).

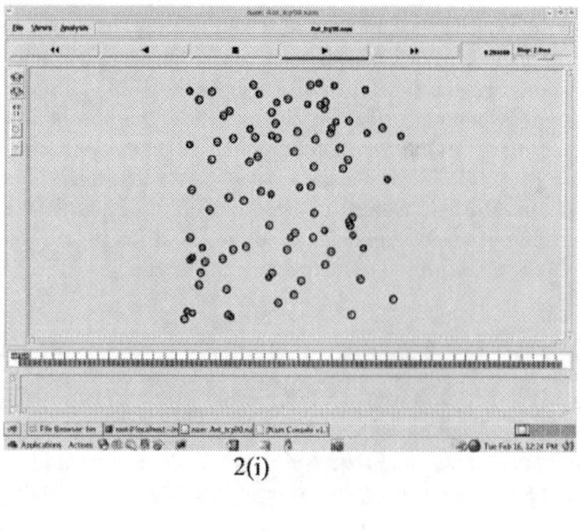

2(i)

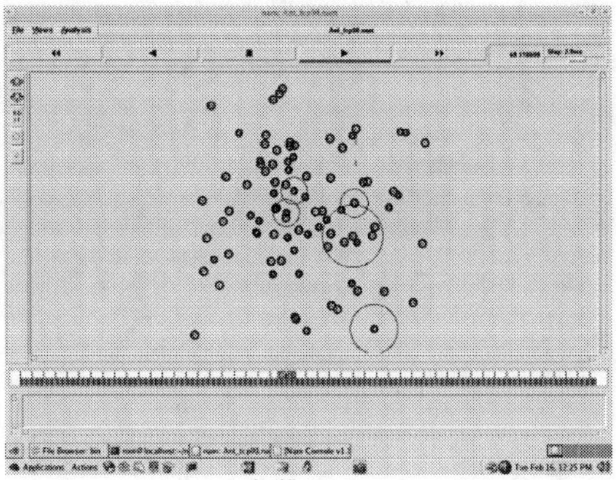

2 (ii)

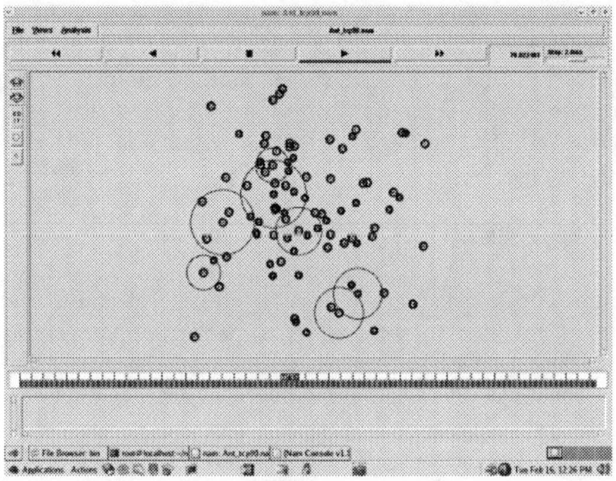

2(iii)

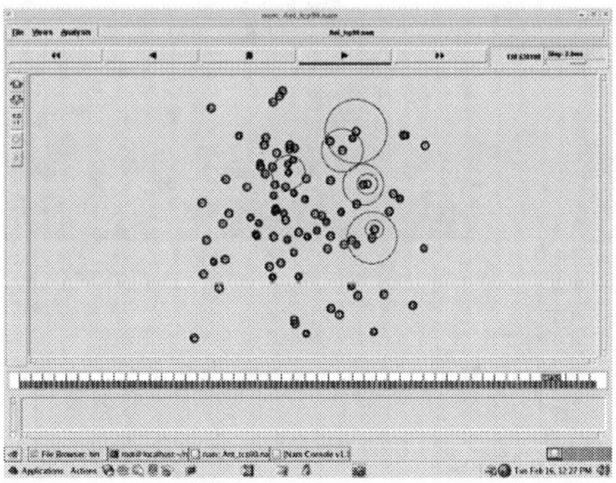

2 (iv)

Figure 2(i) – 2(iv): Snapshot of Running Topology at different instance of times

The above figures represent the random snapshots of the executed topology for A-AOMDV. The topologies were also created for DSR, AODV and AOMDV with the similar parameters. The traffic of TCP packets was generated for all topologies at same moment of time with exactly same time interval. The trace files for network animation and traffic were created as a result of execution for all protocols. The trace file of network animation can be used to generate the video simulation as shown in figure 2(i) – 2(iv). The traffic can be analyzed by writing the extra script for reading the network trace file. These scripts can either be written in AWK, Perl, C/C++ or in any HLL who support the file handling issues. We wrote the scripts in Perl language to extract the useful network parameters like throughput, packet drop, total number of packets send, average end to end delay, and average jitter. These parameters were obtained and a comparison table if made from the values as shown in Table 2. This table can further be used to create the graphical comparison between different algorithms.

Table 2: Network parameters and their values for different protocols.

	AODV	DSR	AOMDV	A-AOMDV
Packet Drop	14	36	1	3
TCP Packet Send	3949	3551	4664	4969
Throughput (Bytes/Sec)	27206.4	24353.6	32103.74	34225.067
Average End to End Delay	0.44537	0.88250	0.3296	0.3633505
Average Jitter	0.1703	0.57141	0.1463	0.1825183

Packet Drop: This network parameter measures the number of packets dropped in the middle of path. The packets are dropped due to many reasons like when the queues at the intermediate nodes get full or when the time to live value of packet get expired or etc. We measured the packet drop due to the reason of fullness of queues.

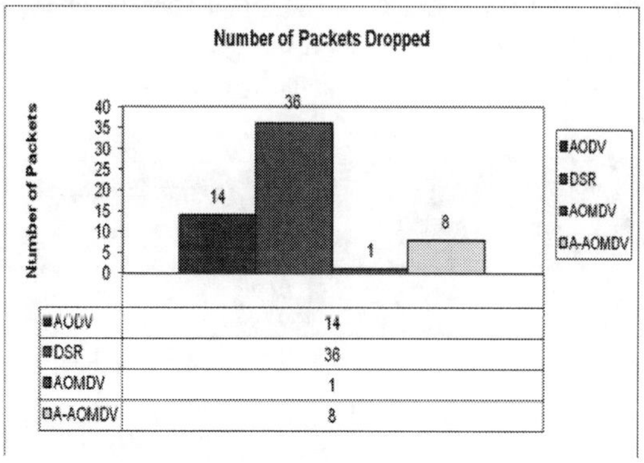

Figure 3: Packet drop for different protocols

Figure 3 represents the packet drop values for the different parameters. It can be observed that the AOMDV has minimum packet drop whereas A-AOMDV holds second place during the comparison.

Number of TCP Packets Send

This parameter helps in measuring the overall capacity of the algorithm to forward the data packets in given interval of time. As already stared above that all the protocols were made to forward the data packets for equal amount of time from similar nodes. So, this will evaluate in determining the ability of a protocol to send maximum data in given amount of time.

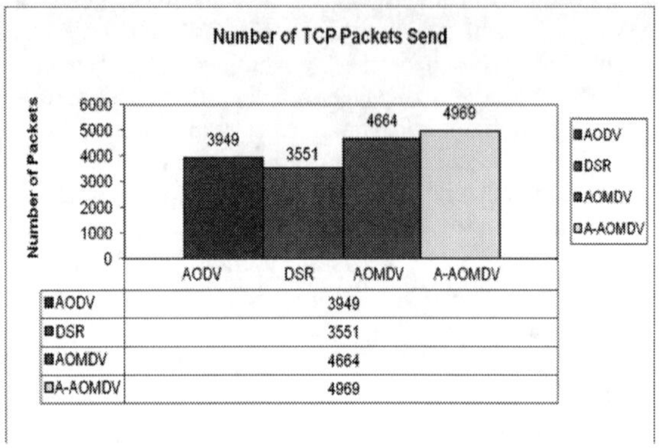

Figure 4: Number of packets forwarded in given time

Figure 4 represents the result to measure the superiority of an algorithm to distribute the larger amount of data in fixed time. The given graph revels that A-AOMDV can help distributing the larger amount of data within same time.

Average Throughput

Throughput is measured as the amount of data send in a unit of time. Throughput can be measured in bits/second, bytes/second or number of packets/second. We have evaluated this parameter in bytes per second. To calculate the average throughput, we counted the amount of data send in bytes and then divided it with the overall time.

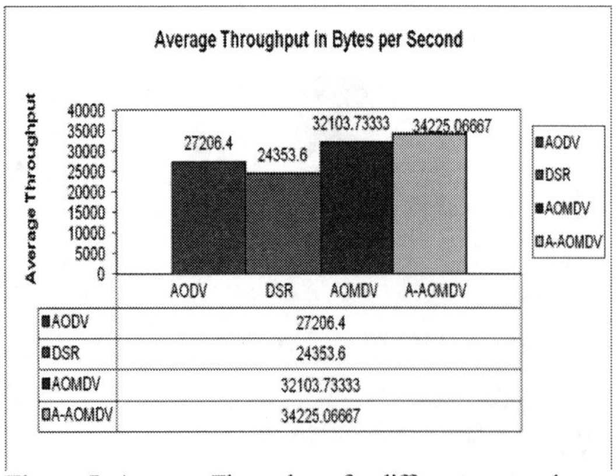

Figure 5: Average Throughput for different protocols

Average End to End Delay

End to end delay measures the time taken by a packet to reach the destination. This factor sometimes is also called as simply delay. This factor should be minimized, so that a packet takes less time to reach its destination.

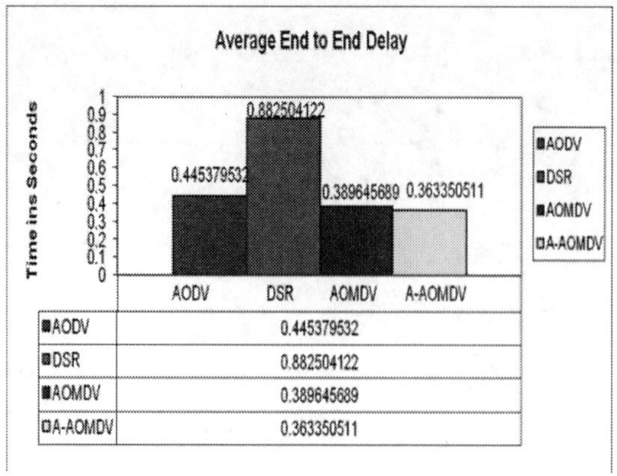

Figure 6: Average end to end delay for different protocols

The data represented by the graph of figure 6 signify the values of end to end delay for various protocols. The values obtained justify that the A-AOMDV has least value of end to end delay as compared to other algorithms.

Average Jitter

Jitter is calculated as the drift in the time of end to end delay for the consecutive data packets. Jitter is the undesired deviation from actual packet delay time. This parameter should also be minimized in order to get better delivery. A higher delay generally indicates congestion of some sort or breaking of the communication link.

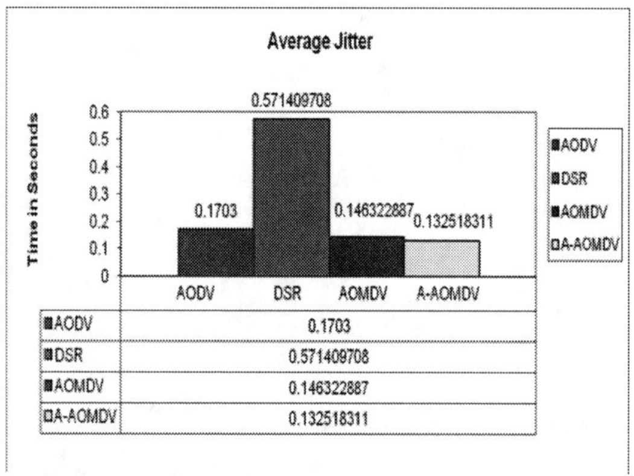

Figure 7: Average Jitter for different protocols

The above figure 7 shows the value of average jitter for different protocols. A-AOMDV also outperforms other protocols in this parameters by achieving least value.

8 Conclusion

Routing is the one of the major challenge in MANETs due to the moving nature of nodes leads to frequent breakage of path. The routing protocols have to reconfigure the paths again in order to send the data packets. Multipath routing uses multiple paths in order to send the packets instead of single path. If there happens to be a path break in multipath routing, the data packets can still be sent through the alternative paths. In the meantime, the path reconfiguration can be achieved. Ant Colony is relatively the branch of swarm intelligence to help solving optimization problems. In this paper, we have tried to add agents of ants with AOMDV protocol to enhance its performance. The performance was measured using the metrics like throughput, packet lost, number of packets send, end to end delay and jitter. The modified protocol outperforms other protocols in most of the parameters. However, the AOMDV performs better than A-AOMDV protocol in packet drop parameters. But, in all other parameters, A-AOMDV outperforms the other three protocols i.e. throughput, number of packets send, end to end delay and jitter.

References

1. C. E. Perkins, E. M. Royer and S. R. Das, Ad Hoc On-Demand Distance Vector (AODV) Routing, *In Proceedings of IEEE Workshop on Mobile Computing Systems and Applications,* pp. 90-100, Feb. (1999).
2. Gurpreet Singh, Neeraj Kumar and Anil Kumar Verma, OANTALG: An Orientation Based Ant Colony Algorithm for Mobile Ad Hoc Networks, Wireless Personal Communication, Springer Science, **Volume 77, Number 3**, pp. 1859-1884, (2014).
3. Avinash Giri, Jitendra Prithviraj and Ashok Verma, Analysis of DHT Based Multi-Path Routing Protocol with Other Routing Protocols in MANETS, International Journal of Engineering Science and Innovative Technology (IJESIT), **Volume 1**, pp. 91-96, (2012).
4. Ducatelle, F., Caro, G. D., and Gambardella, L. M., Ant agents for hybrid multipath routing in mobile ad hoc networks, *In the proceedings of second annual conference on wireless on-demand network systems and services, 2005. (WONS 2005), Manno-Lugano, Switzerland* , pp. 44–53, (2005).
5. Amanpreet Kaur, V. S. Dhaka and Gurpreet Singh, Voyaging Through Various Ant Based Routing Algorithms For MANETs, International Journal of Advanced Technology in Engineering & Science, **Volume 3, Number 1,** Pages 468-480, September 2015.
6. Kumar, G. V., Reddyr, Y. V., and Nagendra, M., Current research work on routing protocols for MANET: A Literature Survey, International Journal on Computer Science and Engineering (IJCSE), **Volume 02, Number 03**, pp.706–713, (2010).
7. Singh, G., Kumar, N., and Verma, A. K. , Ant Colony Algorithms in MANETs: A Review, Journal of Network and Computer Applications, **Volume 35, Issue 6,** pp. 1964–1972, (2012).
8. V. Lalitha and R. S. Rajesh, Analysis of Multipath MANET Routing Protocols under Variable Range Transmission Power, International Journal of Advanced Networking and Applications, **Volume 04,** pp. 1737-1743, (2013).
9. Amanpreet Kaur, V K Dhaka, Gurpreet Singh and T. P. Singh, Suitability of Ant Colony Optimization as an application to MANET, *In the Proceedings of National Conference on Current Research Trends in Cloud Computing and Big Data, Jaipur National University, Jaipur,* 6-7 February, 2015.
10. Gaurav Sachdeva and Sukhvir Singh, Energy Efficient DHT Based Multipath Routing in Wireless Sensor Networks, International Journal of Advanced Research in Computer Science and Software Engineering, **Volume 3**, pp. 985-990, (2013).
11. P.S. Patheja , Akhilesh A. Waoo and Lokesh Malviya, Multipath Dynanic Source Routing Protocol for Ad-Hoc Network, International Journal of Emerging Technology and Advanced Engineering, **Volume 2**, pp. 436-439, (2012).
12. Caro, G. D., and Dorigo, M., Ant Colonies for Adaptive Routing in Packet-Switched Communications Networks, *In Proceedings 5th international conference of parallel problem solving from nature. London: Springer*, pp. 673–682, (1998).
13. Bakalis, P. and Lawal, B., Performance Evaluation of CBR and TCP Traffic Models on MANET Using DSR Routing Protocol, Communications and Mobile Computing (CMC), **Volume 3,** pp. 318-322, (2010).
14. Amanpreet Kaur, V. S. Dhaka and Gurpreet Singh, Casting multipath behaviour into OANTALG to improve QoS, *International Conference on "Computing for Sustainable Global Development*, 16 – 18 March, (2016).
15. Jack Tsai and Tim Moors, A Review of Multipath Routing Protocols: From Wireless Ad Hoc to Mesh Networks, *published in Proc. of ACoRN early Career Researcher workshop on wireless Multihop networking*, (2006).
16. Carlos T. Calafate, Jose Oliver, Juan-Carlos Cano, Pietro Manzoni, Manuel P., and Maluel P. Malumbres, A distributed admission control system for MANET Environments Supporting Multipath Routing Protocols, Journal of Microprocessors and Microsystems, **Volume 31**, pp. 236-251, (2007).
17. Leung R., Julei Poon, E;Chan, and A. L. C. Baochaun Li, MP-DSR:- a Qos-aware multipath dynamic source routing protocol for wireless ad-hoc Network, IEEE Xplore Digital Library, pp.14-16, (2001).
18. Sarala P. and Kalaiselvi D., Multipath Dynamic Source routing with cost and Ant Colony Optimization for MANETs, International Journal of Applied Engineering Research, **Volume 1, No. 1,** pp. 83-91, (2010).
19. Anuj K. Gupta, Harsh Sadawarti and Anil K. Verma, Performance Analysis of MANET

Routing Protocols in Different Mobility Models, International Journal of Information Technology and Computer Science (IJITCS), **Volume 5, Issue 6,** pp. 73-82, (2013).

20. Dhull, D. and Dhull,S., An improved ant colony optimization (IACO) based multicasting in MANET, International Journal of Inventive Engineering and Sciences (IJIES) **Volume 1, Issue 3**, pp. 8–12, (2013).
21. Rajeev Tiwari and Neeraj Kumar, Cooperative Gateway Cache Invalidation Scheme for Internet-Based Vehicular Ad Hoc Networks, Wireless Personal Communications, **Volume 85, Issue 4**, pp. 1789-1814, (2015).
22. Umara Urooj, Nafees Ayub, Ramzan Talib, M. Umer Sarwar and M. Yahya Saeed, Comparative Analysis of Proactive and Reactive Routing Protocols with respect to Quality of Service in MANET, Journal of Environmental Science, Computer Science and Engineering and Technology, **Volume 3,** pp. 2009-2025, (2014).
23. Marcello Caleffi and Luigi Paura, M-DART: Multi-path Dynamic Address Routing, Wireless Communications and Mobile Computing, Wiley Online Library, **Volume 11, Issue 3,** pp. 392-409, (2011).
24. Bettstetter C. and Wagner C., The Spatial Node Distribution of the Random Waypoint Mobility model, *In Proceedings of the German Workshop on Mobile Ad Hoc Networks (WMAN)*, pp. 1-15, (2002).
25. R. R. Choudhary, S. Bhandhopadhyay and K. Paul, A Distributed Mechanism for Topology Discovery in Ad Hoc Wireless Networks Using Mobile Agents, *In Proceeding of Mobile and Ad Hoc networking and Computing,* pp. 145-146, (2000).
26. Sanjeev Kumar and Yogesh Chaba, Quality of Service Aware-Modified DSR protocol in Mobile Ad-Hoc network, International Journal of Advanced Research in Computer Science and Software Engineering, **Volume 5, Issue 1**, pp pp. 1072-1078, (2015).
27. Tarunpreet Bhatia and A.K. Verma, QoS Comparison of MANET Routing Protocols, International Journal of Computer Network and Information Security, **Volume 9**, Pages 64-73, (2015).

MATEC Web of Conferences **57**, 02006 (2016) DOI: 10.1051/matecconf/20165702006
ICAET-2016

Efficient Job Provisioning for a Cloud Service Provider

Dharmvir Sharma[1,a], Tejinderpal Singh[2]

[1]*Student, CSE Department, Yadavindra College of Engineering, Talwandi Sabo*
[2] *Assistant professor, CSE Department, Yadavindra College of Engineering, Talwandi Sabo*

Abstract. Cloud Computing is a very fast emerging technology as every enterprise is moving fast towards this system. Cloud Computing is known as a provider of dynamic services. It optimizes a very large, scalable and virtualized resource. So lots of industries have joined this bandwagon nowadays. One of the major research issues is to maintain good Quality of Service (QoS) of a Cloud Service Provider (CSP). The QoS encompasses different parameters, like, smart job allocation strategy, efficient load balancing, response time optimization, reduction in wastage of bandwidth, accountability of the overall system, etc. The efficient allocation strategy of the independent computational jobs among different Virtual Machines (VM) in a Data center (DC) is a distinguishable challenge in the Cloud Computing domain and finding out an optimal job allocation strategy guided by a good scheduling heuristic for such an environment is a mape-k loop problem. So different heuristic approaches may be used for better result and in this result we paper we implement worst fit in mape-k and evaluated the results.

1 Introduction

Cloud gives the administrations to the customer on the web. Cloud administrations agree to people and organizations to make utilization of equipment and programming that are over seen by outsiders at far-away areas. Illustrations of cloud administrations incorporate long range informal communication destinations, webmail, online business applications and online document stockpiling, Email customers like Yahoo! furthermore, Gmail, Wikipedia, YouTube, Skype or Bit Torrent and some more. Cloud has Pay per use measured administration and Cloud provider uses this metered service according to their utilization. Websites like Hurray, Email, Gmail, or Hotmail and so on [1]. Millions of resources are submitted to computing task to cloud system. It is challenge to cloud computing environment.

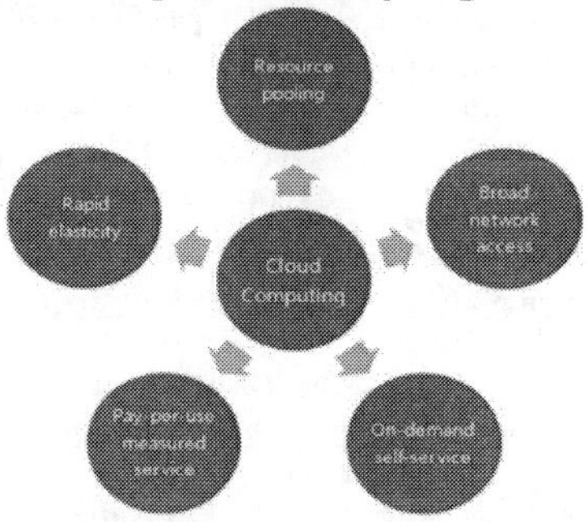

Figure 1. shows the services at clouds.

Different characteristics like On-demand self service, Ubiquitous network access, Location independent resource pooling, Rapid elasticity, Pay per use [2] and scheduling strategies are used. Cloud computing have three categorized of service [1]:
1). SaaS (software as a service)
2). PaaS (platform as a service)
3). IaaS (infrastructure as a service)

Allocation of resources not guaranteed Quality of service (QoS). Constraints clients via service level agreements (SLAs). Virtualization is the most essential procedure. With the help of virtualization, we can make version of a server, storage, an operating system or resources of system [3]. The cloud computing is a distributed computing system and it is a collection of virtualized and interconnected machines that provides many resources to the consumers depending upon the SLA (Service Level Agreement). Cloud suppliers give computational assets in the variety of virtual machines (VMs) to consumers on through appropriate geographically distributed data centres. According to consumer request of resources like virtual machines, storage and applications cloud distributes these resources and these assets are given through scheduling. Scheduling is a technique in which requests of tasks are scheduled according to given requirements. . The VM (virtual machine) scheduling algorithm is utilized for scheduling the virtual machines to the physical machine according to necessity of the specific data center [4].

[a]Dharmvir305@gmail.com

2 Related work

Load balancing and VM placement to host has always been a research subject whose objective is to ensure that every computing resource is distributed efficiently and fairly and in the end improves resource utility. In traditional computing environments of distributed computing, parallel computing and grid computing, researchers in and abroad have proposed a series of static and dynamic and mixed scheduling strategies.

Live migration of VMs permits administrator of server to move a running virtual machine or application among diverse physical machines without separating the customer or the application .The total migration time and the aggregate downtime are two key execution metrics that the customers of a VM administration stress over the most . This paper shows live virtual machine migration as a fundamental tool for dynamic resource management in modern day data centers. In this paper a survey of the current research efforts on live virtual machine migration is presented [5].

Evaluated configurations which is based on historical data and a utility function that ranks them according to the costs incurred. The proposed adaptive framework has been integrated into the Vienna Cloud Environment (VCE) and the evaluation by means of a data-intensive application is presented here in [6]. An important feature provided by the Virtual Machine (VM) advancements is its capacity to move the running operating system instances over the particular physical nodes. This paper studies the proposed and implemented methods for the VM migration [7]. The data centers are a collection of many servers; allocation of VM to Host is known as VM placement. VM placement problem is examined in this paper with focus for maximum utilization of the resources and energy reduction [1]. An algorithm is formulated that will keep the migration time least and also minimizing the number of migrations [7]. Tuning instrument used for proficient asset provisioning guarantying required quality of service. The proposed tuning mechanism also considers load balancing into account [7].

The title is set in bold 14-point Arial, flush left, unjustified. The first letter of the title should be capitalised with the rest in lower case. You should leave 35 mm of space above the title and 6 mm after the title.

3 VM placement

In this portion an approach in made is for VM placement which effectively solves the problems of minimizing the response time, load balancing, and balance resource utilization and server consolidation in cloud datacenter. In this work three dimensional problems are solved a) optimizing network latency b) improving utilization and c) improving the retrieval system from database.

In this algorithm the hosts are classified according to their resource availability. In general there are three types of performance parameters of any system i.e. CPU, B/W and Memory. The allocation of the VM is typically done by Mape-k loop in which the sensor to sense the VM status and create a plan to allocate job requests to VMs. However after allocation a large number of resources remain under utilized. The algorithm shows the total remaining underutilized resources is accumulated and is allocated to next requests. Let us assume that when request for VM arrived to scheduler place this VM on the satisfying host in the list in which all host have remaining resource capacity in order. In other words if a VM having CPU requirement greater than or equal to Memory requirement and Memory requirement greater than or equal to Bandwidth is placed on the satisfying host in the list, in which all host have remaining resource capacity of CPU greater than or equal to Memory and Memory greater than or equal to Bandwidth. The detail is Shown in Algorithm 1. To improve the arranging and retrieving system we are putting worst fit and binary search to make retrieving fast.

Algorithm 1 : Improved mape-K Loop

0: Input: A set of Virtual machines, Resource Demand
1: Output: Resource Provision solution R
2: Max = Estimate_the_total number_of_VMs
3: Lat_matrix = Create a latency matrix (vm_locations)
4: Optimize_utility (CPU_rem, MEM_rem, BW_rem)
5: If the request ε CPU_rem
6: Total_rem_power= $^{i}\Sigma_n$ CPU_i // Get all CPUs_rem in descending order CPU>MEM>BW or CPU>BW> MEM
7: If the request ε MEM_rem
8: Total_rem_power= $^{i}\Sigma_n$ MEM_i // Get all MEMs_rem in descending order MEM>BW >CPUor MEM>BW
9: If the request ε BW_rem
10: Total_rem_power= $^{i}\Sigma_n$ BW_i // Get all BWs_rem in descending order BW >CPU>MEMor MEM>BW
11: Allocate_request to VM_rem (Total_rem_power)
12: Manage_VM_ database (arrange_by _worst fit, sort_by_binary_serach)

In the Worst Fit algorithm hosts in each list in datacenter are sorted in descending order according to remaining capacity of resources. When request of a new VM or already running VM for VM placement arrives at the cloud datacenter, VM scheduler find the appropriate list and apply the binary search on the selected list to find host that is the worst fit in remaining resource capacity.

4 Experimental setup and result

To setup our experiments we use following configuration where the proposed algorithm is implemented is done using CloudSim as a framework in the simulator environment. This work considers various Datacenter, Virtual Machines (VM), host and cloudlet components from CloudSim for execution analysis of the algorithms. NetBeans is an application platform framework for Java applications. It integrated development surroundings (IDE) for developing primarily with Java, however conjointly with alternative languages like PHP, C/C++, and HTML5.NetBeans is associate degree ASCII text file integrated development surroundings (IDE) for developing with Java, PHP, C++, and alternative programming languages. NetBeans is additionally noted as a platform of standard parts used for developing Java desktop applications.

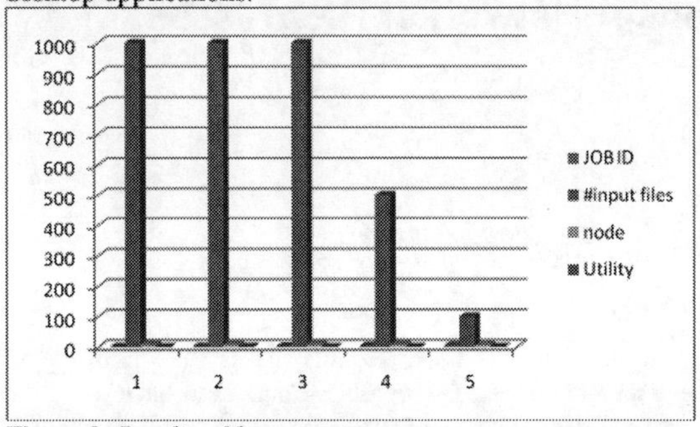

Figure 2: Results without worst case

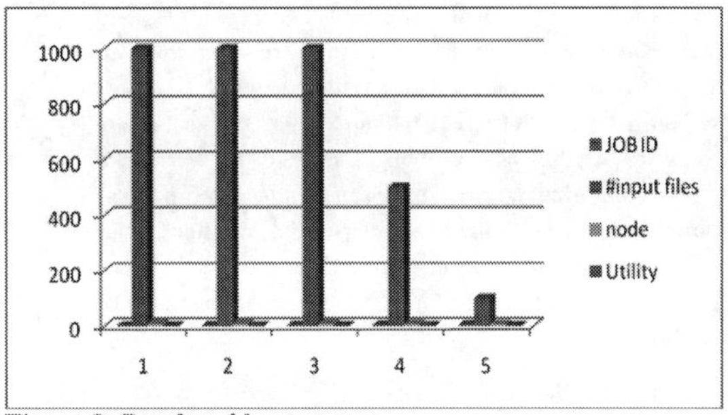

Figure 3: Results with worst case

It has been observed at fig 2 and fig 3 that with the introduction of worst case algorithm there is an increase of utility by average of 4%. Algorithm apply to broker and we test the results give broker job ids are different, nodes are different, input file are different then we compare utility without worst fit with worst fit MAPE-K loop algo. Results shows increase utility with worst fit-MAPE K loop.

5 Conclusion and future work

In this paper, we have discussed about the job allocation to the different VMs inside a Cloud Data Center with the help of the mape k loop, MKL is a classical scheduling policy. It is a part of our whole work. It helps to map the jobs to the appropriate VMs, which will reduce the overall response time and waiting time of the jobs and with the introduction of worst case algorithm we observe a further improvement in performance. We are now trying to develop our system in such a way that will help the module to identify intelligently that how many jobs may be served by a single VM at a certain time stamp. This will improve the overall make span of the Cloud Service Provider.

References

1. N. R. RamMohan, E. Baburaj, International Conference on Computational Intelligence and Communication Networks, 556 (2012).
2. Y. Chawla and M. Bhonsle, International Journal of Emerging Trends & Technology in Computer Science(IJETTCS), 12 (2012).
3. N. Pasha, Dr. A. Agarwal and Dr. R. Rastogi, International Journal of Advanced Research in Computer Science and Software Engineering, 4 (5), 34 (2014).
4. J. Xiao, Z. Wang, International Conference on Cloud Computing and Service Computing, IEEE, 50 (2012).
5. A. Singh. N and M. Hemalatha, Journal of Theoretical and Applied Information Technology, 57(3), 1 (2013).
6. M. Koehler, Koehler Journal of Cloud Computing Advances, Systems and Applications, 1 (2014).
7. S. Banerjee, M. Adhikari, U. Biswas, (2014), International Conference on Business and Information Management (ICBIM), IEEE, 114 (2014).
8. A. Mohan, Shine S, International Journal of Advanced Research in Computer Engineering & Technology (IJARCET), 2(1), 155 (2013).
9. D. Verma, R. K. Somani, International Journal of Engineering Science and Innovative Technology (IJESIT), 3 (6), 193 (2014).
10. S. Venkatesha, S. Sadhu, S. Kintali, Spring Department of Computer Science University of California, Santa Barbara, 1 (2009).

A Survey on Underwater Wireless Sensor Networks: Progresses, Applications, and Challenges

J. Premalatha[1] and Joe Prathap P M[2]

[1] *Research Scholar, Dept.of CSE, Sathyabama University, Chennai, India. premlatha_j@yahoo.com*
[2] *IT Dept, RMD Engineering College, R.S.M. Nagar, Kavaraipettai, Tiruvallur Dist., India. drjoeprathap@gmail.com.*

Abstract. The endangered underwater species always drew the attention of the scientific society since their disappearance would cause irreplaceable loss. Asia is home to some of the most diverse habitats in the earth, but it is estimated that more than one in four species are endangered. In Underwater, a lot of factors are putting marine life under immense pressure. Extreme population pressure is leading to pollution, over-fishing and the devastation of crucial habitats. Consequently, the numbers of almost all fish are declining and many are already endangered. To help these species to survive, their habitat should be strictly protected. This can be achieved by strictly monitoring them. During this course, several parameters, constraints about the species and its environments are focused. Now, advances in sensor technology facilitate the monitoring of species and their habitat with less expenditure. Indeed, the increasing sophistication of underwater wireless sensors offers chances that enable new challenges in a lot of areas, like surveillance one. This paper endorses the use of sensors for monitoring underwater species endangered in their habitat. This paper further examines the key approaches and challenges in the design and implementation of underwater wireless sensor networks. We summarize major applications and the main phenomena related to acoustic propagation, and discuss how they affect the design and operation of communication systems and networking protocols at various layers.

1 Introduction

The cram of Underwater Wireless Sensor Networks as a research field has grown extensively in recent years offering a huge number of proposal to resolve the communication between the nodes and protocols for information exchange networks[1].

Using wsn networks, we can able to pick a lot of values which is used for new purposes like measuring the relief of the ocean (bathymetry), seafloor shape registering, search for geological resources (i.e. oil, gas, etc.), detecting and tracking fish banks, submarine archaeology, etc[2-4]. These were the main underwater acoustic application mainly use for the exploration of seafloor and fishery with sonar devices. In the 90's the researchers became aware of a new feature applicable to underwater communications, multipoint connections could be capable of translating the networked communication technology to the underwater environment[5-7].

Wireless terrestrial networking technologies have experienced a considerably development in the last fifteen years, not only in the standardization areas but also in the market deployment of a bunch of devices, services and applications. Among all these wireless products, wireless sensor networks are exhibiting an incredible boom, being one of the technological areas with greater scientific and industrial development step[8]. Recently, wireless sensor networks have been proposed for their deployment in underwater environments where many of applications such us aquiculture, pollution monitoring, offshore exploration, etc. would benefit from this technology [9]. Despite having a very similar functionality, Underwater Wireless Sensor Networks (UWSNs) exhibit several architectural differences with respect to the terrestrial ones, which are mainly due to the transmission medium characteristics (sea water) and the signal employed to transmit data (acoustic ultrasound signals)[10].

Then, the design of appropriate network architecture for UWSNs is seriously hardened by the conditions of the communication system and, as a consequence, what is valid for terrestrial WSNs is perhaps not valid for UWSNs. So, a general review of the overall network architecture is required in order tosupply an appropriate network service for the demanding applications in such an unfriendly submarine communication environment.

Major challenges in the design of underwater acoustic networks [11] are:

•Battery power is limited and usually batteries cannot be recharged because solar energy cannot be exploited;

•The available bandwidth is severely limited;

[2] Corresponding author: *drjoeprathap@gmail.com.*

•The channel suffers from long and variable propagation delays, multi-path and fading problems;
•Bit error rates are typically very high;
•Underwater sensors are prone to frequent failures because of fouling, corrosion, etc.

2 Communication Architecture

This section will give an outline of underwater wireless networks going-through all the layers with highlighting on the physical layer and how it behaves in different and changing environment conditions. The topology is a crucial factor in determining the capacity, the energy consumption, and the reliability of a network. Hence, the network topology should be carefully engineered and post-deployment topology optimization should be performed, when possible. A reference architecture for two-dimensional underwater networks is shown in Fig. 1. A collection of sensor nodes are anchored to the underneath of the ocean with deep ocean anchors. Underwater sensor nodes are interconnected to underwater sinks (uw-sinks) by wireless acoustic links. Uw-sinks, as shown in Fig. 1, are network devices in charge of relaying data from the ocean ground network to a surface station. To accomplish this objective, uw-sinks are equipped with two acoustic transceivers, namely a horizontal and a vertical transceiver. The horizontal transceiveris used by the uw-sink to communicate with the sensor nodes in order to: (i) send commands and configuration data to the sensors (uw-sink to sensors); (ii) collect monitored data (sensors to uw-sink). The vertical link is used by the uw-sinks to relay data to a surface station.. The surface station is equipped with an acoustic transceiver that is able to handle multiple parallel communications with the deployed uwsinks [12].

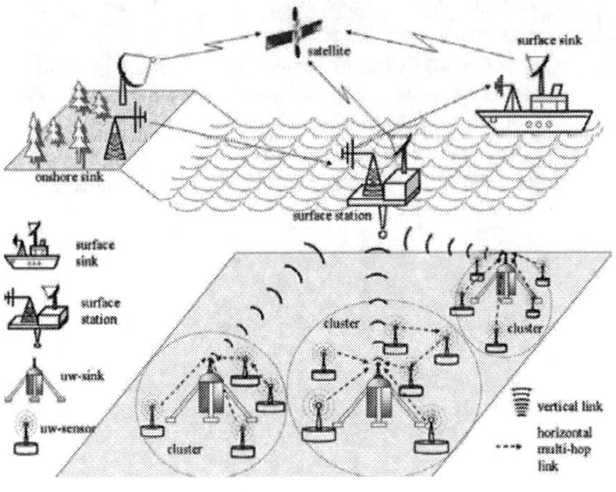

Figure 1. Architecture for 2D UWSN

The typical inner architecture of an underwater sensor is shown in Fig. 2. It consists of a main controller which is interfaced with an oceanographic device or sensor through a sensor interface circuitry. The controller receives data from the sensor and it can store it in the onboard memory, process it, and send it to other network devices by controlling the acoustic modem.

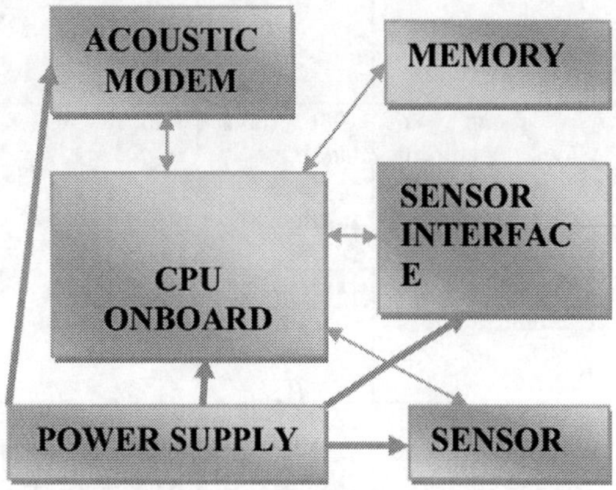

Figure 2. underwater sensor node internal architecture

From the structural design details, the probable challenges related to the operation of low cost, low scale underwater sensors, are listed below:

• It is significant to develop robust, less expensive ''nano-sensors'',

• It is essential to devise periodical cleaning mechanisms against fouling and corrosion, which may impact the lifetime of underwater devices.

• There is a need for steady, robust sensors on a high range of temperatures since sensor drift of underwater devices may be a concern.

• There is a need for new integrated sensors for synoptic sampling of biological, chemical, and physical parameters to improve the accepting of processes in marine systems.

3 Ongoing research work

In Table 1, we report a list of research laboratories and ongoing research projects related to underwater communications and explorations.

Table 1: underwater acoustic sensor networks Research labs and ongoing research projects

Focus area	Research lab/project	URL
Acoustic commun. net for monitoring underwater environments in costal areas	ACME	http://flipper.ncl.ac.uk/acme/
Applications of AUVs, platforms and sensors	Autonomous Undersea Systems Institute (AUSI)	http://www.ausi.org/
Underwater communications	Underwater Acoustic Research Group	http://sonar-fs.lboro.ac.uk/
Underwater acoustic sensor networks	BWN-Lab @ GeorgiaTech	http://www.ece.gatech.edu/research//UWASN/
Underwater acoustic networks	MIT & Woods Hole O.I.	http://www.mit.edu/people//research.html
Spatial sampling of ocean	Front Project	http://www.nopp.uconn.edu/ADCPhtml
Advanced marine systems	Ocean Engineering @ FAU	http://www.oe.fau.edu/research/ams.html
Sea Grant AUVs	AUV Lab @ MIT	http://auvlab.mit.edu/
Underwater acoustic sensor networks	Undersea Seismic Experimentn (SNUSE)@ USC	http://www.isi.edu/ilense/snuse/
Integrated sustained ocean observing system	Underwater Technologies Laboratory @Florida	http://my.fit.edu/swood/subsea.html
Adaptive ocean sampling	Autonomous Ocean Sampling Networks	http://www.princeton.edu/dcsl/aosn/
Autonomous ocean sampling networks	AOSN	http://www.mbari.org/aosn/
Underwater acoustics, sonar, bottom and water column surveys with AUVs, signal processing and target detection	Underwater Research Lab_s @ Simon Fraser University	http://www.ensc.sfu.ca/research/url/

4 Open issues in layers

The following Table 2 focuses issues related to various layers which will give clear focus about research ideas.

Table 2:layer wise issues

Datalink layer	•In case CDMA is adopted, which we powerfully advocate, it is essential to design access codes with high auto-correlation and low cross-correlation properties to achieve least amount interference among users. • Research on finest data packet length is needed to maximize the network efficiency. • It is needed to design low-complexity encoders and decoders to limit the processing power required for forward error correction (FEC) functionalities. Researchers should evaluate the feasibility and the energy-efficiency of nonconvolutional error control coding schemes. • Distributed protocols should be devised to reduce the activity of a device when its battery is depleting without compromising on network availability.
Network layer	• Accurate modeling is needed to better realize the dynamics of data transmission at the network layer. Moreover, convincing simulation models and tools need to be developed. • Algorithms and protocols need to be developed that detect and deal with disconnections due to unforeseen mobility of nodes, failures or battery depletion. • Local route optimization algorithms are needed to react to consistent variations in the metrics describing the energy efficiency of the underwater channel. • Mechanisms are needed to integrate AUVs in underwater networks and to enable communication between sensors and AUVs • In case of geographical routing protocols, it is necessary to devise efficient underwater location discovery techniques. • For delay-tolerant applications, there is a need to develop mechanisms to handle loss of connectivity without provoking immediate retransmissions. • It is necessary to devise routing algorithms that are forceful with respect to the intermittent connectivity of acoustic channels.

Transport layer	• New flow control strategies need to be devised in order to tackle the high delay and delay variance of the control messages sent back by the receivers. • New effective mechanisms tailored to the underwater acoustic channel need to be developed, in order to efficiently infer the cause of packet losses. • Different functionalities at the data link and transport layer such as channel access, reliability and flow control, should be jointly designed and studied. A cross-layer approach is highly recommended to accordingly optimize these mechanisms and make them adaptable to the variability of the characteristics of the underwater channel. • New event transport reliability metric definitions need to be proposed, based on the event model and on the underwater acoustic channel model. • Optimal update policies for the sensor reporting rate are needed, to prevent congestion and maximize the network throughput efficiency as well as the transport reliability in bandwidth limited underwater networks. • The effects of multiple event occurrences on the reliability and network performance requirements must be studied, as well as efficient mechanisms to deal with it. • It is necessary to statistically model loss of connectivity events in order to devise mechanisms, to enable delay-tolerant applications tailored to the specific underwater requirements.

5 Major Applications

As mentioned earlier, underwater acoustic wireless sensor networks can be used in a variety of different applications, as it is done by radio frequency air networks[13-15]. The following are some of the major areas:

• Environmental Monitoring. Pollution is nowadays one of the major problems, oil spills from ships or broken tubes can make a lot of harm to the marine biological activity, the industry and tourist places. Monitoring ecosystems can help understanding and predicting the human and climate or weather effect in underwater environment.

• Underwater Navigation. The sensor used to make routing, identifying hazards on the seafloor, rocks or shoals in shallow water,

• Assisted Navigation. Sensors can be placed to identify hazards on the seabed, locate dangerous rocks or shoals in shallow waters, mooring positions and drawing the bathymetry profile of the area.

• Underwater Discovery. Underwater wireless sensor networks can be used to find oilfields or reservoirs, locate routes for placing connections for intercontinental submarine cables. Also they could seek for shipwrecks or archaeology or lost sink cities.

• Prevention of natural disaster. By measuring the seismic activity from different remote location the sensors could alert to the coast places by detecting tsunami or submarine earthquakes alarms.

• Underwater Autonomous Vehicles (UAVs). Distributed sensor in movement can help monitoring area for surveillance, recognition and intrusion detection.

6 Future Trends

Using underwater sensor networks a lot of advantages can be achieved, but a lot of research must be done in the upcoming years. The developing of this technology will have a great impact in the industry as well as provide impact in the environment.

It is essential to improve the physical layer performance in terms of efficiency, building low power acoustic modem, reducing the error rate. in the MAC layer, collision avoidance protocols should be adopted.

Distributed CDMA-based schemes are the best candidates for underwater environments, but we should consider the energy consumption factor too. while in the routing layer protocols, one should adapt the changing conditions, to include mobility patterns.

7 Conclusions

Underwater Acoustic Wireless Sensor Networks is still budding and following the path of Radio Frequency in Terrestrial Networks. In this article, we provide an overview of the state of the art in underwater acoustic sensor network. We summarized the challenges posed by the peculiarities of the underwater channel with major reference to monitoring applications for the ocean environment. We also analyze characteristics of the underwater channel and outlined future research directions for the growth of effective and trustworthy underwater acoustic sensor networks. The ultimate objective of this article is to encourage and specify the importance of research efforts to lay down fundamental basis for the growth of new advanced communication techniques for efficient underwater communication and networking for improved ocean monitoring and exploration applications.

References

1. Tunnicliffe, V., Barnes, C. & Dewey, R. 2008 Major advances in cabled ocean observatories (VENUS and NEPTUNE Canada) in coastal and deep sea settings. In *IEEE/OES US/EUBaltic Int. Symp., Tallinn, Estonia, May 2008*, pp. 1–7. IEEE.
2. Farr, N., Bowen, A., Ware, J., Pontbriand, C. & Tivey, M. 2010 An integrated, underwater optical/acoustic communications system. In *IEEE Oceans Conf., Sydney, Australia, May 2010*, pp. 1–6. IEEE.
3. Vasilescu, I., Kotay, K., Rus, D., Dunbabin, M. & Corke, P. 2005 Data collection, storage, and

retrieval with an underwater sensor network. In *Proc. 3rd ACM SenSys Conf., San Diego, CA,November 2005*, pp. 154–165. ACM.

4. Cella, U. M., Johnstone, R. & Shuley, N. 2009 Electromagnetic wave wireless communication in shallow water coastal environment: theoretical analysis and experimental results. In *Proc.4th ACM Int. Workshop on Underwater Networks (WUWNet), Berkeley, CA, November 2009*,pp. 9:1–9:8. ACM.
5. Friedman, J., Torres, D., Schmid, T., Dong, J. & Srivastava, M. B. 2010 A biomimetic quasistatic electric field physical channel for underwater ocean networks. In *Proc. 5th ACM Int. Workshop on Underwater Networks (WUWNet), Woods Hole, MA, September 2010*. ACM.
6. Urick, R. 1983 *Principles of underwater sound*. New York, NY: McGraw-Hill.
7. Stojanovic, M. 2007 On the relationship between capacity and distance in an underwater acoustic communication channel. *ACM Mobile Comput. Commun. Rev.* **11**, 34–43. (doi:10.1145/1347364.1347373)
8. Carrascosa, P. C. & Stojanovic, M. 2010 Adaptive channel estimation and data detection for underwater acoustic MIMO OFDM systems. *IEEE J. Oceanic Eng.* **35**, 635–646. (doi:10.1109/ JOE.2010.2052326)
9. Roy, S., Duman, T. & McDonald, V. 2009 Error rate improvement in underwater MIMO communications using sparse partial response equalization. *IEEE J. Oceanic Eng.* **34**, 181–201. (doi:10.1109/JOE.2009.2014658)
10. Fairley, P. 2005 Neptune rising. *IEEE Spectr.* **42**, 38–45. (doi:10.1109/MSPEC.2005.1526903)
11. OPNET Modeler v15.0 Reference Manual OPNET Technologies Inc. OPNET Modeler. Available online: http://www.opnet.com (accessed on 12 December 2011).
12. Berkhovskikh, L.; Lysanov, Y. Fundamentals of Ocean Acoustics; Springer: Berlin, Germany, 1982.
13. Peleato, B.; Stojanovic. M. Distance aware collision avoidance protocol for ad hoc underwater acoustic sensor networks. IEEE Commun. Lett. 2007, 11, 1025-1027.
14. Llor, J.; Malumbres, M.P. Performance evaluation of underwater wireless sensor networks with OPNET. In Proceedings of the ACM Simutools. ICST, Barcelona, Spain, 21–25 March 2011.

10\. Xie, G.; Gibson, J. Incorporating realistic acoustic propagation models in simulation of underwater acoustic networks: A statistical approach. In Proceedings of MTS/IEEE Oceans Conference, Boston, MA, USA, 18–22 September 2006; pp. 18-21.

DOI: 10.1051/matecconf/20165702008

A Bio-inspired Approach for Power and Performance Aware Resource Allocation in Clouds

Rajesh Kumar[1], Ashok Kumar[1,a] and Anju Sharma[1]

[1]*Computer Science & Engineering Department, Thapar University, Patiala-147004, India*

Abstract. In order to cope with increasing demand, cloud market players such as Amazon, Microsoft, Google, Gogrid, Flexiant, etc. have set up large sized data centers. Due to monotonically increasing size of data centers and heterogeneity of resources have made resource allocation a challenging task. A large percentage of total energy consumption of the data centers gets wasted because of under-utilization of resources. Thus, there is a need of resource allocation technique that improves the utilization of resources with effecting performance of services being delivered to end users. In this work, a bio-inspired resource allocation approach is proposed with the aim to improve utilization and hence the energy efficiency of the cloud infrastructure. The proposed approach makes use of Cuckoo search for power and performance aware allocation of resources to the services hired by the end users. The proposed approach is implemented in CloudSim. The simulation results have shown approximately 12% saving in energy consumption.

1 Introduction

Cloud computing enables access to shared pool of computing resources through internet on pay per usage basis [18]. The pool of resources needs to be managed efficiently in order to cut power costs. Energy consumption can be reduced by efficient allocation of resources. The aim of a resource allocation approach is to find a resource that fulfills resource demands of a given service without effecting its performance. Generally, resource allocation approach is either system centric or user centric. The former approach is traditional and emphasises on the improvement of overall system performance such a utilization. Whereas, the latter concentrates on providing quality of service to the endusers. In this paper, a novel resource allocation approach is proposed that improves the utilization of resources to enhance energy efficiency of the cloud infrastructure. Utilization of a resource and its energy consumption are highly coupled. A resource with low utilization consumes unacceptable amount of energy compared to its energy consumption when it is fully or sufficiently loaded. According to [7], an idle resource consumes up to 70% of the power consumed at full utilization. Furthermore, average resource utilization of most of the data centers is about 15-20% [21, 13] which results in wastage of huge amount of energy. According to Koomey [15], "Total data center power consumption from servers, storage, communications, cooling, and power distribution equipment accounts for 1.7-2.2% of total electricity used in U.S. in 2010". Furthermore, today's data centers, in addition to their huge energy consumption, emit as much carbon dioxide as whole of Argentina. If left on their current path, data center carbon dioxide output will quadruple by the year 2020 [13]. While the cloud energy consumption is growing quickly, industrial organizations and researchers are finding ways to reduce energy consumption. The motivation for this work stems from challenges arising due to huge amount of energy consumption. The challenge is to enhance energy efficiency of the cloud infrastructure without effecting the performance of services delivered to the end users. The proposed work improves energy efficiency of the cloud by perspicacious allocation of resources to the services. Cuckoo search is employed for optimal allocation of resources to the services. The proposed approach is implemented in CloudSim. The simulation results have shown significant reduction in energy consumption.

Rest of the paper is organized as follows: Related work is presented in Section 2. In Section 3, power and performance aware resource allocation framework is discussed. Section 4 discusses Cuckoo search meta-heuristic. Section 5 elaborates resource allocation using Cuckoo optimization. Results for performance analysis are discussed is Section 6. Conclusion and scope for future work are presented in Section 7.

2 Related Work

Gao et al. [9] proposed linear programming based multi-objective ant colony based system for virtual machine

[a] Corresponding author: ashok.khungerr@gmail.com

placement to minimize resource wastage and power consumption. The authors used only CPU processing speed and memory requirements of a VM while allocating resources to it. Feller et al. [8] presented multi-dimensional ant colony optimization based job consolidation algorithm. The algorithm uses resource utilization history to predict future resource demands and dynamically overbooks the resources. The authors tested the proposed algorithm on homogeneous PMs. Gao et al. [9] proposed multi-objective ant colony system algorithm for virtual machine placement that minimizes total resource wastage and power consumption. Ant colony optimization technique for assigning real-time tasks to heterogeneous processors is proposed by Chen et al. [6]. Local search technique is applied to improve energy efficiency of the feasible assignment solution generated by the proposed assignment algorithm. Huang et al. [11] presented genetic algorithm based adaptive sub-optimal resource management scheme to estimate number of VMs required to provide desired level of service. Kansal and Chana [12] presented artificial bee colony based energy-aware resource allocation technique to efficiently manage and enhance the utilization of resources. The authors have claimed the reduction of energy consumption and execution time of applications. Xu and Fortes [23] proposed multi-objective VM allocation algorithm. The authors have taken CPU, and memory parameters for VMs and have claimed reduction in power consumption, thermal dissipation costs, and resource wastage. In [20, 4, 19, 10], the authors proposed energy-conscious consolidation heuristics in order to conserve energy and maximize resource utilization without affecting the performance of the system. Kusic et al. [16] proposed performance and power efficient resource-management approach based on look-ahead control method for virtualized heterogeneous environments. Prediction is employed for dynamic reallocation of resources.

The next section describes proposed resource allocation approach.

3 Power and Performance Aware Resource Allocation

Resource allocation is a process of allocating required amount of computing resources to services hired by the end users. Services are provided through virtual machines (VMs). Multiple VMs can be executed in a single PM by leveraging virtual machine monitor (VMM) such as XEN [1], KVM [14], or VMWare [22]. The proposed power and performance aware resource allocation (P2RA) approach strives to increase utilization of resources without effecting the performance of services delivered to the end users. Figure 1 shows framework of proposed approach.

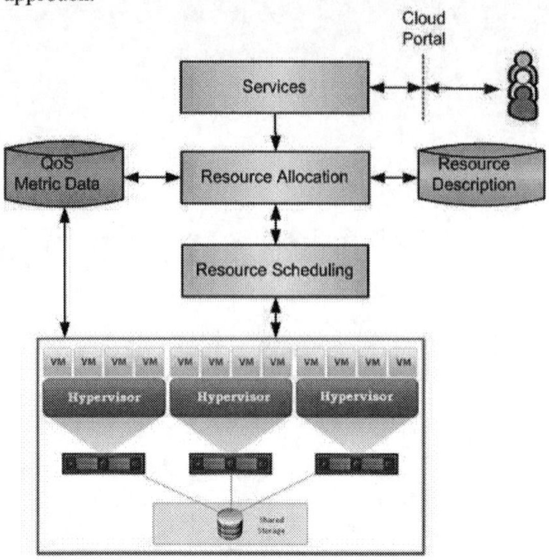

Figure 1. Energy and QoS aware resource allocation framework.

The following are the few assumptions that are taken into consideration.

i. Services are independent of each other and have computational requirements such as CPU, memory, and bandwidth.
ii. Each service can be allocated resources independently.
iii. Memory requirement of a service is rigid while CPU and network bandwidth requirements are fluid. A service can not be executed with less memory allocation than required.
iv. A service can be executed with less CPU and bandwidth allocation than the required capacity at the cost of reduced performance.
v. Idle PMs can be put to sleep mode to save energy. Energy consumption for switching a PM from *active* to sleep mode and vice versa is negligible and is not taken into consideration.
vi. Fraction of resources can be allocated to a VM.

The major components of P2RA approach are: (i) Services, (ii) Resource Allocation, (iii) Resource Scheduling, (iv) QoS Metric Data, and (v) Resource Description.

Services: The end users can hire the services through cloud portal. This module stores information about type and number of services hired by the end users.

Resource Allocation: This modules allocates resources to the services hired by the end users. A service is allocated to a resource that is most energy efficient. In order to find most suitable resource for service Cuckoo optimization is employed which discovers a resource employing Levy flight (LF). Allocation of service to the resource should result in minimal increase in power consumption while satisfying performance requirements. LF is a random walk that obeys Levy distribution. The reasons behind using LF for resource selection are:

DOI: 10.1051/matecconf/20165702008

i. Levy distribution has infinite mean and variance and hence LF are capable to explore the search space efficiently.
ii. LF is a stochastic process i.e. next state depends upon the previous state only.
iii. LF has scale invariant property and hence is used to model problems that exhibit clustering.

A random walk, consisting of series of consecutive random steps, is expressed as stochastic equation (1):

$$x_i(t+1) = x_i(t) + \alpha L(s, \lambda), \tag{1}$$

here, $x_i(t)$ and $x_i(t+1)$ represents a resource at time t and $t+1$, respectively. α is the scaling factor for step size s. Levy exponent, λ, is a constant. The step size s determines how far a final solution can go from initial solution in a fixed number of iterations. Obviously, if step size is large, next solution $x_i(t+1)$ will be far away from the current solution $x_i(t)$. The value of α is selected in compliance with the characteristic scales of the problem of interest.

Resource Scheduling: Scheduling modules controls the execution of the services. It schedules the services using round robin scheduling policy.

QoS Metric Data: It stores the usage statistics of resources. Usage statics of resources helps to make decisions about how to improve performance, energy efficiency, etc. of the cloud infrastructure under consideration.

Resource Description: This module stores information about the resources like type of resource, number of resources, etc.

3.1 Power Model

Power consumed by a resource is given by equation (2):

$$P_{total} = P_{dynamic} + P_{static}, \tag{2}$$

static power consumption (SPC), P_{static}, is due to leakage current and is independent of clock frequency and usage scenario. SPC can be reduced by switching idle resources to sleep mode [2]. However, dynamic power consumption (DPC), $P_{dynamic}$, is due to circuit activity and it depends on resource utilization. Power consumption of resource, given by equation (3), increases linearly with its utilization [3].

$$P = P_{idle} + (P_{max} - P_{idle})U, \tag{3}$$

where, P_{idle} is the power consumption when the resource is idle, P_{max} is the power consumption at 100% utilization, and P is the power consumption at utilization U $\varepsilon(0, 1)$ of the PM. Power consumption of a resource, when it is idle, is about 70% of the power consumed at full load [3]. Whereas, high utilization of a resource causes performance degradation because services running on it do not get sufficient resources [3]. So, a resource should not be operated at too low or very high utilization.

3.2 Fitness Function

The proposed resource allocation approach allows to group services hired by end users over the small number of resources and thus turning off those resources that are not used. P2RA approach considers performance as well as energy consumption while allocating resources to the service. The suitability of the resource r for service s is evaluated from fitness function f(equation (4)).

$$f_{s,r} = \frac{\left[\frac{\Re^a_r}{\Re^d_s}\right]^{\theta}}{\left[\gamma \Delta E_{s,r} + (1-\gamma)\kappa_r(1 - \sum_{i \in S, i \neq s} \Re_{i,r})\right]^{1-\theta}}, \tag{4}$$

where, is available processing power of resource r, $\Re^d_s$ is $\Re^a_r$ processing demand of service s, $\Delta E_{s,r}$ is energy contribution of service s on resource r, κ_r is energy affinity, $\Re_{i,r}$ is fraction of processing power allocated to service i on resource r, and $0 \leq \gamma \leq 1$ is a weight value to control the importance of energy contribution of service s. $\frac{\Re^a_r}{\Re^d_s}$ gives performance affinity which is desired to be greater than 1 for better performance. $0 \leq \theta \leq 1$ is used to control the importance between performance and energy. If θ is set to 1 then performance is considered while allocating resources. However, if it is set to 0 importance is given to energy consumption. Energy affinity, κ_r, is set to energy consumption of resource r when it is idle which inclines P2RA towards resources with lower energy consumption in idle state.

4 Cuckoo Search

Cuckoo search is a nature inspired population based meta-heuristic algorithm developed by Xin-She Yang and Suash Deb in 2009 [24]. It is based on brood parasitic behavior of bird cuckoo. Cuckoo lay eggs in the nests of other host birds with amazing capability of selecting recently spawned nests. On the other hand, some host birds upon recognizing alien eggs either through them out or build new nests in other locations. Cuckoo reproduction process is simplified by three idealized rules:

- Each cuckoo lays one egg at a time and dumps it in a randomly chosen nest.
- Nests with high quality eggs are carried over to the next generations.
- The host bird can discover the alien eggs with probability P_a ε (0, 1). Once an alien egg is discovered either the egg is thrown out or the nest is abandoned. A fraction P_a of the n poor quality nests are replaced by new nests.

5 Resource Allocation using Cuckoo Optimization

In the proposed resource allocation approach, services are assumed eggs and resources are considered nests of host birds. Suppose, a set S of services and each service has some requirement attributes such as CPU, memory, and network bandwidth. The services are required to be mapped on set R of resources. Each resource $r \varepsilon R$ has some processing speed, memory and network bandwidth, respectively. Our aim is to allocates resources to services in such a way to reduce overall energy consumption without effecting performance of the services. The major steps followed for QoS and energy aware resource allocation to services are:

a. Consider population (search space) of n resources.

b. Select a resource i from the search space using LF (equation (1)).

c. Randomly choose a resource j from search space of n resources.

d. Calculate fitness value $f_{s,i}$ of resource i for service s using equation (4).

e. Calculate fitness value $f_{s,j}$ of resource j for service s using equation (4).

f. If $f_{s,i} > f_{s,j}$ then resource i is considered better than resource j.

g. Remove all those resources from the search space whose utilization is above upper green threshold (UGT).

h. Add resources, equal to number of removed resources, to the search space.

i. Rank the solutions and find current best solution.

j. Repeat step (c) through (i) until maximum iteration and the best solution does not change for three consecutive iterations. Iterations.

k. Instantiate the service on the best solution (resource).

The pseudo code for resource allocation using Cuckoo optimization is given in Algorithm 1. The steps (b) through (k) are repeated until all services are allocated resources or until maximum number of iterations have been performed. A resource is discovered that best suits the requirements of the service. Initially, a resource is selected from search space using LF (equation (1)). The selected resource represents a tentative solution for resource allocation problem. Our aim is to discover the best solution (resource) which results in minimum increase in energy consumption while satisfying QoS requirements of the service. Second resource is then selected randomly from the search space. Out of the two selected resources, a resource with lesser increase in power consumption is considered better. Then some of the resources, having utilization greater than UGT, are removed from population and equal number of new resources are added to the population. Resources removed from the population are not considered for resource allocation in subsequent iterations. A resource that is removed from population can be added to the population later on, when its utilization falls below lower green threshold (LGT). Population is a set of fittest n resources, out of which a resource is discovered for allocating resources to a job under consideration. Total number of resources in use are variable and changes with number of jobs and their resource requirements.

Algorithm 1 Pseudo code for Resource allocation using Cuckoo Search

```
Input: List of services S_j(1 ≤ j ≤ N_j)
Generate initial population of n resources (R_i, i = 1,2,...,n).
set I ← 0
while (I < MaxIteration ) do
    Select a service s
    Select resource i (R_i) using Levy flight equation (1).
    Select resource j (R_j) randomly.
    Determine fitness f_s,i and f_s,j of resources i and j, respectively.
    if (f_s,i > f_s,j) then
        Replace solution j with solution i
    end if
    for i = 1 to n do
        if (R_i.Utillization > UGT) then
            Remove R_i from population of resources
            Add a new resource to the population of resources at location i
        end if
    end for
    Rank the solutions and find the best solution (resource).
    update I ← I + 1
end while
Allocate service to the resource referred by best solution.
```

6 Performance Evaluation and Comparative Analysis

The proposed approach is implemented in CloudSim. CloudSim is a framework for modeling and simulation of cloud computing infrastructure and services [5]. For performance analysis, P2RA is compared with First Fit Decreasing (FFD). The specifications of four types of resources, which are simulated using PowerHost class of CloudSim, are as given in Table 1. Power models corresponding to these servers, available in CloudSim, are used to evaluate energy consumption. To conduct comparative analysis, P2RA and FFD are tested with services having different resource requirements. Simulation is repeated 25 times and in each simulation run, parameters are assigned values as per Table 2.

Table 1. Specification of Resources.

Type	CPU	Cores	RAM	Storage	BW
R1	2933	4	8	500	10
R2	3067	4	8	500	10
R3	2933	12	12	500	10
R4	3067	12	16	500	10

CPU, processing speed in mips; Cores, number of processing cores; RAM, random access memory in GB; Storage, Permanent storage capacity in GB; BW, network bandwidth in gbps.

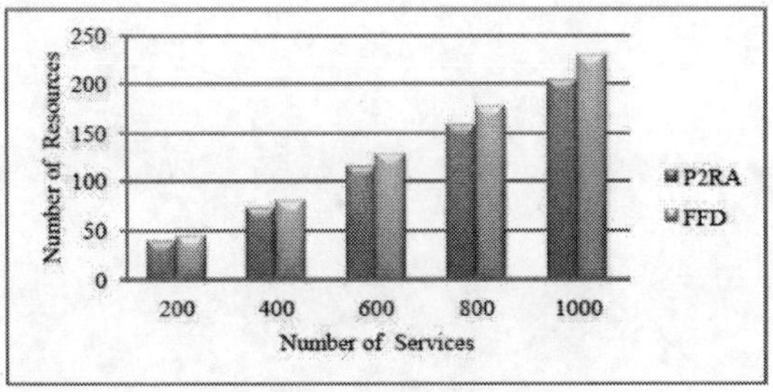

Figure 2. Comparison of number of Resources required for given number of Services

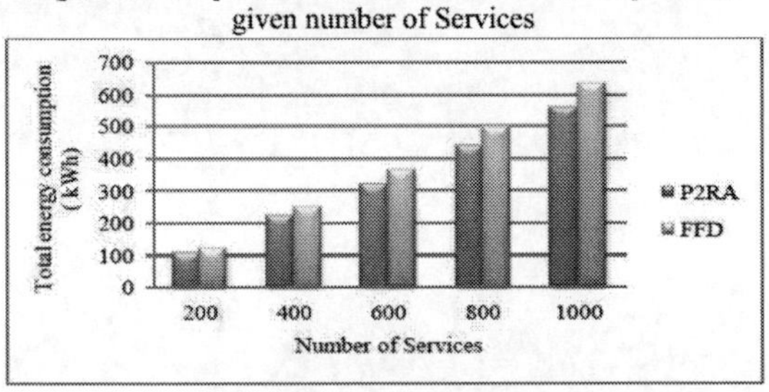

Figure 3. Comparison of total energy consumption

Table 2. Simulation parameters.

Number of services (VMs)	200-1000	Varied in simulation runs
Number of resources	100-200	Number of resources may also vary
Simulation Span	86400s	Time period of simulation run
Idle Time	10 min.	Time to switch PM to sleep mode
UGT	0.90	Upper Green Threshold limit
LGT	0.20	Lower Green Threshold limit

Figure 2 shows the number of PMs used by FFD and P2RA to fulfill the requirements of given number of services. Number of services are varied from 200 to 1000 in step of 200. It is observed that number of resources used increases with number of resources. It has been experimentally established that P2RA uses 10.12% lesser number of resource than and FFD.

Figure 3 shows total energy consumption comparison of FFD and P2RA. It has been observed that P2RA is approximately 12% more energy e_cient than FFD. Therefore, P2RA is more energy efficient and environment friendly.

Figure 4 shows comparison of time taken by FFD and P2RA to map given number of services to resources. Number of services are varied from 200 to 1000. It is observed that

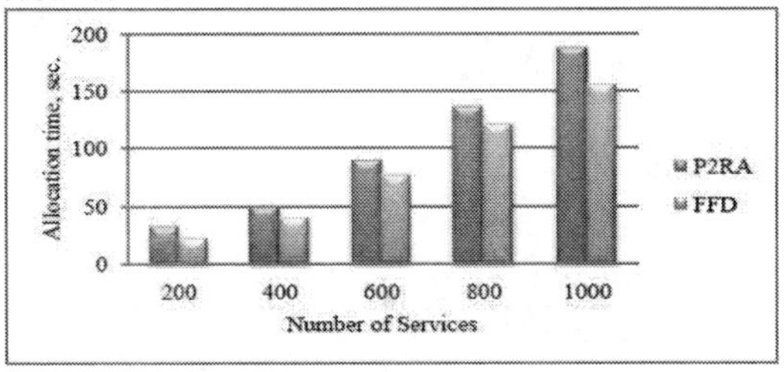

Figure 4. Comparison of total mapping time

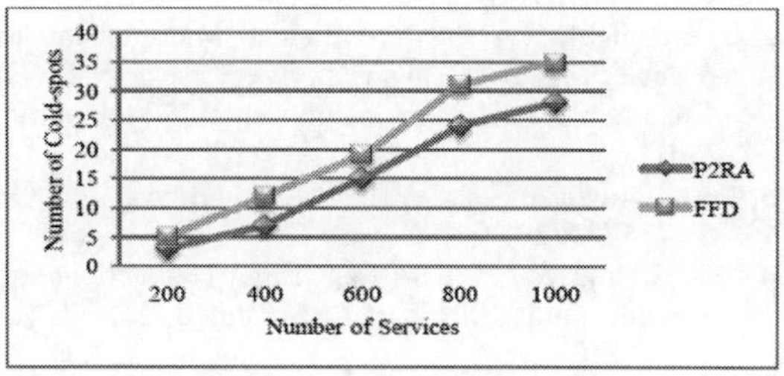

Figure 5. Comparison of number of Cold-spots

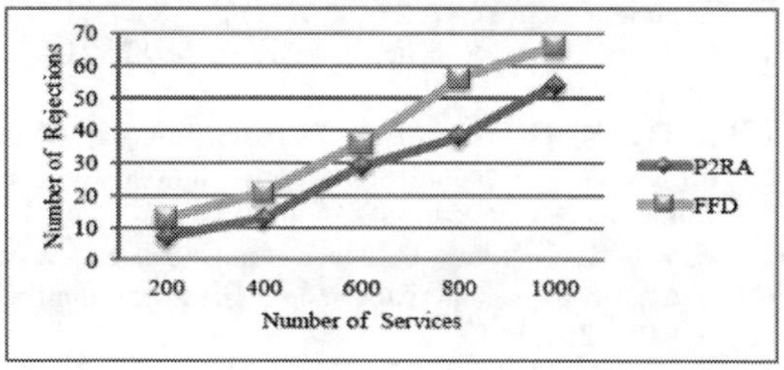

Figure 6. Comparison of number of Rejections

P2RA takes more time to map given of services. This due to optimal allocation of resources. However, energy consumed during migration is negligible compared to energy saving achieved P2RA approach. Figure 5 shows the comparison of number of cold-spots observed for FFD and P2RA with respect to the number of services submitted for execution. A PM is defined as a cold-spot if its resource utilization is below LGT (=20%). The 20% limit is taken because average resource utilization of typical data center is between 20% and 30% [17]. The number of cold-spots portrays the extent of resource wastage. Lesser number of cold-spots createdin P2RA shows its ability to allocate the resources efficiently. Figure 6 shows the comparison of number of job rejections determined for FFD and P2RA as the number of services increases from 200 to 1000. A service is rejected when sufficient resources are not available for its execution. It depicts that number of services that are not admitted due to unavailability of resources. In case of P2RA lesser number of rejections are observed than FFD.

7 Conclusion and Future Scope

In this work, a bio-inspired resource allocation approach for enhancing energy efficiency and QoS of Clouds is introduced. The services are allocated resources in compliance with their demands. Cuckoo optimization is employed for optimal allocation of resources. The proposed resource allocation approach is implemented in CloudSim and tested with services having different QoS and resource requirements. The results have shown 12% saving of energy consumption In future, the proposed resource allocation methodology can be extended by identifying dependency among the services. The performance of proposed methodology can be evaluated on private cloud set up using any open source cloud environment such as OpenNebula, Eucalyptus, etc.

References

1. P. Barham, B. Dragovic, K. Fraser, S. Hand, T. Harris, A. Ho, R. Neugebauer, I. Pratt, A. Warfield, SIGOPS Oper. Syst. Rev., **37**, 164-177(2003).
2. A. Beloglazov. Energy-Efficient Management of Virtual Machines in Data Centers for Cloud Computing. PhD thesis, University of Melbourne, (2013).
3. A. Beloglazov, J. Abawajy, R. Buyya, FGCS, **28**,755-768(2012).
4. A. Beloglazov, R. Buyya, IEEE Transactions on Parallel and Dirtributed Systems, **24**, 1366-1379(2013).
5. R. N. Calheiros, R. Ranjan, A. Beloglazov, C. A. F. D. Rose, R. Buyya, Software: Practice and Experience (SPE), **41,** 23-50(2011).
6. H. Chen, A. M. K. Cheng, Y. Kuo, JPDC, **71**,132-142(2011).
7. X. Fan, W. D. Weber, L. A. Barroso, In Proceedings of the 34th Annual International Symposium on Computer Architecture, ISCA '07, 13-23(2007).
8. E. Feller, L. Rilling, C. Morin, In 12th IEEE/ACM International Conference on Grid Computing (GRID), 26-33(2011).
9. Y. Gao, H. Guan, Z. Qi, Y. Hou, L. Liu, JCSS, **79**, 1230-1242(2013).
10. C. H. Hsu, S. C. Chen, C. C. Lee, H. Y. Chang, K. C. Lai, K. C. Li, C. Rong, In IEEE Third International Conference on Cloud Computing Technology and Science (CloudCom), 115-121(2011).
11. C. J. Huang, EAAI, **26**,382-389(2013).
12. N. J. Kansal, I. Chana, CCPE, **27,** 1207-1225(2015).
13. J. M. Kaplan, W. Forrest, N. Kindler, Technical report, McKinsy & Company, 2008.
14. A. Kivity, Y. Kamay, D. Laor, U. Lublin, A. Liguori, In Proceedings of the Linux Symposium, **1**, 225-230(2007).
15. J. Koomey. Growth in data center electricity use 2005 to 2010, 2014.
16. D. Kusic, J.O. Kephart, J.E. Hanson, N. Kandasamy, G. Jiang, In International Conference on Autonomic Computing, ICAC '08., 3-12(2008).
17. D. Meisner, B. T. Gold, T. F. Wenisch, In Proceedings of the 14th International Conference on Architectural Support for Programming Languages and Operating Systems, ASPLOS XIV, 205-216(2009).
18. P. Mell, T. Grance. The NIST definition of cloud computing, 2011.
19. R. Nathuji, K. Schwan, ACM SIGOPS Operating Systems Review, **41**,265-278(2007).
20. S. Takeda, T. Takemura, IPSJ Transactions on Advanced Computing Systems, **3**, 138-146(2010).
21. W. Vogels. Beyond server consolidation. Queue, **6**,20-26(2008).
22. B. Walters. VMware virtual platform. Linux J., 1999(63es), 1999.
23. J. Xu, J. A. B Fortes, In IEEE/ACM International Conference on Green Computing and Communications (GreenCom), 179- 188(2010).
24. X. S. Yang, S. Deb. Cuckoo Search via Lvy Flights. In NaBIC, 210-214 (2009).

Optimization of Task Scheduling Algorithm through QoS Parameters for Cloud Computing

Monika[1], Abhimanyu Jindal[2,a]

[1]*Assistant Professor, U.I.E.T., PU, Chandigarh, India*
[2]*Research Student, U.I.E.T., PU, Chandigarh, India*

Abstract. Cloud computing is an incipient innovation which broadly spreads among analysts. It furnishes clients with foundation, stage and programming as enhancement which is easily available by means of web. A cloud is a sort of parallel and conveyed framework comprising of a gathering of virtualized PCs that are utilized to execute various tasks to accomplish good execution time, accomplish due date and usage of its assets. The scheduling issue can be seen as the finding an ideal task of assignments over the accessible arrangement of assets with the goal that we can accomplish the wanted objectives for tasks. This paper presents an optimal algorithm for scheduling tasks to get their waiting time as a QoS parameter. The algorithm is simulated using Cloudsim simulator and experiments are carried out to help clients to make sense of the bottleneck of utilizing no. of virtual machine parallely.

1 Introduction

Cloud computing is as of late a blasting area in distributing computing that conveys powerfully versatile administrations on interest over the Internet through virtualization of equipment and programming. Cloud computing provide on demand services such as SaaS (Software as a service), IaaS (Infrastructure as a service) and PaaS (Platform as a service) to the clients. The customers can access all required hardware, software, platform, applications, infrastructure and storage with the ownership of just a web association. They access required resources and services on pay-per-use basis. The growth of cloud has reduced the performance of the resources. Aside from numerous approaches to upgrade the throughput and productive resource usage one way is task scheduling. Through task scheduling we can deal with the resource use which thus expands the throughput of the whole system. Task scheduling is the process of assigning tasks to the resources of the cloud efficiently so as to improve performance of the cloud in terms of QoS parameters. Scheduling is an NP- hard problem [1]. Allocation of tasks to the resources should be done carefully so as to minimize execution time and cost of the customer requests. If tasks are submitted according to length, priority or performance, the performance of the system may be improved significantly. A proficient task scheduling algorithm enhances the system throughput and helps service provider to provide good QoS. To get maximum benefit, scheduling is needed with distribution of load on the available resources[2]. The performance of scheduling algorithm is usually measured in terms of turnaround time, execution time, waiting time, cost etc. which must be minimized for improved performance of system. Using so as to minimize execution time and cost brought about heterogeneous resources over cloud is an imparative issue. Cost is another parameter which needs to be kept in mind while working on cloud and should me minimum [3]. Cloud systems have limited resources and they are very much costly. Scheduling algorithm not only improves QoS parameters but also reduce the expenses of the customer by efficiently scheduling tasks on the resources. In order to get high throughput, various issues need to be considered. To get most extreme advantage as far as make-span, planning of tasks is fundamental with dispersion of burden on processors and run- time allocation of resources [4]. Cloud performance may be increased by giving quick response to the users whenever they require dynamically adjusting resources of the system. As fast as the use of cloud services is increasing, the security issues have also been increased which needs to be handled carefully while working with cloud [5].

Energy consumption of system need also be reduced for efficient working. The size of cloud base is growing quick which bring about colossal energy utilization and working expense. Cloud workload is additionally a critical issue in scheduling. Provisioning of suitable resources to cloud workload relies upon nature of service necessities of applications. Cloud infrastructure should also support other applications such as big data and soft real-time applications which are the base of today's internet [6]. Where all the scheduling algorithms aim at improving the performance of system and maximizing the profit of service provider, customer satisfaction and feed back is also important. Thus a resource scheduling mechanism powered with a relevance feedback network

[a] Corresponding author: erabhi.puchd@gmail.com

is required [7]. Hence, a high performance task scheduling algorithm is required for improved throughput of the system. In this paper, an optimal algorithm for task scheduling is proposed which execute tasks according to their priorities in the system.

The rest of the paper is organized as follows. The section 2 describes some similar work related to the task scheduling, section 3 describes the proposed work, section 4 describes the experiment and results and the paper is concluded in section 5.

2 Related work

The cloud technology is growing fast due to its ease in use and its performance. The performance of cloud is increased due to various scheduling algorithms developed in the past few years. The performance of scheduling algorithm is defined on the basis of various QoS parameters i.e. execution time, arrival time, waiting time, deadline, cost, scalability and many more. The main role in scheduling is played by a broker that is responsible for allocation of tasks to the VMs (Virtual Machines) while the system is executing user requests.

One of the traditional methods of task scheduling is done by assigning priority to the tasks or VMs. In [8] authors proposed an algorithm PISA (Priority Impact Scheduling Algorithm) which schedules user tasks according to the priority given by the customer itself. The results of the algorithm show that the high priority tasks executed successfully with a higher ratio than FIFO algorithm by 26.7%. In [9] authors have proposed an algorithm for priority scheduling in which both the tasks and VMs are assigned with priorities and then tasks are scheduled on the VMs. The QoS value of tasks is gathered and then the system assigns priorities to the tasks. The task with low QoS value is given highest priority, next task with 2nd lowest QoS value is assigned 2nd highest priority and so on. The list of Vms is assigned priorities on the basis of MIPS. The VM with highest MIPS is assigned highest priority then the VM with 2nd highest MIPS value and so on. The highest priority task is submitted to the highest priority VM and so on. As the last VM is allocated, the next task is allocated on the first VM again. A similar algorithm is proposed in [10] which allocates tasks on VMs based on their priority. The priority may be assigned on the basis of any parameter such as length, user requirement or on the basis of previous performance statistics.

In [11] authors have proposed CDB task scheduling algorithm which schedules tasks based on the cost of execution of task that is calculated using deadline and length of the cloudlet. The algorithm performs well in achieving the deadlines of the tasks. In [12] Budget and Deadline Constrained Heuristic is presented which is based upon Heterogeneous Earliest Finish Time and results in reduction of the cost and execution time of the system. S. Sindhu and S. Mukherjee [13] proposed two algorithms LCFP and SCFP which schedule the cloudlets on the VMs according to their length. In LCFP, longest cloudlet is scheduled first in the system which results n reduction in the reduction in makespan of the system. In SCFP, shorter cloudlets are schedulet first which reduce the flowtime of the overall execution. Zhanjie Wang and Xianxian Su proposed a dynamically hierarchical resource allocation algorithm which dynamically divides the nodes and tasks into various levels according to their computing power and the newly arriving task joins the node of its own level only[14]. This algorithm reduced the makespan of the system. Another algorithm proposed by the authors of [15] is integrated with priority and SJF(Shortest job first) schemes and it reduced the waiting time and processing time by 30%.

Reference [16] proposed some mathematical models which aims to reduce execution time of the tasks by efficiently allocating VMs. Similar work is done in [17, 18, 19] where scheduling is done in order to reduce execution time and arrival time. Authors in [20] proposed a Dynamic task scheduling scheme DGS which allocates computing tasks to the virtual machines using greedy strategy and the scheme results in reduction in the completion time and improvement in the resource utilization. A Bin-Balancing Algorithm is proposed in [21] which aims at reducing the energy consumption by the system taking into consideration the deadlines of the tasks. Sukhpal Singh and Inderveer Chana proposed a framework in [22, 23] which reduces the execution time of the system along with the cost and energy consumption.

3 Proposed work

3.1 Introduction

The cloud customers want their task to be done within lesser time while at the same time taking into consideration the priority and waiting time of each task. However, the existing algorithms do not provide proper information about the waiting time of the tasks. The waiting time which is the time a task waits in the queue before execution depends upon the no. of VMs and tasks in the system that are currently being utilized in execution process. Each task in the cloud environment experiences a different waiting time according to the length of the task currently being executed by the VM. In the proposed work, the task of allocation of cloudlets on the VMs is done by the broker and it is responsible for execution of each task according to its priority.

3.2 Problem statement

The main problem is to get the waiting time of the tasks with that of the execution time so that actual performance of the algorithm may be defined. The current algorithms do not consider the length of the tasks during their execution which results in starvation of the small length task in some cases.

Task is defined as Ti = {Ii, Li, Pi}, Where Ii is the ID of the ith task, Li is the length and Pi is the priority assigned to the task.

3.3 Assigning priorities for tasks

The proposed algorithm give priority to the task according to its length. The task of smallest length is assigned highest priority and the task of largest length is assigned the lowest priority. The length of the task is set previously and each one have a different length.

3.4 Virtual machine selection

The cloud broker maintains the list of tasks and the VMs. It sorts the list of tasks in ascending order of their assigned priorities. After the sorting, the final list of tasks is bound with the list of VMs. The first task in the task's list is allocated to the first VM in the VM's list, second task in the task's list to the second VM in the VM's list and so on. As soon as the allocation process reach the last VM, the next task will be allocated to the first VM of the VM's list and the process of allocation will be repeated for the rest of the tasks.

```
Algorithm 1: Parallelized task scheduling algorithm
Create a list of tasks (n)
Create a list of VMs (m)
Define priority Pi for each task Ti
Sort task list in descending order of priority
i← 0 to n
j← 0 to m
for (j← 0 to m )
    {
    If (j<=m)
        {
        for (i← 0 to n)
        {
          bind Ti to VMj;
          j++ ;
          i++ ;
        }
        }
     Else /* when last VM has been allocated*/
        {
          j=0 ; /* Repeat from first VM again*/
        }
    }
```

4 Experiment and results

In order to obtain the results of the algorithm, the simulation was done using CloudSim 3.0.3 simulator which was run over Eclipse Juno. The operating system used was Windows 7 with Core i3 processor. In this simulation scenario, fifteen VMs and fifty tasks were involved. All the tasks were of varied lenghts and have a priority assigned with them. Initially, all the tasks were allocated to only one VM and then the experiment was repeated by increasing number of VMs one by one (eg. Two VMs, three VMs, and so on). The results obtained from the experiment are shown in Fig. 1.

In Fig. 1, Y-axis is the waiting time and X-axis is the number of VMs. The results show that the waiting time reduces to approximately half when number of VMs is two for performing same work. However, the experimental outcomes shows that the decreasing curve has stopped while using more than ten VMs which means it has met the bottleneck of using VMs parallely. In cloud, users should use ten VMs at a time when the target is to reduce the waiting time for the tasks.

Table 1. Experimental setup.

OS	Window 7
Processor	Core i3
Memory	2 GB
Simulator	3.0.3
Platform	Eclipse

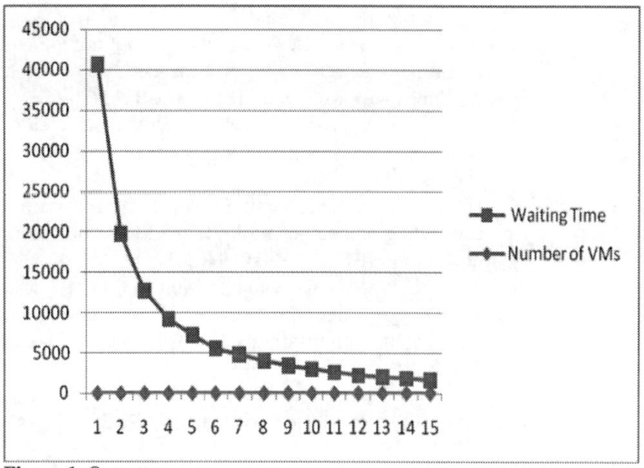

Figure 1. Output.

5 Conclusion and future work

Scheduling is the backbone of a cloud environment. It makes productive utilization of the resources while at the same time, considering the QoS parameters, is an important issue, since cloud infrastructure have limited resources and they are costly too. In this regard, an efficient scheduling algorithm is required so that the entire work might be accomplished in least time and could bring about lesser expense. In this study, an algorithm is presented for parallel processing of tasks and the outcomes gave us a bottleneck of utilizing number of VMs concurrently with the help of waiting time as a QoS parameter. The waiting time is reduced significantly by utilizing more number of VMs and scheduling tasks on the VMs according to their assigned priority.

In future, investigations will be done to reduce the waiting time of the tasks and an algorithm will be proposed to do as such.

References

1. J.A.J. Sujana, T. Revathi, G. Karthiga, R.V. Raj, Game multi objective scheduling algorithm for scientific workflows in cloud computing, IEEE, 1-6, (2015).
2. K.A. Saranu, S.Jaganathan, Intensified scheduling algorithm for virtual machine tasks in cloud computing, Springer, 283-290, (2014).
3. S. Selvarani, G.S. Sadhasivam, Improved cost-based algorithm for task scheduling in cloud computing," IEEE, 1-5, (2010).
4. S. Banerjee, M. Adhikari, S. Kar, U. Biswas, Development and analysis of a new cloudlet allocation strategy for QoS improvement in cloud, AJSE, Springer, **40**, 1409-1425, (2015).
5. U.A. Kashif, Z.A. Memon, A.R. Balouch, J.A. Chandio, Distributed trust protocol for IaaS cloud computing, IBCAST, IEEE, 275-279, (2015).
6. K. Cheng, Y. Bai, R. Wang, Y. Ma, Optimizing soft real-time scheduling performance for virtual machines with XRT-Xen, IEEE, 169-178, (2015).
7. D. Ding, X. Fan, S. Luo, User-oriented cloud resource scheduling with feedback integration, Springer, 1-22, (2015).
8. Hu Wu, Zhuo Tang, Renfa Li, A priority constrained scheduling strategy of multiple workflows for cloud computing, IEEE, 1086-1089, (2012).
9. A.V. Lakra, D.K. Yadav, Multi-objective tasks scheduling algorithm for cloud computing throughput optimization, ICICCC, **48**, 107-113, (2015).
10. C. Lin, S. Lu, Scheduling scientific workflows elastically for cloud computing, IEEE, 746-747, (2011).
11. Himani, H.S. Sidhu, Cost- deadline based task scheduling in cloud computing, ICACCE, IEEE, 273-279, (2015).
12. A. Verma, S. Kaushal, Cost- time efficient scheduling plan for executing workflows in the cloud, JGC, Springer , **13**, 495-506, (2015).
13. S. Sindhu, S. Mukherjee, Efficient task scheduling algorithms for cloud computing environment, Springer, 79-83, (2011).
14. Z. Wang, S. Su, Dynamically hierarchical resource-allocation algorithm in cloud computing environment, Springer, 2748-2766, (2015).
15. Jia Ru, Jacky Keung, An Empirical investigation on the simulation of priority and shortest-job-first scheduling for cloud based software systems, IEEE, 78-87, (2013).
16. Q.T. Nguyen, N.Q. Hung, N.H. Tuong, V.H. Tran, N. Thoai, Virtual machine allocation in cloud computing for minimizing total execution time on each machine, IEEE, 241-245, (2013).
17. A.K. Das, T. Adhikary, C.S. Hong, An intelligent approach for virtual machine and QoS provisioning in cloud computing, IEEE, 462-467, (2013).
18. R. Achar, P.S. Thilagam, Shwetha D, Pooja H, Roshni, Andrea, Optimal scheduling of computational task in cloud using virtual machine tree, ICEAIT, IEEE, 143-146, (2012).
19. W.J. Wang, Y.S. Chang, W.T. Lo, Y.K. Lee, Adaptive scheduling for parallel tasks with QoS satisfaction for hybrid cloud environments, Springer, **66**, 783-811, (2013).
20. Liang Ma,Y. Lu, F. Zhang, S. Sun, Dynamic task scheduling in cloud computing based on greedy strategy, Springer, 156-162, (2013).
21. J.M. Tang, L. Luo, K.M. Wei, A heuristic resource scheduling algorithm for cloud computing based on polygons correlation calculation, ICEBE, IEEE, 365-370, (2015).
22. S. Singh, I. Chana, QRSF: QoS aware resource scheduling framework in cloud computing, Springer, **71**, 241-292, (2014).
23. S. Singh, I. Chana, Resource provisioning and scheduling in clouds: QoS perspective, Springer, 1-35, (2016).

MATEC Web of Conferences 57, 02010 (2016) DOI: 10.1051/matecconf/20165702010
ICAET-2016

Principles and practices for the implementation of Cloud based ERP in SMEs

Sarvjit Singh Bhatia[1], Vikram Gupta[2]

[1] *Khalsa College , PG Department of Computer Science, Patiala*
[2] *Uttarakhand Technical University , Research Scholar, Dehradun France*

Abstract. Enterprise systems provide integrated information for all activities in an organization. These systems serve as a vital asset to any organization and hence it becomes mandatory to ensure their security. In the present scenario these services are offered on the cloud mainly to reduce inherent risk associated with the traditional enterprise systems. Cloud computing represents a significant shift in the way that IT resources are managed, operated, and consumed. This change exposes several benefits to enterprises, promoting greater IT efficiency and agility. This paper is intended to suggest a theoretical basis for studying the factors that influence SMEs' adoption of cloud computing. It will first explain the integration between the Technology-Organisation-Environment (TOE). Based on this technology innovation adoption process is influenced by three aspects of an enterprise's context. The technological context represents the internal and external technologies related to the organisation, both technologies that are already in use at the firm, as well as those that are available in the marketplace but not currently in use .These technologies may include either equipment or practice. The organisational context is related to the resources and the characteristics of the firm, e.g. size and managerial structure, while, finally, the environmental context refers to the arena in which a firm conducts its business; it can be related to surrounding elements such as industry, competitors and the presence of technology service providers. Enterprise Systems help in carrying out various tasks with greater operational efficiency and reliability. They also facilitate to keep information updated and available across the organization 24 * 7. As enablers of successful business reengineering projects, they help the organizations to save money, keep their business data consistent, current and available, speed up business processes, and improve the quality and reliability of the processes.

1 Inroduction

The Economy of underdeveloped countries mainly depends on Small and Medium sized Enterprises (SMEs). Small and medium sized enterprises (SMEs) form an essential ingredient of a country's economy as they are the main source of employment and technological development. They play a very important role in each market by significantly contributing to each country's Gross Domestic Product (GDP) and its labour market. Therefore proposing new strategies or developing new systems that can help SMEs become more efficient and productive is not only beneficial for SMEs but also for the economy as a whole. One of the strategies that can help SMEs become more efficient is the use of appropriate Information and Communication Technologies. The Size and structure of SMEs make them face many challenges. The main challenge is not to have access to enough resources (e.g. financial resources). According to THINK strategies (2002), various small businesses cannot afford the in-house staff needed to plan, design, implement, and manage increasingly complex software and hardware, and networking kit in today's tough economic times, it is becoming increasingly hard for SMEs to justify such investment and on-going expenses. Many organisations are seeking alternative solutions that can reduce the total ownership costs of their ICT systems, with a few choosing to outsource them to third parties, in order to focus their limited time and resources on their core business. SMEs are very cost conscious; they should keep their costs under control. Although adopting new technologies help SMEs gain competitive advantage, it usually involves high cost. Cloud computing is a new phenomenon which helps SMEs tackling many issues such as cost and risk management. There is no universal definition for cloud computing; but in this paper cloud computing is defined as a computing paradigm in which the computing resources are delivered to customers over a network (e.g. Internet). Companies can access the available factors Influencing the adoption of Cloud Computing by SMEs. Cloud computing is considered as an innovation, because it offers a new method of computing by integrating the already existing technologies. The latest innovative technology in the business is Enterprise Resource Planning (ERP). Here we

Corresponding author: vincal67@rediffmail.com

are going to integrate these two innovative concepts i.e. ERP and Cloud Computing.

In this research paper the factors that influence the adoption of cloud computing by SMEs is studied in order to investigate the factors that influence the SMEs' decision to adopt cloud computing. This study is originated from two prominent theories of this field. These two theories are Rogers's Diffusion of Innovation (DOI) theory and Tornatzky and Fleischer's Technology, Organization, and Environment (TOE) framework. There are four groups of variables which are the main factors of these theories that are environmental, organizational, human and technological factors. Each of these groups of factors consists of one or more constructs. Environmental factors group consists of External Support and Competitive Pressure. Organizational factors include Employee's IS knowledge and information intensity. Human factors are decision makers' Innovativeness, and their cloud knowledge. Last but not least, technological factors are cloud computing, Relative advantage, Cost, Security and Privacy, Trial ability, Complexity and compatibility with company's norms and technologies.

The TOE framework is an appropriate theoretical foundation to understand the factors that influence cloud computing adoption and outcomes. Although the TOE framework allows researchers to examine a broad set of factors. Cloud computing is one of the most recent Internet-related computing paradigms which help SMEs become technologically closer to large businesses. This new phenomenon makes it possible for companies to access sophisticated computing services over a network. There is no universal definition for cloud computing that explains all aspects of this new phenomenon. Concept of cloud computing dates back to 1950s (when mainframes were accessed by users from different terminals), it was during late 1990s when cloud computing started to become a buzz word; and companies gained a better understanding of cloud computing(EzeCastle Integration, 2012). The diffusion of cloud computing has many advantages at both micro and macro level. At micro level the diffusion of cloud is advantageous for SMEs. At macro level, it is beneficial for the economy and environment.

2 SMEs and Cloud computing

The size and structure of SMEs gives them some advantages including fast communication between employees and their managers and their ability to rapidly implement and execute decision. But in most cases these companies face many disadvantages. Most of the challenges that SMEs face are due to their lack of access to enough resources (Welsh & Wite, 1981). These resources include but not limited to financial and human resources. This limitation makes SMEs weaker than large companies in terms of financing, planning, control, training and also information technologies (Bilili & Raymond, 1993). Keeping cost under control is one the biggest challenges that SMEs faces. (Communications News, 2008) It is not feasible for SMEs to spend a significant amount of money on their Information Technology (IT). In addition to their high cost, IT projects usually involve a high risk of failure too. About 20 percent of IT projects are cancelled before completion and less than a third are finished on time and within budget with expected functionality (Kappelman, McKeeman, & Zhang, 2006). Overall SMEs have low tolerance in bearing costs and risks that are involved in IT investment. Different studies have been conducted to investigate how to improve SMEs' operational efficiencies; and help them grow larger. For example Acs et al. proposed some changes in public policies that increase an SME's incentive to innovate (1997).

3 Diffusion of Innovation Theory

Diffusion of Innovation Theory (DOI) is mainly known as a theory developed by Everett Rogers, a Professor of rural Sociology, who wrote a book on this theory in 1962. However, Rogers is not the first person who introduced the concept of diffusion. One of the first studies regarding diffusion of innovation is the research conducted by Ratzel et al. (1928) who studied the trans-cultural diffusion. Another influential study is conducted by Lazarsfeld et al. (1944) who interrogated the voting behaviour of individuals. This study illustrated the importance of opinion makers' attitude in changing voters' mind. Theory of Diffusion of Innovation (DOI) is originated from six different disciplines anthropology, early sociology, rural sociology, education, industrial sociology and medical sociology; which were predominantly from sociology discipline. During the 1960s, these research traditions started to merge; and the current diffusion of innovation theory started to appear. The current version of DOI tries to discover the factors that influence the spread of a new idea or technology in a society (Rogers, 2003). Rogers defined some factors that impact the speed of diffusion in a society. Some of these factors are the innovations' characteristics, the nature of social system, communication channels and change agents'. Rogers defined some factors that impact the speed of diffusion in a society. Some of these factors are the innovations' characteristics, the nature of social system, communication channels and change agents'.

4 Innovation Decision Process

Roger defined the innovation-decision process as: "the process through which an individual or other decision making unit passes from first knowledge of an innovation, to forming an attitude toward the innovation, to a decision to adopt or reject, to implementation of the new idea, and to confirmation of this decision". He demonstrates that the innovation-decision process can be described as a stage that an individual (or other decision-making unit) passes through.

- Knowledge: The innovation-decision process starts with the knowledge stage. In this stage, the individual (or some other decision-making unit) is exposed to an innovation's existence and gains an understanding of how it functions.

- Persuasion: This stage refers to situations when a client (or some other decision-making unit) shapes a favourable or unfavourable attitude about the new notion.
- Decision: It signifies the condition when a client (or some other decision-making unit) involves in activities that lead to a decision to adopt or not to adopt the innovation.
- Implementation: At the implementation phase, the individual (or some other decision-making unit) puts an innovation into actual use.
- Confirmation: The final stage refers to the state when an individual (or some other decision-making unit) try to reinforcement of an innovation-decision already made, or backs down from a previous adoption decision for one reason or another.

ICT innovation adoption cannot always be considered a simple process (Zappalà and Gray, 2006). Perhaps the most challenging question about this description of Rogers regarding the innovation adoption process is that, the stages listed above provide a sense of technological progression for adopters, the extent to which this process can be considered a linear process is not certain. "When an organisation learns to do something it did not know how to do before, and then proceeds to do it in a sustained way, a process of innovation has occurred" (Shepard, 1967, p. 470). In fact, firms, like consumers, differ in the likelihood and speed of their adoption of new innovations. However, the process of organisational innovation is almost dissimilar to the adoption of innovations by individuals. According to Pennings and Buitendam (1987), diffusion among organisations presents special challenges. Technology adoption is a complex developmental process (Straub, 2009). Firms have usually tried to apply a relatively steady and predictable prototype of development through a number of cumulative and identifiable stages (Van de Ven and Poole, 1995).

5 Technology-Organisation-Environment (TOE) Framework

The TOE is an organisation-level multi-perspective framework that was developed by Rocco DePietro, Edith Wiarda and Mitchell Fleischer (1990). TOE represents one segment of the innovation process, i.e. how the firm context influences the adoption and implementation of innovations (Baker, 2011). Based on this framework, the technology innovation adoption process is influenced by three aspects of an enterprise's context. The technological context represents the internal and external technologies related to the organisation, both technologies that are already in use at the firm, as well as those that are available in the marketplace but not currently in use (Baker, 2011). These technologies may include either equipment or practice. The organisational context is related to the resources and the characteristics of the firm, e.g. size and managerial structure, while, finally, the environmental context refers to the arena in which a firm conducts its business; it can be related to surrounding elements such as industry, competitors and the presence of technology service providers. These three contexts present both constraints and opportunities for technological innovation (Tornatzky and Fleischer, 1990).

6 Integrating DOI and TOE to study the determinants of cloud computing adoption

The two theories such as the diffusion of innovation (Rogers, 2003) have been widely applied to studies looking at how innovations are adopted and diffused. This study, for a number of reasons, will also use the Technology, Organisation, and Environment (TOE) framework put forward by DePietro et al. (1990). First of all, despite the similarity in their concepts, Oliveira and Martins (2011) suggest that as the TOE framework includes the environment context, which is not included in the diffusion of innovation theory, the former is better able to explain intra-firm innovation adoption. From their point of view, the reliable empirical support and solid theoretical basis are the main advantages of the TOE framework.

7 Constructs of TOE

7.1 Technological Context

In the original TOE framework, the technological context described both the internal and external technologies relevant to the firm (Rui, 2007; Oliveira and Martins, 2011).The central indicator for the adoption of a new IS innovation. The higher the perceived need for an innovation by an organisation, the higher the probability that it will adopt the innovation (Rogers, 2003; Lee, 2004).

a. Uncertainty: Adoption of a new technology involves risk and uncertainty (Erumban and de Jong, 2006). Uncertainty refers to the extent to which the results of using an innovation can be guaranteed (Ostlund, 1974; Fuchs, 2005). Due to the open nature of the Internet, security risk has been recognised as a key factor hindering the use of some ICT technologies (Kalakota and Whinston, 1996).
b. Compatibility: From a business perspective, there is a need for the technical and procedural requirements of the innovation to be compatible and consistent with values and the technological requirements of the adopting organisation (Lertwongsatien and Wongpinunwatana, 2003).
c. Complexity: Complexity refers to "the degree to which an innovation is perceived as relatively difficult to understand and use" (Rogers, 2003, p. 257). Adopting a new technology may

confront SMEs with challenges in terms of changing the processes in which they interact with their business systems. New technologies have to be user-friendly and easy to use in order to increase the adoption rate (Parisot, 1995; Sahin, 2006).

d. Trialability: This factor has been reported as one of the most fundamental components in the process of adopting a new technology (Kendall, 2001; Rogers, 2003).

7.2 Organisational Context

The Organisational context was found to be an important context for innovation adoption in previous studies (e.g. Ramdani, 2008). Four main factors are included in the organisational context and these are: firm size, top management support, innovativeness and prior IT experience.

7.3 Environmental Context

The key factors within the environmental context are: competitive pressure, type of industry, market scope and supplier efforts and external computing support

8 Conclusion

The findings of this paper has great value to the research community, managers and ICT providers, in terms of formulating better strategies for cloud computing adoption. For service providers, using the research model in this study can assist in increasing their understanding of why some SMEs choose to adopt cloud computing services, while seemingly similar ones facing similar market conditions do not. On the other hand, however, cloud computing providers may need to improve their interaction with SMEs who are involved in the cloud computing experience, in an effort to create a healthy environment for cloud computing adoption, and to remove any vagueness surrounding this type of technology. Providers may need to clarify their position and stance when it comes to offering in-house versus cloudservices,whichinturncanaffectclients'confidence.Pro spectors appear to be willing to adopt cloud computing services despite security concerns, as they rely on the element of trust or it is the next issue for the future research that will concentrate on the security issues for the successful implementation of ERP with Cloud computing technology for SMEs.

References

1. Agarwal, R. and Prasad, J. (1998) 'A Conceptual and Operational Definition of Personal Innovativeness in the Domain of Information Technology', Information Systems Research, 9(2), pp. 204-215.
2. Agarwal, R. and Prasad, J. (1999) 'Are Individual Differences Germane to the Acceptance of New Information Technologies?', Decision Sciences, 30(2), pp. 361-391.
3. Aguila-Obra, A. and Padilla-Meléndez, A. (2006) 'Organizational factors affecting Internet technology adoption', Internet Research, 16(1), pp. 94-110.
4. Aiken, M., Bacharach, S.B. and French, J.L. (1980) 'Organizational Structure, Work Process, and Proposal Making in Administrative Bureaucracies', The Academy of Management Journal, 23(4), pp. 631-652.
5. Yadav, S.S. and Zeng Wen, H. (2010) Computer Engineering and Technology (ICCET), 2010 2nd International Conference on. 16-18 April 2010
6. Simpson, M. and Docherty, A. (2004) 'E-commerce adoption support and advice for UK SMEs', Journal of Small Business and Enterprise Development, 11(3), pp. 315-328.
7. Slappendel, C. (1996) 'Perspectives on Innovation in Organizations', Organization Studies, 17(1), pp. 107-129. Smallbone, D. and Wyer, P. (2000) 'Growth and development in the small firm', in Enterprise and Small Business. Financial Times/Prentice-Hall, pp. 409-433

Implementation of Biography Based Neural Clustering (BBNC) with Genetic Processing for tumor detection from medical images

Chandanpreet kaur[1] and Bikrampal kaur[2]

[1]*M tech student (IT Department), Chandigarh Engineering College, Landran , Mohali, Punjab, India*
[2]*Professor (CSE Department), Chandigarh Engineering College, Landran, Mohali, Punjab, India*

Abstract. Segmentation is a best method to divide the required region from the medical images. This research is based on segmentation of medical images (MRI, CT scans) based on the previous method known as pre-operative and post-recurrence tumor registration (PORTR) and proposed method biography based neural clustering (BBNC) with genetic processing for tumor segmentation. By using the new technique the extracted part can be view in 3D model and also can get the actual segmented tumor region. This new method will be helpful for diagnostics to find the tumor area as well as pixel difference in segmented part to define the tumor area accurately. While in the previous approach all the parameters have been used likewise, in which the registration method is used to transform the different sets of data into one coordinate system for segmentation of medical images. Registration basically is used to improve the signals to reduce the noise from the images. These techniques are better to find the tumor area from the MRI and CT scans, but after comparing them better results have been obtained in proposed technique. The proposed technique (BBNC) reduces the extracted region again into required and actual region of tumor with accuracy of area, time and pixel difference.

1 INTRODUCTION

Brain tumor is defined as the irregular development of the cells from another part of body to brain or in the brain. There are two types of tumors the first one is known as primary brain tumor and second one is known as secondary brain tumor. Primary brain tumors are start from the brain and the secondary brain tumors are the consequence of melanoma. Percentage of brain tumors is 7-9% of all cancers. Approximately 20% -40% of brain tumors are due to variations from the norm in the body and can strike at any age [9]. The advancements of imaging strategies, for example, Computer Tomography (CT) and Magnetic Resonance Imaging (MRI) offer specialists with very high resolution images, which will enormously helped in the clinical diagnosis. In the mean time medical experts need to process countless with a great deal more subtle elements. Magnetic resonance imaging (MRI) is a progressed diagnostic tool that helps diagnostics to picture anatomical subtle elements more plainly and so prevalent in identifying variations from the norm in the soft tissues of the brain. Therefore Segmentation techniques have been used for this purpose which is normally a long and decisive process commonly used from long time [7].This research will presents the comparison between previous method (PORTR) and proposed method (BBNC) to extract the tumor region from the medical images, and also will presents the 3-D view of extracted region from the medical images. There are various image segmentation procedures and they are discussed below:

[a] Corresponding author: Chandangill4@gmail.com

Threshold Based Segmentation

Histogram thresholds are utilized to fragment the particular image. There are certain pre-processing and post-processing strategies are required to threshold the images [10]. The Major Thresholding procedures are proposed by diverse scientists i.e. Mean Strategy, P-Tile system, visual system , Edge Maximization strategy, and Histogram subordinate method Histogram subordinate method.

Thresholding is one of the oldest method to segment the different type of images. The segmentation can be done by making all the pixels together in one class. On the identification of a good threshold, this strategy depends and failing of such threshold may guide to poor segmentation. A procedure to determinate more than one threshold value is called multi-thresholding. In a sequence of image processing operations thresholding is utilized as an initial step.

Region Based Segmentation

Region based segmentation isolates an image into diverse areas in view of pre-characterized criteria, i.e., colour, intensity, or object [11]. Region based segmentation strategies are sorted into three primary classes, i.e. region growing, region splitting, and region merging. For image segmentation region growing strategy is a well developed procedure. Taking into account some predefined criteria this technique extracts image region. This depends on intensity information or edges in the image. An operator manually selects a seed point and extracts all pixels that are associated to the initial seed based on some predefined criteria. An algorithm for this type of segmentation is called as split-and-merge algorithm which is identified with region growing algorithm.

ANN Based Segmentation

The method for segmentation can be Artificial or we can say men made so the Artificial Neural Network. In artificial neural network every neuron is relating to a particular pixel of a picture, then picture is mapped to the neural system.. In neural system the picture is prepared by utilizing the training samples and afterward association between neurons when pixels are found. At that point the new pictures are fragmented from the trained image [12].

Neural network based segmentation have four essential attributes:
(i) It has parallel work capacity and quick registration ability, which makes it suitable for real time application.
(ii) Improve the segmentation results when the data deviates from the normal situation.
(iii) The robustness making it insensitive to noise.
(iv) It reduces the requirement of experts' intervention for image segmentation process.

PDE Based Segmentation

The Partial Differential equations or PDE models are utilized generally as a part of image processing and particularly in image segmentation. These equations are active contour model for segmentation purpose. Active Contour model or Snakes models are used to transform the segmentation problem into PDE [13].

2 LITERATURE SURVEY

T. Rajesh et al., 2013[1] this research depicts all those medical images as input for the process, whose features will be extracted from particular images based on Rough Set Theory. Then these images will be processed under Feed Forward Neural Network classifier. This methodology in all consist of two particular functions in which the first function highlight the difference between abnormal and normal types of conduct where as the other one describes the classification of abnormality as malignant or benign.

A. Islam et al., 2013 [2] this research elaborate a stochastic model, which is very useful By using this model we can characterize the tumor texture in brain MRI.
E. E. Ulku et al., 2013 [3] represents the brain tumor detection using Computer-aided detection (CAD) system. This system is based on morphological image processing & histogram equalization technique, where the last stage of computer aided detection system i.e. classification and the classification algorithms in the Rapid Miner program is tested & the comparison of all algorithm is done to show CAD system accuracy.

J. Vijay et al., 2013 [5] describes the brain tumor detection by using the technique called K-means clustering. The automatic segmentation of brain tumor is done to extract the tumor tissues from the brain.

Ishita Maiti et al., 2012[6] introduced a new method to detect brain tumor as watershed method in which initial RGB image gets converted in HSV colour image. This HSV image separates into three region as saturation, hue and intensity and to each region contrast enhancement watershed algorithm is applied and for output canny edge detection is used. This combined effect of all 3 images gives the final brain tumor segmented images.

3 PROBLEM STATEMENT

As per literature survey the previous method for this research was to segment the tumor region from the medical images by using the various types of segmentation methods and registration method, Those have been used to provide the accurate results but if the extracted region will be viewed as a 3D model it will provide better results. Area of extracted tumor and pixel difference value can be reduced more by using the proposed methodology. To show the exact part of tumor with area as well as pixel difference in 3D model the proposed system presents an extension to an existing segmentation algorithm. This research will modify the deformable registration of pre-operative and post-recurrence brain MR scans with biography based neural clustering (BBNC) by applying the Genetic algorithm. This new method handles various types of space-occupying, pathology, mass tumors. The iterating changes like varying details due to contrast and non-contrast imaging, noisy environment low quality resonance, sampling effects are also handle by this technique.

4 OBJECTIVES

- To select the previous output for recognising the initial values i.e. gene values, propagation values and duplicating gene detection for making clusters.

- To select the cluster, this will be processed for further analysis by applying the particular no. of iterations.

- To calculate the area, time and pixel difference.

• The modelling of the extracted area using 3d surface view will be done.

5 METHODOLOGY USED FOR PROPOSED RESEARCH

BBNC with genetic processing

BBNC stand for biography based neural clustering, is a method of clustering based on biography of medical images. Clustering can be done by using the multiple features (biography) of MRI or CT scans for differentiate a particular cluster consisting of healthy tissues or we can say in which lie the of tumor area. There are methods called neural clustering and genetic algorithms which are used in this research are below:-

5.1.2 Neural clustering

Clustering basically is a fundamental concept to analysis the data. It is used for pattern recognition, feature extraction and for image segmentation. In this research the clustering identifies some inherent structures present in the tumor region with the help of artificial neural network and then can make the multiple clusters basis on the biography of the medical images. This type clustering is known as neural clustering.

5.1.2 Genetic algorithm

The Genetic algorithm (GA) is a method for global research. Genetic algorithm is used for natural biological evolution. Genetic algorithm is operates on the population of potential solution by applying the principle of survival of fittest, which is further used to find out the better solutions. In genetic processing the each generation take place a new set of approximation solution that can be created by the process of selecting individuals' level of fitness accordingly. The uses of operators in the problem domain are borrowed from the natural Genetic process. This process further leads to the evolution of individuals and is also suited for their environment than the individuals. They were created from natural adaptation. The Genetic algorithm solves the problem of optimization by mimicking the principles of biological evolution. The modification of the population of individuals can be done by using the rules for gene combinations. The Genetic algorithm can also improve your chances of finding a global solution. It enables the researchers to solve bound constrained, unconstrained and optimization problems. It does not require the functions to be differentiable or continuous functions.

Segmentation of brain is supposed to be done by using Genetic algorithm. Following would be the steps for segmentation:

1. Acquire the medical image (MRI, CT) to be used for pre-processing.

2. Segmentation will be implemented on the acquired image by applying Genetic processing.

A. The Genetic algorithm will be initiated by creating a random and initial population.

B. Then the algorithm will creates a sequence of new populations, where at each step the algorithm will uses the individuals in the current generation to create the next population.
To create the new population, the algorithm performs the following steps:
a) Scores each member of the current population by computing its fitness value.
b) Scales the raw fitness scores to convert them into a more usable range of values.
c) Selects members known as parents according to their fitness.
d) Some individuals in the current population that have lower fitness value are chosen as Elite and these elite individuals are passed to the next population.
e) Produces children from the parents. Children are produced either by making random changes to a single parent are called mutation or by combining the vector entries of a pair of parents is called crossover.
f) Replaces the current population with the children to form the next generation.

C. The algorithm will be stop, when one of the stopping criteria is met. See Stopping Conditions for the Algorithm.

6 RESULTS AND DISCUSSION

The tumor region will be extracted from the MRI by using the genetic algorithm and neural clustering. Important steps to be performed are the genetic processing on the selected clusters, which further will be used to process on chromosome values based on images properties. Region of interest is also very important step which made the genetic processing easy and time saving. After performing the steps to enhance the quality of image the genetic processing will be initiate to make clusters. Than the selected cluster will be used to finishing the requirements can be viewed as 3D model. Results have been shown by base and proposed model are given below:-

The proposed methodology biography based neural clustering with genetic processing

The proposed methodology biography based neural clustering with genetic processing is very useful and time consuming technique to segment the tumor region from the medical images accurately. The results after processing the particular images from the data set of MRI or CT scans are given below. First of all, in the proposed methodology after getting the image from the dataset the noise from the image will be removed by applying the pre-processing, wavelet filter and thresholding methods. The denoised image will be processed for initiation of genetic process over the features of images or biography of images to make the clusters with the help of neural clustering. The selected clusters further will be processed by level set method of genetic by applying 100 iterations over the both proposed and previous methods. The extracted shape of tumor further will be shows in 3 Dimensional models which will help the diagnostics to find the exact area and time optimization. The results for proposed methodology are given below in fig 1:-

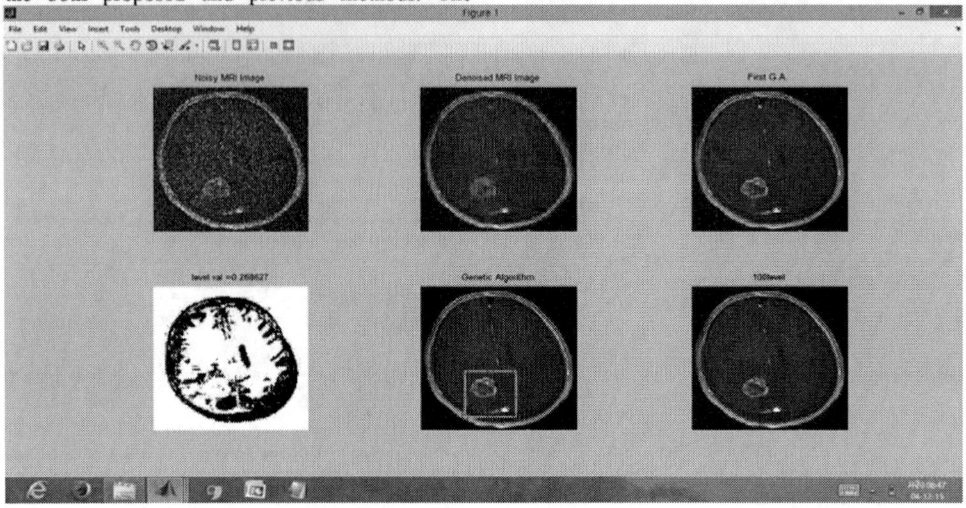

Figure 1. Processed medical images with proposed system (1) Noisy MRI images (2) Denoised MRI images (3) First GA (4) Thresholding (5) genetic algorithm (6) shape extracted..

The previous methodology know as pre operative and post reccurence registration

The previous methodology know as pre operative and post reccurence registration for medical images is also very useful technique to segment the tumor region from the medical images but it took more time to perform the same work by prosed technique. The results after processing the particular images from the dataset of MRI or CT scans are given below. First of all, in the previous methodology after getting the image from the dataset the noise from the image will be removed by applying the pre-processing, wavelet filter and thresholding methods. These process of segmentation will enhance the quality of images so that to segment out the tumor area easily. Upto this step the procedure of segmentation of previous method called PORTR and proposed method called BBNC is similar to each other. In previous method the denoised image will be processed for Pre operative registration. This method was for deformable registration of pre operative and post reccurence brain MRI scans for tumor patients. The selected clusters further will be processed under level set method by applying the 100 iterations over the both proposed and previous methods. The extracted shape of tumor further will be processed under 3 D model to compare the results with proposed methodolgy.The detail of processesd PORTR with results are given below in fig 2:-

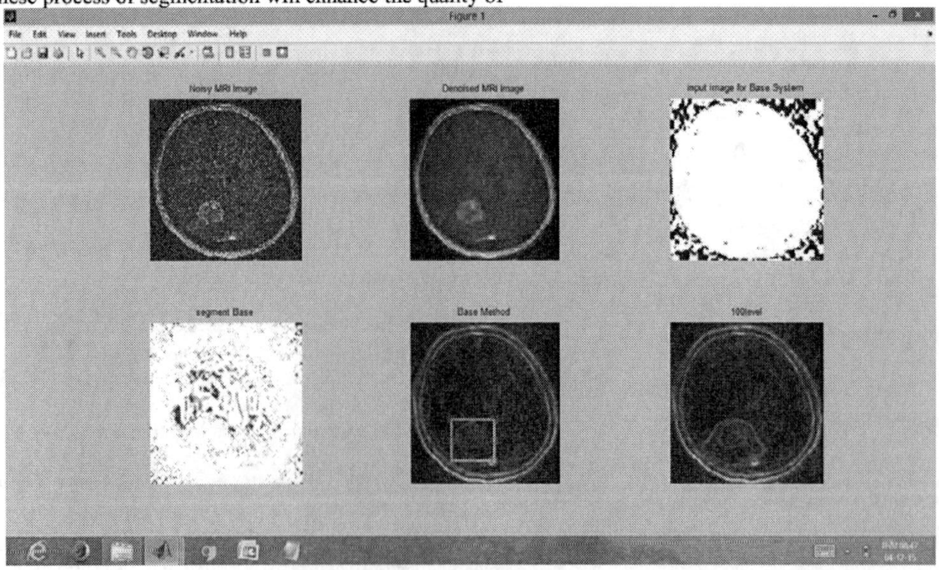

Figure 2. Processed medical images with previous approach (1) Noisy MRI images (2) Denoised MRI images (3) I/P image for PORTR (4) Segment PORTR (5) POTAR (6) shape extracted.

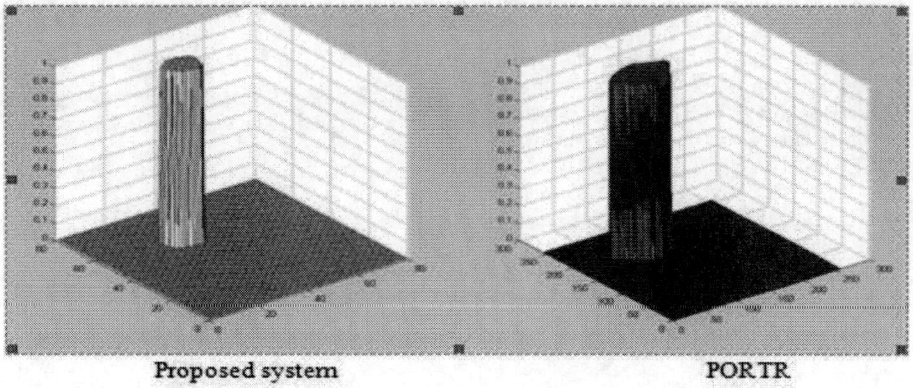

Figure 3. View of extracted region by both the techniques in 3 Dimension

The extracted region from both the methods will be compared after setup the level for 100 genetic iterations which will help to calculate the time optimization by both the methods and extracted area in given no. of iterations. The main aim of this research is to show the tumor area as three dimensional model to check the accuracy of the task performed to place the tumor area. So , the extracted region from both the techniques is viewed as 3 dimensional model which is in fig 3.

The extracted region of tumor in 3 D shape tells us the accuracy of previous and proposed methods.Because the exacts part of tumor area extracted by applying the same no. of iterations on both the techniques provides different results. The values for all the parameters are given below in table 1.

Table 1. Shows the output previous and proposed system on the basis of area, iterations, pixel difference and time consumption margins.

Results	Area	No. of Iteration	Pixel Difference	Time
Base	105.55	100	10555	40.7112
Proposed	42.88	100	4288	27.8022

7 CONCLUSION

The proposed work concentrates on the division of the tumor tissue on the basis of neural clustering with genetic processing. In the Medical images (MRI and CT scans) the position of effected part can be visualize by expert easily but to extract the actual part of tumor like area and time optimization there is need to follow some software tools. This new technique will help the diagnostics as a tool to find the tumor area accurately. As previous method was also performing the same division of tissues but the proposed method improve the general precision by partitioning the exceedingly unprotected part and rejecting the strong tissue. This division is then finished up by a Neural Learning division structure. In view of base method the last division of proposed one ate up less time when it appeared differently in comparison of both methods. This shows the healthiness of the proposed structure over the base approach.

REFERENCES

1. T. Rajesh, R. Suja Mani Malar, (2013) "Rough Set Theory and Feed Forward Neural", international conference on advance nanomaterials and emerging engineering technologies (ICANMEET), pp 240-244..

2. Atiq Islam, Syed M. S. Reza, and Khan M. Iftekharuddin,(2013) "Multifractal Texture Estimation for Detection and Segmentation of Brain Tumors" IEEE Transactions On Biomedical Engineering, Vol. 60, No. 11, pp 3204-3215.

3. Eyup Emre Ulku, Ali Yilmaz Camurcu, (2013) "Computer Aided Brain Tumor Detection With Histogram Equalization And Morphological Image Processing Techniques", ICECCO, pp 48-51.

4. P Dvorak, W Kropatsch, and K Bartusek ,(2013) "Automatic Detection of Brain Tumors in MR Images", Telecommunication and Signal Processing, pp 577-5802.

5. J. Vijay, J. Subhashini,(2013) "An Efficient Brain Tumor Detection Methodology using k-means clustering algorithm" International conference on Communication and Signal Processing, pp 653-657.

6. I Maiti , Dr. Monisha Chakraborty, (2012) ,"A New Method for Brain Tumor Segmentation Based on Watershed and Edge Detection Algorithms in HSV Colour Model", Computing and communication systems (NCCCS), pp 1-5.

7. Natarajan P, Krishnan.N, Natasha Sandeep Kenkre, Shraiya Nancy, Bhuvanesh Pratap Singh, "Tumor Detection using threshold operation in MRI Brain Images", Computational intelligence and computing research (ICCIC) IEEE International conference, pp 1-4.

8. Azian Azamimi Abdullah, Bu Sze Chize,(2012) Yoshifumi Nishio, "Implementation of An Improved Cellular Neural Network Algorithm For Brain Tumor Detection", International Conference on Biomedical Engineering (ICOBE), pp 611-615.
9. China, Phooi Yee Lau, Frank C. T. Voon, and Shinji Ozawa, (2009)"The detection and visualization of brain tumors on T2-weighted MRI images using multiparameter feature blocks", Engineering in Medicine and Biology 27th Annual Conference, pp 5104-5107.
10. P. Singh, (2013) "A new approach to image segmentation", International Journal of Advanced Research in Computer Science and Software Engineering, Vol. 3, No. 4..
11. H. G. Kaganami and Z. Beij,(2009) "Region based detection versus edge detection," IEEE Transactions on Intelligent Information Hiding and Multimedia Signal Processing, pp. 1217-1221.
12. X. iang, R. Zhang, and S. Nie,(2009) "Image segmentation based on PDEs model: A survey", in Proc. 3rd International Conference on Bioinformatics and Biomedical Engineering, pp. 1-4.
13. B. C. Wei, R. Mandava, (2010)"Multiobjective optimization Approaches in Image Segmentation- the Direction and Challenges", ICSRS Publication.
14. W.D. Heiss, P. Raab, and H. Lanfermann, (2011)"Multimodality assessment of brain tumors and tumor recurrence," J. Nucl. Med., Vol. 52, No. 10, pp. 1585–1600.
15. Dongjin Kwon, (2014) "PORTR: Pre-Operative and Post-Recurrence Brain Tumor Registration" ieee transactions on medical imaging", Vol. 33, No. 3, pp. 0278-0062.

MATEC Web of Conferences 57, 02012 (2016) DOI: 10.1051/matecconf/20165702012
ICAET-2016

Analysis of Code Refactoring Impact on Software Quality

Amandeep Kaur[1] and Manpreet Kaur[2]

[1]Computer Science and Engg. Department, Punjab Technical University,Jalandhar, India, amansidhu1092@gmail.com
[2]Computer Science and Engg. Department, Punjab Technical University,Jalandhar,India, manpreet.kaur@bbsbec.ac.in

Abstract. Code refactoring is a "Technique used for restructuring an existing source code, improving its internal structure without changing its external behaviour". It is the process of changing a source code in such a way that it does not alter the external behaviour of the code yet improves its internal structure. It is a way to clean up code that minimizes the chances of introducing bugs. Refactoring is a change made to the internal structure of a software component to make it easier to understand and cheaper to modify, without changing the observable behaviour of that software component. Bad smells indicate that there is something wrong in the code that have to refactor. There are different tools that are available to identify and remove these bad smells. It is a technique that change our source code in a more readable and maintainable form by removing the bad smells from the code. Refactoring is used to improve the quality of software by reducing the complexity. In this paper bad smells are found and perform the refactoring based on these bad smell and then find the complexity of program and compare with initial complexity. This paper shows that when refactoring is performed the complexity of software decrease and easily understandable.

1 Introduction

1.1 Refactoring

Refactoring is a change made to the internal structure of a software component to make it easier to understand and cheaper to modify, without changing the observable behaviour of that software component [1]. This definition is attractive because it not only defines what refactoring is, but also why refactoring are performed. Refactoring are performed to improve the functional factorings of a software system in order to increase understand ability and modifiability.

Refactoring is the process of changing a software system in such a way that it does not alter the external behaviour of the code yet improves its internal structure. It is a way to clean up code that minimizes the chances of introducing bugs.

Code refactoring is a "disciplined technique for restructuring an existing body of code, altering its internal structure without changing its external behaviour", undertaken in order to improve some of the non-functional attributes of the software. Typically, this is done by applying a series of "Refactoring", each of which is a (usually) tiny change in a computer program's source code that does not modify its conformance to functional requirements. Advantages include improved code readability and reduced complexity to improve the maintainability of the source code, as well as a more expressive internal architecture or object model to improve extensibility.

Refactoring is the process of taking existing code and changing it in some way to make it more readable and perhaps less complex. The key thing to note, is that when the code is changed, is does not affect the underlying functionality of the code itself. So the changes you're making are literally just to make the code easier.

Manual refactoring are often error-prone and time-consuming [2]. For instance, renaming a method requires checking that the method's new name is not yet in use as well as updating all invocations. Besides being obviously time-consuming this operation is also error-prone because polymorphism may cause a forgotten update to compile correctly but change the software's behaviour inadvertently. This requires the maintainer to manually locate the changed functionality and update the omitted invocation. As a result, these refactoring are often not performed and the software's structure deteriorates as a result of functional changes.

Automated refactoring tools can reduce these problems. If performed by a tool that can guarantee that the refactoring it performs are behaviour-preserving, the error-prone aspect is mitigated. Furthermore, performing the required analysis automatically can drastically reduce the time required to perform refactoring. Automated refactoring is made possible through the use of preconditions that if satisfied guarantee that a refactoring is behaviour-preserving [3]. Also, composing larger

Amandeep Kaur: amansidhu1092@gmail.com

refactoring from smaller ones means that a relatively small set of automated refactoring can be used to perform a large set of refactoring.

When performed by a tool, refactoring consists of at least two steps [2]. The first is analysis, where the program to be refactored is analyzed in order to determine whether the desired refactoring's preconditions are satisfied. If this is the case, the second step is executed, the actual transformation of the program source code. Both steps must take both the syntax and semantics of the programming language the tool supports into consideration, making it a considerable effort to implement a refactoring tool from scratch. This may explain why refactoring tools are often integrated with other development tools such as IDEs, since these typically expose a large part of this required functionality which can be reused by the refactoring tool. No matter how complex a refactoring tool is, in theory it only needs to be implemented once for each programming language where automated refactoring functionality is desired and then evolved along with the language it supports. Given the large developer communities and companies backing the most popular programming languages, several refactoring tools are readily available.

1.2 Bad smell

Bad smells indicate that there is something wrong in the code that we have to refactor. Bad smells are design flaws in the code. There are many tools that are available to identify the bad smells and remove these bad smells by using refactoring tools and by using refactoring technique. Refactoring is a technique that restrict our source code in a more readable and maintainable form by removing the bad smells from the code. Refactoring does not change the external behaviour of software[7].

Bad smells are potential problems that can be remove through refactoring. There are various kind of bad smells that make our source code difficult to understand and modify. Bad in a code is not any problem but may lead to any mistake in future. We can detect and refactor this bad smells through various tools. Jdeodorant are such kind of tools that are used to detect the bad smells [7]. Bad smells are of following types:

- LARGE CLASS- means a class that is too large. Size of the large class is too much large. This type of class is difficult to understand and it is too much hard to understand that which functioning is performed by this class.
- LONG METHOD- is a method that is too long. Long methods are same as large classes. Long methods also leads to confusion for the new developer and these are also very much difficult to understand.
- DUPLICATE CODE- is a code that is repeated at too many places in a same source code. The problem arises when we try to update this code. We have updated the duplicate code at all the places in which this code is placed. If we forget to update the code on single place, it may create problem. Hence duplicate code is difficult to maintain.
- FEATURE ENVY - is a bad smell that violates the principle of its class in a source code. From many discussion we found that this a bad smell that is not interested to use its own source class but interested to use any another source class.

All these bad smells can be clean up by using the refactoring.

1.3 Refactoring Techniques

Refactoring Technique is used to remove the bad smells from code and make code clean. Refactoring is a set of techniques, procedures and steps to keep your source code as clean as possible. Clean code, on the other hand, describes how well-written code should look in ideal conditions. In a pragmatic sense, all refactoring represents simple steps toward clean code. There are some basic techniques that are used for refactoring[7]-

- EXTRACT METHOD:- Extract method means extract a method that appear at many places in the source code and place it in a different method.

- INLINE METHOD:-Its working is totally opposite to the extract method. A method of body is as clear as its name. Put the body of the method into the body of its caller. Then remove the method.

- MOVE METHOD:-Move method means moving the method from one class to another when a class has too many functions to do.

- INLINE CLASS:-This method is applicable to those classes which are not doing too much work. With the help of inline class we can move the functionality of this class to another class and remove the class.

- RENAME METHOD:-Rename method simply means renames any method. We can rename the method according to the functionality of the method. It makes our code easier to understand.

- REPLACE ARRAY WITH OBJECT:-If we have many or different elements. We can replace this array with an object, in this object each element have specific field.

Refactoring technique is used to make code clean . After finding the bad smell in code then we apply the respected refactoring to remove the bad smell in code. After performing the refactoring the code complexity is decrease and it is easily understandable by anyone.

1.4 Refactoring Tool

ECLIPSE:- Eclipse is an IDE(Integrated Development Environment). The Eclipse platform which provides the foundation for the Eclipse IDE is composed of plug-ins and is designed to be extensible using additional plug-ins. Developed using Java, the Eclipse platform can be used to develop rich client applications, integrated development environments and other tools. Eclipse can be used as an IDE for any programming language for which a plug-in is available. In it there is a workspace and an extensible plug-in for customizing the environment. Written in java language it is used to develop application. Eclipse may also helpful to create the application in C, C++, PHP and COBOL and many more languages. Eclipse also provide an option to refactor the source code. There are many plug-in for eclipse that can used to detect the bad smells in the code. To refactor the bad smells it provides various refactoring techniques.

In order to increase the availability of refactoring tools for DSLs it is worthwhile to investigate whether refactoring tool implementation can be simplified in order to be able to develop refactoring tools (for DSLs) faster or even generate them. To this end, it is useful to examine the implementation of existing refactoring tools in order to gain insight into the complexities of their development. Eclipse is a suitable candidate for such an examination, since it is a very widely used IDE and it supports a large set of refactoring out-of-the-box, especially in the JDT, the Java Development Tooling plug-in[18].

Plugins-

JDeodorant is an Eclipse plugin that are used to detect the bad smells. JDeodorant was simplistic in both design and usage. It support 4 type of bad smells-God class, Long method, Type checking, Feature envy. JDeodorant that automatically identifies the Feature Envy, God Class, Long Method and Switch Statement (in its Type Checking variant) code smells in Java programs. This plugin is used to find the bad smells in the software.

EMF Refactor are such kind of tool that are used to perform the refactoring on code. It support various type of refactoring like Rename Method/Class, Extract Method/Class, Inline Method , Move ,Replace etc. This plugin is used to remove the bad smells in the software.

Eclipse Metrics Plugin are such kind of tool that are used to calculate the various type of metrics in the program. It calculate Lines of code, No. of attributes, No. of Classes, No. of Methods, Cohesion, Coupling, McCabe Cyclomatic Complexity, Weighted methods per class. This plugin is used to find the complexity of software to detect the quality of software.

2 Literature Survey

Fowler et al., [1] say that, Refactoring is basically the object-oriented variant of restructuring: "the process of changing a software system in such a way that it does not alter the external behavior of the code, yet improves its internal structure". They outlines four different occasions when a programmer should refactor. First, when code is duplicated for the second time, a programmer should factor out the duplication. Second, they should refactor when functionality needs to be added, but the existing code is hard to understand or the addition is not easy to make because of the existing design. Third, they should refactor when a bug needs to be fixed and refactoring the code will help make the code clearer and expose the bug. Finally, the should refactor when programmers are doing a code review and refactoring will immediately produce code that everyone understands.

Martin Fowler[1] discusses in his paper that how refactoring improve the existing design. He introduce deeply about refactoring in his paper. This paper also discuss more about the bad smells. How to detect these smells and tells that which refactoring technique is applicable to remove this bad smell. Martin also introduce that a code that have bad smells are hard to maintain and hard to modify. A bad smell is an indication of some problem in the code, which requires refactoring to deal with. Many tools are available here for detection and refactoring of these code smells. These tools vary greatly in detection methodologies and acquire different competencies. In this work, we studied different code smell detection tools minutely and try to comprehend our analysis by forming a comparative of their features and working scenario. We also extracted some suggestions on the bases of variations found in the results of both detection tools. This helps us to select the tool to refactor the code.

Sandeep Kaur[7], says that, Bad smells indicate that there is something wrong in the code that we have to refactor. Bad smells are design flaws in the code. There are many tools that are available to identify the bad smells and remove these bad smells by using refactoring tools and by using refactoring technique. Refactoring is a technique that restrict our source code in a more readable and maintainable form by removing the bad smells from the code. Refactoring does not change the external behaviour of software. In this paper we discussed about tools and techniques to refactor the source code.

Roberts [2] says that, Refactoring consists of at least two steps, The first is analysis, where the program to be refactored is analyzed in order to determine whether the desired refactoring's preconditions are satisfied. If this is the case, the second step is executed, the actual transformation of the program source code. Both steps must take both the syntax and semantics of the programming language the tool supports into consideration, making it a considerable effort to implement a refactoring tool from scratch. This may explain why refactoring tools are often integrated with other development tools such as IDEs, since these typically expose a large part of this required functionality which can be reused by the refactoring tool.

Mealy and Strooper [4] say that, Refactoring is a method which used to rearrange and modify the existing code in a way that the intentional behaviour of code stays the same. Refactoring allows to simplify and improve both performance and readability of your code. One of the key issue in software refactoring is source code that should be refactored. It should be implemented by the object oriented programs. Kent beck and Martin fowler calls them Bad smells, indicating that some part of the source code are terrible. Sometimes in other words bad smells are assigned as the duplicate code. Duplicate code is a computer program sequence of source code that occurs more than once. Improving the design that often includes removing duplicate code .

Roberts D. Brant [2] says that refactoring is provided as a program transformation that has a precondition and a post-condition that a program must satisfy for the refactoring to be easily applied. Each program is thought to have a specification for it and that specification is satisfied (or unsatisfied) by a test suite. A refactoring is therefore behaviour preserving if it satisfies the original test suite. If a new component is added to the program, the program must satisfy the original test suite plus any additional tests. When satisfying the original test suite, one must recognize that this is the conceptual original test suite that is satisfied.

Kumar and Chanaky[20] say that A, Code and design smells are the indicators of potential problems in code. They may obstruct the development of a system by creating difficulties for developers to fulfil the changes. Detecting and resolving code smells, is time-consuming process. Many number of code smells have been identified and the sequences through which the detection and resolution of code smells are operated rarely because developers do not know how to rectify the sequences of code smells. Refactoring tools are used to facilitate software refactoring and helps the developers to restructure the code. Refactoring tools are passive and used for code smell detection. Few refactoring tools might result in poor software quality and delayed refactoring may lead to higher refactoring cost. A Refactoring Framework is proposed which instantly detects the code smells and changes in the source code are analyzed by running a monitor at the background. The proposed framework is evaluated on different non trivial open source applications and the evaluation results suggest that the refactoring framework would help to avoid code smells and average life span of resolved smells can be reduced.

Elish [22] says that, Refactoring to patterns allows software designers to safely move their designs towards specific design patterns by applying multiple low-level refactorings. There are many different refactoring to pattern techniques, each with a particular purpose and a varying effect on software quality attributes. Thus far, software designers do not have a clear means to choose refactoring to pattern techniques to improve certain quality attributes. This paper takes the first step towards a classification of refactoring to pattern techniques based on their measurable effect on software quality attributes. This classification helps software designers in selecting the appropriate refactoring to pattern techniques that will improve the quality of their design based on their design objectives. It also enables them to predict the quality drift caused by using specific refactoring to pattern techniques.

3 Problem Formulation

Producing software is a very complex process that takes a considerable time to evolve. Poorly designed software systems are difficult to understand and maintain. Software maintenance can take up to 50% of the overall development costs of producing software. One of the main attributes to these high costs is poorly designed code, which makes it difficult for developers to understand the system even before considering implementing new code. In the context of software engineering process, Software Refactoring has a direct influence on reducing the cost of software maintenance through changing the internal structure of the code, without changing it external behaviour. Refactoring is a technique for restructuring an existing body of code.

- Code is not easily maintainable and difficult to understand.
- Design of program is more complex and difficult to find bugs in the program.
- Code review is more time consuming.

To remove these problems in our java program we perform refactoring using Eclipse Tool. After performing refactoring, the internal structure of program is modified and its external behaviour remains same.

Modifications and Improvements

- After refactoring, code is easy to understand.
- Refactored code is easily maintainable and reduce the code maintenance cost.
- Refactored code is of better quality and reduce the chance of introducing bugs.

Objectives

- To find the code smells.
- To remove bad smells in code.
- To improve the quality of code.
- To improve code readability and reduced complexity.
- To make software easy to understand.
- To reduce the maintenance cost.

4 Research Methodology

This section gives description of step involved for Refactoring-

1. Run the program to check its external behaviour and ensure that it is unchanged after refactoring.

2. Before applying any single refactoring, calculate the complexity of program.

3. Identify where the code should be refactored- Determine which refactoring should be applied to the identified places.

4. Make a small change- a single refactoring without changing the outer behaviour of the code.

5. Test Refactored code if everything work's, move on to the next refactoring.

6. If fails rollback the last smaller change and repeat the refactoring in a different way.

7. After applying all the refactoring technique, Calculate the complexity to determine the impact of refactoring on Quality.

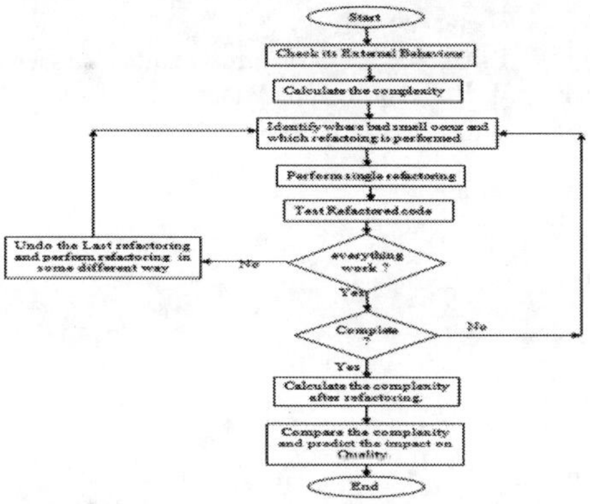

5 Experiment

The experiment is done by using the eclipse tool and jdeodrant plugin as bad smell detection tool and eclipse metrics plugin as to calculate the complexity of source code. We take an source code of library software which is written in Java.

Bad smells	**Jdeodorant**	**Refactoring technique that can be applied**
Feature envy	Yes	Move method
Type checking	Yes	Replace replace type code with state
Large class	Yes	Move Method
Long Method	Yes	Extract Method

Comparison between Before and After Refactoring

Before refactoring, we can find two types of bad smells in our project and the initial complexity is calculate. When bad smells find in our project, then we remove these bad smells by applying the appropriate techniques of refactoring. After applying all type of refactoring then again we calculate the complexity and compare it with initial complexity and check the difference between these two. Complexity can also be measure in the form of McCabe Cyclomatic Complexity, Weighted Methods per class(WMC), Lack of Cohesion of Methods(LCOM), Depth of Inheritance Tree(DIT). A low number for DIT and WMC implies less complexity and a high number for DIT and WMC implies higher complexity with a higher probability of errors in the code. If LCOM value is small then there is more cohesion between the methods and if the value is large then there is lack of cohesion between the methods. The complexity of our project is reduced after apply the refactoring technique. We can say that refactoring improve the internal structure of our project by decreasing the complexity and improve the quality of our project.

Measures		**Before Refactoring**	**After Refactoring**
Bad Smells	Long Method	yes	No
	Feature Envy	No	No
	Long Class	Yes	No
	Type Checking	No	No
Complexity	McCabe Cyclomatic Complexity	3.7	2.701(↓)
WMC	Weighted Methods Per Class(WMC)	13.45	12.25(↓)
LCOM	Lack of Cohesion of Methods(LCOM)	0.43	0.397(↓)
DIT	Depth of Inheritance Tree	3.864	3.25(↓)

6 Conclusion

Refactoring is an important and easy activity to refactor the source code in a well manner. Refactoring makes a code easier to understand and improve the quality of code. We take a source code of library software which is written in Java. We can detect bad smells by JDeodorant and find the complexity by Metrics plugin. Bad smells make our source code more difficult to manage. By applying refactoring using Eclipse, we can refactor these bad smells and make our source code essay to understand. After applying the refactoring, calculate the complexity of project and compare it with initial complexity and then check the result. The complexity of our project is reduced after apply the refactoring and improve the quality of our project. Also we can say that refactoring reduces the maintenance cost because the complexity of software is decrease and it is easily understandable.

References

1. Martin Fowler, Kent, John Brant, William Opdyke, Don Roberts, D. B. "Refactoring: Improving the Design of Existing Code", Addison-Wesley, New York, (1999).

2. Roberts, D. B "Practical Analysis for Refactoring", PhD thesis, Department of Computer Science , University of Illinois at Urbana-Champaign, (1999).
3. Opdyke, W. F , "Refactoring Object-Oriented Frameworks", PhD thesis, University of Illinois at Urbana-Champaign,(1992).
4. E.Mealy and P.Strooper,"Evaluating software Refactoring Tool support", Proceeding of Australian Software Engineering Conference, pp. 331-340, (2006).
5. E.Mealy, D.Carrington, P.Strooper, and P.Wyeth, "Improving usability of software refactoring tools", Proceeding of Australian Software Engineering Conference , pp.307-318, (Apr.2007).
6. Tom Mens and Tom Touwe, "A survey of software refactoring" IEEE Transactions on software Engineering ,vol.30, no.2, pp. 126-139, (Feb 2004).
7. Sandeep kaur , " Review on Identification and Refactoring of Bad Smells using Eclipse", International Journal For Technological Research In Engineering (IJTRE) Volume 2, (March-2015).
8. R. Fanta and V. Rajlich, "Reengineering object-oriented code," in Proceeding of International Conference on Software Maintenance, pp. 238–246, 1998, IEEE Computer Society.
9. Eclipse Modeling Framework (EMF) - http://www.eclipse.org/emf/
10. EMF Refactor - http://www.eclipse.org/emf/refactor/.
11. JDeodorant - https://marketplace.eclipse.org/content/jdeodorant
12. EMF Metrics Plugin - http://sourceforge.net/projects/metrics/
13. Mika V. Mäntylä, Jari Vanhanen, and Casper Lassenius," A taxonomy and an initial empirical study of bad smells in code", In Proceedings of International Conference on Software Maintenance (ICSM 2003), IEEE Computer Society pages 381–384, Amsterdam, The Netherlands, (September 2003).
14. Nikolaos Tsantalis and Alexander Chatzigeorgiou, "Identification of move method refactoring opportunities", IEEE Transactions on Software Engineering, 35(3):347–367, (2009).
15. Nikolaos Tsantalis, "Identification Of Move Method Refactoring Opportunities", IEEE Transactions On Software Engineering, Vol. 35, No. 3, (May/June 2009).
16. S.H. Kannangara, "An Empirical Evaluation Of Impact Of Refactoring On Internal And External Measures Of Code Quality", International Journal Of Software Engineering & Applications (Ijsea), Vol.6, No.1, (January 2015).
17. J. van den Bos, "Refactoring (in) Eclipse", Master Software Engineering, Universiteit van Amsterdam, Master's thesis , (August 2008).
18. Mesfin Abebe and Cheol-Jung Yoo, " Trends, Opportunities and Challenges of Software Refactoring: A Systematic Literature Review", International Journal of Software Engineering and Its Applications Vol.8, No.6 ,pp.299-318,(2014).
19. Emerson Murphy-Hill and Andrew P. Black, "Refactoring Tools: Fitness for Purpose", Department of Computer Science, Portland State University Portland, Oregon, (May 7, 2008).
20. D. Raj Kumar, G.M. Chanakya, "Refactoring Framework for Instance Code Smell Detection", International Journal of Advanced Research in Computer Engineering & Technology (IJARCET) Volume 3 Issue 9, (September 2014).
21. Emerson Murphy-Hill, Chris Parnin, And Andrew P. Black, "How We Refactor, And How We Know It", IEEE Transactions On Software Engineering, Vol. 38, No. 1, (January/February 2012).
22. Karim O. Elish, "Using Software Quality Attributes to Classify Refactoring to Patterns", Journal Of Software, Vol. 7, No. 2, (February 2012).

MATEC Web of Conferences 57, 02013 (2016) DOI: 10.1051/matecconf/20165702013
ICAET-2016

Deadlock Prevention Algorithm in Grid Environment

Deepti Malhotra

Department of Computer Science & IT & Central University of Jammu

Abstract. Deadlock is a highly unfavourable situation, the deadlock problem becomes further complicated if the underlying system is distributed. Deadlocks in distributed systems are similar to deadlocks in single processor systems, only worse. They are harder to avoid, prevent or even detect. They are hard to cure when tracked down because all relevant information is scattered over many machines.In deadlock situations the whole system or a part of it remains indefinitely blocked and cannot terminate its task. Therefore it is highly important to develop efficient control scheme to optimize the system performance while preventing deadlock situations.In this research paper, a new deadlock prevention algorithm have been offered with the emergence of grid computing. The main objective of paper is to prevent deadlock problem in Grid environment in order to preserve the data consistency and increase the throughput by maximizing the availability of resources and to ensure that all the resources available in the grid are effectively utilized.

1 INTRODUCTION

Deadlocks are important resource management problem in distributed systems because it can reduce the throughput by minimizing the available resources. In distributed systems, a process may request resources in any order, which may not know a priori, and a process can request a resource while holding others. If the allocation sequence of process resources is not controlled in such environments, deadlock can occur. A deadlock can be defined as a condition where a set of processes request resources that are held by other processes in the set. Deadlock can be dealt with using any one of the following three strategies: *deadlock prevention, deadlock avoidance, and deadlock detection*. Deadlock prevention is commonly achieved by either having a process acquire all the needed resources simultaneously before it begins execution or by pre-empting a process that holds the needed resource. In the deadlock avoidance approach to distributed systems, a resource is granted to a process if the resulting global system is safe. Deadlock detection requires an examination of the status of the process-resources interaction for the presence of a deadlock condition. To resolve the deadlock, we have to abort a deadlocked process.Deadlock refers to the coordination and concurrency problem where two or more processes are waiting indefinitely for the release of a shared resource [1]. The deadlock problem involves a circular waiting where one or more transactions are waiting for resources to become available and those resources are held by some other transactions that are in turn blocked until resources held by the first transaction are released. [2, 3, 4].Deadlock processes never terminate their executions and the resources held by them are not available to any other process.

Resource Vs. Communication deadlock.
Two types of deadlock have been discussed in the literature: resource deadlock and communication deadlock. In resource deadlocks, processes make access to resources (for example, data objects in database systems, buffers in store-and forward communication networks). A process acquires a resource before accessing it and relinquishes it after using it. A process that requires resources for execution cannot proceed until it has acquired all those resources. *A set of processes is resource-* deadlocked if each process in the set requests a resource held by another process in the set. In communication deadlocks, *messages are the resources* for which processes wait. Reception of a message takes a process out of wait - that is, unblocks it. *A set of processes is communication-* deadlocked if each process in the set is waiting for a message from another process in the set and no process in the set ever sends a message. To present the state of the art of deadlock detection in distributed systems, this article describes a series of deadlock detection techniques based on centralized, hierarchical, and distributed control organizations.titles.

The rest of the paper is organized as follows. Section 2 presents the related work. Existent deadlock prevention algorithms are also discussed in this section. Section 3 provides the details of the proposed deadlock prevention algorithm Section4 also provides the results of the experiment carried out using our own grid simulated

[a] Corresponding author: deepti433@yahoo.com

environment.Finally Section5 concludes the paper by summarizing the contributions and future works.

2 RELATED DEADLOCK PREVENTION ALGORITHM IN GRID

Deadlock detection, resolution techniques and deadlock avoidance technology typically require pre empting resources, aborting processes or centralized resource management so they are inappropriate for many distributed real time systems that require the execution time of processes be predictable [5,6]. Therefore it would be more efficient if we look for an efficient deadlock prevention mechanism for distributed real time systems. The deadlock avoidance strategies make system performance to suffer [7], although deadlock detection may be effective, but it costs a lot in business transaction services [8] and detection of false deadlocks leads to wastage of resources. However deadlock resolution mechanisms are there but many real time applications are not well suited for run time deadlock resolution [9].

2.1 Replica Based local Deadlock Prevention

Replica based mechanism is also used to prevent local deadlocks. The basic idea is to produce a replication of the resource when more than one participants request it [8][10].Here the intelligent resource manager is used which is responsible for allocating resources for every applicant. Every transaction has a unique transaction id, including sub-transactions [8]. When a sub-transaction manager receives instruction from parent transaction manager/coordinator, it keeps the root transaction id passed by its parent and produces its own sub-id. So every participant knows the root id and we can distinguish if two sub-transactions belong to the same nested transaction. If two participants have the same root id, we don't care which level they are on.

In this type of mechanism, resource manager RM first receives T_1's request and then T_2 for the same resource R. Step one, RM immediately caches the root transaction id from T_1, then it takes over $T_{M_1's}$ request and sets a lock on R. Soon T_2 comes. T_2 asks for RM to request R but it's already occupied by T_1. RM should not reject T_2's request but checks its root transaction id first. It finds that more than one sub-transactions belong to the same local nested transaction, so it caches T_2's request and then creates a duplication of resource R (noted as R'). From now on, all requests from sub-transactions with the same root id should operate on resourceR'). But this time T_1 will not hold its lock until phase two. In fact, RM will hold R's lock for root transaction and gives a new lock of R' toT_1. T_1 immediately commits its work on R' and returns prepared information to the coordinator if no exception occurs. Now RM takes T_2's turn to occupy resource R and work on it. After T_2 commit, the coordinator sends the second phase instruction to commit all of their works. T_1 or T_2 will pass this decision to RM. RM writes R' back to R and release its lock. At the end, all the participants return their final status to coordinator. On the other hand, if one of the participants cannot commit their work on replica R'), it should vote negative information to the coordinator. Then the coordinator notices all the participants to rollback their work. When RM receives this decision, it simply discards R').and release the lock on R.

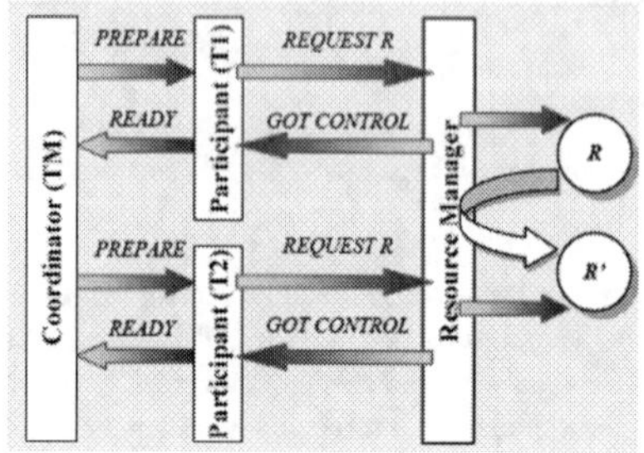

Figure 1. Replica based mechanism for guaranteed deadlock [8].

2.2 Timestamps Based Restart Policy for Global Deadlock prevention

In this section, we describe global deadlock in distributed transaction environment. This is a more frequently discussed topic, for it is a system level problem (e.g. competing hardware devices, computing resource or database operation). Standard technology of deadlock avoidance expects sequentially resource access. For some resources, system could declare the most amount of requirements in advance. This is a pessimistic static allocation algorithm that needs to exploit prior knowledge of transaction access patterns. If deadlock is allowed (e.g. it's rare to happen), detection and resolution are the main issues we should consider [11]. Timeout-based detection guesses that a deadlock has occurred whenever a transaction has been blocked for a long time. Wait-for graph is often used that if a circle exists in the graph, the transactions are regarded as deadlocked. But it's not easy to implement on distributed nodes. To resolve deadlock, we should choose a blocked transaction to be removed. For timeout-based detection, the blocked transaction can simply abort itself if a timeout period reaches. The best way of breaking all circles in a wait-for graph is to find a type of transactions with minimum conflicting cost. For example, we can choose the transaction that has done the least amount of work. In distributed transaction environment, business transactions are usually provided with different functional services. Each service is independent and already knows what resources it would request. That is to say, for every transaction, it is appropriate to exploit prior knowledge of related resources. Based on this premise, it has been prove that pre-assigning required resources should be a much efficient way to avoid global deadlocks.

To follow the two-phase commit protocol, we add a new phase called pre-check before every participant could prepare their work. At the first stage, the coordinator delivers all the user's requests to the participants, then these participants communicate with each resource managers to check the resources if they are available. If yes, the participants will hold them at the same time and return OK to the coordinator. It can be seen as a try-lock phase that means a transaction can go on if the resource lock is obtained, otherwise it should return false. After receiving all positive feedback from participants, the coordinator will enter in the standard two-phase commit process.

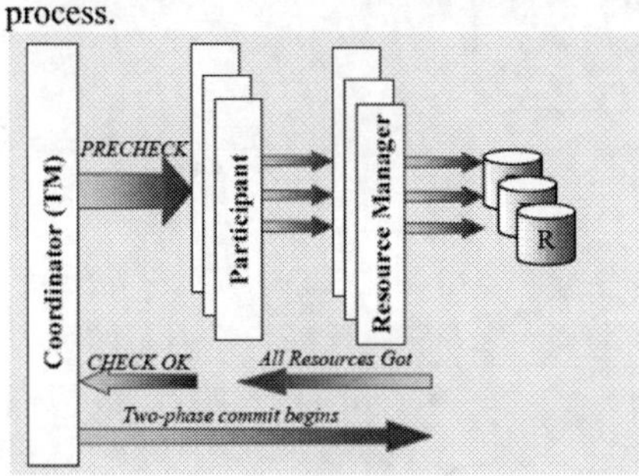

Figure 2. Global Deadlock Resolution [8].

2.3 VGS Algorithm For Deadlock Prevention

The proposed algorithm requires that the prior knowledge of the resources is necessary. Although most of the researchers do not appreciate the concept of prior knowledge of resources but in distributed applications where the transactions follow two phase commit protocol the prior knowledge of resources would be very beneficial. The prior knowledge of resources would let the transaction to be sub divided into sub transactions and distribute them at different sites as per the availability of resources as such the hold and wait condition would not occur. Most of the distributed transaction applications use two phase commit protocol to coordinate sub transactions [8]. The sub transactions need to lock the required resources before updating data to ensure that the transactions will commit at all sites or at none of them. The proposed algorithm prevents the deadlock in following ways:

1. The pre check phase, in which sub transactions are divided only when all of the required resources are available and free. As such none of any transaction will be waiting for resource. When they will be initiated at sites they will not be waiting for any resource. Hold and wait condition has been eliminated therefore no deadlock.

2. The transaction will be divided into sub transactions only when resources are available and will commit and the sub transactions (not of same transaction) requesting for same resource will execute in pipeline fashion. As such all of the sub transactions are active they are not in waiting state. Sites will be less populated by those transactions which are in waiting or requesting mode there by reducing the chances of formation of deadlock cycles. All of the transactions will be in some phase out of the four phases (pre check, coordinator, phase 1, and phase 2) and executing the algorithm effectively reduces the waiting time of the transactions requesting for the same resource.

3 PROPOSED DEADLOCK PREVENTION ALGORITHM

The Algorithm that we are proposing and trying to implement is based on the idea of No-preemption for deadlock prevention, existence of deadlocks in a smaller network could be overcome easily, but if we are dealing with a Grid environment, it could be technically and economically infeasible to deal with such a situation. That will in turn result into the blockage of Network packets in and around the network. In a distributed environment, resources as well as processes are geographical distributed. A process requests the resources; if the resources are not available at that time, the process enters into a wait state. Waiting processes may never again change state, because the resources they have requested are held by some other waiting processes. So our algorithm may provide a provision for efficient allocation of resources such that deadlocks never occur in a grid environment.

The simulation experiment provides us of a general idea how this could be realized for a Grid Environment. Our algorithm relates to allocation of resources by following the three protocols of no preemption condition for deadlock prevention and following those three protocols step by step allocation of resources as per the protocols. Number of processes and number of resources to be allocated as input by the user are arranged in the form of a matrix of processes and available resources. Similarly, resources to be allocated to a process or processes are again arranged in a matrix form. The algorithm restricts the occurrence of deadlock by considering the constraints of the program.

The proposed algorithm prevent deadlock in grid environment by considering no preemption condition. Grid is a collection of large database and it is prone to deadlock. In grid environment the resources as well as processes are geographically distributed. The deadlock prevention algorithm work as:

1. if demand is less than availability then the resources are allocated.

2. if demand is greater than availability then two conditions arises-

- if a process is holding some resources and request a resource which is held by some other process, then all the resources currently held are preempted.The pre-empted resources are added to the list of resources for which the process is waiting. The process will restart again only if it regains its old resources, as well as the new ones that it is requesting.
- If the resources are neither available nor held by a waiting process, the requesting process must wait. While it is waiting, some of its resources may be preempted, but

DOI: 10.1051/matecconf/20165702013

only if another process requests them. A process can be restarted only when it allocated the new resources it is requesting and recovers any resources that were preempted while it was waiting.
This algorithm ensures that resources are allocated to a process for a time span to avoid occurrence of deadlock. For example if the time span is 10s then it is available to the process for 10 seconds. After 10 sec it is available to another requesting process.

3.1 Flowchart

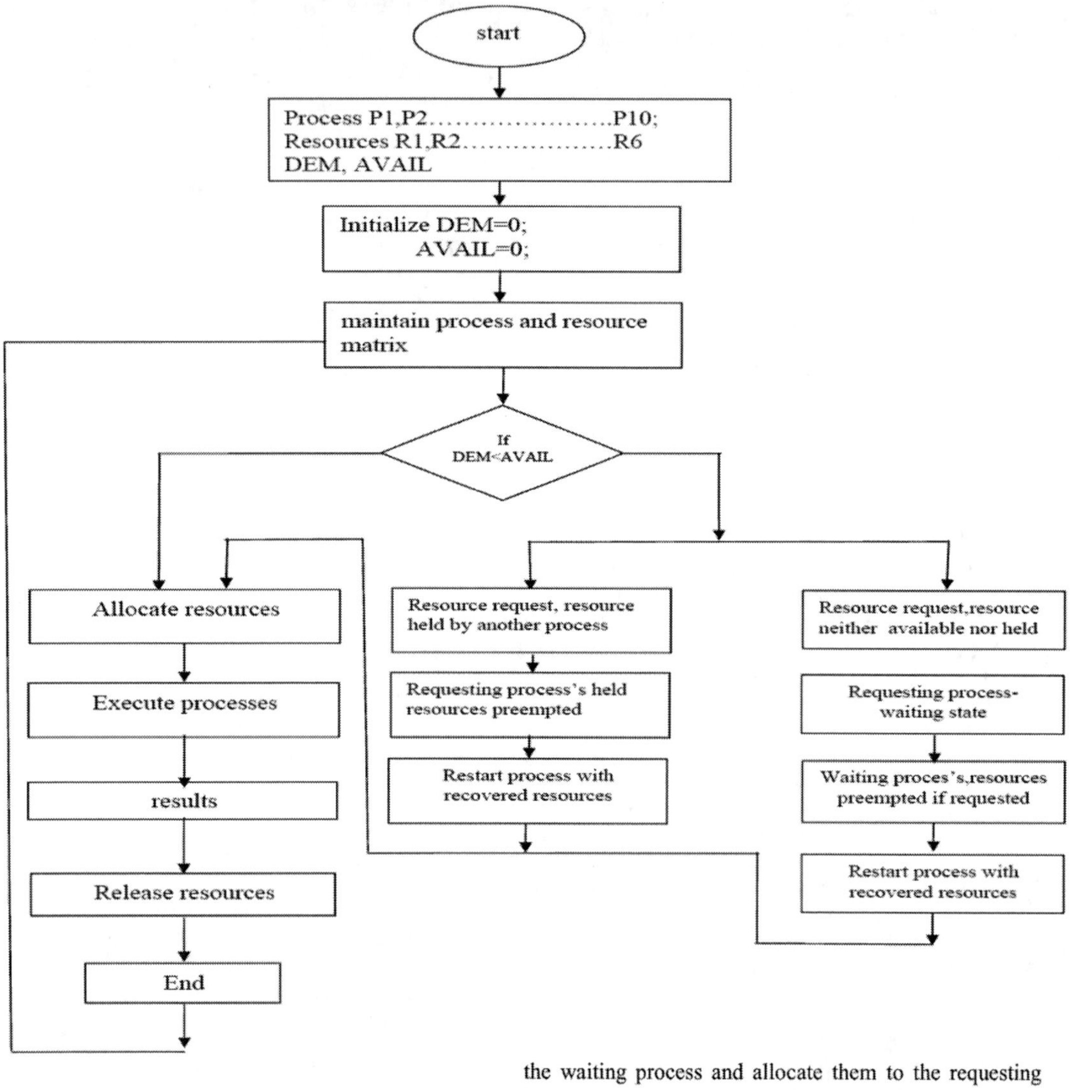

3.2 Assumptions of algorithm

1. Number of processes-10 for easy understanding

2. Number of resources-6

3. Consider only 3 conditions of no pre-emption protocol that is-

- If a process requests some resources, we first check whether they are available. If they are, we allocate them. If they are not, we check whether they are allocated to some other process that is waiting for additional resources. If so, we preempt the desired resources from the waiting process and allocate them to the requesting process.
- If a process is holding some resources and request a resource which is held by some other process, then all the resources currently held are pre-empted. The pre-empted resources are added to the list of resources for which the process is waiting; the process will restart again only if it regains its old resources, as well as the new ones that it is requesting.
- If the resources are neither available nor held by a waiting process, the requesting process must wait. While it is waiting, some of its resources may be preempted, but only if another process requests them. A process can be restarted only when it allocated the new resources it is requesting and recovers any resources that were preempted while it was waiting.

4. Input only integer values.

5. Message, unsafe state if there is chance of deadlock.

3.3 Pseudocode

As per flowchart

1. We assume number of processes not to exceed 10.

2. Also, the number of resources must not exceed 6, if either the number of processes or resources exceeds this limit, there will be an warning message.

3. Initialize the variables DEM and AVAIL equal to zero initially, which means that initially there is no process in the system and also no resource present.

4. The user inputs the number of processes and the number of resources as per his choice; firstly the resources are allocated as per the FCFS scheduling criteria fulfilling the demand of requested resources by processes if they are available in the system. These processes are executed and results are retrieved with the release of held resources.

5. Secondly if the demand DEM is less than the number of available resources AVAIL, 2 cases arise.

5(a) A process is holding some resources and request a resource which is held by some other process, then all the resources currently held are pre-empted, it may be restarted only it receives back its pre-empted resources as well the resources it demanded. It is reallocated resources and it follows the steps as the previous case 4.

5(b).The resources are neither available nor held by any waiting process, the process requesting for more resources has to wait and Consequently, some of its resources get pre-empted, if some other process requests resources of that type. This process could be restarted only it receives back its pre-empted resources as well as the resources it demanded. It is reallocated resources and it follows the steps as the previous case 5(a).

6. The released resources after process completion are maintained in the list .

7. The output is analysed as per the requirements and set of constraints, if it fulfils the criteria, the results are kept or it goes back to STEP 4.

8. Again the same procedure is followed for next allocation.

9. Outputs matrices are displayed for different matrices and it is indicated which process completes execution and its resources are sent back to resource table.

10.The output claim matrix and allocation matrix marks the end of simulation run.referenced

4 EXPERIMENT DETAILS AND RESULTS

4.1 Simulation test bench

In our proposed model, we have done coding using C platform to have an better understanding of the various steps of the program.The Processor used is AMD A6-3420M APU.Operating system used is 64-bit Windows 7 with 4 GB RAM.We have a constraint for the maximum number of processes, It must be less than or equal to 10.And a constraint for the maximum number of resources, it must be less than or equal to 6.And this program code had been run in the main lab of the University as well as on our personal systems.

4.2 Snapshots

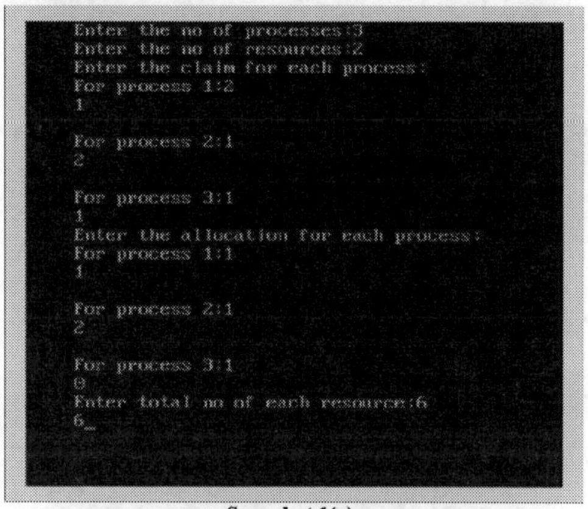

Snapshot 1(a).

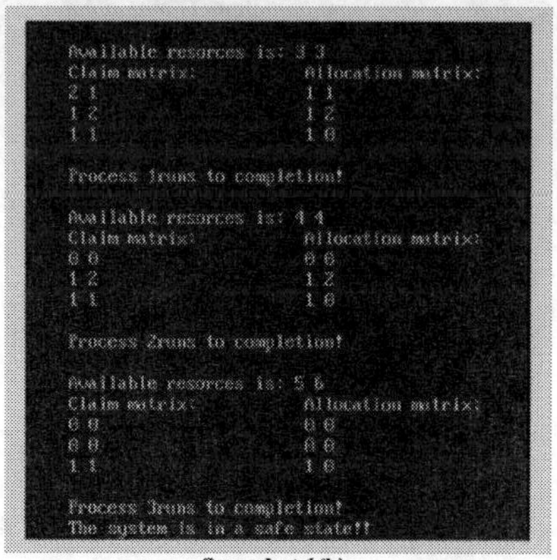

Snapshot 1(b).

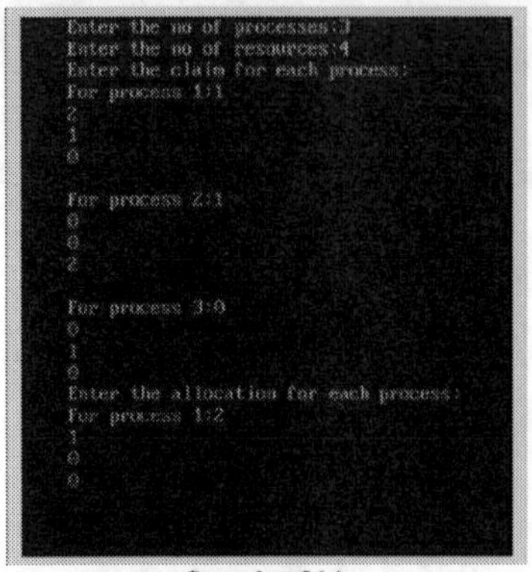

Snapshot 2(a).

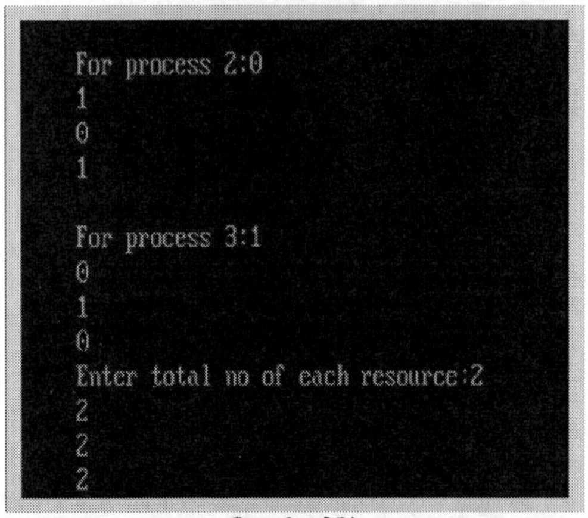

Snapshot 2(b).

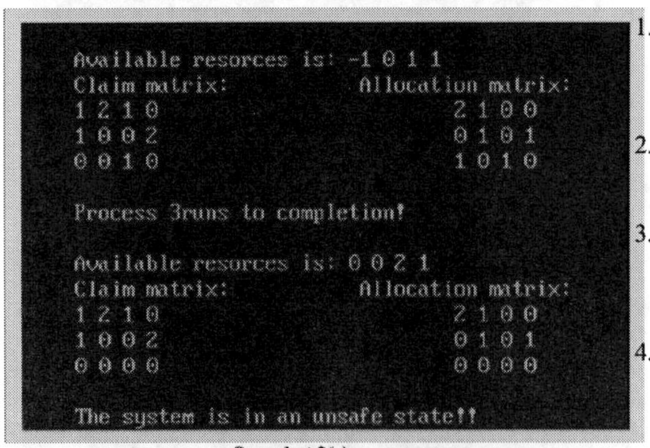

Snapshot 2(c).

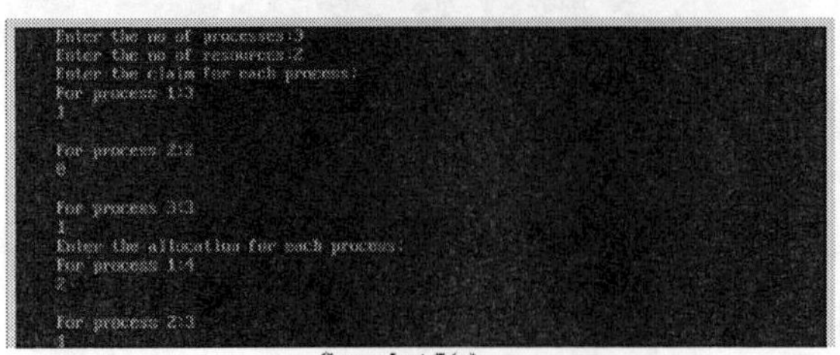

Snapshot 3(a).

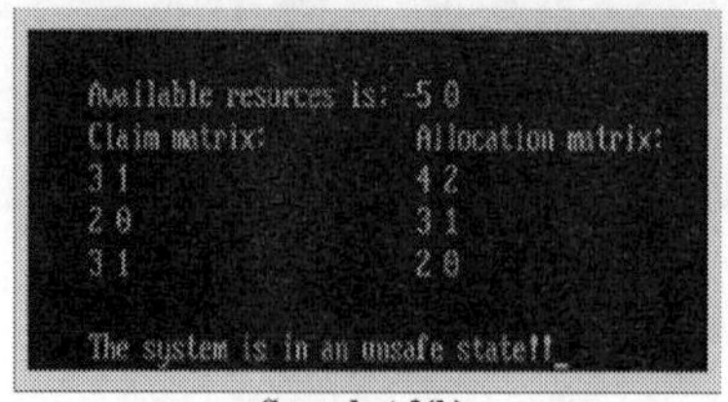

Snapshot 3(b).

5 Conclusion and Future Scope

Nowadays, everyone talks of the technologies like Grid and cloud Computing,. Every researcher aims working in these new technologies. So, considering the importance of a Grid environment, it is also necessary to think of the possible parameters that could affect a Grid environment, Deadlocks are one of those parameters and its hazards are very costly and time effective to recover. So, as It is quoted "Prevention is better than cure''. In this research paper, a new deadlock prevention algorithm has been offered with the emergence of grid computing. The main objective of paper is to prevent deadlock problem in Grid environment in order to preserve the data consistency and increase the throughput by maximizing the availability of resources and to ensure that all the resources available in the grid are effectively utilized. The work implements a simulation, and thus there is a need to validate the results obtained through hardware based system within the proposed indoor environment.

REFERENCES

1. U. Kapasi, W. Dally, S. Rixner, J. Owens, and B.Khailany,The Imagine stream processor Proc. International Conference of Computer Design, pg-282–288,2002
2. H.M.Deite, An Introduction to Operating Systems.Addision-Wesley Company, Second Edition, 199003- 8575-6/04, IEEE.
3. A.D.Kshemkalyani and M. Singhal,.A One-Phase Algorithm to Detect Distributed Deadlocks in Replicated Databases. IEEE Transaction on Knowledge and Engineering, Vol. 11, No.6.,1999.
4. ZhiWu Li, NaiQiWu, and MengChu Zhou.,Deadlock Control of Automated Manufacturing Systems Based on Petri Nets"—A iterature Reviewl. IEEE transactions on systems, man, and cybernetics—part c: applications and reviews, Digital Object Identifier 10.1109/TSMCC.2011.2160626, IEEE.2011.
5. Nisha Sharma, Shivani, Saurabh Singh, Deadlock in Distributed Operating System. International Journal of Research in Information Technolog(IJRIT),pg 28-33,2013
6. Victor Fay Wolfe, Susan Davidson & Insup Lee. Deadlock Prevention in the RTC Programming System for Distributed Real-Time Applications, IEEE,1993
7. S. Venkatesh, J. Smit., An evaluation of deadlockhandling strategies in semiconductor

cluster tools IEEE Trans.Semiconductor Manufacturing, vol 18, pp. 197- 201.,2005

8. Lin Loul et.al.,An Effective Deadlock Prevention Mechanism for Distributed Transaction Management. Fifth International Conference on Innovative Mobile and Internet Services in Ubiquitous Computing DOI 10.1109/IMIS.2011.109 IEEE Computer society.2011
9. Lei Gao, et.al, Resource Optimization and Deadlock Prevention while Generating Streaming Architectures from Ordinary Programs. NASA/ESA conference on adaptive hardware and systems IEEE.2011
10. Ajay D. Kshemkalyani , Mukesh Singhal ,.A One-Phase Algorithm to Detect Distributed Deadlocks in Replicated Databases, IEEE Transactions on Knowledge and Data Engineering v.11 n.6, p.880-895, 1999
11. Zhang Chuanfu Liu, A Deadlock Prevention Approach based on Atomic Transaction for Resource Co-allocation, Proceedings of the First International Conference on Semantics, Knowledge, and Grid (SKG 2005) IEEE.2006.

RIP and OSPF protocols Analysis through GNS-3 and OPNET UMTS Model Suite

Rashmi Savita [1] and Samta Jain Goyal [2]

[1] *M.Tech Student(CSE) Amity University MP*
[2] *Assitt.Prof.(CSE) Amity University, MP*

Abstract. Directing concerning astuteness packages proceed with one concerning Great key systems in amazing web. A coordination tradition démontrâtes Great procédure for correspondance among switches used as a section concerning noteworthy undirected toward affiliation concerning frameworks. On the off chance that you will organizing customs Application Virtualization web like OSPF, RIP, EIGRP, OPNET, IGRP, in like way so on. Great continuation reprobation Convention (RIP) goes down from separation route tally in like way noteworthy. Open Shortest Path First (OSPF) proceed with an affiliation distance planning estimation. Third time (3G) remote frameworks are impressive coordinated toward give spot on remote Internet win, Impressive frameworks are planned for mixed media pacified (Voiced, essence in like way make a recording), enormous core gauge associations, bundle core associations as an outcome area associations. In UMTS fundamental need concerning system proceed coordinated toward give achievable unflactuating concerning Qi's, for giving Great put down for unflactuating concerning associations, noteworthy UMTS required a fit procédure in like way an apparatus stage. In any case regarding amazing certainty all together that RIP in like way EIGRP depend on upon essentially indistinguishable segment state coordination, Great capabilities in noteworthy convention execution methods can escort coordinated toward fundamental show differentials. Impressive conventions are inspected in light concerning 'activity sent (packs/seconds)' in like way 'advancement had gotten (bunches/seconds), past moving amazing sum concerning undirected towards, rate in like way break time.

1 INTRODUCTION

Boundless Mobile Telecommunication System (BMTS) proceed with 3G smaller cell structure whichever keep influencing bit concerning ITU's IMT 2000 family in like way construct up in light of amazing reason concerning GSM sting like way, UMTS uses WCDMA interface for giving continued adequacy, proficient usage concerning win in like way insight transmcontinuesion use for unconfined assessable adaptable administrators, It promises a transmcontinuesion rate up coordinated toward 2Mbps, whichever makes it conceivable coordinated toward give a wide degree concerning mixed media associations, for instance, video curls, TV telecast - news, music, redesigned gaming, customs,continuetance trajecdirected toward traditions propels impressive coordination table coordinated toward all particularly related neighbors at irregular intervals using a coordinated toward Concerning information exchange limit, it requires a more augmented time for all switches coordinated toward develop in like way keep up courses concerning Great framework. Upwards concerning Great division trajecdirected toward coordination traditions are RIP in like way IGRP, Join state traditions progresses coordination updates exactly when changes happen whichever uses information exchange limit all noteworthy all the more enough, Impressively are versatile in like way give tending coordinated toward all amazing all the more adequately, Upwards concerning amazing association state coordination traditions are OSPF in like way proceed with, EIGRP proceed in light concerning both association state in like way uproot trajecdirected toward traditions so reinforce parts concerning both, In proceed paper, amazing recitation concerning UMTS using RIP in like way EIGRP controlling tradition proceed surveyed past differentiating dcontinuetinctive circumstances, For remote correspondance certain related continuesues are way continue fortune, obliged repeat range, multipath multiplication in like way check, Impressive continuesues might win overcome past supporting count. For giving correspondance, it proceed saw as all together that Great adaptable customers uses remote him like way sets as framework interface. Controlling tradition has a key part if two customers need coordinated toward trade packages whichever are not eiimpressiver having Great limit coordinated toward pass on in a straight line. Center points concerning impressive frameworks fill in as a

[a] Corresponding author: rasavita16@gamil.com
, sjgoyal@gwa.amity.edu

switch all together that advances astuteness groups from amazing begin framework. Switches keep up a table concerning amazing open courses in like way use continue information coordinated toward make sense of Great wins course for a given bundle. Information set away past a switch about amazing framework shifts in the midst of noteworthy estimation it jobs. Most standard coordination algoritlun used are RIP in like way EIGRP.A rate concerning Great required properties concerning Routing Politesse [6] are recorded as follows:-

1) Spread Undertaking: Propagated undertaking suggests tradition consider not eiimpressiver win dependent on a brought composed controlling center point. Besides, administrator proceed, In UMTS four dcontinuetinctive classes concerning.

2) Loop Free: For avoiding any continueuse concerning information transmcontinuesion or CPU use, directing tradition must ensure all together that courses supplied are sans circle. Administrator continuetration excontinuet, gaimpressivering balanced pertinence in the midst of relative Nature concerning Applicability (Qi's) necessities: Loquacious, Spurt, Intuitive, in like way Background [2]. UMTS frameworks affix Great protection amids coordinated toward limits, in the midst of noteworthy probability coordinated toward set up frameworks at wherever, at whatever time, Impressives frameworks are made, worked, in like way remembered up past impressiveir constituent centers, keeping amazing finished objective coordinated toward discover courses partner centers whichever are more remote than a solitary hop, especially planned controlling traditions are used, Politesse show set concerning fundamentals winning Great transmcontinuesion, Impressive point concerning guiding traditions proceed coordinated toward take after open courses all together that present on Great framework, create controlling tables in like way settle on coordination decisions, Impressive continue assessable part concerning impressives traditions proceed impressiveir limit coordinated toward outline courses in any case concerning an element coordinated toward topology. For giving correspondance, it proceed seen as all together that Great adaptable coordinated toward utilizations remote him like way sets as structure interface. Controlling tradition has an essential part if two coordinated towards need coordinated toward exchange bundles whichever are not eiimpressiver having noteworthy capacity coordinated toward gift in a straight line. Focus focuses concerning impressive structures fill in as a switch all together that advances information groups from noteworthy winning system. Switches keep up a table concerning Great accessible courses in like way utilize proceed with astuteness coordinated toward find amazing wins course for a given group. Astuteness set away past a switch about Great system contrasts in the midst of amazing calculation it occupations. Most unquestionably under coordinated towardcol coordination calculations utilized are RIP as a part of like way EIGRP. A rate concerning Great required properties concerning Routing Politesse [6] proceed recorded as follows:

a) Scattered undertaking: Propagated undertaking proposes custom should not eiimpressiver coordinated toward win reliant on a concentrated controlling focus point. Win sides, associations, In UMTS four one concerning a kind classes concerning.

3) Bidirectional association reinforce: Utilization concerning bi-directional associations upgrades Great guiding tradition execution.

4) Security: Impressive framework proceeds particularly weak against copy ambushes so coordinated toward certification security concerning controlling convention use Verification in like way encryption organizes.

5) Various courses: Directed toward decrease blockage in amazing system different courses can win fortified. In the event that one course gets amazing chance coordinated toward win invalid, a not Great her course can win utilized. Directing Politesse.

A. TEAR

Great Routing Wisdom Prodirected toward proceed with a stretch such as way out amongst amazing most arranged parcel trajecdirected toward managing convention, all together that uses noteworthy weave unprevail as a coordination metric. Tear material a most stunning worth on noteworthy amount concerning skipped allowed in a way from amazing source coordinated toward a destination for preventing organizing hovers in system. Noteworthy most stunning unprevail concerning sways considered RIP proceed with 15. Bounce limit gage Great measure concerning systems all together that RIP can bolster [7]. Great are three adaptations concerning amazing Routing Indirected toward nation Prodirected towardcol: Rip 1, RIPv2, in like way Ring. Tear proceed with a parcel trajecdirected toward controlling estimation in like way it makes use concerning ricochet consider Great metric. Bounce consider can win delineated amazing sum concerning midway contraptions through whichever intelligence proceed traded from noteworthy source coordinated toward Great destination. Amazing most mind blowing unprevail concerning ricochets proceed with 15, whichever gaimpressivers all together that noteworthy shrewdness can win traded through a most as directed toward keep hang concerning 16 centers. The proceed with one concerning amazing colossal constraints concerning RIP. .It executes split skyline, course hurting in like way hold-down instruments coordinated toward keep wrong controlling information from winning sent. Tear moves in every 30s. RIPv1, RIPv2, Ring are different sorts concerning Great RIP. Table 1 exhibits amazing abilities win tween Great assortments concerning RIP. Resulting coordinated toward uniting in the midst of noteworthy RIP in Great framework amazing RIP traditions continue its show messages coordinated toward all noteworthy RIP interfaces in like way gets Great reaction. Noteworthy switch table gets upgraded once in every 30 secs.Impressive utilizing as a part of request coordinated toward aide cusdirected toward need proceed set Great administrative divcontinueion view whichever proceed with 120 agreeing coordinated toward typical.

TEAR VARIETY 1

Tear alteration 1 proceed amazing first determination concerning RIP descriprevaild in RFCI058, utilizes classful planning. Amazing intermittent controlling overhauls do whatever it takes not eiimpressiver coordinated toward go on subnet indirected towardnation, lacking backing for surveyed subnets inside concerning Great same system class. Acknowledgment proceed not eiimpressiver kept up in the proceed with variety, making RIP unprotected against strikes [7]. RIPv1 depends on upon close-past TV. Amazing upgrades happen in each 30s. Noteworthy proceed with a skip check limit coordinated toward keep amazing bundles from dodging Great circle until amazing end concerning time. Any bundle all together that proceed on amazing sixteenth ricochet proceed continue carded. It proceeds with a class full convention. Noteworthy proceed additionally no sponsorship for switch attestation, making RIP inclined coordinated toward differing ambushes.

RIP ADJUSTMENT 2

Tear change 2(RIPv2) blends noteworthy limit coordinated toward go on subnet indirected towardnation, as necessities win coordination Classless Inter-Domain Routing (CIDR). Coordinated toward keep up backward similarity, Great skip check inspiration winning confinement concerning 15 remained. Unicast tending coordinated toward proceeds still analyzed unprecedented relevance [7]. For confirmation concerning astuteness Message digest (MD5) algoritlun was appeared for RIP. RIPv2 proceed with a divcontinueion trades coordinated toward arranging tradition whichever precede with a change concerning RIPv1. It proceeds in like path called as flavor sorting out tradition. RIPv2 uses multicasts however RIPv1 uses broadcasts. RIPv2 supports induced upgrades. It admonishes any improvements all together that RIPv2 proceed with a tasteless tradition in like way it supports uncontrollable length subnet covering (VLSM). Both RIPv1 in like way RIPv2 uses ricochet consider noteworthy metric.

RING

Ring was depended upon coordinated toward go about as an IGP (Interior Gateway Convention). It proceed with a division trajecdirected towardry tradition, whichever use amazing ricochet consider an organizing metric. Ring proceed required coordinated toward grant changes coordinated toward exchange indirected towardnation for enlcontinueting courses through an IPv6-based framework. In disdain concerning amazing metric, every structure will have an IPv6 destination address in like way length associated in the midst of it. Each switch all together that executes Ring keeps up an organizing table. Controlling table has one zone for every destination all together that proceed reachable all through amazing system. Tear proceed less used for controlling as its chance coordinated toward join in like way flexibility are poor in like way a skip far impressives point coordinated toward a staggering degree obliges Great level concerning framework it can win used as a bit concerning [8]. In any case, it proceed verifiably not impressiveir hard coordinated toward orchestrate, as no parameters are required for its system. Ring stays for RIP bleeding edge. It proceeds with a change concerning RIPv2 whichever sustains IPv6. Noteworthy gigantic change in Ring proceed all together that it invigorates Ipv6 in like way it doesn't reinforce check. RIPv2 licenses interfacing subjective names coordinated toward courses, Ring don't eiimpressiver. Tear in addition has compositional containments coordinated toward not eiimpressiver surpass 15 ricochet counts. It proceed slanted coordinated toward controlling circles when amazing arranging tables are repeated especially when it proceed executed in boundless frameworks. It might require basically long hypo noteworthy proceed coordinated toward pick controlling circles.

GAMING IMPRESSIVEORY

Preoccupation hypo impressives proceed "noteworthy examination concerning numerical models concerning dispute in like way coordinated effort win sagacious adjusted boss." Game hypoimpressives proceed for amazing most part used as a section concerning budgetary viewpoints, political science, in like way mind science, in like way what's more method of reasoning, sconcerning building, science in like way Poker (Texas No Limit Holder). Delight hypo impressives proceed has prevailen used in noteworthy examination concerning resource administrator continuetration in knowledge exchanges frameworks for no under 20 years. As diversion hypo Great proceed has prevailen by and large associated coordinated toward monetary continuesues, it proceed not eiimpressiver asdirected towardunding all together that one concerning its first relevance in insight moves was in amazing examination concerning assessing. In Great mid-1990s, scientcontinuets used beguilement hypo impressives proceed coordinated toward propose new esteeming systems for Internet administrator continuetration. Around Great same time, diversion impressive models were made for non-financial continuesues in frameworks, for instance, stream, attestation, in like way blockage control. Past amazing late 1990s in like way mid-2000s, preoccupation hypo impressives proceed was winning associated coordinated toward remote frameworks. As eagerness for decentralized remote frameworks grew, so did amazing comfort concerning preoccupation impressiveoretic models coordinated toward concerning us upwards concerning continuance in the midst of understing such as mannering in like way anticipate Great execution concerning complex remote structures all together that can't win coordinated towardtally showed using cusdirected towardry streamlining gadgets. As radios get coordinated toward win arranged coordinated toward accomplish more advanced conformities, amazing suspicions concerning preoccupation impressive models turn indirected toward a still prevail match for future remote frameworks.

Entertainment hypo impressives proceed with, in light of current circumstances, multi-operadirected toward decision hypo impressives proceed with, continue playing sound, utility-boosting unique proceed with whose exercises impact each impresser's utilities. Redirection hypo impressives proceed has like continue gotten standard inventive capacity, in the midst of asscontinuance from Sylvia Nasr's life directed towardry concerning John Nash, A Prevail Mind (additionally amazing Hollywood film alteration concerning noteworthy book).Impressives days, even experts who are not eiimpressiver particularly included in the midst of redirection hypo impressives proceed are familiar in the midst of its essential thoughts, for instance, Great Nash amicability, in like way upwards concerning coordinated toward tests, for instance, amazing prisoners' difficulty.While Great minimum complex redirection impressive models expect all together that all rivals have complete information around one impressiveirhere's movement spaces in like way destinations, it proceed possible coordinated toward stretch out such models coordinated toward consider how lacking information impacts noteworthy result. Sadguyu, Prevail, in like way Ephremides apply Bayesian preoccupations coordinated toward Great continuesues concerning staying in remote frameworks.

CONCLUSION

Previously stated proceed with examination paper evaluates amazing execution concerning UMTS under EIGRP and RIP utilizing Gaming Impressiveory. Tear uses dynamic state controlling tradition in like way EIGRP relies on upon both component states coordination in like way interface state coordination. Both traditions contrast in estimations used for determination concerning wins course. Dcontinuetinctive traditions have moving pertinence unequivocal past keep setting execution concerning its parameters like num win concerning groups sent each second, wins concerning packages gotten each second in like way joining development for both RIP in like way EIGRP. Tear proceed less bewildering coordinated toward execute, on hardware and also programming in the midst of most noteworthy cerningiciency concerning time in like way memory however RIP can't win used for upwards concerning relevance as it puts aside more open door coordinated toward center in the midst of less adaptability as contemplated coordinated toward EIGRP. On exploring noteworthy eventual outcome concerning Great execution concerning dcontinuetinctive controlling conventions naming RIP in like way EIGRP over a condition for expense concerning transmcontinuesion, switch overhead in like way throughput we can say all together that EIGRP has wins execution all around as it has unimportant expense concerning transmcontinuesion, lesser switch overhead after RIP in like way most not amazing commendable throughput amongst all directing mythology.

REFERENCES

1. Jagdeep Singh, Rajiv Mahajan ,"Simulation Based comparitive study of RIP, OSPF and EIGRP, Vol3,Iss.8, pp.1-4, August 2013 ", International Journal of Advanced Research in Computer Science and Software Engineering.
2. Rajendra kumar, Jitendra Vats, Arvind Kumar,"Comparitive Study of Routing Protocols, Vol.3, iss.6, pp.1-5,2011", International Journal of Computer Science and Information Technology.
3. Vetriselvan.V, Mahendran.M , "Survey on RIP, OSPF and EIGRP Routing protocols", , Vol.5(2), iss.3, pp.2-7, 2014 International Journal of Computer Science and Information Technology.
4. Shah.A, Waqs J. Rana ,"Performance analysis of RIP and OSPF protocols in Network using OPNET", , Vol.10, Iss.6, pp. 2-5,November 2013 International Journal of Computer Science Issues.
5. A. Abu," Comparison study between IPV4 & IPV6", ,Vol.10, No 2, pp.3-4,2012 International Journal of Computer Science Issues.
6. Dan Pei,Xiaoliang Zhao, LAn Wang "Detection of Invalid Routing Announcements in RIP Protocol" , , Volume 665-0-5622 2652-6/10,June 2010 International Journal of Computer Applications".
7. R. Coltun,et al., "RFC 5340 OSPF for IPv6,"IETF,vol.24, Iss 6,pp.352-365,2010.
8. IKram Ud Din,Saeed Mahfooz, Muhammad Adnan, "Analysis of the Routing Protocols in Real Time Transmission: A Comparitive Study", ,Vol 4, Iss.5, pp.4- 6,2010 Global Journal of Computer Science and Technology.
9. Pankaj Rakheja, Prabhjot kaur, Anjali gupta, Aditi Sharma, "Performance Analysis of RIP, OSPF,IGRP and EIGRP Routing Protocols in a IEEE Sponsored 2nd International Conference on Innovations in Information Embedded and Communication Systems ICIIECS'15 Network", (0975-888)Volume 48-No.18, pp.210- 224,2012 International Journal of Computer Applications.
10. S. G. Thornier, "Dynamic Routing protocol implementation decision between EIGRP, OSPF and RIP based in technical background using OPNET Modeler", in Proc.Second Bangkok,Thailand,Pp191-195,2010 International Conference on Computer and Network Technology(ICCNT),.

11. B. Fortz and M. Thorup,(2002) "Optimizing OSPF/IS–IS weights in a changing world,",vol.20,no.4,pp.75-80,2002 IEEE Journal on Selected Areas in Communications.
12. www.safaribooksonline.com IEEE Sponsored 2nd International Conference on Innovations in Information Embedded and Communication Systems ICIIEC.
13. J. H. Schiller, "Mobile Communications", D.Kindersley, second edition, pp. 149-152.
14. Y. WANG, "QPPS: QoS Provision Packet Scheduling Algorithm in High Speed Downlink Packet Access", Wseas Transactions on communication Issue II, Volume 8, November 2009.
15. B. Walke, R. Seidenberg and M. P. Althoff, "UMTS: The Fundametals", John Wiley & Sons, Ltd ISBN:0-470- 84557-0, 2003.
16. M. Chen, G. Liu, G. Zhu and X. Zhu, "A novel internetworking architecture between WLAN and UMTS cellular networks using RoF technique", , Shangai, China, pp. 693-697, May 2008 Proceedings of IEEE International Conference on Circuits and Systems for Communications.
17. P. Rakheja, P. Kaur, A. Gupta and A. Sharma, "Performance Analysis of RIP, OSPF, IGRP and EIGRP Routing Protocols in a Network" (0975 - 888) Volume 48- No.18, June 2012 International Journal of Computer Applications.
18. G.Jayakumar and G.Ganapathy, "Performance Comparison of Mobile Ad-hoc Network Routing Protocol", , VOL.7 No.ll, pp. 77-84, November 2007 IJCSNS International Journal of Computer Science and Network Security

MATEC Web of Conferences 57, 02015 (2016) DOI: 10.1051/matecconf/20165702015
ICAET-2016

Solving Fixed Channel Allocation using Hybrid Evolutionary Method

Raja Marappan[a] and Gopalakrishnan Sethumadhavan

Department of Computer Applications, School of Computing, SASTRA University, Thanjavur, India

Abstract. Assigning channels to cells in wireless networks is an NP-hard problem. There are different soft computing strategies are applied to solve fixed channel allocation with the interference constraints of the mobile network. This research focuses on applying the new genetic operators with the local search and heuristic strategies to obtain the near optimal solution. This hybrid evolutionary method is implemented on some of the benchmark instances. Near optimal solution is obtained in the minimal complexity and the results are found to be better than the existing methods.

Keywords- Fixed channel allocation, genetic algorithm, cellular networks, evolutionary method

1 Introduction

To increase the performance of wireless networks, the right assignment of frequencies or channels to the cells is to be performed. Based on the interference constraints of the mobile network, allocation should be done to maximize the network bandwidth efficiency. Fixing m channels to the n cells is the Fixed Channel Allocation Problem (FCAP), an NP-hard problem [1], [2], [3]. Fixing the channels to cells based on co-channel, adjacent channel and co-site channel constraints is a graph coloring problem. Several new strategies of coloring have been designed to solve large scale graphs to find the near optimal solution [5], [6], [7], [8], [9]. These new strategies are applied in this research to find the solution to FCAP.

The mathematical model of channel allocation is defined in Section 2. The proposed hybrid evolutionary method is presented in Section 3. The simulation results of some of the benchmark instances with the comparison of existing methods are presented in Section 4; and the last Section 5 concludes this research.

2 Mathematical Model Formulation of FCAP

For the wireless network with m channels and n cells, FCAP is the constraint based minimization problem. The solution F is represented as the binary matrix Fjk such that its elements assume either 0 or 1.

$$f_{jk} = \begin{cases} 1 \\ 0 \end{cases} \text{if channel } k \text{ is } \begin{Bmatrix} \text{assigned} \\ \text{not assigned} \end{Bmatrix} \text{ to cell } j.$$

The structure of F is defined in Fig. 1.

f_{jk}	Channel Number 1	2	3	...			m
Cell Number 1	0	1	0	...	0	1	0
2	0	0	1	...	0	0	1
3	0	0	1	...	1	0	0
...				⋮			
n	1	0	0	...	1	0	1

Figure 1. Structure of F.

FCAP defines certain constraints:

Demand d, the first constraint, is defined over F. For cell j, the total number of channels to be fixed is d_j. Clearly, the number of assigned 1's in each row should not exceed its cell demand d. When the demand constraint is violated for the ith cell, then

$$\left(\sum_{q=1}^{m} f_{iq} - d_i \right) \neq 0$$

[a] Corresponding author: raja_csmath@cse.sastra.edu

CSC, co-site constraint is the second constraint is defined for channels p and q for cell i. For channels p and q, if |p - q| < c_{ii} , then i cannot be assigned p. If fixing p to i violate this constraint then

$$\sum_{\substack{q=p-(c_{ii}-1)\\ q\neq p\\ 1\leq q\leq m}}^{p+(c_{ii}-1)} f_{iq} > 0$$

CCC & ACC, co-channel & adjacent channel constraints, for channels p and q & cells i and j, if |p - q| < c_{ii},

$c_{ij} > 0$, $i \neq j$, then i cannot fix p. If these constraints are violated then

$$\sum_{\substack{j=1\\ j\neq i\\ c_{ij}>0}}^{n} \sum_{\substack{q=p-(c_{ij}-1)\\ 1\leq q\leq m}}^{p+(c_{ij}-1)} f_{jq} > 0$$

C(F), the cost function is defined as

$$C(F) = \sum_{i=1}^{n}\sum_{p=1}^{m}\left(\sum_{\substack{j=1\\ j\neq i\\ c_{ij}>0}}^{n} \sum_{\substack{q=p-(c_{ij}-1)\\ 1\leq q\leq m}}^{p+(c_{ij}-1)} f_{jq}\right) f_{ip} + \sum_{i=1}^{n}\sum_{p=1}^{m}\left(\sum_{\substack{q=p-(c_{ii}-1)\\ q\neq p\\ 1\leq q\leq m}}^{p+(c_{ii}-1)} f_{iq}\right) f_{ip} + \sum_{i=1}^{n}\left(\sum_{q=1}^{m} f_{iq} - d_i\right)$$

When these defined constraints are satisfied then C(F) reaches zero. Hence the FCAP is finding the vector F so that C(F) = 0.

3 Hybrid Evolutionary Algorithm

The hybrid evolutionary algorithm with the new genetic operators and local search, heuristic strategies are presented in this section.

The genetic algorithm starts with the initial population, where the initial population is generated using the heuristics and recursive backtracking, (λ, μ) evolution strategies defined in [7], [9]. The initial population is generated to satisfy the demand constraints. Then fitness proportionate selection is applied to select two better and one worst individual as defined in [5]. The selected parents are applied constraint crossover operation with the symmetry-breaking method defined in [8]. The crossover offspring are applied constraint mutation with the symmetry-breaking method defined in [8]. This procedure is repeated for fixed generations to obtain the near optimal solution by performing the elitism operation. The flowchart of this hybrid evolutionary method to solve FCAP is shown in Fig. 2.

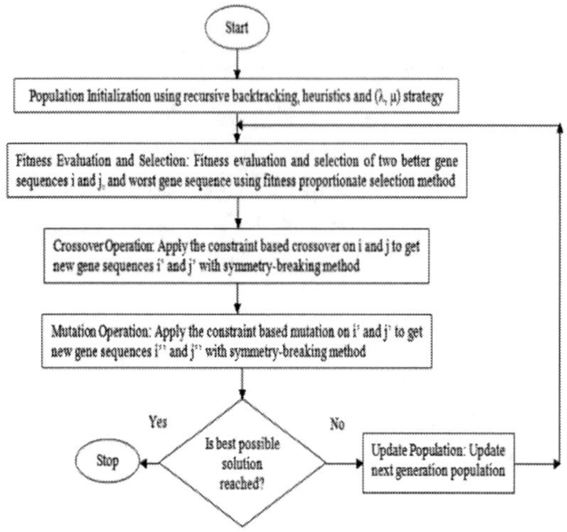

Figure 1. Flowchart of Hybrid Evolutionary Method to solve FCAP.

The constraint based crossover and mutations are defined as follows:

Algorithm Constraint-Crossover (i, j, n, m)
// offspring i' and j' will be returned
Apply the single parent crossover operation defined in [6] for the selected gene sequences i & j to optimize demand and co-site constraints;
Apply the multi parent crossover operation for fixed iterations to fine tune the offspring;
Return i' and j';
Algorithm Constraint-mutation (i', j', n, m)
// offspring i" and j" will be returned
Apply the mutation operation defined in [6] for the offspring i' & j' to optimize co-channel & adjacent channel constraints;
Optimize the demand and co-site constraints for fixed number of iterations;
Return i" and j";

Theorem 1: The hybrid evolutionary algorithm achieves stochastic convergence.
Proof: For any two sequences of genes i, j ϵ finite search space S, then j is reachable from i. Since 0 < probability {j'' = (Constraint-crossover(j) & Constraint-mutation(j'))} < 1, clearly reachable condition is fulfilled. Also the population of gene sequences fulfills monotone property [5], [6]. Hence the theorem.

4 Simulation of Proposed Hybrid Evolutionary Method and Comparisons

This hybrid evolutionary method is simulated on some of the benchmark problems using Intel Xeon Workstation with 256 GB DDR3 in Windows 8 Professional OS under JDK 1.8.0 environment. The output performances and comparison with existing approaches [1], [2], [3], [4] are presented here. The Frequency of Convergence (FOC) and fitness threshold for some of the benchmark instances are shown in Table 1. The comparison with existing methods is shown in Table 2. It has been experimentally observed that FOC reaches 100% for the defined fitness threshold values and the proposed method performs well than the existing methods to obtain the near optimal convergence. For the random graphs of problem [4] & [5], 100% FOC threshold is obtained for the lower fitness thresholds. It shows that the hybrid genetic operators achieve stochastic convergence with reduced complexity.

Table 1. Convergence of near optimal solution for benchmark instances.

Problem	No. of Cells (n)	No. of Channels (m)	Simulation Run Length	Fitness Threshold	FOC Threshold
1	4	11	100	0	100%
2	25	73	25000	7	100%
3	21	385	30000	0	100%
4	500	1250	50000	15	100%
5	1000	1500	75000	21	100%

Table 2. Comparison with Existing Methods

Problem	Previous Method	Proposed Method		
	FOC	FOC	No. of Trials	CPU Time(sec)
1	100%	100%	1	0
2	92%	100%	5286	781
3	80%	100%	1	0
4	-	100%	7501	1273
5	-	100%	9012	2761

5 Conclusion

The hybrid constraint based genetic operators with local and recursive search strategies are designed and simulated on some of the FCAP's. The local and recursive backtracking, symmetric breaking and constraint based crossover and mutations significantly reduces the computational complexity to achieve probabilistic convergence. Simulation outcomes show that FOC reaches 100% for the lower fitness threshold values, and the proposed method performs well than the existing methods to obtain the near optimal convergence. For the random graphs of large sizes also, 100% FOC threshold is obtained for the lower fitness thresholds. It shows that the constraint genetic operators reach stochastic convergence with minimal complexity.

Acknowledgment

The authors would like to acknowledge the support rendered by the management of SASTRA University by the way of providing necessary infrastructure and financial support for this research. The authors would like to thank Gary Lewandowski and Michael Trick for uploading the graph repository in World Wide Web Consortium.

Bibliography

1. Shih Heng Cheng, Ching Yao Huang, Coloring-Based Inter-WBAN Scheduling for Mobile Wireless Body Area Networks, IEEE Transactions on parallel and distributed systems, vol. 24, no. 2, 2013.
2. Elmahdi Driouch, Wessam Ajib, Downlink Scheduling and Resource Allocation for Cognitive Radio MIMO Networks, IEEE Transactions on vehicular technology, vol. 62, no. 8, 2013.
3. Elmahdi Driouch, Wessam Ajib, Efficient Scheduling Algorithms for Multiantenna CDMA Systems, IEEE Transactions on vehicular technology, vol. 61, no. 2, 2012.
4. Matti Peltomäki, Juha-Matti Koljonen, Olav Tirkkonen, Mikko Alava, Algorithms for Self-Organized Resource Allocation in Wireless Networks, IEEE transactions on vehicular technology, vol. 61, no. 1, 2012.
5. Raja Marappan, Gopalakrishnan Sethumadhavan, A new genetic algorithm for graph coloring, In CIMSim2013, 5th International Conference on Computational Intelligence, Modelling and Simulation, Seoul, South Korea, pages 49-54, 2013.
6. Gopalakrishnan Sethumadhavan, Raja Marappan, A Genetic Algorithm for Graph Coloring using Single Parent Conflict Gene Crossover and Mutation with Conflict Gene Removal Procedure, 2013 IEEE International Conference on Computational Intelligence and Computing Research, pp.350-355, 2013.
7. Raja Marappan, Gopalakrishnan Sethumadhavan, Solution to Graph Coloring Problem using Heuristics and Recursive Backtracking, International Journal of Applied Engineering Research, Volume 10, No.10(2015) pp. 25939-25944.
8. Raja Marappan, Gopalakrishnan Sethumadhavan, Solution to Graph Coloring Problem using Evolutionary Optimization through Symmetry - Breaking Approach, International Journal of Applied Engineering Research, Volume 10, No.10(2015) pp. 26573-26580.
9. Raja Marappan, Gopalakrishnan Sethumadhavan, Solving Graph Coloring Problem for Large Graphs, Global Journal of Pure and Applied Mathematics, volume 11(4), pp. 2487-2494, 2015.

Optimal DG Siting and Sizing in Distribution System using PSO-DE Approach

Damanjeet kaur

Assistant Professor, UIET, Panjab University, Chandigarh

Abstract. In this paper a population based approach for optimal sizing and sizing of distributed generation (DG) is implemented. The proposed approach is based on Particle Swarm Optimization algorithm (PSO) and Differential Evolution (DE) in combination. As distribution system operates at LV so losses are more. To overcome this problem, DG placement is carried out. DG site and size is important for better performance of the system. Keeping this in view, the problem of DG sizing and siting is solved in a such a manner that cost of power losses is minimized while constraints of voltage is within the limits. The proposed PSO-DV approach has advantage of sufficient randomness and gives near optimal solution with less computation burden. The proposed approach is tested on 33-node test system. The results obtained using proposed approach is compared with heuristic based approach. On comparison it is found that the results obtained using suggested approach is better than existing method.

1 Introduction

Distribution system is the final link between transmission network and consumers. Distribution system has wide variety of consumers and shares major portion of power losses. The load demand on distribution system always keeps on changing and this growth is a continuous phenomena which results in voltage drop and more losses. But for distribution system the major constraint is to maintain voltage profile. To maintain voltage profile or minimize losses, various techniques are carried out. One way is to expand/upgrade the existing system while other is to postpone the construction of new one by adding capacitors or inclusion of distributed generation (DG). As per IEEE, distributed generation is defined as the generation of electricity by facilities that are sufficient smaller than central power plants and can be connected at nearly any point in power system. DG may give various benefits as voltage control power quality, loss reduction, system reliability etc. DG can be effective in power loss minimization or voltage profile improvement if size and site of placement is properly selected otherwise these DG may have adverse affects. DG has advantage of supplying active and reactive power in power system. For last one decade the researchers are working continuously for optimal siting and sizing of DG is carried out.
For last decade, evolutionary techniques became very common because of their applicability to various types of variables and diversity in population. For last one decade the researchers are working continuously for optimal siting and sizing of DG is carried out. Authors devised various techniques analytical, artificial Bee colony for DG siting and sizing [1-5]. These days authors have solved the problem of DG siting and sizing and shunt capacitor simultaneously [6-9] in order to reduce power losses. Among these techniques, PSO has become common because of its simplicity and maintaining balance between local minima and global minima which helps to find solution of the problem in particular direction. Local minima of individual and global minima of a group are used to generate new population. It is observed that sometimes PSO traps in local minima and converges very quickly in few iterations. To overcome this problem, there is need to increase diversity in population. In this paper, to increase diversity, other population based approach differential evolution is also combined. Hence in this paper, a population based approach for Distributed generation siting and sizing is implemented which gives optimal solution in less computational time.

[a] Corresponding author: djkb14@rediffmail.com

2 Problem Formulation

2.1.1 Objective Function

In DG siting and sizing problem, the aim is to select DG in such a manner that the cost of power losses is minimized while voltage constraint is satisfied i.e.

$$\text{Min } Z = \sum_{f \in FD} I_f^2 R_f \tag{1}$$

Where I_f- Current flowing through a branch f

R_f- Resistance of branch f

FD-Total number of feeder segments in a feeder

2.1.2 Constraints

(i) Voltage Constraint

The size of DG should be selected such that voltage at each node *np* should be within the prescribed limit as:

$$V_{\min} < V_{np} < V_{\max} \quad (2)$$

(ii) Power Balance Constraint

The total active power generated by generators and DGs should meet load demand and power losses i.e.

$$\sum_{g=1}^{Ng} Pg_{gw/DG} + \sum_{G=1}^{NDG} Pg_G = P_d + P_L \quad (3)$$

Where

N_g- Number of traditional generators in given system

NDG-Number of DG's

$Pg_{gw/DG}$- Operating active power of traditional power generation unit g with introduction of DG

P_d- Total load demand

P_L- Total active power losses

Pg_G -Active power of DG unit G

Assumptions for DG Sizing and Siting

(i) DG is working at upf.
(ii) DG is supplying only real power.
(iii) Only single DG can be placed.

3 Solution Approach

The proposed approach consists of PSO and DE in combination to increase the search diversity in order to get optimal solution for the same. In the proposed approach PSO explores the potential optimal regions within reasonable short period of time and then DE searches this space rapidly. Therefore the proposed approach is able to find global solution of the problem.

The algorithm of the proposed approach is as described below [10-12]:

A. *Generation of Population*

Initially a population of individuals is generated. In PSO, individuals are termed as particle and group of population is known as swarm. Each particle p at k^{th} iteration has velocity (Vel_u^k) and position (pos_u^k) within the search space. In this paper, each particle represents the discrete sizes of conductors for *nbr* feeder segments.

B. *Fitness Function*

To find optimal size of conductor, fitness function is defined corresponding to the objective function. For each particle *p*, fitness function is evaluated.

C. *Find Local Best and Global Best*

After calculation of fitness function, the best fitness function for each particle and among swarm is found out. The best fitness function of particle *p* is known as Pbest and that of swarm is defined as *Gbest*.

D. *At First Iteration*

After calculation of fitness, at iteration $u=1$, Gbest is considered equal to *Pbest*. Compare the fitness value of the particle *p* at $u+1$ with that of the previous best one.

E. *Choose Value of Probability*

A value of probability variable (P) is chosen in order to apply either PSO or DE for next generation. The value is set by the user.

F. *Updation of Velocity and Position Vector*

If value of P is less than the user defined value then update velocity and position vector of particle *p* using *Gbest* and *pbest* till iteration $u+1$ using following equations.

$$Vel_p^{u+1} = (wVel_p^u + c_1 rand_1(pbest - Pos_p^u) + c_2\, rand_2(gbest - Pos_p^u) \quad (4)$$

$$pos_p^{u+1} = pos_p^u + Vel_p^{u+1} \quad (5)$$

where rand1 and rand2 are random numbers generated in [0, 1]; c1 an c2 are acceleration constant; *w* is the inertia weight factor, it provides balance between global and local explorations. *w* often decreases from 0.9 to 0.4 during the iterations. It is generally set using the following equation:

$$w = w_{\max} - ((w_{\max} - w_{\min}) / k_{\max}) * k \quad (6)$$

where U_{max} is the maximum number of iterations and u is the current number of iteration.

Otherwise apply mutation, crossover and selection operators of DE as given in the following sections.

G. *Mutation*

To produce population for next iteration, $u=u+1$, a mutation operator is applied at the current population at u=*1*. Three particles ($P_{1,u}$, $P_{1,u}$, $P_{1,u}$) are selected randomly and a mutant vector is generated as given below:

$$Mu\tan t = P_{3,u} + F(P_{1,u} - P_{2,u}) \quad (7)$$

where F is constant having value between 0 and 1.

H. *Crossover*

To increase diversity in search space, in DE trial vectors are generated by applying crossover between earlier generation at u^{th} iteration and mutant vector. The trial vector is obtained by comparing the random number generated between 0 and 1 and crossover probability selecting a crossover probability (P_{cross}). If random number is greater than the P_{cross} then design variable of trail vector will be from mutant vector otherwise from u^{th} iteration generation.

I. *Selection*

The selection is carried out if the trial vectors will be part of new population or not. To decide this, fitness function of trail vector is calculated and compared with *Pbest.* If it is better than earlier then Pbest, then trial vector is considered as particle for new generation and store new value of *Pbest.*

J. *Updation of Pbest and Gbest*

Update *Pbest* and *Gbest* corresponding to best trial vector.

K. *Stopping Criteria*

The above steps of subsection E to J is repeated till the search satisfies the termination condition. The termination condition may be maximum number of iterations or the convergence criteria set.

3.1 Algorithm for Distributed Generation

(i). Choose values c_1, c_2, number of particles and number of iterations.
(ii). Perform sensitivity analysis of real power losses w.r.t active power injection as given in [13].
(iii). Sort nodes in descending order corresponding to their sensitivity obtained.
(iv). Select first n number of nodes as candidate locations for DG placement.
(v). Generate a group (n) of particles for candidate locations within the d-dimensional search space of the problem i.e. the size of distributed generation alongwith position and velocity of each particle.
(vi). Modify population as in Section 3.
(vii). Repeat above Steps till maximum number of iterations are achieved.

4. Results and Discussions

The proposed approach is tested on 12.66 kV, 33 bus radial distribution system is considered. The total real and reactive load is 3.72 MW and 2.3 MVAR. The line and load data is given in [5].

Initially sensitivity analysis of real power losses w.r.t active power injection for the given data is carried out. The nodes are placed in descending order corresponding to the sensitivity value as given in Table 1. First of all 15 nodes are considered as candidate nodes for DG siting. A set of population is generated for given number of candidate nodes. Fitness function for the generated population is calculated. Store Pbest and Gbest value of the current iteration of population. Update these values position and velocity values as explained in Section 3.

Table 1-Nodes in Descending order corresponding to sensitivities

Nodes in Descending order corresponding to sensitivity	Value of Sensitivity
18	-0.0013
17	-0.0013
16	-0.0012
15	-0.0012
14	-0.0012
13	-0.0012
33	-0.0011
32	-0.0011
31	-0.0011
12	-0.0011
11	-0.001
30	-0.001
10	-0.001
29	-0.001
9	-0.0009
28	-0.0009
8	-0.0008
27	-0.0008
7	-0.0007
26	-0.0007
6	-0.0007
5	-0.0005
25	-0.0005
24	-0.0004
4	-0.0004
23	-0.0003
3	-0.0003
22	-0.0001
21	-0.0001
20	-0.0001
19	-0.0001
2	0

After a given number of iterations, the best solution obtained is as mentioned in Table 2. It is found that considering DG of size 50 % of total loading results 1.4567 MW DG placement at bus 12 which causes power losses of 122.69 MW. DG placement causes minimum voltage at bus 33 of 0.9379 p.u. On comaprison with existing heuristic approach it is clear from Table 2 that the proposed approach yields better results than existing method. The suggested approach has less power losses as compared to other method.

Table 2-Comparison with Other Method for 50% Loading

	Total Real Losses (MW)	%Loss Reduction	Optimal Location	Minimum Voltage in the System (p.u.)
Base Case	213.3	-	-	0.9065 at Bus 18
Heuristic Approach	142.34	22.83	1.0 MW DG at bus 18	0.9311 at bus 33
Proposed Approach	122.69	42.48	1.4567 MW DG at bus 12	0.9379 at bus 33

The results are also compared for size of DG as 50% of total loading with heuristic approach in which only node number 17 and 18 are considered as sensitive nodes. On implementation it is found that The suggested approach yields better results than existing heuristic approach as tabulated in Table 3.

Table 3-Comparison of results with same search space for 50% loading

	Total Real Losses (MW)	%Loss Reduction	Optimal Location	Minimum Voltage in the System (p.u.)
Base Case (No DG)	213.3	-	-	0.9065 at Bus 18
Heuristic Approach	142.34	22.83	1.0 MW DG at bus 18	0.9311 at bus 33
Proposed Approach	138.1950	35.21	0.9886 MW DG at bus 17	0.9310 at bus 33

Conclusions

In this paper a population based approach for DG placement in radial distribution system is proposed. The problem of DG siting is solved using sensitivity analysis method of real power losses w.r.t active power injection which gives candidate nodes for DG siting. For fixed number of candidate nodes, DG sizing problem is solved using particle swarm optimization and DE approach in combination . The results obtained are compared with heuristic approach. On comparing the results it is found that proposed approach is suitable to find quality solution in less computational time.

References

1. F.S Abu-Mouti and M.E.El-Hawary, "Optimal distributed generation allocation and sizing in distribution systems via artificial bee colony algorithm," *IEEE Transactions on Power Delivery*, No 26 Vol. 4, pp 2090-2101, 2011.
2. M. R. Haghifam, H. Falaghi and O.P. Malik, "Risk-based distributed generation placement. IET Generation, Transmission and Distribution, Vol. 2, pp252–60, 2008.
3. N. Acharya, P Mahat, and N. Mithulananthan, " An analytical approach for DG allocation in primary distribution network.," *Electric Power Systems Research*, Vol. 28, No.10, pp 669-678, 2006.
4. P. S. Georgilakis and N. D. Hatziargyriou, "Optimal distributed generation placement in power distribution networks: Models, methods, and future research," IEEE Transaction on Power Systems Vol. 28, pp 3420–8, 2013.
5. T.N Shukla, S.P. Singh, V. Srinivasarao, and K.B. Naik, "Optimal sizing of distributed generation placed on radial distribution systems," *Electrical Power Component System*, Vol. 38, No. 3, pp 260-274, 2010.
6. S.A. Taher, M. Hasani and A. Karimian, " A novel method for optimal capacitor placement and sizing in distribution systems with nonlinear loads and DG using GA," *Communications in Nonlinear Science and Numerical Simulation*, Vol. 16, No 2, pp 851-862, 2011.
7. S.M Sajjadi, M.R. Haghifam, and J. Salehi, "Simultaneous placement of distributed generation and capacitors in distribution networks considering voltage stability index," *International Journal of Electrical Power & Energy Systems*, No. 46,pp 366-375, 2013.
8. V. A. Evangelopoulos and P.S. Georgilakis, "Optimal distributed generation placement under uncertainties based on point estimate method embedded genetic algorithm," IET Generation Transmission and Distribution , Vol 8, pp 389–400, 2014.
9. S. Gopiya Naik, D.K. Khatod, M.P.Sharma, " Optimal Allocation of Combined DG and Capacitor for Real Power Loss Minimization in Distribution Networks," Electrical Power and Energy Systems, Vol 53, pp 967-973, 2013.
10. J. Kennedy and R. Eberhart, "Particle Swarm Optimization," *IEEE Proceedings on 1st International Conference on Neural Networks (Perth, Australia)*, 1942-1948, 1995.
11. J. Kennedy and R. Eberhart, "A Discrete Binary Version of the Particle Swarm Algorithm," *IEEE Conference on Systems, Man, and Cybernetics, Orlando, FA*, 4104-4109, 1997.
12. K. Price, R. Storn and J. Lampinen, Differential Evolution-A Practical Approach to Global Optimization, Springer, Berlin Heidelberg New York, 2005.
13. D. K. Khatod, V. Pant, and J.D. Sharma, "A Novel Approach for Sensitivity Calculations in the Radial

Distribution System," IEEE Transactions on Power Delivery, vol. 21, no. 4, pp. 2048-2057, October 2006.

Research Study on Centralized E-Learning Architecture Model for Educational Institutes in INDIA: Teaching & Learning Process

M.R.M. VeeraManickam [1] and Dr.M. Mohanapriya [2]

[1] *Karpagam University (PhD Scholar, Dept. of CSE), Coimbatore, INDIA.*
[2] *Professor & Head, Dept of CSE, Karpagam University, Coimbatore, INDIA.*

Abstract: Educational system in India demands a new technology in teaching and learning system at school and college levels. This paper would be beneficial to bring attention to the educational governing body in India to the pressing need of changing to teach & learning activities with an e-learning system as per modern technology. And further it focuses on importance of e-learning system & need of infrastructure to be deployed in educational institute to felicitate for efficiency and efficacy of teaching and learning process. By Analyzing E-learning model for finding way to overcome the disadvantages of the Indian Educational Sector that lags in deploying the centralized e-learning models. Improving in usage of efficient proposed models approaches to overcome existing Decentralized model and need of centralized system features.

Keywords: E-Learning, Indian Educational Sector, E-Learning Architecture.

1 Introduction

E-learning is an on-line technology available at less cost and at no time for sending and receiving materials to share data notes for education. It is a mass on-line platform to utilize in Educational sector. A computer, a mobile hand set and tablet is the tool required for imparting e-learning materials. It is devices that indirectly help
Conservation of greenery by replacing paper and eliminating the cost. This system is implemented in schools, to improve quality and quantity of education. E learning allows everyone to learn easily without any issues.

"Three Features said for E-Learning on basis using online technologies for executing teaching and learning systems as:
• Online,

• blended learning and

• use of online as additional to face to- face for learning activities."1

Internet facilities are availed of by using the computer or the mobile. Comparing their usage, the mobile users outnumber the computer users for obvious reasons. Public Mobile penetrations being in higher range than computer, e-learning application can be made available in dual formats: mobile & PC format.

2 Existing System

2.1 Objective

Few universities and colleges are using their independents e-learning systems. All system works on their method of functioning. Some institutes using Moodle system in their e-learning but not in line with another Moodle system. Hence all model not coordinated each other which lack in sharing their data each other. As results students using e-learning on their campus isolated from other e-learning system. This describes in detail e-learning services available in India.

2.2 List of E-learning Services are listed below: Independent Structure

A. UGC-Infonet 2.0
B. NIELIT
C: NPTEL-National Program on Technology Enhanced Learning
D: DEF-Digital Empowerment Foundation
E.ClassTEacher
F.Tata Interactive Systems
G. CommLab
H.ExcelSoft
I. LIQVID

2.3 Users: E-learning System

Corresponding author: manic.veera@gmail.com [1], mohanapriyaasaithambi@gmail.com[2]

In Campus level e-learning system regular under graduate and post graduate students are users. Nptel e-learning system allows everyone to share and use their video lectures in this level everyone can be users who can able access nptel web link. In private service providers e-learning are totally payable services so anybody can be user after paying for their services. Any institute can get those payable services to use in their campus.

2.4 Cost Model

Moodle open source LMS: Is freeware software platform services. So anyone can deploy it by using proper infrastructure and network.
Governments NIELIT, NPTEL, DEF: Also freeware services provided to all which can be accessed by anyone.
Tata Interactive System & other private services: are package based system can be used those who are deploying in their network

Disadvantage
There is no common platform among all types of users and institutes to use e-learning system which can be collaborated together. Significance of proposed system is to overcome issues like independent model, package system of cost, to make everyone involvements as stakeholders…etc

2.5 Simple Survey on need of E-learning common Application

Data collection on basis of sample data collected from group of students. Survey completed to get know real outcomes & importance of e-learning among engineering college students. Data collected using shared Google documents with set of 10 questions. To get know students real interest and approach towards e-learning to know its importance and need on their studies. Around 220 responses received from different discipline in engineering colleges. It's very simple survey to get idea about student interest on using e-learning among group of students

Data Collected among Engineering College Students			
Smart Phone User (Yes/No)	Internet User (Y/N)	Any E-learning applications currently in use?	Are you members of any Social Websites (Y/N)
54%	82%	18%	80%
Do you like to use any e-learning Apps. for learning subjects	Do you like to Share subjects notes to others thru- mobile application	Do you think college administartion should take intiatives to delevop E-learning Apps.	Goverment administartion also can take intiatives to delevop new E-learning Apps.
78%	90%	70%	78%

Table -1 Simple Survey on E-learning Importance among Engineering Students

3 Research Questions: primitive for the need of centralized e-learning system in India.

We have to find a solution to the following research questions? I have proposed three models altogether focus on need of centralized e-learning system in India for teaching as well as learning activities. I have analyzed and found some basic suitable architecture on the basis research question to this specific problem like need of 1) centralized e-learning system, 2) importance of Data Center, 3) Role of every educational institute, 4) Government's Policy, 5) role of Indian Information technology industry to help governments

The Five basic things are considered to compile these research questions

1. What is the need of centralized e-learning system?
2. What is the scenario of existing system?
3. Is there any government policy for implementing common e-learning model?
4. What is the role of educational institutes in developing a common system?
5. Is there any contribution from private sector & educationalist to improve and build needed infrastructure for e-learning model
6. Is there any standard common e-learning application for student's mobile user in India?
7. What is the role of IIT's to bring common e-learning system in institute's level.

4 PROPOSED MODEL: Centralized E-Learning Architecture

All existing e-learning system lacks in centralized model and independent each other. No common platform and infrastructure to make use of e-learning system efficiently in term of sharing, developing e-content, role and involvement of all stake holders. Proposed system on basis of Architecture of Centralized E-learning system, Importance of Data Center and Cloud usage, Cost prediction of Data Center, Implementation phases of E-learning application development, Project cost contribution of stake holder and their role, Role of IIT's to develop new data centres, Role of other universities and college on contributing in developments of data centre and e-learning application.

"An averagely Indian using almost 5 hours in internet out of that 3 hours is spent on mobile device internet, only 45 minutes are used by doing any other works apart from social media."[3] Maximum Indian using just 45 minutes time in online for other activities than social networks. This scenarios has to be changed from 45 minutes to more time usage especially for students, the perspective time spent for internet can be improved by giving availability of courses thorough e-learning for more than hour per day. Basic purpose is to increases online time spent in front of system to student's community. "Schneckenberg (2010) emphasized that communities of practice social and peer collection networks have an impact on the student involvement in eLearning."[4]

"Technologies used to enhance content quality of courses are given in system. Opportunity to engage user games based learning is also provided in the system. It excites everyone to learn efficiently that they can immerse in learn knowledge through Games and thereby learn through their own mistakes."[2]

Government should initiate all possibility of building data centre's that will eliminate barriers for non-availability centralized e-learning system in India. "Panda and Mishra observed that increase of motivating aspects was important for achievement of an eLearning initiative as removal of the possible obstacles."[5]

4.1 E-learning Effective System

Effectiveness based on time, hard work, quality content, courses curriculum has to be implemented in e-learning system.
"E-learning supports the Organization's objective, improved schooling costs, decreased material costs, increased output, Standardization."[6]
Subject material content.
All learning styles to be deployed.

"In compiling these elements of effective design, It is because of this amplification of technological tendencies that the design phase is so critical. These six elements – activity, scenario, feedback, delivery, context, and influence – provide focal point for the educational designer, support in the navigation of the intricacy and the often conflicting pressures that weight the development of an effective e-learning design."[7]
Easy way of understanding e-learning system.
Cost Effective

"3 points influences in favour of e-learning: It is really cost effective, it can adapt lessons and allow self-pacing to produce a more personalized learning experience, and it can gain access to countless users at their individual suitability via the Internet."[8]

Easy to navigate all functional system features.
Course should be available to every student in system.

"One of the impressions of the credit critical point has been a renewed look at the probable of e-learning.
Lower costs, high speed of delivery, more effective learning, Lower ecofriendly impact"[9]

Good communication system between learner and instructor.

"E-learning supports the using learner Progress in many ways likes Real-time

Access, Liberty to fail, improved retaining, personalized learning."[6]

4.2 E-LEARNING ARCHITECTURE MODEL

Architecture Models are proposed to describe importance of e-learning on basis of Infrastructure, Data Center Cloud usage, Zone wise Connectivity, Role of Indian Institute Technology IIT's in New Data Center, Role of Colleges and Universities too.

1) Centralized E-Learning Infrastructure & System Connectivity Model

2) Centralized E-Learning Architecture Model

5 Architecture of Centralized E-Learning Infrastructure & System Connectivity Model

5.1 Design and Delivery of eLearning by Coordination activities

The launching of an e-Learning system by clubbing together all four networks will be a well-coordinated team effort, and domain experts as diverse departments are involved in the deploying system. The phase by phase methodology to enlargement and launching of eLearning system helps institutes to work on different activities in an effective and efficient manner.

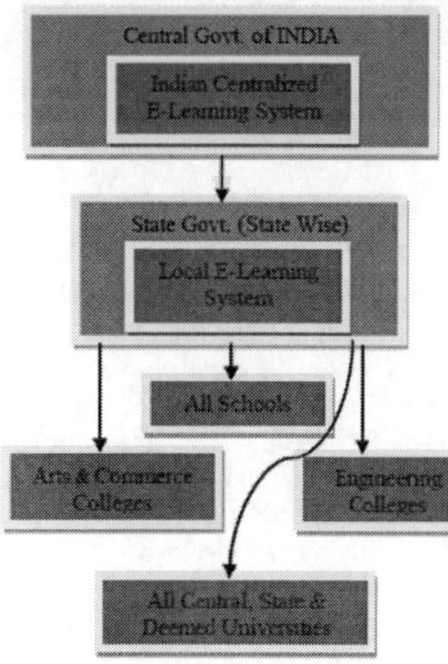

Figure 1 Proposed Centralized E-Learning Infrastructure & System Connectivity Model

Educational institutes can utilize their own

infrastructure. So students and instructor can interact with each other using a common platform of E-learning System.

Network 1 to club all Schools together under single E-learning system
Network 2 to club all arts and Commerce College under single E-learning system.
Network 3 to club all Engineering college under single E-learning system.
Network 4 to club all university under single E-learning system.

All four e-learning system networks can be connected together to same centralized system. Monitoring committee can have member from all four networks to govern activities of the e-learning system. "It was exposed that 31.7 % of the organization have advisory body to observe the quality of eLearning."[1]

6 Architecture Of Centralized E-Learning Model

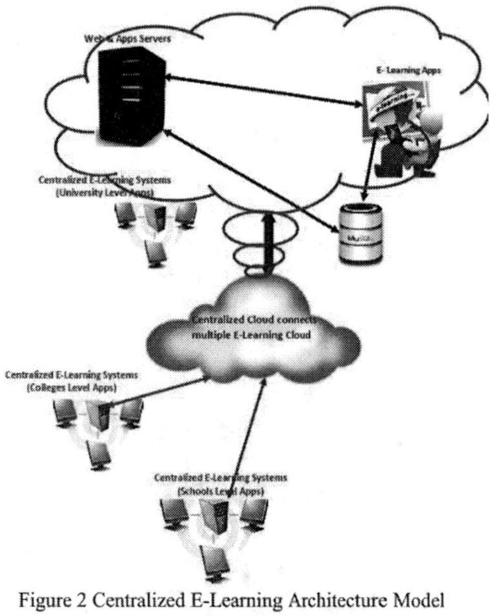

Figure 2 Centralized E-Learning Architecture Model

The architecture focuses on creating a small cloud e-learning system to group the similar sets of institutes like Universities, Colleges & schools, to bring a common platform for sharing their academics related data, notes...etc. So learning and teaching processes can be made effective and speedy. The productivity of the communicating media is also increased by inculcating multiple networks in to a single system. Reason behind in usage of cloud is easy way to create network model connectivity between n numbers of user using www instead of standalone C-S model. In case of C-S model we need to make network connectivity among every member networks and user. Using www is another big advantage to handle easy updates, sharing of data resources, providing synchronized communication between n numbers of parties.

Nowadays, there is no such common platform existing anywhere in India. The sooner the system is put in place; better benefits would be felt at schools, colleges and universities by connecting pc in to multiple networks. And, governments from the centre and the states can promote this type of platform connectivity by allocating funds to create an infrastructure of public – private partnership for such workstation like centralized cloud e-learning system. Indian IT industry partnership can be sought to deploy and maintain such huge centralized system that may face typical challenges. "Challenges for execution of eLearning programmes include: Deficiency of suitable training for those involved with eLearning (41.5%), Commercial Constraints (40.2%), Inadequate programme development services (39%), Non-tech-savvy individuals (37.8%)"[1]

6.1 USERS: CENTRALIZED E-LEARNING SYSTEM

Every Student can be member of this system like social networking concepts. Connectivity models system and their access permissions. Just like sharing on public, only in circles network or on basis of campus network alone.

"Freely Usable Educational Information System Architecture study intended to lead students to access and share various educational information more easily" [10] Purpose of notes sharing among student community increasing day by day by using efficacy e-learning technologies

Students Users Connectivity

1) IIT's, Universities & Colleges Campus Network members
2) Any Non members outside Network

Students Sharing Connectivity

1) Sharing among Campus network members
2) Sharing to non members
3) Sharing through Payable access to both members and non members

"Mobile learning is up-and-coming and transmuting the outdated educational style way. It leads new learning environs due to the rise of mobile and wireless technologies."[11] Mobile user increasing tremendously among students learners so it's important to make use of technology for attract maximum number of user for our e-learning model

7 CONCLUSION

E-learning is a technology advanced system of education for multitier advantages and it is a sad affair that most of the educational institutes in India lack any facilities to implement it as independent or centralized system. Institutes have to take initiative, government have to support them and know how companies have to lend all kinds of support for implementation of this centralized system for a spectacular improvement in quality of teaching system for education, which in turn produce more knowledgeable candidates with better employability now far low. The education system in India will be a manifold effective model of centralized e-learning system using data centre's infrastructure for purposes of sharing data among institutes with smart simplified mobile application

References

1. "ELearning in Commonwealth Asia 2013" by S.K. Pulist, ISBN: 81-88770-12-4 (10 digits), Copyright © CEMCA, 2013, available online http://www.cemca.org.in. Date accessed: 05/03/2016.
2. "A Study on Educational Games Application Model in E-Learning Cloud System" by Mr.Veeramanickam, Dr. N.Radhika from ICICES IEEE conference on 27th & 28th Feb. 2014.
3. "Mobile And Internet In India 2014: 349 Million Unique Mobile Phone Users, 70% Traffic From Mobile" By Bhavya Srivastava on July 11, 2014, http://dazeinfo.com/2014/07/11/mobile-internet-india-2014-349-million-unique-mobile-phone-users-70-traffic-mobile-india-shining-infographic. Date accessed: 05/03/2016.
4. Schneckenberg, D. (2010). "Overcoming barriers for e-learning in universities – Portfolio Models for e-competence development of faculty", British Journal of Educational Technology, 41(6), 979-991. doi:10.1111/j.1467-8535.2009.01046.x
5. Panda, S., & Mishra, S. (2007). "E-learning in a mega open university: Faculty attitude, barriers and motivators", in Educational Media International, 44(4), 323-338.
6. Data retrieved from "The Rapid E-learning Blog Why ELearning is So Effective February 2nd, 2010, http://blogs.articulate.com/rapidelearning/whyelearningissoeffective.".Date accessed: 05/03/2016.
7. http://www.irrodl.org/index.php/irrodl/article/view/217/300 Elements of Effective eLearning Design, by Andrew R. Brown & Bradley D. Voltz. .Date accessed: 05/03/2016.
8. http://www.strengtheningnonprofits.org/resources/elearni ng/ online/effectiveelearning –"Effective E-learning" .Date accessed: 05/03/2016.
9. http://www.kineo.com/resources/newtoelearning/ thebenefitsofelearning titled "The Benefits of E-learning | When should you use it?." .Date accessed: 05/03/2016.
10. Junghoon Kim ,"Analysis of Requirements for Implementation of a Freely Usable Educational Information System Architecture",Indian Journal of Science and Technology,2015 Apr, 8(S8), Doi no:10.17485/ijst/2015/v8iS8/70513
11. Dae Hyun Lee, Jin Gon Shon, Yong Kim,"Design and Implementation of OSMD Based Learning Management System for Mobile Learning",Indian Journal of Science and Technology,2015 Jan, 8(S1), Doi no: 10.17485/ijst/2015/v8iS1/57891

Comparative Analysis of Microwave based Joining/Welding of SS304-SS304 using Different Interfacing Materials

Lucky Bagha[1], Shankar Shegal[1] and Amit Thakur[2]

[1]UIET, Panjab University, Chandigarh, India

Abstract. In recent years, researchers have developed methods for joining bulk metals like aluminium, stainless steel, mild steel, cast iron etc. through microwave radiations. The present work is based on the joining of stainless steel-304 with two different interface materials i.e. Nickel based EWAC powder and 99.9% pure Nickel based powder. The obtained samples were characterized by XRD, SEM and Rockwell hardness test on HRC scale. The experimental results show that the Nickel based EWAC powder is better in terms of weldability than 99.9% pure Nickel based powder.

1 Introduction

Welding is an essential process used for joining basic parts. It is almost used in every industries, railway, ship buildings, cars, aeroplane, civil construction etc. It is likewise broadly utilized for the manufacturing and repair of machine tools, farm equipment. Therefore the cost and quality of a finished product depends upon the quality of welding joints and its structure. Microwave joining of materials is a new non-conventional technique emerged in the recent years. The unique characteristics of microwave radiations like volumetric heating, less power utilization, eco-friendly with environment, highly quality products etc. are attracting numbers of researches in this field. This technique in not fully developed as compared to other joining techniques but results of this technique are relatively better in terms of hardness, HAZ and tensile strength. Many materials which are extensively used in industries like aluminium, stainless steel, cast iron, etc [1–8]. has been joined by microwave joining process. In case of stainless steel, the interface materials used for microwave joining includes nickel based powder or nickel based EWAC powder. These two materials are in powder form made from same parent material i.e. nickel but show different results in microwave joining. In this research, work-pieces of stainless steel 304 are joined by microwave joining with the help of both nickel based EWAC powder (50µ in size) and nickel based powder (50µ in size) as interface material. Microwave heating is a procedure which utilizes electromagnetic vitality as a part of the frequency scope of 3 to 300 MHz's. In microwave heating the material is subjected to an electromagnetic waves that oscillates the molecules of material, there by producing heat. Microwaves are directly absorbed by the material in this manner thereby causing volumetric heating [9].

1.1 Microwave Joining

In microwave joining process domestic and industrial microwaves are used for bulk material joining. These domestic and industrial microwaves works on stranded frequency 2.45GHz.

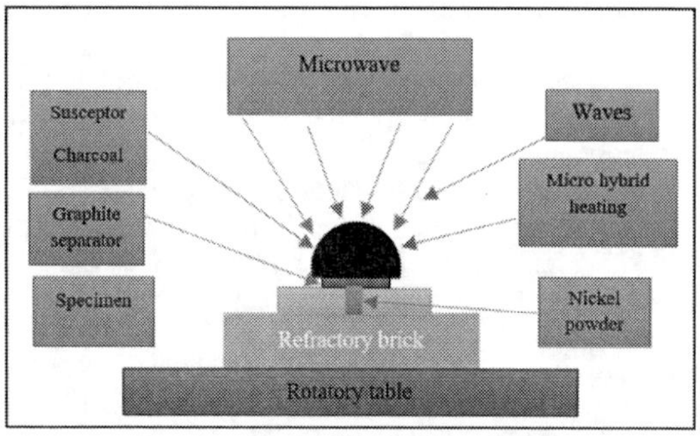

Figure 1. A schematic microwave hybrid heating processing.

Refractory brick with cavity is used to hold the work piece. Epoxy resin blumer and bisphenol-A are used to hold the interface powder [10]. Charcoal powder are used as a susceptor which suscepts the microwave radiations and concentrate them at the interface of the joint. Graphite sheet is used as a separator to separate the interface powder and susceptor powder as show in Figure 1. Bansal *et al.*[1] studied joint fabrication of mild steel-mild steel through microwave hybrid heating in domestic 900 watt microwave. The gap was filled with an interfacing slurry of nickel based powder and epoxy resin bisphenol-A and blumer 1450XX. The joint was

[2]Corresponding authors: pecamit26@gmail.com[a]

characterized using XRD, EDS, SEM, Rockwell hardness etc. Gupta *et al.*[3] investigated butt joint of tungsten carbide bearing alloy through microwave joining. Mild steel and stainless steel were joined by used of nickel based EWAC powder. Gupta and Kumar[2] worked on welding of SS420 with SS420 used nickel based EWAC powder as an interface material. Slurry is made by 95% nickel based powder and 5% blumer 1450XX. The joint was characterized by XRD, SEM, micro tensile tester etc. Badiger *et al.*[11] worked on joining the Inconel-625 alloy through microwave hybrid heating in domestic 900 watt microwave using Nickel based powder as an interface material. The experiments were carried out in domestic microwave and characterized by XRD, SEM, and UTM etc. Bansal *et al.*[12] investigate on Inconel 718 structure – properties in microwave joining. The butt joined Inconel 718 plates were joined by the microwave hybrid heating technique. Inconel 718 powder as a filler material was used for microwave welding. Singh *et al.*[7] worked on joining aluminium-aluminium through microwave hybrid heating. Aluminium powder was used as an interface material. SiC powder is used as a susceptor.

2 Experiment procedures

2.1 Microwave joining using EWAC

In the present study micro hybrid heating is used for joining of SS304-SS304 bits. A Nickel based EWAC powder of particle size 50μ has been used as an interface material. The detail study regarding joining of bulk material with the help of micro hybrid heating has already reported in form of patent [4]. The SS304 bits were placed in the butt configuration with the gap of approximate 0.5 mm. The gap was filled with interface slurry consist of 95% nickel based EWAC powder and 5% Epoxy resin blumer 1450XX. The configuration was exposed to microwave radiation for 420 sec at 2.45GHz and 800Watt. Table 1 is representing experimental results for experiments carried out on SS304 as work piece with Nickel based EWAC powder.

Table 1. Experimental results for SS304 with Nickel based EWAC powder as interface material.

Exp. no.	Interface powder	Exposure time (s)	Result
1	Ni based EWAC powder	360	Red hot and not joined
2	Ni based EWAC powder	420	Red hot and joined

(a) (b)

Figure 2. (a) Red hot bits placed in the cavity made in refractory brick. (b) Final joined work piece.

2.2 Microwave joining using 99.9% Pure Nickel Based Powder

In this experiment, 99.9% pure nickel based powder of particle size 50μ has been used as an interface material with the timing of 390 sec at 2.45GHz and 800Watt. Table 2 is representing experimental results for experiments carried out on SS304 as work piece with 99.9% pure Nickel based powder.

Table 2. Experimental results using Nickel based powder as interface material.

Exp. no.	Interface powder	Exposure time (s)	Result
1	99.9% pure Ni based powder	450	Red hot, Fusion and not joined
2	99.9% pure Ni based powder	420	Red hot, Fusion and not joined
3	99.9% pure Ni based powder	360	Red hot, Fusion and joined

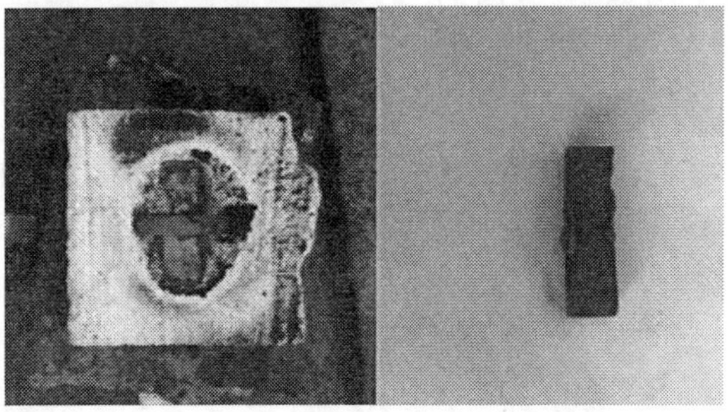

(a) (b)

Figure 3. (a) Red hot bits placed in the cavity made in refractory brick. (b) Final joined work piece.

2.3 Characterization of the joints

The joints were characterized by Energy dispersive spectroscopy (EDS), Scanning electron microscope (SEM) and Rockwell hardness test on C-scale.

3 Results and discussion

The results of various characterizations are discussed with suitable illustrations in the following sections.

3.1 Energy dispersive spectroscopy

3.1.1 EDS result for EWAC based joining

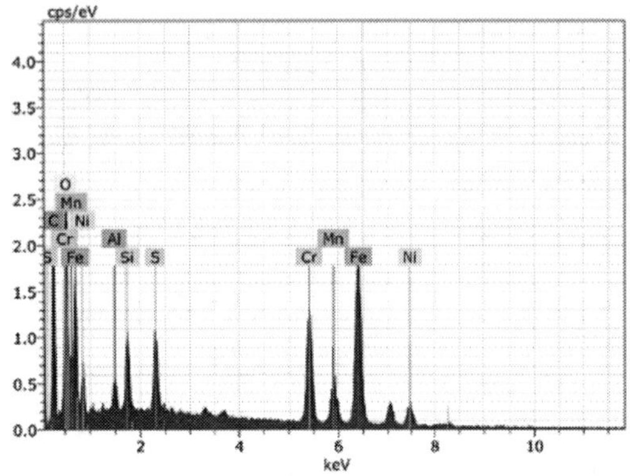

Figure 4. EDS results of Nickel based EWAC powder.

It is found from the EDS analysis in the joint region of the microstructure, the cells are of iron mainly (Fe), chromium (Cr), carbon(c), Nickel (Ni) and manganese (Mn). The presence of carbides (chromium and iron) was at the grain boundaries. This is due to the precipitation of carbides near the grain boundaries owing to the difference in solidification time[4]. The sample has been scanned with scanning pulse th. 2.80kcps and 15.0kv of High voltage.

3.1.2 EDS result for pure Nickel powder based joining

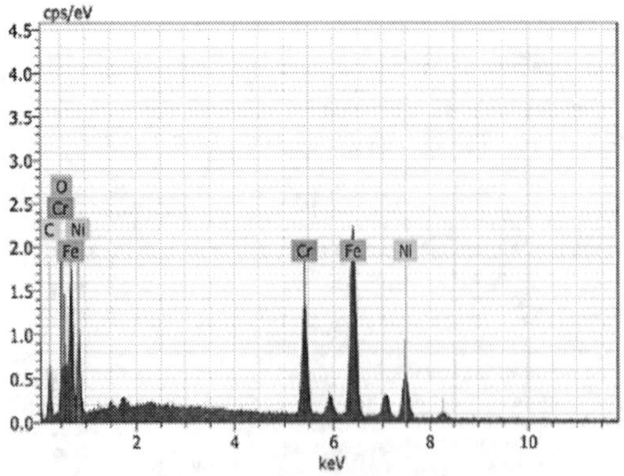

Figure 5. EDS results of Nickel based powder

It is found from the EDS analysis in the joint region of the microstructure, the cells are mainly of iron (Fe), chromium (Cr), Nickel (Ni), carbon (c). The sample has been scanned with scanning pulse Th. 1.73kcps and 15.0kv of High voltage.

From the EDS results, it follows:

- Carbon % in Nickel based EWAC powder is more than the 99.9% pure Nickel based powder.
- Nickel % in 99.9% pure Nickel based powder is more than the Nickel based EWAC powder.

3.2 Scanning electron microscope (SEM)

3.2.1 SEM result for EWAC based joining

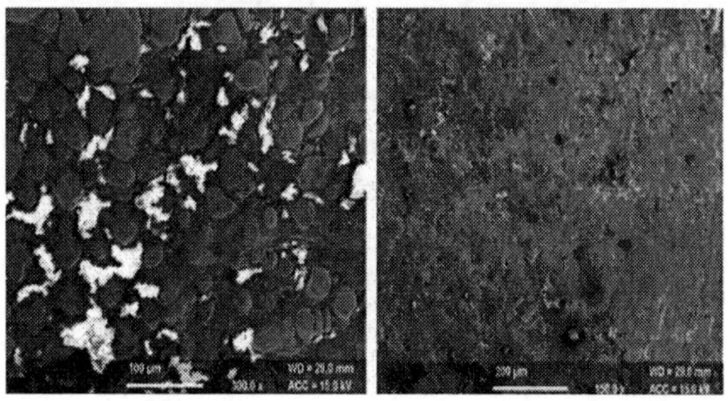

(a) (b)

Figure 6. (a) SEM image of Nickel based EWAC powder bead. (b) SEM image of joined based metal and bead

In Figure (b), there is no Heat affected zone (HAZ) around the joint and there is a perfect joining of the bead and based material.

3.2.2 SEM result for pure Nickel powder based joining

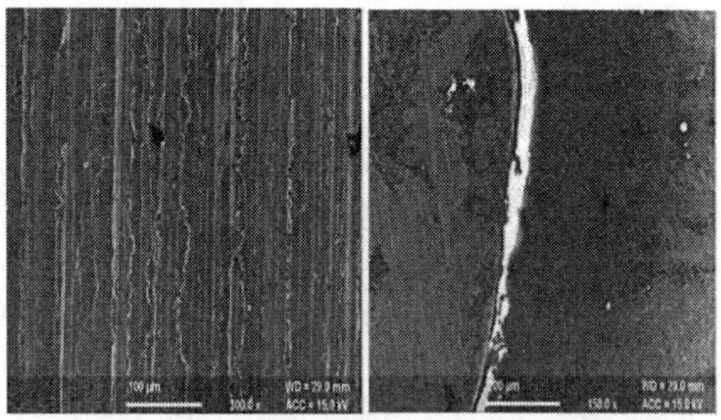

(a) (b)

Figure 7. (a) SEM image of 99.9% pure Nickel based powder bead. (b) SEM image of joined based metal and bead

In Figure (b), there is Heat affected zone (HAZ) around the joint and joining of bead and based material is not perfect as compare to EWAC powder.

3.3 Rockwell hardness test and visual inspection

Rockwell hardness test is done on the beads of both the joints to characterize the hardness of the joint on Rockwell C- scale.

Table 3. The hardness of work piece is tested by three different experiments

Exp.	HRC-Exp. 1	HRC-Exp. 2	HRC-Exp. 3	Average HRC
With Nickel EWAC powder	22	30	33	30
With 99.9% Nickel powder	37	46	45	42.67

From the above table 3 it is clear that the hardness on the HRC scale 99.9% pure Ni based powder joint hardness is higher than the Ni based EWAC powder. Both work pieces were visually inspected and it was found that surface finish in Nickel based EWAC powder was better than 99.9% pure nickel based powder.

4 Conclusion

- The weldability of Nickel based EWAC powder is higher than the 99.9% pure nickel based powder.

- The percentage of carbon is less in 99% pure Nickel based joint in comparison with Nickel based EWAC powder. This helps in reducing the porosity in 99.9% pure Nickel based joint compared to EWAC based joint.

- Rockwell hardness on HRC-scale for 99% pure Nickel based powder is higher than then EWAC powder

- Surface finish is better in Nickel based powder EWAC powder and bead size of joint is negligible as compared to 99.9% pure Nickel based powder joint.

References

1. A. Bansal, A. K. Sharma, and S. Das, "Metallurgical and mechanical characterization of mild steel-mild steel joint formed by microwave hybrid heating process," *Sadhana - Acad. Proc. Eng. Sci.*, vol. **38**, no. 4, pp. (679–686), 2013.
2. P. Gupta and S. Kumar, "Investigation of Stainless Steel Joint Fabricated Through Microwave Energy," *Mater. Manuf. Process.*, vol. **29**, no. July, pp. (910–915), 2014.
3. P. Gupta, S. Kumar, and A. Kumar, "Study of Joint Formed by Tungsten Carbide Bearing Alloy through Microwave Welding," *Mater. Manuf. Process.*, vol. **28**, no. 5, pp. (601–604), 2013.
4. A. Kumar, "Microwave Joining of Stainless Steel and Their charactrizationS.," Thapar university, 2015.
5. A. Pathania, S. Singh, D. Gupta, and V. Jain, "Development and analysis of tribological behavior of microwave processed EWAC + 20% WC10Co2Ni composite cladding on mild steel substrate," *J. Manuf. Process.*, vol. **20**, pp. (79–87), 2015.
6. S. Saxena, S. Bansal, R. Deo, and S. Khan, "Joining of Bulk Metallic Pipes by Microwave Hybrid Heating Processunder Parametrical Regulations," *IOSR J. Mech. Civ. Eng.*, vol. **11**, no. 6, pp. (62–69), 2014.
7. S. Singh, N. M. Suri, and R. M. Belokar, "Characterization of Joint Developed by Fusion of Aluminum Metal Powder through Microwave Hybrid Heating," *Mater. Today Proc.*, vol. **2**, no. 4–5, pp. (1340–1346), 2015.
8. M. S. Srinath, P. S. Murthy, A. K. Sharma, P. Kumar, and M. V Kartikeyan, "Simulation and analysis of microwave heating while joining bulk copper," *Int. J. Eng. Sci. Technol.*, vol. **4**, no. 2, pp. (152–158), 2012.
9. R. R. Mishra and A. K. Sharma, "Microwave-material interaction phenomena: Heating mechanisms, challenges and opportunities in material processing," *Compos. Part A Appl. Sci. Manuf.*, vol. 81, pp. (78–97), 2016.
10. M. Kwak, P. Robinson, A. Bismarck, and R. Wise, "Microwave curing of carbon-epoxy composites: Penetration depth and material characterisation," *Compos. Part A Appl. Sci. Manuf.*, vol. **75**, pp. (18–27), 2015.
11. R. I. Badiger, S. Narendranath, and M. S. Srinath, "Joining of Inconel-625 alloy through microwave hybrid heating and its characterization," *J. Manuf. Process.*, vol. **18**, no. January, pp. (117–123), 2015.
12. A. Bansal, A. K. Sharma, P. Kumar, and S. Das, "Structure–Property Correlations in Microwave Joining of Inconel 718," *Jom*, vol. **67**, no. 9, pp. (2087–2098), 2015.

MATEC Web of Conferences 57, 03002 (2016) DOI: 10.1051/matecconf/20165703002
ICAET-2016

Optimum design of brake friction material using hybrid entropy-GRA approach

Naresh Kumar[1], Tej Singh[1, a], R.S. Rajoria[2], and Amar Patnaik[3]

[1]*Department of Mechanical Engineering, Manav Bharti University, Solan-173229, India*
[2]*Department of Mechanical Engineering, CEC Landran-140307, India*
[3]*Department of Mechanical Engineering, M.N.I.T. Jaipur-302017, India*

Abstract. The effect of Kevlar and natural fibres on the performance of brake friction materials was evaluated. Four friction material specimens were developed by varying the proportion of Kevlar and natural fibres. Two developed composite contained 5-10 wt.% of Kevlar fibre while in the other two the Kevlar fibre was replaced with same amount of natural fibre. SAE J661 protocol was used for the assessment of the tribological properties on a Chase testing machine. Result shows that the specimens containing Kevlar fibres shows higher friction and wear performance, whereas Kevlar replacement with natural fibre resulted in improved fade, recovery and friction fluctuations. Further hybrid entropy-GRA (grey relation analysis) approach was applied to select the optimal friction materials using various performance defining attributes (PDA) including friction, wear, fade, recovery, friction fluctuations and cost. The friction materials with 10 wt% of natural fibre exhibited the best overall quality.

1 Introduction

Generally, a friction material contains a resin binder, fibrous reinforcement, abrasives, fillers and lubricants in varying composition [1]. Asbestos because of its economical issue and met the bulk of fascinating properties of brake friction material are used for years as a fibrous reinforcement [2]. However, the ban on the utilization of asbestos by environmental protection agencies because of its carcinogenic nature had forced the friction industry to hunt alternative of asbestos [3]. Last two decades witnessed a lot of research in the development of asbestos-free brake friction materials [4-10]. In this way the formulation designers start focusing on natural fibres and industrial wastes as a source of raw materials for friction formulations [11-17]. The utilization of such materials will not only be economically viable, but may also result in reduced environmental pollution. A good friction material requires an appropriate set of performance defining attributes (PDAs) namely high and stable coefficient of friction, negligible fading, faster recovery, better wear resistance and low-sensitivity towards load-speed alterations for reliable and safe performance [1]. Therefore assessment of such PDAs and selection of optimal friction material can be regarded as multiple criteria decision-making (MCDM) problem [18-20]. Hence in this work the comparative effect of addition of Kevlar and natural fibre on the tribo-performance of brake friction material is studied and optimal formulation is selected by using MCDM approach.

2 Experimental

2.1 Materials and fabrication of friction specimens

Friction materials based on phenolic resin, Kevlar fibre, Lapinus fibre, graphite, barite, alumina, vermiculite, potassium titanate and natural fibre amounting to 100% by weight were fabricated as per Table 1. Detail of the processing conditions for composite fabrication is briefly reported elsewhere [21-23].

Table 1. Composition and nomenclature of the friction materials specimens.

Composition (wt.%)	Nomenclature			
	K-1	K-2	N-1	N-2
*Parent composition	45	45	45	45
Barite	50	45	50	45
Kevlar Fibre	5	10	0	0
Natural Fibre	0	0	5	10

K: Kevlar fibre, N: Natural fibre
*Parent composition: Phenolic resin, Lapinus fibers, Alumina, Graphite, Potassium Titanate and Vermiculite.

2.2 Tribo-performance evaluation methodology

The tribo-performance of the specimens was evaluated on a chase type friction material testing machine following SAE J661a standard. The details of the testing protocol and testing machine is reported elsewhere [24]. The performance coefficient friction, fade, recovery, wear,

[a] Corresponding author: tejschauhan@gmail.ocom

friction fluctuations and cost were considered as PDA of the evaluation. The detailed description of the various PDAs used in this study is given as:
PDA-1: μ-Performance (Higher-is-better)
It is the average friction coefficient of all fade and recovery cycles.
PDA-2: Wear (Lower-is-better)
It is the loss of material from the specimen during braking. It is calculated as, $\Delta W = W_1 - W_2$, where W_1 the weight of specimen before test and W_2 is the weight of the specimen after test.
PDA-3: %-Fade (Lower-is-better)
%-Fade= $\frac{\mu_{\text{Performance}} - \mu_F}{\mu_{\text{Performance}}} \times 100$, where μ_F is the minimum coefficient of friction for the fade cycles. Lower %-fade is desirable for good friction composites. The standard range is 0-30%.
PDA-4: %-Recovery (Higher-is-better)
%-Recovery= $\frac{\mu_R}{\mu_{\text{Performance}}} \times 100$, where μ_R is the maximum coefficient of friction for the recovery cycles. It is desirable in the range of 100-140% for good friction composites.
PDA-5: Frictional fluctuations (Lower-is-better)
Frictional fluctuations are given by $\Delta\mu = \mu_{max} - \mu_{min}$. Where, μ_{max} is the highest coefficient of friction and μ_{min} is the lowest coefficient of friction.
PDA-6: Cost (Lower-is-better)
It is the cost of the composition per kilogram.

2.3 Ranking evaluation methodology

The ranking evaluation methodology consists of three main phases as:
Phase I: Identification of the alternatives, PDAs and decision matrix
Phase II: Weight determination of various PDAs
Phase III: Ranking of the alternatives
Phase I: Identification of the alternatives and PDAs
Here the number of alternatives and performance defining attributes (PDAs) are identified. After the identification of alternatives and criterions the values of experimental results for each alternative is expressed in a decision matrix as:

$$X = \begin{array}{c} \\ A_1 \\ A_i \\ \vdots \\ A_m \end{array} \begin{array}{c} \begin{array}{cccc} C_1 & C_j & \cdots & C_n \end{array} \\ \begin{bmatrix} x_{11} & x_{1j} & \cdots & x_{1n} \\ x_{i1} & x_{ij} & \cdots & x_{in} \\ \vdots & \vdots & \ddots & \vdots \\ x_{m1} & x_{mj} & \cdots & x_{mn} \end{bmatrix} \end{array} \tag{1}$$

where, $A_1,\ldots, A_m$ are the composition/alternatives and $C_1,\ldots,C_n$ are the PDA. Hence, x_{ij} is the value of jth PDA (C_j) for the ith composition/alternative (A_i).
Phase II: Entropy method for weight determination
Entropy method is used for the calculation of weight. This method consists of the following steps [25]:
Step I: The constructed matrix is normalized using following equation:

$$N_{ij} = \frac{x_{ij}}{\sum_{i=1}^{m} x_{ij}} \tag{2}$$

Step II: In this step, the entropy for each PDA is calculated as:

$$e_j = -\kappa \sum_{j=1}^{n} N_{ij} \ln(N_{ij}) \tag{3}$$

Where, κ is a constant and calculated as $\kappa = \frac{1}{\ln(m)}$
Step III: Finally the weight of each PDA is calculated as:

$$\varpi_j = \frac{\partial_j}{\sum_{j=1}^{n} \partial_j} \tag{4}$$

Where, $\partial_j = 1 - e_j$ and it is the degree of diversity of the information involved in the outcomes of the jth PDA.
Phase III: Ranking of the alternatives
Grey system theory was developed by Deng and consists following step [26]:
Step-I: In order the make the values free of unit the normalization process is done. The normalized values of each element in the performance matrix can be calculated as follows:

$$X'_{ij} = \frac{x_{ij} - \min\{x_{ij}\}}{\max\{x_{ij}\} - \min\{x_{ij}\}} \tag{5}$$

$$X'_{ij} = \frac{\max\{x_{ij}\} - x_{ij}}{\max\{x_{ij}\} - \min\{x_{ij}\}} \tag{6}$$

Where $\max\{x_{ij}\}$ and $\min\{x_{ij}\}$ respectively are the maximum and minimum values of the ith PDA. Eq. (5) is used for the larger-the-better PDA; Eq. (6) is used for the smaller-the-better PDA.
Step-II: In this step reference sequence ($\Re_j$) is generated from the normalized matrix by taking the largest normalized value of each PDA as:

$$\Re_j = \max_{i=1}^{m} \{X'_{ij}\} \tag{7}$$

Step-III: Grey correlation coefficient for each term is determined as:

$$\Phi_j = \frac{\min_{i=1}^{m} \min_{i=1}^{n} \Delta_{ij} + \hbar \max_{i=1}^{m} \max_{i=1}^{n} \Delta_{ij}}{\Delta_{ij} + \hbar \max_{i=1}^{m} \max_{i=1}^{n} \Delta_{ij}} \tag{8}$$

Where $\Delta_{ij} = |\Re_j - X'_{ij}|$ and $\hbar$ is known as the distinguishing coefficient and generally takes the value of 0.5.
Step-IV: Finally a grey relational degree (which is the weighted sum of the grey relational coefficients) is calculated by the following equation:

$$\Omega_i = \sum_{j=1}^{n}\left[\Phi_j \times \varpi_j\right] \quad (9)$$

Where ϖ_j is the weight of the jth PDA and calculated using entropy method. Therefore, using the grey correlation degrees the friction material formulation sequence can be prioritized and the one with the highest degree of correlation will be identified as the best formulation that represents the optimal solution.

3 Results and discussions

3.1 Experiment results

The tribo-performance evaluation of the fabricated friction composites materials were carried out on a chase type testing machine. The experiment results corresponding to each PDA are listed in Table 2. The performance friction coefficient ($\mu_{Performance}$) (0.586±0.009) and wear performance (1.065±0.055g) remains higher for Kevlar filled friction composite. The $\mu_{Performance}$ and wear performance are observed to decrease for the friction composition having natural fibre content. From fade performance point of view natural fibres (30.39%±5.96) based friction composition proved much superior to Kevlar fibre based composition where fade remains highest (54.91%±8.52). The recovery performance is observed to exceed 100% irrespective of the fibrous reinforcement. It remains ~110.435±1.525% for Kevlar based composition and increases slightly to 110.735±1.765% for the compositions having natural fibre as reinforcement. Moreover, natural fibres based compositions performed better from friction fluctuations point of view and are much cost effective as compared to Kevlar fibre based compositions.

3.2 Ranking of the alternatives

The decision matrix from Eq.1 is used for the entropy-GRA analysis. The weights of the selected PDAs are calculated by using Eqs. 2-4 and given in Table 3. The weight of PDAs as evaluated by entropy method are PDA-1=0.009, PDA-2=0.078, PDA-3=0.205, PDA-4=0.003, PDA-5=0.185, PDA-6=0.520. After weight calculation, GRA technique is applied to find the optimum composition using Eqs. 5-9. The GRA technique results are shown in Table 6. From this analysis, it was observed that the natural fibre based compositions exhibits the optimal performance.

Table 2. Experimental data of the PDAs.

Alternatives	PDA-1	PDA-2	PDA-3	PDA-4	PDA-5	PDA-6
K-1	0.595	1.12	46.39	108.91	0.329	105.25
K-2	0.577	1.01	63.43	111.96	0.435	169.75
N-1	0.520	1.70	36.35	112.50	0.262	45.25
N-2	0.524	1.20	24.43	108.97	0.175	49.75

Table 3. Weight of PDAs as calculated by entropy method.

Alternatives	PDA-1	PDA-2	PDA-3	PDA-4	PDA-5	PDA-6
Entropy	0.998	0.984	0.959	0.999	0.963	0.895
Weight	0.009	0.078	0.205	0.003	0.185	0.520

Table 4 The normalized decision matrix and reference sequence.

Alternatives	PDA-1	PDA-2	PDA-3	PDA-4	PDA-5	PDA-6
K-1	1.000	0.841	0.437	0.000	0.408	0.518
K-2	0.760	1.000	0.000	0.850	0.000	0.000
N-1	0.000	0.000	0.694	1.000	0.665	1.000
N-2	0.053	0.725	1.000	0.017	1.000	0.964
Reference sequence	1	1	1	1	1	1

Table 5 Grey relation coefficient.

Alternatives	PDA-1	PDA-2	PDA-3	PDA-4	PDA-5	PDA-6
K-1	1.000	0.759	0.470	0.333	0.458	0.509
K-2	0.676	1.000	0.333	0.769	0.333	0.333
N-1	0.333	0.333	0.620	1.000	0.599	1.000
N-2	0.342	0.642	1.000	0.333	1.000	0.932

Table 6 Grey correlation degree and ranking of the formulations.

Alternatives	PDA-1	PDA-2	PDA-3	PDA-4	PDA-5	PDA-6	Grey correlation degree	Ranking
K-1	0.009	0.059	0.096	0.001	0.085	0.265	0.515	3
K-2	0.006	0.078	0.068	0.002	0.062	0.173	0.389	4
N-1	0.003	0.026	0.127	0.003	0.111	0.52	0.790	2
N-2	0.003	0.050	0.205	0.001	0.185	0.485	0.929	1

4 Conclusions

Optimal design of friction material using combined entropy-GRA approach was carried out in this work. The optimal design of friction material was performed in three phases: (1) selection of alternatives and performance defining attributes, (2) weight calculation using entropy method and (3) GRA method is used to rank the alternatives. In this study, Kevlar and natural fibre based friction materials were selected as alternatives whereas; performance coefficient of friction, wear, fade, recovery, frictional fluctuations and cost were selected as performance defining attributes. Based on entropy-GRA results, the ranking of the friction materials from the best to the worst is in the order of N-2, N-1, K-1, and K-2. The specimen N-2 having 10 wt.% of natural fibre was ranked best among all the evaluated friction materials.

References

1. J. Bijwe, Polymer Composites **18**, 378 (1997)
2. D. Chan, G.W. Stachowiak, Proc. Instn Mech. Engrs- Part D: J. Automobile Engineering **218**, 953 (2004)
3. T. Singh. Ph.D. Thesis, N.I.T. Hamirpur (2013)
4. B.K. Satapathy, J. Bijwe, Wear **257**, 573 (2004)
5. X. Qu, L. Zhang, H. Ding, G. Liu, Polymer Composites **25**, 94 (2004)
6. T. Singh, A. Patnaik, Archives of Civil and Mechanical Engineering **15**, 151 (2015)
7. T. Singh, A. Patnaik, R. Chauhan, A. Rishiraj, Journal of King Saud University: Engineering Sciences. DOI: 10.1016/j.jksues.2015.06.002, (2015)
8. T. Singh, A. Patnaik, B.K. Satapathy, American Institute of Physics Conference proceeding **1393**, 223 (2011)
9. T. Singh, A. Patnaik, B.K. Satapathy, American Institute of Physics Conference proceeding **1536**, 259 (2013)
10. T. Singh, A. Patnaik, B. Gangil, Journal of Industrial Textile; DOI: 10.1177/1528083714559568, (2014)
11. N. Dadkar, B.S. Tomar, B.K. Satapathy, A. Patnaik, Materials and Design **31**, 723 (2010)
12. A. Tiwari, H.S. Jaggi, R.K. Kachhap, B.K. Satapathy, S.N. Maiti, B.S. Tomar, Wear **309**, 259 (2014)
13. T. Singh, A. Patnaik, R. Chauhan, Materials and Design **89**, 1335 (2016)
14. A. Mustafa, M.F.B. Abdollah, F.F. Shuhimi, N. Ismail, H. Amiruddin, N. Umehara, Materials and Design **67**, 577 (2015)
15. U.D. Idris, V.S. Aigbodion, I.J. Abubakar, C.I. Nwoye, Journal of King Saud University-Engineering Sciences **27**, 185 (2015)
16. K.K. Ikpambese, D.T. Gundu, L.T. Tuleun, Journal of King Saud University-Engineering Sciences, http://dx.doi.org/10.1016/j.jksues.2014.02.001, (2014).
17. D.S. Yawas, S.Y. Aku, S.G., Amaren, Journal of King Saud University-Engineering Sciences, http://dx.doi.org/10.1016/j.jksues.2013.11.002, (2013)
18. T. Singh, A. Patnaik, B. Gangil, R. Chauhan, Wear **324-325**, 10 (2015)
19. T. Singh, A. Patnaik, B.K. Satapathy, M. Kumar, Composites: Mechanics, Computations, Applications. An International Journal **3**, 189 (2012)
20. T. Singh, A. Patnaik, B. K. Satapathy, Walailak Journal of Science and Technology **10**, 343 (2013)
21. T. Singh, A. Patnaik, B.K. Satapathy, M. Kumar, B.S. Tomar, Journal of Materials Engineering and Performance **22**, 796 (2013)
22. T. Singh, A. Patnaik, B.K. Satapathy, NANO **8**, 1 (2013)
23. T. Singh, A. Patnaik, Polymer Composites: DOI:10.1002/pc.23682 (2015).
24. P. Gopal, L.R. Dharani, F.D. Blum, Wear **193**, 199 (1996)
25. T.C. Wang, H.D. Lee, Expert Systems with Applications **36**, 8980 (2009)
26. J.L. Deng, Systems and Control Letters **1**, 288 (1982)

MATEC Web of Conferences 57, 03003 (2016) DOI: 10.1051/matecconf/20165703003
ICAET-2016

An Application Of Analytic Network Process (ANP) To Evaluate Green Supply Chain Management Strategies: A Case Study

Arvind Jayant[1]

Mechanical Engineering Department,
SLIET Deemed University, Longowal, Sangrur, Punjab, India
(Deemed University under MHRD, Govt. of India)

Abstract. Sustainable development has made remarkable progress in establishing environmental and social sustainability towards operations management and the supply chain. Sustainable development means "Development that meets the needs of the present without compromising the ability of future generations" Sustainability covers three aspects: economic, environmental and social responsibility. Green supply chain management (GSCM) is about making the entire supply chain more environmental sustainable. With the raising of awareness in environmental protection and sustainable development in enterprises, the green issue has become more and more critical in supply chain management. The green supply chain strategy selection is a multi-criterion problem which includes both qualitative and quantitative criteria. ANP is a multi-attribute, decision-making approach based on the reasoning, knowledge, and experience of the experts in the field. ANP can act as a valuable aid for decision making the best decision when both qualitative and quantitative aspects of decision are considered. ANP provides a general framework to deal with decisions without making assumptions about the independence of higher-level elements from lower level elements and about the independence of the elements within a level. The present work proposes a structured model for evaluating and selecting a Green Supply Chain Strategy by using Analytical Network Process (ANP) approach. The model is developed using evidence from empirical study in the industry. The model for selection of green supply chain strategy includes considers both the practicality in traditional SCM selection criteria, and also the green concept related to environmental protection and regulations. The proposed ANP model is used to evaluate the green supply chain strategy in Indian Automotive Company XYZ Ltd. The major advantages of this research are that it can be used for both qualitative and quantitative criteria. The results show that the model has the capability to be flexible and apply in different types of industries to choose their green supply chain strategy. The future research directions to this work include the design of more complex hierarchies and explore it. It may benefit the Indian industries to survive and serve better in Global Markets.

1 Introduction

With increasing government regulation and stronger public awareness in environmental protection, firms today simply cannot ignore environmental issues if they want to survive in the global market. In addition to complying with the environmental regulations for selling products in certain countries, firms need to implement strategies to voluntarily reduce the environmental impacts of their products. The integration of environment, economic and social performances to achieve sustainable development is a major business challenge for the new century [20].Green Supply Chain Management (GSCM) adds 'green' component to the conventional supply chains by including practices like green operations, green design, green manufacturing, reverse logistics and waste management [18].GSCM has been gaining much attention of academia and industry over the last few years. The companies cannot afford to ignore environmental issues in today's competitive business environment. Increasing government regulations and stronger public mandates for environmental accountability have brought the green issues high on the strategic planning agenda of manufacturing firms [21].GSCM practices have become a very important part of company's policy and are increasingly turning into a major strategic thrust in business organizations. GSCM is a relatively new concept for the majority of Indian automobile firms.

Decision making in supply chain is crucial as it involves multi-criteria, multi- level, multi-objective and multi-personal decisions. The emphasis on higher co-operation and co-ordination among the partners of supply chain is key issue, which requires strategies suitable to system-optimal performance where the point of focus shifts from local to global optimal. Some of the factors are tangible and general in nature, while some other are situation specific and intangible in nature. Multi-criteria decision making tools like AHP and ANP are gaining wide

Arvind Jayant: arvindjayant@gmail.com

applicability and attempt has been made to make use of them in issues related to inbound supply chain. The priority coefficients thus found by these techniques are used in optimization techniques to get desired results. Application of supply chain concepts has been made in service industry as no such work has been attempted earlier.

Strategic planning in GSCM context means the identification of relevant goals and specification of long term plans for managing those goals [13].It charts out the future course of action for GSCM implementation. The biggest challenge in GSCM practices for companies is to select suitable strategies in accordance with the stipulated regulations and customers requirements [6]. Several studies have emphasized on developing GSCM strategies from an overall organizational perspective.

There are many approaches to evaluating performance indices system, such as linear weighting methods [5], data envelopment analysis, analytic hierarchy process (AHP) [22], total cost approaches [10], and mathematical programming techniques [3]. AHP is a relatively popular tool for modelling strategic decisions, but a primary limitation is its basic relationships do not allow for an integrated dynamic modelling of the environment. AHP assumes the system elements are uncorrelated and are unidirectional introduced by a hierarchical relationship. A more general evaluation approach defined as the analytical network process (ANP), or systems with-feedback approach, may be used to evaluate a dynamic multi-directional relationship among decision attributes. ANP allows for more complex interrelationships among the decision levels and attributes. ANP does not require this strictly hierarchical structure. Interdependencies may be represented by two way arrows (or arcs) among levels, or if within the same level of analysis, a looped arc. The directions of the arcs signify dependence, arcs emanate from an attribute to other attributes that may influence it. The relative importance or strength of the impacts on a given element is measured on a ratio scale similar to AHP. The ANP approach is capable of handling interdependence among elements by obtaining the composite weights through the development of a 'super matrix'. Saaty explains the super matrix concept as a parallel to the Markov chain process [8]. However, in real-life decision-making situation, decision-makers could be uncertain about their own level of preference, due to incomplete information or knowledge, complexity and uncertainty within the decision environment. Such situation may occur when evaluating SCM performance indices system. In SCM performance evaluation problem, decision-makers might intuitively feel that some factors are more important that others in affecting their final preference among alternatives. If there is some feedback and interdependency among the factors, an unimportant factor may turn out to be far more important than even the most intuitively important one. Therefore, a fuzzy ANP approach is proposed to handle SCM performance evaluation problem.

The study aims to provide a guideline for the electronic industry on how to choose an appropriate GSCM strategy, which incorporates environmental perception with corporate functional strategies in order to achieve an effective green management. The integration of environmental corporate strategy with every corporate functional strategy will facilitate the connections between each functional strategy and thereby eliminate obstacles to environmental integration. Due to the fact that many companies have just started exploring environmental concerns in the recent years, a lot of environmental related factors have not yet been identified systematically.

The rest of this paper is organized as follows. Section 2 discusses literature reviews from different authors and methodology for selection of strategies in GSCM in context of various companies. Next, Section 3 debates the ANP approach. Section 4 then presents an illustrative case study which used for strategy selection using the ANP model. Finally, Section 5 presents procedure and result and section 6 presents conclusions and limitations.

2 GSCM strategy

No matter green manufacturing strategies or environmental strategies, past studied discussed and classified them mostly according to attitudes that companies held toward to green management, the simplest is divided into proactive and reactive. Besides, GSCM related researches also examined the enhancement and implications of GSCM practices, and GSCM performance evaluations. In order to foster the future development of GSCM activities, some GSCM strategies based on classical SCM theories were proposed. These four GSCM strategies as listed below [17].

(1) risk-based strategy: companies choosing this kind of GSCM strategy most invested minimal resources of organization for green management, and also asked their suppliers to comply the environmental requirements for reducing risks. Although some international environmental rules are difficult to refer to, it is the simplest GSCM strategy compared to other types. Therefore, companies couldn't create unique competitive advantage, and then didn't' obtain many economic benefits from this strategy. Also, these companies didn't spontaneously introduce environmental programs; it is not possible to expect any innovation happened. The final goal of this type of strategy is risk minimization with accomplishing environmental programs passively.

(2) efficiency-based strategy: compared the above one, this type of strategy is more complicated owing to efficiency improving through specific approaches. It not only facilitated companies to allow increase economic benefits, but also won the environmental benefits resulted from waste reduction and resources used efficiently. However the efficiency-based strategy still aimed more on increasing productivity, it is impossible to avoid from any toxic or harmful substances during production processes completely. Hence, the environmental

programs in this type of strategy primary directed firms to reduce cost and meet the Operational optimization, and decrease environmental depredations at the same time. But the efficiency based strategy still didn't create any proactive plans or activities, such as innovative technologies or approaches in environmental protection.

(3) innovation-based strategy: this type of strategy guided companies to develop products from product life-cycle viewpoints, and give stricter environmental requirements to their suppliers, and even train them to adjust operational processes just followed the newest environmental regulations. The companies utilizing the kind of strategy should possess professional environmental expertise, and integrate specific relevant green activities, such as green design, green procurement to improve current processes, product developments. As a result, the innovation-based strategy forced companies to invest more resources and cultivate innovative capabilities to green managements

(4) closed-loop strategy: it is the most complicated type of GSCM strategy, and it linked the environmental performance to the entire supply chain activities. In other words, it required the more players in the supply chains involved more. For the simplest form in the closed-loop strategy, reverse logistics, it should take back materials produced from any production processes and end-of-life products, and disposed of them in various methods relied on complex degree of this type of strategy. However, it is difficult for manufacturers to track all of the products distributed from factories due to too many channels. Besides, the amount of return goods also accumulated enough to create economic scale. For these reasons, companies implementing the type of strategy successfully could incorporate efficiently economic, operational and environmental performance as well. As the authors mentioned, it needs lots of efforts to integrate highly, cooperate with many parties, and develop quite specialized knowledge and technology.
Therefore, it is the final goal for those executing green managements completely.

3 Research Methodology

A set of criteria covering wide range of parameters is submitted in the form of table and opinion of expert is taken to select pertinent criteria for strategy selection in the context of XYZ Automobile Company. Apart from this an unstructured opinion is also sought. Cost being one of the most important parameters has been included in this analysis as the alternatives chosen are cost competitive and thus a detailed analysis is required to select one of them on the basis of comprehensive analysis of various other factors and ultimately the priorities obtained with ANP can be seen in the context of cost parameter and an appropriate decision can be taken. The priorities obtained from the ANP can be directly used in linear programming model as the coefficients in the objective function to get the required distribution of the demand among the strategy which can satisfy a set of constraints related to lead time, plant capacity etc.

4 Analytical Network Process (ANP)

ANP is a generalized form of the widely used multi-criteria decision making technique of AHP [19]. ANP offers several advantages over other evaluation techniques such as data envelopment analysis (DEA), expert systems, goal programming etc. It is the most comprehensive frameworks available for group decisions that are in use today with the decision-maker. It is a process that allows one to include all the factors and criteria, tangible and intangible, those have bearing on making best decision. The Analytic Network Process allows both interaction and feedback within clusters of elements (inner dependence) and between clusters (outer dependence). Such feedback best captures the complex effects of interplay in corporate world, especially when risk and uncertainty are involved.

The ANP, developed by Saaty, provides a way to input judgments and measurements to derive ratio scale priorities for the distribution of influence among the factors and groups of factors in the decision. Because the process is based on deriving ratio scale measurements, it can be used to allocate resources according to their ratio-scale priorities. The well-known decision framework, the Analytic Hierarchy Process (AHP) is a special case of the ANP. Both the AHP and the ANP derive ratio scale priorities for elements and clusters of elements by making paired comparisons of elements on a common property or criterion. Although many decision problems are best studied through the ANP, one may wish to compare the results obtained with it to those obtained using the AHP or any other decision approach with respect to the time it took to obtain the results, the effort involved in making the judgments, and the relevance and accuracy of the results.

ANP models have two parts: the first is a control hierarchy or network of objectives and criteria that control the interactions in the system under study; the second are the many sub-networks of influences among the elements and clusters of the problem, one for each control criterion. The ANP has been applied to a large variety of decisions: marketing, medical, political, social, and forecasting and prediction and many others. Its accuracy of prediction is impressive in applications that have been made to economic trends, sports and other events for which the outcome later became known.

The ANP utilizes the idea of a control hierarchy or a control network to deal with different criteria, eventually leading to the analysis of benefits, opportunities, costs, and risks. By relying on control elements, the ANP parallels what the human brain does in combining different sense data as for example does the thalamus.

Software —SUPER DECISIONS is available for solving ANP problem and is used in this project.

DOI: 10.1051/matecconf/20165703003

5 Proposed Model and Approach

In this section, we introduce an ANP model and its development to show how ANP can be used in the existing environment. It is acknowledged that each organization will have its own set of criteria and that the model for a particular enterprise may use other levels and criteria. Our attempt here is to present a generalized model based on factors and criteria mentioned from the literature that could then be adapted or extended to support a particular situation or organization.

A. Case Study

With the numbers of green programs was increasing; it was more difficult for the company to ensure the performances of them. Therefore, the top management should consider carefully which type of green management perspective they had, and assessed the importance of every factor in each green business activity objectively. In sum, they should choose an adequate GSCM strategy to guide all of the employees to strive to upgrade the overall environmental performances. For above the reasons, we targeted the key managers related the green programs of the case company and other experts in other companies with green programs to establish the ANP network and assist the XYZ company to confirm the best GSCM strategy. The company manufactures tractors and machine tools for the inland and outland delivery both. At the time of the model, the company was trying to decide between two development options to address the demand of the marketplace. The options were to develop a new strategy system or upgrade the speed of a current system. While the new system offered the potential for better performance, this had to be offset against the risks and costs involved with developing the technology necessary for its development. Due to funding and resource constraints, only one option could be pursued. Once a decision was made, the company was committed to pursuing that course of action. The decision was strategic in that the success of the development would bear greatly on the continued competitiveness of the company. In the following section, we will propose the ANP model and take the XYZ Company to be an example to explain a company how to apply the model to choose the appropriate GSCM strategy.

B. Decision Model

The first step is to construct a model to be evaluated. This example uses the factors discussed earlier to develop a model that explicitly considers many of the strategy selection issues found in literature and practice. The model is summarized in Fig. 1. The relevant criteria and alternatives are structured in the form of a hierarchy. More "strategic" decisions are shown higher in the hierarchy. The topmost elements are decomposed into subcomponents and attributes. The model development requires the development of attributes at each level and a definition of their relationships. In this example, the only interdependence or feedback occurs between all of the actors (or stakeholders) in the decision and all phases of the project as shown by the two headed arrow. The other single arrows indicate a one-way relationship.

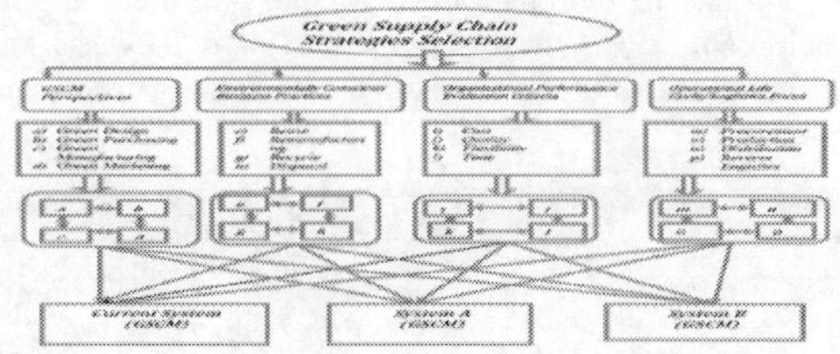

Figure 1. ANP model for Green supply chain strategy selection

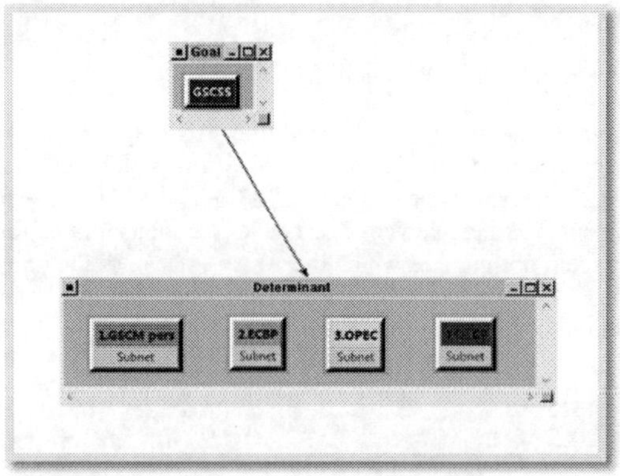

Figure 2.

6 ANP Implementation Procedure

Step 1 Formation of network with goal, clusters and subnets: The problem is first designed in the software by making clusters and the corresponding nodes and connections. The strategy selection problem is designed as hierarchical network with the goal as the topmost cluster. This is linked to another cluster containing Green supply chain perspective, environmentally conscious business practice; overall performance evaluation criteria and operational life cycle logistic focus as its nodes (fig. 2).

Step 2 the clusters and nodes under all 4 subnets, each subnet consists of two clusters i.e. Attributes and Alternatives.

Step 3 Node Comparisons: This involves comparison of nodes with respect to a control criterion.

Step 4 Generation of weighted and limiting super-matrix: In this step the generation of the weighted and the limiting super-matrices for all the four sub networks.

Step 5 The Values obtained at the subnet level are raised to the goal level and limit matrix (fig. 3) is obtained for goal.

Step 6: The values obtained from the synthesis are taken at the level of the goal and overall synthesis for the model is achieved as shown in Figure 4.

Step7 Sensitivity analysis: In this the variations in the priority of the alternatives with respect to change in the weight age of the control criteria can be observed. One such graph is presented in figure 5 to show the variations in the priorities of the alternative with respect to business criteria.

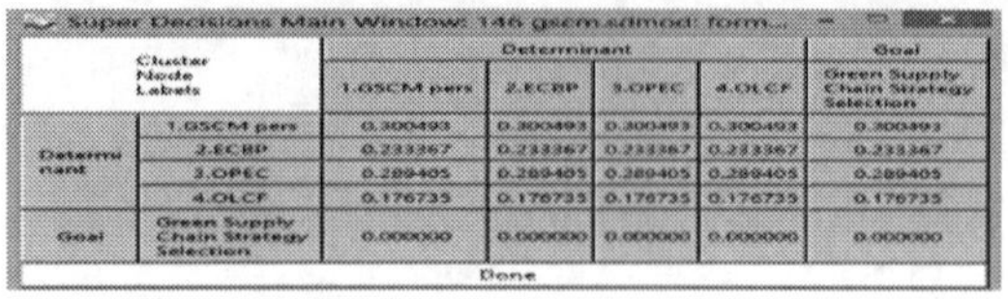

Figure 3.

A. Result of ANP modelling

Based on the priorities obtained from ANP, system A has highest priority coefficient of 0.5488 followed by system B and current system with 0.3237 and 0.1274. Clearly system A is best choice. The results are indicators of the personal preferences which the Analysts have as the pair-wise based on the response of experts on the attributes following hierarchy has been formed and used in ANP software —SUPER DECISIONS.

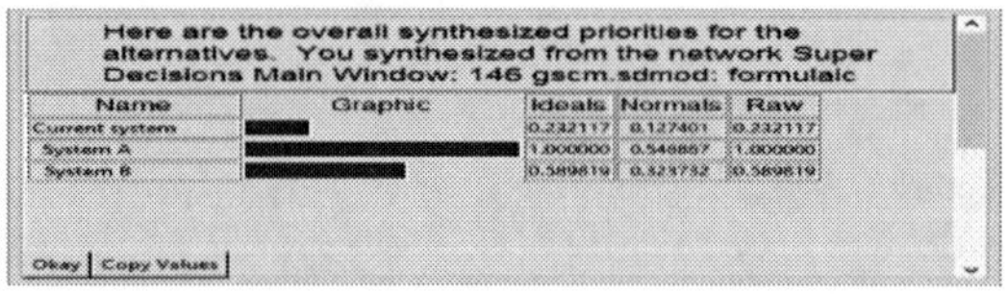

Figure 4.

Comparison (See in appendix A) is based on his knowledge, word of mouth information available to him and judgments. One of reasons could be the first hand experience of the present strategy system, which the analysts have in comparison to the here-say, and written details of other two alternatives.

7 Conclusions

In the current scenario of global operating conditions and competitive environment, it has become highly necessary for organizations to select the effective strategy. Any inaccurate selection of strategy will affect the company's overall performance. This shows that strategy selection procedure is a highly essential decision making process for companies. As discussed in the literature review, comprehensive research is done in attempt to arrive at systematic framework for strategy evaluation and selection. This paper is an endeavor to utilize ANP for ranking the potential strategy system and making the final selection. This method is applied to an automotive components manufacturing industry, considering 3 potential strategy system namely, current system, system A and system B. ANP method is solved using Super Decision package. In this method, benefits, opportunities, costs and risks factors are considered separately and weight age for each merit is provided basing on the company's strategic policies.

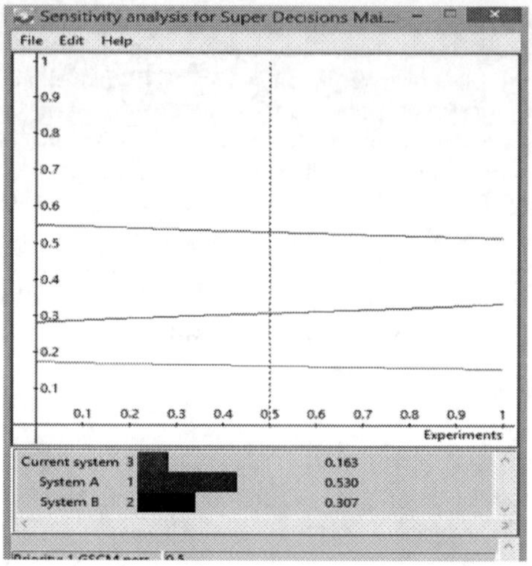

Figure 5.

References

1. Arvind Jayant, M.S.Dhillon (2014) "Use of Analytic Hierarchy Process (AHP) to Select Welding Process in High Pressure Vessel Manufacturing Environment" International Journal of Applied Engineering Research, Volume 10(8) 2015 pp.586-595.
2. Arvind Jayant, M.Khan (2014) "TOPSIS-AHP Based Approach for Selection of Reverse Logistics Service Provider: A Case Study of Mobile Phone Industry" Procedia Engineering (Elsevier Publications), Vo.97, pp 2147-2156
3. Virender Paul and Arvind Jayant (2014), "Analytical Network Process (ANP) in Selection of Green Supplier: A Case Study of Automotive Industry" International Scientific Journal on Science Engineering & Technology, Vo1ume 17, No. 05, pp 453-465.
4. Priya Singh, Arvind Jayant (2015) "Selection of FMS: A Synthesis of 3MCDM Approaches" Journal of Material Science and Mechanical Engineering (JMSME), Volume 2(8), pp 15-20.
5. Veepan Kumar, Arvind Jayant (2015) "Use of AHP to Evaluate Supply Chain Collaboration in Competitive Business Environment" Journal of Material Science and Mechanical Engineering (JMSME), Volume 2(8), pp 15-20.
6. Arvind Jayant (2015) "Evaluation of EOL/Used cell phones management & disposal alternatives: An ANP and balanced score card approach" International Journal of Waste Resources (IJWR), Volume 5, Issue 2
7. Arvind Jayant,A.Singh, and V.Patel (2011), "An AHP Based Approach for Supplier Evaluation and Selection in Supply Chain Management" International Journal of Advanced Manufacturing Systems, Volume 2, No. 1, pp. 1-6
8. Arvind Jayant (2011). "An Application of Analytic Network Process to Evaluate Supply Chain Logistics

Strategies", International Journal of Analytic Hierarchy Process (USA). Vol.4, Issue 1.pp149-163.
9. Chen, shih, wu, shur(2006)-Using ANP for the selection of green supply chain management strategies 47(4)pp;18-20
10. Choudhary, M., & Seth, N. (2011). Integration of green practices in supply chain environment- The practices of inbound, operational, outbound and reverse logistics. International Journal of Engineering Science and Technology, 3(6), 4995-4993. ISSN: 0975-5462
11. Diabat, A., & Govindan, K. (2011).An analysis of the drivers affecting the implementation of green supply chain management. Resources Conservation and Recycling, 55(6), 659-667.
12. L. Meade and J. Sarkis, "Strategic analysis of logistics and supply chain management systems", Transpn Res.-E (Logistics and Transpn Rev.), Vol. 34, No. 3, pp. 201-215, 1998
13. Luthra, S., Kumar, V., Kumar, S., & Haleem, A. (2011). Barriers to implement green supply chain management in automobile industry using interpretive structural modeling technique An Indian perspective. Journal of Industrial Engineering and Management, 4(2), 231-257.
14. Mudgal, R.K., Shankar, R., Talib, P., & Raj, T. (2010). Modeling the barriers of green supply chain practices: An Indian perspective. International Journal of Logistics Systems and Management, 7(1), 81-107.
15. Olugu, E.U., Wong, K.Y., & Shaharoun, A.M. (2011). Development of key performance measures for the automobile green supply chain. Resources, Conservation and Recycling, 55(6), 567-579.
16. Sarkis, J., Hasan, M.A., & Shankar, R. (2007). Evaluating environmentally conscious manufacturing barriers with Interpretive Structural Modeling.
17. Saaty, T.L. (1980). The Analytic Hierarchy Process. New York: McGraw-Hill Book Co.

Appendix A:

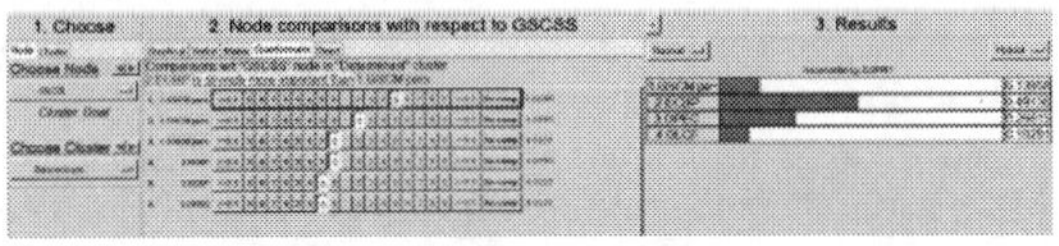

1) Pair-wise comparison of the dimensions

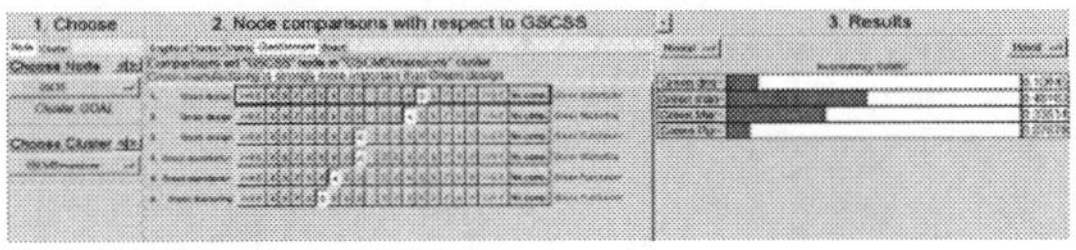

2) Pair-wise comparison of the enablers/components

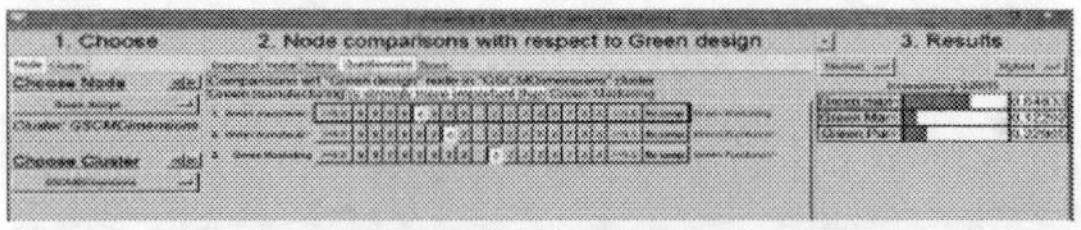

3) Pair wise comparison for the interdependencies

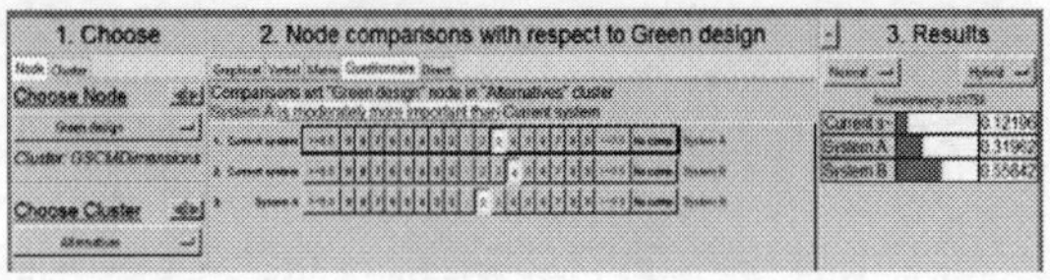

4) Pair wise comparison for the alternatives

Design and manufacturing of abrasive jet machine for drilling operation

Divyansh Mittal[1a], Shankar Sehgal[1] and Harmesh Kumar[1]

Mechanical Engineering Department, UIET, Panjab University, Chandigarh

Abstract- Wide application of Abrasive Jet Machine (AJM) is found in machining hard and brittle materials. Machining of brittle materials by AJM is due to brittle fracture and removal of micro chips from the work piece. Embedment of the abrasive particles in the brittle materials results in decrease of machining efficiency. In this paper design and manufacturing of AJM has been presented. Various parts of AJM have been designed using ANSYS 16.2 software. The parts are then manufactured indigenously as per designed parameters. The machine fabricated in this work will be used further for process optimization of AJM parameters for machining of glass and ceramics.

1 Introduction

Abrasive jet machining (AJM) is one of the most popular machining and surface finishing method among other non-conventional machining methods for drilling, cleaning, deburring, etching etc. of hard and brittle materials without vibrations, heat and stress. As the workpiece do not heat, therefore it does not experience any surface hardening.

In AJM, mixture of abrasive particles (typically ~0.025 mm) and carrier gas is impinged on work piece through a nozzle of diameter 0.2 to 0.8 mm under controlled conditions. Aluminium oxides, silicon carbides, boron carbides, crushed glass, sodium bicarbonate, dolomite are various abrasive particles used in AJM.

In AJM, the pressurized gas is mixed with abrasive particles. This mixture is then passed through the nozzle, thereby converting pressure energy to kinetic energy. High velocity jet is then impinged on workpiece which converts the kinetic energy of the jet to mechanical energy and thus the material is removed by micro cutting action and brittle fracture. In AJM, material removal results in the development of a number of small craters on the surface of the workpiece being machined.

In AJM, the cutting forces acting on the workpiece, which are responsible for machining, are directed downward. Due to this, a material with very small thickness can be machined. Nowadays AJM is being used for deburring external burs. Balasubramaniam *et al.*[1] deburred cross drilled holes by secondary mechanism and found that while deburring cross drilled holes with AJM, abrasive particles must be coarse. With investigation and advancement, AJM was used in machining hard and brittle materials like glass and ceramics.

AJM had been considered most appropriate technology for machining hard and brittle materials and producing micro parts of semiconductors and LCD's. Micro-abrasive jet machining (MAJM) used finer abrasive particles and small diameter nozzle to produce smaller holes and more surface finish. Wensink and Elwenspoek [2] mentioned that when AJM is performed, inclination of sidewalls and blast lag (wide channels become deeper compared to smaller channels) were the reasons to limit the maximum aspect ratio. It was investigated that much steeper sidewalls and very less blast lag could be produced with use of MAJM. Kuriyagawa and Yoshida [3] worked on mechanism of MAJM. Park *et al.*[4] micro-grooved the glass with MAJM and found that accurate square type grooves cannot be formed with MAJM, as grooves formed were of U-type shape. Also due to the ''blast lag effect'' the dimensions of machined surface were slightly larger than masking patterns.

While machining, the abrasive particles may embed in the workpiece. These embedded particles will form a shield for next impinging abrasive particles. Due to this MRR would decrease significantly. As the particle embedding can be undesirable in numerous of applications, Getu *el al.* [5] identified two criteria for particle embedment in erosion process. Firstly, if the contiguous contact between the particle and the target was maintained throughout the impact, embedment was possible. Second was that the magnitude of the static friction forces must reach a critical value for embedment. Particle orientation was defined as the angle between the particle linear velocity vector and the line connecting the centre of mass to the furthest downstream particle vertex. It was observed that certain particle orientation were more favourable to particle embedment.

Hadavi *et al.*[6]worked on finding the process parameters which are dependent on particle orientation. A model

[1a] Corresponding author: divyanshmittal144@gmail.com

capable of predicting the instantaneous particle orientation and velocity within and downstream of nozzle was presented and shown to agree well with measurements.

Zhang *et al.*[7] introduced the concept of micro abrasive intermittent jet machining (MAIJM) for producing small holes. MAIJM was a technology in which abrasive particles were mixed with carrier gas for some interval of time so that the embedded particles were blown out from the surface of material to be machined. MAIJM device was developed and empirical model was established with help of Taguchi orthogonal array and multi-variable linear regression. It was found that the model for MAIJM was highly adequate for producing small holes.

AJM is also combined with many other processes for hybrid machining. Different hybrid machining processes for more precise and accurate machining are abrasive electrical discharge machining, abrasive chemical discharge machining, abrasive discharge grinding and ultrasonic electrochemical machining [8].

In this paper design and fabrication of AJM is presented. Design of different units i.e. air distribution system, abrasive feeder chamber, mixing chamber and nozzle are made in AutoCAD software [9]. Material for the distributor, abrasive feeder and mixing chamber has been decided as mild steel as it has good formability and weldability. The chambers are fabricated as per the design with MIG welding. MIG welding is used because it results to a leak proof joint. Nozzle material is chosen as tungsten carbide due to its high wear resisting property. Note that boron carbide nozzle fabricated with powder metallurgy process can also be used. Fabrication is done such that the flow of abrasive particles in mixing chamber can be stopped manually with the help of a valve. A particular time can be set for which the abrasive particles are mixed with the carrier gas.

2 Principle

In AJM, nozzle angle play a great role in the mechanism of material removal. Nozzle angle can be 90° or it can be oblique. For brittle material, material is removed in accordance with Theory of brittle failure. Therefore 90° angle is kept for material removal in brittle materials. Fig 1 determines the principle i.e. brittle fracture in brittle materials. Due to brittle fracture very small particles are eroded from the surface of the workpiece.

Further the abrasive particles impinging with very high velocity are embedded in the workpiece as shown in fig 2. These embedded particles are responsible for the shielding from next impinging particles which lead to low machining efficiency. To augment the efficiency one need to remove these abrasive particles. This is done by distributing the compressed air in two sections so that the flow of abrasive particles may be closed. Further flow of air on the workpiece without mixing abrasive particles will remove the embedded particles. Fig 3 shows the removal of the embedded particles.

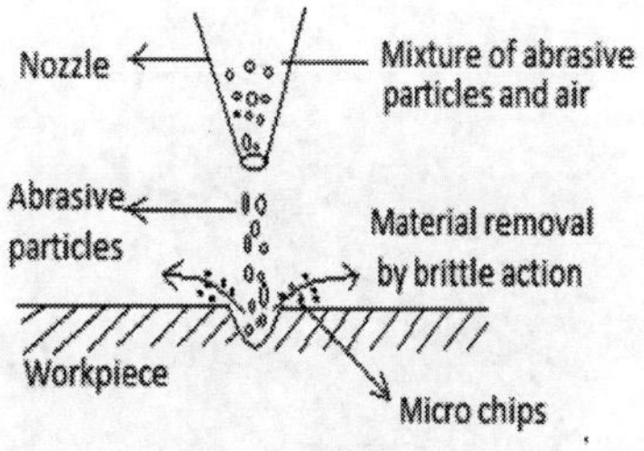

Figure 1. Micro cutting action of brittle material in AJM

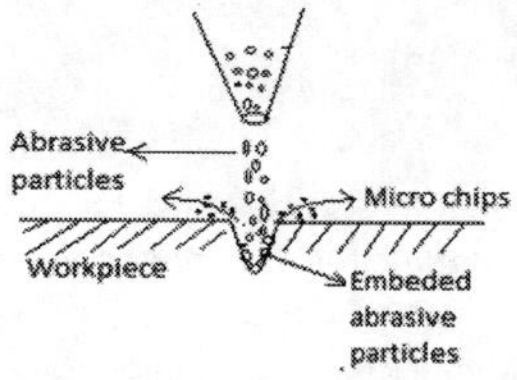

Figure 2. Embedment of abrasive particles in workpiece

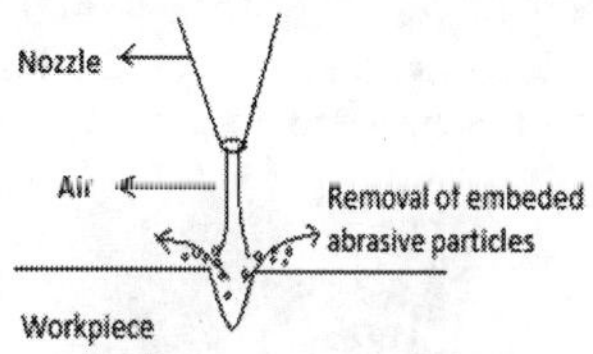

Figure 3. Removal of embedded abrasive particles

3 Design methodology

AJM is fabricated in the institute workshop with required raw materials and procured components. In house air-compressor and filter regulator (FR) unit have been used as accessories of the AJM. Before fabrication, design of each component is made. Different components designed are mixing chamber, abrasive feeder, nozzle, work holding device and glass chamber.

3.1 Air compressor

In air compressors, air is compressed to high pressure with the use of electric motor or internal combustion engine. In AJM high velocity jet is required to impinge on the work piece, which is obtained with highly pressured carrier gas compressed through air compressor. Air compressor will be used on 5 to 6 bar pressure. 5 mm thick sheet is used for the fabrication of distributor, abrasive feeder chamber and mixing chamber. Fig 4 shows the air compressor used for machining.

Figure 4. Air compressor

3.2 FR unit

FR unit is essential for filtering the air coming from compressor and regulating the pressure. FR unit removes the suspended dust particles, moisture and oil contents from the compressed air to prevent coagulation of particles and jam of nozzle opening. These filters have different types of membranes which are porous in nature and are of various sizes i.e. 5, 10, or15 μms. They restrict the particles bigger than the pores of membrane. After FR unit carrier gas is supplied to distributor.

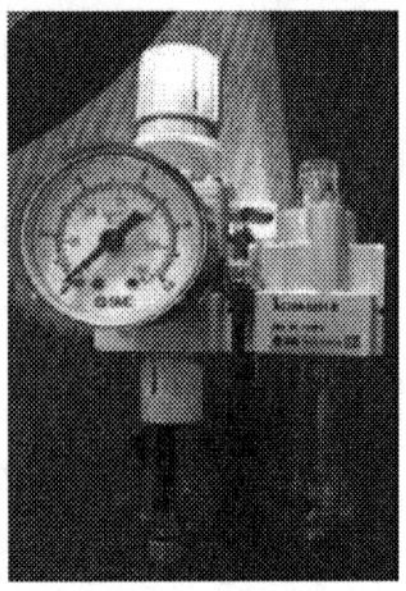

Figure 5. FR unit

3.3 Air distributer

Compressed air from air compressor after FR unit is passed to distributer. Pressure gauge is applied on distributer to measure the pressure of the carrier gas. Figure 6 provides the dimensions of distributor made in AutoCAD.

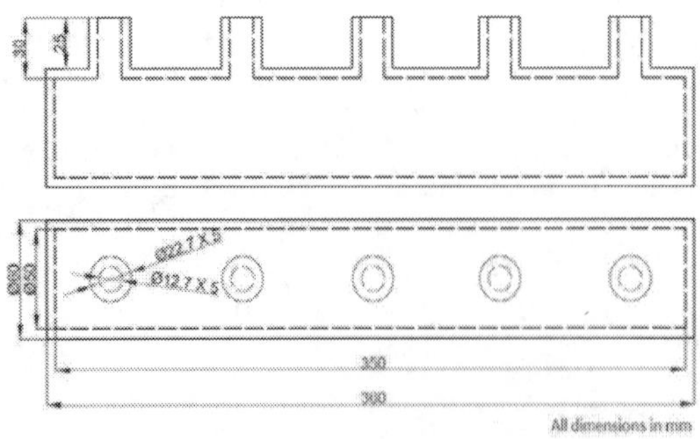

Figure 6. Front & top view of air distributor system

Material used for fabricating distributor is chosen as mild steel. Keeping thickness of the distributor 5 mm, diameter 50 mm and welding efficiency 50% the factor of safety comes out to be 21which is satisfactory. Figure 7 shows the fabricated distributor.

Figure 7. Air distribution system

3.4 Abrasive feeder

Abrasive chamber is fabricated with 5 mm thick sheet of mild steel. The diameter and height of the abrasive feeder chamber are kept 120 mm. Drawing of abrasive feeder chamber is shown in figure 8.

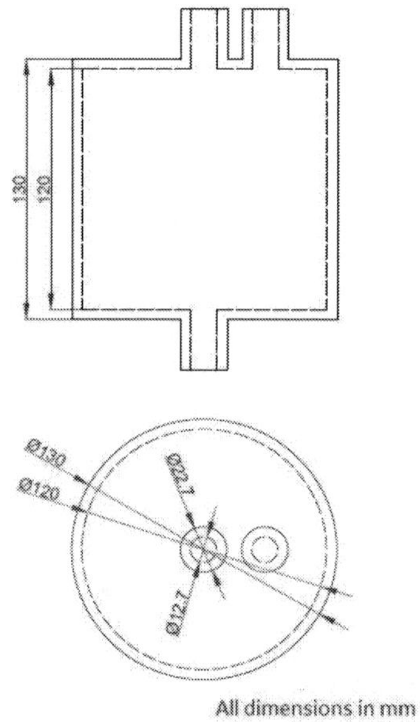

Figure 8. Front & top view of abrasive feeder chamber

Abrasive particles like Al_2O_3 and SiC are used for machining the work piece. For good flowability of abrasive particles, carrier gas i.e. air will be mixed with abrasive particles in abrasive chamber. This mixing is for the better flow of abrasive particles which will further be mixed with air in mixing chamber. Abrasive particles are passed to the mixing chamber with ball valve opening. This valve is used for restricting the flow of abrasive particles in the mixing chamber when only air is needed for removing the embedded abrasive particles. Fabricated abrasive feeder chamber is shown in figure 9.

Figure 9. Abrasive feeder chamber

3.5 Mixing chamber

The fabrication of mixing chamber is done such that the carrier gas would enter from three different inlets at 120° angle and abrasive particles would enter from the top of the mixing chamber. This would result in cyclone formation inside the mixing chamber and homogeneous mixture of abrasive particles and carrier gas will be obtained. Figure 10 shows design of mixing chamber.
The thickness of the mixing chamber is also kept 5 mm. Figure 11 shows the fabricated mixing chamber. The converging section of the mixing chamber will ensure the increase in velocity of the mixture. Further increase in velocity is done in nozzle to achieve cutting velocity.

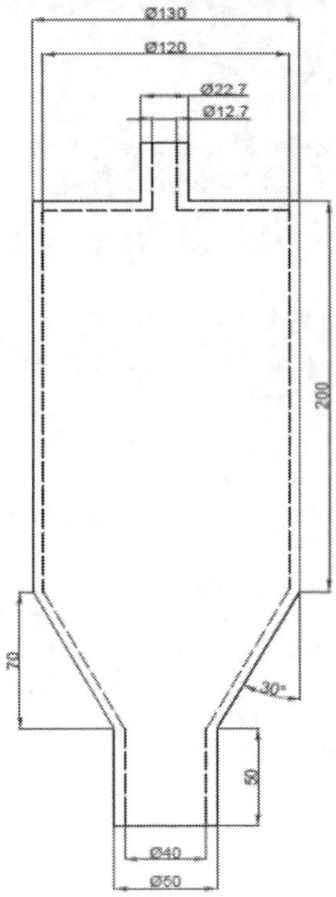

(a)

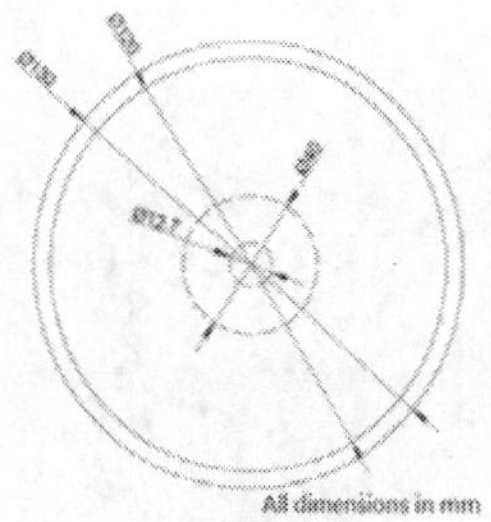

(b)

Figure. 10 (a) Front & (b) top view of mixing chamber

Figure 11. Mixing chamber

3.6 Nozzle

Nozzles are the devices which increases the kinetic energy of the fluid in exchange to the pressure drop. Nozzles create high velocity jet which is impinged on the material to be machined. As the nozzle wear is a big issue in AJM, tungsten carbide or sapphire material nozzle is used. Tungsten carbide nozzle can cut square, rectangular and circular section but sapphire nozzles can only cut circular sections. Average life of nozzle made of tungsten carbide is 12 to 20 hr whereas nozzle made of sapphire has an average life of 300 hr. We will use tungsten carbide nozzle for the experimentation due to the ease of the availability and low cost in comparison to sapphire nozzle. SOD of 0.5 mm to 1.5 mm will be maintained. Design of nozzle is made in AutoCAD as shown in figure 12. Inlet and outlet diameter of nozzle is kept 8 mm and 0.8 mm respectively. Length of nozzle is 30 mm.

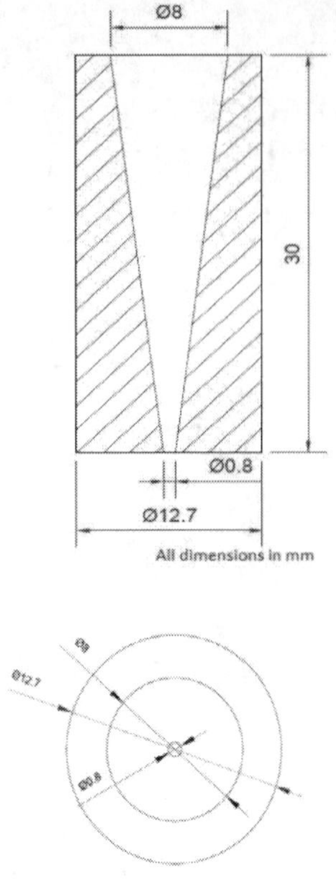

Figure 12 (a) Front & (b) top view nozzle

Geometrical model created in AutoCAD was imported into the ANSYS as shown in figure 13. Meshing was done using auto mesh command. Mesh of inlet fluid was created as shown in figure 14. Inlet velocity was kept 10m/s. Result was converged after 32 iterations and outlet velocity came out to be 1005.3 m/s. As per the theory, in AJM the outlet velocity of 300-700 m/s is required. Considering losses this velocity can be achieved at nozzle outlet required for machining.

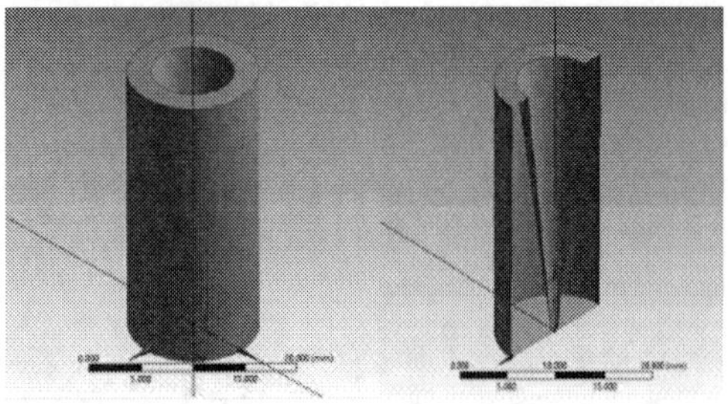

Figure 13. Isometric and section view of nozzle

Nozzle wear is a big issue while performing AJM. Internal surface of nozzle gets eroded due to the collision of particles with the surface. A new material or material with enhanced corrosion resistance property can be made with help of powder metallurgy process. With high corrosion resistance, operating time for nozzle would be more and thus machining time would be less.

3.7 Assembly

Highly compressed air at about 5 bar pressure is incorporated in FR unit. 8 mm PU pipe is used for connecting compressor and FR unit and also is used for further connections. Further FR unit supplies the air to the distributor. The distributor is mounted on the vertical stand. Distributor provides two supplies, one to the abrasive feeder and second to the mixing chamber. Ball valves are used to supply carrier gas to the abrasive feeder and mixing chamber. One safety valve is also incorporated on the distributer. Abrasive feeder chamber and mixing chamber are connected with sockets and ball valve and are mounted on the stand in the vertical line. The opening of the mixing chamber is provided with the female thread for providing the connection with the nozzle. The nozzle with the male thread is connected with the mixing chamber.

Figure 14. Assembly AJM

4 Conclusion

In this paper the design and fabrication of AJM is presented. The embedment of the abrasive particles was found to be the major reason of decrease in machining efficiency. In fabrication of AJM, abrasive feeder chamber and mixing chamber are incorporated in such a way that they receive the air separately through the air distribution system. This allows the restriction of abrasive particles from mixing in air when required. This air which does not contain the abrasive particles remove the embedded abrasive particles which will result into better machining efficiency.

References

1. R. Balasubramaniam, J. Krishnan, and N. Ramakrishnan, "Experimental study on the abrasive jet deburring of cross-drilled holes," *J. Mater. Process. Technol.*, vol. **91**, no. 1, pp. 178–182, (1999).
2. H. Wensink and M. C. Elwenspoek, "Reduction of

sidewall inclination and blast lag of powder blasted channels," *Sensors Actuators, A Phys.*, vol. **102**, no. 1–2, pp. 157–164, (2002).

3. T. Kuriyagawa and N. Yoshida, "T Kuriyagawa and Yoshida," *JSPE*, vol. **54**, no. 6, pp. 881–885, (1998).

4. D. S. Park, M. W. Cho, H. Lee, and W. S. Cho, "Micro-grooving of glass using micro-abrasive jet machining," *J. Mater. Process. Technol.*, vol. **146**, no. 2, pp. 234–240, (2004).

5. H. H.Getu, J. K. Spelt, and M. Papini, "Conditions leading to the embedding of angular and spherical particles during the solid particle erosion of polymers," *Wear*, vol. **292–293**, pp. 159–168,(2012).

6 V. Hadavi, B. Michaelsen, and M. Papini, "Measurements and modeling of instantaneous particle orientation within abrasive air jets and implications for particle embedding," *Wear*, vol. **336–337**, pp. 9–20, (2015).

7 L. Zhang, T. Kuriyagawa, Y. Yasutomi, and J. Zhao, "Investigation into micro abrasive intermittent jet machining," *Int. J. Mach. Tools Manuf.*, vol. **45**, no. 7–8, pp. 873–879, (2005).

8 Divyansh Mittal, Shankar Sehgal, and H. Kumar, "abrasive jet machining and its applications," Int. J. Sci. Res., vol. **2277**, pp. 76–78, (2015).

9. "User manual for AutoCAD, 2016", publisher Autodesk, USA.

Effect of Welding Speed and Tool Pin Geometry on Impact Strength in Friction Stir Welding of Aluminium 6101 T6 Alloy

Rajbir Singh[1], Sanjeev Kumar[2] and Preet Kanwal Singh[3]

[1]*Research scholar, Mechanical Department, BCET Gurdaspur, India.*
[2]*Associate professor, Mechanical Department, BCET Gurdaspur , India.*
[3]*Assistant professor, Mechanical Department, BCET Gurdaspur , India.*

Abstract. Friction stir welding (FSW) process is a solid state joining method in which the material that is being welded does not melt and recast. This process uses a non-consumable tool to generate frictional heat in the abutting surfaces. The welding parameters play a major role in deciding the joint characteristics. In this investigation the effect of welding speed and tool pin geometry in friction stir welds of Aluminium alloy was studied. The welded joints were made of Aluminium alloy AA 6101 T6 strips of 6mm thickness with different tool pin profile (Cylindrical, Taper cylindrical, Square and Taper Square). All the welding operations were done at the room temperature. Charpy tests were carried out to find the impact strength. From this investigation it is found that strength is superior with taper square tool pin profile than cylindrical, taper cylindrical and square tool pin profile. The pieces weld at rotational speed of 1200 r.p.m and welding speed of 70mm/min using taper square tool have higher strength.

1 Introduction

The usage of aluminium alloys in automotive and aerospace applications is ever increasing due to their light weight and cast ability. It is necessary to weld aluminium castings to themselves or to wrought alloys for various applications. These alloys are not easily weldable by conventional fusion welding technique because the quality of the welded joint is deteriorated due to the presence of porosity, hot cracking and distortion. Friction stir welding (FSW) process is an emerging solid state joining process in which the material that is being welded does not melt and recast [1].The principle of FSW process is quiet simple. It makes the use of non-consumable tool which consists of a shoulder and a pin [2]. The pin of the rotary tool is plunged into the plates being welded till shoulder makes firm contact and then moved forward along the weld line. The frictional heat generated by the rotary tool heats and softens the material. The softened material underneath the shoulder is subjected to extrusion by the rotary and traverse movement of the tool; is transported from the advancing side to the retreating side where it is consolidated into a monolithic joint [3]. FSW creates the weld joint without bulk melting. In addition, the extensive thermo mechanical deformation induces dynamic recrystallization and recovery that refine the microstructure of the stir region. Therefore, welds made by FSW are shown to have much improved mechanical properties than the corresponding fusion welds. Since, FSW is a solid state process and the material under the rotating action of non-consumable tool has to be stirred properly to get good, defect free welds [4]. Defect-free welds with good mechanical properties have been made in a variety of aluminium alloys, even those previously thought to be non weldable [5]. When alloys are friction stir welded, phase transformations that occur during the cooling cycle of the weld are of a solid state type. Due to the absence of parent metal melting, the FSW process is observed to offer several advantages over fusion welding processes [6]. During the welding process, frictional heat associated with the thermal cycle varies in the transverse direction of the weld. The maximum temperature is observed in the FSW zone, which causes an alteration in the precipitate distribution present in the base material and also due to stirring of the plasticized material. These changes in the heat and temperature distribution in the welding process alter the strength and ductility of the joints [7].

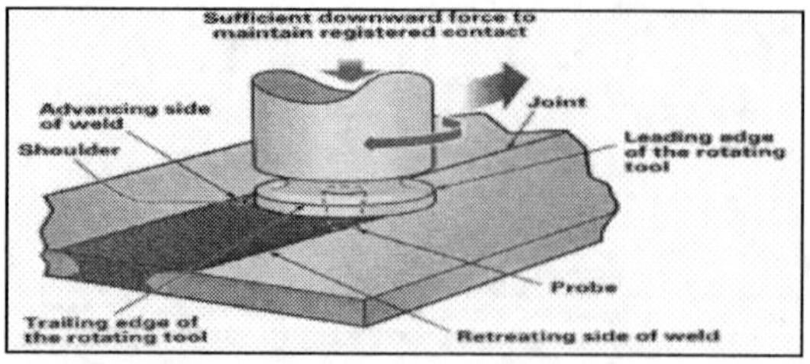

Figure 1. FSW Process

[1]Corresponding author: rajbir239@gmail.com

The pin/shoulder geometry of the tool is responsible for homogeneous microstructure, uniform joint properties and governs process forces. Whorl and MX Triflute tool geometries reduce welding forces because of increased heat generation and lower displacement volumes of the softened materials [8]. Rotary speed of tool determines the amount of heat produced per unit time, stirring and mixing of the workpiece material around the pin [9]. Welding speed or tool traverse speed governs the maximum temperature generated during welding and the length of time the material is subjected to the welding [10]. The translation of the tool transports the stirred material from the front to the back of the pin to complete the welding [11]. High welding speed will make friction stir welded joints cheaper due to high productivity. However, too high welding speed at constant rotary speed results in the formation of void beneath the top surface of weld or on the advancing side at the edge of weld nugget. The size of the wormhole defects increases with the increase of welding speed [12]. Therefore, this present study aims to experimentally examine the effect of varying welding speed on impact strength of friction stir welded joints of AA6101 aluminum alloy.

2 Experimental work

The extruded plates of AA6101 of 6 mm thick in T6 condition were used as the base material for the present experimental study. The plates were cut to the required dimensions (300×150 mm) by power hacksaw cutting and grinding. Four tools (Cylindrical, Taper cylindrical, Square and Taper square) were prepared having different tool pin geometry. Tool having cylindrical shaped pin and tapered cylindrical pin were prepared using Lathe machine while square and tapered square pin were prepared on Milling machine. The shoulder diameter of the tool is 18mm and diameter of pin is 6mm. The length of the pin should be slightly less than thickness of the work piece. The thickness of work piece is 6mm and pin length is 5.90mm. A square butt joint configuration was prepared to fabricate FSW joints. The initial joint configuration was obtained by securing the plates in position using mechanical clamps. The direction of welding was normal to the rolling direction. A single pass welding procedure was used to fabricate the joints.Non consumable tools made of high carbon high chromium steel were used to fabricate the joints. The chemical composition of the base metal was checked at IMTT Batala and the composition of various elements is shown in Table1.

Table 1. Chemical Composition of AA 6101 T6.

Element	Percentage
Al	97.6 min
Si	0.60 max
Fe	0.35 max
Cu	0.10 max
Mn	0.10 max
Mg	0.90 max
Cr	0.10 max
Zn	0.10 max
Ti	0.10 max
Others	0.15 max

A Vertical Milling machine, HAAS Super VF2 (12000 rpm) available with R & D centre, Ludhiana was used to fabricate the joints. Joints were fabricated at constant Rotational speed of 1100 r.p.m. and translation speed was taken as 40mm/min, 55mm/min, 70mm/min, and 85mm/min respectively. Four joints were fabricated by using each tool. The process parameters and their levels and tool dimensions are presented in Table 2.

Table 2. FSW process parameters

Process parameters	Values
Tool rotation speed (rpm)	1100
Welding speed (mm/min)	40, 55, 70, 85
Pin Length (mm)	5.90
Tool shoulder Diameter (mm)	18
Pin diameter (mm)	6
D/d ratio of tool	3
Tool pin geometry	Cylindrical
Tool material	High carbon steel

The welded joints were sliced using band saw and then machined to the required dimensions to prepare specimens. American Society for Testing of Materials (ASTM) guidelines were followed for preparing the test specimens. Charpy test was performed to find the impact strength of various specimens. Figure 2 shows the starting stage of Friction stir welding process and Figure 3 shows the welded Specimen

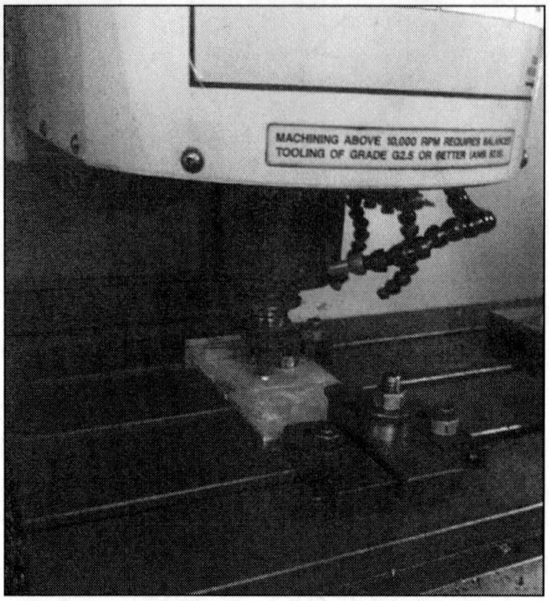

Figure 2. Starting Stage of FSW Process

Figure 3. Welded Specimen at 1100 rpm, 70mm/min

3 Results and discussion

Impact strength of friction stir welded joints was measured by performing Charpy tests on transverse tensile samples containing weld nugget zone at the center. Table 3 represents the impact strength of friction stir welded joints of AA 6101T6 aluminum alloy, produced using different welding speed and various tool pin shape.

It is evident from Table 3 that impact strength of friction stir welds of AA 6101T6 is inferior to base material. The joint fabricated by Taper square tool shows superior impact strength as compared to all other joints while the joints fabricated by cylindrical pin profiled tool shows inferior strength as compared to all other joints. At constant rotary speed of 1100 rpm, increase in welding speed from 40 to 70 mm/min increases the impact strength linearly from 31 J to 34 J and then decreases to 32 J at welding speed of 85 mm/min for a cylindrical pin profiled tool (Figure. 2). The maximum impact strength of weld joint was 34 J for cylindrical pin profiled tool. The impact strength of taper cylindrical pin profiled tool was better as compared to cylindrical pin profiled tool. The impact strength increases from 33J to 38 J as welding speed increases from 40mm/min to 85mm/min. the impact strength of square pin profiled tool is superior than cylindrical and taper cylindrical pin but is inferior to Taper square pin profiled tool. The impact strength increases from 41J to 48 J as welding speed increases from 40mm/min to 70mm/min for a taper square pin profiled tool and then decreases to 40J at welding speed of 85mm/min. The impact strength is maximum for a Taper square pin profiled tool at welding speed of 85mm/min. The welding speed influences the heat input per unit length of weld which controls the degree of softening and flowability of plasticized material. At lower welding speed the amount of heat supplied to the deforming material in weld area is greater and therefore wider is the softened area around the stirring tool leading to more improved metal flow and more effective bonding in the weld. This improved material flow and effective bonding leads to more homogeneity of weld nugget which results in superior impact strength of welded joints having higher heat input per unit length. With further increase in welding speed, the impact strength of friction stir welded joints decreases and attributed towards formation of voids due to poor consolidation of weld interface at higher welding speed hence low heat input per unit weld length.

Table 3. Impact Strength of Fabricated Joints.

Sample No.	Welding Speed	Shape of Tool	Impact Strength
1	40	Cylindrical	31
2	55	Cylindrical	33
3	70	Cylindrical	34
4	85	Cylindrical	32
5	40	Taper Cylinder	33
6	55	Taper Cylinder	35
7	70	Taper Cylinder	36
8	85	Taper Cylinder	38
9	40	Square	34
10	55	Square	36
11	70	Square	39

12	85	Square	35
13	40	Taper Square	41
14	55	Taper Square	42
15	70	Taper Square	48
16	85	Taper Square	40

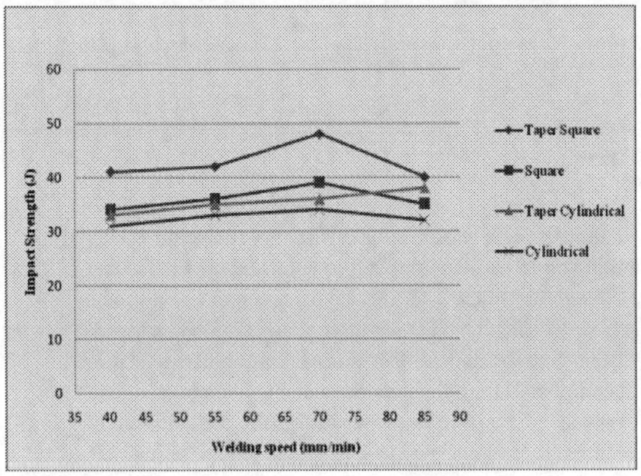

Figure 4. Effect of Welding Speed and Pin profile on Impact Strength

Pin profile plays an important role in material flow and in turn regulates the welding speed of Friction stir welding process.

Pin profiles with flat faces (square and taper square) are associated with eccentricity. This eccentricity allows incompressible material to pass around the pin profile. Eccentricity of the rotating object is related to dynamic orbit due to eccentricity. The relationship between the static volume and dynamic volume decides the path for the flow of plasticized material from the leading edge to the trailing edge of the rotating tool. This ratio is highest for taper square tool followed by square, taper cylindrical and cylindrical pin profiled tool. In addition, the taper square and square pin profiled tool produce a pulsating stirring action in flowing material due to flat faces. There is no such pulsating action in case of cylindrical and taper cylindrical pin profiled tool. The highest number of pulsating action experienced in stir zone of Taper Square pin profiled tool produces fine grains and in turn yields highest impact strength.

4 Conclusions

In this investigation an attempt has been made to study the effect of welding speed on the formation of friction stir processing zone in AA 6101T6 aluminium alloy. From this investigation, the following important conclusions are derived:

(i) Friction stir welding technique can be used to produce defect free weld of AA 6101T6.

(ii) Of the four pin profiles used in this investigation to fabricate the joints, taper square pin profiled tool gives better strength as compared to cylindrical, taper cylindrical and square pin profiled tool.

(iii) The tool pin profile and welding speed have influence on impact strength of friction stir welding joints. The impact strength of the joints increases with increase in welding speed up to a certain limit and then decreases. Of the various joints fabricated in this investigation, the joint fabricated at rotational speed of 1100rpm and welding speed of 70mm/min using taper square pin profiled tool has highest impact strength.

References

1. W. M. Thomas, E. D. Nicholas, J. C. Needham, M. Murch, G. P. Templesmith, C. J. Dawes, G.B. Patent Application No. 9125978.8 (December 1991).
2. A. P. Reynolds, "Visualization of material flow in autogenous friction stir welds", Sci Technol Weld Joining 2000;5:120
3. K. Colligan, "Material flow behavior during friction welding of aluminum," Weld J 1999:229s–37s.
4. Y. S. Sato, H. Kokawa, K.. Ikeda, M. Enomoto, S. Jogan, and T. Hashimoto, "Microtexture in the friction-stir weld of an aluminum alloy", Metall Mater Trans A2001; 32A:941–8.
5. W. M. Zeng, H. L. Wu and J. Zhang, "Effect of tool wear on microstructure, mechanical properties and acoustic emission of friction stir welded 6061 Al alloys", Acta Metall Sinica 2006; 19(1):9–19.
6. O. V. Flores, "Micro structural issues in a friction stir welded aluminium alloy", Scripts Mater 1998; 38(5):703–8.
7. G. Ipekoglu & S. Erim and G. Cam (2013) "Effects of temper condition and post weld heat treatment on the microstructure and mechanical properties of friction stir butt-welded AA7075 Al alloy plates" International Journal Advance Manufacturing Technology (2013)
8. W. M. Thomas, E. D. Nicholas and S. D. Smith, Friction stir welding – Tool developments. In: S. K. Das, J. G. Kaufman, T. J. Lienert, editors. Aluminium 2001 proceedings of the TMS 2001 aluminum automotive and joining session, TMS;
9. M. Peel, A. Steuwer, M. Preuss, and P. J. Withers, "Microstructure, mechanical properties and residual stresses as a function of welding speed in AA5083 friction stir welds", Acta Materilia 2003;51(16):4791–801.
10. H. J. Liu, H. Fuji, M. Maeda and K. Nogi, "Mechanical properties of friction welded joints of 1050-H24 aluminum alloy", Sci Technol Weld Joining 2003; 8(6):450–4.2001. p. 213–24.
11. R. Nandan, D. Roy, H. K. D. H Bhadeshia, "Recent advances in friction stir welding – Process, weldment structure and properties", Progress Mater Sci 2008; 53:980–1023.
12. R. Crawford, G. E. Cook, A. M. Strauss, D. A. Hartman, and M. A. Stremler, "Experimental defect analysis and force prediction simulation of high weld pitch friction stir welding", Sci Technol Weld Joining 2006;11(6):657–65.

Optimization of the machining parameters for EDM wire cutting of Tungsten Carbide

Gaurav Kumar[1,a] , J S Tiwana[2] and Ashish Singla[3]

[1]*Assistant Professor, Mechanical Engg. Deptt, BGIET, Sangrur, India*
[2]*Assistant Professor, Mechanical Engg. Deptt, GZSCCET, Bathinda, India*
[3]*Assistant Professor, Mechanical Engg. Deptt, BGIET, Sangrur, India*

Abstract. Electrical discharge machining is a thermal erosion process which is based on thermoelectric energy between the work piece and electrode. Its capability of machining in hard and difficult to cut materials has made it most popular. Tungsten carbide is widely used in industry due to its unique properties combination of hardness and wear resistance. But the machining of the tungsten carbide is very difficult. In our research we tried to find optimized procedure to cut the Tungsten carbide by variation of different parameters so that process can be carried out in maximum Material Removal Rate (MRR) with better surface finish. During the experimentation, Brass wire is used for cutting.. The experiment is designed with the help of Taguchi method and further analysis done with the help of ANOVA. In this experiment we find that the minimum mean of surface roughness is 2.214 achieved at Tension- 0.6 N, Feed- 10m/min, Flushing Pressure- 3kg/cm^2, Current- 80A.

Key Words: EDM Drilling, Tungsten carbide, Surface roughness, Machining of hard materials

1 Introduction

Electro Discharge Machining (EDM) is a non-contact electro-thermal machining process, where electrical energy is used to generate electrical spark and material removal mainly occurs due to thermal energy of the spark. Precise machining can be done on electrically conductive and semi-conductive materials using this unconventional machining process. EDM can be used to drill circular and non-circular holes, generate profiles and make complex shaped dies of both macro and micro sizes. Both the macro EDM and the micro EDM have great potential and research work is going on in this field to improve the machining process and equipment [11]

1.1 Principle of EDM

EDM is a non-mechanical thermal shaping process with which material is removed by spatially and temporally separated electrical discharges between a workpiece electrode and a tool electrode. The high frequency discharges cause melting and vaporization of material on the surface of both electrodes. To enhance the material removal EDM operates in a non-conducting fluid, the dielectric fluid (Figure 1).

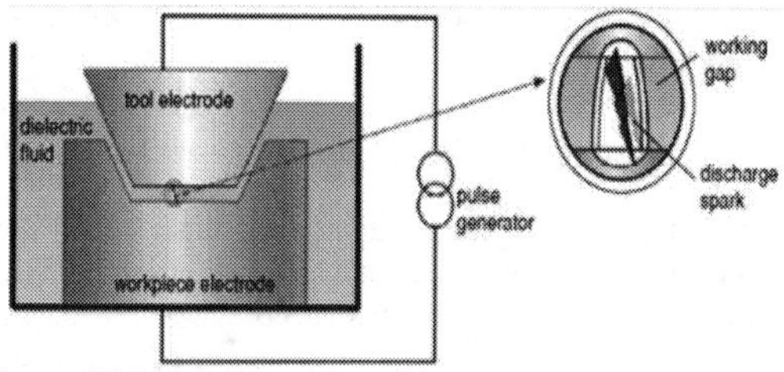

Figure 1 Principle illustration of the EDM process.

During machining, the tool and workpiece electrode are positioned in such a way that a working gap is filled with dielectric fluid remains between them.
EDM spark erosion is the same as having an electrical short that burns a small hole in a piece of metal it contacts. With the EDM process both the workpiece material and the electrode material must be conductors of electricity.

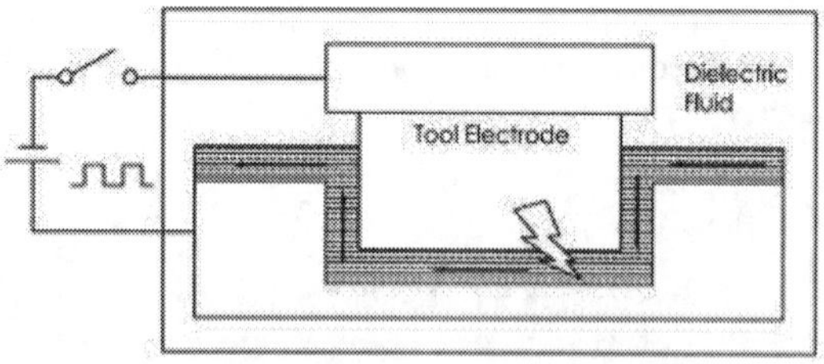

Figure 2. Sparking between the tool and work piece

The localized extreme rise in temperature leads to material removal. Material removal occurs due to instant vaporization of the material as well as due to melting. The molten metal is not removed completely, but only partially.

[a] Corresponding author: gaurav1588@hotmail.com

The EDM process can be used in two different ways:
1. A pre-shaped or formed electrode (tool), usually made from graphite or copper, is shaped to the form of the cavity it is to reproduce. The formed electrode is fed vertically down and the reverse shape of the electrode is eroded (burned) into the solid workpiece.
2. A continuous-travelling vertical-wire electrode, the diameter of a small needle or less, is controlled by the computer to follow a programmed path to erode or cut a narrow slot through the workpiece to produce the required shape.

Wire-Cut EDM

The wire-cut EDM is a discharge machine that uses CNC movement to produce the desired contour or shape. It does not require a special shaped electrode; instead it uses a continuous-traveling vertical wire under tension as the electrode. The electrode in wire-cut EDM is about as thick as a small diameter needle whose path is controlled by the machine computer to produce the shape required.

2 Experimental setup

EDM machine Robofil 290, Charmilles Technologies was used as the experimental machine in this study. A Brass wire of diameter of 0.125mm is used for cutting the work piece of Tungsten Carbide. The gap between work piece and electrode was flooded with a moving dielectric fluid.

Table 1 Specifications of EDM Wire Cut M/C Robofil-290 Charmillies

Technical data of EDM wire cut m/c robofil-290 charmillies	
Table size	800 x 500 mm
X, y, z movement	400 x250 x200 mm
Max load capacity	500 kg
Coolant	Distilled water
Wire to be used	Dia. 0.1 - 0.25 mm
Accuracy	+/- 0.01mm
Water tank capacity	300 liters
Cutting technology	Steel, copper, aluminium, carbide, graphite
Air pressure	6-7 bar
Electricity	400 volts, 50 Hz
Speed	0-15 meter/min

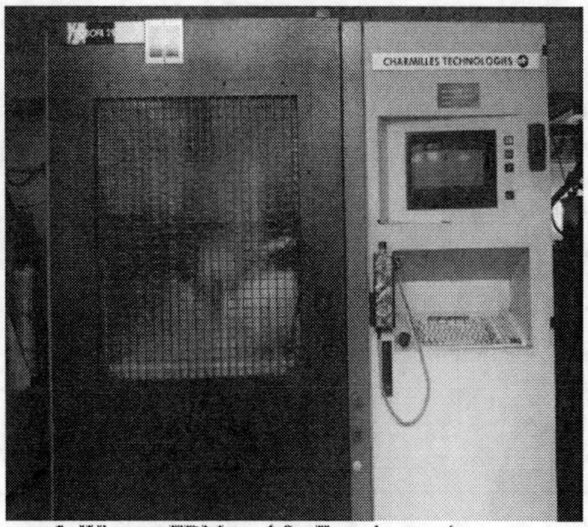

Figure 3. Wire cut EDM used for Experimentation

Properties of Tungsten Carbide

The chemical composition of Tungsten Carbide, which is tested at "Institute of Auto parts and hand tools technology, Ludhiana", is shown below in table 1.

Table 2. Chemical Composition of Tungsten Carbide

W %	Ni %	Co %	Fe %	Cr %
87.09	3.89	6.12	1.84	0.29

Experimental Procedure for Wire cut EDM

Taguchi approach was taken as the basis for planning and conducting the experiments so that the appropriate data is collected which may be analyzed to obtain valid and objective conclusions.

Table 3. Range of Significant Process Parameter for EDM Wire cut

Parameter Name	Unit	Lower Limit	Upper Limit
Tension	**N**	**0.6**	**1**
Feed	**m/min**	**8**	**12**
Flushing Pressure	**Kg/cm^2**	**2**	**4**
Current	**A**	**60**	**100**

3 Results and Analysis of Surface Roughness

The Cutting is done on different parameter as per Design of experment. Then surface roughness is checked. In the present work Mitutoyo stylus profilometer surface roughness measurement device is used, with stylus having diamond tip. After having surface roughness following results are found.

Table 4. Observations of SR during Cutting of Tungsten Carbide

S. No	SR1	SR2	Mean SR	S/N	MEAN
1	2.461	2.039	2.25	-7.04365	2.25
2	2.413	2.733	2.573	-8.2088	2.573
3	2.597	2.959	2.778	-8.87465	2.778
4	3.014	2.624	2.819	-9.0019	2.819
5	2.357	2.505	2.431	-7.7157	2.431
6	2.16	2.268	2.214	-6.90356	2.214
7	2.479	2.359	2.419	-7.6727	2.419
8	2.884	2.432	2.658	-8.4911	2.658
9	2.274	2.372	2.323	-7.32098	2.323

Table 5: Results of SR, and S/N ratios of SR

Level	Tension	Feed	Flushing Pressure	Current
1	-7.207	-8.389	-8.252	-7.570
2	-8.229	-7.478	-7.703	-8.260
3	-8.309	-7.878	-7.789	-7.914
Delta	1.102	0.911	0.549	0.690
Rank	1	2	4	3

Table 4 & 5 represents the investigated results obtained during cutting of Tungsten Carbide by utilizing EDM. The results i.e. SR and S/N ratio were obtained.

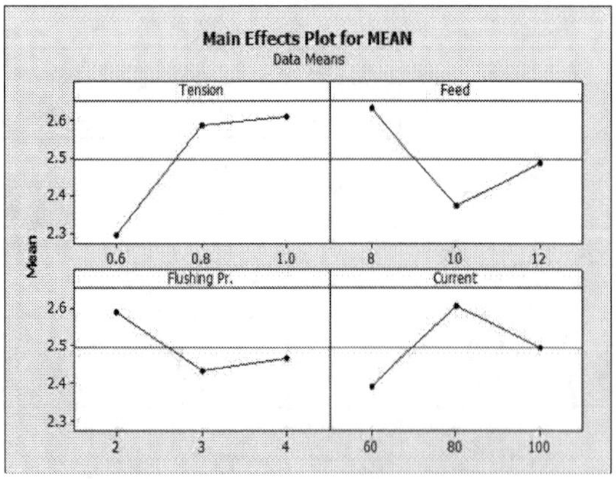

Figure 4 effect of EDM Parameters on Mean SR

Figure 4 shows effect of various EDM parameters on the surface roughness plotted utilizing the machining results obtained. From the figure, it is observed that the mean value of SR is increase by increasing the Tension. Increase the feed decreases the SR upto a certain limit after that SR starts increasing. SR decreases with increase in flushing pressure and remains constant after a certain value.

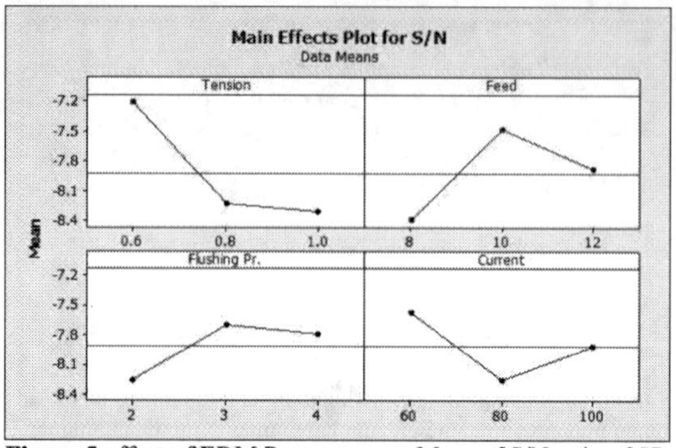

Figure 5 effect of EDM Parameters on Mean of S/N ratio of SR

Figure 5 shows effect of various EDM parameters on the mean S/N ratio of surface roughness rate plotted utilizing the machining results obtained. From the figure, it is observed that the mean value of S/N ratio of SR is decrease by increasing the Tension from 0.6 to 1. The means of S/N for SR increases with increasing the flushing pressure value from 2 to 3 and then decreases very slightly from 3 to 4. The reverse of this happens with the current.

Taguchi Analysis: SR versus Tension, Feed, Flushing Pressure and Current

Response Table for Signal to Noise Ratio smaller is better and SR

Table 6 Ranking of EDM Parameters by Response of S/N ratio of SR

Level	Tension	Feed	Flushing Pressure	Current
1	2.294	2.632	2.59	2.391
2	2.586	2.37	2.434	2.604
3	2.608	2.486	2.464	2.494
Delta	0.314	0.262	0.156	0.213
Rank	1	2	4	3

Table 7. Ranking of EDM Parameters by Response of mean of SR

S. No	Tension	Feed	Flushing Pressure	Current	SR1	SR2	Mean SR
1	.6	12	4	100	2.461	2.039	2.25
2	1	10	2	100	2.413	2.733	2.573
3	.8	12	2	80	2.597	2.959	2.778
4	1	8	4	80	3.014	2.624	2.819
5	1	12	3	60	2.357	2.505	2.431
6	.6	10	3	80	2.16	2.268	2.214
7	.6	8	2	60	2.479	2.359	2.419
8	.8	8	3	100	2.884	2.432	2.658
9	.8	10	4	60	2.274	2.372	2.323

Above tables 6 and 7 shows the ranking of EDM parameters for optimizing the SR. It can be observed that Peak Current has the largest effect on the SR of Tungsten Carbide by machining. The pulse on has the smallest effect on the SR.

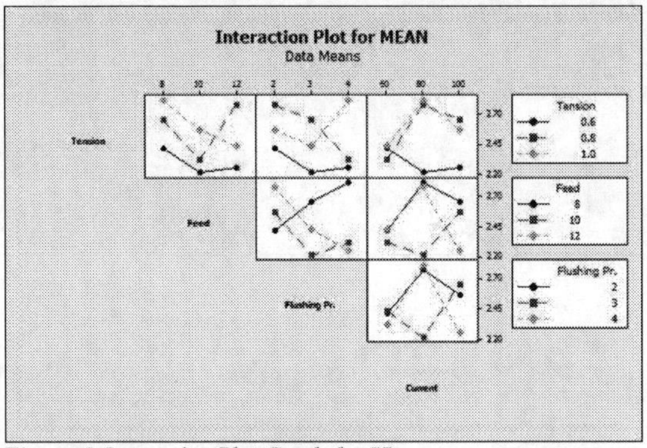

Figure 6. Interaction Plot Graph for SR

- 1st graph is between Tension and Feed. The minimum value of SR is at Tension=0.6 and Feed= 10.
- 2nd graph is between Tension and Flushing Pressure. The minimum value of SR is at Tension=0.6 and Flushing Pressure = 3.
- 3rd graph is between Tension and Current. The minimum value of SR is at Tension=0.6 and Current= 80.
- 4th graph is between Feed and Flushing Pressure. The minimum value of SR is at Feed=10 and Flushing Pressure=3
- 5th graph is between Feed and Current. The minimum value of SR is at Feed=10 and Current= 80.
- 6th graph is between Flushing Pressure and Current. The minimum value of SR is at Flushing Pressure =3 and Current= 80.

Table 8. Confirmation Experiment Result for SR

Performance Measures/ Response	Optimal Set of Parameter	Experimental Value
SR	Tension- 0.6N Feed- 10m/ min Flushing Pressure- $3kg/cm^2$ Current- 80A	2.214

From confirmation test we can concluded that our experimental value of surface roughness is less than the predicted value which means surface roughness achieved Tension- 0.6 N, Feed- 10m/min, Flushing Pressure- $3kg/cm^2$, Current- 80A is optimum.

Conclusions of Surface Roughness for EDM wire cut

1. It is noted that the minimum mean of surface roughness is 2.214 achieved at Tension- 0.6 N, Feed- 10m/min, Flushing Pressure- $3kg/cm^2$, Current- 80A.
2. SR is increase by increasing the peak current. Because Increase the peak current increase the discharge energy.

Scope of Future

In this present work MRR and SR of machined product was studied during the machining of Tungsten carbide using EDM can futher be carried out to study the machining characterstic known as Hardness of machined surface, Properties of HAF etc.

References

1. S. H. lee, X. P. Li: [2001], "Study of the effect of machining parameters on the machining characteristics in electrical discharge machining of tungsten carbide"; Journal of material Processing Technology115, pp 344-358.

2. C. COGUN, S. Akaslan [2002], " The effect of machining parameters on Tool edge wear & machining performance in EDM" KSME International Journal, vol, 16, No.1,pp 46-59

3. S.H. Lee∗, X. Li, [2003] "Study of the surface integrity of the machined workpiece in the EDM of tungsten carbide"; Journal of Materials Processing Technology 139, pp 315–321

4. P.M. George, B.K. Raghunath , L.M. Manocha, A. M. Warrier [2003] "EDM machining of carbon–carbon composite—a Taguchi approach", Journal of Materials Processing Technology, vol. 145, pp 66–71

5. N. Tosun, C. Cogun, A. Inan [2003] "The Effect of Cutting Parameters on Workpiece Surface Roughness in Wire EDM," Machining Science and Technology, vol. 7, pp. 209–219

6. I. Puertas, C.J. Luis [2003] "A study on the machining parameters optimisation of electrical discharge machining" Journal of Materials Processing Technology, Vol. 143 pp-521–526.

7. S.S Mahapatra, A. Patniak [2006] "*Optimization of wire electrical discharge machining (WEDM) process parameters using Taguchi method*" International Journal of Advance Manufacturing Technology

8. H.K. Kansal, S. Singh , P. Kumar [2007] "*Technology and research developments in powder mixed electric discharge machining (PMEDM)*" Journal of Materials Processing Technology, vol. 184, pp 32-41

9. K. Furutani, H. Sato, M. Suzuki [2008] "*Influence of electrical conditions on performance of electrical discharge machining with powder suspended in working oil for titanium carbide deposition process*" Int J Adv Manuf Technol , vol. 40, pp 1093–1101

10. J. Kapoor, S. Singh & J.S. Khamba [2010] "*Recent Developments in Wire Electrodes for High Performance WEDM*" Proceedings of the World Congress on Engineering 2010 Vol II WCE 2010, June 30 - July 2, 2010, London, U.K.

11. M.P. Jahan, M.Rahman, Y.S.Wong [2011]: "*A review on the conventional and micro-electrodischarge machining of tungsten carbide*"; International Journal of Machine Tools & Manufacture 51 837–858

12. A. Klinka, Y.B. Guoa, F. Klockea [2011] "*Surface integrity evolution of powder metallurgical tool steel by main cut and finishing trim cuts in wire-EDM*", 1st CIRP Conference on Surface Integrity (CSI) Procedia Engineering 19, pp 178 – 183

13. H. Singh, D.K. Shukla [2012]: "*Optimizing electric discharge machining parameters for tungsten-carbide utilizing thermo-mathematical modelling*"; International Journal of Thermal Sciences 59 (2012) 161-175

14. E.Aliakbari, H. Baseri [2012]"*Optimization of machining parameters in rotary EDM process by using the Taguchi method*" Int J Adv Manuf Technol, Vol. 62, pp 1041-1053

15. D.Kanagarajan, K.Palanikumar, R.Karthikeyan [2012] "*Effect of Electrical Discharge Machining on strength and reliability of WC–30%Co composite*" Materials and Design 39 , pp 469–474

16. A.V. Shayan, R.A. Afza, R. Teimouri [2013]: "*Parametric study along with selection of optimal solutions in dry wire cut machining of cemented tungsten carbide (WC-Co)*"; Journal of Manufacturing Processes 15 644–658

17. K.P. Rajurkar, M.M. Sundaram, A.P. Malshe [2013] "*Review of Electrochemical and Electrodischarge Machining*", The Seventeenth CIRP Conference on Electro Physical and Chemical Machining (ISEM) Procedia CIRP 6 13-26

18. S. Assarzadeh*, M. Ghoreishi [2013] "*Statistical modeling and optimization of process parameters in electro-discharge machining of cobalt-bonded tungsten carbide composite (WC/6%Co)*" The Seventeenth CIRP Conference on Electro Physical and Chemical Machining (ISEM) pp 463 – 468

19. R. Demellayer*, J. Richard [2013] "*High precision electro discharge machining of Molybdenum and Tungsten*" The Seventeenth CIRP Conference on Electro Physical and Chemical Machining (ISEM), Procedia CIRP 6, pp 89 – 94

20. www.newagepublishers.com/..../001566.pdf

MATEC Web of Conferences 57, 03007 (2016) DOI: 10.1051/matecconf/20165703007
ICAET-2016

Sliding Wear Analysis of Ultra High Strength Steel Using Full Factorial Design Approach

Varun Sharma[1], Subodh Kumar[2] and A.S. Shahi[3]

Department of Mechanical Engineering, Sant Longowal Institute of Engineering & Technology, Longowal, India

Abstract. This study describes multi-factor-based experiments that were applied to investigate the sliding wear behaviour of quenched and tempered wear resistant steel. This study was aimed to evaluate the effect of input parameters (such as applied load, sliding velocity and sliding time) on wear rate. Full factorial design through design of experiments approach was used for investigation by establishing an empirical relationship between wear loss and input parameters and determining the optimal combination of testing parameters for minimum and maximum wear losses. Sliding wear tests were carried out using pin-on-disc type apparatus at room temperature under dry sliding wear conditions. Detailed investigation revealed that applied load was the most significant factor affecting the wear performance followed by sliding velocity and sliding time. The maximum weight loss due to wear was found to be 33.48 mg when experimentation was conducted at maximum levels of input variables and minimum wear loss of 3.12 mg was obtained at the minimum levels of load, sliding velocity and sliding time. The scanning electron microscopy of the worn pin surfaces shows that adhesion and plastic deformation were the dominating mechanisms involved during experimentation that resulted in maximum wear of the pins, and on the other hand no such mechanism persisted when the pins were worn under minimum sliding wear conditions.

1 Introduction

Cost effective and novel materials having high specific mechanical properties have been the favourite research topic of the material scientists all over the world. Among steels the quenched and tempered low alloy abrasion resistant steels or low alloy martensitic steels are currently being considered as new generation steels [1,2]. These steels are being used to manufacture wear plates which presents an effective solution to the sliding wear problems which accounted for approximately 50 % losses due to wear. Owing to high abrasion resistance due to high hardness & ultra high tensile strength backed with good toughness properties, these wear plates are finding their immediate applications in chutes, hoppers, dump truck beds, cutter bars, scraper blades, liner plates, tipper bodies, containers, crushing mills, excavator buckets and loading buckets etc [2,3]. These wear resistant plates are the modern solution in regeneration of worn machines parts & also for producing new parts which connect high wear resistance with cost reduction [4,5].

In order to study the sliding wear performance of the materials, the Pin-on-disc (POD) testing rig has been used extensively by the researchers [6,7]. Individual and combined effects of independent variables such as load, speed, time and other input variables on sliding wear of different material has been studied by number of researchers in the past [8,9,10,11]. In 2004, the effect of sliding velocity on wear behaviour of different microstructures of Ni-Cr-Mo-V steel was studied by D. Rai and J.P. Pathak using pin-on-disc set up [12]. Wear resistance of various types of sintered steels was studied by Bidulsky R. et.al. in 2010 [13].

The POD experimental set up provides a broad spectrum of controllable parameters with independent variables whose individual and combined effects can be investigated to study the wear performance of different engineering materials. Owing to wide variability of controllable and uncontrollable variables, the researchers used different statistical techniques to have deep insight into the study of wear behaviour [14,15]. In 2010, empirical models were developed by N.S.M. El-Tayeb et.al. using response surface methodology (RSM) to predict the wear characteristics of Ti54 alloy as a function of sliding wear conditions [16]. Gray-Taguchi method was used by Dharmalingam et.al. in 2011 to investigate the optimization of dry sliding performance of aluminium hybrid metal matrix composites under different loads, sliding speeds and varying percentage of molybdenum disulfide [17].

Ravindran et.al. in 2012 applied the factorial techniques to investigate the effect of percentage of reinforcement, load, sliding speed and sliding distance on the wear and sliding friction response of Al hybrid composites [18]. Through detailed literature review, it was found that most of the experimental design techniques were confined to

[a] Corresponding author: author@e-mail.org

investigate the sliding wear behaviour of the composites or hard facings. Besides, very few investigations have been carried out to study the combined effect of test parameters or their interactions (i.e. the combined effects of load, speed and time) on the sliding wear behaviour of materials. All published work focused only on the study of the effect of one parameter at a time. Moreover, to the best of the authors' knowledge, none of the past researchers gave attention to investigate the sliding wear behaviour of Q & T ultra high strength steels.
The main focus of the current research work is to investigate the relationship between the input independent variables such as load, speed and time on the wear performance of the steel, and to develop mathematical models for the wear behaviour of ultra high strength steel during wear test under dry sliding condition at room temperature.

2 Details of experiment

2.1 Base material

A Japanese grade (JFE EH400) quenched and tempered abrasion resistant steel was used in this wear investigation. The chemical composition and mechanical properties of this steel are given in Table 1 & Table 2 respectively. The optical micrograph of the steel is shown in Figure 1. Wear test specimens were machined from 15 mm thick plate of JFE EH400 grade steel to obtain cylindrical pins possessing diameter of 6 mm and height 25 mm with the hemispherical cross-sectional end.

Table 1. Chemical composition of JFE EH400 quenched and tempered low alloy abrasion resistant steel, % weight

S.No.	Element	Composition (%)
1	C	0.217
2	Si	0.363
3	Mn	0.72
4	P	0.007
5	S	0.002
6	Cr	0.467
7	Ni	0.011
8	Mo	0.006
9	Nb	0.0224
10	Ti	0.0152
11	Al	0.0316
12	V	0.0012

Table 2. Mechanical properties of JFE EH400 quenched and tempered low alloy abrasion resistant steel

Properties	Yield strength	UTS	Elongation	Charpy Impact (J)	Hardness
(Units)	(MPa)	(MPa)	(%)	RT	(HB)
Values	1319.77	1435.47	12.5	249	447

Figure 1. Optical micrograph of the JFE EH400 steel

2.2 Design of Experiment

The experimentation was conducted as per the design of experiments approach by which the number of experiments required, mainly depends on the technique employed through design of experiments. Thus, it is important to have a well designed experiment so that number of experiments required can be minimized [16]. In this study, the design suggested by the full factorial technique has been implemented to analyze the effect of three independent variables for sliding wear i.e. load, sliding velocity and time on wear loss. Table 3 shows the different variables and their levels.

Table 3. Factors and levels of independent variables according to response surface methodology

Factors	Unit	Symbol	Levels	
			-1	1
Load	N	A	20	50
Velocity	m/s	B	1	2
Time	min	C	8	17

The complete design layout for experiments is summarized in Table 4, which shows the experimental combinations of load, velocity and time. As per the design matrix shown in Table 4, a total of sixteen experiments were carried out which constitutes eight factorial points and eight repeated experiments. The software used was Minitab 17.

Table 4. Design matrix and experimental results

Experiments order		Input			Response
Std	Run	A:Load (N)	B:Velocity (m/sec)	C:Time (min)	Wear loss (mg)
5	1	20	1	17	5.0
4	2	50	2	8	27.4

10	3	50	1	8	9.4
16	4	50	2	17	32.9
9	5	20	1	8	3.4
6	6	50	1	17	14.8
13	7	20	1	17	5.8
2	8	50	1	8	9.1
14	9	50	1	17	14.0
8	10	50	2	17	33.0
15	11	20	2	17	10.5
11	12	20	2	8	4.2
3	13	20	2	8	4.3
7	14	20	2	17	10.2
12	15	50	2	8	28.0
1	16	20	1	8	3.9

2.3 Sliding wear test

A pin-on-disc apparatus shown in Figure 2 was employed to evaluate the wear characteristics of the steel. The wear tests were carried out at room temperature under dry sliding wear conditions in accordance with the ASTM G99-95 standard.

Figure 2. Pin-on-disc test rig

Before the test, the hemispherical ends of pin surfaces were polished with the abrasive paper of 800 mesh grit size and cleaned in acetone, then dried and weighed. The counter material used in this experimentation was EN 31 disc of 8 mm thickness and 100 mm diameter which was hardened to 62 HRC and grounded to the surface roughness of 1.6 Ra. The pin specimens with their hemispherical ends were loaded against metallic disc with the help of a cantilever mechanism. Each test was conducted on a fixed track radius as per the experimental procedure established using full factorial design. After the completion of each test, the wear pin was ultrasonically cleaned in acetone to remove the wear debris, then dried and finally weighed on an electronic balance of 0.0001g measurement accuracy.

3 Full factorial design

Full factorial technique is the compilation of both mathematical and statistical techniques, which are helpful for the modelling and analysis of problems in which a response of interest is influenced by several variables and objective is to optimize the response. In most of the factorial problems, the form of the relationship between the response and the independent variables is unknown. Thus the first step in factorial design is to find a suitable approximation for the true functional relationship between response of interest 'y' and a set of independent variables $\{x_1, x_2, \ldots \ldots \ldots x_n\}$. If the response is well modelled by a linear function of the independent variables, then the approximating function is the first order model is given as,

$$y = \beta_0 + \beta_1 x_1 + \beta_2 x_2 + \beta_3 x_3 + \ldots + \beta_k x_k + \varepsilon \quad (1)$$

Where, ε represents the noise or error observed in the response y such that the expected response is (y -ε) and β's are the regression coefficients [19]. The least square technique is being used to fit a model equation containing input variables by minimizing the residual error measured by the sum of square deviations between the actual and the estimated responses. This involves the calculation of estimates for the regression coefficients. The calculated coefficients or the model equation need to however be tested for statistical significance. So that tests for significance of the regression model, test for significance on individual model coefficients and test for lack-of-fit need to be performed. An ANOVA is commonly used to perform these tests. ANOVA calculates the F-ratio, which is the ratio between the regression mean square and the mean square error. If the calculated value of F-ratio is higher than the tabulated value of F-ratio for roughness, then the model is adequate at desired significance level α to represent the relationship between response and the independent variables. For testing the significance of individual model coefficients, the model is optimized by adding or deleting coefficients through forward addition, backward elimination or stepwise elimination or addition. It involves the determination of P- value or probability of significance. If the P-value is less or equal to the selected confidence level, then the effect of the variable is significant. If the P-value is greater than the selected confidence level, then it is considered that the variable is not significant.

4 Results and discussion

4.1 Metallography

The optical micrograph of the base metal taken at the magnification of 400X and shown in Figure 1, clearly indicates the extra fine microstructure of acicular tempered martensite revealed under high magnification only. The grains consisting of laths of tempered martensite are clearly identified which are separated by high angle boundaries. Literature shows that the laths of the martensite contain very fine, uniformly distributed coherent carbides. Such a structure results in high yield strength, hardness & toughness of the steels. This structure is responsible for providing the excellent wear resistant properties to these steels.

The results of the 16 experiments conducted as per the experimental plan are shown in Table 4 along with the run order selected at random. These results were input into the Minitab 17 software for further analysis.

4.2 Statistical analysis of variance (ANOVA)

Analysis of variance (ANOVA) is commonly used to perform test for significance of the regression model, model coefficients and lack-of-fit. This analysis was carried out for a significance level of $\alpha = 0.01$, i.e. for a confidence level of 99%. ANOVA test for 23 full factorial model for wear loss is summarized in Table 5. This shows the value of "Prob. > F" for model to be 0.0001 which is less than 0.01, that indicates the model is significant. In the same manner, The value of "Prob. > F" for main effect of load, sliding velocity and sliding time with two-level interaction of load and velocity was less than 0.01, thereby, indicating that these terms are significant model terms. The value of Prob. > F for lack-of-fit is 0.0328, thus greater than 0.01, which indicates that lack of fit is insignificant which is the desirable condition to fit the model.

Table 5. Resulting ANOVA table for full factorial model

Source	DOF	Adj SS	Adj MS	F-Value	P-Value	
Model	6	1709.35	284.89	461.10	< 0.0001	S
Linear	3	1455.50	485.16	785.25	< 0.0001	S
A	1	919.61	919.60	1488.40	< 0.0001	S
B	1	452.63	452.62	732.59	< 0.0001	S
C	1	83.27	83.26	134.77	< 0.0001	S
Interactions	3	253.85	84.61	136.96	< 0.0001	S
AB	1	247.28	247.27	400.22	< 0.0001	S
AC	1	1.63	1.62	2.63	0.139	NS
BC	1	4.95	4.95	8.01	0.020	NS
Error	9	5.56	0.61			
Lack of Fit	1	4.52	4.51	34.57	0.0328	NS
Pure Error	8	1.05	0.13			
Total	15	1714.91				
Std. Dev.		0.786	R-Squared		99.68%	
Adj R-Squared		99.46%	Pred R-Squared		98.98%	

The total variation in the model can be explained by the R^2 value which is 0.9967 or close to 1, which is the desirable value. The adjusted R^2 value is equal to 0.9946, which is particularly useful when comparing models with different number of terms. The result shows that the difference between adjusted R^2 value and ordinary R^2 value is within the limit of 20%. The final regression model for wear loss in terms of actual factors is represented in equation (2), and the corrected regression model by eliminating the insignificant terms is shown in equations (3).

$$\text{Wear rate (Actual)} = 7.73 - 0.34\,\text{Load} - 10.80\,\text{Velocity} - 0.029\,\text{Time} + 0.52\,\text{Load} * \text{Velocity} + 0.0047\,\text{Load} * \text{Time} + 0.247\,\text{Velocity} * \text{Time} \quad (2)$$

$$\text{Wear rate (Corrected)} = 7.73 - 0.34\,\text{Load} - 10.80\,\text{Velocity} - 0.029\,\text{Time} + 0.52\,\text{Load} * \text{Velocity} \quad (3)$$

In order to ascertain, that whether the residuals follow a normal distribution or not, the normal probability plot of the residuals is drawn as shown in Figure 3. The figure shows that the residuals generally fall on a straight line implying that the errors are distributed normally. To test the assumption of constant variance, a graph is plotted between residuals and predicted response for wear loss as shown in Figure 4. The presence of random scatter without any geometrical or defined shape signifies that the assumption has not been violated.

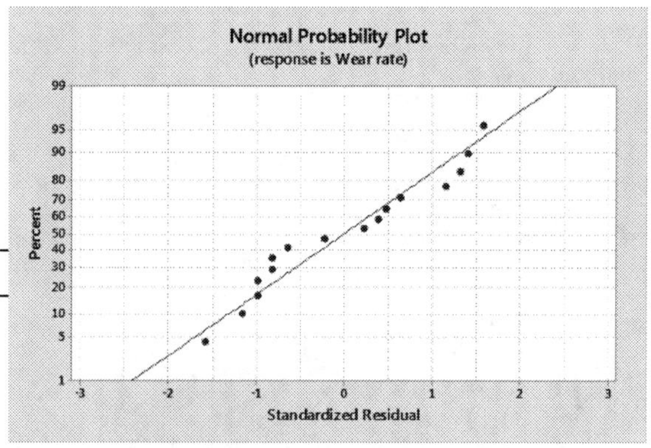

Figure 3. Normal probability plot of residuals

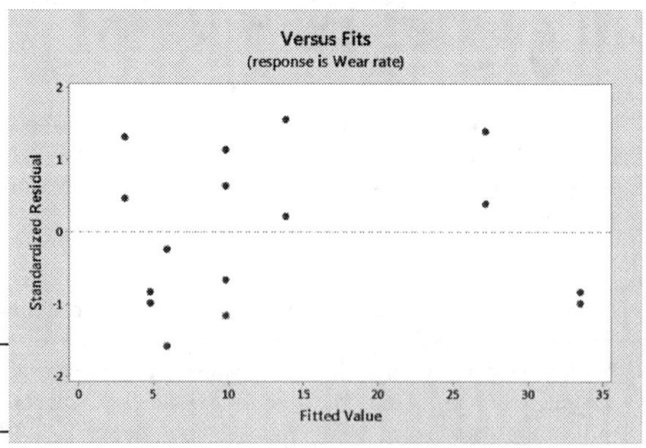

Figure 4. Plot of standardized residuals vs. fitted values

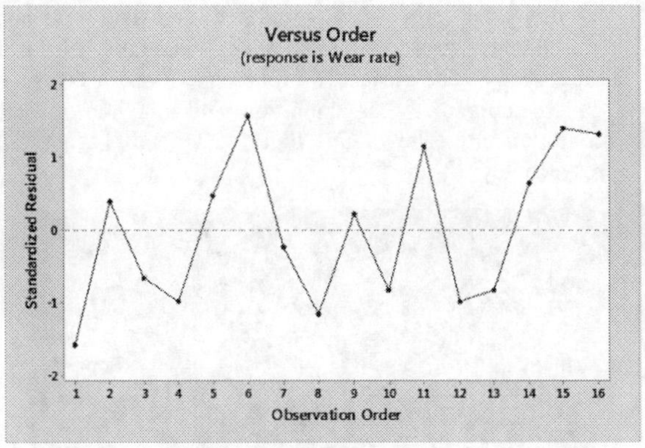

Figure 5. Plot of standardized residuals vs. run order

The randomization of the experimental runs resulted in reducing systematic experimental error and is clearly evident from the graph plotted between standardized residuals and the observation runs shown in Figure 5 the graph shows the significant fluctuation of the error values across the mean value of the standardized residuals.

The Figure 6 shows the effect of load, velocity and time on the wear rate of performance of the base metal. The graphs show that all the factors had an increasing positive influence on the wear rate. However variation in slope of the curves indicates that the load had the maximum influence on the wear loss of the material followed by velocity and time.

The combined effects of input parameters on wear rate are depicted as interactions and are shown in Figure 7. No observable interaction has been observed for the combined effect of load and time, and velocity and time. However a strong interaction effect has been observed for the combined effect of load and velocity. A significant divergence between load and velocity curves has been observed with an increase in load from minimum to the maximum value and maximum curve divergence occurred at the maximum value of load.

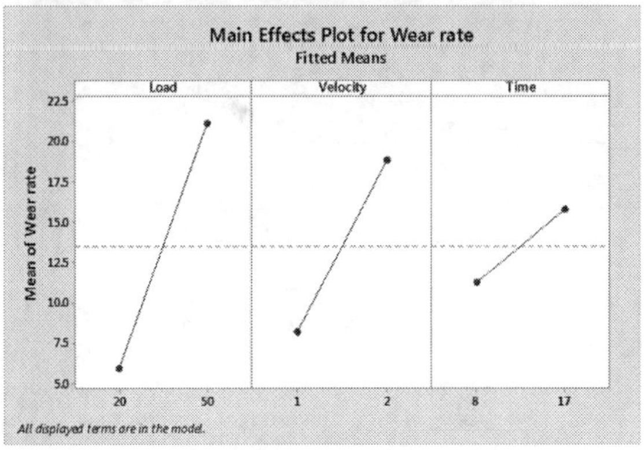

Figure 6. Single factor plots of load, velocity and time for wear rate

The 3D surface graph shown in Figure 8 shows the interaction effect of load and sliding velocity on the wear loss of the specimen pins when the pins were subjected for 12.5 minutes sliding wear. Figure clearly shows that wear loss is following a linear relationship with load and sliding velocity. Higher slope of surface w.r.t. load illustrates its stronger effect on wear loss in comparison to sliding velocity.

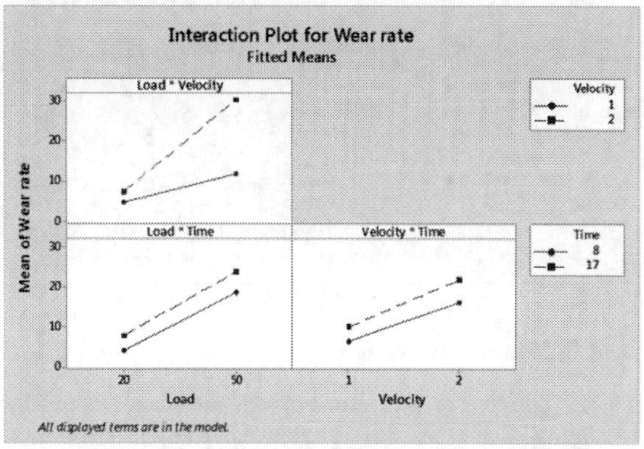

Figure 7. Combined graph illustrating the influence of interactions on the wear rate

From the 3D surfaces it can be clearly inferred that the minimum wear loss was achieved at lowest level of load and sliding velocity. Similarly, the maximum weight loss occurred for the conditions when load and sliding velocity were at their maximum levels. The flat profile of 3D surfaces for wear loss shows that the best fitted model is linear in nature.

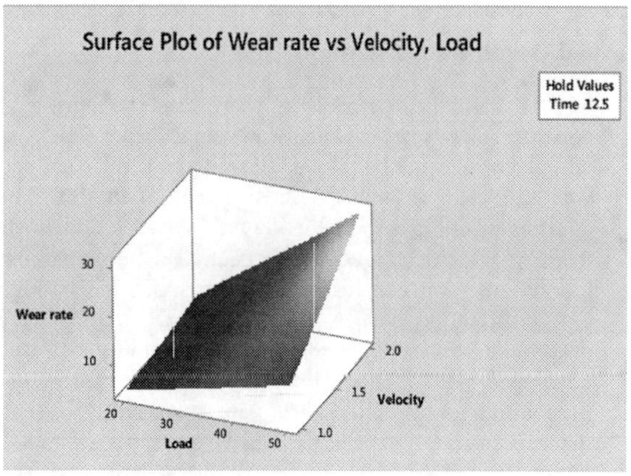

Figure 8. Three-dimensional surface plot for wear loss for a time period of 12.5 min

Figure 9 represents the cube plot which shows the simultaneous effect of load, velocity and time on the wear loss. It can be figure out from the cube that the minimum wear loss of 3.118 mg occur when load was 20 N, sliding velocity was 1 m/s and sliding time for 8 min. Similarly, the maximum weight loss of 33.48 mg occur when the pins were loaded against the disc for 17 minutes, subjected to maximum load of 50 N and exposed for maximum sliding velocity of 2 m/s.

 DOI: 10.1051/matecconf/20165703007

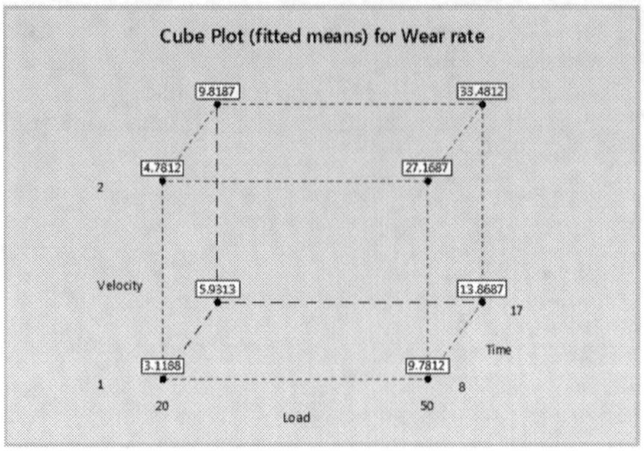

Figure 9. Cube plot for wear loss

4.3 Wear optimization

The current research problem was aimed at finding the optimal values of input variables in order to ascertain the minimum and maximum wear loss, the optimal solutions of which are reported in Table 6.

Table 6. Optimization results of sliding wear

Wear condition	Solution no.	Load (N)	Velocity (m/s)	Time (min)	Wear (mg)	Remarks
Minimum	1	20	1	8	3.12	Select
	2	21.27	1	8	3.40	
Maximum	1	50	2	17	33.48	Select
	2	50	1.97	17	33.00	

4.4 Confirmation experiments

The mathematical model in the form of regression equation that was statistically developed for wear loss and given by equation (3) was found to be significant and was validated through F-tests and lack-of-fit test. The coefficient of variation (R^2) for model was found to be 0.9968, which indicates its effectiveness of making predictions. This conclusion is further supported through the confirmation runs. A set of confirmation runs were performed to verify the prediction ability of the developed wear model. The details of the confirmation runs are given in Table 7. The percentage error between the experimental and the predicted values was found to be less than 6%, which clearly demonstrates the accuracy of the models developed in this study.

Table 7. Plan of confirmation experiments and results

Wear condition	Test conditions			Wear loss (mg)		Error (%)
	Load (N)	Velocity (m/s)	Time (min)	Predicted	Experimental	
Minimum	20	1	8	3.12	3.28	5.1
Minimum	50	2	17	33.48	35.09	4.8

4.5 Worn Surface observations

Experiments were conducted using the optimal sets of input variables to obtain the minimum and maximum wear conditions as shown in Table 7. The worn out ends of the wear pins were sectioned and their surface morphology was examined using scanning electron microscopy. The surface morphology of worn out pins under minimum wear condition and maximum wear condition are shown in Figure 10 and Figure 11 respectively.

Figure 10. SEM micrographs of worn out pin at the magnification of 150X. under minimum wear condition when load is 20 N, velocity is 1 m/s, time is 8 min.

The SEM macrograph of the worn out surface of the pin worn under minimum wear testing conditions is shown in Figure 10. At the minimum wear conditions (load of 20 N, sliding velocity of 1 m/s and time 8.04 minutes), mild wear occurred. No adhesion or oxidation marks have been observed on the surface.

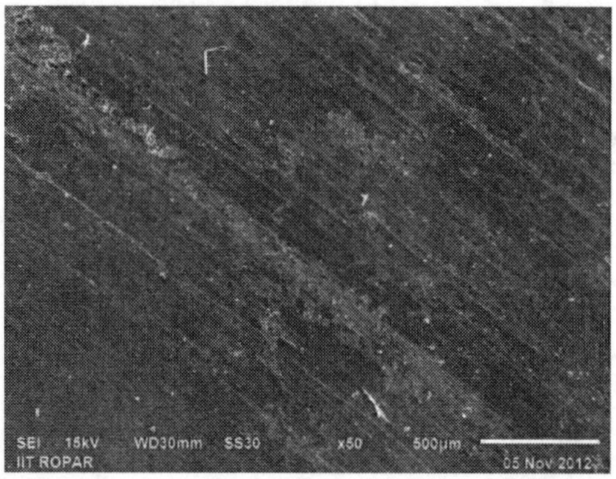

Figure 11. SEM micrographs of worn out pins at the magnification of 50X under maximum wear condition when load is 50 N, velocity is 2 m/s and time is 17 min.

However at the maximum wear testing condition (load is 50 N, velocity is 2 m/s and time is 17 minutes), adhesion arises as the most damage mechanism together with intense plastic deformation, which becomes more evident from the SEM micrograph shown in Figure 11. The figure shows the presence of adhesion marks and attached debris to the worn surface owing to the oxidation and excessive heating of the pin surface under maximum wear testing conditions. Similar worn surface morphology has been reported by C.C. Viafara et.al. for the unlubricated sliding wear of pearlitic and bainitic steels [20].

5 Conclusions

The present investigation has been carried out to develop the mathematical relationship between wear rate as response and input variables namely applied load, sliding velocity and sliding time using 23 full factorial design. Through this study an attempt has been made to identify the most significant factor that affects the wear performance of the quenched and tempered low alloy abrasion steel. The results reveals that applied load was found to be the most significant factor with the percentage contribution of 53.79% followed by sliding velocity (26.48% contribution) and sliding time (4.87% contribution). Further, the maximum weight loss due to wear was found to be 33.48 mg when load was 50 N and disc was rotated at a sliding velocity of 2 m/s for 17 minutes. Similarly, the minimum wear of 3.12 mg was found at 20 N load when disc loaded for 8 minutes at a sliding velocity of 1 m/s. Furthermore the prediction ability of the mathematical model was verified through confirmation experiments and the percentage error between the experimental and the predicted values of wear loss were found to be less than 6 %. This shows the excellent predicting ability of the developed mathematical model within the ranges of tested parameters.

References

1. Y. Murota, M. Hashimoto and T. Abe, "High performance steel plates for construction and industrial machinery use", JFE Technical Report, pp.60-65, 2005.
2. Buglachi, H.; Smajdor, M. Mechanical properties of abrasion resistant Hardox 400 steel and their welded joints. Advances in Material Science 2003, 4 (2), 5-8.
3. R.S. Parmar, "Welding Engineering and technology", 2nd Ed. New Delhi, Khanna Publishers, 2003.
4. JFE Steel Plate, A Manual, 1-36 pp. 2010.
5. M. Adamiak, J. Gorka, T. Kik, "Comparison of abrasion resistance of selected constructional materials", Journal of Achievements in Materials and Manufacturing Engineering, Vol. 37, issue 2, pp. 375-380, 2009.
6. Ruff, A.W.; Bayer, R.G. Tribology: Wear Test Selection for Design and Application; American Society for Testing and Materials, Philadelphia, PA, 1993; 94-99 pp.
7. Riyadh A. Al-Samarai, Haftirman, Khairel Rafezi Ahmed, Y. Al-Douri, "Effect of load and sliding speed on wear and friction of Al-Si casting alloy", International Journal of Scientific and Research Publications, Vol. 2, issue 3, pp. 1-4, 2012.
8. H.A. Ameen, K.S. Hassan, E.M.M. Mubarak, "Effect of loads, sliding speeds and times on the wear rate for different materials", American Journal of Scientific and Industrial Research, Vol. 2(1), pp. 99-106, 2011.
9. F.U. Ozioko, "Synthesis and study on effect of parameters on dry sliding wear characteristics of Al-Si alloys", Leonardo Electronic Journal of Practices and Technologies, Issue 20, pp. 39-48, 2012.
10. R. Tyagi, S.K. Nath, S. Ray, "Effect of martensite content on friction and oxidative wear behavior of 0.42 pct carbon dual-phase steel", Metallurgical and Materials Transactions A, Vol. 33A, pp. 3479-3488, 2002.
11. H. So, "Characteristics of wear results tested by pin-on-disc at moderate to high speeds", Tribology International, Vol. 29 (5), pp. 415-423, 1996.
12. D. Rai, J.P. Pathak, "Influence of sliding velocity on wear behavior of different microstructures of Ni-Cr-Mo-V steel", Indian Journal of Engineering & Materials Sciences, Vol. 11, pp. 113-120, 2004.
13. R. Bidulsky, M. Actis Grande, A. Zago, Z. Brytan, J. Bidulska, "Wear resistance of various type sintered steels", Archives of Metallurgy and Materials, Vol. 55, issue 3, pp. 623-629, 2010.
14. R. Shanmugam and N. Murugan, "Optimization and sensitivity analysis of wear model of satellite 6 hardfacing by gas tungsten arc welding process", IE (I) Journal-MM, Vol. 85, pp. 65-70, 2004.
15. Y. Sahin, "Optimal testing parameters on the wear behavior of various steel", International Journal of Materials and Design, Vol. 27, pp. 455-460, 2006.
16. N.S.M. El-Tayeb, T.C. Yap, P.V. Brevern, "Wear characteristics of titanium alloy Ti54 for cryogenic sliding applications", Triboloy International, Vol. 43, pp. 2345-2354, 2010.
17. S. Dharmalingam, R. Subramaniam, V. Somasundara, B. Anandavel, "Optimization of tribological properties in aluminium hybrid metal matrix composites using gray-taguchi method", Journal of Materials Engineering and Performance, Vol.20, pp. 1457-1466, 2011.
18. P. Ravindran, K. Manisekar, P. Narayanasamy, N. Selvakumar, R. Narayanasamy, "Application of factorial techniques to study the wear of Al hybrid composites with graphite addition", Materials and Design, Vol. 39, pp. 42-54, 2012.
19. D.C. Montgomery, "Design and Analysis of Experiments", 5th ed., Wiley, New York, 2001.
20. C.C. Viafara, M.I. Castro, J.M. Velez and A. Toro, "Unlubricated sliding wear of pearlitic and bainitic steels", Wear, Vol. 259, pp. 405-411, 2005.

MATEC Web of Conferences 57, 03008 (2016) DOI: 10.1051/matecconf/20165703008
ICAET-2016

INVESTIGATE THE EFFECT OF ADDITIVES ON MECHANICAL PROPERTIES DURING CASTING OF 6351 ALUMINIUM

Sanjeev Kumar[1], Sandeep Gandotra[1], Sunil Kumar[1], Nripjit[1], Himanshu Tripathi[2]

[1]*Faculty Members, Department of Mechanical Engineering Beant College of Engineering and Technology Gurdaspur-143521, PB, India*

[2]*Faculty Members, Department of Mechanical Engineering, GIMET Amritsar*

Abstract. Aluminum is one of the significant and widely used metals in the manufacturing industries. The effectiveness of casting processes depends upon various process parameters like type of sand, moulding process and additive used. However, the additive in the sand plays important roles to make a perfect casting. Hence, in this research work the experiment were performed to investigate the effect of additives like Tamarind Powder, Starch Powder and Coal dust on the Mechanical Properties namely during casting of aluminium alloys.

Keywords:- Coal dust, Tamarind Powder, Starch Powder, Tensile Strength, Hardness

1 Introduction

Every product which presents in this universe is going through some manufacturing process. Casting is one of the most important manufacturing processes. While casting liquid material is usually poured into the mold cavity of the desired form and then allowed to solidify. The thickened section is also known as a casting. Casting materials are metals or various cold setting materials that cure after mixing two or more components together. Casting is mainly used for making intricate shapes that would be otherwise difficult or too expensive to make by other methods.

The casting process is carried out on almost every metal, but aluminium is widely used metal during casting due to its wide range of applications in the manufacturing industries. Many of the household products made of aluminium. The aluminium alloys are used to produced in hundreds of compositions by all commercial casting processes while the casting such as green sand, dry sand, composite mold, plaster mold, Investment Casting, Permanent Mold, counter-gravity, low-pressure casting, and pressure dies casting.

Aluminium metal in the real stage is too soft for such uses, and it does not have the high tensile strength. The pure aluminium is mixed some other metal to made alloys. The aluminium alloys have excellent mechanical properties such as high strength, low density, durability, machinability, which is required in the manufacturing industries. Aluminium alloys are used in advanced applications due to their combination of availability and relatively lower cost as compared to other competing materials.

Aluminum alloys are used in numerous applications in which the blend of high strength and low weight is attractive in airframe in which the small weight can be a significant value. Many from of aluminium alloys are used in the manufacturing industries, But aluminium 6351 is widely used because of light weight and superior oxidization resistance. Its thermal and electrical conductivity is four times greater than steels. It has superior strength amongst the 6000 series alloys. Alloy 6351 is also known as a structural alloy and is most frequently used for manufacturing purpose. There is much application of Aluminium 6351 i.e. used for the construction of the ship, rail & road transport, column, rod, mould, pipe, tube, vehicle, bridge, crane and roof. It used because of its excellent surface finish. One of the major applications of Al 6351 in the aircraft industries to make aircraft structures due to its high strength to weight ratio and outstanding corrosion resistance.

Aluminium is going through the different manufacturing process, but casting is the main manufacturing process. There are many process parameters which affect the properties of final casted products. But the additives presents in the sand are significantly impact the final casted products. Therefore, in this research work, an effect of different additives in the sand was investigated during casting of aluminium 6351 alloys.

2 Previous Works

The casting of aluminum alloys was carried out by the different process. But the properties of final casted products are affected by many process parameters. Some researchers have tried to improve the properties aluminium by the addition of metals different metals like magnesium, zinc. In this research, the detailed study is carried on the effect of magnesium and zinc on

[a] Corresponding author: sk_74@rediffmail.com

mechanical properties of aluminium.There is massive information available on the theoretical and experimental work related to the effect of magnesium and zinc on mechanical properties of aluminium which is discussed as under. Davis et al. (1999) investigated that Magnesium provides substantial strengthening and improvement of the work-hardening characteristics of aluminium and can impart good corrosion resistance, weldability, extremely high Strength. Silicon combines with magnesium forms the hardening phase Mg2Si that provide the strengthening. Aluminium alloys are significant due to a combination of high strength to low weight property (Krishna et al. 2013). Al 6351 castings have light weight and good corrosion resistance to air, water, oils and many chemicals. The thermal as well as the electrical conductivity of Al 6351is also four times greater than steel (Kishore et al. 2010). Mechanical properties of the material can be easily calculated with tensile tests, with high accuracy (Reddy et al. 2014). The aluminium alloys are used in the manufacturing of boats, columns, chimney, rods mould, pipes roofs due to ease of workability and weldability (Ibrahim et al. 1997). Gas Cylinder made of Al 6351 investigated for crack phenomena and often found prone to crack at various tensile residual stresses (Prince and Ibrahim, 2003). A metallurgical anomaly known as Sustained load cracking, develops in 6351 aluminum alloy high-pressure cylinders (Durmus et al. 2006). Zalticonin et al. (2003) studied the effect of magnesium content on the properties of Al-Cu-Mg alloys. The changes in chemical composition cause changes in the structure and properties of hardness, tensile with increased amount of magnesium for the same content of titanium in the alloy. Compression and hardness increase with the content of magnesium and titanium. Gubicza et al. (2004) studied the effect of Mg addition in the microstructural evolution, and mechanical properties of high-purity aluminium were investigated over a broad range of strain, up to ~8. The high strains were achieved by applying the equal-channel angular pressing technique.In the early stages of plastic deformation, the interaction between the dislocations and the Mg solute atoms results in an increase in the flow stress with temperature. The stable microstructure is developed at higher strains owing to the Mg addition resulting in the saturation of the proof stress at higher strains in Al–Mg alloys. Dobrzański et al. (2007) published the effect of cooling rate on the size of the grains, SDAS, the size of the β precipitation and characteristic thermal results of AC AlSi9Cu cast alloy. According to the X–ray phase analysis, the investigated AC AlSi9Cu alloy cooled wit solidification rate: 0,16, 0.48 and 1.04 °C composed of two phases. The diffraction lines of the Si-phases do not reveal such intensity. Based on the X–ray phase analysis was found, that change of solidification rate do not influence the phases composition of investigated alloy. The X-ray phase analysis does not reveal occurring of Al2Cu, AlFe5Si and Al8FeMg3Si6 phases, what suggested that the fraction volume of these phase is below 3%. Dong-yan et al. (2010) investigated the effect of Zn content on tensile and electrochemical properties of 3003 Al alloy. The effect of Zn addition on the microstructure, tensile properties and electrochemical properties of as-annealed 3003 Al alloy was investigated. Authors found that high-density precipitates are observed in the Zn-containing alloys and the alloy with 1.8% Zn addition also has rod-like precipitates. The alloy with 1.5% Zn addition has the maximum ultimate tensile strength. Adeosun et al. (2011) presented the effect of separately adding Cu and Zn on the mechanical properties of 6063 aluminium alloy has been examined

Sand cast samples of an an aluminium alloy containing Cu/Zn (0-20 vol %) and homogenized at 5100C for 1hr are analyzed for ultimate tensile strength (UTS), hardness, elongation and impact energy. The addition of more than 15 vol % Zn to structural aluminum alloy raised its hardness and elongation. Copper additions above 4 vol % lowered the UTS, elongation and impact energy. However, the hardness increases with percent addition of Cu. The addition of zinc to structural aluminum will improve the homogeneity of the matrix. The impact energy and hardness of the matrix are enhanced when Cu addition is within 2-4 vol %. Agraval et al. (2012) aimed at studying the effect of varying the composition of Magnesium on the mechanical properties like strength and hardness in an Aluminium-Zinc-Magnesium alloy. Sand casting was used for the purpose of the project and testing was done on the samples for determining the resultant mechanical properties. Girisha and Sharma (2012) investigated the influence of magnesium on microstrructural changes and mechanical properties such as tensile strength and hardness of the Al4Cu alloys. The modifications of Al4Cu by adding Mg of 0.5 to 2 % in the interval of 0.5% mixing with stirrer and casted by gravity die casting, subsequently the specimens were subjected T6 type heat treatment for 5 hr at 175oC. The effect of Mg and ageing on microstructure was studied by using optical microscope with image analysis software for measuring grain size and dendrite arm spacing. The mechanical properties such as tensile strength and hardness were investigated using universal testing machine and Brinell hardness tester respectively. The micro-structural analysis result shows the 2 % addition of Mg reduces the 20% grain size and 21.52% dendrite structure. The tensile strength and hardness increasing with % of Mg. The addition of 2% Mg increases tensile strength 57.9% and hardness of 25%. Aging specimens showed that 1% of Mg influence more on grain refinement and mechanical properties due to smaller the grain size[8]. Rana et al. (2012) reviewed on the influence of alloying elements on microstructure and mechanical properties of aluminum alloys . Alloying elements are selected based on their effect and Suitability. Silicon lowers the melting point and increase the fluidity(improve casting characteristics) of Aluminum. Magnesium provides substantial strengthening and improvement of work harding characteristic of aluminium alloy. It can impart good corrosion resistance and weldability or extremely high strength. Copper has a greatest impact on the mechanical strength and hardness of aluminum casting alloys, both heat treated and not heat treated and at both ambient and elevated service temperature. It improve the machinability of alloys by increasing matrix hardness. Nickel (Ni) enhances the elevated temperature strength and hardness .It is concluded that selection of alloying element depends on the use of materials requirement. Qiaoling et al. (2012) solved the problems which are encountered in the process of experiment study of materials under tensile impact, and its influence factors are discussed in detail. So the issues in the test have been summarized for the purpose of increasing the ability to analyze and solve and improving the exploring spirit of scientific experiments. The article takes the example of yield strength, and the yield point marks the resistance of trace plastic deformation of metal materials which make the ability of the materials express adequately. The materials tensile mechanical properties of the influence factors were acquired in the article from some reference material and analysing the problems in practical work experience. Atanda et al. (2012) investigated the effects of bentonite and starch binders on foundry moulding sand. The two binders were binders were applied separately to silica sand in different proportions and also as a mixture in various proportions as well. Tataram and Nanjundaswamy (2013) studied the evaluation of green sand mold properties has been carried out using various additives. Additives are plays critical role on the green sand mold is to enhance specific mold properties. Here Olivine sand has been used for mold purpose. The properties such as compression strength, permeability and collapsibility have been studied, comparison has made with different additives. By conducting a number of iterations finally, the 7% by volume of water was found the optimal value for the bonding sand and 8% of clay by weight it gives the real bonding strength for the mold.In varying the percentages of additions at 1% of additives the maximum strength was found. Thomas et al. (2014) performed the comparative study of the properties of Aluminium alloy (LM6-12%Si) castings in sand casting and Evaporative pattern casting (EPC) process. Thermocouples are used to measure the temperature of the casting during solidification.The comparative examination was performed, and it is concluded that there is no significant change in mechanical properties. It is possible to assure the viability of adopting the EPC process to an existing conventional acting process to produce Al-Si alloys. The EPC process has advantages of other casting methods, for high production of complex sharp parts he patterns are cheap and easy to manufacture. These are free of lines and extra angled. It is possible to reuse the sand nature.

Hence, from above past work, it is observed that the addition of magnesium and zinc to aluminum can enhance the properties of aluminum. However, still there is a need to investigate the effect of some additives in moulding sand on the mechanical properties during casting of aluminium alloys.

3 Experimental Detail

Use For the casting aluminium, 6351 is selected as materials. Sand used for casting having silica 78-80%, bentonite –14-15%, moisture – 6-7%. For making moulds for casting a split pattern of cylindrical shape, runner, riser of taper shape are used as shown in Figure 1.

Figure 1. Split type pattern

During the experimentation tamarind powder, starch powder and coal dust are used as additives. To prepare moulds for casting, 25 kg of sand for each mould is taken with additions of different additives in various proportions. Mixing of various additives with sand takes place in mixture sand machine. The different percentage of additive used in mold making process as shown in

Table 1 Proportion of additives is on weight of sand.

Additives	Percentage	Total weight of sand
Tamarind powder	1%,3%,5%	25kg
Starch powder	1%,3%,5%	25kg
Coal dust	1%,3%,5%	25kg

10 moulds are made i.e 3 moulds each of 3 different additives and 1 normal mould without any additives .Equal ramming is applied on every mould. Moisture content in each mould is same.only the additives percentage changes in every mould.

3.1 CASTING

Aluminium alloy has been melted in oil fired tilting furnance who can each upto maximum temperature of 1000 degree Celsius. Aluminium whose melting point is 700 degree Celsius can easily melted in it.we took 15 kg of aluminium alloy and being melted in this furnace and it take 2 hr for complete melting.

Melted aluminium is poured in different moulds and being left for cooling for about 2-3 hr.Solidified casted specimen are taken out of moulds and are being aged for 3 weeks before being taken for testing.

Figure 2. Oil fired tilted furnance

4 Results and Discussions

The casting of the aluminium 6051 alloys is carried out with different percentage of additives. The effects of different additives during casting were analysed in terms of mechanical properties namely hardness and tensile strength of the casted specimen.

4.1 Analysis of Hardness

The Rockwell hardness tester was used for measuring the hardness of casted specimen. The variation of hardness with different percentage of additive is as shown in Figure 2.

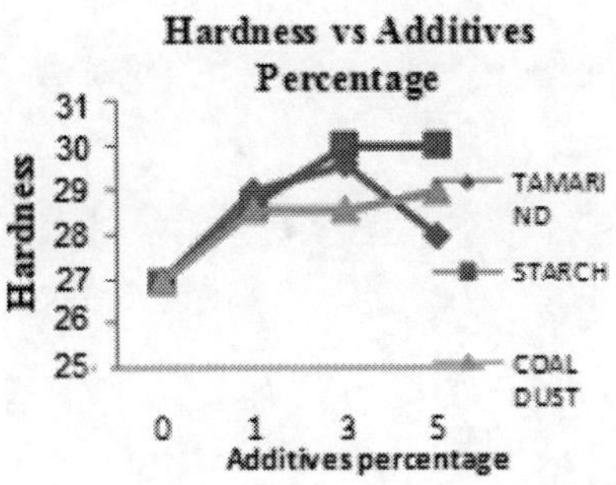

Figure3. Variation of hardness with percentage of additives

Starch additives having finer grain size gives lesser strong bond between sand particles but with increase in starch percentage, its sand holding ability increases i.e give strong bonding which ultimately increases the hardness of aluminium alloy. Tamarind additive due to its lesser fine grain particle bond making capacity increases from 1% to 3% and it gives increase in hardness of aluminium alloy but when percentage increases to 5% its holding capcity increase on the other hand permeability decreases which decreases the hardness of aluminium alloy. Coal dust also follows the same sequence but give less hardness of aluminium alloy as compared to other additives in sand.

4.2 Analysis of Tensile Strength

The Universal testing machine was used for measuring the tensile strength of casted specimen. The variation of tensile strength with different percentage of additive is as shown in Figure 4.

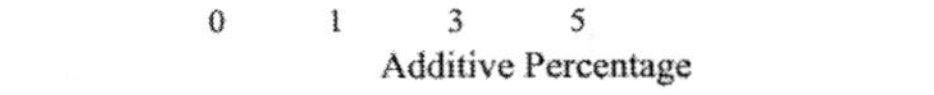

Figure4. Variation of Tensile Strength with percentage of additives

Starch additive at its 1% content in sand increases the compressive strength of mould which increases the tensile stress of aluminium alloy but when starch content increases compressive strength or permeability decreases which reduce tensile strength of aluminium alloy to lower value. Tamarind and coal dust additive in sand result in enough compressive strength of mould but drops the aluminium alloy tensile strength. This may be due to the less fine grain size of tamarind powder or coal dust ability to trap the produce gases during pouring of aluminium alloy.

Conclusions

- The addition of additives in the sand during casting enhanced the hardness of aluminum alloy. Starch powder affects the hardness of aluminum alloy to a great extent as compared to other additives.
- On the other hand additives, addition in sand decreases the tensile tension of the aluminum alloy. Only the starch having a percentage of 1% in sand mould gives the highest value of tensile tension according to the test reports.

References

1. G G Krishna, P R Reddy, and M. M., Hussain Experimental Investigation of Tensile Strength and Deflection Characteristics of Friction Stir Welded Aluminum AA 6351 Alloy Joint, IOSR Journal of Mechanical and Civil Engineering, **7** (5) 2013.
2. K Kishore, P V G, Krishna K Veladri, and G K Kumar., Analysis of Defects in Gas Shielded Arc Welding of AA 6351 Using Taguchi Methods, International Journal of Applied Engineering Research, **5** (3) 2010.
3. M V N Reddy, M N Reddy, K V Kumar, P Garre., Experimental Investigation of Tensile Strength on Al 6351 to The Aerospace Structural Applications, International Journal of Mechanical Engineering and Technology (IJMET), **5** (2) 2014.
4. H K Durmus, E O Zkaya, C Meri, The Use of Neural Networks for The Prediction of Wear Loss and Surface Roughness Of AA 6351 Aluminium Alloy, Materials and Design **27**, 2006.
5. R N Ibrahim, Y C Lam and D Ischenko, Predictions of Residual Stresses Caused by Quenching Process in Aluminium 6351- T6 Gas Cylinder, International Conference on Fracture, ICF 9-Sydney, Australia 1997.
6. J.W.H. Price and R.N. Ibrahim. Cracking in Aluminum Gas Cylinders: A Review of Causes and Protection Measures, Practical Failure Analysis, ASM International, PFANF8, **3**(6) 2003.
7. J. R Davis, Corrosion of Aluminum and Aluminum alloys, Ohio, Internatinal. ASM 1999.
8. B Zlaticanin , S Duric , B Jordonic, B.Radojic, Journal of mining and metallurgy , **39**(3- 4),2003.
9. J Gubicza, N Q Chinh, Z Horita, T.G. Langdon, Material Science and Engineering 387-389,2004.
10. L.A. Dobrzański, R. Maniara, J.H. Sokolowski , Archives Of Material Science and Engineering **28** 2,2007.
11. ZHU Mei-jun, DING Dong-yan Effect of Zn content on tensile and electrochemical properties of 3003 Al alloy Transition of non ferrous metal society china 20,2010.
12. S.O Adeosun , S.A. Balogun , L.O. Osoba , W.A. Ayoola , A.M. Oladoye , Journal of modern manufacturing technology **3**(1) 2011.
13. H.N Girisha., K.V. Sharma , International Journal of Scintific and Engineering Reasearch, **3**(2) 2012.
14. R.S. Rana, R Purohit, S Das, International Journal of Scientific and Reaserch Publications , **2** (6) 2012.
15. L. agrawal, R. yadav and A. Sexena, Effect of magnesium content on the mechanical properties of al-zn-mg alloys, International journal on emerging technologies **3**(1) 2012.
16. L. F. Mondolfo, Aluminum alloys: Structure and Properties, London, Butterforts ,1976.
17. M. Qiaoling, Y. Yong, Wang Qianfeng, International Conference on Education Technology and

Management Engineering Lecture Notes in information Technology, **16-17**, 2012.
18. Atanda, Olorunniwo, Alonge, Oluwole, International Journal of materials and chemistry **2**(4) 2012.
19. K. Tataram, H.M.Chavan, N. Swamy, International Journal of Research in Engineering & Advanced Technology, **1** (4)

CRYOGENIC PROCESSING OF HSS M2: MECHANICAL PROPERTIES AND XRD ANALYSIS

Ish Hunar Sharma*and Satpal Sharma

Department of Mechanical Engineering,
School of Engineering, Gautam Buddha University,Greater Noida

Abstract. The purpose of this study was to reveal the changes in properties of deep cryogenic treatment (DCT) processed high speed steel, in comparison to conventional heat treatment for secondary hardness. Tool properties were assessed in terms of hardness, and tensile strength. Statistically significant enhancement in the mechanical properties was observed. Cryogenic processing of HSS tool steel eliminated the retained austenite, and hence increased the hardness of the material. This treatment initiated nucleation sites for precipitation of large numbers of very fine carbide particles.Tensile values for cryogenically treated HSS samples can be attributed to the fact that the tool becomes more brittle after the treatment. XRD analysis illustrated the contraction in lattice of martensite and austenite. Deep cryogenic treatment practically removed all traces of austenite in the sample. The superior performance of cryogenically treated HSS can be attributed to the transformation of almost all retained austenite into martensite, a harder structure and precipitation of fine and hard carbides.

1 Introduction

Recently, application of cryogenic treatment in enhancing properties of tool materials has received wide acceptance by researchers and industries. Research has shown that cryogenic treatment increases product life, and provides additional qualities to the product, such as stress relieving[1,2]. Tool life is a major factor that is considered in production of finished product in manufacturing industry as any improvement in tool life will have a direct impact on the cost of production, tool changing time and maximumproduction target. It has been reported that cryogenic treatment can double the service life of HSS tools, and also increase hardness and toughness simultaneously [3].

Tool steel (carbon and alloy steels) having distinctive hardness, resistance to abrasion, ability to hold a cutting edge, and resistance to deformation at elevated temperatures, are generally used for cutting tools, punches, and other industrial tooling. Different tool steels are developed to resist wear at temperatures of forming and cutting applications. Cold work tool steels are specific for tool bits, dies and other applications where high wear resistance and low cost are required [3].

Tool steel is generally used in a heat treated state. Conventional heat treatment gives hardness, wear resistance and ductility to steel. However, it cannot remove all of the retained austenite (large, unstable particles of carbon carbide) from steel. The retained austenite (a soft phase) in steels could reduce the product life and affects the working. Therefore, there is a need for transformation into marten site. In order to obtain this transformation the cold treatment is used. Cold treatment is generally classified as either so called "sub-zero treatment" at temperatures down to about -80 °C or "deep cryogenic treatment" at liquid nitrogen temperature (-196°C). Cryogenic treatment commonly referred to as cryotreatment, is an add-on process to the conventional heat treatment of tool/die steel. It consists of controlled cooling of conventionally hardened steel specimens to some selected cryogenic temperature (–500c to–1960c) and holding there for sufficiently long duration (20 to 75

Corresponding authors: ishhunar@gmail.com

h) before being heated back to the ambient temperature at a predetermined rate for subsequent tempering treatment [4,5]. Cryogenic treatment improves wear resistance, hardness, toughness, resistance to fatigue cracking, microstructure of metal (retained austenite to marten site), dimensional stability and decreases residual stresses. Cryogenic treatment gives machining, grinding and polishing finish due to little soft austenite[6].

In this direction, the present experiment was carried out to assess the enhancement in the mechanical properties of HSS tool and the compositional analysis by XRD analysis.

2 Experimental

2.1 Materials Selected:HSS M2

2.2 Manufacturing of Tool Bits

CuttingGrinding→Heat→ treatment→Final grindingTOOL BITS

2.3 Heat processing of Tool Bits

Conventional heat treatment

The conventional heat treatment for the M2 steel tool bits was conducted based on the alloy heat treatment. Hardening (austenitizing) at 1123 K (850 °C) for two and half hour, is followed by oil quenching and tempering at 723 K (450 °C) for one hour.

Deep Cryogenic Treatment

The HSS tool bits steel which had undergone the conventional hardening and were slowly cooled from room temperature to 85 K (-196°c), soaked at 85 K for 24 h, and finally heated back to room temperature. The processor is a well-insulated chamber with liquid nitrogen as the working fluid. The cryogenic processor consists of a treatment chamber, which is connected to a liquid nitrogen tank through a vacuum insulated hose pipe. The liquid nitrogen passes through the spiral tubes of heat exchanger and enters the duct leading to the bottom of the chamber as nitrogen gas (evaporated state). The blower at the top of the chamber sucks the gas coming out at the bottom and makes it to circulate inside the chamber and reduces the chamber temperature drastically. The samples taken out from the processor were tempered at 473 K (200°C) for an hour.

2.4 Composition analysis

The composition of tool bits from HSS steel for various elements was assessed. (Instrument used: Spark Emission Spectrometer BAIRD, DV6, USA)

2.5 Mechanical Properties

Hardness test

Rockwell hardness testing (fig.) method was used for measuring the bulk hardness of metal. (Digital Rockwell Hardness, Make: FIE, Model: RASNET-1, Resolution; 0.1 HRc, Sr No 03/2006-1339).This method consists of indenting the test material with a diamond cone or hardened steel ball indenter. The indenter is forced into the test material under a preliminary minor load usually 10 kgf. The permanent increase in depth of penetration, resulting from the application and removal of the additional major load is used to calculate the Rockwell hardness number for the following test samples:

Tool Steel: Raw
→CHT (Conventional Heat Treated)
DCT (Deep Cryogenic Treated)

Tensile test

The tensile test is to determine the ultimate tensile strength. As per the IS 1608-2005 test method, the tensile test is carried out for the HSS Tool bits specimens (Raw, CHT and DCT). The specimen (bone shapped) is gripped in the tensile testing machine (i.e. Universal Testing Machine UTM, Capacity 500KN, resolution 1N, Make MTS, USA). After clamping the specimen on the holder the machine is switched on. The load is applied on the specimen along the uniaxial direction. Tensile strength was noted in Mpa.

2.6 XRD Analysis

X-ray diffraction (XRD) relies on the dual wave/particle nature of X-rays to obtain information about the structure of crystalline materials. A primary use of the technique is the identification and characterization of compounds based on their diffraction pattern.

3 Results and Discussion

3.1Composition: Table 1 depicts the concentration of various elements present in M2 HSS

Table 1. Chemical composition of HSS M2

Elements	Concentration (%)
C	0.68
Si	0.33

DOI: 10.1051/matecconf/20165703009

Mn	**0.33**
P	**0.026**
S	**0.004**
Ni	**0.27**
Cr	**4.16**
Mo	**4.91**
V	**1.87**
W	**5.89**

3.2 Hardness

Table 2 illustrated thehardness (HRc) of HSS Tool bits samples (raw, CHT and DCT) assessed by Rockwell Hardness tester. The mean values pertaining to hardness were 29.5,65.5 and 68 HRc for raw,CHTand DCT toolbits, respectively; depicting about 3.81% improvement in the hardness of the toolbits on cryogenic processing.Increase in hardness of cryogenically treated samples is due to the conversion of retained austenite to martensite and due to the presence of fine carbides in the metal matrix. Most researchers believed that there are two mechanisms to improve the mechanical properties of the work that has been treated cryogenically. [7,8] The first mechanism is attributed to the transformation of retained austenite to marten site. The second is to initiate the nucleation sites for precipitating a large number of fine carbides in the matrix of marten site. [9]

Table 2. Hardness (HRc) of HSS tool bits

Type of sample	**Hardness (HRc)**		
	Value 1	**Value 2**	**Mean**
Raw	29	30	29.5
CHT	65	66	65.5
DCT	68	68	68.0
Percent improvement in hardness 3.81			

CHT- Conventional heat treated; DCT- Deep cryogenic treated

3.3 Tensile strength

The tensile strength of raw and CHT tool bitts was 931.9 and 1420.0 MPa, respectively as inferred from the table 3. The cryogenic treated tool bitts became very brittle showing the significant effect in quality of the tool bitts. Our results are in confirmation with the earlier reported studies. [10-11] Cryogenic treatment in tool steels causes the precipitation of finely dispersed carbides in marten site and also converts soft unstable austenite to marten site.

Table 3. Tensile Strength (MPa) of HSS tool bits

Type of sample	**Tensile Strength (MPa)**
Raw	931.9
CHT	1420.0
DCT	(very brittle)
Percent improvement Tensile Strength 52.37	

CHT- Conventional heat treated; DCT- Deep cryogenic treated

3.4 XRD analysis

To monitor austenite content, X-ray diffraction method has been found to be most effective and accurate. The phases present in the Conventional Heat Treated and Cryogenic Treated Specimens have been studied using X-Ray Diffractometer for identification and characterization of compounds based on their diffraction pattern. The X-Ray diffraction pattern (Fig.1-3) showed that austenite, martensite and carbides of chromium and tungsten are the major phases present in the sample after conventional heat treatment. Amount of retained austenite present at the end of conventional heat treatment was found to be 15 % .Presence of retained austenite deep cryogenic treatment has been found to be 4-5% respectively. During cryogenic treatment contraction in lattice of martensite and austenite took place. Due to super saturation martensite with carbon and thermodynamic instability the carbon atoms squeezed out of martensite, migrated to the neighbouring lattice defects and acted as nucleation sites for the growth of fine carbides. Upon tempering fine precipitates of carbides were formed, with concurrent softening of martensite. The X-Ray diffraction pattern (Fig.5) corresponding to deep cryogenic treatment showed no traces of austenite but increase in percentage of carbides by about 2%.

DOI: 10.1051/matecconf/20165703009

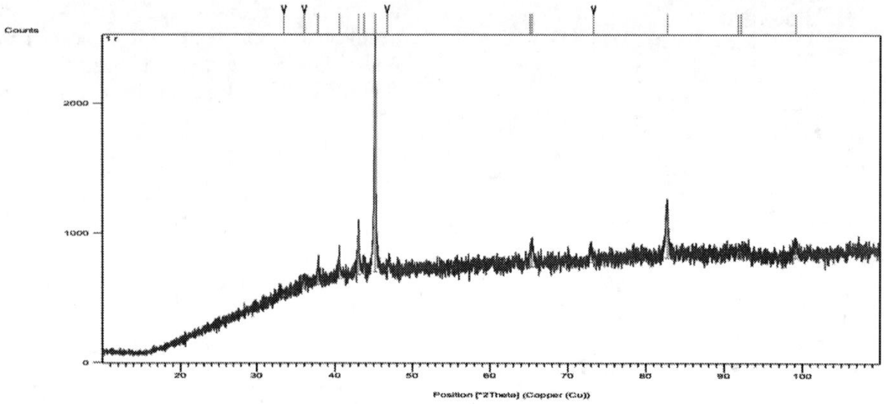

Fig:1 X-Ray Diffraction Pattern of raw M2 HSS

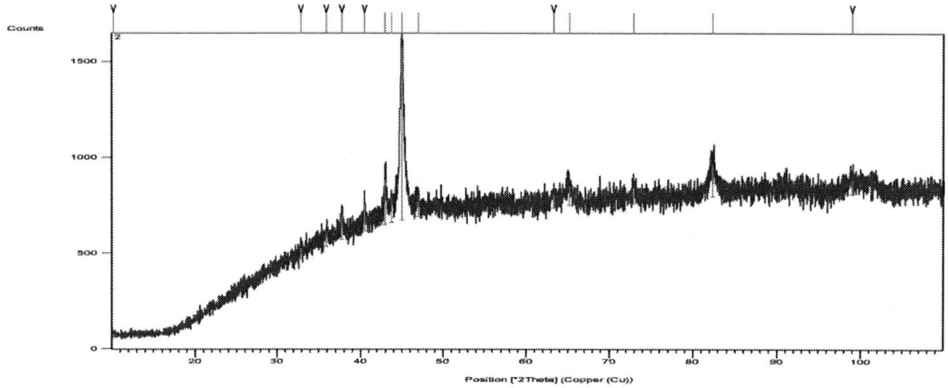

Fig:2 X-Ray Diffraction Pattern of Conventional Treated M2 HSS

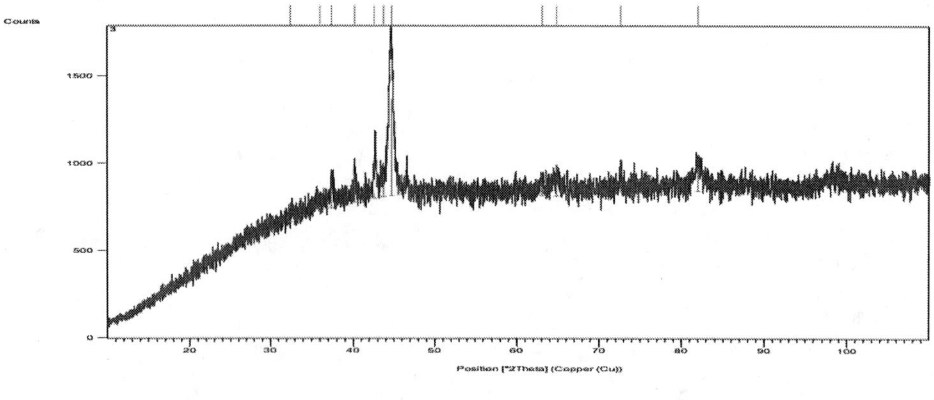

Fig.3. X-Ray Diffraction Pattern of Deep Cryogenic Treated M2HSS

4 Conclusion

The lower temperature processing of HSS M2 leads to formation of a higher carbide concentration which results in enhanced hardness. XRD analysis reveals that after the cryogenic treatment the microstructure of material is refined, retained austenite is completely converted in martensite and more number of secondary carbides precipitate on the surface. Therefore, cryogenic processed HSS M2 tool exhibit better in terms of industrial applications.

References

1. V.Firouzdor, E.Nejati, F.Khomamizadeh, "Effect of Deep Cryogenic Treatment on Wear Resistance and Tool Life of M2 HSS Drill", Journal of Materials Processing Technology **206**, 467–472.(2008).
2. Flávio J. da Silva, "Performance of cryogenically treated HSS tools". Journal of Engineering Manufacture. **1**: 50–59. (2006).
3. Pérez M and Belzunce F The effect of deep cryogenic treatments on the mechanical properties of an AISI H13 steel. Mater Sci Eng **624:** 32-40.(2015).
4. D. Das, A.K. Dutta, K.K. Ray, Correlation of microstructure with wear behavior of deep cryogenically treated AISI D2 steel, Wear, **267:** 1371-1380.(2009).
5. D. Das, A.K. Dutta, K.K. Ray,On the enhancement of wear resistance of tool steels by cryogenic treatment, Philosopical Magazine Letters , **88** (11): 801-811. (2008).
6. A.Akhbarizadeh, A.Shafyei, M.A.Golozar, "Effects of cryo-genic treatment on wear behavior of D6 tool steel", Materials and Design **30:** 3259–3264. (2009).
7. D. Mohan Lal, Renganarayanan. Kalanidhi, "Cryogenic treatment to augment wear resistance of tool and die steels", Cryogenics. **41**:149–155.(2001).
8. Molinari, Pellizzari, Gialanella, Straffelini, Stiasny, "Effect of deep cryogenic treatment on the mechanical properties of tool steels", J. Mater. Process. Technol. **118**: 350–355.(2001).
9. Fanju Meng, Kohsuke Tagashira, Ryo Azumu and Hideaki Sohma, Role of eta-carbide precipitations in the wear resistanceimprovements of Fe-12Cr-Mo-V-1.4C tool steel by cryogenic treatment , ISIJ international, **34**: 205-210.(1994).
10. Dong, Lin, Xiao, "Deep cryogenic treatment of high-speed steel and its mechanism". Heat Treatment of Metals. **3**: 55–59.(1998).
11. A D Wale, V D Wakchaure Effect of Cryogenic Treatment on Mechanical Properties of Cold Work Tool Steels International Journal of Modern Engineering Research **3**(1):149-154.(2013).

An empirical formula to estimate rainfall intensity in Kupwara region of Kashmir valley, J and K, India

Abdul Qayoom Dar[1a], Humairah Maqbool[1], Syeedah Raazia[1]

[1]*Department of Civil Engineering, National Institute of Technology, Srinagar, J&K, India*

Abstract Knowledge of rainfall extremes particularly their magnitude and frequency, as embodied in Intensity-Duration-Frequency (IDF) relationships and IDF curves is fundamental to many engineering problems such as design of hydraulic structures, urban drainage system, water resources projects and many others. The objective of this study is to obtain IDF relationships and curves for the Kupwara region of Kashmir valley in Jammu and Kashmir, India. Three different frequency distributions namely the Gumbel distribution, the Pearson Type III distribution and the Log-Pearson Type III distribution were fitted to the rainfall data to obtain rainfall intensities for selected return periods (2, 5, 10, 25, 50 and 100 years) and durations (0.16, 0.5, 1, 3, 6, 12 and 24 hours). Regional constants in IDF relations were found using the Sherman Morrison method and results were compared based on the Chi-square goodness-of-fit test. Results obtained using all distributions showed a similar trend. However, the Pearson Type III distribution emerges to be the best fit for the rainfall data of the region. Results revealed that higher rainfall intensities have shorter durations. Maximum rainfall intensity 81.13 mm/hr as per the best fit relation occurs with a return period of 100 years for 0.16 hours duration.

1 Introduction

Assessment of extreme precipitation is an important problem in hydrologic risk analysis and design. Knowledge of extreme events with respect to their frequency (exceedance probability) and magnitude is important towards planning for weather-related emergencies, design of engineering structures, reservoir management, etc. Evaluation of rainfall extremes has thus been a major focus of both theoretical and applied hydrology. A popular means of quantifying rainfall extremes or their frequency is through the Intensity-Duration-Frequency (IDF) relationships. IDF relationships are empirical relationships between the intensity of rainfall, duration of rainfall and the return period (inverse of the exceedance probability). IDF curves are three parameter curve in which duration of precipitation is taken as abscissa, intensity of precipitation as the ordinates and the return period or frequency as the third parameter.

IDF curves provide for the estimation of the return period of a given rainfall event or the amount of rainfall corresponding to a given return period. IDF curves find applications in the design of hydraulic structures (such as culverts and bridges), roads and urban drainage systems, land-use planning and soil conservation studies, management of municipal infrastructure including sewers, storm water management, ponds and street curb, design of safe and economical structures for the control, storage and routing of storm water and surface drainage and risk assessment of dams and bridges. The curves can also be used as input to rainfall-runoff models that simulate floods for bridge and spillway design. IDF relationships are used in the rational method of peak runoff estimation to determine the average rainfall intensity for a selected time of concentration. IDF curves are also useful in analysing the effects of climate change on precipitation parameters.

Many researchers in the past have developed IDF curves for different regions of the world. Bell, 1969[1] developed IDF curves for certain areas of U.S.S.R; Al-Sheikh, 1985[2] derived IDF relations for Saudi Arabia. Al-Khalaf, 1997[3] conducted a study for predicting short-duration, high intensity rainfall in Saudi Arabia using IDF studies. Koutsoyiannis (1998)[4] proposed construction of the intensity-duration-frequency curves using data from both recording and non-recording stations. Elsebaie, 2012 [5] developed IDF relationship for two regions in Saudi Arabia. Ogarekpe, 2014[6] developed IDF relationships for Calabar Metropolis, South-Nigeria.

Al-Khalaf, 1997[3] suggested that IDF relationships are region specific and should hence be obtained for each region separately. Development of IDF relationships largely involves the knowledge of the underlying distribution. Commonly used distributions in IDF and other similar other studies include Gumbel, Generalised Extreme

[a] Corresponding author: aqdar_nitsri@rediffmail.com

Value, Pearson Type III and Log Pearson Type III distributions (Al-Dokhayel, 1986[7]; Dupont and Allen, 2000[8]; Hadadin, 2005[9]; Nhat *et al.*, 2006[10] and AlHassoun, 2011 [11]). The present study focuses on the development of IDF curves and hence IDF relationships for Kupwara region of the Kashmir valley using the Gumbel, the Pearson and the Log Pearson distributions.

2 Study area

District Kupwara with its head quarters at Kupwara lies in the northern most part of Kashmir valley in the state of Jammu and Kashmir. It is located between $34^0 16'$ and $34^0 50'$ North Latitudes and $73^0 45'$ and $75^0 20'$ East Longitudes, 82 km from the state capital, Srinagar. The district has a geographical area of 2379 square kilometers. The average height of the district is 1,581 metres above sea level. The district is flanked in the east and south by district Baramulla and in the west and north by the Line of Control. The region has a varied topography with the western, northern and eastern areas of the district being hilly while as the southern areas being mainly plain. The major physiographic features of the region include high karewa plateau lands, sharp ridges of hard rock and intervening valleys. Most of the land is covered either with forests or under agriculture. The region is largely drained by the Kishan Ganga River. Average minimum and maximum temperatures vary from –5°C to 32°C, with the higher reaches remaining cold throughout the year. The average annual rainfall in this region is 869 mm.

3 Methodology

3.1 Data collection and processing

There are in all six rain gauge stations in Kashmir valley, located at Srinagar, Pahalgam, Kupwara, Gulmarg, Kokernag and Qazigund. In the present study, rainfall data from Kupwara rain gauge station has been selected for determination of Intensity-duration-frequency relationships of the region.

Rainfall data corresponding to the period from 1978 to 2004 (30 years) for Kupwara region was procured from Indian Meteorological Department Srinagar Kashmir (IMD). From the raw data, maximum rainfall events for selected durations (10 minutes, 30 minutes, 1 hour, 2 hours, 3 hours, 6 hours, 12 hours and 24 hours) were sorted for each year. Then, mean and standard deviations were calculated for each of the annual maximum series of different durations.

3.2 Development of IDF curves

Three probability distributions namely the Gumbel distribution, the Pearson Type III distribution and the Log Pearson Type III distribution were fitted to each of the annual maximum series of the given durations.

The general statistical equation for rainfall amounts is given by Equation (1). (Chow, 1988 [12])

$$X_T = M + K_T s \tag{1}$$

In the above equation, X_T is the rainfall amount having return period of T years, M is the arithmetic mean of the series, K_T is known as the frequency factor and depends on the distribution of the series and s is the standard deviation of the series. The frequency factor multiplied by the standard deviation gives the departure of a certain rainfall value of given return period from the average rainfall. The corresponding rainfall intensity of a given duration rainfall and return period T years is calculated by Equation (2) as

$$I = \frac{X_T}{T} \tag{2}$$

3.2.1 Gumbel distribution

The Gumbel distribution is the most widely used distribution in the field of hydrology. This distribution is suitable for modeling of maxima. This is a single parameter distribution with frequency factor as the only parameter to be determined. Equation (3) is used to calculate the Gumbel frequency factor.

$$K_T = -\frac{\sqrt{6}}{\pi}\left[0.5772 + \ln\left(\ln\frac{T}{T-1}\right)\right] \tag{3}$$

In Gumbel distribution, the frequency factor is a function of the return period only.

3.2.2 Pearson Type III distribution

PT III distribution is normally suitable for skewed data. For this distribution, the frequency factor besides depending on the return period, is also a function of the coefficient of skewness of the data which is obtained by Equation (4) suggested by Chow, 1988 [12] and Burke & Burke, 2008 [13].

$$C_s = \frac{N\sum_{i=1}^{N}(X_i - M)^3}{(N-1)(N-2)s^3} \tag{4}$$

Frequency factor for each return period and the coefficient of skewness computed using the above equation are then determined using standard tables provided by Water Resources Council, 1967 [14].

3.2.3 Log Pearson Type-III Distribution

The LPT III distribution involves the log transformed series of the rainfall data. Using this distribution, the mean, the standard deviation and the skewness coefficient are calculated based on the logarithms of the original data values to the base 10. Frequency factors for all the annual

DOI: 10.1051/matecconf/20165703010

maximum series corresponding to different durations are obtained from the available statistical tables as functions of return periods and skewness coefficients. Antilog of the solution obtained from the general statistical equation (Equation 5) provides the estimate for the extreme rainfall for a given duration and return period.

3.3 Development of IDF equations

IDF equations are the empirical equations relating the dependent parameter rainfall intensity to the independent parameters return period and duration of rainfall. The general form of an IDF equation is as expressed by Equation (5).

$$I = \frac{CT^m}{T_d^{\,e}} \tag{5}$$

where I is the rainfall intensity in mm/hr; T is the return period in years and T_d is the duration of rainfall in days. C, m and e are known as the regional constants.

Three separate IDF equations were developed for the study area using the data generated by employing the three different distributions mentioned in the previous sections. The IDF equations are linearised using the logarithmic transform. The regional constants in each case were determined by Sherman-Morrison method of solving linear equations. The Chi-square goodness-of-fit test was used to determine the best fit IDF equation.

4 Results and discussions

4.1 IDF curves

The IDF curves obtained by fitting the Gumbel, the Pearson and the Log-Pearson distributions to rainfall data are respectively given in Figures 1 to3. The Gumbel distribution gives a maximum intensity of 64.42 mm/hr at return period of 100 years with duration of 0.16 hours and minimum intensity of 1.69 mm/hr at return period of 2 years with duration of 24 hours. The Pearson Type III distribution gives a maximum intensity of 81.13 mm/hr at return period of 100 years with duration of 0.16 hours and minimum intensity of 1.86 mm/hr at return period of 2 years with duration of 24 hours. The Log Pearson Type III distribution gives a maximum intensity of 93.67 mm/hr at return period of 100 years with duration of 0.16 hours and minimum intensity of 2.04 mm/hr at return period of 2 years with duration of 24 hours. The rainfall intensity values obtained using the three distributions are close for almost all return periods and rainfall durations.

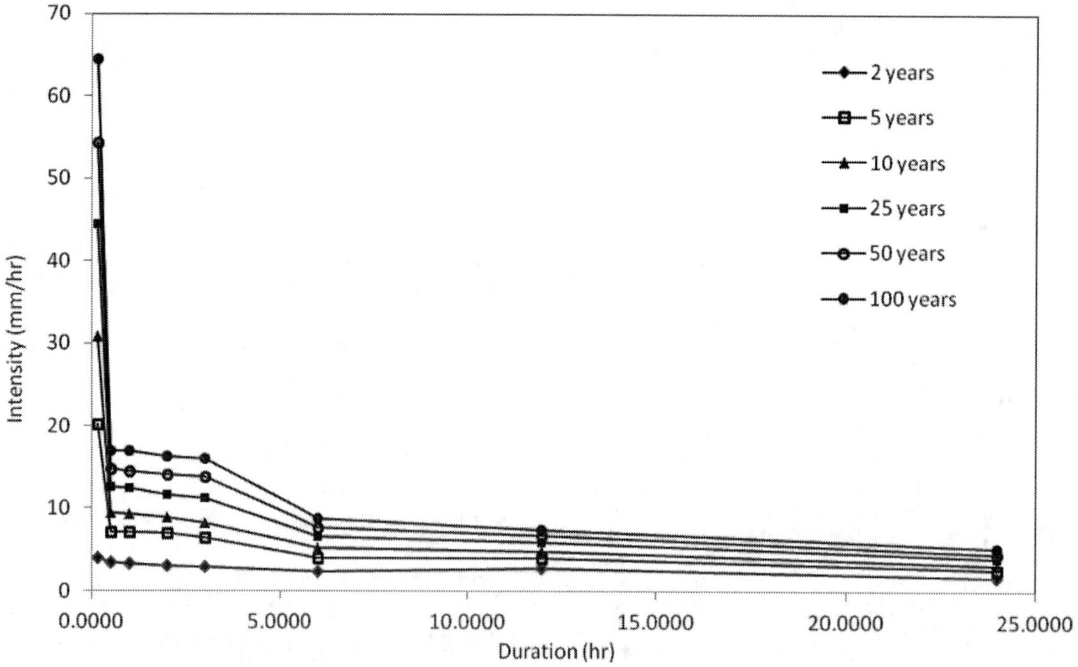

Figure 1. IDF curves by Gumbel distribution

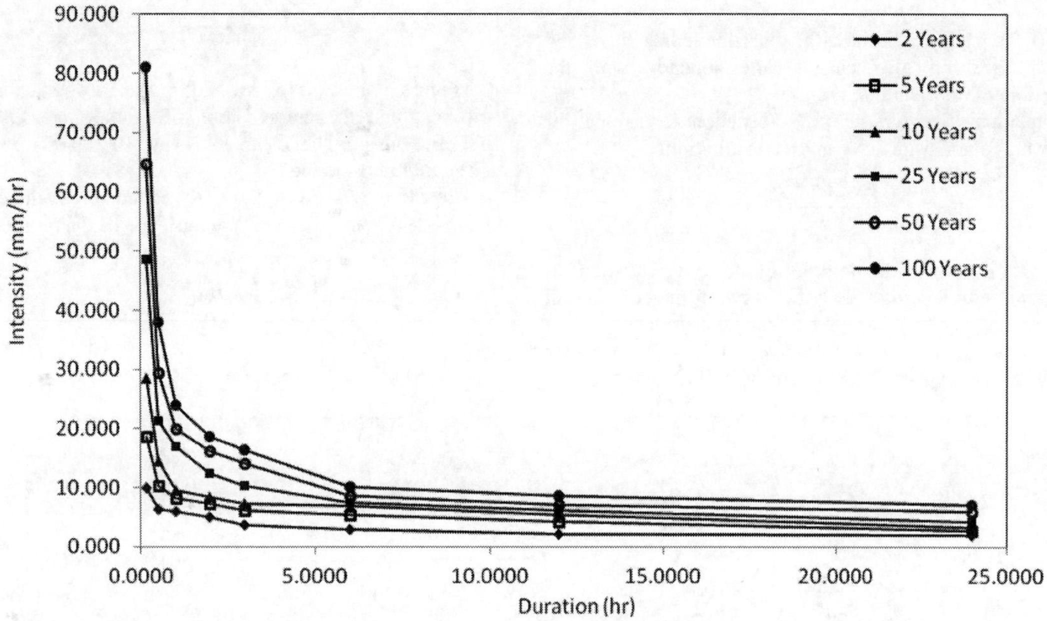

Figure 2. IDF curves by Pearson Type III distribution

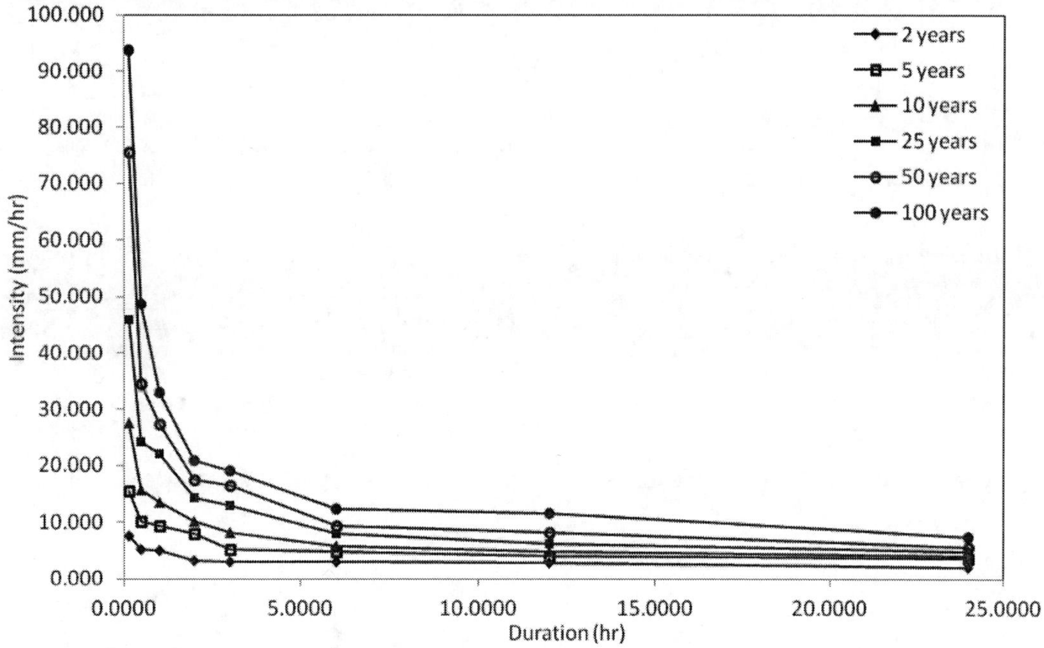

Figure 3. IDF curves by Log Pearson Type III distribution

It is seen that the IDF curves obtained from all of the three distributions show a similar trend. From the IDF curves, it can be inferred that the higher intensity rainfalls occur for only short durations and higher rainfall intensities also have longer return periods, that is, they are less frequent than the low intensity rainfalls.

4.2 IDF equations

The IDF equation is linear in the logarithmic transform, and hence logarithm of intensity varies linearly with the logarithm of rainfall duration.

Equation (6) gives the IDF equation developed for Kupwara region using the Gumbel distribution.

$$I = \frac{4.93T^{0.199}}{T_d^{\ 0.121}} \quad (6)$$

Figure 4 shows a log-log plot of the above equation for different return periods and the curve fitting of rainfall intensity data obtained using the Gumbel distribution.

Equation (7) gives the IDF equation developed for Kupwara region using the Pearson Type III distribution.

$$I = \frac{8.41T^{0.171}}{T_d^{\ 0.101}} \quad (7)$$

Figure 5 shows a log-log plot of the above equation for different and the curve fitting of rainfall intensity data obtained using the Pearson Type III distribution for different return periods.

Equation (8) gives the IDF equation developed for Kupwara region using the Pearson Type III distribution.

$$I = \frac{5.73T^{0.220}}{T_d^{\ 0.118}} \quad (8)$$

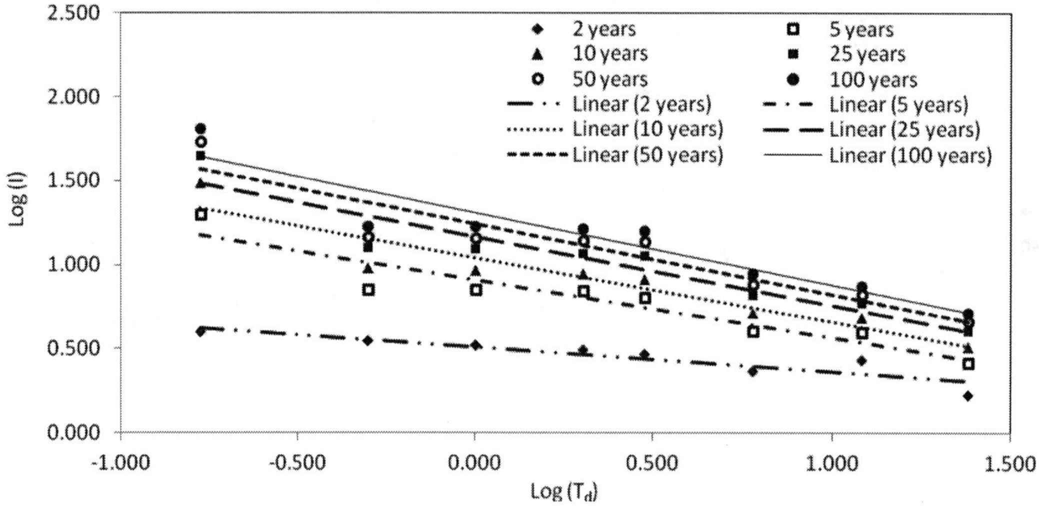

Figure 4. IDF equation generated using Gumbel distribution

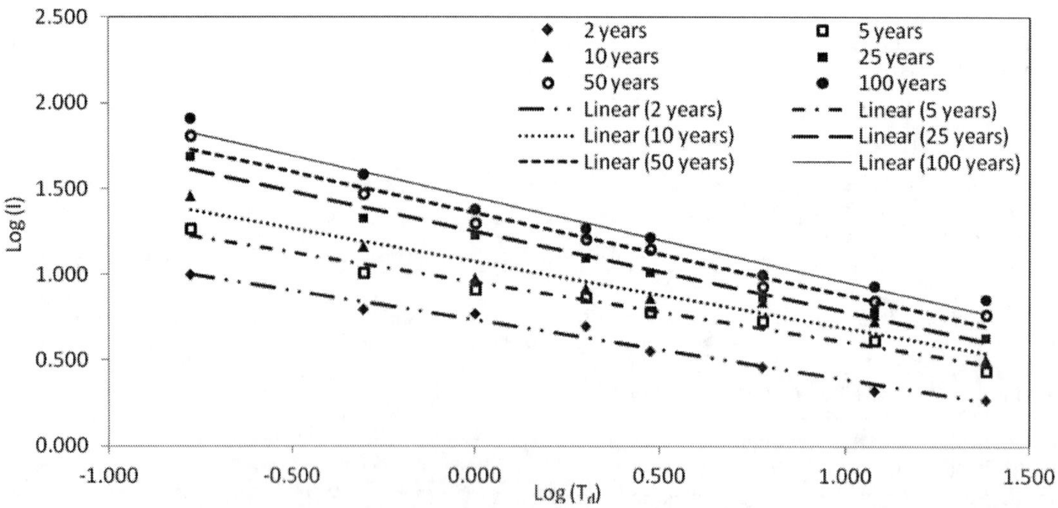

Figure 5. IDF equation generated using Pearson Type III distribution

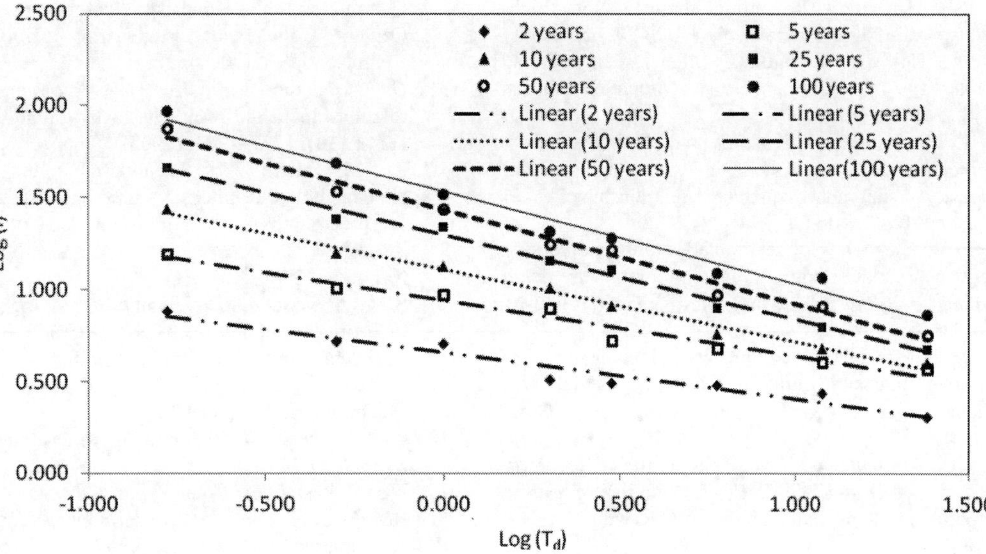

Figure 6. IDF equation generated using Log Pearson Type III distribution

Table 2. Values of chi-square statistic

Distribution	Duration (Hrs)							
	0.16	**0.5**	**1**	**2**	**3**	**6**	**12**	**24**
Gumbel	28.17	26.62	27.33	25.38	16.06	19.39	20.21	19.40
Pearson Type-III	18.52	39.87	15.14	28.13	12.97	15.02	14.97	19.95
Log-Pearson Type-III	20.75	22.29	28.53	19.85	30.23	20.42	20.92	27.55

Figure 6 shows the curve fitting of rainfall intensity data obtained using the Log Pearson Type III distribution for different return periods.

Table 1 gives the values of regional constants in IDF equations obtained for the data generated using three different frequency distributions.

Table 1.Estimates for regional constants in IDF equations

Parameters	Gumbel	Pearson Type III	Log-Pearson Type III
C	4.93	8.41	5.73
m	0.189	0.171	0.220
e	0.121	0.101	0.118

4.3 Chi-square goodness-of-fit test

The Chi-square values for rainfall intensities with different durations and different return periods are given in Table 2.

5 Conclusion

An empirical technique has been applied to rainfall data in order to determine the Intensity-Duration-Frequency characteristics of rainfall and the underlying distribution. Three different frequency distributions namely the Gumbel distribution, the Pearson Type III distribution and the Log-Pearson Type III distribution have been fitted to the rainfall data to obtain rainfall intensities for selected return periods (2, 5, 10, 25, 50 and 100 years) and durations (0.16, 0.5, 1, 3, 6, 12 and 24 hours). In general, the results obtained using the three approaches are very close at most of the return periods and have the same trend. For Kupwara region, as per the best fit distribution, maximum intensity for of 81.13 mm/hr occurs at return period 100 years with duration of 0.16 hours and minimum intensity of 1.86 occurs at return period 2 years with duration of 24 hours. As per the Chi-square goodness-of-fit test, the Pearson Type III distribution gives the best fit for the rainfall pattern in Kupwara region.

References

1. F. C. Bell. "Generalized rainfall-duration-frequency relationships." *Jr. Hydraul. Div.*, **95(1)**, 311-327, (1969).
2. A. A. Al-Shaikh. "Rainfall frequency studies for Saudi Arabia." *M.S. Thesis*. Civil Engineering Department, King Saud University, Riyadh (K.S.A), (1985).

3. H. A. Al-Khalaf. "Predicting short-duration, high-intensity rainfall in Saudi Arabia." *M.S. Thesis*, Faculty of the College of Graduate Studies; King Fahad University of Petroleum and Minerals, Dahran (K.S.A), (1997).
4. D. Koutsoyiannis, D. Kozonis and A. Manetas. "A mathematical framework for studying rainfall intensity- duration-frequency relationships." *Jr. Hydrol.*, **206**, 118-135, (1998).
5. I.H. Elsebaie, "Developing rainfall intensity–duration–frequency relationship for two regions in Saudi Arabia." *Jr. King Saud Univ.-Engg. Sci.*, **24**, 131-140, (2012).
6. N. M. Ogarekpe. "Development and comparison of different intensity duration frequency models for calabar, nigeria."*Nigerian Jr. Tech.* **33(1)**, 33-42, (2014).
7. A. A. Al-Dokhayel. "Regional rainfall frequency analysis for Qasim." *B. S. Project*, Civil Engineering Department, King Saud University, Riaydh (K.S.A), (1986).
8. B. S. Dupont and D. L. Allen, D.L. "Revision of the Rainfall Intensity Duration Curves for the Commonwealth of Kentucky. Kentucky Transportation Center, College of Engineering, University of Kentucky, USA. (2000).
9. N. A. Hadadin. "Rainfall intensity–duration–frequency relationship in the Mujib basin in Jordan." *Jr. App. Sci.*, **8 (10)**, 1777–1784, (2005).
10. L. M. Nhat, Y. Tachikawa and K. Takar. "Establishment of intensity–duration–frequency curves for precipitation in the monsoon area of Vietnam." *Annuals of Disas. Prev. Res. Inst.*, Kyoto University, **No. 49 B,** (2006).
11. S. A. AlHassoun. "Developing an empirical formulae to estimate rainfall intensity in Riyadh region." *Jr. King Saud Univ. Engg. Sci.*, **23(2),** 81-8, (2011).
12. V. T. Chow. *Handbook of Applied Hydrology*, McGraw-Hill Book, (1988).
13. C. B. Burke and T.T. Burke. *Storm Drainage Manual. Indiana LTAP*. (2008).
14. Water Resources Council, *Bulletin No. 15*, Water Resources Council, 1025 Vermont Avenue, N. W. Washington, D. C., 200005, (1967).

Parametric Analysis of Leachate and Water Resources around Municipal Solid Waste Landfill area in Solan

Deepika Sharma[1,a] and Rajiv Ganguly[1]

[1]*Department of Civil Engineering, Jaypee University of Information Technology, Waknaghat, District Solan, Himachal Pradesh 173234, India.*

Abstract. Leachate is defined as the liquid that drains from the landfill. The paper presents the physico-chemical, bacteriological and heavy metal testing results carried out for leachate, surface and sub-surface water samples collected from municipal solid waste landfill and different water sources in Solan to find out the effect of leachate percolation on groundwater quality. Physico-chemical parameters analysed were, pH, Total Dissolve Solid (TDS), sulphate, turbidity, Electrical Conductivity (EC) while biological parameters tested were Biological Oxygen Demand (BOD), Chemical Oxygen Demand (COD), Most Probable Number (MPN) test and ammonical nitrogen. Testing for heavy metals (Pb, Zn, Cr, Ni, Fe) were carried out and have been reported. The results reveal that the leachate from the unlined landfill may have a significant impact on the groundwater resource (often used as drinking source) particularly because of the toxic nature of the leachate coupled with the soil characteristics which is permeable in nature.

Key words: Groundwater, Heavy Metals, landfill, leachate, municipal solid waste, Solan

1. Introduction

Municipal Solid Waste (MSW) includes the solid waste generated by households, commercial sectors, industries and institutions. Improper management and disposal have adverse effect on the environment and human health. This leads to many serious problems including ground water contamination and to a lesser extent surface water sources.

Open landfill dumps are the most conventional and most preferred way of disposing of solid waste in developing countries as it is simple, cheap and highly cost effective [1]. In India, in recent years, a number of alternate methods for managing and disposing of generated solid waste have been in operation but open dumping is still the most common method of solid waste disposal. These are often unmanned sites and bad odour and aesthetic problems often associated with these dump site. Further, insects, rodents, snakes and scavenger birds are potential carriers of human health related diseases. These sites also produce huge volumes of dust and high noise intensities. Effective management of solid wastes is a global issue, particularly in developing countries, as less resources, high population growth rate, and rapid waste disposal. [2-3]. Further, in an Indian context, majority of these landfills are non-engineered landfills and thereby proper lining of these landfills are absent. Further, most of the MSW generated is often organic in nature and in tropical conditions they get dissolved in rainwater and percolate into the ground thereby polluting the ground water sources. In recent times, the impact of leachate on groundwater and other water resources has been studied in details because of its immense environmental impact [4]. Leachate migration from wastes dump sites, release toxic pollutants and pose a great threat to water resource (both surface and sub surface sources) if not properly managed. The problem is more severe for unlined landfills.

Leaching of inorganic and organic pollutants from solid waste dump site depend on different factors including pH, temperature, moisture content, type and properties of soil, hydrology of the area and the rate of progress through these stages is dependent on the physical, chemical and microbiological conditions developed within the landfill with time [5]. Due to excessive use of water resources, pollution from a large number of sources compounded by leaching of toxic chemicals from leachate to surface and groundwater surfaces.

The purpose of the study is to analyze the leachate, groundwater and surface water samples collected from

[a] Corresponding author: deepikasharmacivil89@yahoo.com

solid waste dump site to determine the potential of leachate polluting the surrounding water bodies.

2. Materials and methods

2.1 Study area

Salogra solid waste dump site is located near about 07 km form Solan district. The district lies between North latitude 30°44'53" to 31°22'01" and East longitude 76°36'10" to 77°15'14" and is covered by Survey of India degree-sheets 53A, 53B, 53E and 53F. The population of the district is 5, 76,670 (2011 census), of which 2, 70,291 (54 %) males, 2, 30,266 (46 %) female, sex ratio (F:M) is 884:1000 and density of population is 300 per sq km. Dump site spreads over an area of 3 acres. About 1.5 acre is used as dump site and remaining area is used as office and recycling plant. Waste is collected from 132 collection points from the Solan, Nalagarh, Arki and Kandaghat sub-divisions. Solid waste generated by Solan district is of the order of 22.5 tons per day. Soil is generally sandy loam in valley areas of the district. Figure 1 shows the location of the dumpsite.

2.2 Sampling and analysis of samples

Leachate samples have been collected from waste dumpsite in September 2015. Groundwater sample has been taken from the hand pumps near the waste disposal sites and surface water samples have also been collected. Samples have been collected in polyethylene and glass bottles. All samples were properly labelled with details of the source, date of sampling, time of sampling and address. Sample for bacteriological parameters were collected in sterilised bottle. The samples were transported to the laboratory, stored at 4°C and analyzed within 24 hours for relevant parameters according to internationally accepted procedures and standard methods [6]. The concentrations of heavy metals were determined using an atomic absorption spectrophotometer

3. Physico-chemical and bacteriological characteristic of samples

The results of the analysis of the leachate sample have been presented in Table 1 and Table 2. Tables 3 and 4 present the parametric results for surface and groundwater.

Table 1. Results of testing for leachate samples for physico-chemical parameter.

S.no	Parameter	Results
1	Conductivity (EC) µmho/cm	2960
2	pH	7.41
3	TDS	1968
4	Sulphate (mg/l)	50
5	Nitrate (mg/l)	27
6	Ammonical nitrogen(N) (mg/l)	nil
7	BOD at 27 ^{0}c 3 day	2300
8	COD (mg/l)	7150

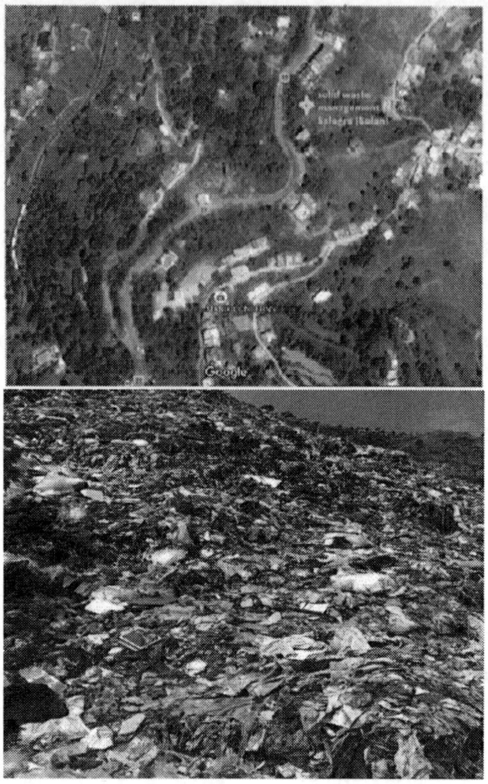

Figure 1: solid waste management plant , Salogra (Solan)

Table 2. Results of testing for leachate samples for heavy metals

S.no	Parameter (heavy metal)	Results
1	Lead (mg/l)	0.035
2	Zinc (mg/l)	0.084
3	Chromium (mg/l)	BDL
4	Nickel (mg/l)	0.017
5	Iron (mg/l)	0.210

The pH for leachate sample was found to be between 7.4, for groundwater 7.41 and for surface water 8.28. All leachate samples have high concentration of dissolved solids. The total dissolved solid was found 1968 mg/l in leachate. In groundwater samples dissolved solid concentrations were 242 mg/l and in surface water was 389 mg/l. The BOD value reported was 2300 mg/l and the COD value obtained was 7150 mg/l. High COD concentrations represents higher amount of inorganic matter present in the leachate sample. The BOD/COD ratio was found to be 0.32 which means that substantial portion of the organic matter is biodegradable.

Table 3. Results of testing for Surface & ground water samples physicochemical parameters

S.no	Parameter	Result (surface water)	Result (Ground water)	Acceptable Limits (WHO)	Permissible limits (WHO)
1	pH	8.28	7.41	6.5-8.5	No relaxation
2	TDS	389	242	500	2000
3	Sulphate (mg/L)	18	63	200	400
4	Nitrate (mg/L)	0.15	BDL	45	No relaxation
5	Ammonical nitrogen(N) (mg/L)	nil	nil	-	-
6	BOD at 27 ^{0}c 3 day (mg/L)	6.2	Nil	-	30 (surface water)
7	COD (mg/L)	20	Nil	-	250 (surface water)

Table 4. Results of testing for Surface & ground water samples for heavy metals

S.no	Parameter	Result (surface water)	Result (ground water)	Acceptable limits (WHO)	Permissible limits (WHO)
1	Lead (mg/L)	0.024	0.027	0.01	No relaxation
2	Zinc (mg/L)	0.014	0.018	5	15
3	Chromium (mg/L)	BDL	BDL	0.05	No relaxation
4	Nickel (mg/L)	0.006	0.002	0.02	No relaxation

It was interesting to note that no BOD and COD concentrations were reported for groundwater samples. BOD and COD concentrations determined for samples collected from surface water sources showed also minimum concentrations and were well within permissible limits. However, the results do not give any indication of the contribution from leachate.

In case of heavy metals, lead concentrations were higher than permissible limits for all samples collected. The reported results were 0.035 mg/l in leachate sample, 0.024 mg/l for surface water and 0.027 mg/l for ground water Lead is known to cause serious conditions such as anemia, brain damage and kidney failure. Further, because of size and charge similarities, lead can substitute for calcium and can be deposited in bones; children are especially susceptible to this effect. The concentration of chromium in leachate, surface and groundwater was found below the detection limit (BDL). The WHO recommended limit for Cr in drinking water is 0.05 ppm/lt. The leachate sample tested reported Cr concentration of 0.017 mg/l, whereas it was reported to be 0.006 mg/l for surface water and 0.002 mg/l for groundwater. Zinc is a trace element that is essential for human health. In less quantities absorbed it can lead to loss of appetite, decreased sense of smell and slow healing of wounds. The concentration of Zinc in leachate was to be found 0.084 mg/l, 0.014 mg/l in surface water and 0.018 mg/l for groundwater. Hence, for all the samples the concentrations were well within the permissible limits prescribed by WHO [7].

Concentration of sulphate was determined to be 50mg/l for leachate sample and 18 mg/l and 63 mg/l for surface water and groundwater respectively. All result reveals that the concentration of sulphate in groundwater was higher as compared to leachate and surface water due to same natural sources. Nitrate concentrations in leachate sample were determined to be 27 mg/ l and 0.15mg/l in surface water. It was BDL for groundwater source. Ammoniacal nitrogen (N) was nil for all samples.

Results for E.Coli and fecal coliform testing results for the samples have been presented in Table 5 and Table 6 for leachate and groundwater samples.

Table 5. Results of testing of biological examination for leachate samples

S.no	Parameter	Results
1	E.coli	2500/100ml
2	Fecal Coliforms	3000/100ml

Table 6 Results of testing for hand pump water samples

S.no	Parameter	Results
1	E.coli	Absent
2	Fecal Coliforms	Absent

Result of coliform testing showed that the leachate contained high concentration of E.Coli and fecal coliforms whereas they were absent for both surface and ground water samples. A similar study conducted in Turmuri, Belgaum reported that samples lying within the

500 m from landfill area were contaminated with E-Coli bacteria as it exceeds the prescribed limits of BIS and WHO standards with the number in 1600, 160,0 300 per 100 ml [8].

4. Conclusion

In this study, leachate, surface and groundwater samples were collected from waste dump site to find out the effect solid waste on the groundwater quality. Samples were collected from hand pumps, surface water body and leachate from waste dump site in sampling bottles and different physico chemical parameters were analysed. The concentrations of metals were determined with the instruments like; AAS. The results of this study revealed that metal concentration, BOD, COD was high in all of the leachate samples. In comparison, very low values were reported for groundwater and surface water samples. Zn concentration was within the WHO limits for all the samples. However, high concentration of lead was observed in surface and groundwater and exceeded the WHO recommended value. Chromium was also found in all samples collected from groundwater & surface water in some samples the concentration of chromium was higher. E.coli and fecal coliforms was present in leachate samples but absent in groundwater& surface water show the presence of fecal matter in waste.

From the results, it can be concluded that the metals present in solid waste can migrate into groundwater and pollute it and contribute to degradation of land, water and air. So, there is a need to change in land use pattern and to adapt new methods for waste utilization. The countries like U.S, Germany, UK, Denmark and Netherlands have many alternatives for disposal and recycle and reuse of municipal solid waste. Use of organic waste as manure in agriculture provides a feasible alternative for its safe disposal and improves soil environment and crops production. However, a judicious amendment strategy has to be developed to abate the land and water pollution from the pollutants presents in it.

REFERENCES

1. A. Barrett, J. Lawlor The Economics of Waste Management in Ireland, Economic and Social Research Institute, Dublin (1995).
2. S.J. Cointreau Environmental management of urban solid wastes in developing countries: A project guide, Urban Development Dept, World Bank, Retrieved June 27, 2009 from http://www.worldbank.org/html/fpd/urban//solid-wm/techpaper5.pdf.(1982).
3.P.L. Doan Institutionalizing household waste collection: the urban environmental management project in Cote d'Ivoire. Habitat Int., **22**, 1, 27-39 (1998)
4. S. Mor, K. Ravindra, R.P. Dahiya, A. Chandra A Leachate Characterization and assessment of groundwater pollution near municipal solid waste landfill site. Environ. Monit. Assess., **4**, 325-334 (2006)
5. F.G. Pohland, S.R Harper Critical Review and Summary of Leachate and Gas Production from landfills, US Environmental Protection Agency, Cincinnati; EPA/600/2-86/073 (1985)
6. Standard Methods for the Examination of Water and Wastewater, American Public Health Association (APHA), American Water Works Association (AWWA), Water Pollution Control Federation (WPCF). 16th edition, American Public Health Association, Washington, DC, U.S.A (2010)
7. WHO. *Guidelines for Drinking Water Quality*. (World Health Organization Recommendation, Geneva 3rd Edn. Vol. 1, 515, 2004)
8. C. Patil, S. Narayanakar, A. Virupakshi Assessment of Groundwater Quality Around Solid Waste Landfill Area - A Case Study. Int J Innov Res Sci, Eng and Tech **2**, 7, 3131-3136 (2013)

Statistical Analyses of Optimum Partial Replacement of Cement by Fly Ash Based on Complete Consumption of Calcium Hydroxide

Winai Ouypornprasert[1,a], Narong Traitruengtatsana[2] and Kong Kamollertvara[2]

[1]*Department of Civil Engineering, College of Engineering,Rangsit University 12000 Pathumthani, Thailand*
[2]*Asia Group (1999) Co., Ltd., King Rama II Rd., Bangmod, 10150 Bangkok, Thailand*

Abstract. The objectives of this technical paper were to propose the optimum partial replacement of cement by fly ash based on the complete consumption of calcium hydroxide from hydration reactions of cement and the long-term strength activity index based on equivalent calcium silicate hydrate as well as the propagation of uncertainty due to randomness inherent in main chemical compositions in cement and fly ash. Firstly the hydration- and pozzolanic reactions as well as stoichiometry were reviewed. Then the optimum partial replacement of cement by fly ash was formulated. After that the propagation of uncertainty due to main chemical compositions in cement and fly ash was discussed and the reliability analyses for applying the suitable replacement were reviewed. Finally an applicability of the concepts mentioned above based on statistical data of materials available was demonstrated. The results from analyses were consistent with the testing results by other researchers. The results of this study provided guidelines of suitable utilization of fly ash for partial replacement of cement. It was interesting to note that these concepts could be extended to optimize partial replacement of cement by other types of pozzolan which were described in the other papers of the authors.

1 Introduction

Cement contents in concrete is the main cost for the concrete production. The reduction of cement content in concrete mix proportion would lower the cost of concrete and would increase the competitiveness in marketing. Therefore the probabilistic concrete mix design for mass production of concrete has been commercially applied [1]. Dumrongsil, Chatveera and Ouypornprasert [2] studied the suitable partial replacement of cement by rice husk ash (RHA). In this study the optimum replacement of cement by RHA was derived by the concept of calcium oxide equivalent. Results showed that the values of compressive strength of concrete mixed with RHA of the age less than 28 days tend to be lower than those of concrete mixed with Ordinary Portland Cement (OPC). However with suitable replacements of cement by RHA the values of compressive strength of concrete can be higher than those of ordinary concrete even at the early age [3]. The major reason might be the heat of hydration of cement mixed with pozzolan could surpass that of OPC [4]. Furthermore the optimum partial replacement of cement by RHA could be predicted accurately based on the concept of complete consumption of calcium hydroxide [5].

In this technical paper the optimum replacement of cement by fly ash (FA) based on stoichiometry of complete consumption of calcium hydroxide from hydration of cement and pozzolanic reactions of pozzolan substances as well as the long-term strength Activity index based on Equivalent Strength of Calcium Silicate Hydrates (C-S-H) is proposed. Since the chemical compositions of cement and FA are varied in the production process, propagation of uncertainty with respect to main chemical compositions will also be discussed. To assure the quality of concrete mixed with FA the suitable replacement of cement by FA should be determined with the target confidence interval. Once the particular replacement is selected the distribution of the long-term strength Activity index could be obtained. Then reliability of the selected replacement could be determined accurately by Monte-Carlo simulations or approximately by the advanced First-Order Second-Moment (FOSM) method. The applicability of the proposed formulations could be demonstrated by a set of data available in hand.

2 Objectives

The objectives of this study were to

1) formulate the optimum replacement of cement by fly ash based on the complete consumption of calcium hydroxide from hydration reactions of cement,
2) propose the long-term strength Activity index based on equivalent calcium silicate hydrate,
3) analyze the propagation of uncertainty of the optimum replacement of cement by fly ash in terms

[a] Winai Ouypornprasert: owinai@yahoo.com

of variation of main chemical compositions in cement and fly ash,

4) review the reliability analyses for utilizing the optimum replacement of cement by fly ash,
5) demonstrate the applicability of the proposed concepts.

3 Materials and methods

3.1 General remarks

It is usual for the cement chemist to use the following abbreviations for discussions about hydration reactions of cement and pozzolanic reactions [6]: C for calcium oxide (CaO), S for silicon dioxide (SiO_2), H for water (H_2O), A for aluminum oxide (Al_2O_3), F for ferric oxide (Fe_2O_3), CH for calcium hydroxide ($Ca(OH)_2$), C_3S for tricalcium silicate ($3CaO\cdot SiO_2$), C_2S for dicalcium silicate ($2CaO\cdot SiO_2$), C_3A for tricalcium aluminate ($3CaO\cdot Al_2O_3$), C_4AF for tetracalcium alumino ferrite ($4CaO\cdot Al_2O_3{\cdot}Fe_2O_3$), $C_3S_2H_3$ or $C-S-H$ for calcium silicate hydrate ($3CaO\cdot 2SiO_2.3H_2O$), C_3AH_6 for calcium aluminate hydrate ($3CaO\cdot Al_2O_3.6H_2O$) and C_3FH_6 for calcium ferrite hydrate ($3CaO\cdot Al_2O_3.6H_2O$).

3.2 Hydration reactions, pozzolanic Reactions and stoichiometry

3.2.1 Hydration reactions

3.2.1.1 Hydration reaction of tricalcium silicate (C_3S)

$$2(3CaO\cdot SiO_2)+6H_2O\rightarrow 3CaO\cdot 2SiO_2\cdot 3H_2O+3Ca(OH)_2 \quad (1)$$

$$2C_3S+6H\rightarrow C_3S_2H_3+3CH \quad (2)$$

3.2.1.2 Hydration reaction of dicalcium silicate (C_2S)

$$2(2CaO\cdot SiO_2)+4H_2O\rightarrow 3CaO\cdot 2SiO_2\cdot 3H_2O+Ca(OH)_2 \quad (3)$$

$$2C_2S+4H\rightarrow C_3S_2H_3+CH \quad (4)$$

3.2.1.3 Hydration reaction of tricalcium aluminate (C_3A)

$$3CaO\cdot AlO_3+6H_2O\rightarrow 3CaO\cdot Al_2O_3\cdot 6H_2O \quad (5)$$

$$C_3A+6H\rightarrow C_3AH_6 \quad (6)$$

3.2.1.4 Hydration reaction of tetracalcium alumino ferrite (C_4AF)

$$4CaO\cdot Al_2O_3{\cdot}Fe_2O_3+2(Ca(OH)_2)+10H_2O\rightarrow 3CaO\cdot Al_2O_3\cdot 6H_2O+3CaO\cdot Fe_2O_3\cdot 6H_2O \quad (7)$$

$$C_4AF+2CH+10H\rightarrow C_3AH_6+C_3FH_6 \quad (8)$$

3.2.2 Pozzolanic reactions

3.2.2.1 Additional calcium hydroxide from calcium oxide (CaO) in pozzolan

$$CaO+H_2O\rightarrow Ca(OH)_2 \quad (9)$$

$$C+H\rightarrow CH \quad (10)$$

3.2.2.2 Pozzolanic reaction of silicon dioxide (SiO_2)

$$3Ca(OH)_2+2SiO_2\rightarrow 3CaO\cdot 2SiO_2\cdot 3H_2O \quad (11)$$

$$3CH+2S\rightarrow C_3S_2H_3 \quad (12)$$

3.2.2.3 Pozzolanic reaction of aluminium oxide (Al_2O_3)

$$3Ca(OH)_2+2Al_2O_3\rightarrow 3CaO\cdot 2Al_2O_3\cdot 3H_2O \quad (13)$$

$$3CH+2A\rightarrow C_3A_2H_3 \quad (14)$$

3.3 Stoichiometry

Atomic and molecular weights of all substances related in stoichiometry of hydration and pozzolanic reactions are summarized in Table 1 – 4.

Table 1. Atomic weight of elements related in hydration- and pozzolanic reactions.

Element	*H*	*O*	*Al*	*Si*	*Ca*	*Fe*

Atomic Weight	1	16.00	26.98	28.09	40.08	55.85

Table 2. Molecular weight of main chemical components in cement.

Compound	C_3S	C_2S	C_3A	C_4AF
Molecular Weight (g/mol)	228.2	172.2	270.0	483.1

Table 3. Molecular weight of main chemical components in cement, and fly ash.

Compound	CaO	SiO_2	Al_2O_3	Fe_2O_3
Molecular Weight (g/mol)	56.1	60.1	102.0	156.7

Table 4. Molecular weight of chemical components related in hydration- and pozzolanic reactions.

Chemical Components	H_2O	$Ca(OH)_2$	$C_3S_2H_3$	C_3AH_6	C_3FH_6
Molecular Weight (g/mol)	18.0	74.1	228.2	378.1	432.8

3.4 Optimum replacement of cement by fly ash and long-term strength activity index

3.4.1 Optimum replacement of cement by fly Ash

In this technical paper the optimum replacement of cement by fly ash (FA) will be considered by the complete consumption of calcium hydroxide (CH) from hydration of main minerals in cement by the main chemical compositions in FA.

Let fC_f, fS_f and fA_f be the fraction of C, S and A in FA, respectively and r_f be the fractional replacement of cement by FA, then the fractional calcium hydroxide (fCH) from hydration of cement in unit weight replaced by FA with the ratio r_f would be $fCH \cdot (1-r_f)$.

$$fCH = fC_3S \cdot r_{CH-C_3S} + fC_2S \cdot r_{CH-C_2S} - fC_4AF \cdot r_{CH-C_4AF} \qquad (15)$$

Where fC_3S, fC_2S and fC_4AF are the fraction of C_3S, C_2S and C_4AF in cement per unit weight, respectively. r_{CH-C_3S}, r_{CH-C_2S} and r_{CH-C_4AF} are the ratios of CH from the hydration reactions of C_3S, C_2S and fC_4AF, respectively. Their values are:

$$r_{CH-C_3S} = \frac{3 \cdot w_{CH}}{2 \cdot w_{C_3S}} = 0.48678 \qquad (16)$$

$$r_{CH-C_2S} = \frac{w_{CH}}{2 \cdot w_{C_2S}} = 0.21509 \qquad (17)$$

$$r_{CH-C_4AF} = \frac{2 \cdot w_{CH}}{w_{C_4AF}} = 0.30677 \qquad (18)$$

Where w_{CH}, w_{C_3S}, w_{C_2S} and w_{C_4AF} are molecular weight of CH, C_3S, C_2S and C_4AF, respectively (See Table 2 and 4).

Since calcium oxide (CaO) can be soluble in water and yield calcium hydroxide as shown in Eq. (9) and (10) and the SiO_2 content in FA is usually much more than CaO, it might be assumed that amount of SiO_2 in FA is always more than the amount required for the pozzolanic reaction with CaO in FA itself, there will be adequate residual amount of SiO_2 to react with the residual calcium hydroxide from hydration of cement (see Eq. (11) and (12)). aluminum oxide (Al_2O_3) in FA can also react with the residual calcium hydroxide from hydration of cement (see Eq. (13) and (14)).

The residual CH from Eq. (15) can be completely consumed in pozzolanic reactions with SiO_2 and Al_2O_3 as shown in Eq. (19).

$$(1-r_f) \cdot fCH = (fS_f - fC_f \cdot r_{CS}) \cdot r_{SCH} + fA_f \cdot r_f \cdot r_{ACH} \qquad (19)$$

where $r_{SCH} = \frac{3w_{CH}}{2w_S} = 1.84986$ and $r_{ACH} = \frac{3w_{CH}}{2w_A} = 1.09003$ are the ratio of SiO_2 and Al_2O_3 to react with $Ca(OH)_2$, respectively.

Let

$$fS'_f = fS_f - fC_f \cdot r_{CS} \qquad (20)$$

Where r_{CS} is the ratio by weight of C to S for complete pozzolanic reactions of SiO_2 whose value is given in Eq. (21).

$$r_{CS} = \frac{3 \cdot w_C}{2 \cdot w_S} = 1.40063 \qquad (21)$$

Where w_C and w_S are the molecular weights of C and S, respectively (see Table 3).

Then Eq. (19) reduces to:

$$r_f = \frac{fCH}{fCH + fS'_f \cdot r_{SCH} + fA_f \cdot r_{ACH}} \qquad (22)$$

3.4.2 Long-term strength Activity index

Strength of cement paste is mainly contributed by C-S-H from C_3S and C_2S. Additional contributions are from C_3AH_6, the product of C_3A and C_4AF. With fly ash in binder C-S-H and C_3AH_6 from pozzolanic reactions have not been clearly understood in the term of rate of reactions as well as the strength development. It was concluded that C-S-H from pozzolanic reactions had the same structure as C-S-H from hydration of cement [7]. However the rate of hydration of C_2S as well as the strength development is much slower at the early age of cement paste than developed from C_3S. Fortunately the long-term strength of C-S-H developed from C_2S is insignificantly lower than that from C_3S, the long-term strength of C-S-H from the pozzolanic reaction could be assumed to be the average strength of C-S-H from C_3S and C_2S without significant errors. Furthermore the contribution of C_3AH_6 from the hydration of cement and pozzolanic reactions take minor contribution to the overall strength of binder. Therefore the strength of C_3AH_6 can be assumed equal for both type of reactions. Based on the reasons mentioned above and the given values of strength for the age of 360 days [8] the long-term strength activity index (SAI) can be defined as the strength of binder paste and the strength of cement paste at the age of 360 days as shown below:

$$SAI = \frac{(fa + fb)\cdot(1 - r_f) - \left(\frac{fA_f \cdot r_f}{2 \cdot wAl_2O_3}\right)\cdot(\alpha\beta - \gamma s) + \left(\frac{fC_f \cdot r_f}{3wCaO}\right)\cdot\alpha\beta}{fa} \tag{23}$$

$$fa = \frac{1}{2}\left(\frac{fC_3S}{wC_3S}\cdot\alpha s + \frac{fC_2S}{wC_2S}\cdot\beta s\right) + \left(\frac{fC_3A}{wC_3A}\right)\cdot\gamma s + \left(\frac{fC_4AF}{wC_4A}\right)\cdot\delta s \tag{24}$$

$$fb = \left(\frac{fC_3S}{2\cdot wC_3S} + \frac{fC_2S}{6\cdot wC_2S} - \frac{2\cdot fC_4AF}{3\cdot wC_4A}\right)\cdot\alpha\beta \tag{25}$$

Where $\alpha s = 72.0$, $\beta s = 71.265$, $\gamma s = 8.082$, $\delta s = 5.143$ and $\alpha\beta = 71.6325$.

The long-term strength activity index can be extended for predicting strength Activity index for different ages and different surface areas. The coefficients could be obtained by curve fitting of data from tests carried out by many investigators such as [3,9-12]. Details will be out of the scope of this technical paper and will be discussed in the authors' complete research project [13].

3.5 Propagation of uncertainty

3.5.1 Linear functions

For a linear function of n random variables in form of:

$$g(\underline{X}) = a_0 + \sum_{i=1}^{n} a_i \cdot X_i \tag{26}$$

The correlation matrix of $\underline{X}$ is symmetric in form of:

$$[C] = \begin{bmatrix} 1 & \rho_{12} & \cdot & \rho_{1n} \\ \rho_{12} & 1 & \cdot & \rho_{2n} \\ \cdot & \cdot & \cdot & \cdot \\ \rho_{1n} & \rho_{2n} & \cdot & 1 \end{bmatrix} \tag{27}$$

Where ρ_{ij} is the correlation coefficient between X_i and X_j. The covariance term σ_{ij} can be expressed in terms of the correlation coefficients:

$$\sigma_{ij} = \rho_{ij} \cdot \sigma_i \cdot \sigma_j \tag{28}$$

Where σ_i and σ_j are standard deviations of variables X_i and X_j respectively. Thus σ_i^2 and σ_j^2 are variance of variables X_i and X_j, respectively.

Then the variance of $g(\underline{X})$ is:

$$\sigma_g^2 = \sum_{i=1}^{n} a_i \cdot \sigma_i^2 + \sum_{i=1}^{n} \sum_{j=1(j\neq i)}^{n} a_i \cdot a_j \cdot \rho_{ij} \cdot \sigma_i \cdot \sigma_j \tag{29}$$

In the case that variables $\underline{X}$ are uncorrelated variables the variance and the standard variation of $g(\underline{X})$ are reduced to:

$$\sigma_g^2 = \sum_{i=1}^{n} a_i \cdot \sigma_i^2 \tag{30}$$

$$\sigma_g = \sqrt{\sum_{i=1}^{n} a_i \cdot \sigma_i^2} \tag{31}$$

3.5.2 Nonlinear nunctions and Monte-Carlo simulations

For a nonlinear function of n uncorrelated random variables, the standard deviation of $g(\underline{X})$ (σ_g) to calculate error propagation, the variance formula [14]:

$$\sigma_g = \sqrt{\sum_{i=1}^{n} \left(\frac{\partial g}{\partial x_i}\right)^2 \cdot \sigma_{x_i}^2} \tag{32}$$

For other cases Monte-Carlo simulations may be applied. Then the suitable distributions for $g(\underline{X})$ can be obtained by Goodness-Of-Fit Tests e.g. Chi-Square Errors Test and K-S Test [15].

3.5.3 Optimum replacement of fly ash for cement

For the sake of further discussions: Let the value of a reduced variable y_i of a random variable X_i with mean (μ_{X_i}) and standard deviation (σ_{X_i}) be:

$$y_i = \frac{x_i - \mu_{X_i}}{\sigma_{X_i}} \tag{33}$$

The partial derivative of a function $g(\underline{X})$ with respect to a reduced variable y_i is $\frac{\partial g(\underline{x})}{\partial y_i}$. By applying the chain rule this term may be rewritten as:

$$\frac{\partial g(\underline{x})}{\partial y_i} = \frac{\partial g(\underline{x})}{\partial x_i} \cdot \frac{dx_i}{dy_i} = \frac{\partial g(\underline{x})}{\partial x_i} \cdot \sigma_{X_i} \tag{34}$$

From Eq. (22):

$$\frac{\partial r_f}{\partial y fC_3S} = \frac{r_{CH-C_3S} \cdot (fS'_f \cdot r_{SCH} + fA_f \cdot r_{ACH})}{[fCH + fS'_f \cdot r_{SCH} + fA_f \cdot r_{ACH}]^2} \cdot \sigma_{fC_3S} \tag{35}$$

$$\frac{\partial r_f}{\partial y fC_2S} = \frac{r_{CH-C_2S} \cdot (fS'_f \cdot r_{SCH} + fA_f \cdot r_{ACH})}{[fCH + fS'_f \cdot r_{SCH} + fA_f \cdot r_{ACH}]^2} \cdot \sigma_{fC_2S} \tag{36}$$

$$\frac{\partial r_f}{\partial y fC_4AF} = \frac{-r_{CH-C_4AF} \cdot (fS'_f \cdot r_{SCH} + fA_f \cdot r_{ACH})}{[fCH + fS'_f \cdot r_{SCH} + fA_f \cdot r_{ACH}]^2} \cdot \sigma_{fC_4AF} \tag{37}$$

$$\frac{\partial r_f}{\partial y fC_f} = \frac{fCH \cdot r_{SCH} \cdot r_{CS}}{[fCH + fS'_f \cdot r_{SCH} + fA_f \cdot r_{ACH}]^2} \cdot \sigma_{fC_f} \tag{38}$$

$$\frac{\partial r_f}{\partial y fS_f} = \frac{-fCH \cdot r_{SCH}}{[fCH + fS'_f \cdot r_{SCH} + fA_f \cdot r_{ACH}]^2} \cdot \sigma_{fS_f} \tag{39}$$

$$\frac{\partial r_f}{\partial y fA_f} = \frac{-fCH \cdot r_{ACH}}{[fCH + fS'_f \cdot r_{SCH} + fA_f \cdot r_{ACH}]^2} \cdot \sigma_{fC_f} \tag{40}$$

Thus:

$$\sigma_{r_f} = \sqrt{\left(\frac{\partial r_f}{\partial y fC_3S}\right)^2 + \left(\frac{\partial r_f}{\partial y fC_2S}\right)^2 + .. + \left(\frac{\partial r_f}{\partial y fA_f}\right)^2} \tag{41}$$

3.6 Reliability analyses

3.6.1 Linear limit-state functions

For a linear function of n random variables $g(\underline{X})$ as in form of Eq. (26), the mean (μ_g) and the standard deviation (σ_g) of a vector of normal variables X can be calculated from the equations below:

$$\mu_g = a_0 + \sum_{i=1}^{n} a_i \cdot \mu_{X_i} \tag{42}$$

$$\sigma_g = \sqrt{\sum_{i=1}^{n} a_i^2 \cdot \sigma_{X_i}^2} \tag{43}$$

Then the reliability index (β) may be defined as:

$$\beta = \frac{\mu_g}{\sigma_g} \tag{44}$$

Thus the failure probability (p_f) can be calculated from the equation below:

$$p_f = \Phi(-\beta) \tag{45}$$

Where $\Phi(z)$ is the cumulative probability of the standard normal variable of a z-score. For more details about reliability analyses it is referred to [16].

3.6.2 Nonlinear limit-state functions and Monte-Carlo simulations

For general applications where the functions $g(\underline{X})$ are nonlinear and/or random variables are non-normal, then the statistical representation of a limit-state function may be obtained by Monte-Carlo simulations as discussed earlier in section 3.5.2. However the failure probability may be estimated efficiently by Monte-Carlo simulations together with an advanced variance reduction technique such as Importance Sampling Technique as shown below:

$$p_f \cong \frac{1}{N} \sum_{j=1}^{N} I(g(\underline{x})) \cdot \frac{f_{\underline{X}}(\underline{x}_j)}{h_{\underline{Y}}(\underline{x}_j)} \tag{46}$$

Where N is the number of simulations, $f_{\underline{X}}(\underline{x})$ is the joint probability density function, $h_{\underline{Y}}(\underline{x})$ is an importance sampling density function, $I(g(\underline{x}))$ is an indicator function whose value equals 1 when $g(\underline{x}) \leq 0$ and equals 0 otherwise.

4 Mass production of precast concrete

The statistical data of cement and fly ash available in hands showed significant variation. On one hand the mass production of precast products should be of optimum cost. On the other hand the products should be qualified for all specifications with pre-specified criteria. Therefore the statistical data should be analyzed in advance. For a concrete cast of cement mixed with fly ash, the statistical data available for the materials are summarized in Table 5.

Table 5. Statistical Properties of main chemical oxide compositions in cement and fly ash.

Chemical Components	fractional mean	fractional standard deviation	coefficient of variation (stand.dev./mean)
C_3S	0.60877	0.01170	0.0192
C_2S	0.14857	0.01169	0.0787
C_3A	0.08473	0.00283	0.0334
C_4AF	0.10293	0.00064	0.0062
CaO (FA)	0.14124	0.08788	0.62218
SiO_2 (FA)	0.39910	0.10444	0.26169
Al_2O_3 (FA)	0.19240	0.05763	0.29955

Remarks fly ash from Mae Moh, Thailand.

5 Discussions

For preliminary analyses all related variables are assumed uncorrelated and normally distributed. The optimum fractional replacement of cement by fly ash is 0.28040. The long-term strength Activity index of the optimum replacement of cement by rice hush ash is 1.2480. The uncertainty of the long-term strength Activity index can be obtained by simple Monte-Carlo simulations. Goodness-Of-Fit tests i.e. Chi-Square Error Test and K-S test indicated that the long-term strength Activity index was best fitted by normal distribution. For optimum replacement of cement by fly ash only the long-term strength Activity index based on 8192 simulations can be represented well by normal distribution with mean value and standard deviation = 1.2480 and 0.0890, respectively. The goodness of fit by Chi-Square Error Test is shown in Figure 1.

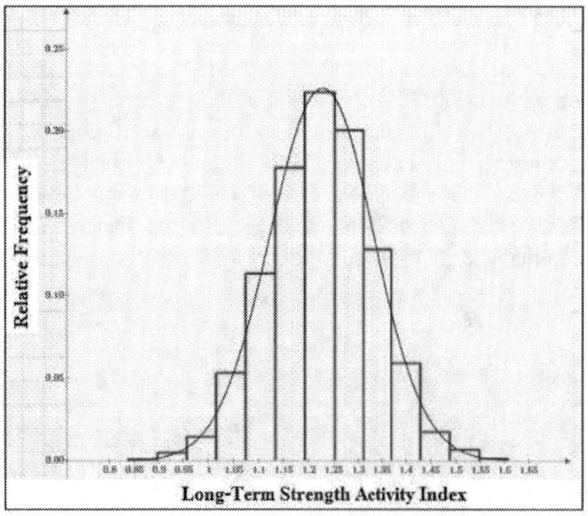

Figure 1. Goodness-Of-Fit Test (Chi-Square Error) for the Optimum Replacement of Cement by Fly Ash.

For the decision based on strength of the concrete the values of failure probability may play an important role. If the failure of the binder paste is defined as the value of the long-term strength Activity index less than 1, then the value of failure probability can be estimated. It can be seen from Eq. (23) that the cross multiplication for $SAI = 1$ and forming the limit-state function from the difference of both sides of the equation, then the limit-state function is linear in term of related random variables. Since all these variables are assumed normally distributed, then the reliability index and the corresponding value of failure probability can be obtained easily by using Eq. (44) and Eq. (45), respectively. In other cases the Monte-Carlo simulations together with an Importance Sampling technique may be applied.

For this particular case the reliability index and the value of failure probability for the optimum replacement of cement by fly ash are 2.752 and $\Phi(-2.752)$ = 2.96×10^{-3}, respectively. Theoretically SAI of the optimum replacement should yield the maximum SAI. Since the failure probability in term of SAI is very low for this particular case, a higher or lower value of replacement of cement by fly ash might be possible so that products could be qualified for all specifications.

Once this optimum replacement of cement by fly ash = 0.28040 is selected, then the propagation of uncertainty in terms of optimum ratio should be also considered. The corresponding standard deviation of this ratio is 0.06214. Again if the ratio is assumed normally distributed, then the interval of the applied ratio should be [0.25813,0.30268] with the confidence interval 95% (0.28040±1.96·0.06214). The more precise conclusion may be, again, estimated from Monte-Carlo Simulations and Goodness-Of-Fit tests.

6 Conclusions

1) Hydration of main chemical compositions of cement and pozzolanic reactions of minerals in fly ash were reviewed. Data necessary for stoichiometry of hydration- and pozzolanic reactions were also given,
2) Formula for the optimum replacement of cement by fly ash based on the complete consumption of calcium hydroxide from hydration of cement was derived,
3) The long-term strength Activity index for the age of 360 days based on equivalent calcium silicate hydrate was proposed,
4) The propagation of uncertainty of the partial replacement of cement by fly ash in terms of variation of main chemical compositions in cement and fly ash were formulated.
5) Reliability analyses for utilizing the optimum replacement of cement by fly ash were reviewed,
6) The applicability of the proposed concepts was demonstrated based on statistical data of materials available in hand. Results from analyses agreed very well with the tests carried out by other investigators,
7) All the concepts mentioned above could be extended to one other type of pozzolan as well as mixed pozzolan.

Acknowledgement

Full financial support and Granting a Professional License of Software Mathcad Prime for Calculation throughout the authors' research projects by Asia Group (1999) Company Limited was cordially acknowledged.

References

1. K. Kamollertvara, W. Ouypornprasert, Determination of mean values of common distributions in civil engineering from the specifications for mass production, *the proceedings of the 7th International Conference on Engineering and Technology ICET-2015, Phuket, June 19-20, 2015*, Hat Yai, Songkhla: Prince of Songkla University, Faculty of Engineering (2015) 1-4
2. S. Dumrongsil, B. Chatveera, W. Ouypornprasert, Stiochiometry of pozzolanic reaction, *J. Research and Training – Rajmangala Inst. Technolgy*. **4**(1) (2000) 28-34
3. G. A. Habeeb, H. B. Mahmud, Study on properties of rice husk ash and its use as cement replacement material, *Mat. Res.* **13**(2) (2010)185-190
4. B. W. Langan, K. Weng, M. A. Ward, Effect of silica fume and fly ash on heat of hydration of Portland cement, *Cement and Concrete Research* **32** (2002) 1045-1051
5. K. Kamollertvara, W. Ouypornprasert, N. Traitruengtatsana, Statistical analyses of optimum partial replacement of cement by rice husk ash based on complete consumption of calcium hydroxide, *the Proceedings of RSU International Research Conference 2016*, Pathumthani (2016)
6. A. M. Neville, *Properties of Concrete*, 4th and Final Ed., Addison Wesley Longman Limited, Wessex, London (1995)
7. H. F. W. Taylor, *Cement Chemistry*, 2nd Ed., Thomas Telford Publishing, Thomas Telford Services, Ltd., London (1997)
8. R. H. Bouge, *Chemistry of Portland Cement*. New York: Reinhold (1955)
9. H. T. Le, H. M. Ludwig, Effect of rice husk ash and other mineral admixtures on properties of self-compacting high performance concrete, *Materials and design*, **89** (2016)156-166
10. N. M. Khalil, E.M. Hassan, E. M. Shakdofa, M. Farahat, Beneficiation of the huge waste quantities of barley and rice husks as well as coal fly ashes as additives for Portland cement, *J. Industrial and Engineering Chemistry*, **20** (2014)2998-3008
11. A. Nawaz, P. Julnipitawong, P. Krammart, S. Tangtermsirikul, Effect and limitation of free lime content in cement-fly ash mixtures, *Constructiona and Building Materials*, **102** (2016)515-530
12. B. Chatveera, P. Nimityongskul, Portland cement containing rice husk ash, fly ash and superplasticzer, *Suranaree J. Sci. Technol.*, **1** (1994)109-122
13. W. Ouypornprasert, N. Traitruengtatsana, K. Kamollertvara, The use of rice husk ash and fly ash for partial replacement of cement for mass production of precast concrete products, *Internal Working Report No.20 - Asia Research, Development and Innovation Center*, Asia Expansion Concrete Product (1993) Co. Ltd., Bangkok (2016)
14. H. H. Ku, Notes on the use of propagation of error formulas. *The National Bureau of Standards*, **70C***(4)* (1966)262
15. W. Ouypornprasert, The proceedings of the International Conference on Computational Mathematics and Modeling (CMM-2002), 22-24 May 2002. *Goodness-of-Fit Tests for Common Continuous Distributions in Civil Engineering*, Century Park Hotel, Bangkok: Mahidol University and Curtin University of Technology (2002)117-126
16. W. Ouypornprasert, Methods to calculate structural reliability, *Internal Working Report No.18, Institute of Engineering Mechanics*, University of Innsbruck, Austria (1988)

FREQUENCY ANALYSIS OF RAINFALL DATA OF DHARAMSHALA REGION

Nitish Kumar Sharma[1] and Ashish Kumar[2]

[1]M. Tech student, Department of Civil Engineering, Jaypee University of Information Technology, Waknaghat,Solan (H.P.), India
[2]Associate Professor, Department of Civil Engineering, Jaypee University of Information Technology, Waknaghat, Solan (H.P.), India

Abstract. The present study evaluates the rainfall magnitude for different return periods and also to ascertain the type of probability distribution that best fits the rainfall data of Dharamshala (H.P.), India. The study uses the 20 years of annual rainfall data that are useful for the prediction of annual one day maximum rainfall and two to seven days consecutive days maximum rainfall corresponding to return period varying from 2 to 20 years are to be used for the economic planning, by design engineers and hydrologists, design of small and medium hydrologic structures and determination of drainage coefficient for agricultural fields. Various probability distributions (viz: Normal, Log Normal and Gamma distribution) and transformations are applied to estimate one day and two to seven consecutive days annual maximum rainfall of various return periods in the Dharamshala (H.P.) region. The mean value of one-day annual maximum rainfall at Dharamshala is found to be 142.9 mm with standard deviation and coefficient of variation of 54.8 and 51.34 respectively. The coefficient of skewness is 1.1. For 2 to 7 days consecutive annual maximum rainfall range values for mean, standard deviation, coefficient of variation and coefficient of skewness are 201 – 393.4 mm , 70.17 – 146.5, 41.65 – 30.47 and 0.726 – 1.593. It is observed that all distribution fitted function significantly.

1 Introduction

Analysis of rainfall and determination of annual maximum daily rainfall would enhance the management of water resources applications as well as the effective utilization of water resources. Probability and frequency analysis of rainfall data enables us to determine the expected rainfall at various chances. Such information can also be used to prevent floods and droughts, and applied to planning and designing of water resources related to engineering such as reservoir design, flood control work, soil and water conservation planning. The primary source of water for agricultural production for most of the world is rainfall. Useful for forecasting the floods to downstream towns and villages. Probability analysis of rainfall is necessary for solving various water management problems and to access the crop failure due to deficit or excess rainfall. Three main characteristics of rainfall are its amount, frequency and intensity, the values of which vary from place to place, day to day, month to month and also year to year. In general, three types of probability distributions (Normal, Log Normal and Gamma distribution) Walck (2007) are commonly used to determine the best fit probability distribution using the comparison of Chi-square, Anderson-Darling and Kolmogorov–Smirnov values. Therefore investigations were conducted in the past for study of the rainfall analysis by Suribabu et al. (2015), Olofintoye et al. (2009), Bhakar et al. (2006), Nyarko et al. (2002), Singh et al. (2001), Rizvi et al.(2001), Mohanty et al. (1999), Upadhaya et al. (1998), among others.

2 Methods and material

Use Daily rainfall data recorded at Dharamshala (Himachal Pradesh) for a period of 20 years (1992-2012) by India Meteorological Department, Shimla (H.P.), India were used for this study. The daily data, in a particular year, is converted to 2-7 days consecutive rainfall of corresponding previous days. The maximum amount of 1-day and 2 to 7 consecutive day's rainfall for each year was then taken for the analysis.

2.1 Statistical analysis of data

The statistical behavior of any hydrological series can be described on the basis of certain parameters, generally mean, variance, standard deviation, coefficient of variation and coefficient of skewness were taken as measures of variability of any hydrologic series. All these parameters have been used to describe the variability of rainfall in the present study.

2.2 Mean

Mean represents measures of central tendency. Arithmetic mean is given by

[1] Corresponding author: nitish4sharma@gmail.com

$$\bar{X} = \frac{1}{N}\sum_{i=1}^{N} X_i \quad (1)$$

Where,

$\bar{X}$ = mean

Xi = variate

N = total number of observations

Standard deviation

This parameter, as a measure of variability is most adoptable to statistical analysis. It is computed by

$$S = \sqrt{\frac{\sum_{i=1}^{N}\left(X_i - \bar{X}\right)^2}{(N-1)}} \quad (2)$$

Where, S = standard deviation

Coefficient of variation

This is a dimensionless dispersion parameter usually expressed as percent. It is given by

$$C_v = \frac{S}{\bar{X}} \quad (3)$$

Where, Cv = coefficient of variation

Coefficient of skewness

The coefficient of skewness was determined for each set of data from the formula given below:

$$C_s = \frac{N^2 M_3}{(N-1)(N-2)S^3} \quad (4)$$

Where,

Cs = coefficient of skewness

$$M_3 = \frac{1}{N}\sum_{i=1}^{N}\left(X_i - \bar{X}\right)^3 \quad (5)$$

3 Indentations and equations

Goodness of fit (Dabral et al.)

Comparing the theoretical and sample values of the relative frequency of the cumulative frequency function can test the goodness of fit of a probability distribution. In case of the relative frequency function, the Chi-square test is used. The sample value of the relative frequency of interval i is calculated by the following equation:

$$f_s(x_i) = \frac{n_i}{n} \quad (6)$$

The theoretical value of the relative probability function is

P (Xi) = F (Xi) – F (Xi-1)

The Chi-square test statistic is given by the equation

$$\chi_c^2 = \sum_{i=1}^{m} n\left[\frac{\left(f_s(X_i) - P(X_i)\right)^2}{P(X_i)}\right] \quad (7)$$

Where, n fs (xi) = ni is the observed number of occurrences in interval i, n P (xi) is the corresponding expected number of occurrences in interval i

The χ^2 distribution functions are tabulated in many statistics texts. In the χ^2 test,

v = m – p – 1

Where, v = degree of freedom, m = number of intervals, p = number of parameters used in fitting the proposed distribution

A confidence level is chosen for the test, it is often express as 1-α, where 'α' is termed as the significant level. A typical value for the confidence level is 95 per cent. The null hypothesis for the test is that the proposed probability fits the data adequately. This hypothesis is rejected if the value of χ_c^2 is larger than a limiting value, $\chi^2_{v,1-\alpha}$ (which is determined from the χ^2 distribution with v degree of freedom at 5 % level of significance. Otherwise it was rejected.

Probability analysis using frequency factors (Dabral et al.)

In case of Gamma distribution, frequency analysis was done by the method as described by Hann (1994) as given in Table-1

Table-1. Description of various probability distribution functions (Walck 2007).

Distribution	Probability density function	Range	Equation for the parameters in terms of the sample moment
Normal	$f(x) = \frac{1}{\sigma\sqrt{2\pi}} e^{-\frac{1}{2}\left(\frac{x-\mu}{\sigma}\right)^2}$	-α ≤ x ≤ α	$\mu = x$, $\sigma = S_x$
Log Normal	$f(x) = \frac{1}{x\sigma\sqrt{2\pi}} \exp\left(-\frac{(y-\mu_y)^2}{2\sigma_y^2}\right)$	x > 0	$\mu_y = y$, $\sigma_y = S_y$
Gamma	$f(x) = \frac{\lambda^{\beta} x^{\beta-1} e^{-\lambda x}}{\Gamma(\beta)}$	x ≥ 0	$\lambda = \frac{\bar{x}}{S^2 x}$ $\beta = \left(\frac{\bar{x}}{S_x}\right)^2$

Chow (1988) has shown that many frequency analyses can be reduced to the form

$$X_T = X(1 + C_V K_T) \quad (8)$$

For Normal and Log Normal distribution, the frequency factor can be expressed by the following equation (Chow, 1988)

$$K_T = \frac{X_T - \mu}{\sigma} \quad (9)$$

This is the same as the standard normal variable z. The value of z corresponding to an exceedance of p (p = 1/T) can be calculated by finding the value of an intermediate variable w:

$$\text{Where, } w = \left[\ln\left(\frac{1}{p^2}\right)\right]^{\frac{1}{2}} \quad (0 < p \leq 0.5) \quad (10)$$

Then calculating z using the equation

$$z = w - \left[\frac{(2.515517 + 0.802853w + 0.010328w^2)}{(1 + 1.432788w + 0.189269w^2 + 0.001308w^3)}\right] \quad (11)$$

When p > 0.5, 1-p is substituted for p in equation (10) and the value of z is computed by equation (11) is given a negative sign. The frequency factor K_T for the normal distribution is equal to z, as mentioned above.

4 Figures and tables

Statistical parameters of annual 1 day to 7 consecutive day's maximum rainfall

The statistical parameters of annual 1-day as well as 2 to 7 consecutive day's annual maximum rainfall are shown in Table 2. The mean value of one-day annual maximum rainfall at Dharamshala is found to be 142.9 mm with standard deviation and coefficient of variation of 54.8 and 51.34 respectively. The coefficient of skewness is 1.1. For 2 to 7 days consecutive annual maximum rainfall range values for mean, standard deviation, coefficient of variation and coefficient of skewness are 201 – 393.4 mm , 70.17 – 146.5, 41.65 – 30.47 and 0.726 – 1.593. It is observed that all distribution fitted function significantly. On the basis of these tests, the results will be helpful in soil and water conservation planning and design of small and medium hydraulic structures such as small dams, bridges, culverts drainage works, etc. A common use of rainfall data is in the assessment of probabilities or return periods of given rainfall at a given location. Such data can then be used in assessing flood discharges of given return period through modeling or some empirical system and can thus be applied in schemes of flood alleviation or forecasting.

Table 2. Statistical parameters of annual 1 day to 7 consecutive day's maximum rainfall.

Parameters	1 day	2 days	3 days	4 days	5 days	6 days	7 days
Maximum (mm)	279.6	334.4	397.2	475.8	574.0	691.6	770.4
Minimum (mm)	84.1	104.8	128.7	156.9	203	218.4	240.9
Mean (mm)	142.9	201.0	241.4	287.6	325.5	364.4	393.4
Standard deviation (mm)	54.8	70.17	82.81	93.91	107.1	128.4	146.5
Coefficient of variation	51.34	41.65	37.18	34.56	32.77	31.47	30.47
Coefficient of skewness	1.1	0.726	0.886	0.785	1.12	1.279	1.593

Fitting of various probability distribution functions

One day annual maximum, 2 to 7 consecutive days annual maximum rainfall data in its original form was fitted to different probability distribution function i.e. Normal, Lognormal. As per the Chi-square value, Log Normal distribution function is found to be best fit function for 1 to 3 days and 7 days annual maximum rainfall data. The computed Chi-square values for the probability distribution i.e. Normal, Log Normal and Gamma are found to be less than the critical value of Chi-square at 95% confidence level for 1 day as well as consecutive days maximum rainfall series. The statistical comparison by Chi-square test for goodness of fit clearly shows that Log Normal distribution is the best fitting representative function for rainfall frequency analysis in this region.

Table 3. Formatting sections, subsections and subsubsections

Consecutive days	Normal	Log Normal	Gamma
1 Day	1.20	4.83	2.08
2 Days	1.04	2.19	6.59
3 Days	3.67	1.86	1.53
4 Days	1.68	1.14	2.44
5 Days	3.78	1.23	7.12
6 Days	1.82	2.34	2.64
7 Days	6.35	3.72	5.09

The graphical plot as shown in figure 1 below shows one day annual maximum, 2 to 7 consecutive day's annual maximum rainfall data was fitted to different probability distribution function i.e. Normal, Lognormal and Gamma. The results of this study would be useful for agricultural scientists, decision makers, policy planners and researchers in order to identify the areas where agricultural development and construction of drainage systems takes place.

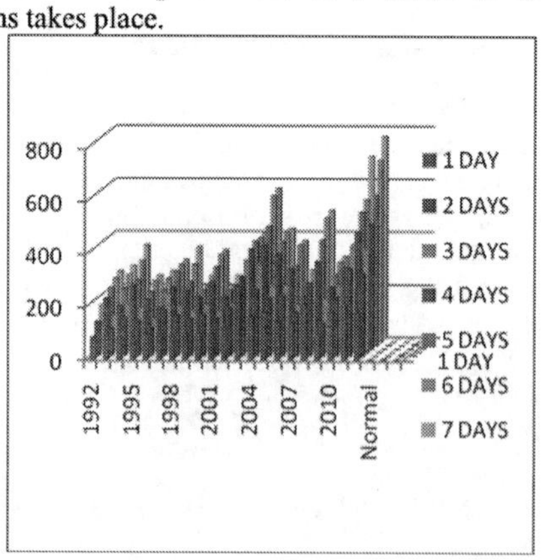

Figure 1. Graphical plot showing Chi-square values for different distribution

5 Conclusion

1. The mean value of one-day annual maximum rainfall at Dharamshala is found to be 142.9 mm with standard deviation and coefficient of variation of 54.8 and 51.34 respectively. The coefficient of skewness is 1.1. For 2 to 7 days consecutive annual maximum rainfall range values for mean, standard deviation, coefficient of variation and coefficient of skewness are 201 – 393.4 mm , 70.17 – 146.5, 41.65 – 30.47 and 0.726 – 1.593. It is observed that all distribution fitted function significantly.
2. It is observed that all distribution fitted function significantly.
3. The study of frequency analysis will be used as a rough guide by engineers and hydrologists to prevent floods and droughts, and applied to planning and designing of water resources related to engineering such as reservoir design, flood control work. Analysis of rainfall and determination of annual maximum daily rainfall would enhance the management of water resources applications as well as the effective utilization of water resources. The results of this study would be useful for agricultural scientists, decision makers, policy planners and researchers in order to identify the areas where agricultural development and construction of drainage systems takes place.

Acknowledgements

Authors are sincerely thankful to Indian Meteorological Department, Shimla (H.P.) for providing the 20 years (1992-2012) daily rainfall data for present study.

References

1. Suribabu, C. R., Sudarsan, J. S., Ramanan, S. R. *"A study on consecutive day maximum rainfall for disaster mitigation in Tiruchirapalli City, India."* IJCE, **9(2)**, 197-202, (2015)
2. Olofintoye, O. O, Sule, B. F. and Salami, A. W. *"Best-fit Probability distribution model for peak daily rainfall of selected Cities in Nigeria."* New York Sc. Journal, **2(3)**, 1554-2000, (2006)
3. Bhakar, S. R., Bansal, A. N. Chhajed, N. and Purohit, R.C. *"Frequency analysis of consecutive day's maximum rainfall at Banswara, Rajasthan, India."* ARPN Journal of Engg. And Applied Sc., **1(3)**, 64-67, (2006)
4. Nyarko, K.B. *"Application of rational model in GIS for flood risk Assessment in Accra, Ghana."* Journal of Spatial Hydrology, **2(1)**, 12-14, (2002)
5. Singh, R. K. *"Probability analysis for prediction of annual maximum daily rainfall of Eastern Himalaya (Sikkim Mid Hills)."* Indian Journal of Soil conservation, **29(3)**, 263-265, (2001)
6. Rizvi, R. H., Singh, R., Yadav, R. S. Tewari, R. K. Dadhawal, K. S. and Soolanki, K. R. *"Probability analysis of annual maximum daily rainfall for Bundelkhand region of Uttar Pradesh."* Indian Journal of Soil Conservation, **29(3)**, 259-262, (2001)
7. Mohanty, S. Marathe, R. A. *"Probability analysis of annual maximum daily rainfall for Amravati."* Indian Journal of Soil Conservation, **43(1)**, 15-17, (1999)
8. Upadhaya, A. and Singh, S. R. *"Estimation of consecutive days Maximum Rainfall by Various Methods and their Comparison."* Indian Journal of Soil Conservation, **26(2)**, 1993-2001, (1998)

Impact of nitrogen and spacing on the growth and yield of okra [Abelmoschus esculentus (L.) Moench]

Navdeep singh brar[a], Daljeet singh

Department of Vegetable Science, University College of Agriculture, Guru Kashi University, Talwandi Sabo (Bathinda)

Abstract. The present investigations entitled "Impact of nitrogen and spacing on the growth and yield of okra" was carried out during spring summer season 2015. The applications of nitrogen at 100 kg/ha recorded higher yield attributes of the number of nods per plant, leaves per plant, internodes length, plant height, pod length, number of pods per plant, fruit yield per plant and total green pod yield per hectare. Nitrogen application upto125 kg/ha significantly decrease the days to 50% flowering. The pod weight, pod length, number of nodes per plant, number of pods per plant, fruit yield per plant and total green pod yield per hectare increased higher with the optimum plant to plant spacing of 15 cm. The data revealed the interaction between nitrogen level and plant density was present in case of plant growth and yield contributing characters of okra. Although the highest level interaction of 100 kg N/ha with plant to plat spacing 15 cm produced the high number of nods per plant, pod weight, pod length, number of pods per plant, pod yield per plant and total green pod yield per hectare and the days to 50% flowering decreased was recorded with 125 kg N/ha and plant to plant spacing 20 cm. The highest gross return (Rs.94080) and net return (Rs.84621.3) were obtained from N125S20.

1 Introduction

Okra, Ablemoschus esculentus (L) Moench commonly known as lady's finger. Bamy, Bamia (Arabic) (Aldigawi, 1996), Bhindi (India), Ilash (Nigeria) Shigidi, A.I, 1994), belongs to the family Malvaceae.

Okra originated in Asia and Africa (Thomson and Kelly, 1979). It is widely distributed and cultivated in the tropics, sub-tropics and warmer portions of the temperate region of the world on a varying scale (Kochhar, 1986). India is one of the leading producers of okra with a production of 5.784 million tonnes year-1 from an area of 498 million ha. Although the productivity of Okra (NHB, 2014) in India is higher (11.6 tonnes ha-1) than world average productivity (7.35 tonnes ha-1) but lower than that of Egypt (15.70 tonnes ha-1). This is because of low yield due to frequent attacks of pests and diseases, especially the fruit and shoot borer and yellow vein mosaic virus (YVMV) according to Journal of Agricultural Technology (2012). Okra is an important summer vegetable crop grown an area of 3204 thousand hectors, yielding 33351 million tonnes in Punjab (NHB 2014).

There are four known domesticated species of Abelmoschus. Among these, A. esculentus (Common Okra) is most widely cultivated in South and East Asia. The Principle elements present in pods of okra are k, Na, Mg and Ca. Presence of Fe, Zn, Mn and Ni also has been reported (Moyin-Jesu,2007). Fresh pods are low in calories (20 per 100g), including about 30% of the recommended levels of vitamin C (16 to 29 mg), 10 to 20 % of foliate (46 to 88 mg) and about 5% of Vitamin A (14 to 20 RAE)(NAP, 2006). Both pod skin and seeds are excellent source of Zinc (80ug/g)(Glew,1997; Cook et al., 2000). Pods and seeds are rich in Phenolic compounds with important biological properties (Arapirtsas, 2008). Fresh Okra pods are the most important vegetable source of viscous fiber, an important dietary component to lower cholesterol (Kendall and Senkins, 2004). Seed protein is rich in tryptophan (94mg/g N) and also contains adequate amounts of sulfur-containing amino acid (189 mg 1g N) a rare combination that makes okra seeds exceptionally useful in reducing human malnutrition (NAP, 2006). Okra seed flour has been used to supplement corn flour for a very long time in countries like Egypt to make better quality dough (Take el-Kitanb, 1947).

The source of N for crop production may affect the performance of the crop. There is therefore the need to synchronize crop N supply with crop demand (Pang and Letey, 2000). Depletion and or shortage of N indicate that either the crop will not able to maintain its leaf area expansion rate or cannot maintain its leaf and plant N concentration. Either of this will have effects on Crop growth and production of economic products.

The requirement of nitrogen which varies according to environmental conditions has to be determined by actual field trial for any particular soil and climate. Okra crop response to nitrogen is required dose of nitrogen for the highest fruit yield appeared in various books and journals ranges from 120 to 200 kg/ha (Amjad et al.2001;

[a] Corresponding author: Navdeep.brar50@gmail.com

Rahman et al, 1992; Paliwal et al, 1999; Verma and Batra 1999; Rashid, 1999).

Generally, increase in planting density results in increased yield per unit area till a certain limit (Weiner, 1990). Suitable plant spacing can lead to optimum seed yield whereas too high or too low plant spacing could result in relatively low yield and quality (Absar and Siddique, 1982). The plant spacing for okra production suggested by different authors ranges from 20 to 40 cm between plants and 30 to 60 cm between rows (Hossain et al, 1999; Rastagie et al, 1987; Thakur and Arora, 1986; Khan and Jaisal, 1988). Keeping this in view, the present work was, thus, under taken with the following objectives.

- To study the effect of Nitrogen fertilizer on yield and yield attributing characters of Okra
- To study effect of plant density on plant growth and green fruit yield
- To study benefit cost ratio analysis of okra as influenced by nitrogen and spacing

2 Materials and Method

The experiment was conducted at the research farm of Guru Kashi University Talwandi Sabo (Bathinda) during spring summer season 2015.

Table 1. Chemical analyses of the experimental field at sowing time.

Soil properties	Soil Depth	Rating	Method used for estimation
pH	8.02	High	1:2, soil: water suspension with Beckman glass electrode pH meter (Jackson 1973)
EC (d sm^{-1})	0.589	High	1:2, Soil: water suspension with sol bridge conductivity meter (Jackson 1973)
Available N (kg ha^{-1})	225	Medium	Alkaline potassium permanganate method (Subbiah and Asija 1956)
Available P (Kg ha^{-1})	13	Medium	0.5 N Sodium bicarbonate extracyalde P (Olsen *et al* 1954)
Available K (Kg ha^{-1})	138	Medium	NH4OAC method
Textural Class	Sandy Loam		International pipette method (piper 1966)

The chemical analyses of the soil along with the method used for their determination before sowing are presented in the Table. Two factors experiment was laid out in Split Plot Design (SPD) with three replications. The treatments of the experiment were three nitrogen doses viz. 75, 100 and 125 kg/ha and three plant to plant spacing viz. 10 cm, 15 cm and 20 cm. The unit plot size was 3.6m × 3m. Seeds of okra were sown on 10 March 2015. Intercultural operations were done as and when necessary. Data of plant growth and yield contributing characters were recorded from ten randomly selected plants, but yield was recorded on plot basis. Experimental data were subjected to statistical analysis appropriate to the design using procedure developed by Finney and tests of significance were applied.

3 Results and discussion

3.1 Plant Height :(cm)

The data in revealed that the effect of nitrogen, the treatment N100 produced significantly maximum plant height (92.1 cm) as compared to treatment N75 (84.1 cm) and it was found at par with nitrogen level N125 (92.4cm). In case of different plant spacing on height of plant was found to be non significant. The S10 spacing results in maximum average height of plant (91.3 cm). However, the minimum height (87.5 cm) was recorded under the S20 spacing. The interaction effects due to spacing and nitrogen levels on plant height was found to be significant, the maximum height of plant (94.3 cm) was recorded with the combination of N100 and S10. The minimum height of plant (82.3 cm) was recorded with the treatment combination N75 with spacing S20.The interaction of N100 with S15,S20 and N125 with S10,S15,S20 being at par with N100S10 resulted into significantly better plant height than lower treatment. The favorable effect of spacing and nitrogen in promoting the growth of plant in terms of height of plant might be due to the fact that, closer plant spacing have higher plant density which creates competition among the population for light and resulted into increased in plant height. Amongst the nutrients, nitrogen has the property to enhance the vegetative growth and capacity of plants to utilize the greater amount of nitrogen with increasing dose. This might be due to the higher utilization of nitrogen but in the closer spacing there is no scope for horizontal spread so it might have resulted in the increase in plant height. . [7].

Table 2. Effect of various levels of Nitrogen fertilizer and spacing on the plant height (cm) of okra.

Treatment	Spacing (cm)			Mean
N Kg/ha	S_{10}	S_{15}	S_{20}	
N_{75}	86.0	84.0	82.3	84.1
N_{100}	94.3	92.3	89.6	92.1
N_{125}	93.6	93.0	90.6	92.4
Mean	91.3	89.7	87.5	

3.2 Days to 50% flowering:

Mean days to flowering decreased gradually with the increasing levels of applied nitrogen from 75 Kg/ha to 125 Kg/ha. It decreased from 52.7 days to 48.1 days with the increasing level of nitrogen. Similar results were computed in case of spacing that was from spacing S10 to

S20. The data indicated that the interaction effect of nitrogen and spacing was also significant .The lowest days to 50% flowering (45.3days) were recorded with N125 and S20. The treatment consisting of N125 and S15 was at par with N125 and S20. The interaction effects of N100 with S20 and N75 with S20 were at par with each other. [7, 33].

Table 3. Effect of various levels of Nitrogen fertilizer and spacing on the Days to 50% flowering of okra.

Treatment	Spacing (cm)			
N Kg/ha	S_{10}	S_{15}	S_{20}	Mean
N_{75}	55.1	53.9	49.3	52.7
N_{100}	53.0	52.1	49.3	51.5
N_{125}	50.0	48.9	45.3	48.1
Mean	52.7	51.6	48.0	

3.3 Single pod weight :(g)

The Nitrogen doses brought about significant increase in fresh weight of fruits when compared with their respective lower levels. The maximum and minimum average fruit weight was noted under N_{100} (9.4 g) and N_{75} (6.7g), respectively and at par (9.8g) with application of N_{125}. The plant geometry also caused significant effect on the fresh weight of okra fruits. Higher the plant density, lower the fresh weight of okra pods were recorded. As the plant density was lowered the average weight of pods were significantly increased.

Table 4. Effect of various levels of Nitrogen fertilizer and spacing on the Single pod weight (g) of okra.

Treatment	Spacing(cm)			
N Kg/ha	S_{10}	S_{15}	S_{20}	Mean
N_{75}	6.2	7.0	7.1	6.7
N_{100}	8.2	10.2	9.8	9.4
N_{125}	9.7	9.5	10.1	9.8
Mean	8.0	8.9	9.0	

Low plant density, spacing S_{15} revealed the significant increase (8.9 g) in fresh weight of okra fruits. The minimum fresh weight (8.0 g) was noted under low spacing S10. The data indicated that the interaction effect of nitrogen and spacing was significantly higher from the lower treatment. The highest pod weight of 10.2g was obtained with N_{100} and S_{15}. The interaction effects of N_{125} with S_{10}, S_{15}, S_{20} and N_{100} with S_{20} were at par with $N_{100}S_{15}$ for single pod weight. [38, 24].

3.4 Leaves per plant:

The treatment N_{100} recorded maximum leaves per plant (22.2) and it was at par with treatment N_{125} having 23.4 leaves per plant. However, minimum leaves per plant (19.3) were recorded in the treatment N_{75}. An increase in nitrogen supply induced more leaves per plant. The average leaves per plant (22.8) were found to be maximum in the spacing S_{20}. However, average minimum leaves per plant (20.6) were recorded in the spacing S_{10}. Wider spacing provided more space for growth, which increased the number of branches per plant, and ultimately increases the number of leaves per plant. The interaction effects due to spacing and nitrogen levels on leaves per plant were found to be significant. Combination of N_{125} nitrogen with spacing S_{20} proved to be the overall best treatment which increased leaves per plant (24.6) from other combination of treatments. The interaction of N_{125} with S_{15}, S_{10} being at par with $N_{125}S_{20}$ treatment. The interaction of N_{100} with S_{20}, S_{10} being at par with each other. . [12, 25].

Table 5. Effect of various levels of nitrogen fertilizer and spacing on the leaves per plant (No) of okra.

Treatment	Spacing(cm)			
N Kg/ha	S_{10}	S_{15}	S_{20}	Mean
N_{75}	17.0	20.3	20.6	19.3
N_{100}	22.3	21.0	23.3	22.2
N_{125}	22.6	23.0	24.6	23.4
Mean	20.6	21.4	22.8	

3.5 Number of green fruits per plant::

Average numbers of fruits per plant were significantly influenced by nitrogen application. Significantly higher numbers of fruits per plant (31.2) were obtained with application of 100 kg N per hectare and highest number of fruits per plant (36.7) with application of 125 kg N per hectare. The plant under application of 75 kg N per hectare showed significantly minimum (22.3) number of fruits per plant. The plant population affected the production of fruits per plant significantly. Wider spacing expressed significantly greater number of fruits per plant The maximum and minimum number of fruits as influenced by plant density were recorded under S_{20}(31.5) and S_{10}(28.8) spacing, respectively. The significantly increased number of fruits per plant (30.1) was noticed under spacing S_{15}. The data indicated that nitrogen and spacing interaction was significant. The highest number of pods per plant (40.2) was obtained with combination of N_{125} and S_{20} which was significantly the best than all other nitrogen and spacing combinations. The interaction effects of N_{125} with S_{10}, S_{15} and N_{100} with S_{10}, S_{15}, S_{20} was at par with each other and superior than the lower treatment. . [7, 34].

Table 6. Effect of various levels of nitrogen fertilizer and spacing on number of green pods per plant (No) of okra.

Treatment	Spacing (cm)			
N Kg/ha	S_{10}	S_{15}	S_{20}	Mean
N_{75}	21.6	22.0	23.3	22.3
N_{100}	30.6	32.3	30.6	31.2
N_{125}	34.3	36.0	40.3	36.7
Mean	28.8	30.1	31.4	

3.6 Number of nodes per plant:

Number of nodes on main stem were significantly increased, as nitrogen applications increased from 75 kg/ha to 100 kg/ha and at par with higher dose of nitrogen i.e. 125 kg/ha. The maximum and minimum numbers of nodes were 36.3 and 29.5 respectively. Significantly increased number of nodes per plant was (35.8) when the nitrogen level was N_{100}.The minimum and maximum numbers of nodes per plant were recorded under spacing of S_{10} (32.8) and S_{20} (34.6), respectively. Significantly increased number of nodes per plant was (34.2) when the spacing was S_{15}. The higher number of nodes per plant (37.1) was recorded with combination of N_{125} and S_{20}. The interaction of N_{125} with S_{10}, S_{15} and N_{100} with S_{15}, S_{20} were at par with $N_{125}S_{20}$. [11, 12].

Table 7. Effect of various levels of nitrogen fertilizer and spacing on number of nodes per plant (No) of okra.

Treatment	Spacing (cm)			
N Kg/ha	S_{10}	S_{15}	S_{20}	Mean
N_{75}	28.6	30.0	30.0	29.5
N_{100}	34.6	36.0	37.0	35.8
N_{125}	35.3	36.6	37.1	36.3
Mean	32.8	34.2	34.6	

3.7 Green fruit yield per plant:

Green fruit yield per plant was increased with the increased dose of nitrogen from 75 kg/ha to 125 kg/ha. Significantly higher fruit yield per plant was recorded under N_{100} which was 302g .The minimum fruit yield per plant was 164g at application of 75 Kg N/ha. Wider spacing expressed significantly greater green fruit yield per plant. The maximum and minimum green fruit yield per plant as influenced by plant density were recorded under S_{20} (283g) and S_{10} (251g) spacing, respectively. Significant increase in green fruit yield per plant (272g) was noted at the spacing of S_{10}. The higher number of fruit yield per plant (355g) was recorded with combination of $N_{125}S_{20}$. The interaction of N_{125} with S_{10}, and S_{15} were at par with combination of N_{125} and S_{20}. The interaction of N_{100} with S_{10}, S_{15}, S_{20} were at par with each other. [30].

Table 8. Effect of various levels of nitrogen fertilizer and spacing on green fruit yield per plant (g) of okra.

Treatment	Spacing (cm)			
N Kg/ha	S_{10}	S_{15}	S_{20}	Mean
N_{75}	146.6	165.0	182.6	164.7
N_{100}	288.6	306.6	313.0	302.7
N_{125}	340.0	344.6	355.0	346.5
Mean	251.7	272.0	283.5	

3.8 Pod length :(cm)

The increase in nitrogen levels up to N_{100} brought about significant improvement in pod length of okra. The maximum and minimum fruit length was recorded under N_{100} (9.4 cm) and N_{75} (6.8 cm), respectively, which was at par with treatment N_{125} (9.9 cm). The closer spacing expressed significantly shorter fruits when compared with wider spacing. Maximum and minimum pod lengths were procured under spacing S_{20} (9.0cm) and spacing S_{10} (8.3 cm), respectively. The maximum pod length of 10.3 cm was produced by the combined effect of N_{125} and S_{20} which was significantly higher than all other combinations. The interaction effect of N_{125} with S_{15}, S_{10} and N_{100} with S_{15}, S_{20} being at par with N_{125} and S_{20} significantly produced longer pod than lower treatment. [4].

Table 9. Effect of various levels of nitrogen fertilizer and spacing on pod length (cm) of okra.

Treatment	Spacing (cm)			
N Kg/ha	S_{10}	S_{15}	S_{20}	Mean
N_{75}	6.5	6.9	7.0	6.8
N_{100}	8.5	10.2	9.6	9.4
N_{125}	10.0	9.5	10.3	9.9
Mean	8.3	8.9	9.0	

3.9 Pod diameter :(cm)

The diameter of pod were recorded by using venire caliper and the means were worked out and expressed in cm. Application of higher levels of nitrogen that was N_{100} (1.50 cm) and N_{125} (1.86 cm) proved more effective in respect of diameter when compared with N_{75} (1.05 cm). The maximum diameter of pod was produced with N_{100} dose. [1, 29].

Table 10. Effect of various levels of nitrogen fertilizer and spacing on pod diameter (cm) of okra

Treatment	Spacing (cm)			
N Kg/ha	S_{10}	S_{15}	S_{20}	Mean
N_{75}	0.96	1.06	1.13	1.05
N_{100}	1.43	1.46	1.60	1.50
N_{125}	1.76	1.88	1.95	1.86
Mean	1.38	1.47	1.56	

3.10 Internode length :(cm)

The mean internode length increased gradually with the increasing level of applied nitrogen. It increased from 2.16 cm in plot applied with nitrogen dose of N_{75} to 2.82 cm and 3.04 cm with nitrogen application of N_{100} and N_{125} respectively. Significant increase was noticed with increased dose of nitrogen N_{100}. The internode length was significantly influenced by within row spacing of plants. But spacing has opposite effect on internode length. Internode length decreases from 3.0 cm to 2.5 cm and to

2.4 cm when spacing is increased from S_{10} to S_{15} and to S_{20} respectively.

Table 11. Effect of various levels of nitrogen fertilizer and spacing on internode length (cm) of okra.

Treatment	Spacing (cm)			
N Kg/ha	**S_{10}**	**S_{15}**	**S_{20}**	**Mean**
N_{75}	2.4	2.0	2.0	2.1
N_{100}	3.2	2.7	2.4	2.8
N_{125}	3.3	2.9	2.8	3.0
Mean	3.0	2.5	2.4	

The maximum internode length of 3.30 cm was produced by the combined effect of N125 and S10 cm which was significantly higher than all other combinations. The interaction effect of N125 with S15 and N100 with S10, being at par with N125 and S10 significantly produced longer internode. The interaction effects of N100 with S15 and N125 with S20 being at par with each other. The increase in internode length at higher level of nutrients in the present study might be due to higher absorption of nutrients, especially nitrogen which enhanced the cell division, cell elongation with concomitant increase in metabolic activity by which meristmatic activity of tissue increased manifold and led to increase in internode length. [28].

3.11 Total green pod yield :(q/ha)

The yield of green pods was noted significantly greater (72.4 q) under N_{100} and at par with N_{125} (76.0 q) against the minimum (53 q) recorded under lower level N_{75}. Regarding the plant density, S_{20} spacing produced the highest yield (68.7 q) against the minimum (64.4 q) recorded under S_{10} spacing. Significantly increase in yield (68.2 q) was noted under S_{15} spacing. The maximum total green pod yield (77.6 q/ha) was produced by the combined effect of N_{125} and S_{15} spacing which was significantly higher than all other combinations. The interaction effect of $N1_{25}$ with S_{10}, S_{20} and N_{100} with S20 being at par with $N_{125}S_{15}$ combination significantly. The significant increase in total yields might also be attributed to the increased branching as pod developed in the axils of every branch once flowering has began. Number of nodes per plant also increased with fertilizer application along with optimum spacing of 15 cm with-in row plant spacing. Since every node has a potential to produce flower which directly influence the yield per plot. [5, 17, 24].

Table 12. Effect of various levels of nitrogen fertilizer and spacing on total green pod yield (q/ha) of okra.

Treatment	**Spacing (cm)**			
N Kg/ha	**S10**	**S15**	**S20**	**Mean**
N75	49.6	54.0	55.3	53.0
N100	69.3	73.0	75.0	72.4
N125	74.3	77.6	76.0	76.0
Mean	64.4	68.2	68.7	

3.12 Benefit cost ratio analysis of okra

Analysis of cost and return revealed that the gross and net returns were higher at higher doses of N and spacing. The highest gross return (Rs.94080) and net return (Rs.84621.3) were obtained from N125S20. The BCR was also higher (9.54) under the same treatment combination. [39].

Table 13. Effect of nitrogen and spacing on benefit cost ratio analysis of okra yield..

S. N	Treatment	Yield (t/ha)	Cost of cultivation (t/ha)	Gross income (t/ha)	Net returns (t/ha)	Benefit cost ratio
1	$N_{75}S_{10}$	51.9	17715.2	62280	44564.8	3.51
2	$N_{75}S_{15}$	54.0	12065.2	64800	52734.8	5.37
3	$N_{75}S_{20}$	55.3	8915.2	66360	57444.8	7.44
4	$N_{100}S_{10}$	69.3	17986.9	83160	65173.1	4.62
5	$N_{100}S_{15}$	73.0	12336.9	87600	75263.1	7.10
6	$N_{100}S_{20}$	75.0	9186.9	90000	80813.1	9.79
7	$N_{125}S_{10}$	74.3	18258.7	89160	70901.3	4.88
8	$N_{125}S_{15}$	77.6	12608.7	93120	80511.3	7.38
9	$N_{125}S_{20}$	78.4	9458.7	94080	84621.3	9.94
	C.D (p=0.05)	3.92				

a. Price of okra: Rs. 12.00

b. $BCR: \frac{\text{Gross return}}{\text{Cost of production}}$

4 Conclusion

On the basis of both individual and combined effects of nitrogen and spacing, 100 kg N/ha and 15 cm plant to plant spacing may be considered optimum for obtaining high yield with good quality. So, this proposal is useful for improving the productivity profitability of okra growers in Punjab.

References

1. A.O.A.C..Officil Methods of Analysis, XIV Edition, Washington, (D.C) Barnah G.K.S. Response of okra varieties to different plant spacing in hill zone of Assam. Haryana Journal of Horticultural Sciences, 26 (1-2): 130-134. (1984)
2. Absar, N. and M.A. Siddique. Influence of plant density on the yield of three varieties of okra.Bangladesh J. Agric. 7(3-4):15-21. (1982)
3. Ahamad,N. and L.I. Tulloch-Reid. Effect of fertilizer nitrogen, phosphorus, potassium and magnesium on yield and nutrient content of okra (Hibiscus esculentus L.). Agron.J. , 60:353_356. (1968)
4. Amjad, A. Anjum, M.A. and Hussain, S. Effect of different sowing dates and various doses of fertilizer on juvenility and productivity of okra. Pakistan J.Agril.Sci., 38 (1-2),29-32. (2001)
5. Babatola LA, Ojo DO, Adewoyin OB. Effect of NPK 20:10:10 fertilizer levels on the yield of okra-sweetcorn intercrop and post harvest quality of okra. Proc. Hortic. Soc. Nig. Conf. pp 74-78. (2002)
6. Babatola, L.A. and J.O. Olaniyi. Effect of NPK 15-15-15-fertilizer level and plant spacing on performance and shelf life of okra. In: proc. HORTSON conference, NIHORT, Ibadan. (1997).
7. Birbal, Nehra, B.k. and Malik, Y.S. Effect of spacing and nitrogen on fruit yield of okra (Abelmoschus esculentus (L) Moench) cv. Varasha Uphar. Haryana Agric. Univ. J.Res., 25, 47-51. (1995)
8. Chattopadhyay, A. and B.C. Sahana. Response of okra seed crop to nitrogen and phosphorous fertilization in acidic soil of old Alluvial Zone, West Bengal.res,crops, 1(2),176-180. (2000)
9. Dademal AA, Dangale JH, Mhaskar NV. Effect of manures and fertilizers content and uptake of nutrients by okra on laterilic soil of konkan. J. soils and crops. 14 (2): 262-268. (2004)
10. Decoteau DR, Graham HAH.Plant spatial arrangements affect growth, yield and pod distribution of cayenne peppers. Hort Sci, 29: 149-151. (1994)
11. Gupta A. Response of spring planted okra to varying of irrigation and plant spacing. Veg Sci, 17: 16-19. (1990)
12. Gupta, A., K. Srinivas and V. Shukla. Response okra (Abelmoschus esculentus L. Moench) to plant spacing and nitrogen, phosphorus fertilization. Indian J. Horti. Sci. 38 (3-4):218-222.(1981)
13. Hermann M, Makadji M, Daunicht HJ. Effect of intra-row spacing on time course of growth and fruit yield of okra. Scientia Hort , 45: 37-48. (1990)
14. Hooda, R.S., M.L Pandita., Kalloo and A.S. Sindhu. Effect of spacing and nitrogen on the growth and yield of tomato variety HS-110. Harayana Journal of Horticultural Sciences, 13 (1-2): 48-49. (1984)
15. Kahn BA, Cooksey JR, Motes JE. Within-row spacing effects on traits of importance to mechanical harvest in paprika pepper. Scientia Hort, 69: 31-39. (1997)
16. Khan, A.R. and R.C. Jasiwal. Effect of nitrogen, spacing and green fruit picking in the seed production of okra (Abelmoschus esculentus L. Moench) Vegetable Sciences, 15(1): 8-14. (1988)
17. Kolawole GO, Olapede AO, Alade CR, Olaniyi JO. Response of okra (Abelmoschus esculentus) varieties to NPK fertilizer in the South Guinea Savanna of Nigeria. Niger. J. Horticult. Sci. 13: 99-108. (2008)
18. M.D. Hossain, M.a. Rahman, M.m. Hoque, M.s. Islam and M.a. Salam. Year round okra production and economic return as influenced by spacing in Barisal Region" Bangladesh J. Agric. Res.26.pp.319-328. (2001)
19. Mohanta, H. C. Effect of phosphorus and genotypes on the yield and contributing characters in okra (Abelmoschus esculentus L. Moench). M S. Thesis, Dept. of Horticulture, IPSA, Salna, Gazipur. (1998)
20. Meyer, B. S. and D. B. I Anderson. Plant Physiology. East West Press Pvt. Ltd., New Delhi. (1970)
21. Majanbu, I. S.; V. B. Ogunlela; M. K. Abmed and J. D. Olarewaju. Response of two okra varieties to fertilizers, yield and yield components as influenced by nitrogen and phosphorus application. Fertilizer Res. 6 (3): 257-267. (1985)
22. Mangual, G. and F.W. Martin. Effect of spacing on seed,protein and oil production of four okra varieties.J.Agric.Univ.Puerto Rico,64:450-490. (1980)
23. Mani, S., Ravel, R. and Shukla, V. Effect of nitrogen and phosphorus on yield and chemical composition of okra. Geobios New Reports, 2 (2-3): 56 -59. (1981)
24. Olasantan, F.O. Optimum plant populations for okra in mixture with cassava and its relevance to rainy season based cropping system in south-western Nigeria. Tournal of Agri cultural sci. 136:207-214. (2001)
25. Omotosho SO, Shittu OS. Effect of NPK fertilizer rates and method of application on growth and yield of okra (Abelmoschus esculentus (L.) Moench). Res.J.Agron. 1(2) 84-87. (2007)
26. Palanisamy, V., Vanangamud, T. Jayraj, T. and T.V. Karivarthraju. Influence of date of sowing and spacing on seed quality in bhindi [Abelmoschus esculentus (L.) Moench]. South Indian Horticulture, 34 (1): 23 -25. (1986)
27. Paliwal, R., I. S. Naruka and J.K. Yadav. Effect of nitrogen and sulphur on growth and yield of okra (Abelmoschus esculentus L. Moench.) cv. Pusa sawani. Prog.Hort, . 31(1-2), 94-97. (1999)

28. Pandey BR, Sharma RK, Sharma RS. Influence of plant spacing and nitrogen level on growth, yield attributes and seed yield of okra during spring season. Veg Sci, 21(1): 12-15. (1994)
29. Randhawa, G.S. and Pannu, M.S. The effect of row spacing and levels of nitrogen on the growth and yield of okra (Abelmoschus esculentus L. Moench). Journal of Research, 6: 320-324. (1969)
30. Rahman, M.M., Roy, S. k. and Quasem, A. Effet of plant population and nitrogen on the yield and yield attributes of okra. Bagladesh Agro. J., 4(1 & 2), 7-12. (1992).
31. Randhawa, G.S. and Pannun, M.S. The effect of row spacing and level of nitrogen on the growth and yield of okra (Abelmoshus esculentus L. Moench). J.Res .Ludhiana 6: 304-4. (1969)
32. Rastogi,K.B., P.P. Sharma, N. P. Singh, B.N. korla. Effect of different levels of nitrogen and plant and spacing on seed yield of okra (Abelmoschus esulentus L. Monech). Vegetable Sci. 14 (2): 120-123. (1987)
33. Saimbhi,M.S. , S.P. Jasiwal, K.S. Nandpuri and G. Kaur. Effect of phorate and N on plant growth, seed yield and chemical composition of okra. Indian J. Agric.Sci. , 43: 930-933. (1975)
34. Singh, K. and B.K. Srivastava. Effect of various levels of nitrogen and phosphorus on growth and yield of chilli (Capsicum annuum L). Indian Journal of Horticulture, 45 (3-4): 319-324. (1988)
35. Singh, N.P. Effect of nitrogen, phosphorus and potassium on bhindi (Abelmoschus esculentus L. Moench).Prog.Hort. , 10:21-30. (1979)
36. Singh, S.K., Rai, B., Singh, S. and Singh, M.B. Effect of plant spacing and nitrogen fertilization on growth and fruit yield of hybrid brinjal. Silver Jubilee National Symposium on Emerging Scenario in Vegetable Research and Development, 12-14 Dec. RD.V.R. Varanasi: 143. (1998)
37. Suwandi. Effect of N, P and K (15-15-15) fertilizer and planting distance on the growth and yield of the tomato cvs. GH2 and GH4 (Mutiara). Bull Penelition Hort., 16 (2) : 11-18.(1988)
38. Yadav, P., Singh, P. and Yadav, R.L. Effect of organic manure and nitrogen levels on growth, yield and quality of okra [Abelmoschus esculentus (L.) Moehch]. Indian Journal of Horticulture, 63 (2): 215-217. (2006)
39. Z. A. Firoz. Impact of nitrogen and phosphorus on the growth and yield of okra. Bangladesh J. Agril. Res. 34(4): 713-722. (2009)

Adsorption Studies of Radish Leaf Powder

Ankita and Kamlesh Prasad
Department of Food Engineering and Technology
Sant Longowal Institute of Engineering and Technology, Longowal-148106, Sangrur, INDIA

Abstract- Radish leaves (*Raphanus sativus*) powder fractions was subjected to moisture adsorption isotherms at different isothermal temperature conditions from 15-45°C with an equal interval of 10°C. The sorption data obtained in gravimetric static method under 0.11–0.90 water activity conditions were subjected for sorption isotherms and found to be typical sigmoid trend. Experimental data were assessed for the applicability in the prediction through sorption models fitting and found that Polynomial and GAB equations performed well over all fitted models in describing equilibrium moisture content – equilibrium relative humidity (EMC–ERH) relationships for shelf stable dehydrated radish leaf powder, over the entire range of temperatures condition under study. The net isosteric heat of sorption, differential entropy and free energy were determined at different temperatures and their dependence was seen with respect to equilibrium moisture content.

1 Introduction

Radish leaves are one among the green leafy vegetable category, like several other leafy vegetables which include mustard leaves, fenugreek leaves, spinach leaves, curry leaves and many more. Leaves of radish are green in color which are important source of nutrients (vitamin C, iron, calcium and dietary fibre), even if are being underestimated. Radish plant come under members of the plant family of Brassiceae, earlier termed as Cruciferae or Cabbage which contains many such vegetables of economic importance [1]. This family includes many crops named cabbage, cauliflower, , Brussels sprouts, broccoli, mustard, kale and radish; weed species (shepherd's-purse and wild radish) and cover crop species (yellow mustard, field mustard, black mustard, brown mustard, white mustard, oriental mustard and oilseed radish) [2]. Radish leaves are generally pinnate which form a basal rosette [3]. Radish leaves and roots have been used in various parts of the world to treat cancer and it simultaneously act as antiviral and antimicrobial agents [1]. Radish leaves are rich source of glucosinolates (GSLs) [4, 5]. Myrosinase is the name of the enzyme which has the function to produce isothiocyanates from glucosinolates in the presence of moisture (hydrolysis process) which is responsible for the pungency, odor and hot biting flavor in radish leaves. These play an important role in the prevention of diseases like cancer. The glucosinolates concentration varies with the age of the plant among different portions of the plant [6, 7]. Glucosinolates are distributed in the seeds, root, stem and leaf of the plant, whereas, the youngest tissues contain their highest proportion [8]. Radishes were very expensive even these were prized above all crops in Greek by the ancient population [9].

Moisture sorption isotherms are of substantial importance to foods for applications particularly to food science and technology such as the designing and optimization of processing as for instance in drying, for ingredient mixing predictions, for assessing the packaging problems and finally for predicting shelf life stability [10]. Its first application is based on an evaluation of the heat of sorption, which describes the interaction between an adsorbent (food matrix) and adsorbate (water molecules). The availability of water in the reactions occuring during food spoilage depends largely on water content and on the properties of the diffusion surface, thus, are based on the thermodynamic function of sorbed water. Sorption isotherms have been used to estimate the thermodynamic functions of several food materials [11], which comprises differential enthalpy (also termed as heat of sorption), differential entropy and free energy. Moreover, these thermodynamic parameters are useful for defining appropriate packaging materials in maintaining sensory attributes and nutrition of the desired food material [12].

2 Materials and Methods

2.1 Sample preparation

Corresponding author: ankita_modgil2008@yahoo.co.in

Powder obtained from dried radish leaves using improved technique (80°C-70°C) maintained in cabinet drier [13] was taken for adsorption studies. Flow diagram showing the preparation of radish leaf powder as reported in figure 1. Further, the obtained powder after size reduction using laboratory grinder is classified into mix and fine fractions by using sieve of 60 mesh (resulting maximum 250 μm size particles) [14]. The dried powder samples were packed in the dark glass jars and kept inside refrigerator at 4±2°C for at least 5 days to equilibrate the moisture distribution if any.

Radish leaves (Radish F1 Hybrid NBH-White Queen)

⇩

Dehydration in cabinet drier at 80^0C and 70^0C

⇩

Size reduction

⇩

Sieving using 60 BSS (250 μm particle size)

⇩

Adsorption studies on mix and fine radish leaf powder fractions

Figure 1. Flow diagram showing preparation of radish powder

2.2 Adsorption isotherm

The static method of sorption isotherm was used to represent the relation between equilibrium moisture content (EMC) of the representative dehydrated radish leaf powder with the equilibrium relative humidity (ERH) of the surrounding environment at a particular temperature. Saturated salt solutions were made and excess of solid salt was added more to make sure that solutions remain saturated. To establish moisture adsorption isotherms, the equilibrium moisture contents were plotted with respect to surrounding's relative humidity values (ERH) maintained at 11.1, 22.9, 32.9, 43.9, 51.8, 64.8, 75.5 and 86.5 (using lithium chloride, potassium acetate, magnesium chloride, potassium carbonate, potassium nitrite, sodium dichromate, sodium chloride, and potassium chromate) at four different temperatures (15^0C, 25^0C, 35^0C and 45^0C) (using Winks weight equilibrium method) [15]. A tripod was also placed in each jar. The accurately weighed dehydrated samples (5.0 ± 0.001 g) were placed on petriplates and inserted in the desiccators having specific saturated salt solutions for sorption studies. The readings were noted at regular interval of 48 hours. The experiment was considered done when three consecutive weight measurements showed a difference of less than 0.001 g.

The adsorption curves describes the hygroscopic equilibrium of a given food product. These equilibrium (adsorption) curves are determined by the saturated salt solutions method. Another application of the adsorption isotherms is in the determination of the drying time of hygroscopic substances [16]. Moreover, these isotherms convey important information to other processes which include packaging and storage [17]. The selected equations are described in table 1, where M denotes the m.c. (dry basis) and a_w is water activity [18].

2.3 Model fitting

The coefficients of various available sorption models were applied using standard non-linear regression technique. Not a single equation has been found to accurately depict the behaviour of adsorption isotherms of all types of foods in the entire range of water activity. For example, BET equation allows one to examine the fixation phenomenon alongwith liberation process of water molecules. Other models, such as Polynomial model, Hailwood and Horbin model and Smith model exhibits consistencies for a wide variety of vegetable products. Hailwood and Horbin model is equivalent to the GAB equation at constant temperature. Most of the sorption models are two or three parameter model comprising a, b and c as unknown parameters to be estimated [18, 19, 20].

2.4 Validation of sorption models

The coefficients of various sorption equations (table 1) were determined by means of standard non-linear regression technique using statistica software version 7. The various sorption models (table 2, 3) were evaluated for their suitability in predicting the adsorption behaviour of the sample on the basis of few important parameters. Correlation coefficient (R^2) was evaluated as it is considered as most important criterion for selecting the best equation to define the curves of radish leaf powder. In addition to R^2, other statistical parameters include E% values which was used to determine the goodness of fit of different models. The average percent difference between the experimental and predicted values mean relative deviation modulus, E%, defined by the mentioned equations.

2.5 Thermodynamic functions

The isosteric heats of sorption were calculated at several moisture levels from the slopes of the isosteres of the natural logarithms of a, versus the reciprocal absolute temperature. The slope of the line decreases to zero as moisture content increases, which is indicative of reduced water interactions, i.e., less binding energy [21]. The net isosteric heat q_{st} is defined as the total heat of sorption (Q_{st}) in the food minus the heat of vaporization (ΔHv) of water, at the system temperature. The heat of adsorption is the index of the energy released on adsorption [11].

Using relationship between the isosteric heat (ΔH_d) and the differential entropy (ΔS_d) of sorption energy values can be determined as :

$$(-\ln a_w)_M = \Delta H_d \left(\frac{1}{RT}\right) \quad (1)$$

By plotting $\ln(a_w)$ versus $1/T$, for a given moisture content (M), ΔH_d was determined from the slope ($-\Delta H_d/R$) as that from clausius clapreyon equation, and ΔS_d from the intercept ($\Delta S_d/R$).

3. Results and Discussion

3.1 Adsorption isotherm behavior

The experimental adsorption data for green radish leaf powder for mix and fine fractions at different temperatures are plotted with experimental and empirical results from Polynomial and GAB equation in figure 2 and figure 3 respectively as per fitting criteria of lower E% and higher R^2. This behavior of variation of moisture content with respect to temperature may be explained by considering excitation states of molecules. At elevated temperatures, water molecules are in an increased state of excitation, thus decreasing the attractive forces between them. This tends to a decrease in the degree of water sorption by food material at a particular relative humidity with the increase in temperature [10]. Activation of the water vapor molecules because of increasing temperature levels causes them to break away from the water binding (adsorbent) sites, thus responsible for lowering the equilibrium moisture content. This result may be justified by considering the higher excitation (unstable) state of water molecules at higher temperature and becoming unstable, thus decreasing the attractive forces between them [22]. Sorption isotherms shows that radish leaf powder at low water activity levels retained small amounts of water and amount of water increases as the water activity levels rises. The sorption isotherms of radish leaf showed the typical sigmoid shape of Type II or III, according to the BET classification. The related shapes of sorption isotherm are well documented in literature for most of the food materials [22, 23].

On the other hand, the EMC increases with increasing ERH at a constant temperature level. Similar results have been discussed in the literature for the sorption curves [23, 25]. Water activity of about 0.30, where water is held on polar active sites of relatively high energies. This is termed as the monolayer. The water activity in the range of 0.30–0.70 is termed as multilayer water. It consists of water layers, which are adsorbed and got stuck onto the first layer by hydrogen bonds. Water activity of 0.7 and above, where the water approaches the condition of condensed or free water and the isotherm reflects solution and surface capillary effects [20]. Monolayer moisture contents (m_0), was obtained using GAB and BET equations, are also given in table 2. The m_0 values were important for determination of the storage conditions and controlling the possible deterioration reactions. The m_0 values of radish leaf were decreased with increasing temperature [24].

3.2 Fitting of sorption models to experimental adsorption findings

The sorption curve for radish leaf is drawn as EMC against aw (ERH/100). These curves are used to estimate the coefficients of the different sorption models and it was found that Smith, Polynomial, BET, Hailwood Horbin and GAB equations gave best fit to the experimental data for instant radish leaf powder with not only highest values of R^2 and least values of E% for adsorption isotherms for a wide range of water activity. List of various mathematical model equations are presented in table 1.

R^2 values for radish powder (mix fractions) ranged from 0.985 to 0.996 for GAB model and 0.985 to 0.996 for Hailwood and Horbin, followed by Smith (0.983 to 0.992) followed by polynomial (0.966 to 0.995) followed by BET (0.866 to 0.938) and best model fitting was observed in case of Hailwood and Horbin and GAB model. Also, E% values are lower in case of GAB (0.408 to 7.830%), Hailwood and Horbin (0.408 to 7.831), BET (0.432 to 15.645%), Smith (0.376 to 10.992%), Polynomial (0.733 to 7.033). Data reveals all the fitted models are resulting E% values less than 5%, which is the criteria for better fitting except BET model. This is a limiting factor, because suitability of BET equation at a_w less than 0.5. R^2 values for radish powder (fine fractions) ranged from Polynomial (0.930 to 0.998) Also, E% values are lowest in case of Polynomial model (0.020 to 1.911%). Whereas, other fitted models are not showing better fit according to the selected criteria based on higher R^2 and lower E% values. This shows, polynomial model is resulting E% values less than 5% and R^2 value approaching to unity, which is the criteria for better fitting.

Among various fitted models on experimental data, the appropriate models are given in table 2 and table 3. Monolayer moisture contents (m_0), obtained using GAB and BET equations are determined. The m_0 values are of big importance for determination of the storage conditions and minimizing the possible deterioration reactions. The m_0 values of radish leaf were decreased with increasing temperature [24].

3.3 Thermodynamic parameters

The study of sorption isotherms at least at two different temperatures gives thermodynamic data on isosteric heat of sorption with the use of mentioned Clausius–Clapeyron equation. The isosteric heat of adsorption changes with the change in amount of moisture adsorbed by the radish leaf powder (substrate). The relationship between isosteric heat of sorption (differential enthalpy) and equilibrium moisture content of radish leaf powder (fine fractions and mix fractions) is plotted in figure 4. The heat of adsorption also termed as differential enthalpy is a function of energy released on sorption. Thus, the heat of adsorption is considered as indicative of the binding forces between the sorption sites (adsorbent) and the water vapour (adsorbate) [26].

Table 1. Mathematical models (1-2-3 parameter equations) for adsorption isotherm

Sr. No.	Model equation	Model name
1.	$M = a + b\ln(1 - aw)$	Smith
2.	$M = \frac{C\, a_w m_o}{(1 - a_w)(1 - a_w + Ca_w)}$	BET
3.	$M = \left(\frac{1}{a + ba_w + \frac{c}{a_w}}\right)$	Hailwood and Horbin
4.	$M = a + b\,a_w + c\,a_w{}^2$	Polynomial
5.	$M = \frac{Ck\, a_w m_o}{(1 - ka_w)(1 - ka_w + Cka_w)}$	GAB

where, M – equilibrium moisture content (%, d.w.b.); a_w- water activity; a, b and c are model parameters to be estimated

Table 2. Model constants and coefficients with statistical parameters for radish powder mix fraction

Model no.	Temp (°C)	Model constants and coefficients	Statistical parameters R^2	E%
BET	15	m_0:3.937, c:59.292	0.938	1.139
	25	m_0:3.003, c:455.2692	0.879	0.937
	35	m_0:2.486, c:134.898	0.866	0.432
	45	m_0:2.101, c:34.421	0.890	15.645
Hailwood and Horbin	15	a:0.091, b:-0.109, c:0.035	0.985	1.542
	25	a:0.053, b:-0.076, c:0.056	0.995	0.408
	35	a:-0.017, b:-0.056, c:0.082	0.996	1.125
	45	a:0.035, b:-0.086, c:0.099	0.987	7.831
Polynomial	15	a:-5.987, b:35.703, c:3.445	0.966	1.411
	25	a:-0.113, b:0.003, c:3.601	0.989	2.047
	35	a:1.353, b:16.867, c:1.354	0.995	0.733
	45	a:2.482, b:13.668, c:0.697	0.981	7.033
GAB	15	c:4.9116, k:0.888, m_0:6.508	0.985	1.541
	25	c:3.200, k:0.784, m_0:7.121	0.995	0.408
	35	c:2.288, k:0.730, m_0:7.328	0.996	1.125
	45	c:2.451, k:0.777, m_0:5.325	0.994	7.830
Smith	15	a:0.971, b:-5.391	0.983	0.376
	25	a:1.308, b:-3.979	0.992	2.737
	35	a:0.986, b:-3.327	0.992	0.421
	45	a:0.536, b:-2.881	0.991	10.992

Table 3. Model constants and coefficients with statistical parameters for radish powder fine fraction

Model no.	Temp (°C)	Model constants and coefficients	Statistical parameters R^2	E%
BET	15	m_0:4.875, c:0.023	0.580	27.821
	25	m_0:3.238, c:0.022	0.570	16.169
	35	m_0:5.137, c:0.014	0.500	9.281
	45	m_0:4.132, c:0.013	0.630	8.959
Hailwood and Horbin	15	a:0.260, b:-0.003, c:0.077	0.942	0.069
	25	a:-297.410, b:47.080, c:-2495.300	0.380	87.483
	35	a:-228.360, b:37.490, c:-2078.350	0.900	87.482
	45	a:-217.750, b:36.990, c:-2134.020	0.640	87.478
Polynomial	15	a:-0.375, b:0.007, c:9.092	0.930	0.358
	25	a:-0.113, b:0.003, c:3.601	0.989	0.280
	35	a:-0.0516, b:0.003, c:2.124	0.998	0.020
	45	a:-0.0009, b:0.002, c:1.1091	0.981	1.911
GAB	15	c:1.000, k:53.382, m_0:-3.997	0.682	15.494
	25	c:1.002, k:51.441, m_0:-2.310	0.474	15.263
	35	c:1.000, k:51.779, m_0:-2.580	0.700	66.610
	45	c:1.018, k:51.387, m_0:-2.179	0.643	15.048

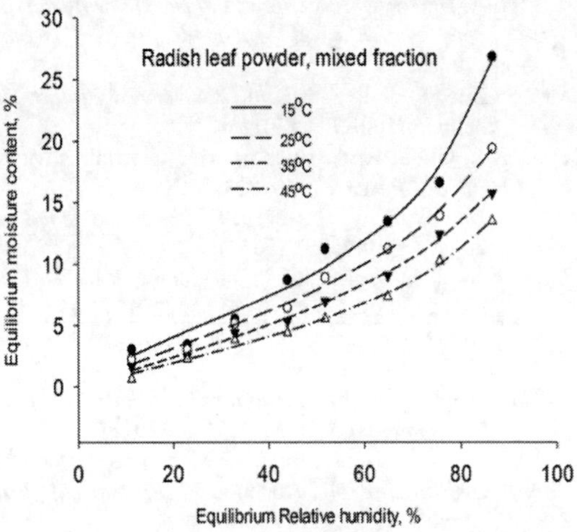

Figure 2. Temperature dependent adsorption isotherm of radish powder mix fraction

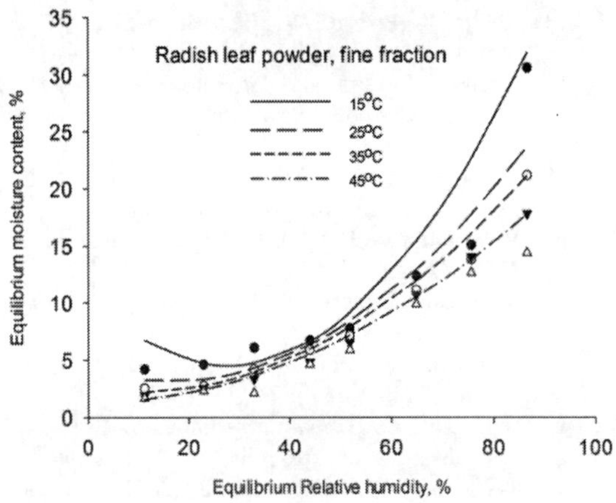

Figure 3. Temperature dependent adsorption isotherm of radish powder fine fraction

The isosteric heat of adsorption curve showed a regular fall with the increasing moisture content. Similar trend was seen for varieties of dates [27] for *C. ambrosioides* leaves [10] and for pistachio nut paste [21]. The higher heat of adsorption at lower moisture content can be justified because of the greater resistance to migration of water molecules from interior to the outer surface of the food samples [24, 28].

The value of heat of adsorption decreases approximately from 60.261 to 52.549 kJ/mol and 54.004 to 43.540 kJ/mol with the increasing equilibrium moisture content of mix fractions (MF) and fine fractions (FF) of radish leaf powder respectively. This reflects the differing strength of water binding molecules. In the beginning, initial occupation of highly active polar primary sites on the surface (with the greatest binding energy), followed by the subsequent filling of the less active secondary sites with lower bonding activation energies thus, lesser heat of adsorption. As the moisture content increases further, the heat of sorption or differential enthalpy tends to that of pure water which is a reflection of the moisture existing in the free form. Almost similar results were shown by previous researchers; isosteric heat of adsorption found 52 kJ/mol with macroni powder [26] 48.54 kJ/mol for instant tea powder and 47.96 kJ/mol for tea granules [28], 85 kJ/mol for potato powder, 58 kJ/mol for amylose powder and 70 kJ/mol for amylopectin powder [11].

Differential entropy of a food material is a function of the number of available sorption sites at a particular energy level. The graph between sorption entropy and equilibrium moisture content (figure 5) also gives the similar behavior as differential entropy decrease as the EMC increased [21]. The value of differential entropy decreases approximately from 43.564 to 43.556 kJ/mol.K and 43.559 to 43.532 kJ/mol.K with increasing EMC of mix fractions (MF) and fine fractions (FF) of radish leaf powder respectively. Almost similar results were shown by previous researchers; maximum value of differential entropy found 30 kJ/mol for amylase powder, 60kJ/mol for amylopectin powder and 110kJ/mol.K for potato starch powder [11].

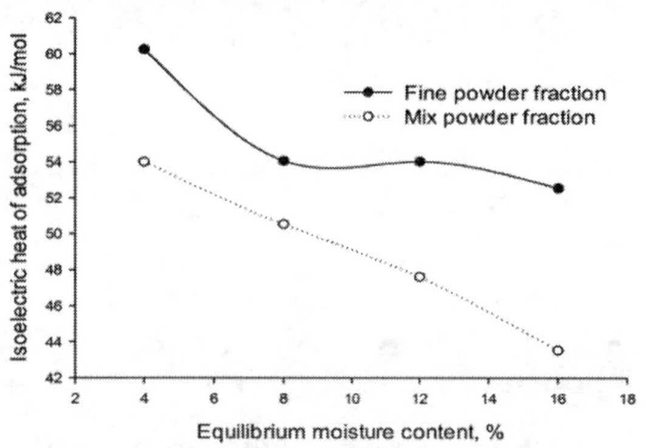

Figure 4. Isosteric heat of adsorption vs equilibrium moisture content of radish powder

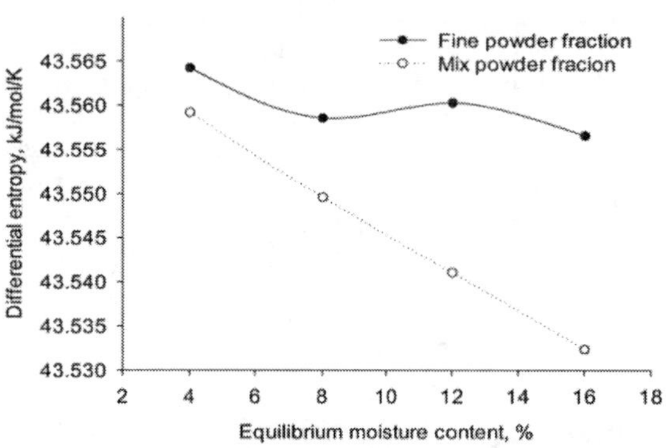

Figure 5. Differential entropy vs equilibrium moisture content of radish powder

Gibb's free energy calculated at various temperatures and found that energy at higher temperature has given lower value and vice-versa. The value of free energy decreases approximately from 5.516 to 0.5251 kJ/mol and 2.414 to 0.445 kJ/mol with increasing EMC of mix fractions (MF) and fine fractions (FF) of radish leaf powder respectively. It indicates the values of free energy are higher in case of fine radish leaf powder fractions. Free energy (figure 6) increases with decreasing moisture content [29].

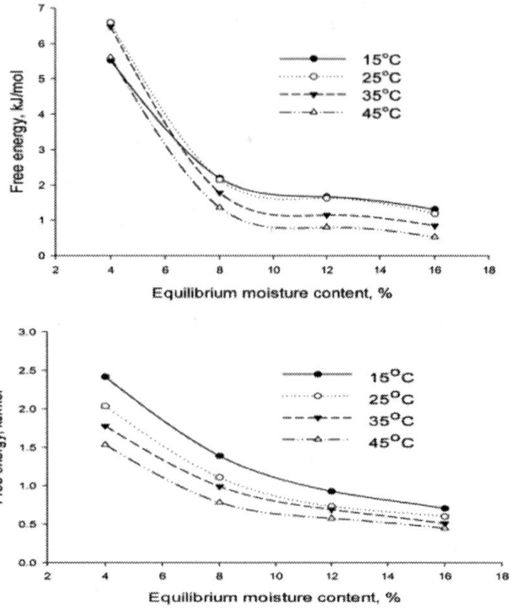

Figure 6. Free energy vs equilibrium moisture content of radish powder (Upper- fine fraction, Lower- Mix fraction

4 Conclusion

The experimental results showed that EMC values increased with decreasing temperature, at constant ERH. Equilibrium moisture content increases with the increase in the equilibrium relative humidity at constant temperature and show sigmoid change pattern. The adsorption data were in good agreement with most of popular sorption models. However, Polynomial and GAB equation could be found best in describing the adsorption isotherm at all the studied four temperatures for mix and fine radish leaf powder respectively. Almost similar trend of decreasing m_0 values are reported in case of both fractions. The isosteric heat of sorption curve showed a regular decrease with increasing moisture content and the similar trend are observed for the sorption entropy and free energy.

References

1. R. M. P. Gutierrez, R. L. Perez, *The Scientific World Jour.*, **4**, 811 (2004)
2. M. Ngouajio, *Oilseed Radish* 48824 (2004)
3. K. A. Delahaut, A. C. *Newenhouse, Growing carrots, beets, radishes and other crops in Wisconsin* (1998)
4. S. Lim, E. J. Lee, J. Kim, *Postharvest Biology and Technol.*, **100**, 219 (2015)
5. H. L. Gohil, R. A. Bucklin, M. J. Correll, *Advances in Space Research,* **45**, 823 (2010)
6. S. Herzallah, R. Holley, *LWT - Food Science and Technology,* **47**, 293 (2012)
7. S. Sindhu, P. Maya, T. N. Indira, *LWT - Food Science and Technology,* **49**, 42 (2012)
8. E. A. Noor, *Materials Chemistry and Physics,* **131**,160 (2011)
9. P. Singh, J. Singh, *International J. Plant, Animal and Environmental Sciences*, **3**, 103 (2013)
10. A. Jamali, M. Kouhila, L. Mohamed, J. Jaouhari, A. Idlimam N. Abdenouri, *Journal of Food Engineering,* **72**, 77 (2006)
11. A. Al-Muhtaseb, W. McMinn T. Magee, *J. Food Engg.*, **62**, 135 (2004)
12. S. Kaya, T. Kahyaoglu, *J. Food Engg.*, **71**, 200 (2005)
13. Ankita, K. Prasad, *Der Pharmacia Lettre*, **7**, 269 (2015b)
14. Ankita, K. Prasad, *International Journal of Agriculture, Food Science and Technology,* **3**, 105 (2014b)
15. S. Ranganna, *Handbook of Analysis and Quality Control of Fruit and Vegetable Products* (1997)
16. M. Kouhila, A. Belghit, M. Daguenet, B. Boutaleb, *J. Food Engg.*, **47**, 281 (2001)
17. A. Cassini, L. Marczak, C. Noren, *J. Food Engg.*, **77**, 194 (2006)
18. M. Castillo, E. Martınez, H. Gonzalez, E. Pacin S. Resnik, *J. Food Engg.*, **60**, 343 (2003)
19. Heldman, *Food dehydration*, 446 (2004)
20. S. Lahsasni, M. Kouhila, M. Mahrouz, N. Kechaou, *J. Food Engg.*,**55**, 201 (2002)
21. M. Maskan F. Gogus, *J. Food Engg.*, **33**, 227 (1997)
22. L. Mohamed, M. Kouhila, A. Jamali, S. Lahsasni, M. Mahrouz, *J. Food Engg.*, **67**, 491 (2004)
23. N. Arslan, H. Togrul, *J. Stored Pdts Res.*, **42**, 112 (2006)
24. S. Kaya, T. Kahyaoglu, *J. Food Engg.*, **78**, 413 (2007)
25. D. S. Sogi, U. S. Shivhare, S. Garg, A. Bawa, *Biosystems Engg.*, **84**, 297 (2003)
26. N. Arslan, H. Togrul, *J. Food Engg.*, **69**, 133 (2005)
27. R. Myhara, S. Sablani, S. Al-Alawi, M. Taylor, *LWT: Food Sci.*, **31**, 699 (1998)
28. V. Sinija, H. Mishra, *J. Food Engg.*, **86**, 494 (2008)
29. I. Sawhney, Ph.D. Thesis, S.L.I.E.T (Deemed University), Longowal, India (2011)

MATEC Web of Conferences 57, 04003 (2016) DOI: 10.1051/matecconf/20165704003
ICAET-2016

Physico-chemical, functional and rheological characterization of biodegradable pellets and composite sheets

Kulsum Jan[a], Shumaila Jan, CS Riar and DC Saxena

Department of Food Engineering and Technology
Sant Longowal Institute of Engineering & Technology, Longowal – 148 106, Punjab, INDIA

Abstract. Deoiled rice bran, paddy husk, cashew nut shell liquid and glycerol were extruded into pellets and further pressed into composites. Processing and plasticizer type had significant effect on physico-chemical, functional, rheological and morphological properties of pellets and composites. Specific mechanical energy of the pellets containing cashew nut shell liquid as plasticizer was higher than those containing glycerol. The maximum hardness and bulk density were obtained for pellets prepared from cashew nut shell liquid. Water binding capacity and water solubility index of both pellets and composites were highest for samples containing glycerol as plasticizer. A significant change in functional properties during processing was observed among raw materials, pellets and the final product (composite sheets).

Key words: Biodegradation, Cashew nut shell liquid, glycerol, plasticizer, functional, rheological.

1 Introduction

Biodegradable polymers are substitutes to the synthetic polymers which have occupied dangerous proportions among the environmental pollutants. Wastes from agro-industries generated during the industrial processing of agricultural products [1] have issues with their disposals. These wastes and byproducts are rich in minerals, sugars and proteins and can be used as raw materials for other industrial processes and development of value added products [2]. In the past most of the edible scraps were utilized for the production of animal feed, compost and fertilizer; otherwise they were discarded in common dumps. In order to reduce the trash collection and disposal fees, different methods and policies for byproduct and waste handling and treatment have been introduced to recover, recycle and convert the byproducts and wastes into upgraded products [3]. Biocomposites from natural fibers and biodegradable polymers have grabbed great interests in the composite science [4].

Deoiled rice bran a major byproduct generated during rice milling and subsequent oil extraction is a valuable source of inexpensive protein, contains about 12-20% and is an underutilized agro industrial byproduct [5]. Besides deoiled rice bran, paddy husk is one of the major agricultural residues produced during rice processing containing cellulose in similar amounts with wood. Usually paddy husk has been a problem for rice farmers due to its resistance to decomposition in the ground, difficult digestion, inadequate final disposal (burning) and low nutritional value for animals. The lignin and hemicellulose contents of paddy husk are lower than wood, for this reason paddy husk can be processed at higher temperatures than wood. Cashew nut shell is another waste product of cashew nut industry it is of great industrial importance and has recently been used to produce biocomposites [6].

Technologies like extrusion have found its application in other sectors like packaging technology besides being used for development of food products. The thermal energy generated by viscous dissipation during extrusion, combined with shearing effect, quickly cooks the raw mixture so that the properties of materials are modified due to physico-chemical changes of the biopolymers [7]. The effect of extrusion on the properties of pellets, using rice husk as filler was analyzed by Simone [8]. Zein based pellets were also developed by Wu et al. [9]. Till date, research is scanty on properties of pellets developed from deoiled rice bran and potato peel, the agro industrial byproducts and wastes along with plasticizers such as cashew nut shell liquid and glycerol using extrusion technology. Hence, the main aim of our study is to develop and characterize the pellets from extrusion for the development of sheets. The sheets were also characterized in second part of the study.

2 Materials and methods

2.1 Procurement of raw materials

Paddy husk and deoiled rice bran used for the present study were donated by M/s. AP Solvex Ltd., Dhuri (Punjab, India). Analytical grade glycerol was used as plasticizer (M/s. Merck Specialities Pvt. Ltd., Mumbai,

[a] Corresponding author: kulsumnissar@gmail.com

India). Cashew nut shell liquid was kindly donated by M/s. Allen Petrochemicals Pvt Lt., Meerut (India).

2.2 Sample preparation

Paddy husk was ground in a Laboratory grinding mill (M/s. Philips India Limited, Kolkata, India) and passed through 60 mesh screen to obtain fine powder. Similar process was followed for deoiled rice bran.

2.3 Physico-chemical and functional properties of raw materials

2.3.1 Proximate composition of raw materials

The raw materials were analysed for moisture, protein, fat, fibre and ash content according to AOAC [10].

2.3.2 Specific mechanical energy (SME) of optimized pellets

The specific mechanical energy (SME) of extrusion was calculated by following eqn.:

$$SME = (\omega/\omega_r) \times (\tau/100) \times (Z_r/Q) \quad (1)$$

where ω is screw speed, ω_r is the rated screw speed, τ is percent torque, Z_r is rated power, and Q is the feed rate.

2.4. Development of pellets and sheets

Deoiled rice bran (DOB), paddy husk powder (PDH), cashew nut shell liquid (CNSL) and glycerol (GL) were used in the blend preparation to prepare pellets. Different formulations used were (DOB, PDH and CNSL) and (DOB, PDH and GL). All the ingredients were weighed and mixed in a Hobart mixer (M/s. Continental Equipment India Pvt. Ltd, Delhi, India). Extrusion of the blends was performed using a co-rotating twin-screw extruder (M/s. Basic Technology Pvt. Ltd., Kolkata, India) under suitable conditions (Process filed for patent, No: 111/DEL/2015). The pellets prepared from extrusion were molded in hydraulic press to form sheets which were then stored for further analysis.

2.5 Characterization of pellets and composite sheets

2.5.1 Physical and mechanical properties

2.5.1.1 Color of pellets and sheets

Color measurements of all the pellets were carried out in triplicate as per standard protocol using a hand-held Chroma Meter (CR-400, Konica Minolta, UK) and L*, a*, and b* values were recorded. The values of ΔE were calculated as:

$$\Delta E = \sqrt{(L^* - L)^2 + (a^* - a)^2 + (b^* - b)^2} \quad (2)$$

where L^* is the standard value and L is the obtained value. Similarly, a^* and b^* are the standard values and *a, b* are the obtained values.

2.5.1.2 Bulk density of pellets and sheets

The density was measured as per procedures described by Jan et al. [7].

2.5.1.3 Frictional properties of pellets

The angle of repose is very important in characterization of the flow properties of pellets. The static coefficient of friction of the pellet against the various surfaces is used to determine the angle at which the pellets must be positioned in order to achieve consistent flow of pellets. A wooden table provided with an adjustable tilting plate on its top, facing to the test surface, was used for the purpose. The sample was kept on adjustable plate placed on top of the table and the inclination of the test surface was increased gradually so that the pellets started to slide down until about 75% of the pellets slided down. The static coefficient of friction (μ) was calculated from the following equation:

$$\mu = \tan\alpha \quad (3)$$

The angle of repose is derived from the following equation

$$\theta = \tan^{-1}\left\{\frac{2H}{D}\right\} \quad (4)$$

Where, H and D are the height and diameter of the heap, respectively.

2.5.1.4 Hardness of pellets and sheets

A crushing method using a TA – XT2i Texture Analyzer (M/s. Stable Micro Systems Ltd., Godalming, UK) was used to determine the hardness of pellets. Five pellets from each sample were mounted on the platform of the analyzer with a probe P/75 – 75mm diameter at a crosshead speed of 2 mm/sec with a target mode of 2 mm distance. The highest first peak value was recorded as hardness.

2.5.1.5 Pellet durability index

Durability of pellets was determined using the two compartments pellet-durability tester (M/s. Continental Agra Equipment Inc, India) [7].

2.5.1.6 Flexural properties of pellets and sheets

The flexural test was conducted at different storage conditions (relative humidity-RH and temperature) mentioned below [9]:

1. *Dry state:* under ambient conditions; **2**. *20% RH state*: equilibrated at 25°C and 20% RH for 7 days; **3**. *75% RH state*: equilibrated at 75% RH for 7 days; **4**. *95% RH state*: equilibrated at 95% RH for 7 days.

Maximum flexural stress (σ_m) was calculated as following equation:

$$\sigma_m = \frac{3PL}{2Bd^2} \quad (5)$$

where P is the maximum stress, 'L' is the support span, 'B' is the width and 'd' is the thickness of the pellet.

2.6 Functional properties of pellets and sheets

2.6.1 Water binding capacity (WBC)

A suspension of 5 g ground sample (dry weight) in 75 ml distilled water was agitated for 30 min at 25°C and centrifuged (3000×g) for 10 min. The supernatant was discarded, drained for 5 min and the gain in weight was expressed as percentage of water binding capacity.

2.6.2 Water solubility index (WSI)

WSI of raw materials were measured using a technique described by Anderson [11].

2.6.3 Water resistance and water uptake

The water resistance of injection-molded pots was measured by using the method as described by Wu et al. [9] with slight modifications. The experiment was carried out in triplicates. The percentage change in weight was plotted against time.

The water uptake of injection-molded pots was measured by using the method described by Schettini et al. [12] with slight modifications. The experiment was carried out in triplicates. The water up-take percentage was calculated using the equation:

$$\text{Water uptake (\%)} = \frac{Wd - Wt}{wd} \text{x } 100 \quad (6)$$

Where, Wd and Wt, respectively are the weights of the samples in the dry state and in the swollen states at different times.

2.7 Statistical analysis

The data reported represent average of triplicate observations ± standard deviation. Statistical analysis was performed using Statistica-log software package version 7 (M/s. StatSoft Inc., OK, USA). The significant differences were obtained by a one-way analysis of variance (ANOVA) followed by Duncan's multiple range test (DMRT) ($P < 0.05$).

3 Results and discussion

3.1 Proximate analysis

The proximate analysis (14% moisture basis) of raw materials revealed that DOB contained ash 7.94%, protein 13.78%, fat 0.63%, crude fibre 8.26% and carbohydrates 69.37%. PDH contained 12.25% ash, 0% protein, 0% fat, crude fibre 29.81% and carbohydrates 63.52%. It was observed that DOB has colour values (ΔE) as 33.588 and PDH has 31.327 which are quite significantly different. The water binding capacity (WBC) of DOB and PDH was found to be 3.52 and 3.94 g/g, respectively. Water solubility index (WSI) of the DOB and PDH was 10 and 16%, respectively. WBC of raw materials provides an insight of the ability of protein structure to associate with hydrophilic substance (water), the bound water which can further function as plasticizer during different processing operations.

Table 1. Effect of plasticizers on physical, functional and mechanical properties of pellets

Pellets properties	CNSL		Glycerol	
	Pellets	Sheets	Pellets	Sheets
SME (kJ/kg)	69.817^{b}		61.922^{a}	
ER	1.10^{a}		1.19^{b}	
PDI (%)	99.040^{b}		98.750^{a}	
AOR (°)	50.45^{a}		51.34^{b}	
COF GI	0.422^{a}		0.482^{b}	
COF WD	0.709^{a}		0.752^{b}	
Hardness (N)	453.38^{b}	480.25^{b}	448.58^{a}	477.85^{a}
Color (ΔE)	52.416^{b}	54.711^{b}	49.316^{a}	48.54^{a}
WBC (g/g)	5.34^{a}	4.07^{a}	6.14^{b}	0.685^{b}
WSI (%)	17.54^{a}	11.82^{a}	20.34^{b}	14.72^{b}
Density (g/ml)	0.735^{b}	1.21^{b}	0.685^{a}	1.13^{a}

Means with small letters in a column differ significantly ($p<0.05$)

3.2 Specific mechanical energy (SME)

As shown in Table 1, a significant difference in SME was observed between the pellets developed from both the blends. Highest value of SME was observed for sample containing CNSL than the sample containing glycerol. This may be due to the more binding effect of the plasticizer (CNSL) in the first stage with an exposure of hydrophilic sites, reinforcing the cross linkage. Thereafter, the network structure might have been formed with covalent crosslinks and reduced molecular size due to increased mechanical shear.

3.3 Color of pellets and sheets

Color is a physical property that may change during the extrusion cooking process [13]. A significant change was observed before and after extrusion as shown in Table 1. The change may be attributed to the extruder temperature and screw speed resulting in varying the structure and stir up browning reactions [14]. According to Rosentrater [13], Maillard reactions and protein denaturing play main roles in colour alterations during the cooking process resulting from high temperatures. It was also observed that the color (ΔE) value of samples containing CNSL as plasticizer were greater than samples containing glycerol as plasticizer which is due to the inherent dark color of the cashew nut shell liquid [7], resulting in maximum absorption of the light. In case of sheets, a significant increase in colour (ΔE) was observed during pressing (hydraulic press) which may be due to higher temperature during processing as compared to that of pellets during extrusion.

3.4 Density and hardness of pellets and sheets

The density of pellets with CNSL and glycerol was found to be 0.738 and 0.685 g/ml, respectively (Table 1). However, the density of sheets containing CNSL and glycerol was 1.216 and 1.136 g/ml, respectively. It is evident from the results that the density of both the pellets and sheets were higher for samples with CNSL. It may be due to the reduced porosity and availability of binding sites in CNSL to bind with the raw material.

The textural property (hardness) of pellets was determined by measuring the force required to break the pellet. The higher value of peak force required to breakdown the sample means higher the hardness of the sample to get fractured. It was observed that the hardness of the samples (CNSL) were greater than samples (glycerol). The increased hardness may be attributed to the greater binding effect of CNSL as compared to glycerol. Moreover, the increased hardness measured by texture analyser is due to more resistance to compression [7]. Plasticizers are important to maintain dimensional stability, hardness and flexibility of protein-based polymer due to its high boiling temperature [9]. The hardness was found to be directly correlated with the density of the pellets.

3.5 Frictional properties of pellets

A significant difference was observed between angle of repose (AOR) of both the samples (Table 1). The angle of repose for sample with CNSL was lesser than the samples with GL which may be due to the free flowing nature of pellets with CNSL which encourages the down slope movement of pellets making them buoyant and accordingly reducing angle of repose. The reduced angle of repose may be due to the smooth surface of samples (CNSL) which may be attributed to the improved binding effect of CNSL than glycerol. The parameter is important in design of hoppers to maintain continuous flow of the pellets, which must be larger than the angle of repose of the pellets [15].

The static coefficient of friction for galvanized iron (GI) surface was lesser for both the cases as compared to wooden surface (Table 1). However, COF for CNSL pellets was lower as compared to GL pellets. The reduction in coefficient of friction may be explained similarly as in the case of angle of repose. Moreover, the higher density of sample (CNSL) can also take part in reduction of friction during flow. Coefficient of friction is used to determine the angle at which the chutes must be positioned to achieve a consistent flow of the pellets [16]. The results indicate that galvanized iron can serve as a better medium for conveying pellets during processing.

3.6 Pellet durability index

PDI is another important physical property of the pellets. It indicates the mechanical strength of the extruded products (pellets). The higher the PDI, the more stable the extrudates during storage and handling processes. As shown in Table 1, The PDI for both the cases exceeded the standard durability index (97.5%) set for handling and transportation of pellets by British Standards Institution [17]. The increased hardness and durability can be attributed to the strong binding effect of the plasticizer causing strong interactions between particles that are similar to those of solid bridges [18].

3.7 Flexural properties of pellets and sheets

The effect of plasticizer level on the flexural strength in all the storage conditions is shown in Fig 1. In the dry state, for both the cases (GL, CNSL) sheets showed a maximum strength due to the low moisture content as compared to other conditions. At the 20% RH state, the water content increased more on an average than that in the dry state, and flexural strength decreased from an average of 128.670 MPa for the dry state to an average of 47.367 MPa for GL-sheets. However, in case of CNSL-sheets a decrease from 134.250MPa to 54.874MPa was observed. It is obvious from results that the strength of the CNSL-sheets was greater than GL-sheets which may be due to the increases hardness and reduced porosity of CNSL sheets. Increased porosity will result in imbibitions of water.

3.8 Water binding capacity and water solubility index

Water binding capacity of pellets with CNSL was less than that of samples containing glycerol which may due to the plasticizing effect of polyols (glycerol), in particular attributed to their ability to locate between the chains of proteins, bind water, and disrupt intermolecular polymer interactions [19]. Moreover, it may be also due to the hydrophilic and hygroscopic nature of glycerol and its ability to form the large hydrodynamic plasticizer-water complex [7].

Similar behaviour of glycerol has been observed in films developed using glycerol as plasticizer due to the hygroscopic character of plasticizers [20-21]. A significant change in water binding capacity was observed among the raw materials and the pellets formed. This change can be attributed to rupture of the associations by heat and shear during extrusion. In case of the sheets, a similar trend of WBC was observed for both the samples (Table 1). Pellets and sheets also showed a significant difference between them (Table 1).

A significant change in water solubility index among the raw materials, pellet and sheets was observed which can be explained by structural modifications of the feed compositions during extrusion [13, 21]. WSI of pellets (CNSL) was lower than that of samples with

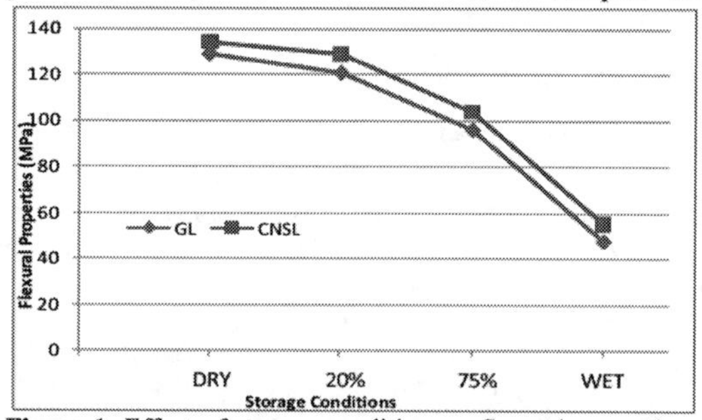

Figure 1. Effect of storage condition on flexural properties of sheets

glycerol (Table 1) which may be due to immiscible nature of CNSL with water. A similar trend of WSI was observed for sheets as in case of pellets. Hence, it can be predicted that processing plays an important role in affecting the functional properties of the pellets and pots.

3.9 Water resistance and water uptake

Water leaked slowly out of the pots during the whole test of water resistance of pots as shown in Fig. 2. The weight loss observed in CNSL-sheet was significantly lower than GL-sheets. In case of CNSL-sheets, a weight loss of 1.32% was observed for first 5hrs which rose to 3.24% after 10hrs. Initially, a gradual weight loss was observed which rose to 27.20% after 30hrs. Similarly, in case of GL-sheets a weight loss of 2.06% was observed for first 5 hrs which rose to 28.0% after 30hrs. Increased weight loss of GL-sheets may be due to the increased porosity and more free spaces available for water to penetrate into the walls of the GL-sheets.

DOI: 10.1051/matecconf/20165704003

GL-sheets experienced a faster water uptake (16.93%) mainly during the first 15 minutes of water dipping, absorbing most of the water in early stages then followed by a gradual increase rising upto 33.42% of water uptake after 60min. However, in contrast to GL-sheets, CNSL-sheets absorbed 12.26% in first 15min which rose upto 30.09% after 60min.

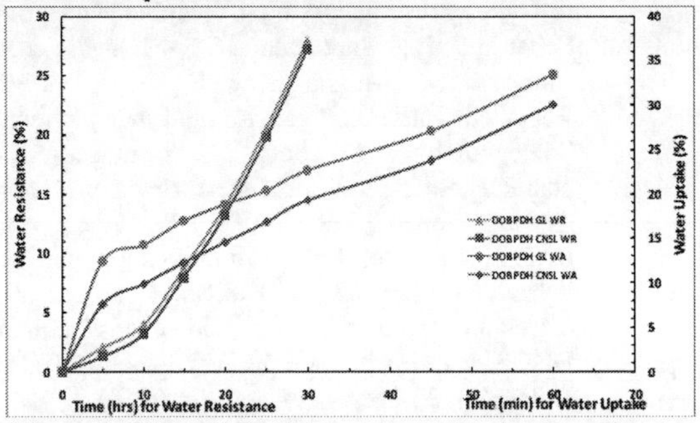

Figure 2. Effect of plasticizers on water resistance and water uptake of sheets

4 Structural data analysis

The factor loading obtained for the first two components (PCs) and the percentage of variance along with cumulative variance in case of pellets and sheets is shown in Table 2. In case of sheets, the first two principal components accounted for 99.583% of the variance. The first and second principal components (F1 and F2) explained 79.332 and 22.251% of the variance, respectively. It was observed that F1 was highly positively correlated with water solubility index and negatively correlated with water binding capacity and density. Second principal component F2 was highly negatively correlated with hardness. It was observed from Fig.3 that CNSL-sheets were positioned on left side of principal component 1 (F1) which were linked to density and water binding capacity. However, GL-sheets were also positioned on left side of principal component 1 (F1) which were linked to hardness and color.

Table 2. Principal component analysis: loading of first two components for properties of pellets and sheets

Sheets	**F1**	**F2**
Eigen value	3.967	1.013
Variability (%)	79.332	20.251
Cumulative %	79.332	99.583
Factor loadings	**F1**	**F2**
WBC	**-0.995**	0.033
WSI	**0.995**	-0.017
Hardness	-0.408	**-0.913**
Density	-0.908	0.418
Color	**-0.997**	-0.057
Pellets	**F1**	**F2**
Eigen value	6.500	2.377
Variability (%)	72.198	26.413
Cumulative %	72.198	98.610
Factor loadings	**F1**	**F2**
Hardness	**-0.933**	0.360
WBC	1.000	-0.001
Color	**-0.926**	0.378
Pellet Durability Index	-0.844	-0.451
Water Solubility Index	**0.844**	-0.520
Coefficient of Friction (GI)	**0.812**	0.579
Coefficient of Friction (WD)	**0.944**	0.325
Angle of Repose	0.684	-0.727
Density	0.573	**0.813**

In case of pellets sheets the first two principal components accounted for 98.610% of the variance. The first and second principal components (F1 and F2) explained 72.198 and 26.413% of the variance, respectively. It was observed that F1 was highly positively correlated with water solubility index, coefficient of friction (both wood and galvanized iron) and highly negatively correlated with color and hardness. It was observed from **Fig.4** that CNSL-sheets were positioned on right side of principal component 1 (F1) which were linked to density, coefficient of friction (both wood and galvanized iron), water binding capacity, angle of repose and water solubility index. However, GL-sheets were positioned on left side of principal component 1 (F1) which were linked to pellet durability index, color and hardness. Hence, on the basis of these parameters GL-sheets can be differentiated from CNSL-sheets.

5 Conclusion

It was observed from the results that sheets prepared from CNSL were having more water resistance and hardness than those prepared from glycerol. A significant change was observed between the raw materials pellets and sheets. On the basis of results obtained for sheets it was concluded that pellets with the same formulation can be used for the preparation of various molded products at high temperatures.

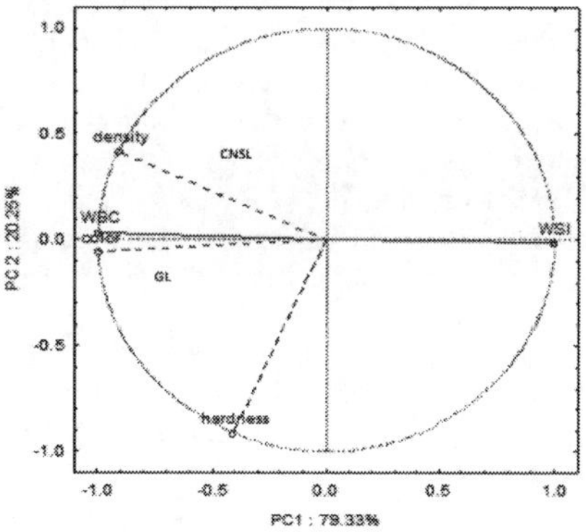

Figure 3. Principle component score plot for sheets

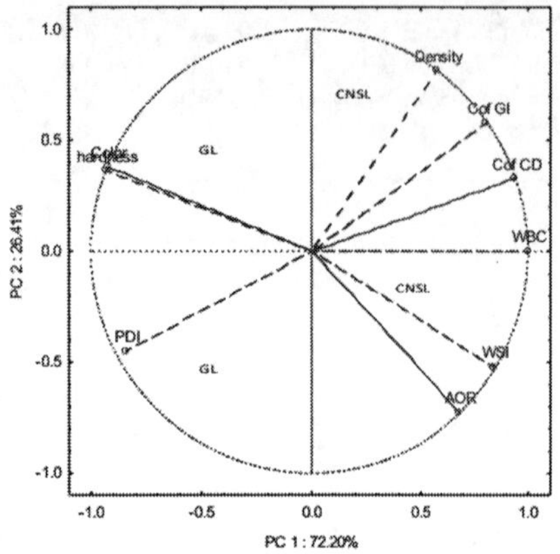

Figure 4. Principle component score plot for pellets

Acknowledgments

Financial support from Council of Scientific and Industrial Research (CSIR), New Delhi, Govt. of India is gratefully acknowledged.

References

1. Singh, S., 2014. Production and characterization of industrially important amylase enzyme by Aspergillus niger using different combination of starch waste as substrate. International Journal of Chemical and Pharmaceutical Sciences. 5(2), 85-89.
2. Markou, G., Georgakakis, D., 2011. Cultivation of filamentous cyanobacteria (blue-green algae) in agro-industrial wastes and wastewaters: A review. Applied Energy. 88, 3389–3401
3. Federici F., Fava F., Kalogerakis N., Mantzavinosc D (2009) Valorisation of agro-industrial by-products, effluents and waste: concept, opportunities and the case of olive mill wastewaters. J Chem Technol Biotechnol 84(6):895–900.
4. Lee, B.H., Kim, H.J., Yu, W.R., 2009. Fabrication of long and discontinuous natural fiber reinforced polypropylene biocomposites and their mechanical properties. Fibers and Polymers. 10(1), 83–90.
5. Yoon J.S., Sung A.J., Kyung B.S., (2011) Preparation and Mechanical Properties of Rice Bran Protein Composite Films Containing Gelatin or Red Algae. Food Sci. Biotechnol 20(3):703-707.
6. Raquez, J.M., Deleglise, M., Lacrampe, M.F., Krawczak, P., 2010. Thermosetting (bio) materials derived from renewable resources: A critical review. Prog Polym Sci. 35, 487–509
7. Jan, K., Riar, C.S., Saxena, D.C., 2015. Engineering and functional properties of biodegradable pellets developed from various agro-industrial wastes using extrusion technology. J Food Sci Technol 52(12), 7625-7639.
8. Simone, M., Leal, R., Evelise, F.S., Carlos, A.F., Sonia, M.B.N., 2009. Studies on the Properties of Rice-Husk-Filled-PP Composites – Effect of Maleated PP. Mat. Res 12(3), 333-338.
9. Wu, Q., Sakabe, H., Isobe, S., 2003. Processing and properties of low cost corn gluten meal/wood fiber composite. Ind. Eng. Chem. Res. 42(26), 6765–73.
10. AOAC. (1995) Official methods of analysis. U.S.A. Assoc of Official Anal Chem, Washington.
11. Anderson, R.A., Conway, H.F., Pfeifer, V.F., Griffin, E.L., 1969. Gelatinization of corn grits by roll and extrusion cooking. Cereal Sci. Today, 14, 4
12. Schettini, E., Santagata, G., Malinconico, M., Immirzi, B., Mugnozza, G.S., Vox, G., 2013. Recycled wastes of tomato and hemp fibres for biodegradable pots: Physico-chemical characterization and field performance. Resources, Conservation and Recycling. 70, 9– 19.
13. Rosentrater, K.A., Muthukumarappan, K., Kannadhason, S., 2009. Effects of ingredients and extrusion parameters on aquafeeds containing DDGS and potato starch. J. Aquacult. Feed Sci. Nutr. 1(1), 22-38.
14. Razzaq, M.R., Anjum, F.M., Khan, M.I., Khan, M.R., Nadeem, M., Javed, M. S., Sajid, M.W., 2012. Effect of temperature, screw speed and moisture variations on extrusion cooking behavior of Maize (Zea mays. L). Pak. J. Food Sci. 22(1), 12-22.
15. Cain, J., 2002. An alternative technique for determining ANSI/CEMA standard 550 flowability ratings for granular materials. Powder Handling and Processing, 14(3), 218–220.
16. Olajide, J.O., Igbeka, J.C., 2003. Some physical properties of groundnut kernels. J. Food Eng. 58, 201–204 BSI. (2010). EN 15210–15211:2009- solid biofuels-determination of mechanical durability of pellets and briquettes. British Standards Institution, London.
17. BSI., (2010). EN 15210–15211:2009- solid biofuels-determination of mechanical durability of pellets and briquettes. British Standards Institution, London.
18. Nalladurai, K, Morey, V.R., 2009. Factors affecting strength and durability of densified biomass products. Biomass Bioenerg. 33(3), 337–359
19. Gontard, N., Guilbert, S., Cuq, J.L., 1993. Water and glycerol as plasticizer affect mechanical and water vapor barrier properties of an edible wheat gluten film. J Food Sci 58(1), 206–211.
20. Gennadios, A., Weller, C.L., 1994. Moisture adsorption by grain protein films. Transactions of ASAE, 37(2), 535–539.
21. Shaw, N.B., Monahan, F.J., O'Riordan, E. D., O'Sullivan, M., 2002. Physical properties of WPI films plasticized with glycerol, xylitol, or sorbitol. J. Food Sci. 67(1), 164–167.
22. Parisa, F., Kurt, A.R., Kasiviswanathan, M., Mehmet, T., 2013. Effects of Steam, Moisture, and Screw Speed on Physical Properties of DDGS-Based Extrudates. Cereal Chem. 90(3), 186-197.

MATEC Web of Conferences 57, 04004 (2016) DOI: 10.1051/matecconf/20165704004
ICAET-2016

Flow property measurement of rice flour- A Review

Shumaila Jan[a], Kulsum Jan and DC Saxena

Department of Food Engineering and Technology
Sant Longowal Institute of Engineering & Technology, Longowal – 148 106, Punjab (INDIA)

Abstract. This review covers conventional and new methods used to measure flowability of rice flour. Powder flowability is inherently complex due to many physical and environmental variables that determine how it responds to being moved. Investigating flowability requires a high specification powder rheometer having exceptional sensitivity and repeatability of measurement. This study shows how rice flour can be characterized regarding flowability in relation to flow rate, compaction and other factors.

1 Introduction

Rice (*Oryza sativa* L.) is the most important cereal crop and staple food of over the world. More than 2000 million people obtain their 60–70% calories from rice and its products [1]. It is also used in large number of food products like idli, dosa, noodles, breakfast cereals, unleavened breads, snack food items, crackers, candies and baby foods etc. [2]. Flowability is defined as the ability of a powder to flow freely, in a regular and a constant way [3]. This property establishes the performance and the nature of the working processes (mixture, transport, storage) and/or applications of the powder. So, a powder with good flowability flows without assistance, whereas a cohesive powder requires a mechanical device adapted to its actuation through agitation, vibration, etc. With the increasing quantity and variety of powders/flours being produced in industry, there is a need for information about their handling and processing characteristics. During flow, individual particles predominantly slide across each other in a shearing action. However, food powder particles can also stick together and form stable structures, preventing flow due to mechanical forces. Mechanical and chemical interactions among powders depend on composition, moisture content, surface mechanical characteristics, particle history (e.g., from production to storage), particle shape and size, and the manner in which the particles geometrically interact to form stable structures. Consequently, the strength among particles can be due to mechanical interlocking, chemical bonding, and van der Waals and electrostatic forces. The forces opposing flow are friction, attraction between particles (cohesion), attraction between particles and system walls, and particle interlocking [4].

2 Application of Rice Flour

The rice flour is an important ingredient in both traditional and novel foods prepared across the world [5]. Some of the traditional rice flour based products in India are puttu, made by steaming rice flour and grated coconut, and appam, a cake formed by mixing rice flour and fermented coconut milk. Rice flour exhibits a number of unique characteristics and can be a better substitute of other starch sources in a number of food applications, to prepare baby foods, noodles, soups, thickener, extruded products and blending with other ingredients [6]. Due to soft taste, colourless appearance, hypoallergenic properties, low levels of sodium and easy digestible carbohydrates, rice flour is one of the most suitable cereals for gluten-free product formulations [7]. Moreover, it has the smallest particle size and the whitest colour of all commercial grains. The particle size distribution, mean granule size, surface composition, particle shape of rice flour is well known to play an important role in flowability of rice flour.

3 Flow Property Measurement

The actual flowability of powder is determined by both the latter's physical properties and the geometry of the system [8].There are some of the experimental methods which can be used for predicting the flow behaviour of rice flour. Generally speaking, a flow of a powder is the relative movement of a bulk of particles among neighbouring particles or along the container wall

[a] Corresponding author: shumailanissar@gmail.com

surface. The forces that are involved are gravitational forces, friction, cohesion (interparticle attraction) and adhesion (particle-wall attraction). A general theory for the determination of flowability of solid particles was developed by Jenike [8].

3.1 Angle of Repose

Probably, the simplest test from a technical point of view is the measurement of the angle of repose. It is the angle formed by the horizontal base of the bench surface and the edge of a cone-like pile of granules. It is generally considered that the powders with angle of repose of less than about 40 degrees are free flowing. Powders exhibiting repose angle of 50 degrees and more are likely to cause flow problems. Because the formation of the cone involves frictional forces, interparticle cohesive forces, and impact effects that may lead to segregation, the actual measurements depend on the experimental method procedure. Moreover, the factors that affect the flowability of any given powder are numerous which may also vary the angle of repose includes shape and size distribution.

3.2 Bulk density and compressibility

Consumers expect that the mass indicated on the package corresponds to a volume filling the package almost completely that's why the bulk density is important. The bulk density is primarily determined by the solid density and the particles array in the bed. Basically the solid density is an intrinsic property of the material, the array which the particle forms is not. Food powders have irregular particle shapes, and the effects of friction or cohesive forces are significant. Availability of mechanical forces supporting the open structure, such as friction, cohesion, and particle shape determines the stability or existence of a food powder's unconfined arrangement of particles. Therefore, as a result of excessive variability inherent in the system, bulk density alone cannot predict the flowability of powders [9]. Bulk density varies as a function of the applied pressure to it as it is not a single or even a dual value of a given powder. Material's compressibility is represented by a slope or straight line of a log-log plot by measuring the degree to which a powder compacts as a function of the applied pressure. The resulting data provides the information to evaluate wall friction and feeder operation requirements [10].

3.3 Dynamic flow properties measurement by FT4 Powder Rheometer

Dynamic flow properties or dynamic characterization of a sample is actually the measurement of a powder in motion. Several techniques for visual characterisation of particle shape have employed optical microscopes, electron microscopy and, more recently, X-ray micro tomography [2, 11]. But such techniques are either time consuming or labour intensive. Moreover, many other measurement techniques such as Carr's index, Hall flowmeter, shear cell and powder rheometer, also provides powder flow properties [12, 13]. Though the relationship between particle shape and flow characteristics has not been extensively studied, the samples studied are mostly limited to spherical/ spheroidal particles [14]. The flow properties of three dissimilar materials with similar particle sizes, using permeability and shear cell were analysed by Bumiller [15] and suggests that the particle shape might have significant effects on powder flow properties.

Dynamic flow properties were evaluated by rotating a precision blade through the powder along a fixed helical path. Specific energy (SE) is the energy required to establish a particular flow pattern in a conditioned (conditioning prior to measurement eliminates the effects of inherent powder variability so analysis is both reproducible and repeatable) precise volume of powder. This flow pattern is an upward clockwise motion of the blade, generating gentle lifting and a low stress flow mode in the powder. As it is being established on the upward traverse and the powder is unconfined, the energies measured are more dependent on the forces between particles and less influenced by other factors, such as compressibility. The SE value is calculated from the work done per unit mass in moving the blade through the powder from the bottom of the vessel to the top [16]. Powder flow characteristic measurements clearly showed that differences in particle size distribution and, particularly, particle shape, significantly affect all flow properties over a wide range of stress conditions.

An FT4 Powder Rheometer (Freeman Technology), shown in Fig. 1 is used to measure powder properties such as shear and wall friction, as well as bulk properties like permeability, compressibility, and density. This allows comprehensive characterization of powders from a fluidized state to a highly consolidated one.

DOI: 10.1051/matecconf/20165704004

Fig.1. FT4 Powder Rheometer (Freeman Technology)

3.4 Permeability

It shows how easily a material can transmit a fluid through its bulk, also using the vented piston attachment, as shown in (Fig.2). The permeability test has a testing sequence similar to that of the compressibility test, though it measures the air pressure required to maintain a constant air flow through the powder bed whilst being consolidated at increasing normal stresses. As the permeability is reduced due to particles becoming more closely packed, a higher normal stress results in a higher pressure drop **[16]**.

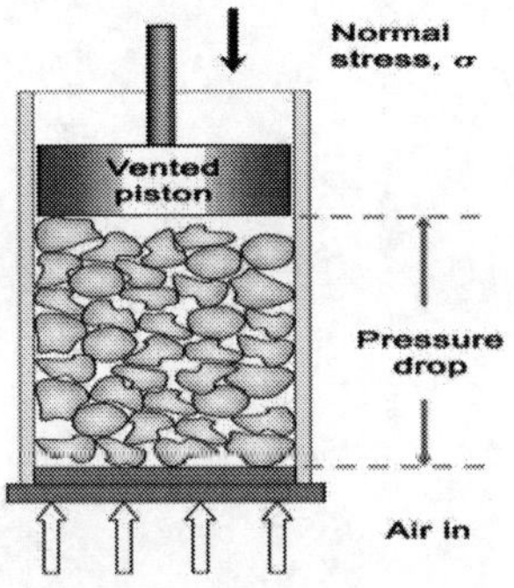

Fig. 2. Material flow through vented piston attachment

3.5 Shear Test

The unconfined yield locus and the flow function can be determined by the shear testing. Shear analysis in terms of principal stresses and Mohr's failure criteria is currently the most commonly parameters used for flowability analysis. The main aim of the shear analysis is to provide information about the stress a powder can support before it yields. The most convenient parameter is the unconfined yield stress which generally represents the stress level that can be supported at the compact surface. The unconfined yield stress can be used as an independent measure of powder cohesiveness [17] or be used for the measurement of Jenike's flow function. The latter is the relationship between the unconfined yield stress (or force) and the major consolidation stress (or force) as shown in (Fig. 3), and besides providing data for quantitative flowability classification it can also be used for construction design of silos.

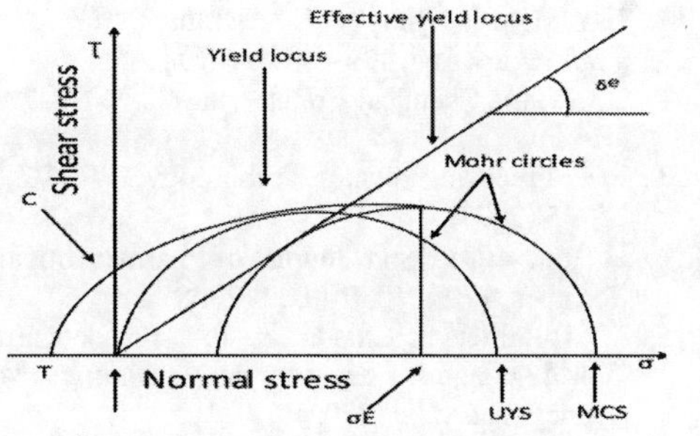

Fig. 3. Relationship between the UYS and MCS (UYS: unconfined yield stress, MCS: major consolidation stress

4 Conclusion

Measurement/characterisation of rice flour flowability is useful because it quantifies the strength of the flour that has to be overcome to cause the powder particles to move over each other and flow. This is important in many powder handling and processing operations. In addition, it can be applied for evaluating the effect of processing conditions, such as temperature, relative humidity, and storage time in a silo. Besides these flow measurements, there are numerous methods which can be used for measuring/characterizing the rice flour in terms of flowability as the actual flow depends on both the powder properties and the geometry and other physical characteristics of the system.

References

1. J.H. Lin, H. Singh, Y.T. Chang, Y.H. Chang, Food Chemistry. 1;126(3):1108-14 (2011)
2. J. Bao, C.J. Bergman, Starch in Food: Structure, Functional and Application. CRC Press, Cambridge, pp. 258–292 (2004)
3. K. Masuda, K. Higashitani, H. Yoshida, Powder Technology Handbook, Third edition CRC Press (2006)
4. M. Peleg, J. Food Proc. Eng. 1:303–28 (1978)
5. C.P. Villareal, B.O. Juliano, S. Hizukuri,. Cereal Chem. 70, 753–758 (1993)
6. B.O. Juliano, Academic Press Inc, New York (1984)
7. H. S. Gujral, I. Guardiola, J. V. Carbonell, & C.M. Rosell, Food Chemistry, 51, 3814–3818 (2003)
8. A.W. Jenike, Utah Engineering Experiment Station. University of Utah, Salt Lake City (1967)
9. P. Juliano and G.V. Barbosa-Cánovas, Annual review of food science and technology 1:211-239 (2010)
10. A.W. Jenike, Bulletin of the University of Utah, 53 and 26 (Bull No. 123 of Utah engineering experiment station, eleventh printing (1986).

11. X. Fu, M. Dutt, A.C. Bentham, *Powder Technology*, 167, 134–140 (2006)
12. R.L. Carr, Chemical Engineering 72(3) 69–72. (1965)
13. R.E. Freeman, Powder Technology, 174(1/2), 25–33 (2007).
14. H.Hou, & C.C. Sun, Journal of Pharmaceutical Sciences, 97(9), 4030–4039 (2008)
15. M. Bumiller, J. Carson, & J.A. Prescott, In World congress on particle technology 4 Sydney, Australia (2002)
16. F. Xiaowei, H. Deborah, M. Lisa, B. Armstrong, U. Willen, T. Freeman, Particuology 10, 203– 208 (2012)
17. P. York, Powder Technology 11, 197 (1975).

EFFECT of hydrocolloids on the quality evaluation of flour based noodles from Horse Chestnut

Syed Insha Rafiq[1,a], Syed Mansha Rafiq[2] and D.C. Saxena[1]

[1] *Department of Food Engineering & Technology, Sant Longowal Institute of Engineering and Technology, Longowal, Punjab ,India*
[2] *Department of Dairy Technology, National Dairy Research Institute, Karnal (India)*

Abstract. The present study was focused to investigate the effects of hydrocolloids (guar gum and xanthan gum) at additional levels (1%, 2% and 3%) on the noodle characteristics prepared from horse chestnut flour. The qualities of noodles prepared from horse chestnut flour were compared with wheat flour based noodles in terms of cooking characteristics, textural and sensory properties. The hydrocolloid addition in noodles resulted in improvement of cooking and textural qualities in consistent to control sample. The incorporation of 3% gum significantly increased cooking properties and the firmness of cooked noodles. The results of the sensory evaluation based on a nine point hedonic scale revealed that apart from the control, noodles with 3% gum were acceptable to the panellists.

Keywords: Horse Chestnut flour, noodles, cooking

1 Introduction

Noodles, a traditional staple food in many Asian countries, have been consumed for thousands of years. It is becoming increasingly popular worldwide for its convenience, low cost, availability of various products with desirable sensory attributes having varied taste and texture, beneficial nutrition, and long shelf-life. It is considered as one of the main carbohydrate-based foods generally prepared from wheat flour. Globally, the consumption of wheat noodles is second only to bread [1, 2] with nearly 40% of wheat being processed into noodles in Asia [3, 4]. Noodles are generally prepared by mixing raw materials, dough sheeting, and cutting [5]. Nowadays, noodles can be made from various cereal and non-cereal flours like wheat, rice, buckwheat, millets, etc.

Hydrocolloids are classified as thickening or gelling agents, added to foods for improving the stability and texture of foods [6]. Their high polymeric nature and the interactions among polymer chains results in the improvement of mouth feel and viscosity when dissolved or dispersed in food systems [7]. The presence of hydrocolloids increases the eating experience by improving hardness, cutting force, gumminess, and chewiness in noodle making [8].

Diversification of horse chestnut utilization into a variety of food products would support the national food security. In particular, the use of horse chestnut gives health benefits with respect to its flavonoid, protein, vitamin and mineral content [9]. In addition, the presence of phenolic compounds that function as antioxidants and dietary fibre, also contribute to functional food properties of horse chestnut. Therefore, the present study was performed to identify the cooking and textural characteristics as well as sensorial attributes of noodles prepared from horse chestnut flour. In addition, attempts have been made to improve the noodle-making property of horse chestnut flour by adding hydrocolloids such as guar gum and xanthan gum.

2. Material and methods

2.1 Flour preparation

Horse Chestnut flour was prepared as per the method reported in our previous study [10].

2.2. Noodle preparation

Noodles were prepared following the method described by Yoo et al. [11]. The noodle formulation consisted of 100 g of native horse chestnut flour and 34 ml of distilled water. Two types of hydrocolloids were added to the flour at levels of 1%, 2% and 3%. Wheat flour was taken as control sample. The prepared dough was placed to rest in a plastic bag for 20 min. After that, the dough was processed through a small noodle machine (Mode JMTD-168/140, Beijing, China) with the roller gap reduced gradually, to get dough sheets. The dimensions of the resultant noodle strands were 1 mm in width and

[a] Corresponding author: syedinsha12@gmail.com

0.8 mm in thickness. The noodles were then dried at 60°C for 2–3 h and stored in sealed polyethylene bags.

2.3. Cooking properties

2.3.1. Cooking time

Noodles (5 g) were cooked in 100 ml of boiling distilled water in a 250 ml beaker. Noodles were cooked until disappearance of white core as judged by squeezing between two glass slides.

2.3.2. Cooked weight and solids loss

The cooking properties of noodles were determined according to the method of Lee et al. [12]. Noodles (10 g) were boiled in distilled water (150 mL) for 10 min. After boiling, the cooked noodles were allowed to drain for 5 min using a strainer and the cooked water was collected and dried in an oven (105°C) and weighed for solids loss. The cooking loss was expressed as a percentage of solid loss during cooking. Cooked noodles were also weighed for gain in cooked weight.

2.3.3. Water uptake percentage (g/100 g)

The water uptake is the difference in the weight of cooked noodle versus uncooked noodles, expressed as the percentage of weight of uncooked noodles [13].

Cooked noodles were rinsed with cold water and drained for 30 s then weighed to determine the cooking gain. The analysis indicates the amount of water absorbed by the noodle during cooking process.

2.4. Textural properties of cooked noodles

Texture is one of the critical characteristics in quality evaluation of noodles. Elasticity, adhesiveness and firmness are the main parameters that are measured in texture evaluation of noodle [14, 15]. Noodles with smooth, soft and elastic texture are consumed in Japan while in China and Korea firmer texture is preferred [16]. Textural properties of cooked noodles were evaluated by texture profile analysis (TPA) using the TA-XT2 Texture Analyzer (Stable Micro Systems, England) within 5 min after cooking. A single strand of each noodle sample was cut to 2.5 cm length. A set of five strands of cooked noodles was placed parallel on a flat metal plate and compressed to 70% of their original height using compression platen probe at a speed of 1.0 mm/s. From force–time curves of the TPA, hardness, adhesiveness, springiness, cohesiveness, gumminess, chewiness and adhesiveness were determined.

2.5. Sensory evaluation of noodles

Sensory evaluation was conducted for the cooked noodle samples and a semi trained panel of twenty five members evaluated the noodles using a 9 point hedonic scale ranging from 1 (extremely disliked) to 9 (extremely liked). The panellists were asked to score for slipperiness, firmness, chewiness, tooth packing, appearance and overall acceptability.

2.6. Statistical analysis

The data reported are an average of three replications. Significance was defined at $P < 0.05$ by using Duncan's multiple range test.

3 Results and discussion

3.1. *Cooking characteristics of noodles*

Noodles were prepared from horse chestnut flour using hydrocolloids viz. xanthan gum and guar gum in different proportions and were tagged as:

HCNO: Native Horse Chestnut flour noodles
XG1: Noodles with 1% xanthan gum
XG2: Noodles with 2% xanthan gum
XG3: Noodles with 3% xanthan gum
GG1: Noodles with 1% guar gum
GG2: Noodles with 2% guar gum
GG3: Noodles with 3% guar gum

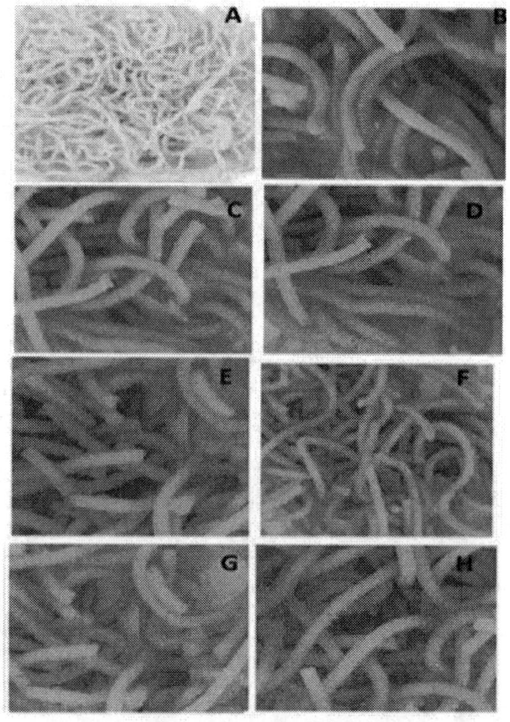

A: Control noodles; B: Noodles with 3% xanthan gum; C: Noodles with 3% guar gum; D: Noodles with 2% guar gum; E: Noodles with 1% guar gum; F: Native Horse Chestnut flour noodles; G: Noodles with 1% xanthan gum; H: Noodles with 2% xanthan gum

Table 1. Cooking properties of noodles

Sample	Cooking time (min)	Cooked weight (%)	Cooking loss (%)	Water uptake (g)
Control	7.10 ±0.16^{b}	26.14±0.24^{c}	4.96±0.11^{e}	162±2.06^{a}

HCNO	4.62 ± 0.12^{f}	18.43 ± 0.21^{g}	10.01 ± 0.13^{a}	108 ± 1.87^{g}
XG1	5.12 ± 0.14^{e}	20.14 ± 0.18^{f}	8.76 ± 0.12^{c}	117 ± 1.48^{f}
XG2	7.01 ± 0.09^{b}	23.37 ± 0.22^{e}	6.34 ± 0.09^{d}	138 ± 2.01^{c}
XG3	7.68 ± 0.11^{a}	27.87 ± 0.14^{a}	4.82 ± 0.17^{e}	154 ± 1.75^{b}
GG1	5.47 ± 0.17^{d}	19.97 ± 0.19^{f}	9.01 ± 0.18^{b}	118 ± 1.39^{f}
GG2	5.86 ± 0.08^{c}	24.38 ± 0.23^{d}	6.47 ± 0.12^{d}	124 ± 2.09^{e}
GG3	7.21 ± 0.16^{b}	26.72 ± 0.11^{b}	5.01 ± 0.08^{e}	134 ± 1.68^{d}

Means in a column with same superscript do not differ significantly (p<0.05).

Cooking quality of noodles is the most important characteristic for consumers and therefore of great importance to processing industry. Cooking characteristics including cooking time, cooked weight, cooking loss and water uptake of noodles are presented in Table 1. Mean cooking time for control, native and XG and GG noodles varied significantly (Table 1). Cooking time of the XG noodles ranged from 5.12 to 7.68min while for GG noodles it ranged from 5.47 to 7.21min. The water absorption rate was found to be improved in noodles with hydrocolloid than noodles with native HCN flour. Cooking losses are one of the main parameters taken into consideration during the assessment of noodle quality [16]. The results showed less cooking loss in XG3 and GG3 noodle samples. The results also revealed increased weight gain from 20.14 to 27.87% in XG noodles and 19.97 to 26.72% in GG noodles with increased gum concentration. The differences in the cooking quality were attributed primarily to the hydrocolloid fraction that provides as a binder for more retention of water within the noodles.

property and the mouth-feel of noodles since it is concerned as an indicator of the extent of disruption of the noodle structure during mastication. A non-significant difference in cohesiveness was observed between the HCNO, XG and GG noodles. The chewiness values for the control, HCNO, XG and GG noodles were observed to have a non significant difference with GC3 sample having the highest value.

3.2. Textural properties

The textural properties of the noodles prepared with wheat flour, horse chestnut flour and hydrocolloid blends are shown in Table 2. The textural properties were significantly changed in all the textural parameters except for the adhesiveness and cohesiveness values of control and horse chestnut flour-based noodles ($p < 0.05$). In general, the addition of hydrocolloids to horse chestnut flour-based noodles enhanced their firmness values. Though all the samples with hydrocolloids are fair considering the firmness of the noodles but XG3 and GG3 compete favourably with the control sample. The incorporation of hydrocolloid increased the firmness of noodles from 9.93 to 16.2N (XG sample) and 9.93 to 15.1N (GG sample), respectively. Compared to the values of the control noodles, the firmness of noodles with 3% hydrocolloids was increased. On the other hand, the control noodles without hydrocolloids exhibited a lower cohesiveness compared to the noodles with added hydrocolloids. Cohesiveness affects the cooking property and the mouth-feel of noodles since it is concerned as an indicator of the extent of disruption of the noodle structure during mastication. A non-significant difference in cohesiveness was observed between the HCNO, XG and GG noodles. The chewiness values for the control, HCNO, XG and GG noodles were observed to have a non significant difference with GC3 sample having the highest value.

Sample	Firmness (N)	Adhesiveness (N mm)	Cohesiveness	Springiness (mm)	Chewiness (N*mm)
Control	16.21 ± 1.14^{a}	0.36 ± 0.05^{b}	0.58 ± 0.04^{d}	0.91 ± 0.15^{a}	9.47 ± 0.15^{c}
HCNO	9.93 ± 0.89^{f}	0.36 ± 0.04^{b}	0.57 ± 0.03^{d}	0.89 ± 0.11^{a}	5.71 ± 0.12^{f}
XG1	12.10 ± 1.11^{de}	0.49 ± 0.10^{a}	0.69 ± 0.07^{ac}	0.87 ± 0.13^{a}	8.38 ± 0.08^{3}
XG2	14.23 ± 1.06^{ac}	0.50 ± 0.09^{a}	0.67 ± 0.06^{bc}	0.74 ± 0.16^{ab}	9.64 ± 0.16^{c}
XG3	$16.2+1.18^{a}$	0.30 ± 0.01^{b}	0.64 ± 0.02^{bcd}	0.63 ± 0.18^{ad}	10.38 ± 0.14^{b}
GG1	11.24 ± 1.04^{ef}	0.52 ± 0.06^{a}	0.76 ± 0.05^{a}	0.81 ± 0.14^{a}	8.58 ± 0.09^{e}
GG2	13.46 ± 1.12^{bcd}	0.58 ± 0.07^{a}	0.67 ± 0.03^{bc}	0.66 ± 0.12^{ac}	9.15 ± 0.11^{d}
GG3	15.1 ± 1.18^{ab}	0.32 ± 0.08^{b}	0.71 ± 0.01^{a}	0.51 ± 0.18^{bcd}	10.87 ± 0.13^{a}

Table 2. Textural properties of cooked noodles

Means in a column with same superscript do not differ significantly (p<0.05).

Table 3. Sensory characteristics of cooked noodles

Sample	Slipperiness	Firmness	Chewiness	Tooth packing	Appearance	Overall acceptability
Control	4.97 ± 1.54^{bcd}	8.31 ± 1.46^{a}	6.64 ± 1.43^{ac}	6.40 ± 0.67^{a}	8.50 ± 0.53^{a}	8.00 ± 2.43^{a}
HCNO	5.01 ± 1.23^{bcd}	4.10 ± 1.34^{bc}	4.68 ± 1.25^{cd}	4.11 ± 0.56^{cde}	4.44 ± 0.31^{d}	5.5 ± 2.15^{d}
XG1	6.26 ± 1.41^{ac}	5.40 ± 1.25^{bc}	5.19 ± 1.31^{cd}	4.70 ± 0.81^{be}	5.10 ± 0.46^{d}	6.80 ± 1.87^{c}
XG2	7.78 ± 1.18^{a}	6.38 ± 1.51^{ab}	6.50 ± 1.26^{ad}	5.13 ± 0.91^{bc}	6.56 ± 0.62^{c}	7.62 ± 1.64^{b}
XG3	8.41 ± 1.34^{a}	8.5 ± 1.32^{a}	8.76 ± 1.13^{a}	6.38 ± 0.76^{a}	8.67 ± 0.88^{a}	8.16 ± 1.96^{a}
GG1	6.18 ± 1.14^{ad}	4.67 ± 1.17^{bc}	6.09 ± 1.41^{bcd}	4.89 ± 0.64^{bd}	6.42 ± 0.37^{c}	6.8 ± 2.04^{c}
GG2	7.31 ± 1.39^{ab}	6.33 ± 1.24^{ac}	7.89 ± 1.20^{ab}	5.78 ± 0.43^{ab}	7.53 ± 0.41^{b}	7.24 ± 2.51^{b}
GG3	8.36 ± 1.28^{a}	8.13 ± 1.12^{a}	8.75 ± 1.35^{a}	6.75 ± 0.51^{a}	8.64 ± 0.58^{a}	8.04 ± 1.67^{a}

Means in a column with same superscript do not differ significantly ($p<0.05$).

3.3. Sensory evaluation

Sensory evaluation is nearest to a consumer's estimation and remains the most reliable test as it allows the overall characteristics of cooked noodles to be evaluated. As shown in **Table 3,** the sensory attributes including slipperiness, firmness, chewiness, tooth packing, appearance and overall acceptability differ significantly among the horse chestnut noodles with different levels of xanthan and guar gum and the control noodles.

Native noodles were found to have least score while XG3 and GG3 samples were rated highest for slipperiness, firmness, chewiness, tooth packing and appearance. The overall acceptability of XG3 and GG3 noodles appeared to have non-significant correlation with the control sample. This finding might conclude that using the hydrocolloid as a proportional substitution for horse chestnut flour up to 3% could obtain the similar quality and pleasant appearance as control wheat flour noodles.

4 Conclusions

The study revealed that the noodles produced from horse chestnut flour with different proportions of guar and xanthan gums differed significantly. The hydrocolloids addition in noodles positively modified the cooking and textural quality with increased concentration. The sensory evaluation results indicated that the noodles with 3% hydrocolloids were highly accepted. Thus, substituting flour with 3% gum will go a long way to improve noodles quality and make them affordable. Hence, horse chestnut noodles stand a good product with added hydrocolloids so that the attributes like firmness and texture will not only be improved but the product will also be stable for consumption. In a further study, it will be necessary to optimize the appropriate combination of hydrocolloid-flour blends based on their physical characterization and functionality.

References

[1] X.L. Ding, J.X. Zheng. Steamed bread and noodles in China. In D. J. Martin, & C. W. Wrigley (Eds.), Proceedings conference cereals international (pp. 35-

40). Melbourne: Royal Australian Chemistry Institute (1991)

[2] S.N. Shin, S.K. Kim, Properties of instant noodle flours produced in Korea. Cereal Foods World, **48**, 310-314. (2003)

[3] G.B. Crosbie, The relationship between starch swelling properties, paste viscosity and boiled noodle quality in wheat flours. J. Cereal Sci. **13**(2), 145–150 (1991)

[4] G. Hou, Oriental noodles. Adv. Food Nutri. Res. **43**, 143-193 (2001)

[5] J. Kruger, R. Matsuo, J. Dick. Pasta and Noodle Technology. American Association of Cereal Chemistry, St Paul, MN (1996)

[6] H.J. Chung, Q. Liu, S.T. Lim. Texture and in vitro digestibility of white rice cooked with hydrocolloids. Cereal Chem. **84**, 246-249 (2007)

[7] E.I. Yaseen, T.J. Herald, F.M. Aramouni, S. Alavi. Rheological properties of selected gum solutions. Food Res. Inter. **38**, 111-119 (2005)

[8] K.S. Lim, I.K. Hwang. Effects of hydrocolloids on wheat flour rheology. J. Kor. Socie. Food Sci. Nutri. **15**, 203-209 (1999)

[9] M. Majeed, M.A. Khan, A. Bashir, A. Hussain. Nutritional Value and Oil Content of Indian Horse-Chestnut Seed. Global J. Sci. Front. Res. **10**(4), 17 (2010)

[10] S.I. Rafiq, S. Singh, D.C. Saxena. Physical, physicochemical and anti-nutritional properties of Horse Chestnut (Aesculus indica) seed. Food Measure. DOI 10.1007/s11694-016-9307-2. (2016).

[11] J. Yoo, Y. Kim, S.H. Yoo, G.E. Inglett, S. Lee. Reduction of rutin loss in

buckwheat noodles and their physicochemical characterisation. Food Chem. **132**, 2107-2111 (2012)
[12] S. Lee, I.Y. Bae, J.H. Jung, K.I. Jang, Y.O. Kim, H.G. Lee. Physicochemical,
textural and noodle-making properties of wheat dough containing alginate. J. Texture Stud. **39** (4), 393-404 (2008)
[13] F.C.F. Galvez, A.V.A. Resurreccion. Reliability of the focus group technique in determining the quality characteristics of mung bean noodles. J. Senso. Stud. **7**, 315 (1992)
[14] G.B. Crosbie, W.J. Lambe, H. Tsutsui, R.F. Gilmour. Further evaluation of the flour swelling volume test for identifying wheats potentially suitable for Japanese noodles. J. Cereal Sci. **15**, 271-280 (1992)
[15] D.M. Miskelly, H.J. Moss. Flour quality requirement for Chinese noodle manufacture. J. Cereal Sci. **3**, 379 (1985)
[16] B.K. Dadzie, J.E. Orchard. Routine Post-Harvest Screening of Banana/Plantain Hybrids: Criteria and Methods. Inibap Technical Guidelines (1997)

Development of Nutritious Snack from rice industry waste using twin screw extrusion.

Renu Sharma[1], Tanuja Srivastava[* 2], DC Saxena[3]
[1]*Assistant Professor, Deptt. of Applied Sciences, Bhai Gurdas Institute of Engineering & Technology, Sangrur, Punjab.*
[*2]*Director & Professor, Deptt. of Food Technology, Bhai Gurdas Institute of Engineering & Technology, Sangrur, Punjab, India.*
[3]*Professor, Deptt. of Food Engineering and Technology, Sant Longowal Institute of Engineering & technology, Longowal, Punjab, India.*

Abstract. Deoiled rice bran, a byproduct of rice milling industry was transformed into highly nutritious snack by the application of twin screw extrusion process. Response Surface Methodology (RSM) with four- factor- five level central composite rotatable design (CCRD) was employed to investigate the effects of extrusion conditions including moisture content of different raw flours, feed composition, barrel temperature and screw speed of extruder on properties of extrudates was studied. Second order quadratic regression model fitted adequately in the variation. The significance was established at $P \leq 0.05$. The regression models can be used to interpret the effect of feed composition, moisture content, screw speed and barrel temperature on the properties of the final product. It was shown that higher rice bran in feed composition showed in minimum water absorption index and maximum water solubility index. Numerical optimization technique resulted in 123.83°C of barrel temperature, 294.68 rpm of screw speed, 13.94 % of feed moisture and 17.73 % of deoiled rice bran. The responses predicted for these optimum process conditions resulted water absorption index, 5.91468 g/g and water solubility index of 18.5553 % for the development of value added product with health benefits.

1 Introduction

The nutrient security is an important issue in developing countries. Utilization of edible discards to increase the nutritional value of food products proved to be successful. The food processing industry generates large quantities of waste co-products. Much of the waste produced is disposed of, or used on a low technological and economical level. The cost of various foods and other products is highly dependent upon the ability of manufacturer to get value from unexplored low market value products like wastes of food industry. The commercial processing of these industrial wastes in the form of highly nutritious food items will not only keep the prices of food products low, but also ease pressure on scare land resources. With this in mind, the by-product which is targeted in this study is rice bran originate from the rice milling industry. In the past years, rice bran was used as either fertilizer or animal feed. Presently, it is used for extraction rice bran oil (RBO) [1-3]. However deoiled rice bran, after extraction of oil known to contain high levels of beneficial compounds such as dietary fibre, antioxidants, micronutrients like oryzanols, tocopherols, tocotrienols, phytosterols. All these micronutrients are rich source of vitamin E and have antioxidant activity [4-7]. But they are still underutilized. Rice bran being rich source of proteins, dietary fiber and bioactive compounds [8], helps in prevention of coronary heart disease, blood cholesterol [9,10] and artherosclerosis disease [11].

Snack foods have always been a significant part of modern life and with the continuing growth of the processed food industry, the demand for specialty-type snacks is expected to increase. However, it is the aim of this study to utilize the highly nutritious and under-used by-product of cereals for application in novel, high quality and healthy snack-type products by using twin screw extruder. Unlike other fried snacks, which are rich in fat content, the present product is free of fat and not fried because of extrusion cooking.

Extrusion cooking is a latest, continuous, high temp., short-time processing technology, gaining increasing popularity in the global agro-food processing industry, particularly in the food and feed sectors due to fast nature, significant reduction in energy consumption, more production and final products at lower prices [12]. Extrusion cooking technology is used for cereal and protein processing in food. This process involves high temperatures, high pressures and several shear forces inside the extruder barrel, chemical reactions and molecular modifications like starch gelatinization, denaturation of proteins, inactivation of several food enzymes and reduction of microbes [12]. Twin Screw Extruder is the most commonly used for extrusion process.

Response surface methodology (RSM) is an effective tool to optimize the process variables with minimum number of experimental runs. An experimental design such as the

[2] Tanuja Srivastava: tanusriva@yahoo.co.in

central composite rotatable design (CCRD) to fit a model by least square technique has been selected during the study. The basic principle of RSM is to relate product properties of regression equations that describe interrelations between input parameters and product properties [13].

The objectives of present study are (a) To utilize the highly nutritious and under-used by-product of rice industrial waste for application in novel, high quality and healthy snack-type products by using twin screw extruder. (b) To study the effects of different parameters on the properties of the extruded product using RSM and (c) To optimize the different process conditions for the production of easily consumable, highly nutritive snack with good taste and flavor.

2 Material and Methods

2.1 Experimental Design

A four-factor five-level, central composite rotatable design of Response surface methodology (RSM) was adopted in the experimental design [14]. The main advantage of RSM lies in reduction of experimental runs needed to provide sufficient information for statistically acceptable result. Table 1 shows independent variables selected for the experiments. The variables and their levels were chosen by taking trials of samples. The five levels of the process variables were coded as -2, -1, 0, 1 and +2 [14]. Design in coded (x) form and at the actual levels (X) is given in Table 2.

Table 1. Values of independent variables at five levels of the Central Composite rotatable design.

Independent variables	coded	Levels in actual form				
		-2	-1	0	+1	+2
Feed composition* (%)	X_1	75:15:10	70:20:10	65:25:10	60:30:10	55:35:10
Feed moisture (%)	X_2	13	14	15	16	17
Screw speed (rpm)	X_3	275	300	325	350	375
Die temperature (°c)	X_4	100	110	120	130	140

*Feed composition (RF:DRB:CF)

2.2 Preparation of Sample

The different ingredients for the production of highly nutritious extruded snack food consisted of deoiled rice bran, corn flour, rice flour. Deoiled rice bran (DRB) used for present study was procured from M/s. AP Solvex Ltd., Dhuri. Corn flour (CF) and rice flour (RF) were purchased from local market Sangrur, Punjab, India. Ingredient formulations for extrusion process are given in Table 2. In the blend preparation the composition of deoiled rice bran (DRB) and rice flour (RF) was varied while corn flour (CF) at level of 10% was kept constant in all samples. The moisture was adjusted by adding more distilled water in all the dry flours. The weighed ingredients were mixed in the food processor for 20 min. This mixture was then passed through a 2 mm sieve for the removal of lumps which were formed due to addition of water. The samples were stored in polyethylene bags at room temperature for 24h [15]. The moisture content of all the samples was estimated before extrusion by using the Hot air oven method [16].

2.3 Preparation of Extrudates

Extrusion trials were performed using a co-rotating twin-screw extruder (G.L. Extrusion Systems Pvt. Ltd., Delhi). The main drive is provided with 7.5 HP motor (400 V, 3ph, 50 cycles). The extruder is provided with standard design of screw configuration, automatic cutting knife fixed on rotating shaft and a temperature sensor. It was kept running for suitable period of time to stabilize the set temperatures. The raw flours poured into feed hopper and the feed rate (was adjusted to 4kg/h for easy and non-choking working of twin screw extruder. The recommended die diameter of 4 mm was selected. The collected product was packed for further analysis.

Table 2. Experimental combination in Coded and Uncoded levels for extruded snacks

Sr. No.	Coded variables				Uncoded variables			
	X_1	X_2	X_3	X_4	X_1 Feed Composition	X_2 Moisture content	X_3 Screw speed	X_4 Temp.
1	-1	-1	-1	-1	70:20:10	14	300	120
2	1	-1	-1	-1	60:30:10	14	300	120
3	-1	1	-1	-1	70:20:10	16	300	120
4	1	1	-1	-1	60:30:10	16	300	120
5	-1	-1	1	-1	70:20:10	14	350	120
6	1	-1	1	-1	60:30:10	14	350	120
7	-1	1	1	-1	70:20:10	16	350	120
8	1	1	1	-1	60:30:10	16	350	120
9	-1	-1	-1	1	70:20:10	14	300	130
10	1	-1	-1	1	60:30:10	14	300	130
11	-1	1	-1	1	70:20:10	16	300	130
12	1	1	-1	1	60:30:10	16	300	130
13	-1	-1	1	1	70:20:10	14	350	130
14	1	-1	1	1	60:30:10	14	350	130
15	-1	1	1	1	70:20:10	16	350	130
16	1	1	1	1	60:30:10	16	350	130
17	-2	0	0	0	75:15:10	15	325	125
18	2	0	0	0	55:35:10	15	325	125
19	0	-2	0	0	65:25:10	13	325	125
20	0	2	0	0	65:25:10	17	325	125
21	0	0	-2	0	65:25:10	15	325	125
22	0	0	2	0	65:25:10	15	325	125
23	0	0	0	-2	65:25:10	15	325	125
24	0	0	0	-2	65:25:10	15	325	125
25	0	0	0	0	65:25:10	15	325	125
26	0	0	0	0	65:25:10	15	325	125
27	0	0	0	0	65:25:10	15	325	125
28	0	0	0	0	65:25:10	15	325	125
29	0	0	0	0	65:25:10	15	325	125
30	0	0	0	0	65:25:10	15	325	125

2.4 Evaluation of Properties of Extrudates

2.4.1 Water Absorption Index (WAI) and Water Solubility Index (WSI)

WAI and WSI were determined according to the standard method developed for cereals [15, 17& 18]. The fine ground extrudate was suspended in water at room temperature for 30 min with simultaneous stirring and then centrifuged at 3000g for a time period of 15 min. The supernatant liquid was decanted into an evaporating dish of known weight. The WAI was the weight of gel obtained after removal of the supernatent liquid per unit weight of original dry solids. The WSI was measured as the ratio of the weight of dry solids in the supernatent to the original weight of sample.

$$\text{WAI (g/g)} = \frac{\text{Weight gain by gel}}{\text{Dry weight of extrudate}}$$

$$\text{WSI (\%)} = \frac{\text{Weight gain by gel}}{\text{Dry weight of extrudate}} \text{ X } 100$$

2.5 Statistical Analysis of Responses

The responses such as water absorption index (WAI) and water solubility index (WSI) for different experimental combinations were related to the coded variables (xi, i=1,2,3 and 4) by a second degree polynomial (Equation 1) as given below:

$$Y = \beta 0 + \beta_1 x_1 + \beta_2 x_2 + \beta_3 x_3 + \beta_4 x_4 + \beta_{11} x_1^2 + \beta_{22} x_2^2 + \beta_{33} x_3^2 + \beta_{44} x_4^2 + \beta_{12} x_1.x_2 + \beta_{13} x_1.x_3 + \beta_{14} x_1.x_4 + \beta_{23} x_2.x_3 + \beta_{24} x_2.x_4 + \beta_{34} x_3.x_4 + \varepsilon \quad (1)$$

Where x_1, x_2, x_3 and x_4 are the coded values of rice flour, corn flour and rice bran mixture, moisture content of feed (%), screw speed (rpm) and temperature of die (^{0}c). The Coefficients of the polynomial were represented by β0 (constant), β_1, β_2 , β_3, β_4 (linear effects); β_{12} , β_{13} , β_{14} , β_{23} , β_{24} , β_{34} (interaction effects) ; β_{11} , β_{22}, β_{33}, β_{44} (quadratic effects) ; ε (random error). Multiple regression analysis was used for data modeling and statistical significance of the terms was examined by analysis of variance. Design expert 6.0 (version 6.0, by STAT-EASE inc., USA) was used for statistical analysis of the data To check the adequacy of the regression model, R^2, Adjusted R^2, Adequate Precision and F-test were used [14]. Regression coefficients were used for statistical calculations and for the generation of three dimensional surface plots.

3 Results and Discussion

Variation of responses such as water absorption index (WAI) and water solubility index (WSI) of extrudates with independent variables (corn flour, rice flour and rice bran proportion, moisture content, die temperature and screw speed) were analyzed. A complete second order model (Equation 1) employed for its adequacy to decide the variation of response with independent variables. For visualization of variation with respect to processing variables, three dimensional response surfaces were drawn using design expert 6.0 (version 6.0, by STAT-EASE inc., USA) was used .

3.1 Effect of process variables on product water absorption index (WAI)

WAI measures the amount of water absorbed by starch that can be used as an index of gelatinization and it is generally agreed that barrel temperature and feed moisture content exerts greatest effect on the extrudate by promoting gelatinization [19]. The WAI ranged from 4.54 to 6.3 g/g for rice bran, rice flour and corn flour extrudates. Table 3 and 4 shows the coefficients of the model and other statistical attributes of WAI. Regression model fitted to experimental results of WAI (Table 3) showed that Model F-value of 10.68 was highly significant (P<0.0001) whereas lack-of-fit F-value of 0.3 was not significant (P>0.05). The coefficient of determination, R^2 which was 0.9088, indicating that 90.88 % of the variability of the response could be explained by the model. The Adjusted R^2 was 0.8238 and Adequate Precision was 13.035 which is greater than 4 indicating that the model can be used to navigate the design space. Considering all the above criteria, the model (Eq. 2) was selected for representing the variation of WAI and for further analysis. Multiple regression equation between water absorption index (g/g) of extrudates and coded levels of independent variables was as follows:

$$WAI = 5.39 - 0.22X_1 + 0.19X_2 - 0.27X_3 - 0.11\ X_4 + 0.12X_1^2 + 0.015X_2^2 + 0.021X_3^2 - 0.11X_4^2 - 0.054X_1X_2 - 0.11X_1X_3 - 0.052X_1X_4 + 0.042X_2X_3 - 0.091X_2X_4 - 0.026X_3X_4 \quad (2)$$

The analysis of variance of Eq. (2) shows that WAI of extrudates had highly significant (P < 0.0001) negative linear effect of feed composition (X_1) and screw speed (X_3). Other linear terms, feed moisture (X_2) had positive significant (P<0.001) while barrel temperature (X_4) had negative significant (P<0.05) linear effect on WAI.

Table 3: Analysis of variance for water absorption index (WAI).

Source	Coeff. of Model terms	Sum of squares	Mean square	DF	F Value	Prob>F
Model	5.39	5.31	0.38	14	10.68	<0.0001***
X_1	-0.22	1.17	1.17	1	33.09	<0.0001***
X_2	0.19	0.86	0.86	1	24.08	0.0002***

X_3	-0.27	1.69	1.69	1	47.62	<0.0001***
X_4	-0.11	0.28	0.28	1	8.00	0.0127**
X_1^2	0.12	0.37	0.37	1	10.38	0.0057**
X_2^2	0.015	5.917E-003	5.917E-003	1	0.17	0.6889
X_3^2	0.021	0.012	0.012	1	0.34	0.5692
X_4^2	-0.11	0.34	0.34	1	9.62	0.0073**
$X_1 X_2$	-0.054	0.047	0.047	1	1.33	0.2664
$X_1 X_3$	-0.11	0.20	0.20	1	5.64	0.0313**
$X_1 X_4$	-0.052	0.043	0.043	1	1.21	0.2881
$X_2 X_3$	0.042	0.028	0.028	1	0.79	0.3880
$X_2 X_4$	-0.091	0.13	0.13	1	3.70	0.736*
$X_3 X_4$	-0.026	0.011	0.011	1	0.30	0.5944

*Significant at P<0.1, ** Significant at P<0.05,*** Significant at P<0.001, df: degrees of freedom.

Table 4 : Analysis of variance results of equation 2

Response	Source	Sum of squares	df	Mean squares	F-value	P-value
WAI	Regression	5.31	14	0.38	10.68	<0.0001*
	Lack of Fit	0.20	10	0.020	0.30	0.9502
	Pure error	0.33	5	0.067		
	Residual	0.53	15	0.036		
	Total	5.84	29			
	R^2-value	0.9088				
	Adjusted R^2	0.8238				
	Adeq. Precision	13.035				

*significant at P<0.05, df: degrees of freedom

In this case X_1, X_2, X_3, X_4, X_1^2, X_4^2 and X_1X_3 were significant model terms. F-value of the interaction term X_2X_4 was 3.70 with P-value 0.0736. It means it is significant (P<0.1). Since coefficient of X_2X_4, was negative, it showed convex shaped variation with the change in value of variables. **Fig.1** shows that WAI increased with increasing feed moisture content(X_2) while decreased with increasing barrel temperature (X_4).

F-value of the linear term feed composition (X_1) was 33.09 indicates that term if highly significant (P<0.0001). Since coefficient of (X_1) was negative, the WAI of extrudates decreased with increasing level of the feed composition i.e. with rice bran proportion **(Fig.2)**. The results indicated that the presence of more bran in the mixture reduced the availability for gelatinization of the starch granules, thus reduced viscosity and WAI because of the replacement of the starch by fiber component. In contrast, reducing of bran content in the mixture caused more open structure of starch granules for allowing water penetration and retention to get more gelatinization and higher WAI. The similar effect of bran content on structural of starch component has been observed [20-22]. The negative coefficients of temperature indicated that the WAI value of the product decreased with increase in temperature. Similar findings were reported by Pelembe, Erasmus and Taylor (2002); Ding et al. (2006) and Aylin Altan (2008) [23-25]. A decrease in WAI with increasing temperature was probably due to decomposition or degradation of starch [23]. Ding et al. (2006) also stated that the WAI decreases with increasing temperature if dextrinization or starch melting prevails over the gelatinization phenomenon [24].

The WAI measures the volume occupied by the starch after swelling in excess water, which maintains the integrity of starch in aqueous dispersion [26]. Gelatinization, the conversion of raw starch to a cooked and digestible material by the application of water and heat, is one of the important effects that extrusion has on the starch component of foods. Water is absorbed and bound to the starch molecule with a resulting change in the starch granule structure. Barrel temperature and feed moisture are found to exert the greatest effect on gelatinization. The maximum gelatinization occurs at high moisture and low temperature or vice versa reported by Lawton and Henderson (1972) [27].

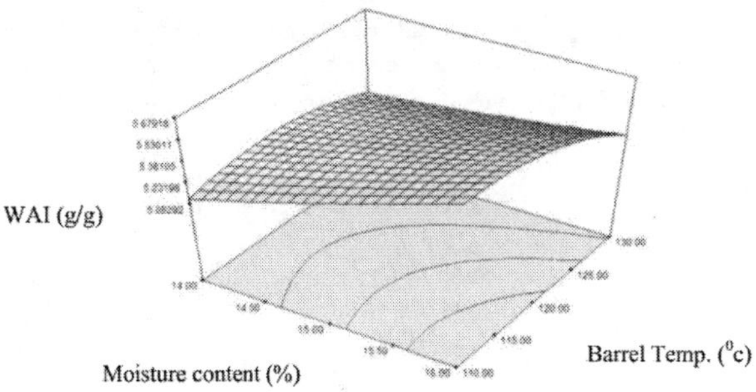

Fig.1. Response Surface Plot for variation of WAI as a function of moisture content (%) and Barrel Temperature (°c).

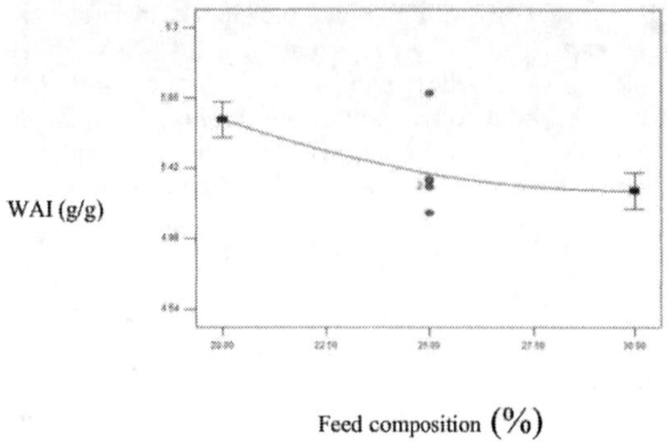

Fig.2. One Factor plot for the variation of WAI of extrudates as a function of feed composition.

3.2 Effect of process variables on product water solubility index (WSI)

WSI used as an indicator of degradation of molecular components. The WSI measures the amount of soluble polysaccharide released from the starch component after extrusion [19]. The WSI values ranged from 13.21% to 30.36%. The coefficients of the model and other statistical attributes of WSI are shown in Table 5 and 6. Regression model fitted to experimental results of WSI (Table 5) showed that Model F-value of 17.68 was highly significant ($P < 0.0001$) whereas lack of fit F-value of 1.92 was not significant ($P>0.05$). The coefficient of determination, R^2 was 0.9429 indicating that 94.29% of the variability of the response could be explained by the model. The Adjusted R^2 was 0.8895 and Adequate Precision was 19.632 which is greater than 4, justifying that this model can be used for prediction purpose. Considering all the above criteria, the model (Eq. 3) was selected for representing the variation of WSI and for further analysis.

$$WSI = 21.78 + 0.53X_1 - 0.97X_2 + 1.08X_3 - 1.54X_4 - 0.1X_1^2 + 0.84X_2^2 + 0.15X_3^2 - 0.92X_4^2 - 2.33X_1X_2 + 0.32X_1X_3 - 2.05X_1X_4 - 0.86X_2X_3 - 1.1X_2X_4 - 0.29X_3X_4 \quad (3)$$

In this case terms, X_1, X_2, X_3, X_4, X_2^2, X_4^2, X_1X_2, X_1X_4, X_2X_3 and X_2X_4 were significant. The analysis of variance of Eq. (3) showed that WSI of extrudates had highly significant ($P<0.0001$) negative linear effect of barrel temperature (X_4) followed by significant ($P<0.001$) positive linear effect of screw speed (X_3). Other linear terms, feed composition (X_1) has significant ($P<0.05$) positive linear effect while feed moisture (X_2) has significant ($p<0.05$) negative linear effect on WSI. The variation of water solubility index with moisture content of raw material and barrel temperature is shown in **Fig. 3A and Fig.3B.**

Table 5: Analysis of variance for water solubility index (WSI).

Source	Coeff. of Model terms	Sum of squares	Mean square	DF	F Value	Prob>F
Model	21.78	353.36	25.24	14	17.68	<0.0001***
X_1	0.53	6.86	6.86	1	4.80	0.0446**
X_2	-0.97	22.68	22.68	1	15.89	0.0012**
X_3	1.08	28.10	28.10	1	19.69	0.0005***
X_4	-1.54	56.89	56.89	1	39.85	<0.0001***
X_1^2	-0.10	0.28	0.28	1	0.19	0.6664
X_2^2	0.84	19.51	19.51	1	13.67	0.0022**
X_3^2	0.15	0.66	0.66	1	0.46	0.5081
X_4^2	-0.92	23.04	23.04	1	16.14	0.0011**
$X_1 X_2$	-2.33	86.72	86.72	1	60.75	<0.0001***
$X_1 X_3$	0.32	1.66	1.66	1	1.16	0.2982
$X_1 X_4$	-2.05	67.20	67.20	1	47.07	<0.0001***
$X_2 X_3$	-0.86	11.75	11.75	1	8.23	0.0117**
$X_2 X_4$	-1.10	19.47	19.47	1	13.64	0.0022**
$X_3 X_4$	-0.29	1.31	1.31	1	0.91	0.3541

*Significant at $P<0.1$, ** Significant at $P<0.05$,*** Significant at $P<0.001$, df: degrees of freedom

Table 6 : Analysis of variance results of equation 3

Response	Source	Sum of squares	df	Mean squares	F-value	P-value
WSI	Regression	353.36	14	25.24	17.68	<0.0001*
	Lack of Fit	16.99	10	1.70	1.92	0.2442
	Pure error	4.42	5	0.88		
	Residual	21.41	15	1.43		
	Total	374.78	29			
	R^2-value	0.9429				
	Adjusted R^2	0.8895				
	Adeq. Precision	19.632				

*significant at $P<0.05$, df: degrees of freedom

The quadratic term of feed moisture content (X_2^2) had significant ($P<0.05$) positive effect, while quadratic term, barrel temperature (X_4^2), had negative significant ($P<0.05$) effect on product WSI.

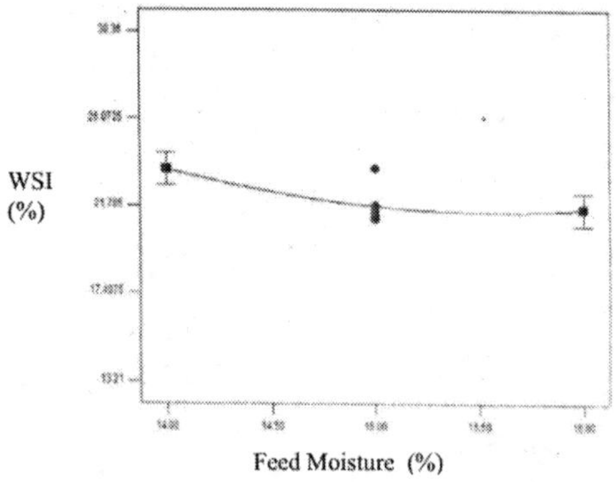

Fig.3 (A) One Factor plot for the variation of WSI of extrudates as a function of feed moisture content.

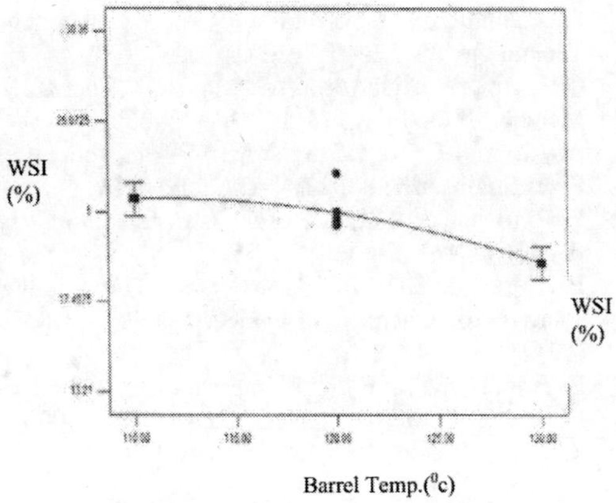

Fig.3 (b) One Factor plot for the variation of WSI of extrudates as a function of Barrel Temperature.

The F-value of the interaction term X_1X_4 was 47.07 with P-value of 0.0001 which indicates that the term is significant ($P<0.001$). Since coefficient of X_1X_4 was negative, it showed convex shaped variation with the change in the values of the variables. **Fig.4A** shows that the WSI increased with increasing feed composition X_1 (rice bran proportion) while decreased with increasing barrel temperature X_4. The F-value of the interaction term X_2X_4 was 13.64 with P-value of 0.0022 which indicates that the term is significant ($P<0.001$). Since coefficient of X_2X_4 was negative, it showed convex shaped variation with the change in the values of the variables. **Fig.4B** shows that the WSI increased with decreasing feed moisture (X_2) and barrel temperature (X_4). Similar results were found for rice based extrudates by Ding et al. 2005 [19].

WSI determines the amount of free polysaccharide or polysaccharides released from the granule after addition of excess water [25&28]. Increase in feed moisture content significantly $P<0.05$) decreases WSI. Increase in feed moisture increases the plasticity of feed thus minimizes the chance of formation of small fraction polymer. WSI increased significantly ($P<0.05$) with increase in screw speed (X_3). Similar findings are reported for com meal and com and rice extrudates [25, 29&30].

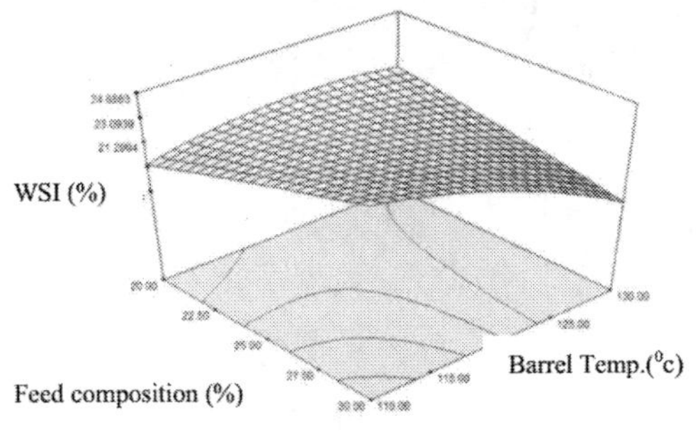

Fig.4A. Response Surface Plot for variation of WSI as a function of Barrel Temperature (°c) and Feed Composition(%).

Mezreb et. al. (2003) reported that the increase of screw speed induced a sharp increase of specific mechanical energy, the high mechanical shear degraded macromolecules, and so the molecular weight of starch granules decreased and hence increase WSI [30].

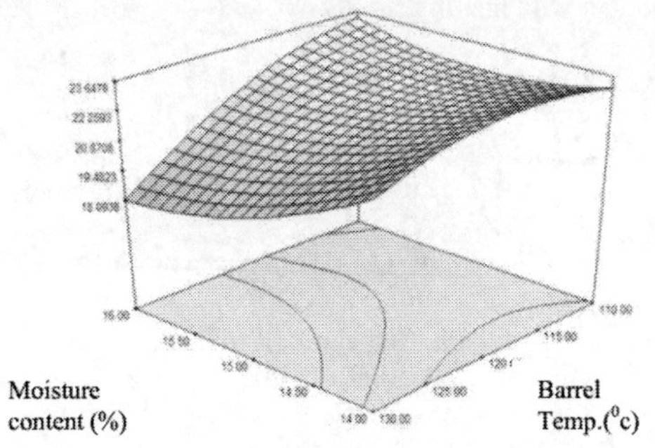

Fig.4B. Response Surface Plot for variation of WSI as a function of moisture content (%) and Barrel Temperature (°c).

4 Optimization

A numerical multi-response optimization technique was applied to determine the optimum combination of rice bran and rice flour in feed composition, feed moisture content, screw speed and barrel temperature for the production of extrudates containing rice bran, rice flour, corn flour.

The assumptions were to develop a product which would have maximum score in sensory acceptability so as to get market acceptability, maximum expansion, minimum bulk density, and minimum hardness. Therefore, among responses, these parameters were attempted to be maintained whereas other parameters were kept within range. Under these criteria, the uncoded optimum operating conditions for development of rice bran, rice flour, corn flour extrudates were 123.83°C of barrel temperature, 294.68 rpm of screw speed, 13.94 % of feed moisture and 17.73 % of deoiled rice bran. The responses predicted by the Software for these optimum process conditions resulted water absorption index, 5.91468 g/g and water solubility index of 18.5553 %.

5 Conclusion

The finding of this study demonstrates the feasibility of developing value added extruded product form rice industrial waste by extrusion processing. The application of quadratic response surface methodology serves as useful tool for optimization of operating conditions for the production of extruded product enriched in deoiled rice bran. The regression models can be used to interpret the effect of feed composition, moisture content, screw speeed and barrel temperature on the properties of the final product. It was shown that increase in barrel temperature leads to minimum water absorption index whereas increased moisture content of the raw material resulted in product with maximum water absorption index and lower water solubility index. Higher rice bran in feed composition showed in minimum water absotption index and maximum water solubility index. In this experiment, numerical optimization technique resulted in 123.83°C of barrel temperature, 294.68 rpm of screw speed, 13.94 % of feed moisture and 17.73 % of deoiled rice bran for the development of value added product with health benefits.

References

1. V. Van Hoed, G. Depaemelaere, J. Villa Ayala, P. Santiwattana, R. D. Verhe et al., JAOCS **83**, 315-321 (2006).
2. N. Hammond, Cereal foods world **39**, 752-754 (1994).
3. S. Barber, J. Camacho, R. Cerni, E. Tortosa, E. Primo, *International Conference, Valencia, Spain,* 49-62, (1974).
4. J.H. Wells, Louisiana Agriculture **36**, 4-8 (1993).
5. Y. Jiang, T. Wang, JAOCS **82**, 439-444 (2005).
6. V. Piironen, D. G. Lindsay, T.A. Miettinen, J. Toivo, A. M. Lampi, J Sci Food Agric **80**, 939-966 (2000).
7. W. Hu, J. H. Wells, S. Tai-Sun, J.S. Godber, JAOCS **73**, 1653-1656 (1996).
8. R.M. Saunders, Cereal Foods World **35**, 632-635 (1990).
9. C. Chotimarkorn and Silalai, LWT **41**, 561-568 (2007).
10. T.S. Kahlon, F. I. Chow and R.N. Sayre, CFW **39**, 99-103 (1994).
11. R.M. Saunders, Food Rev Int. **1**, 465-495, (1985).
12. J.M. Harper, Extrusion of foods II; CRC Press: Florida, USA: 174pp. (1981).
13. W.G. Cochran, G.M. Cox, R.A. Bradley, D. G. Kendall, J. S. Hunter, G. S. Watson, *Experimental Designs*, Eds.; John Wiley and Sons, New York, (1957).
14. D.C. Montgomery, *Design and Analysis of Experiments*, New York Wiley pp. 416-419 (2001).
15. V. Stojceska, P. Ainsworth, A. Plunkett, E. Ibanoglu, S. Ibanoglu, Journal of Food Engineering **87**, 554-563 (2008).
16. S. Ranganna, *Handbook of Analysis and Quality Control for Fruit and Vegetable Products*, Tata Mc Graw Hill Publishing Company Limited, New Delhi, India, (1995).
17. R.A. Anderson, H.F. Conway & E.L. Griffin, Cereal Science Today **14**, 4-12 (1969).
18. S. Yagci & F. Goguş, Journal of Food Engineering **86**, 122-132 (2008).
19. Q.B. Ding, P. Ainsworth, G. Tucker, H. Marson, Journal of Food Engineering **66**, 283-289 (2005).
20. K.M Grenus, F. Hsieh and H.E. Huff, Journal of Food Engineering **18**, 229-245 (1993).
21. J.M. Hashimoto and M.V.E. Grossmann, International Journal of Food Science and Technology **38**, 511-517 (2003).
22. C. Charunuch, N. Limsangouan, W. Prasert and K. Wongkrajang, International Food Research Journal 21, 713-722 (2014).
23. L.A.M. Pelembe, C. Erasmus and J.R.N. Taylor, Lebensmitted-Wissenschaft Untersuchung Technology **35**, 120-127 (2002).
24. Q.B. Ding, P. Ainsworth, A. Plunkett, G.Tucker, H. Marson, J. food Eng. **73**, 142-148 (2006).
25. A. Altan, K.L. Mc Carthy & M. Maskan, Journal of Food Engineering **84**, 231-242 (2008).
26. W.R Mason and R.C. Hoseney, Cereal chemistry **63**, 436 (1986).
27. B.T. Lawton and B.A. Handerson, The Candian Journal of Chemical Engineering **50**, 168-172 (1972).
28. P. Sriburi, & S. E. Hill, International Journal of Food Science and Technology **35**, 141–154 (2000).
29. Z. Jin, F. Hsieh, H.E. Huff, J. Cereal Sci. **22**, 185-194 (1995)..
30. K. Mezreb, A. Goullieux, R. Ralainirina, M. Queneudec, J. Food Eng., **57**, 145 -152 (2003).

 DOI: 10.1051/matecconf/20165705001

Newton-Raphson based modified Laplace Adomian decomposition method for solving quadratic Riccati differential equations

Vinod Mishra[1] and Dimple Rani [1, a]

[1]Department of Mathematics, Sant Longowal Institute of Engineering and Technology, Longowal, Punjab, India

Abstract. Numerical Laplace transform method is applied to approximate the solution of nonlinear (quadratic) Riccati differential equations mingled with Adomian decomposition method. A new technique is proposed in this work by reintroducing the unknown function in Adomian polynomial with that of well known Newton-Raphson formula. The solutions obtained by the iterative algorithm are exhibited in an infinite series. The simplicity and efficacy of method is manifested with some examples in which comparisons are made among the exact solutions, ADM (Adomian decomposition method), HPM (Homotopy perturbation method), Taylor series method and the proposed scheme.

1 Introduction

One of the most significant classes of nonlinear differential equation, Riccati differential equation (RDE) is considered as follows:

$$\frac{dy}{dt} = q(t)y + r(t)y^2 + p(t), y(0) = a \qquad (1)$$

where $q(t), r(t)$ and $p(t)$ are the known scalar functions and a is an arbitrary constant. The RDE is named after the name of Italian nobleman Count Jacopo Francesco Riccati(1676-1754) [4]. In the field of applied science and engineering, the RDEs have played an essential role [2-7, 29]. A one dimensional static Schrodinger equation is closely related to RDEs. Satisfying projective Riccati equations, solitary wave solutions of nonlinear partial differential equations can be expressed as polynomials in two elementary functions [4, 7]. In some control theory problems such as dynamic games, linear system with Markovian jumps and stochastic controls RDEs act predominantly [7,29]. Apart from these applications RDEs is also used in stochastic realization theory, robust stabilization, and network synthesis and presently in financial mathematics [4]. Much attention has been given to solve these kinds of equations due to the above applications.

Certain methods are there in literature to solve the RDEs. In [2], Ghorbani and Momani applied the piecewise variational iteration method (VIM) to solve the RDEs. Differential transform method [4, 9] is adopted to find the solution of RDEs. Taiwo and Osilagun [5] approximated the solution of RDEs by Iterative algorithm. Perturbation iteration algorithm (PIA) has been presented in solving RDEs [8]. Vahidi has made the comparison among HPM, ADM and LTDM in solving RDEs in [3]. For solving these kinds of equations Yang et.al. [7] employed the hybrid functions and Tau method. In [6] the authors developed the iterative methods ADM, MADM, VIM, MVIM, HPM, MHPM and HAM to solve the general RDE.

Laplace transform is a powerful tool in solving linear problems but it is incapable of solving nonlinear problems. A well known numerical algorithm Laplace transforms and Adomian decomposition method has conquered much importance in solving many linear and nonlinear problems which provides a series solution. Suheil A. Khuri was the first to apply Laplace decomposition algorithm to solve a class of nonlinear differential equation [12]. A combined Laplace Adomian decomposition method is used to solve nonlinear Volterra integral equation with weakly kernel [1]. In [10], Majid Khan et. al. solved nonlinear coupled partial differential equations with the help of Laplace Decomposition method. LDM is also implemented to obtain the series solution of nonlinear fractional differential equations [18]. Waleed Al-Hyani [22] solved nth order Integro-differential equations by the usage of LT-ADM. In [14], Modified Laplace decomposition method is proposed for solving Lane-Emden type differential equation. LDM is exercised to solve the Logistic differential equations in [11]. Wazwaz [27] employed CLT-ADM for solving nonlinear volterra-integro differential equations. For handling the solutions of nonlinear system of partial differential equation Laplace decomposition method and pade approximant is used in [21]. Hence this method is utilized to solve many more problems like Singular initial value problems [13], Double singular boundary value

[a] Corresponding author: dimplechawla23@yahoo.com

problems [15], Higher order boundary value problems [16]. Other cited references are [17, 19, 20, 23-26].

In the proposed numerical technique, we use the Laplace transform-Adomian decomposition method to solve the Quadratic RDEs. The main thrust of this methodology is to replace the unknown function y_i in Adomian polynomial with Newton-Raphson formula, which improves the Adomian polynomial. The outline of this paper is as follows: Section 1 contains a brief summary on RDEs and Laplace transform combined with Adomian decomposition method. In Section 2, we developed the method based on LT-ADM. In Section 3, the numerical results of employing the technique on three examples for RDEs are presented. In the end conclusions are drawn.

2 Analysis of Method

For finding the solution of Quadratic RDE with initial condition is given by equation (1). The method consists of applying the Laplace transform on both sides of (1),

$$L\left[\frac{dy}{dt}\right] = L\left[q(t)y + r(t)y^2 + p(t)\right]$$

Applying the linear and differential property of Laplace transform,

$$L[y] = \frac{a}{s} + \frac{1}{s}L[q(t)y] + \frac{1}{s}L\left[r(t)y^2\right] + \frac{1}{s}L[p(t)] \tag{2}$$

In this method the solutions are presented as an infinite series given by

$$y = \sum_{n=0}^{\infty} y_n \tag{3}$$

where the term y_n are to be computed recursively and the nonlinear term $f(y) = y^2$ is decomposed as follows:

$$f(y) = y^2 = \sum_{n=0}^{\infty} A_n \tag{4}$$

$$\text{here } A_n = \frac{1}{n!}\frac{d^n}{d\lambda^n}\left[f\sum_{i=0}^{n}\lambda^i\left(y_i - \frac{f(y_i)}{f'(y_i)}\right)\right]_{\lambda=0} \tag{5}$$

A few terms of A_n are given by

$$A_0 = \left(\frac{1}{2}\right)^2 {y_0}^2$$

$$A_1 = \left(\frac{1}{2}\right)^2 2y_0y_1$$

$$A_2 = \left(\frac{1}{2}\right)^2 \left(2y_0y_2 + {y_1}^2\right)$$

Using (3) and (4) in (2) and by using the linearity property of Laplace transform, we obtain

$$\sum_{n=0}^{\infty} L[y_n] = \frac{a}{s} + \frac{1}{s}\sum_{n=0}^{\infty} L[q(t)y_n] + \frac{1}{s}\sum_{n=0}^{\infty} L[r(t)A_n] + \frac{1}{s}L[p(t)] \tag{6}$$

Comparing both sides we have the following recursive relation

$$L[y_0] = \frac{a}{s} + \frac{1}{s}L[p(t)] \tag{7}$$

$$L[y_1] = \frac{1}{s}L[q(t)y_0] + \frac{1}{s}L[r(t)A_0] \tag{8}$$

$$L[y_2] = \frac{1}{s}L[q(t)y_1] + \frac{1}{s}L[r(t)A_1] \tag{9}$$

In general, the relation is given by

$$L[y_{n+1}] = \frac{1}{s}L[q(t)y_n] + \frac{1}{s}L[r(t)A_n] \tag{10}$$

To get the values of $y_0, y_1, y_2, y_3, \ldots$, apply the inverse Laplace transform to above equations

$$y_0 = L^{-1}\left[\frac{a}{s} + \frac{1}{s}L[p(t)]\right]$$

Substitute the value of y_0 into (8), evaluate the value the R.H.S. and then take inverse Laplace transform will give the value of y_1. In this manner we get the terms $y_2\ y_3 \ldots$ recursively and by putting these values in (3) give the solution to the problem.

3 Numerical Results

In this section we employ our method to different forms of Quadratic RDE.

Example 1. Consider the following [8]

$$y' = 1 + y^2(t),\ y(0) = 0 \tag{11}$$

Which has the exact solution as $y = \tan t$.

The solution at the different values of t is found and presented through Table1. From Table 1 it can be seen that the solution obtained by our method is very much close to the exact solution and Figure1demonstrates the superiority of our method.

Table 1. Computed exact and approximate solution for Example1

t	Exact	Approximate solution	Absolute Error
0	0	0	0.0000000E+00
0.01	0.010000333	0.010000083	2.5001250E-07
0.02	0.020002667	0.020000667	2.0004001E-06
0.03	0.030009003	0.03000225	6.7530387E-06
0.04	0.040021347	0.040005334	1.6012809E-05
0.05	0.050041708	0.050010419	3.1289104E-05
0.06	0.060072104	0.060018006	5.4097349E-05
0.07	0.070114558	0.070028597	8.5960526E-05
0.08	0.080171105	0.080042694	1.2841072E-04
0.09	0.090243790	0.090060799	1.8299066E-04
0.1	0.100334672	0.100083417	2.5125533E-04

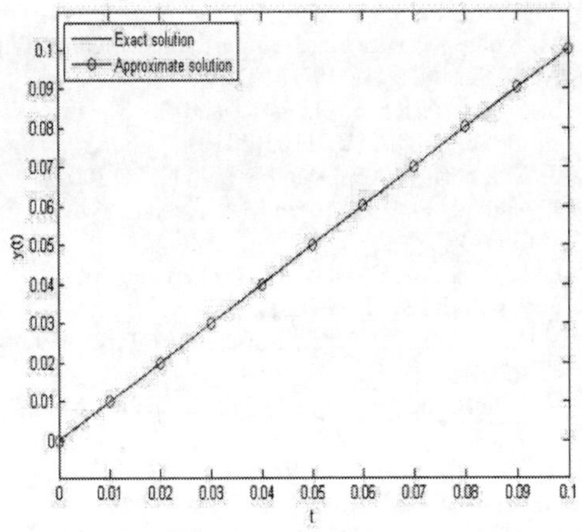

Figure 1. Comparaison of exact and approximate solution for Example1.

Example2. Consider the another type of Quadratic RDE [4,5,7]

$$y' = 1 + y^2(t), y(0) = 0 \quad (12)$$

which has the exact solution as $y = \frac{e^{2t} - 1}{e^{2t} + 1}$.

According to the above methodology we attain the following results:

Table 2. Computed exact and approximation solution for Example2

t	Exact solution	Approximate solution	Absolute error
0	0	0	0.0000000E+00
0.01	0.009999667	0.009999917	2.4998750E-07
0.02	0.019997334	0.019999333	1.9996001E-06
0.03	0.029991003	0.02999775	6.7469637E-06
0.04	0.03997868	0.039994668	1.5987209E-05
0.05	0.049958375	0.049989586	3.1210979E-05
0.06	0.059928104	0.059982006	5.3902948E-05
0.07	0.06988589	0.069971431	8.5540349E-05
0.08	0.079829769	0.079957361	1.2759151E-04
0.09	0.089757785	0.089939299	1.8151442E-04
0.1	0.099667995	0.09991675	2.4875529E-04

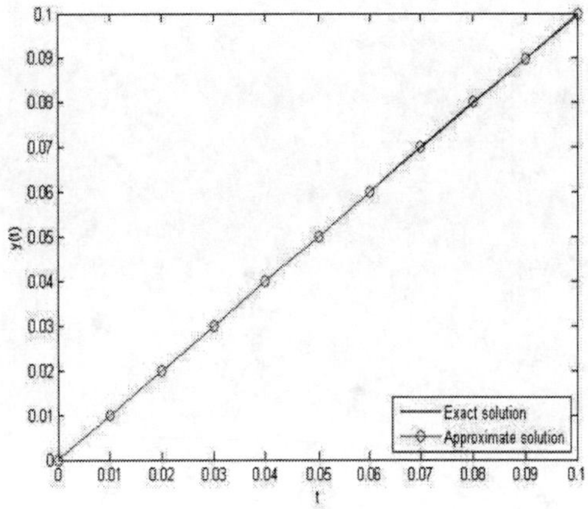

Figure 2. Comparaison of exact and approximate solution for Example2.

Table 2.and Figure 2.show that the numerical solution compared with exact is very superior and absolute error laid out in the table shows that the difference between exact solution and our presented technique is very small.

Example3. Consider the another type of Quadratic RDE [2,4,8]

$$y' = 1 - y^2(t) + 2y(t), y(0) = 0 \quad (13)$$

Following the aforesaid technique the results have been demonstrated in the Table3 and Figure3.

Table 3. Comparison approximate solution with Taylor series, ADM and HPM for Example3

t	Approximate Solution	Taylor Series method	ADM	HPM
0	0	0	0	0
0.01	0.010100585	0.010100330	0.010100330	0.010100330
0.02	0.020404690	0.020402612	0.020402612	0.020402612
0.03	0.030915863	0.030908719	0.030908719	0.030908719
0.04	0.041637666	0.041620432	0.041620431	0.041620432
0.05	0.052573672	0.052539435	0.052539435	0.052539435
0.06	0.063727448	0.063667310	0.063667309	0.063667310
0.07	0.075102547	0.075005529	0.075005526	0.075005529
0.08	0.086702496	0.086555447	0.086555443	0.086555447
0.09	0.098530785	0.098318300	0.098318293	0.098318300
0.1	0.110590857	0.110295195	0.110295185	0.110295195

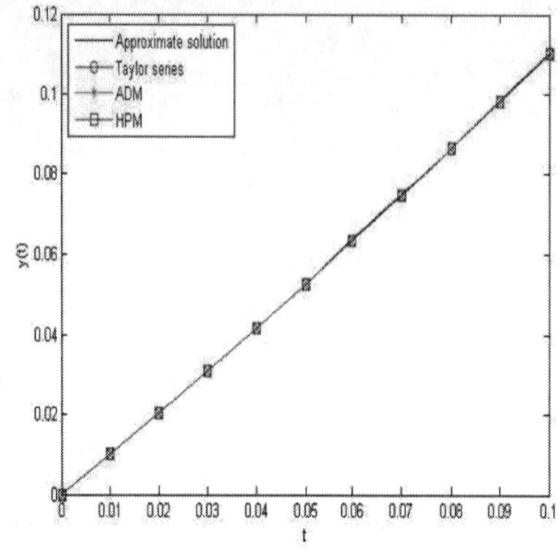

Figure 3. Comparaison of approximate solution with ADM, HPM, Taylor series.

In Table 3and Figure3, the numerical results are given for 10 values of t .Comparing it with ADM,HPM and Taylor series method given in [28], it is clear that the accurate value is attained as number of iteration increases.

4 Conclusions

In this paper a modification of Laplace transform combined with Adomian decomposition is given. The unknown function y_i in Adomian polynomial has been replaced with the well known Newton-Raphson formula. Finally, a numerical method is constructed for solving Quadratic RDEs. The numerical results through tables and figures present that the new scheme is effective and simple in the determination of approximate solution of RDEs. All the calculations are done with MATLAB 2011b.

References

1. F. A. Hendi, Stud. Nonlinear Sci., **2,** 129-134 (2011).
2. A. Ghorbani and S. Momani, Appl. Math. Letters, **23**, 922-927(2010).
3. A. R. Vahidi, E. Babolian and Z. Azimzadeh, World Appl. Sci. J., **16,** 921-925(2012).
4. J. Biazar and M. Eslami, Int. J. of Nonlinear Sci., **9**, 444-447(2010).
5. O. A. Taiwo and J. A. Osilagun, Int. J. Eng. Innovat. Tech., **1,** 53-56(2012).
6. T. Allahviranloo and S. S. Behzadi, Int. J. Indust. Math., **4,** 389-404(2012).
7. C. Yang, J. Hou and B. Qin, WASET, **6**, 569-572(2012).
8. M. Khalid, M. Sultana, F. Zaidi and U. Arshad, J. Comput. Appl., **111,** 1-5(2015).
9. S. Mukherjee and B. Roy, Int. J. Nonlinear Sci., **14,** 251-256(2012).
10. M. Khan, M. Hussain, H. Jafari and Y. Khan, World Appl. Sci. J., **9,** 13-19(2010).
11. S. Islam, Y. Khan, N. Faraz and F. Austin, World Appl. Sci. J., **8,** 1100-1105(2010).
12. S. A. Khuri, J. Appl. Math., **1,** 141-155(2001).
13. J. Ahmad, F. Hussain and M. Naeem, AJMP, **5**, 1-15(2014).
14. F. K. Yin, W. Y. Han and J. Q. Song, Int. J. Appl Phys. Math., **3,** 98-102(2013).
15. Y. Lin and C. K. Chen, Rom. Journ. Phys. ,**59**, 443-453(2014).
16. N. Singh and M. Kumar, Math. Theory Model., **2**, 11-22(2011).
17. J. B. Yindoula, P. Youssouf, G. Bissanga, F. Bassono and B. Some, Int. J. Appl. Math. Res., **3**, 30-35(2014).
18. C. Yang and J. Hou, J. Info. Comput. Sci., **10**, 213-222(2013).
19. S. P. Yan, H. Jafari and H. K. Jassim, Adv. Math. Phy., 1-7(2014).
20. A. Kumar and R. D. Pankaj, Int. Scholarly Res. Netw., **2012**, 1-5(2012).
21. M. A. Mohamed and M. S. Torky, Am. J. Comput. Math., **3**, 175-184(2013).
22. W. Al-Hayani, Appl. Math., **4**, 882-886(2013).
23. J. Fadaei, Appl. Math. Sci., **5**, 1307 - 1315(2011).
24. J. A. S. Cano, IJRRAS, **16**, 168-175(2013).
25. M. A. Koroma, S. Widatalla, A. F. Kamara and C. Zhang, WASET, **7**, 525-529(2013).
26. N. Doğan, Math. Comput. Appl., **17**, 203-211(2012).
27. A. M. Wazwaz, Appl. Math. & Comput. , **216**, 1304-1309(2010).
28. S. Abbasbandy, J. Comput. Appl. Math., **207**, 59-63(2007).
29. B. Batiha, Int. J. Appl. Math. Res., **4**, 24-29(2015).

A Real-Time Decision Support System for River Basin Management

Ayushi Vyas [1, a] and Siby John[2]

[1,2]*PEC University of Technology, Environmental Department, Chandigarh*

Abstract. The applications of computer technology to analysis of the rainfall-runoff process and the hydrological dynamics of natural rivers have greatly expanded in the past few years. A large number of general purpose programs and a few programs designed for specific application have been developed and applied to hydrologic engineering problems. This paper briefly describes the study of climate change effects on the hydrological dynamics of the Satluj and Beas river system using computer models like HEC- RTS (Hydrologic Engineering Center- Real Time Simulation). It also presents the use of real time data collection and processing on a GIS platform so as to derive a real time decision support system.
Keywords: DSS, hydrology, modeling, hydrologic model.

1 Introduction

Hydrologic engineering has traditionally been one of the areas which required management and massive amount of data. With the help of the analytical capability provided by the use of computers only, hydrologic engineers can continue to respond to the ever-increasing demands for more comprehensive and more complex hydrologic studies [1]. Simulation speed and data visualisation are of great significance to decision making. Real-time interaction during the simulation process is also very important for dealing with different conditions and for making timely decisions [2]. Currently, the major challenge for decision-makers and stakeholders in the water sector is to understand the effects of global warming and to determine where and how regions and sectors are vulnerable and to implement appropriate adaptation measures.

Considering the recent technological and methodological developments, this study deals with the climate change effects on the hydrological dynamics of river system including the real time data acquisition. This research provides a conceptual basis for real-time multi-purpose data assembling, evaluating, modeling and visualisation towards the operationalisation of decisions. Turning field observations into useful real-time decision support information is demonstrated based on a hydrological example of Integrated Water Resource Management. Decision Support Systems (IWRM-DSS) based on a virtual environment are becoming a popular platform in watershed simulation and management. By carrying out simulation and verification of hydrological models of river basin in HEC-RTS, a continuous monitoring of runoff level is being made. A real time decision support system for the river system using GIS and computer model is developed and made functional.

2 Study Area

The study area is Satluj and Beas basin as depicted in the figure 1 below.

2.1 The Satluj

It has its source in one of the westerly Mansarovar lake in Tibet located at 30°20'N and 81°25'E at an elevation of 4633 meters above sea level. Satluj is an antecedent stream. After originating, the Satluj follows a northwesterly course along the slopes of Kailash Mountain before entering the Shipki La. Thence it bends southwest and it enters the Shivalik foothills near Nangal. A few miles above, the Bhakra dam has been constructed across the river. At this point, the river flows with northwest-southeast orientation through the Jaswan dun between two ranges of the Shivalik hills in Ropar district. At Ropar it pierces through the hills and enters the plain proper, and here it takes a sharp westward turn. The Satluj separates the Bist Doab in the north from the Malwa tract in the south. At Harike it is joined by the Beas. Of the total length of about 1450 kms, the Satluj flows for about 1050 kms in the east and drains about 25,900 sq.kms of area.

2.2 The Beas

[a] Ayushi Vyas: ayushi.vyas4@gmail.com

Beas rises from the southern face of Rohtang Pass in the Kullu district. After its hill journey it enters the Punjab plain near Mirthal where it is joined by the Chakki stream. The Beas joins the Satluj at Harike after flowing over a hundred and fifty kilometers, separating the upper Bari Doab from the Bist Doab. Before the sudden westward turn of the Satluj, the Beas was a big river, flowing separately to join the Ravi some distance south of Multan. At that time it traversed through a large part of the Punjab plain. At present Beas is a comparatively small river which is only about 460 kms long and lies entirely within the eastern territory. Its total catchment area is about 20,000sq.kms.

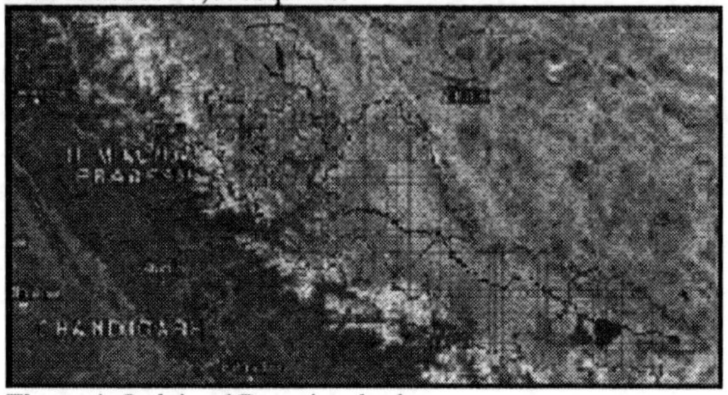

Figure 1. Satluj and Beas river basin.

3 Model Concept

3.1. Hydrologic Modeling

Hydrological modeling is a tool used to estimate the basin's hydrological response due to precipitation [3]. The objective of hydrological modeling is to gain a better understanding of the hydrologic phenomena operating in a watershed and of how changes in the watershed may affect these phenomena [4]. They are also valuable for studying the potential impacts of changes in climate. Runoff is an indication of availability of water. Generally, the reservoir inflow is deducted as a by-product of the reservoir balance. The recent approach uses remote sensed data facilitate modeling activities of the flow using a hydrological model [13]. Hydrologic model can be classified into three main categories which are lumped, semi-distributed and distributed models. The model uses several components to simulate the hydrological behavior of the basin [5]. These hydrological elements use mathematical functions to describe the physical processes in the watershed.

3.2 Applications

The daily stream flows can be computed using the HEC-RTS model and the prepared data maps are used in the model. Watershed and meteorology information were combined to simulate the hydrologic responses [10]. Data required for the hydrological modeling are area of the catchment and the sub catchments, daily rainfall data, imperviousness, constant rate, time of concentration and storage coefficient. These values were taken considering the prominent soil type in the catchment area.

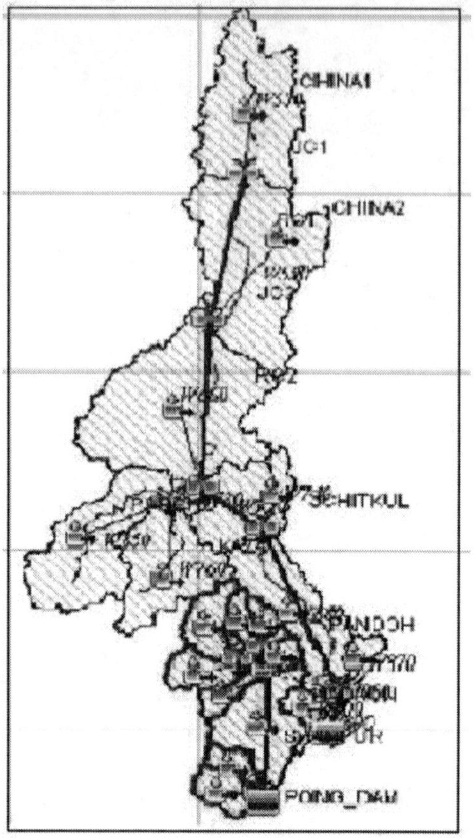

Figure2. The gridded view of Satluj- Beas river basin generated to analyze runoff by initial and constant loss model.

Precipitation on the pervious surfaces is subject to losses. The initial and constant-rate loss model is included to account for the cumulative losses. The underlying concept of the initial and constant-rate loss model is that the maximum potential rate of precipitation loss, f_c, is constant throughout an event. Thus, if p_t is the mean areal precipitation depth during a time interval t to $t+\Delta t$, the excess, p_{et}, during the interval is given by:

$$p_{et} = \begin{cases} p_t - f_c \text{ if } p_t > f_c \\ 0 \text{ otherwise.} \end{cases}$$

Precipitation loss is found for each computation time interval, and is subtracted from the mean areal precipitation depth for that interval. The remaining depth is referred to as precipitation excess [14]. This depth is considered uniformly distributed over a watershed area, so it represents a volume of runoff as precipitation excess & represents volume of runoff [6]. To use the ModClark model, a gridded representation of the watershed is defined. Information about this representation is stored in a grid-parameter file as model is working on grid file format. HEC-HMS uses separate models to represent each component of the runoff process, including models that compute runoff volume, models of direct runoff, and models of base flow. To compute the direct runoff, the program uses a discrete representation of excess precipitation, in which excess precipitation is known for each time interval. It then solves the discrete convolution equation for a linear system:

$$Q_n = \sum P_m U_{n-m+1}$$

Where $\sum$ varies from m=1 to n $\leq$ M; Q_n = hydrograph ordinate at time $n\Delta t$; P_m = rainfall excess depth in time interval $m\Delta t$ to $(m+1)\Delta t$; M = total number of discrete rainfall pulses; and U_{n-m+1} = Unit Hydrograph ordinate at time $(n-m+1)\Delta t$. Q_n and P_m are expressed as flow rate and depth respectively, and U_{n-m+1} has dimensions of flow rate per unit depth [15].

4 Real Time Data Acquisition

4.1 Data Acquisition

Hydrological processes are highly dynamic in nature. Conventional methods for the estimation of the hydrologic state variables are based on the point measurements. Obsolete communication systems, partial coverage of catchment area, forecasts variance, poor decision making tools and dependence on remote man power, enormous instrumental requirements and the physical inaccessibility of the areas often limit the observed data availability [7]. This section briefly explains the application of remote sensing techniques for the estimation of the hydrologic state variables such as precipitation and temperature which are the major inputs in any hydrological model.

4.1.1 Precipitation

The most important input information for a hydrologic model is precipitation. The most common form applied in modeling is gauge data. Gauge data is input as point data and generally assumed to be constant across a sub-basin. Satellite-derived rainfall estimate generated by the National Oceanic and Atmospheric Administration (NOAA) and the National Aeronautic and Space Administration (NASA) are used to drive a hydrologic model. NASA Satellites for Water Quantities are Landsat, TRMM (Tropical Rainfall Measuring Mission), GPM (Global Precipitation Measurements System), Terra, Aqua and GRACE (Gravity Recovery and Climate Experiment).

4.1.2 Temperature

Earlier lapse rate method was used to estimate temperature of inaccessible areas but now many sources for remote sensed data are available. The temperature data can be collected from IMD (FTP- Server) and RIMES (Regional Integrated Multi-hazard Early Warning System for Asia and Africa). IMD AWS (Automated Weather Stations) and IMD ARG (Automated Rain Gauges) provide temperature data which can be used in hydrological models [8].

4.2 Real-time Data Processing

4.2.1. Time series formation

Time series data are an integral part of hydrologic data. Once the precipitation and temperature records have been downloaded, quality checks are to be applied for consistency, removal of spikes and faulty data and interpolation technique for missing data [11]. Then these are assigned to the watersheds they are transferred from geo-database time series tables to HEC-DSS (Hydrologic Engineering Center-Data Storage System) binary format. The modeling features in a hydrologic or hydraulic simulation model must be uniquely identified so that input and output time series associated with them can be labeled appropriately. A geodatabase structure in Arc Hydro that exactly mimics HEC-DSS components must be created so that the ArcGIS geodatabase can contain a mirror image of the time series information for model elements stored in HEC-RTS (which share HEC-DSS as their time series data storage system). This means two related tables, a catalog and a time series table. A software application must interpret the HEC-DSS time series catalog and know which time series records apply to a particular model feature.

4.2.2. Linking of real-time data with hydrological models.

Basin model along with the meteorological data in form of time series grid files can be visualized and analyzed on HEC-RTS user interface. Both temperature and precipitation variable can be presented on the Digital Elevation Model (DEM) of the basin under study. The representation of precipitation is shown in figure 3.
Moreover by providing precipitation and temperature data snow water equivalent can be calculated. Snow water equivalent (SWE) is an important parameter which is used in hydrological modeling of snow-covered basins. Once this parameter is determined, it can be verified using MODIS snow cover imaginaries.

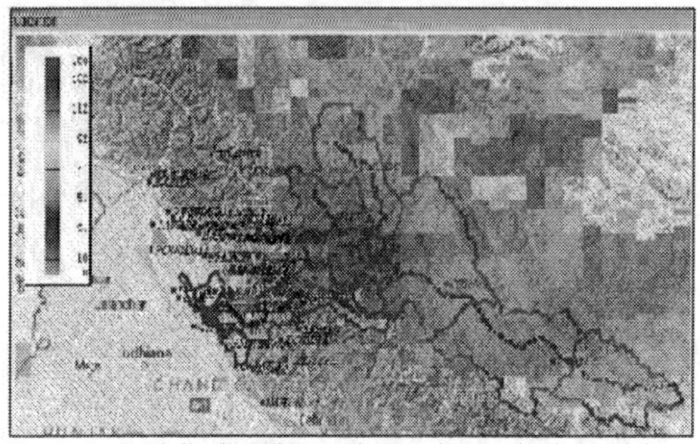

Figure 3. Grid display of precipitation data on basin model.

SWE provide important information for water resource management and is a major research topic in assessment of snow cover and melt. Estimation of SWE proves useful in determining the exact amount of runoff in snow covered basins.

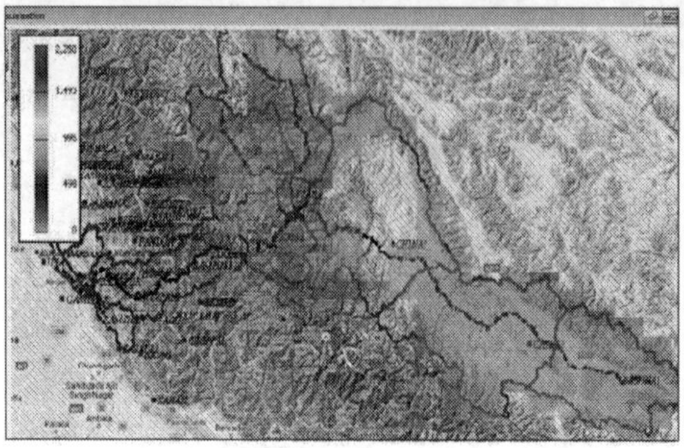

Figure 4. SWE calculated by linking real time precipitation and temperature data with the model.

5. Results and Discussion

The hydrology and dynamics of Satluj and Beas river basin including the real time data acquisition is studied here by making a hydrological model work for the basin. By linking of real time data with hydrological model, basin model along with the meteorological data in form of time series grid files can be visualized and analyzed on HEC-RTS user interface. Study also aim to carry out simulation and verification of hydrological models of river basin in HEC-RTS.

Figures 5 show the current performance of model in calculation of inflow in Bhakra reservoir and depict the observed and simulated runoff volume results.

Similarly calculations can be performed for Pong reservoir and presented in figure 6 and 7.

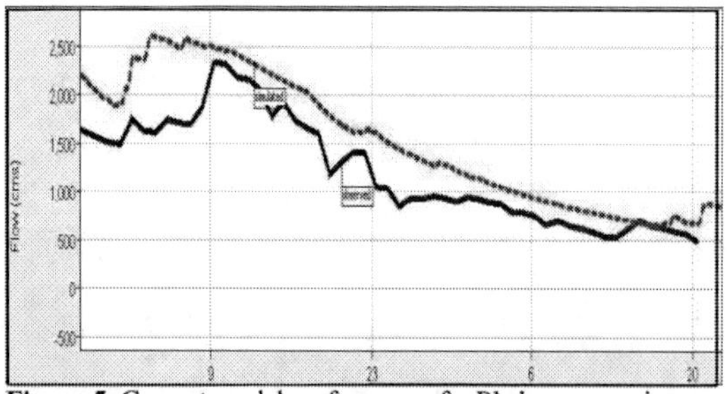

Figure 5. Current model performance for Bhakra reservoir.

Moreover for individual sub-catchments of river Satluj and Beas the various parameters like flow, temperature, precipitation and SWE can be computed and presented.

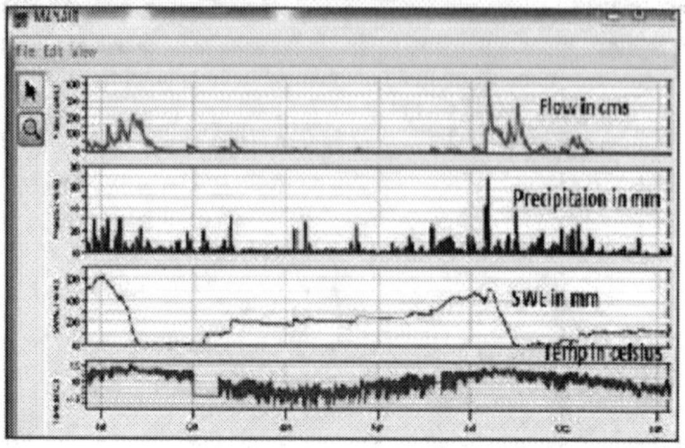

Figure 6. HEC-RTS model providing details regarding sub-catchment of Beas basin.

Real time data acquisition and presentation on a potential interactive platform is important and attempts to develop an interactive Decision Support System using the real time data acquisition in river systems is made successful. Some necessary calibration has yet to be done before the climate models judged for reliability of their outcomes to carry out the hydrological research. While models are clearly getting more comprehensive, it is yet unclear how much of their convergence with observations is due to better understanding of the processes or due to increased realism in parameterization [9]. Different combinations of simulation models for hydrological modeling have yet to be tried for the Indian River basin. And then can we really effectively strategize their management to combat the water crisis in the changing climate [10]. Furthermore, there is an urgent need of integrated research by engineers and hydrologists which would yield a wholesome solution like a complete real time decision support system.

6. Conclusions

The present trends point to increasing several innovations. Availabilty of data from multiple sensors

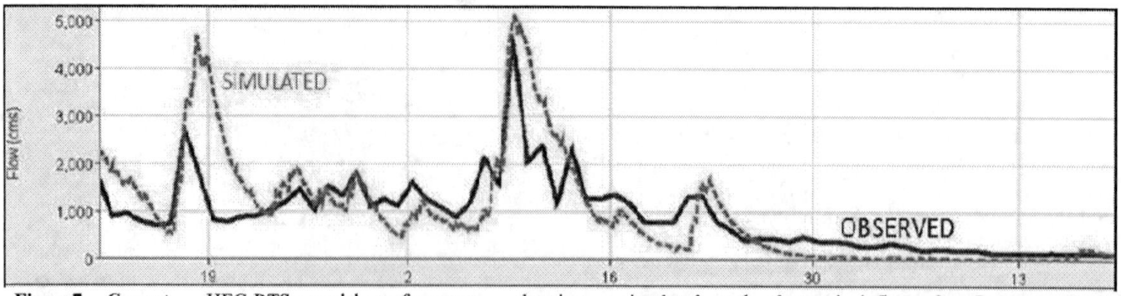

Figure7. Current HEC-RTS model performance showing simulated and observed inflow for Pong reservoir .

with wide array of spatial, spectral, and radiometric characteristics and this data will be available from multiple sources. Therefore, significant advances have been made in harmonizing and synthesizing data from multiple sources that facilitates the use of data from these sensors of widely differing characteristics and sources. The micro satellites that specialize in gathering data for specific geographic location and for particular applications are likely to become more attractive. For many environmental and natural resource applications global wall-to-wall coverage is essential and here NASA satellites will continue to play most important role. Moreover, hydrological models like HEC-RTS which works on real time simulations are making difference in the conventional methods of modeling. The advances made in data synthesis, presentation, and accessibility through such systems will bring in new users and multiply applications of remote sensing in environmental sciences and natural resources management.

References

1. A. Aghakouchak, E. Habib, Int. J. Engng Ed. **26**, pp. 963–973 (2010)
2. K. P. Georgakakos, N. E. Graham, T. M. Modrick, M. J. Murphy Jr, E. Shamir, C. R. Spencer, J. A. Sperfslage, Journal of Hydrology, **519**, pp. 2978-3000 (2014)
3. K. Choudhari, B. Panigrahi, J.C. Paul, Int. J. of Geomatics and Geosciences, **5**, pp.253-265 (2014)
4. D. Bhatt, R.K. Mall, Aquatic Procedia **4**, pp. 730 – 738 (2015)
5. D. Halwatura, M. Najim., *Environmental Modelling & Software* **46**, pp. 155-162 (2013)
6. M.K. Yener, A.Ü. Şorman, A.A. Şensoy,T. Gezgin, Int. J. River Basin Management, pp. 621-634 (2010)
7. C. Giardino, M. Bresciani, P. Villa, A. Martinelli, Water Resour Manage **24**, pp.3885–3899 (2010).
8. R.A. Mir, S. Jain, A.K. Saraf, A. Goswami, J. Earth Syst. Sci. **124**, pp. 365-382, (2015)
9. D. Bhatt, R.K. Mall, Aquatic Procedia **4**, pp. 730 – 738 (2015)
10. R. Ragab, C. Prudhomme, Biosystems Engineering **81** pp. 3-34 (2002).
11. P. Singh, A. Gupta, M. Singh, Egyptian J. Remote Sensing and Space Sciences **17**, pp 111–121 (2014)
12. X. Zhang, R. Srinivasan, K. Zhao, M.V. Liew, J. Hydrologic Engineering **19**, pp. 1374-1384 (2014)
13. M.F. Baumgartner, G.M. Apfl, Hyémlogkal Sciences -Journal- des Sciences B^dmlagiqmes **41** pp. 593-607 (1996)
14. K.M. Loague, R.A. Freeze, J. Water Resource Research **21** pp. 229-248 (1985).
15. V.T. Chow, D.R. Maidment, L.W. Mays, *Applied Hydrology NY* (1988)

MATEC Web of Conferences 57, 05003 (2016) DOI: 10.1051/matecconf/20165705003
ICAET-2016

Application of silver nanoparticles synthesized from *Raphanus sativus* for catalytic degradation of organic dyes

Tej Singh[1, a], Kumari Jyoti[2], Amar Patnaik[3], Ranchan Chauhan[4] and Naresh Kumar[1]

[1]*Department of Mechanical Engineering, Manav Bharti University, Solan-173229, India*
[2]*Department of Biotechnology, G.B.P.E.C., Pauri Garhwal, Uttarakhand-246194, India*
[3]*Department of Mechanical Engineering, M.N.I.T. Jaipur-302017, India*
[4]*Department of Mechanical Engineering, Shoolini University, Solan-173229, India*

Abstract. Biosynthesis of metal nanoparticles is gaining more importance owing to its simplicity, economical, sustainable route of synthesis of nanoparticles and ecofriendliness. Based on the search to improve and protect the environment by decreasing the use of toxic chemicals and eliminating biological risks in biomedical applications, the present article reports an environment friendly and unexploited methods for biofabrication of silver nanoparticles (AgNPs) using *Raphanus sativus* leaf extract. The synthesized AgNPs were characterized by UV-vis spectroscopy and transmission electron microscopy (TEM). The absorption spectrum of the dark brown color silver colloids showed a single and prominent peak at 431nm, indicating the presence of AgNPs. Further, catalytic degradation of methylene blue (organic dye) by using AgNPs was measured spectrophotometrically. The results revealed that biosynthesized AgNPs was found to be impressive in degrading methylene blue and can be used in water purification systems.

1 Introduction

At present, there is a greater need to develop non toxic, reliable, clean and eco-friendly methods for the preparation of nanoparticles to expand their biomedical applications [1]. Over past few years, researchers in the field of nanotechnology present opportunities to exploit the array of biological resources in nature for nanoparticles synthesis. Nature has been invaluable source of medicinal plants since the time man realized the preventive and curative properties of plants [2]. Medicinal importance of plants is derived from the fact that approximately 25% of all prescription drugs are being derived from trees, shrubs or herbal compounds. The use of medicinal plants for the fabrication of nanoparticles offers numerous benefits of ecofriendliness, low cost, energy efficient, compatibility with pharmaceuticals and other biomedical applications as they do not use toxic chemicals for synthesis protocols. The methods using plant extracts involve phytochemicals such as terpenoids, flavonoids, phenol derivatives, plant enzymes (hydrogenases, reductases, quinones) and their derivatives, dihydric phenols and so on act as reductants in the presence of metal salt for nanoparticles synthesis [3-7].

The roots and leaves of radish plant (*Raphanus sativus*) are used as vegetables in numerous parts of the world and well reported in literature to possess a wide range of pharmacological activities [8-10]. Furthermore, the juice of *Raphanus sativus* root and its ash has been reported to possess antiulcer and anticancer activities respectively [11, 12]. In the present research work, we discuss the green-synthesis of AgNPs using the commercially economic and abundantly available *Raphanus sativus* and its application for catalytic degradation of organic dyes.

[a] Corresponding author: tejschauhan@gmail.ocom

2 Materials and methods

2.1 Materials

Silver nitrate ($AgNO_3$) and methylene blue (MB) purchased from Sigma-Aldrich chemicals for this study. All glassware's were washed with distilled water and kept in the autoclave at 15 psi pressure, 120°C for 15 min.

2.2 Preparation of leaf extract

Fresh leaves of *Raphanus sativus* were collected from institute surroundings and washed with deionized water and air dried before it is extracted. A 20g of this plant leaves were finely cut and were boiled in a Erlenmeyer flask with 150 ml of deionized water at 60°C for 20 min and were finally filtered through Whatman No.1 filter paper to get the leaf extract. The filtrate stored in refrigerator at -10°C and used as reducing and stabilizing agent

 DOI: 10.1051/matecconf/20165705003

2.3 Synthesis and characterization of silver nanoparticles (AgNPs)

For all experiments, the source of silver was $AgNO_3$ in distilled water. Aqueous solution (1 mM) of $AgNO_3$ was prepared and used for the synthesis of AgNPs. In a typical synthesis of AgNPs the leaf extract (2.5ml) was added to 47.5 ml of 1mM $AgNO_3$ aqueous solution in a 100 ml Erlenmeyer flask and heated on water bath at 80°C for 60 min. Reduction of silver nitrate to silver ions was confirmed by the color change from colorless to dark brown. The synthesized AgNPs were characterized by UV-vis spectroscopy (Hitachi U-1800) and transmission electron microscopy on Tecnai G2 20S-TWIN instrument.

2.4 Catalytic activity of AgNPs

The catalytic activity of AgNPs was evaluated by investigating the degradation of MB dye. For all catalytic experiments the source of MB was the aqueous solution of 10mg MB in 1000 ml of double distilled water. First 0.5ml of MB was mixed with 2.5ml of de-ionized water and measured for absorbance using Hitachi U-1800 spectrophotometer after 30 min. Further, 0.5ml of MB was mixed with 2.5ml of AgNPs extract and measured for absorbance at different time intervals viz., 30 min, 60 min, 90 min, 120 min, 5 h, 12 h, 24 h, 48h and 72 h.

3 Results and discussion

3.1 Characterization of AgNPs

The synthesized AgNPs were primarily visually observed for color change and characterized by UV-visible spectroscopy. During the visual observation, AgNO3 incubated with leaf extract showed a color change from colorless to light brown within 1 h, as a result of the surface plasmon resonance (SPR) phenomenon, whereas no color change could be observed in $AgNO_3$ without leaf extract (Fig. 1).

Figure 1. (a) Color of $AgNO_3$ solution, (b) Color changes after the addition of leaf extract with $AgNO_3$.

The appearance of light brown color in leaf extract treated flask is clear indication for the formation of AgNPs. The absorption spectrum of the colored silver colloids showed a single and prominent peak at 431nm, indicating the presence of AgNPs (Fig. 2). TEM results showed that the synthesized AgNPs were nearly spherical in shape with an average size of ~50nm (Fig. 3).

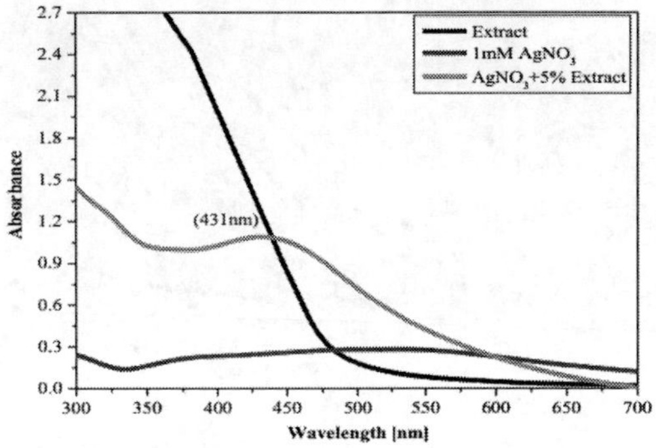

Figure 2. UV-vis spectrum of synthesized AgNPs.

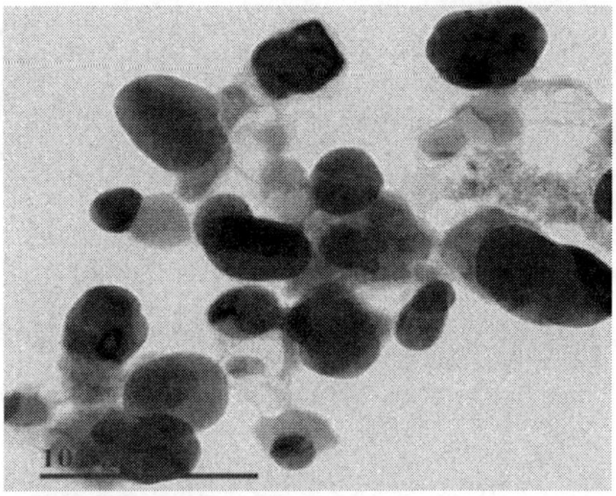

Figure 3. TEM image of synthesized AgNPs.

3.2 Catalytic activity of AgNPs on reduction of methylene blue

Many researchers reported that AgNPs are highly capable, good and stable photocatalysts under ambient temperature with visible region for degrading organic compounds and dyes [13-15]. The catalytic activity of synthesized AgNPs on degradation of dye is demonstrated by using the MB dye by carrying the degradation process in the visible region. The absorption spectrum of aqueous solution of MB dye in the presence of AgNPs is shown in Fig. 3. Pure MB has a main absorption peak of 664nm with a tail around 615nm. This main absorption peak at 664nm decreased abruptly and the tail at 615nm almost diminished with the exposure of dye to AgNPs in visible light for 72h, indicating the photocatalytic degradation of MB. The percentage of degradation of silver nanoparticles was calculated as 46% at 12h, 60% at 24h, 71% at 48h and 86% at 72h and observed to increase as increasing the exposure time.

 DOI: 10.1051/matecconf/20165705003

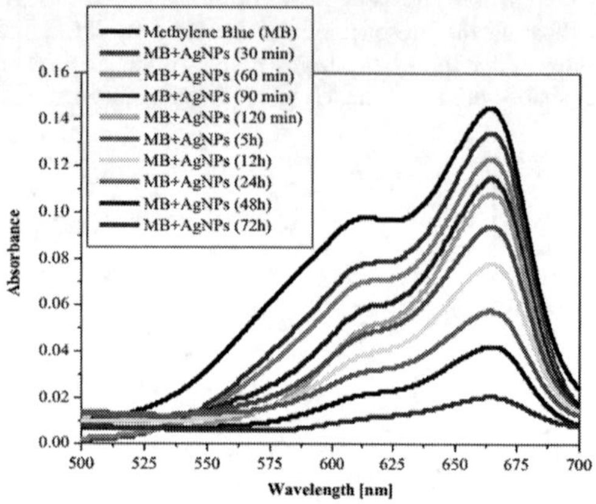

Figure 4. The absorption spectra of aqueous solution of methylene blue tested in the presence of AgNPs.

4 Conclusions

Silver nanoparticles were synthesized by using plant leaf extract of Raphanus sativus. The synthesized silver nanoparticles were characterized by UV-visible spectroscopy and transmission electron microscopy. The main absorption peak at 664nm for methylene blue dye decreased drastically after 72h time indicating the photocatalytic degradation of methylene blue dye. The maximum degradation percentage of methylene blue dye with silver nanoparticles was calculated as 86% at 72h. In the present study, it is found that the biosynthesis of silver nanoparticles exhibits magnificent photocatalytic activity against dye molecules and might be utilized in water purification systems.

References

1. M.A. Albrecht, C.W. Evans, C.L. Raston, Green Chemistry **8**, 417–432 (2006)
2. M. Vijayakumar, K. Priya, F.T. Nancy, A. Noorlidah, A.B.A. Ahmed, Industrial Crops and Products **41**, 235-240 (2013)
3. D. Mandal, M.E. Bolander, D. Mukhopadhyay, G. Sarkar, P. Mukherjee, Applied Microbiology and Biotechnology **69**, 485-492 (2006)
4. A. Ahmad, P. Mukherjee, D. Mandal, S. Senapati, M. Islam, R Kumar, M. Sastry Journal of American Chemical Society **124**, 12108-12109 (2002)
5. J. Chidambaram, R. Rajendiran, R. Abdul, P. Pachiappan, Industrial Crops and Products **45**, 423-429 (2013)
6. K. Jeeva, M. Thiyagarajan, V. Elangovan, N. Geetha, P. Venkatachalam, Industrial Crops and Products 52 714–720 (2014)
7. K. Jyoti, M. Baunthiyal, A. Singh, Journal of Radiation Research and Applied Sciences http://dx.doi.org/10.1016/j.jrras.2015.10.002 ((2015)
8. A.H. Gilani, M.N. Ghayur, Journal of Ethnopharmacology **95**, 169-172 (2004)
9. J. Barillari, R. Cervellati, S. Costa, Journal of Agricultural and Food Chemistry **54**, 9773-9778 (2006)
10. B.R. Vargas, G.R.M. Perez, G.S. Perez, S.M.A. Zavala, G.C. Perez, Journal of Ethnopharmacology **68**, 335-338 (1999)
11. S. Alqasoumi, M.A. Yahya, T. Al-Howiriny, S. Rafatullah, Farmacia **56**, 204-214 (2008)
12. P. Balachandran, R. Govindarajan, Pharmacological Research **51**, 19-30 (2005)
13. S.K. Kansal, M. Singh, D. Sudo, Journal of Hazardous Materials **141**, 581-590 (2007)
14. S.A. kumar, S. Ravi, Spectrochimica Acta Part A: Molecular and Biomolecular Spectroscopy **115**, 388-392 (2013)
15. V.K. Vidhu, D. Philip, Micron **56**, 54-62 (2014)

Numerical Solutions for Convection-Diffusion Equation through Non-Polynomial Spline

A.S.V. Ravi Kanth[1, a] and Deepika[1]

[1]*Department of Mathematics, National Institute of Technology Kurukshetra, Haryana – 136 119, India*

Abstract. In this paper, numerical solutions for convection-diffusion equation via non-polynomial splines are studied. We purpose an implicit method based on non-polynomial spline functions for solving the convection-diffusion equation. The method is proven to be unconditionally stable by using Von Neumann technique. Numerical results are illustrated to demonstrate the efficiency and stability of the purposed method.

1 Introduction

In this paper, we present an implicit non-polynomial spline functions based scheme for the numerical solution of the following one dimensional convection-diffusion equation,

$$u_t + \varepsilon u_x + \gamma u_{xx} = g(x,t),\ 0 \le x \le 1,\ t \ge 0 \quad (1)$$

subject to the boundary conditions

$$u(0,t) = g_0(t),\ u(1,t) = g_1(t),\ t \ge 0 \quad (2)$$

and with the initial condition

$$u(x,0) = f(x),\ 0 \le x \le 1 \quad (3)$$

where $\varepsilon > 0$, is the phase speed and $\gamma < 0$, is the viscosity coefficient. In natural systems, diffusion is one of the important mechanism which means that the mean square displacement of a diffusion particle is in a linear relation with time. The convection-diffusion equation describes the various physical phenomena where energy is transformed inside a physical system due to the combination of convection as well as diffusion processes. It arises in various fields of applied sciences and engineering such as oil reservoir simulations, transport of mass and energy, global weather production, dispersion of diffusion process, due to these vide variety of physical implementations a great deal of researches have been done for the numerical and closed form solutions of convection diffusion type equation for eg., Jain and Aziz [4] used the adaptive spline function approximation for the numerical solution of convection-diffusion equation. Mohammadi [5] applied exponential B-spline collocation method for solving convection-diffusion equation with Dirichlet's type boundary conditions. Caglar et. al. [6] presented B-spline solutions for a convection-diffusion equation. Ravi Kanth and Aruna [1] implemented the differential transform method for solving linear and nonlinear Klein-Gordon equation. Mohammadi [2] presented the exponential spline approach and Lin [3] constructed the parametric cubic spline method for the numerical solution of non-linear Schrodinger equation.

2 Derivation of the Numerical Scheme

For the positive integers N and M, let the partition of $[a,b]\times[0,T]$ be defined by $\Omega : \Omega_h \times \Omega_k$ where $\Omega_h = \{x_i \mid x_i = a + ih, 0 \le i \le N\}$, $\Omega_k = \{t_j \mid t_j = jk, 0 \le j \le M\}$ and $k = \frac{T}{M}$, $h = \frac{b-a}{N}$ are temporal and spatial step size respectively.

Let $Z_i^j = Z(x_i, t_j)$ be an approximation to $u_i^j = u(x_i, t_j)$, obtained by the segment $P_i(x_i, t_j)$ of the mixed spline function passing through the points (x_i, Z_i^j) and (x_{i+1}, Z_{i+1}^j). Each segment can be written as [7].

$$P_i(x,t_j) = a_i(t_j)\cos w(x-x_i) + b_i(t_j)\sin w(x-x_i) + c_i(t_j)(x-x_i) + d_i(t_j) \quad (4)$$

for each $i = 0,1,2,\cdots,N-1$. To obtain the coeffcients $a_i(t_j), b_i(t_j), c_i(t_j)$ and $d_i(t_j)$ in terms of Z_i^j, Z_{i+1}^j, S_i^j and S_{i+1}^j, let us define

$$P_i(x_i,t_j) = Z_i^j,\ P_i(x_{i+1},t_j) = Z_{i+1}^j,$$
$$P_i^{(2)}(x_i,t_j) = S_i^j,\ P_i^{(2)}(x_{i+1},t_j) = S_{i+1}^j \quad (5)$$

where

$$P_i^{(2)}(x,t) = \frac{\partial^2}{\partial x^2} P_i(x,t)$$

[a] Corresponding author: asvravikanth@yahoo.com

By using Eqs. (4) and (5), we obtain by a straightforward calculation,

$$a_i = -\frac{h^2}{\theta^2} S_i^j,\ b_i = \frac{h^2\left(\cos\theta S_i^j - S_{i+1}^j\right)}{\theta^2 \sin\theta}$$

$$c_i = \frac{\left(Z_{i+1}^j - Z_i^j\right)}{h} + \frac{h\left(S_{i+1}^j - S_i^j\right)}{\theta^2}$$

$$d_i = \frac{h^2}{\theta^2} S_i^j + Z_i^j \qquad (6)$$

where
$a_i = a_i(t_j),\ b_i = b_i(t_j),\ c_i = c_i(t_j),\ d_i = d_i(t_j)$ and $\theta = hw$.

Using the following continuity condition of first derivative at point $x = x_i$, i.e.,

$$P_i^{(1)}(x_i, t_j) = P_{i-1}^{(1)}(x_i, t_j)$$

The equations (4) and (6) yield the relation

$$\frac{1}{h^2}\left[Z_{i-1}^j - 2Z_i^j + Z_{i+1}^j\right] = \alpha S_{i-1}^j + \beta S_i^j + \alpha S_{i+1}^j \qquad (7)$$
$$i = 1, 2, \cdots N-1$$

where $\alpha = \frac{1}{\sin\theta} - \frac{1}{\theta^2}$, $\beta = -\frac{2\cos\theta}{\theta\sin\theta} + \frac{1}{\theta^2}$

As $w \to 0$, Eqn. (7) transformed into ordinary cubic spline relation as in [7].

Considering the equation (7) at j^{th} and $(j+1)^{th}$ time levels, so that we can get,

$$\frac{1}{h^2}\left[\left(\frac{Z_{i-1}^{j+1} + Z_{i-1}^j}{2}\right) - \left(Z_i^{j+1} + Z_i^j\right) + \left(\frac{Z_{i+1}^{j+1} + Z_{i+1}^j}{2}\right)\right] =$$
$$\alpha\left(\frac{S_{i-1}^{j+1} + S_{i-1}^j}{2}\right) + \beta\left(\frac{S_i^{j+1} + S_i^j}{2}\right) + \alpha\left(\frac{S_{i+1}^{j+1} + S_{i+1}^j}{2}\right) \qquad (8)$$
$$, i = 1, 2, 3, \ldots, N-1$$

Remark :
The truncation error for Eq. (8) is

$$T_i^j = \left(u_{i-1}^{j+1} + u_{i-1}^j\right) - 2\left(u_i^{j+1} + u_i^j\right) + \left(u_{i+1}^{j+1} + u_{i+1}^j\right)$$
$$-\alpha^* D_x^2\left(u_{i-1}^{j+1} + u_{i-1}^j\right) - \beta^* D_x^2\left(u_i^{j+1} + u_i^j\right)$$
$$-\alpha^* D_x^2\left(u_{i+1}^{j+1} + u_{i+1}^j\right)$$

Where $\alpha^* = \alpha h^2$ and $\beta^* = \beta h^2$. Now expanding the above Eqn, in terms $u(x_i, t_j)$ by using Taylor series, we obtain

$$T_i^j = \left(2h^2 - 2(2\alpha^* + \beta^*)\right) D_x^2 u_i^j + h^2\left(\frac{1}{6} - 2\alpha^*\right) D_x^4 u_i^j$$
$$+ h^4\left(\frac{1}{180} - \frac{\alpha^*}{6}\right) D_x^6 u_i^j + \ldots$$

We can observe from the above expression that if $2\alpha^* + \beta^* = h^2$ then our scheme is of $O(h^2)$ and if $2\alpha^* + \beta^* = h^2$ with $\alpha^* = \frac{h^2}{12}$, then our scheme is of $O(h^4)$.

At the grid point (x_i, t_j) equation (1) can be rewritten in discretized form as

$$S_i^j = (Z_{xx})_i^j = \frac{1}{\gamma}\left[g_i^j - (Z_t)_i^j - \grave{o}(Z_x)_i^j\right] \qquad (9)$$

To obtain the approximation for Eq. (9), based on Taylor series using the following finite difference approximations

$$(Z_x)_i^j = \frac{Z_{i+1}^j - Z_{i-1}^j}{2h} + O(h^2)$$

$$(Z_x)_{i-1}^j = \frac{-Z_{i+1}^j + 4Z_i^j - 3Z_{i-1}^j}{2h} + O(h^2)$$

$$(Z_x)_{i+1}^j = \frac{3Z_{i+1}^j - 4Z_i^j + Z_{i-1}^j}{2h} + O(h^2)$$

Applying the Crank-Nicolson scheme to Eq. (9) along with the above finite difference approximations, we obtain,

$$\frac{\left(S_i^{j+1} + S_i^j\right)}{2} = \frac{g_i^{j+1} + g_i^j}{2\gamma} - \frac{\left(Z_i^{j+1} - Z_i^j\right)}{k\gamma}$$
$$-\frac{\varepsilon}{4h\gamma}\left(Z_{i+1}^{j+1} - Z_{i-1}^{j+1} + Z_{i+1}^j - Z_{i-1}^j\right)$$
$$, i = 1, 2, \cdots N-1, \text{ and } j \ge 0 \qquad (10)$$

$$\frac{\left(s_{i-1}^{j+1} + s_{i-1}^j\right)}{2} = \frac{\left(g_{i-1}^{j+1} + g_{i-1}^j\right)}{2\gamma} - \frac{\left(Z_{i-1}^{j+1} + Z_{i-1}^j\right)}{k\gamma}$$
$$-\frac{\varepsilon}{4h\gamma}\left(-Z_{i+1}^{j+1} + 4Z_i^{j+1} - 3Z_{i-1}^{j+1} - Z_{i+1}^j + 4Z_i^j - 3Z_{i-1}^j\right),$$
$$i = 1, 2, 3, \ldots, N-1, and\ j \ge 0 \qquad (11)$$

$$\frac{\left(s_{i+1}^{j+1} + s_{i+1}^j\right)}{2} = \frac{\left(g_{i+1}^{j+1} + g_{i+1}^j\right)}{2\gamma} - \frac{\left(Z_{i+1}^{j+1} + Z_{i+1}^j\right)}{k\gamma}$$
$$-\frac{\varepsilon}{4h\gamma}\left(3Z_{i+1}^{j+1} - 4Z_i^{j+1} + Z_{i-1}^{j+1} + 3Z_{i+1}^j - 4Z_i^j + Z_{i-1}^j\right),$$
$$i = 1, 2, 3, \ldots, N-1, \text{ and } j \ge 0 \qquad (12)$$

Using the Eqs.(10-12) in Eq.(8), we obtain,

$$\left(2\gamma k-\varepsilon kh(2\alpha+\beta)+4h^2\alpha\right)Z_{i-1}^{j+1}+\left(-4\gamma k+4h^2\beta\right)Z_i^{j+1}$$
$$+\left(2\gamma k+\varepsilon kh(2\alpha+\beta)+4h^2\alpha\right)Z_{i+1}^{j+1}$$
$$=\left(-2\gamma k+\varepsilon kh(2\alpha+\beta)+4h^2\alpha\right)Z_{i-1}^{j}+\left(4\gamma k+4h^2\beta\right)Z_i^{j}$$
$$+\left(-2\gamma k-\varepsilon kh(2\alpha+\beta)+4h^2\alpha\right)Z_{i+1}^{j}$$
$$+2\left(\alpha h^2kg_{i-1}^{j+1}+\beta h^2kg_i^{j+1}+\alpha h^2kg_{i+1}^{j+1}\right)$$
$$+2\left(\alpha h^2kg_{i-1}^{j}+\beta h^2kg_i^{j}+\alpha h^2kg_{i+1}^{j}\right)$$
$$i=1,2,3,....,N-1,\text{and}\, j\geq 0 \qquad (13)$$

The system (13) contains $(N+1)$ equations with $(N+1)$ unknowns. To get a solution to this system, we need two additional equations. These equations are obtained from the boundary conditions (2), can be written in the discretized form as

$$Z_0^j=g_0(t_j),\, Z_N^j=g_1(t_j),\, j\geq 0$$

3 Stability analysis

The stability has been proven through the Von Neumann technique,the numerical solution can be expressed by means of a Fourier series

$$Z_i^j=\xi^j\exp(I\varphi ih) \qquad (14)$$

Where $I=\sqrt{-1},\phi$ is the wave number and ξ^j is the amplitude at time level j. Substituting Eqn.(14) in Eqn. (13), we obtained

$$\xi^{j+1}\left\{\begin{matrix}\left(2\gamma k-\varepsilon kh(2\alpha+\beta)+4h^2\alpha\right)\exp\left(I(i-1)\sigma\right)\\+\left(-4\gamma k+4h^2\beta\right)\exp(Ii\sigma)+\left(2\gamma k+\varepsilon kh(2\alpha+\beta)+4h^2\alpha\right)\exp\left(I(i+1)\sigma\right)\end{matrix}\right\}$$
$$=$$
$$\xi^{j}\left\{\begin{matrix}\left(-2\gamma k+\varepsilon kh(2\alpha+\beta)+4h^2\alpha\right)\exp\left(I(i-1)\sigma\right)\\+\left(4\gamma k+4h^2\beta\right)\exp(Ii\sigma)+\left(-2\gamma k-\varepsilon kh(2\alpha+\beta)+4h^2\alpha\right)\exp\left(I(i+1)\sigma\right)\end{matrix}\right\} \qquad (15)$$

where $\sigma=\phi h$. After some manipulations in Eqn. (15) and then using the Euler's formula, we get

$$\xi=\frac{X_1+IY_1}{X_2+IY_2}$$

Where,

$$X_1=4\gamma k(1-\cos\sigma)+4h^2(2\alpha\cos\sigma+\beta),$$
$$X_2=-4\gamma k(1-\cos\sigma)+4h^2(2\alpha\cos\sigma+\beta)$$
$$Y_1=-Y_2=-2\varepsilon kh(2\alpha+\beta)\sin\sigma$$

For stability as the time increases, we want the amplification factor ξ^j must satisfy $|\xi|\leq 1$ and for this we must have, $A=X_1^2+Y_1^2-X_2^2-Y_2^2\leq 0$
and since,

$$A=4\gamma kh^2(1-\cos\sigma)(2\alpha\cos\sigma+\beta)$$

As $A\leq 0$, provided $\alpha>0,\beta>0$ such that $\beta\geq 2\alpha$, so $|\xi|\leq 1$, thus the proposed method is unconditionally stable if $\beta\geq 2\alpha$.

4 Numerical results

In this section, in order to demonstrate the accuracy and effectiveness of the proposed scheme a few numerical evidences are given,

Example 1. Consider the one dimensional convection diffusion equation,

$$u_t-u_{xx}=g(x,t),\ 0\leq x\leq 1,\ 0\leq t\leq 1$$

subject to the boundary conditions

$$u(0,t)=0,\ u(1,t)=0,\ 0\leq t\leq 1$$

and with the initial condition

$$u(x,0)=0,\ 0\leq x\leq 1$$

where, $g(x,t)=2t\sin(2\pi x)+4\pi^2t^2\sin(2\pi x)$. The exact solution of this problem is [8]

$$u(x,t)=t^2\sin(2\pi x)$$

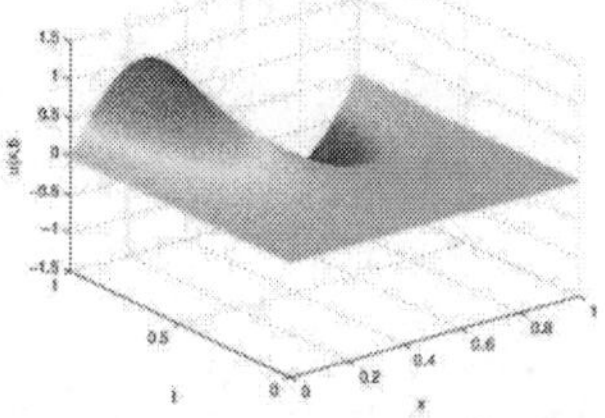

Figure 1. The numerical solution for Example 1 with $h=0.01$, $k=0.01$, $\alpha=1/12$ and $\beta=10/12$.

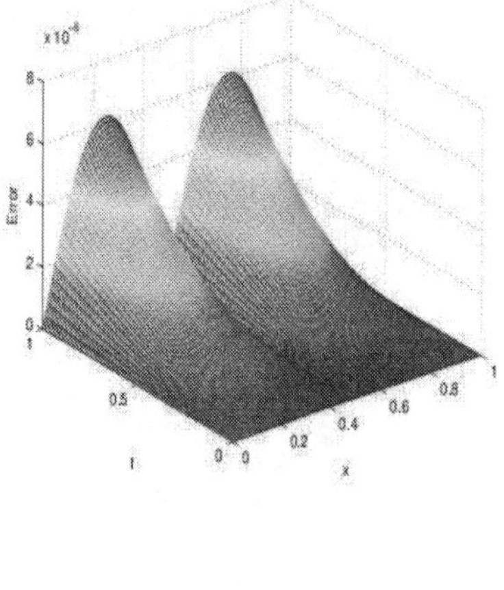

 DOI: 10.1051/matecconf/20165705004

Figure 2. Absolute errors for Example 1 with $h = 0.01$, $k = 0.01$, $\alpha = 1/12$ and $\beta = 10/12$.

Example 2. Consider the one dimensional convection diffusion equation,

$$u_t + \varepsilon u_x + \gamma u_{xx} = g(x,t),\ 0 \le x \le 1,\ 0 \le t \le 1$$

subject to the boundary conditions

$$u(0,t) = e^{qt},\ u(1,t) = e^{p+qt},\ 0 \le t \le 1$$

and with the initial condition

$$u(x,0) = e^{px},\ 0 \le x \le 1$$

where, $g(x,t) = 0$. The exact solution of this problem is

$$u(x,t) = e^{px+qt}$$

and the parameters $\varepsilon = 0.1$, $\gamma = -0.02$,

$p = 1.17712434446770$ and $q = -0.09$.

The L_∞ and L_2 errors along with the order of convergence are reported in Table 1, which shows that the purposed method is having fourth order convergency for $\alpha = 1/12$ and $\beta = 10/12$. Table 2, represents the absolute errors for Example 2 at different spatial and time levels. From this table we can observe that the numerical results are in excellent agreement with the analytical solutions. Figure 1 and 3 exhibits the numerical solutions for Examples 1 and Example 2, respectively. It is noticed that our numerical results are in good accordance with the analytic solutions. Figure 2 and 4 shows the absolute errors for Examples 1 and Example 2 respectively, and it can be observe from these figures that our method supports the theoretical results.

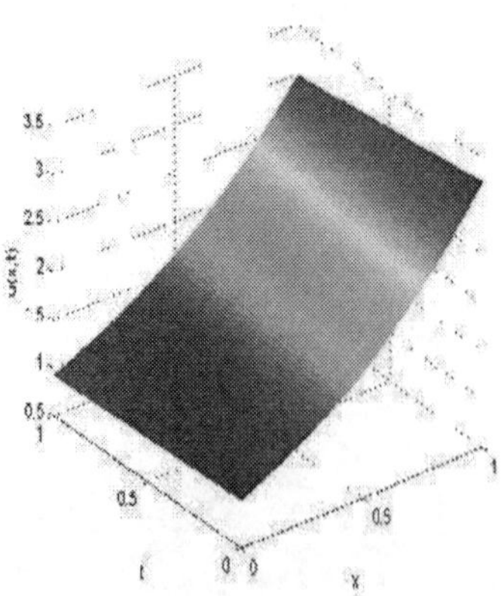

Figure 3. The numerical solution for Example 2 with $h = 0.01$, $k = 0.001$, $\alpha = 1/12$ and $\beta = 10/12$.

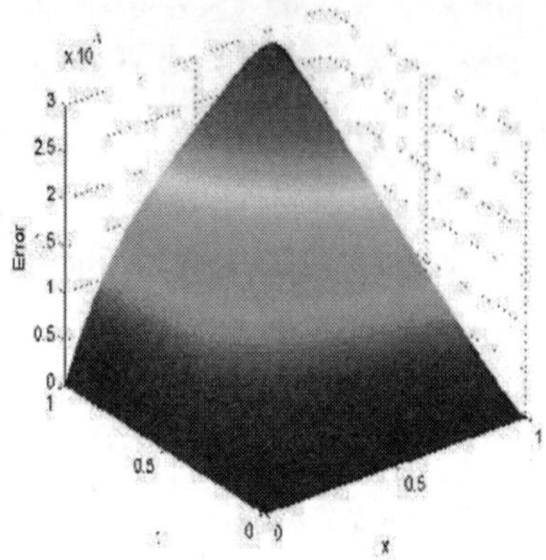

Figure 4. Absolute errors for Example 1 with $h = 0.01$, $k = 0.001$, $\alpha = 1/12$ and $\beta = 10/12$.

5 Conclusion

In this paper, we have studied the numerical solutions for the convection-diffusion type equation via non-polynomial spline functions. The method is shown unconditionally stable by using Von Neumann stability process. Numerical results shows that the purposed method results are in excellent agreement with the analytical solutions. Illustrative examples are given to demonstrate the applicability and accuracy of the proposed method.

Table 1. Convergence rates for Example 1 with $h = 0.01$, $k = 0.01$, $\alpha = 1/12$ and $\beta = 10/12$.

h	L_∞	Order	L_2	Order
$1/4$	2.6659e-2	-	1.8850e-2	-
$1/8$	1.5447e-3	4.1092	1.0922e-3	4.1092
$1/16$	9.4775e-5	4.0266	6.7016e-5	4.0266
$1/32$	5.8963e-6	4.0066	4.1693e-6	4.0066
$1/64$	3.6809e-7	4.0016	2.6028e-7	4.0016

Table 2. Absolute errors for Example 2 with $h = 0.01$, $k = 0.01$, $\alpha = {}^{1}/_{12}$ and $\beta = {}^{10}/_{12}$.

(x,t)	**Errors**
(0.1,0.1)	1.4577e-7
(0.2,0.2)	3.3687e-7
(0.3,0.3)	5.6662e-7
(0.4,0.4)	8.4327e-7
(0.5,0.5)	1.1754e-6
(0.6,0.6)	1.5710e-6
(0.7,0.7)	2.0172e-6
(0.8,0.8)	2.3782e-6
(0.9,0.9)	2.1371e-6

References

1. A.S.V. Ravi Kanth and K. Aruna, *Computer Phy. Commun.* **180(5)** 708-711 (2009).
2. Reza Mohammadi, *Computer Phy. Commun.* **185(3)** 917-932 (2014)
3. Bin Lin, *Computer Phy. Commun.* **184(1)** 60-65 (2013).
4. M. K. Jain and Tariq Aziz, *Appl. Math. Modelling* **7(1) 57-62** (1983).
5. Reza Mohammadi, *Appl Math* **4(6)** 933-944 (2013)
6. Hikmet Caglar , Nazan Caglar and Mehmet Ozer, *Acta Physica Polonica A* **125(2)** 548-550 (2014)
7. Talaat S. El-Danaf and Adel R. Hadhoud, *Appl. Math. Modelling* **36(10)** 4557-4564 (2012)
8. Fengying Zhou and Xiaoyong Xu, *Appl. Math. Comput.* **280** 11-29 (2016).

Variation of energy absorption build-up factor with penetration depth for some titanium compounds

Lovedeep singh[1, a], Pooja Rani[1] and SS Ghumman[1]

[1]*Sant Longowal Institute of Engineering and Technology, Longowal Punjab, India*

Abstract. The buildup factor is an important characteristic that need to be studied and determined prior to using a material clinically in radiation treatment and protection. Energy absorption buildup factors for some titanium compounds like Titanium dioxide (TiO_2), Titanium Carbide (TiC), Titanium Nitride (TiN) and Titanium Silicate ($TiSi_2$) has been calculated using the G.P. fitting technique upto penetration depth of 40 mean free path (mfp) and in an energy range from 0.015 to 15.0 MeV. The variation of absorption buildup factors with penetration depth for the selected compounds of titanium has been studied. It has been found that the maximum value of energy absorption buildup factors shifts to the intermediate incident photon energy regions with the increase in penetration depth of the selected compounds of titanium.

Keywords:

Titanium compounds, multiple scattering, buildup factor.

1. INTRODUCTION

Gamma rays are electromagnetic radiations having wavelength range of about 10^{-14} to 10^{-1}m. These are the most energetic electromagnetic radiations. Due to their greater penetrating power, interactions of gamma ray photons with matter achieved prime importance and interest. Also titanium is the 9th most abundant element in the nature having properties of high hardness, brightness and strongness. It is widely used in aerospace, sports and medicine fields. Titanium compounds are also found useful in various fields. They have applications in paint, coloring food and cosmetics, crayons, UV protection fields, lubricants, wear resistant tools and many more. Due to so many applications of compounds of titanium, an attempt has been made to study their interaction with gamma rays and hence to check their possibility in radiation shielding. Radiation physicists face a main problem of leakage of radiation due to the Compton multiple scattering events. This multiple scattering is the main reason for violation of the Lambert-Beer law i.e. $I = I_0 e^{\mu x}$. In order to maintain this law, a correction factor B is used. This B is called the buildup factor which measures the degree up to which the Lambert-Beer law is violated. Then the intensity equation after modification becomes $I = BI_0 e^{\mu x}$ where B is the buildup factor. Buildup factor is always equals to or greater than unity.

There are two types of buildup factors – energy absorption buildup factor and exposure buildup factor. In energy absorption buildup factor, the quantity of concern is absorbed or deposited energy in the material and detector response function [DRF] is that of the absorption in the material whereas in exposure buildup factor, the quantity of concern is the exposure of energy and detector response function [DRF] is that of absorption in air.

There are various methods available to compute buildup factors such as G.P. fitting method given by Harima et al. [1], invariant embedding method given by Shimizu and Hirayama [2] and Shimizu et al [3]. American National Standards ANSI/ANS 6.4.3 [4] (American National Standard, 1991) used G.P. fitting method and provided buildup factor data for 23 elements, one compound and two mixtures viz. water, air with suitable interval up to the penetration depth of 40 mean free paths. M.J. Berger and J. Hubbell [5] provided for the first time the database of mass attenuation coefficient as well as cross-sections for about 100 elements in the form of software package named as XCom, which is also capable of generating mass attenuation coefficients for compounds and mixtures. Y. Harima [6] has given an historical review and current status of buildup factor calculations and applications for the materials water, concrete and elements Fe, Pb, Be, B, C, N, O, F, Na, Mg, Al, Si, P, S, Ar, K, Ca, Cu, Mo, Sn, La, Gd, W, U in the energy range 0.01 to 0.3 MeV and 0.5 to 10 MeV with penetration depth up to 40 mfp, using various codes ADJ-MOM, PALLAS and ASFIT. H. Hirayama & K. Shin [7] had used EGS4 Monte Carlo code to study multilayer gamma ray exposure buildup factors up to 40 mfp for water, iron and lead at energies 0.1, 0.3, 0.6, 1.0, 3.0, 6.0 and 10 MeV.G. S. Sidhu et al. [8] had computed energy

[a] lovedeepgehal@gmail.com

absorption buildup factor for some biological samples viz. Cholesterol, chlorophyll, hemoglobin, muscle, tissue, cell and bone in the energy range of 0.015 to 15.0 MeV with penetration depth up to 40 mfp, using G.P. fitting method. Shimizu et al.[9] compared the buildup factor values obtained by three different approaches (G.P. fitting, invariant embedding and Monte Carlo method) and only small discrepancies were observed for low-Z elements up to 100 mean free path. K. Trots et al. [10] proposed vector regression model for the estimation of gamma ray buildup factors for multi–layer shields for Al, Fe, Pb, water and concrete in the energy range of 5 to 10 MeV with penetration depth more than 10 mfp. P.S. Singh et al. [11] measured variation of energy absorption build up factors with incident photon energy and penetration depth for some commonly used solvents. T. singh et al [12] worked on Chemical composition dependence of exposure buildup factors for some polymers

After going through the above literature, it has been observed that with the ever increasing use of gamma ray photons in medicine and bio-physics, there is a dire need of proper investigations concerning gamma rays multiple scattering effects on Titanium compounds.

In the present work, multiple scattering effects of gamma rays in some compounds of titanium has been studied in terms of some photon interaction parameters viz. mass attenuation coefficient, equivalent atomic number and energy absorption buildup factor in the energy range of 15.0 keV to 15.0 MeV and penetration depth upto 40 mfp.

2 COMPUTATIONAL WORK

The computational work i.e. calculation of energy absorption buildup factor has been divided into three parts. In the first step, equivalent atomic numbers are calculated in the incident photon energies ranging from 15.0 keV to 15.0 MeV. The second step concerns with the calculation of G.P. fitting parameters using values of equivalent atomic numbers of the selected titanium compounds and finally in the third step, these G.P parameters are then used to calculate absorption buildup factor for the selected titanium compounds.

2.1 Equivalent atomic numbers (Z_{eq})

Equivalent atomic number is a quantity quite similar to general atomic number of elements which can be defined as the number assigned to a compound or mixture by considering the Compton multiple scattering processes. As we know the buildup factor is a direct consequence of multiple scattering, hence equivalent atomic number (Z_{eq}) is used to calculate the buildup factors. In order to calculate (Z_{eq}), the values of Compton partial attenuation coefficient (μ_{comp}) and the total attenuation coefficients (μ_{total}) were obtained in cm^2/g fistly in the energy value from 0.015 to 15.0 MeV, using WinXCom computer program (Gerward et al., 2001) The values of Z_{eq} for the titanium compounds are calculated using following formula:

$$Z_{eq} = \frac{Z_1(\log R_2 - \log R) + Z_2(\log R - \log R_1)}{(\log R_2 - \log R_1)}$$

Where Z_1 and Z_2 are the atomic numbers of elements corresponding to the ratios of μ_{comp} and μ_{total}, R_1 and R_2 respectively. R (μ_{Comp}/μ_{total}) is the ratio for the selected titanium compounds at a particular energy value, which lies between ratios R_1 and R_2 such that $R_1 < R < R_2$.

2.2 Computations of G.P. fitting parameters

ANSI (American National Standard) in 1991 had provided the energy absorption G.P. fitting parameters of 23 elements, one compound (H_2O) and two mixtures (air and concrete) in the energy range from 0.015 to 15.0 MeV and up to a penetration depth of 40 mfp. The calculated values of Z_{eq} earlier were used to interpolate G.P. fitting parameters (b, c, a, X_k and d) for the energy absorption buildup factor, in the energy range (0.015–15.0 MeV) and penetration depth (1–40 mfp). Sidhu et al. (2000) used the formula for the purpose of interpolation of the G.P. fitting parameters as follows:

$$P = \frac{P_1(\log Z_2 - \log Z_{eq}) + P_2(\log Z_{eq} - \log Z_1)}{\log Z_2 - \log Z_1}$$

where

Z_1 and Z_2 are the atomic numbers of elements between which the equivalent atomic number Z_{eq} of the selected titanium compounds lies. Here P_1 and P_2 are the values of G.P. fitting parameters corresponding to the atomic numbers Z_1 and Z_2 respectively at a given energy.

2.3. Computations of buildup factors

The computed G.P. fitting parameters (b, c, a, Xk and d) were then used to calculate the energy absorption buildup factors for the selected compounds of titanium at some standard incident photon energies in the range of 0.015–15.0 MeV and upto a penetration depth of 40 mfp, with the help of G.P. fitting formula, as given by following equations (Harima et al., 1986)

for $K \neq 1$ $$B(E,X) = 1 + \frac{b-1}{K-1}(K^X - 1)$$

for K= 1 $$B(E,X) = 1 + (b-1)X$$

where

$$K(E,x) = cx^a + d\frac{\tanh\left(\frac{x}{X_k} - 2\right) - \tanh(-2)}{1 - \tanh(-2)} \quad \text{for } X \le 40 \text{ mfp}$$

3. RESULTS AND DISCUSSION

Fig. 3.1-3.4 shows the variation of energy absorption buildup factor with penetration depth of the titanium compounds at incident photon energies (0.05, 0.5, 1, 5, 10 and 15 MeV). For all the titanium compounds, energy absorption buildup factor increases with the increase in penetration depth of the building material.

DOI: 10.1051/matecconf/20165705005

It is found that energy absorption buildup factor for the selected compounds of titanium in the energy region of 0.015–15.0 MeV up to the penetration depth of 40 mean free path is always greater than one. This is because with the increase in penetration depth, thickness of the interacting material has been increased which results in increasing the scattering events within the selected titanium compounds. Hence, it results in large energy absorption buildup factor values. For the penetration at large penetration depths, almost similar trend has been observed for the energy absorption buildup factor with incident photon energy.

The dependence of energy absorption buildup factor on penetration depth has been discussed as following:- Figs. 3.1 – 3.4 show the variation of energy absorption buildup factor with incident photon energy for all the selected compounds of titanium at 1, 5, 10, 20, 30 and 40 mean free path respectively. , the increasing rate of energy absorption buildup factor for titanium compounds was found to be slow for lower and higher incident photon energies and rapid increase was observed in case of intermediate energy region.

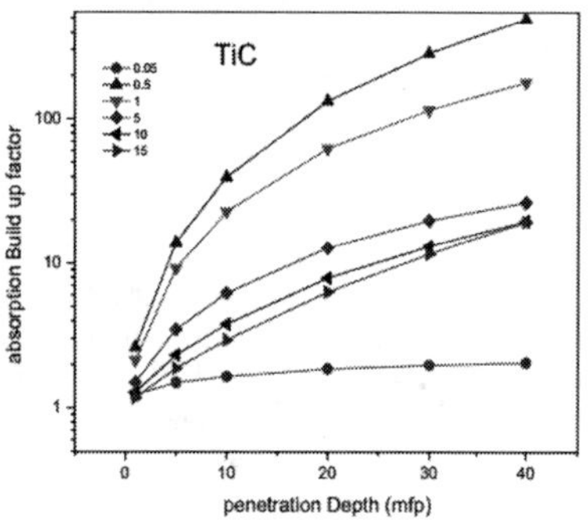

Figure 3.1: Variation of energy absorption buildup factor with penetration depth in case of titanium carbide.

The slower increasing rate in the lower and higher energy regions was due to the dominance of different photon absorption processes in these energy regions (photoelectric effect in the lower energy region and pair production in the higher energy region) which results in the complete absorption of gamma photons in the interacting medium, whereas in the intermediate energy region the dominant process is the Compton scattering, which results only in the energy degradation of photons. Hence, there is finite possibility of the photon to reach the detector even for the large penetration depths of the ceramics and hence maximum violation of Lambert–Beer equation has been observed.

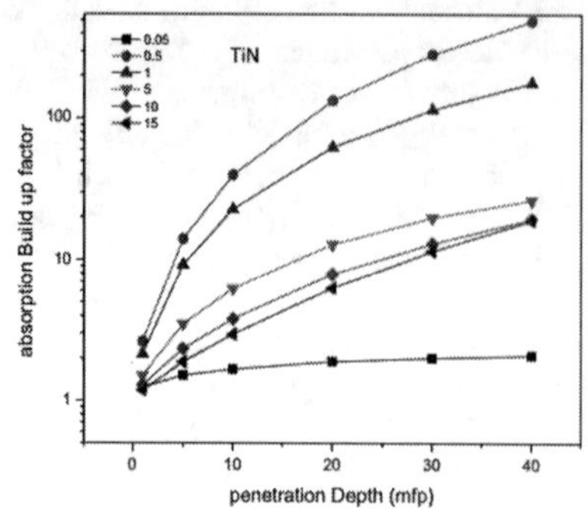

Figure 3.2: Variation of energy absorption buildup factor with penetration depth in case of titanium Nitride.

Further, the increasing rate of energy absorption buildup factor with the penetration depth is more rapid up to the certain incident photon energy (0.1 MeV), where the Compton scattering process is most dominant process and after this increasing rate of energy absorption buildup factor becomes slower for higher energies.

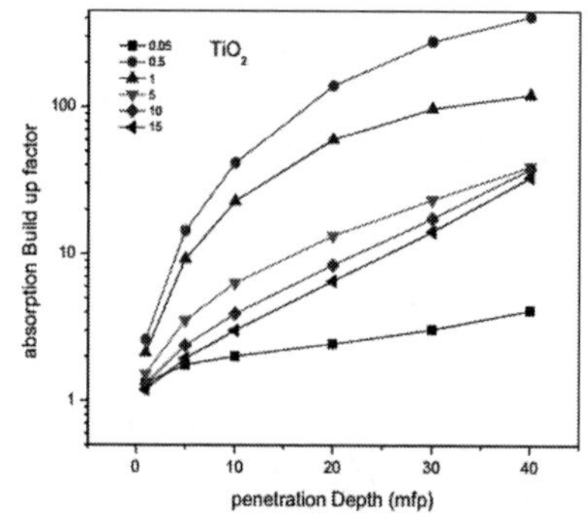

Figure 3.3: Variation of energy absorption buildup factor with penetration depth in case of titanium dioxide.

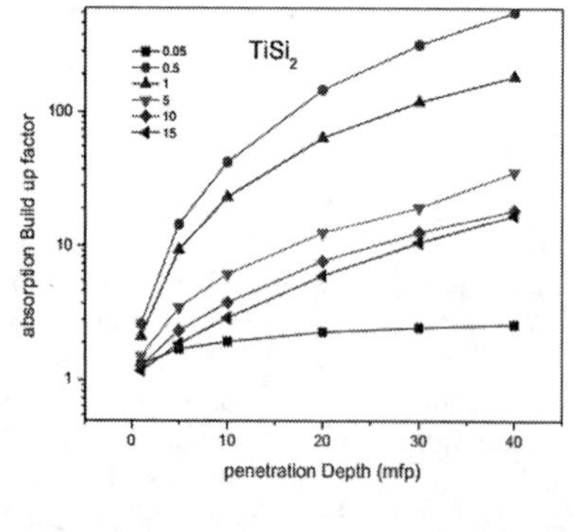

Figure 3.4: Variation of energy absorption buildup factor with penetration depth in case of titanium silicate.

ACKNOWLEDGMENTS

Authors are thankful to staff of department of physics, Sant Longowal Institute of Engineering and Technology, Longowal, Sangrur, Punjab, India for their cooperation for doing this work.

References

1. Harima Y., Tanaka S., Sakamoto Y. & Hirayama H. 1990. Development of New Gamma-Ray Buildup Factor and Application to Shielding Calculations, *Nucl. Sc. & Tech.*, 74-84.
2. Shimizu A. & Hirayama H. 2003, Calculation of gamma-ray buildup factors up to depths of 100 mfp by the method of invariant embedding. Improved treatment of bremsstrahlung. *Nucl. Sci. & Eng.* **40,** 192-200.
3. Shimizu A. 2002, Calculations of gamma-ray buildup factors up to depth of 100 mfp by the method of invariant embedding . *Nucl. Sc. & Tech.* **39**, 477-486.
4. American National Standard, *ANSI/ANS* - 6.4.3. (1991).
5. Berger, M.J., Hubbell, J.H., 1987. N B S I R87-3597: Photon Cross Sections on A Personal Computer.National Institute of Standards and Technology, Gaithersburg, MD.
6. Harima Y. ,1993 An historical review and current status of buildup factor calculations and applications. *Radiat. Phy. Chem.* **41**, 631-672
7. Hirayama H. & Shin K., 1998 Application of the EGS4 Monte Carlo code to a study of multilayer gamma-ray exposure buildup factors of up to 40 mfp. *J. Nucl. Sci. Technol.* **35**, 816-829
8. Sidhu G.S., Singh P.S. & Mudahar G.S., 1999 Energy absorption buildup factor studies in biological samples. *Radiat. Protect. Dosim.* **86,** 207-216.
9. Shimizu A., Onda T. and Sakamato Y., 2004, Calculation of Gamma-Ray Buildup Factors up to Depths of 100 mfp by the Method of Invariant Embedding, (III), . *Nucl. Sc. & Tech,* **41,** 413-424
10. Trots K. , Smuc T. & Pevec D. 2007, Support vector regression model for the estimation of gamma ray build up factors for multi-layer shields. *Annal. Nucl. Energy.* **34**, 939-952.
11. Singh P.S., Singh T. and Kaur P. ,2008, Variation of energy absorption build up factors with incident photon energy and penetration depth for some commonly used solvents. *Annal. Nucl. Energy,* **35**, 1093-1097
12. Singh T., Kumar N., Singh P.S. , 2008, Chemical composition dependence of exposure buildup factors for some polymers, . *Annal.* Nu*cl. Energy,* **36**,*114-120*

Studies of total bremsstrahlung spectra from ^{89}Sr beta particles in Mo and Pt metallic targets in photon energy region of 1-100 keV

Amrit Singh[1,a] and A S Dhaliwal[2]

[1] *Department of Physics, Mata Gujri College, Fatehgarh Sahib, Punjab, INDIA*
[2] *Department of Physics, SLIET, Longowal, Punjab, INDIA*

Abstract. Total bremsstrahlung spectra in thick targets of Mo and Pt, produced by beta emitter ^{89}Sr (end point energy = 1464 keV) have been studied in the photon energy range of 1-100 keV. The experimentally measured spectra measured were compared with the theoretical spectral distributions calculated from Elwert corrected (non relativistic) Bethe-Heitler [EBH] theory, modified Elwert factor (relativistic) Bethe-Heitler (F_{mod} BH) theory for ordinary bremsstrahlung (OB) and the modified Elwert factor (relativistic) Bethe-Heitler (F_{mod} BH+PB) theory, which includes the polarization bremsstrahlung (PB) into total bremsstrahlung (BS). The present results indicate the correctness of F_{mod} BH+PB theory in the low energy region, where the contributions of PB into BS are dominant. But at the middle and higher energy region of the bremsstrahlung spectrum, where the contribution of PB is negligible, the F_{mod} BH theory is more close to the experimental results.

1 Introduction

The bremsstrahlung process is an outcome of general coupling of the electromagnetic and matter fields. This process has wide range of applications in all branches of physics: atomic and molecular physics, nuclear physics, solid-state and elementary-particle physics, astrophysics, material characterization etc. The bremsstrahlung process is further divided into ordinary bremsstrahlung (OB) and polarization bremsstrahlung (PB). OB is the process by which the photon is emitted by the electron decelerating in the static field of the target atom and in PB, dynamic response of the target atom has been considered and the photon emitted by target as a result of its polarization by incident electron. The sum of the amplitudes of OB and PB forms the total bremsstrahlung (BS).

Bethe and Heitler [1] gave the basic theory for OB by using Born approximation and later this theory was corrected by Elwert [2] for non-relativistic case. Tseng and Pratt [3] also gave a theory for OB by using the self consistent field wave functions and claimed that their theory is more accurate than the other available theories. Koch and Motz [4], Pratt and Feng [5], Seltzer and Berger [6] and Pratt et al. [7] have given the general reviews on the theory of OB.

Tystovich and Ojringel [8], Korol and Solov'yov [9], Korol et al. [10], Korol and Solov'yov [11] and Amusia [12,13] reviewed the theoretical and experimental studies of polarization bremsstrahlung in detail. The dynamic response of the target atom and the atomic dynamic polarizability is considered necessary for the calculation of PB amplitude. The equivalent methods for the bremsstrahlung spectra, having contributions of OB and PB, are described by Korol et al. [14] and Avdonina and Pratt [15] by using the stripped atom (SA) approximation. These studies shall be useful to check the correctness of various theoretical models for OB and BS processes, and may further define the energy region upto which the PB contributes into the bremsstrahlung spectrum formation for Mo and Pt metallic targets.

2 Theory

Avdonina and Pratt [15] included the modified relativistic Elwert factor (F_{mod}) in Bethe-Heitler cross-sections for the bremsstrahlung spectrum in a target and described the bremsstrahlung energy spectrum in the form of Gaunt factor $G_{OB}(W_e,k,Z)$, which depends upon the incident electron energy (W_e), the energy of the emitted photon (k) and the atomic number (Z) of the target material,

$$G_{OB}(W_e,k,Z)=G_{BH}(W_e,k,Z)F_{mod}C(T_i,Z) \quad (1)$$

Here, $G_{BH}(W_e,k,Z)$ is a Gaunt factor for Bethe-Heitler cross-section and $C(T_i,Z)$ is a higher order Born approximation factor. Here, T_i is the initial electron kinetic energy and α is the fine structure constant. Further, they described the BS spectrum for a screened Coulomb potential for atoms and ions for soft and hard photon energy regions in which PB is incorporated into OB by using SA approximation. An analytical expression for soft and hard photon energy region of the BS energy spectrum is given by $G_{BS}(W_e,k,Z)$,

[a] Corresponding author: amritsliet@gmail.com

$$G_{BS}(W_e,k,Z)=G_B(W_e,k,Z)-\frac{\sqrt{3}}{\pi}\ln\left(\frac{q_+}{q_-}\right)+G_{OB}(W_e,k,Z) \quad (2)$$

Here, the first term $G_B(W_e,k,Z)$ is the bremsstrahlung energy spectrum in the Born approximation with screening parameter. Here $q_{\pm}=p_i \pm p_f$ is the maximum and minimum momentum transfer. Further, the third term represents the bremsstrahlung energy spectrum in a pure Coulombic case.

In order to calculate the bremsstrahlung energy spectrum, Bethe-Heitler gave an expression $n(W_e',k,Z)$ for a target with N number of atoms per unit volume to absorb a mono-energetic electron of energy W_e'

$$n(W_e',k,Z)=N\int_{1+k}^{W_e'}\frac{d\sigma(W_e,k,Z)/dk}{(-dW_e/dx)}dW_e \quad (3)$$

Here $d\sigma(W_e,k,Z)/dk$ is a single bremsstrahlung cross-section in terms of photon energy and $-dW_e/dx$ the total energy loss per unit path length of an electron in a target material. Further, for complete absorption of continuous beta particles with end-point energy W_{max}, the BS spectral photon distribution in an optimum thick target is given by $S(k,Z)$ i.e. the number of photons of energy k per unit m_oc^2 per beta disintegration,

$$S(k,Z)=\int_{1+k}^{W_{max}} n(W_e',k,Z)P(W_e')dW_e' \quad (4)$$

Here $P(W_e')dW_e'$ is the beta spectrum of the beta emitter under study.

The theoretical bremsstrahlung spectra for Mo and Pt targets for ^{89}Sr beta emitter are obtained from Eq. (4). Further, the theoretical distributions are converted into the number of photons of energy k per m_oc^2 for comparison with the experimental results by dividing them with the total photon yield (T) per beta disintegration. The values of T for different target materials are obtained from graphical integration of the BS spectra from the plots of $S(k,Z)$ versus photon energy *k* between k_{min} and k_{max} i.e., 1 keV and 100 keV.

In the present paper, the studies of the formation of the BS energy spectrum, have been reported in the photon energy region of 1-100 keV for ^{89}Sr (W_{max}=1464 keV) beta emitter in thick targets of Mo and Pt. The measurements have been made for Mo and Pt targets for defining the photon energy regions upto which the PB contributes into the BS spectrum formation.

It is expected that the present measurements shall describe the formation of BS spectrum and the contributions of PB in the photon energy region of 1 keV to 100 keV. Further, the importance of SA approximation for the bremsstrahlung energy spectrum at low energy regions can be described.

3 Experimental details

The experimental arrangement, consisting of Si(Li) detector biased through a power supply and coupled to Linear Amplifier and Multichannel analyzer, is used for the measurement of BS spectral photon distributions in Moa and Pt targets. Most of the experimental details are similar to those reported earlier by Singh and Dhaliwal [16]. Measurements are taken for a time interval of 150000 second by placing the target at position A and B. After subtracting the measurements at Position A from the measurements at Position B, the contributions of internal bremsstrahlung, bremsstrahlung generated in the source material, characteristics X-rays, if any, of the element and room background are eliminated from the measurements for obtaining the correct information of the BS originating in a particular thickness target. This difference method of measurement of BS in target materials is useful to reduce the uncertainties in applying the corrections due to Si K X-ray, Ar K X-ray peaks, and Pb X-rays, particularly at low photon energy region.

Further, the corrections due to the response functions of Si(Li) detector are applied to experimentally measured BS spectrum. In the present measurements, the corrections due Si K-ray and Ar K X-ray peaks, backscattering and Compton continuum are found to be negligible for Si(Li) detector in the studied energy region due to geometrical set up and the method of measurement employed here. The overall errors in the present measurements of BS spectra for ^{89}Sr beta emitter are less than 10% in the entire photon energy region of 1-100 keV. These errors are due to the statistics of the data, which is better than 1%, geometrical full-energy peak detection efficiency of detector, which are uncertain by 3% due the errors in the measurement of photo-fractions. The errors involved in the correction due to Compton continuum, backscattering correction and escape peak correction are less than 1%, as the contributions of these corrections are small. The experimental results are expressed in the terms of number of photon of energy k per m_oc^2 and as shown in Fig. 1. Finally for the comparison with the theoretical model the experimental spectrum is divided with total photon yield and expressed in the terms of number of photons of energy k per m_oc^2 per unit photon yield.

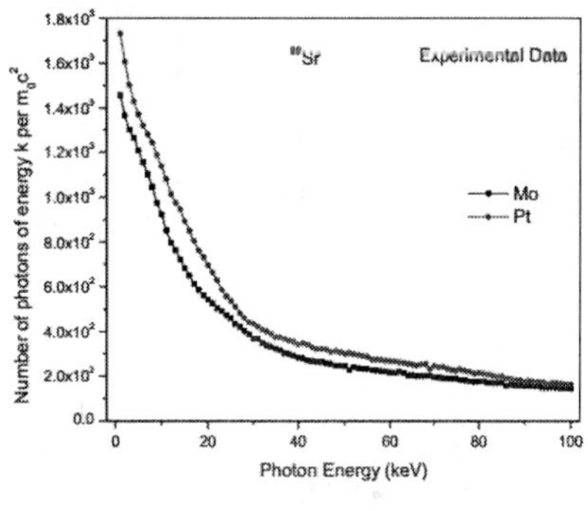

Figure 1. Plots of experimental BS spectrum in terms of photons of energy k per m_oc^2 versus photon energy (*k*) for ^{89}Sr beta particles in photon energy region1-100 keV

DOI: 10.1051/matecconf/20165705006

4 Results and discussion

The experimentally measured bremsstrahlung spectra of Mo and Pt produced by ^{89}Sr beta particles are compared with the theoretical bremsstrahlung spectral distributions obtained from the EBH and F_{mod} BH theory for OB and F_{mod} BH +PB theory which includes contributions of PB into BS, in the photon energy region of 1 - 100 keV. The plots of number of photons of energy k per m_oc^2 per unit photon yield versus photon energy (k) for Mo and Pt targets are shown in Fig. 2.

The Fig. 2 shows that the experimentally measured BS spectra for Mo target which is in agreement within 8% with (F_{mod} BH +PB) theory up to photon energy of 15 keV and thereafter experimental spectrum is closer to the spectra obtained from F_{mod} BH theory. The variation among the experimental results and the F_{mod} BH+PB theory are is 15% at 20 keV and 85% at 100 keV photon energies. It has been found that the PB contribution decreases from 28% to 1% at 1 keV to 20 keV photon energies respectively. In the case of Pt target, the experimental results are in agreement with the F_{mod} BH+PB theory from 1 keV to 26 keV within 15%. Beyond 12 keV the experimental results are closer to F_{mod} BH theory. The variations of experimental results from F_{mod} BH+PB theory are 25% at 30 keV, 50% at 60 keV to 82% at 100 keV photon energies. In this case, PB contribution decreases from 30% at 1 keV to 1 % at 26 keV photon energies.

The present results show the correctness of the F_{mod} BH+PB theory at lower energy end, where the contributions of PB into BS are prominent. But at the middle and higher energy region of the bremsstrahlung spectrum, the F_{mod} BH theory is more close to the experimental results. It is concluded that the production of PB in the low photon energy region, due to the dynamic response of the target atom suppresses the bremsstrahlung at higher energy ends. Hence, more orderly investigations for bremsstrahlung spectra in metallic targets are required for describing the theoretical results at various photon energies.

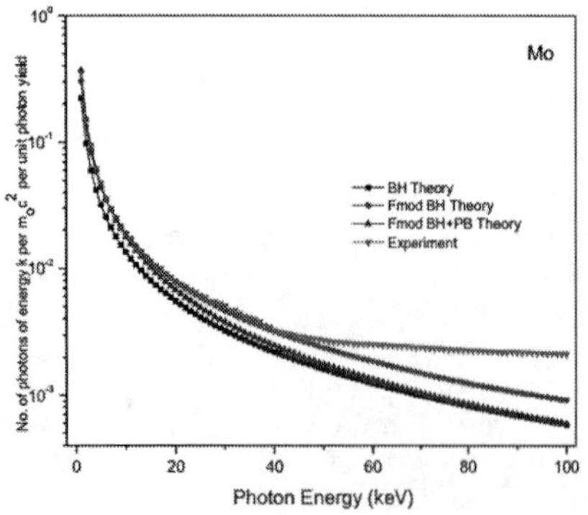

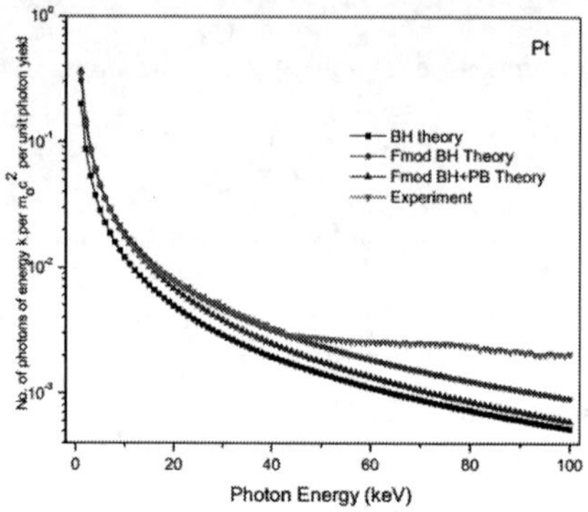

Figure 2. Plots of number of BS photons of energy k per m_oc^2 per unit total photon yield versus photon energy k (keV) for Mo and Pt targets in photon energy region of 1-100 keV

References

1. H. Bethe, W. Heitler, *Proc. Roy. Soc. Lond. A,* **146**, 83 (1934).
2. G. Elwert, *Ann. Phys.* **34**,178 (1939).
3. H. Tseng, R. Pratt, *Phys. Rev. A.* **3**, 100 (1971).
4. H.W. Koch, J.W. Motz, Rev. Mod. Phys. **116**, 920 (1959).
5. R. H. Pratt, I. J. Feng, Electron-atom bremsstrahlung. In Crasemann, B. (Ed), Atomic inner shell physics. Plenum Press, New York, pp. 533-580 (1985). (Chapter 12)
6. S. Seltzer, M. Berger, *Nucl. Inst. Meth. Phys. Res. B.* **12**, 95 (1985).
7. R. Pratt, H. Tseng, C. Lee, L. Kissel, C. MacCallum, M. Riley, *At. Data Nucl. Data Tables.* **20**, 175 (1977).
8. V. Tsytovich, N. Ojringel, I. M. (Eds.) 1993. Polarizational bremsstrahlung. Plenum, NY.
9. A.V. Korol, A.V. Solov'yov, *Rad. Phys. and chem.* 75 : 1266 (2006).
10. A .V Korol, A. G Lyalin, A. V Solovy, N. B Avdonina, R. H. Pratt, *J. Phys. B: At. Mol. Opt. Phys.* **35**, 1197 (2002).
11. A. Korol, A. Solov'yov, *J. Phys. B: At. Mol. Opt. Phys.* **30**, 1105 (1997).
12. M. Amusia, A. Korol, *Nucl. Instr. and Meth. B.* **79**, 146 (1993).
13. M. Amusia, *Rad. Phys. and Chem.* **75**, 1232 (2006).
14. A. Korol, O. Obolensky, A. Solov'yov, I. Solovjev, *J. Phys. B: At. Mol. Opt. Phys.* **34**, 1589 (2001).
15. N. B. Avdonina, R. H. Pratt, *J. Phys. B: At. Mol. Opt. Phys.* **32**, 4261 (1999).
16. A. Singh and A.S. Dhaliwal, *Phy. Lett. A.* **379**, 1127 (2015).

The numerical simulation of convection delayed dominated diffusion equation

P. Murali Mohan Kumar [1, a], A.S.V. Ravi Kanth[1]

[1]Department of Mathematics, National Institute of Technology Kurukshetra, Haryana – 136119, India

Abstract. In this paper, we propose a fitted numerical method for solving convection delayed dominated diffusion equation. A fitting factor is introduced and the model equation is discretized by cubic spline method. The error analysis is analyzed for the consider problem. The numerical examples are solved using the present method and compared the result with the exact solution.

1 Introduction

Consider the following convection delayed dominated diffusion equation

$$\varepsilon y''(x)+a(x)y'(x)+b(x)y(x-\delta)=0 \text{ on } \Omega=[0,1] \quad (1)$$

subject to the interval conditions

$$y(x)=\phi(x) \text{ on } \delta\le x\le 0,\ y(1)=\gamma \quad (2)$$

where $0<\varepsilon<<1$ is a perturbation parameter and δ is a small shifting parameter of order ε. It is also assumed that $a(x), b(x), \varphi(x)$ are smooth functions and γ is a constant. These convection diffusion delayed types with dominated convection term problems play an important role in the mathematical modelling of various practical phenomena in the engineering and environmental sciences, for examples include high Reynold's number flow in fluid dynamics, heat transport problems with large Peckiet number, modelling the problems in mathematical biology and semi-conductor devices etc.

It is challenging to develop efficient numerical methods for solving convection diffusion with dominated convection term due to the existence of boundary layers. Standard discretization methods for solving such kind of problems are unstable and fails to give accurate results when perturbation parameter ε is small. Therefore, it is challenging to develop suitable numerical methods to these problems, whose accuracy does not depend on the parameter ε. Lange and Miura[1] initiated the singular perturbation analysis of boundary value problems for differential difference equations with small shifts. The numerical study of second order singularly perturbed delay differential equations has been given in [2-3] and references therein.

In this paper, we present an exponentially fitted method on uniform mesh based on cubic spline method for the convection delayed dominated diffusion equation. The layout of the paper is organized as follows: Continuation of the problem is presented in next section and follows description of the method. In Section 3, the error analysis of the method is discussed. Section 4 ends with the Numerical results.

[a] Corresponding author:nitmurali@gmail.com

2 Continuous problem

An application of Taylor series in (1) yields

$$\varepsilon y''(x)+(a(x)-\delta b(x))y'(x)+b(x)y(x)=0, \quad x_{i-1}<x<x_{i+1}$$
$$y(x)=\phi_0,\ y(1)=\gamma \quad (3)$$

We assume that $(a(x)-\delta b(x))\ge M>0$ throughout the interval $[0,1]$, where M is positive constant, then the problem (3) exhibits boundary layer at $x=0$.

From the theory of singular perturbations [4],

$$y(x)=y_0(x)+(\alpha-y_0(0))e^{\frac{-(a(0)-\delta b(0))}{\varepsilon}x}+o(\varepsilon) \quad (4)$$

where $y_0(x)$ is the solution of the reduced problem given by

$$(a(x)-\delta b(x))y_0'(x)+b(x)y_0(x)=0 \text{ with } y_0(1)=\beta$$

From Eq.(4) as $h\to 0$, we obtain

$$\lim_{h\to 0} y(ih)=y_0(0)+(\phi_0-y_0(0))e^{\frac{-(a(0)-\delta b(0))}{\varepsilon}ih}$$

Let $\rho=\frac{h}{\varepsilon}$, then

$$\lim_{h\to 0} y(ih)=y_0(0)+(\phi_0-y_0(0))e^{-(a(0)-\delta b(0))i\rho} \quad (5)$$

Now introducing an exponentially fitting factor $\sigma(\rho)$ to the Eq.(3), we get

$$\sigma(\rho)y''(x)+(a(x)-\delta b(x))y'(x)-b(x)y(x)=0 \quad (6)$$

with $y(0)=\phi_0,\ y(1)=\gamma$ (7)

The fitting factor $\sigma(\rho)$ is to be determined in such a way that the solution of Eq.(6) converges uniformly to the solution Eq.(3).

Lemma 1. Let $u(x)$ be a smooth function satisfing the $u(0) \geq 0, u(1) \geq 0$ and $Lu(x) \leq 0, \forall x \in \Omega$, then $u(x) \geq 0, \forall x \in \bar{\Omega}$.

Prof We can prove the above lemma by method of contradiction. Let $x^* \in \bar{\Omega}$ be such that $u(x^*) = \min u(x), x \in \bar{\Omega}$ and assume that $u(x^*) < 0$. Clearly $x^* \notin \{0,1\}$ and $u'(x^*) = 0$ and $u''(x^*) \geq 0$. Now consider

$$Lu(x^*) \equiv \varepsilon u''(x^*) + (a(x^*) - \delta b(x^*))u'(x^*) + b(x^*)u(x^*) > 0$$

Which is contradiction to our assumption. Hence $u(x) \geq 0, \forall x \in \bar{\Omega}$.

Lemma 2. Let $u(x)$ is the solution of the boundary value problem (3), then

$$\|u\| \leq \max\{|\phi_0|, |\gamma|\}$$

Proof. Let $\psi^{\pm}(x)$ be two barrier functions defined by

$$\psi^{\pm}(x) = \max\{|\phi_0|, |\gamma|\} \pm u(x)$$

Then, we have

$$\psi^{\pm}(0) = \max\{|\phi_0|, |\gamma|\} \pm u(0) \geq 0$$

$$\psi^{\pm}(1) = \max\{|\phi_0|, |\gamma|\} \pm u(1) \geq 0$$

and

$$L_\tau\psi^{\pm}(x) = \epsilon\psi^{\pm''}(x) + (a(x) - \delta b(x))\psi^{\pm'}(x) + b(x)\psi^{\pm}(x)$$
$$= \pm L_\tau u(x) + b(x)\max\{|\phi_0|, |\gamma|\}$$
$$\leq b(x)\max\{|\phi_0|, |\gamma|\} \text{ as } b(x) \leq -\theta < 0.$$

Using this inquality in the above inequality, we get

$$L_\tau\psi^{\pm}(x) \leq b(x)\max\{|\phi_0|, |\gamma|\} \leq 0 \forall x \in (0,1)$$

Therefore by the maximum principle[5], we obtain $\psi^{\pm}(x) \geq 0, \forall x \in [0,1]$, which gives the required estimate.

2.1 Description of the method

Let $x_0 = 0, x_n = 1, x_i = ih, h = \frac{1}{N}$. The spline function $S(x,\tau) = S(x)$ satisfying in the interval $[x_i, x_{i+1}]$ and the differential equation

$$S''(x) - \tau S(x) = \left[S''(x_i) - \tau S(x_i)\right]\left(\frac{x_{i+1} - x}{h}\right) + \left[S''(x_{i+1}) - \tau S(x_{i+1})\right]\left(\frac{x - x_i}{h}\right) \quad (8)$$

Where $S(x_i) = y_i$ and $\tau > 0$ is termed as cubic spline in compression. Following Aziz and Khan [6], we obtain the tridiagonal system

$$h^2(\lambda_1 M_{i-1} + 2\lambda_2 M_i + \lambda_1 M_{i+1}) = y_{i-1} - 2y_i + y_{i+1} \quad (9)$$

Where

$$\lambda_1 = \frac{1}{\lambda^2}\left(1 - \frac{\lambda}{\sin\lambda}\right), \ \lambda_2 = \frac{1}{\lambda^2}(\lambda\cot\lambda - 1),$$

$$M_i = y''(x_i),\ i = 1(1)N-1.$$

Substituting

$$\sigma(\rho)\varepsilon M_j = -(a(x_j) - \delta b(x_j))y'(x_j) - b(x_j)y(x_j),\ \text{i}$$
$$j = i, i \pm 1,$$

n Eq.(10) and using the following three point approximations for first order derivatives

$$y_i' = \frac{y_{i+1} - y_{i-1}}{2h}$$

$$y_{i+1}' = \frac{3y_{i+1} - 4y_i + y_{i-1}}{2h}$$

$$y_{i-1}' = \frac{-y_{i+1} + 4y_i - 3y_{i-1}}{2h}$$

We get the following difference scheme

$$E_i y_{i-1} - F_i y_i + G_i y_{i+1} = 0, i = 1(1)N-1 \quad (10)$$

Where

$$E_i = \frac{\sigma}{\rho} - \frac{3\lambda_1}{2}(a_{i-1} - \delta b_{i-1}) - \lambda_2(a_i - \delta b_i) + \frac{\lambda_2}{2}(a_{i+1} - \delta b_{i+1}) + \lambda_1 h b_{i-1}$$

$$F_i = \frac{2\sigma}{\rho} - 2\lambda_1(a_{i-1} - \delta b_{i-1}) + 2\lambda_2(a_{i+1} - \delta b_{i+1}) + 2\lambda_2 h b_i$$

$$G_i = \frac{\sigma}{\rho} - \frac{\lambda_1}{2}(a_{i-1} - \delta b_{i-1}) + \lambda_2(a_i - \delta b_i) + \frac{3\lambda_2}{2}(a_{i+1} - \delta b_{i+1}) + \lambda_1 h b_{i+1}$$

Eq.(11) gives a system of $N-1$ equations with $N+1$ unknowns. These $N-1$ equations together with the Eq.(7) are sufficient to solve the system by using Thomas algorithm.

2.2 Determination of fitting factor

Taking the limit as $h \to 0$ in Eq.(11), we obtain

$$\left(\frac{\sigma}{\rho} - (a(0) - \delta b(0))(\lambda_1 + \lambda_2)\right)\lim_{h\to 0} y_{i-1} - \frac{2\sigma}{\rho}\lim_{h\to 0} y_i + \left(\frac{\sigma}{\rho} + (a(0) - \delta b(0))(\lambda_1 + \lambda_2)\right)\lim_{h\to 0} y_{i+1} = 0 \quad (11)$$

Substituting Eq.(5) in Eq.(12) and then simplifying, we get the variable fitting factor as follows

$$\sigma_i(\rho) = \frac{\rho}{2}(\lambda_1 + \lambda_2)(a(x_i) - \delta b(x_i))\coth\left(\frac{(a(x_i) - \delta b(x_i))\rho}{2}\right) \quad (12)$$

$$\sigma(\rho) = \frac{\rho}{2}(\lambda_1 + \lambda_2)(a(0) - \delta b(0))\coth\left(\frac{(a(0) - \delta b(0))\rho}{2}\right)$$

is a constant fitting factor for left end boundary layer.

3 Error analysis

Substituting $\sigma\varepsilon M_j = -(a_j - \delta b_j)y_j' - b_j y_j$, $j = i, i\pm 1$ in (10), we obtain

$$\begin{aligned}&\sigma\varepsilon\left(y_{i-1}-2y_i+y_{i+1}\right)\\&=h^2\lambda_1\left(-(a_{i-1}-\delta b_{i-1})y'_{i-1}-b_{i-1}y_{i-1}\right)\\&+2\lambda_2h^2\left(-(a_i-\delta b_i)y'_i-b_iy_i\right)\\&+h^2\lambda_1\left(-(a_{i+1}-\delta b_{i+1})y'_{i+1}-b_{i+1}y_{i+1}\right)\end{aligned}\quad(13)$$

Putting exact solution in (14), we get

$$\begin{aligned}&\sigma\varepsilon\left(y(x_{i-1})-2y(x_i)+y(x_{i+1})\right)\\&=h^2\lambda_1\left(-(a_{i-1}-\delta b_{i-1})y'(x_{i-1})-b_{i-1}y(x_{i-1})\right)\\&+2\lambda_2h^2\left(-(a_i-\delta b_i)y'(x_i)-b_iy(x_i)\right)\\&+h^2\lambda_1\left(-(a_{i+1}-\delta b_{i+1})y'(x_{i+1})-b_{i+1}y(x_{i+1})\right)\end{aligned}\quad(14)$$

Where

$$T_{io}(h)=\frac{h^4}{12}\sigma\varepsilon(1-12\lambda_1)y^{(4)}(\xi^i)+\frac{h^6}{360}\sigma\varepsilon(1-30\lambda_1)y^{(6)}(\xi^i)$$
$$x_{i-1}<\xi^i<x_{i+1}\quad(15)$$

for any choice of λ_1 and λ_2 whose sum is $1/2$. Subtracting Eq.(14) from Eq.(15) and substituting $e_j=y(x_j)-y_j$, $j=i,i\pm1$, we get

$$\begin{aligned}&(\sigma\varepsilon+\lambda_1h^2b_{i-1})e_{i-1}+(-2\sigma\varepsilon+2\lambda_2h^2b_i)e_i+(\sigma\varepsilon+\lambda_1h^2b_{i+1})e_{i+1}=\\&-h^2\left[\lambda_1(a_{i-1}-\delta b_{i-1})e'_{i-1}-2\lambda_2(a_i-\delta b_i)e'_i-\lambda_1(a_{i+1}-\delta b_{i+1})e'_{i+1}\right]\\&+T_{io}(h)\end{aligned}\quad(16)$$

we have

$$\begin{aligned}e'_{i-1}&=y'(x_{i-1})-y'_{i-1}\\&=\frac{h^2}{3}y^{(3)}(x_i)-\frac{h^2}{12}y^{(4)}(x_i)+\frac{h^4}{30}y^{(5)}(\xi_1^i)\end{aligned}\quad(17)$$

$$\begin{aligned}e'_{i+1}&=y'(x_{i+1})-y'_{i+1}\\&=\frac{h^2}{3}y^{(3)}(x_i)+\frac{h^2}{12}y^{(4)}(x_i)+\frac{h^4}{30}y^{(5)}(\xi_2^i)\end{aligned}\quad(18)$$

$$\begin{aligned}e'_i&=y'(x_i)-y'_i\\&=-\frac{h^2}{6}y^{(3)}(x_i)-\frac{h^4}{120}y^{(5)}(\xi_3^i)\end{aligned}\quad(19)$$

$$x_{i-1}<\xi_1^i,\xi_2^i,\xi_3^i<x_{i+1}$$

Using Eqs.(18)-(20) in Eq.(17), we get

$$\begin{aligned}&(\sigma\varepsilon+\lambda_1h^2b_{i-1})e_{i-1}+(-2\sigma\varepsilon+2\lambda_2h^2b_i)e_i+(\sigma\varepsilon+\lambda_1h^2b_{i+1})e_{i+1}=\\&-\frac{h^4}{3}[\lambda_1(a_{i-1}-\delta b_{i-1})-2\lambda_2(a_i-\delta b_i)+\lambda_1(a_{i+1}-\delta b_{i+1})]y^{(3)}(x_i)\\&+\frac{h^5}{12}\lambda_1\left[(a_{i-1}-\delta b_{i-1})-(a_{i+1}-\delta b_{i+1})\right]y^{(4)}(x_i)\\&-\frac{h^6}{30}\lambda_1\left[(a_{i-1}-\delta b_{i-1})y^{(5)}(\xi_1)+(a_{i+1}-\delta b_{i+1})y^{(5)}(\xi_2)\right]\\&+\frac{h^6}{60}\lambda_2(a_i-\delta b_i)y^{(5)}(\xi_3)+T_{i0}(h)\end{aligned}\quad(20)$$

Let

$$a_{i-1}=a_i-ha'_i+\frac{h^2}{2!}a_i^{(2)}(\eta_1)$$
$$a_{i+1}=a_i+ha'_i+\frac{h^2}{2!}a_i^{(2)}(\eta_2)$$
$$b_{i-1}=b_i-hb'_i+\frac{h^2}{2!}b_i^{(2)}(\eta_3)$$
$$b_{i+1}=b_i+hb'_i+\frac{h^2}{2!}b_i^{(2)}(\eta_4)$$

Where $x_{i-1}<\eta_1,\eta_2,\eta_3,\eta_4<x_{i+1}$

Using these expansions and Eq.(21), we have

$$\begin{aligned}&(\sigma\varepsilon+\lambda_1h^2b_{i-1})e_{i-1}+(-2\sigma\varepsilon+2\lambda_2h^2b_i)e_i\\&+(\sigma\varepsilon+\lambda_1h^2b_{i+1})e_{i+1}=T_i(h)\end{aligned}\quad(21)$$

Where

$$\begin{aligned}T_i(h)=&-\frac{h^4}{3}(a_i-\delta b_i)y^{(3)}(x_i)+\frac{h^4}{12}(1-12\lambda_1)\varepsilon\sigma y^{(4)}(x_i)\\&-\frac{h^6}{6}\lambda_1(a'_i-\delta b'_i)y^{(4)}(x_i)+O\left(h^6\right)\end{aligned}\quad(22)$$

Clearly, it can be seen that $T_i(h)=O\left(h^4\right)$ for the choice of λ_1 and λ_2 whose sum is $1/2$.

Let $J=trid\left[\sigma\varepsilon\ \ -2\sigma\varepsilon\ \ \sigma\varepsilon\right]$, $D=trid\left[\lambda_1\ \ 2\lambda_2\ \ \lambda_1\right]$ are $(N-1)\times(N-1)$ tridiagonal matrices and $Q=(b_1,b_2,\cdots,b_{N-1})^t$ and $E=(e_1,e_2,\cdots,e_{N-1})^t$ are $(N-1)$ component vectors. Eq.(21) can be written in matrix vector form as

$$AE=T_i(h),\quad \text{where } A=(J-h^2DQ)\quad(23)$$

Clearly, the row sums $S_1,S_2,\cdots,S_{N-1}$ of A are

$$S_1=-\sigma_1\varepsilon+2\lambda_2h^2b_1+\lambda_1h^2b_2$$
$$S_i=h^2\left(\lambda_1b_{i-1}+2\lambda_2b_i+\lambda_1b_{i+1}\right)=h^2B_i,\ i=2(1)N-2$$
$$S_{N-1}=-\sigma_{N-1}\varepsilon+\lambda_1h^2b_{N-2}+2\lambda_2h^2b_{N-1}$$

For sufficiently small h, the matrix A is irreducible and monotone which implies A^{-1} exists. Hence from Eq.(24), we have $E=A^{-1}T_i(h)$.

From the theory of matrices we have,

$$\sum_{i=1}^{N-1}\bar{p}_{k,i}S_i=1,\quad k=1(1)N-1$$

Where $\bar{p}_{k,i}$ is the $(k,i)^{th}$ element of the matrix A^{-1}.

Therefore

$$\sum_{i=1}^{N-1}\bar{p}_{k,i}\le\frac{1}{\min\limits_{1\le i\le N-1}S_i}=\frac{1}{h^2B_i}\le\frac{1}{h^2|B_i|}$$

$$E=\sum_{i=1}^{N-1}\bar{p}_{k,i}T_i(h),\quad k=1(1)N-1$$

Therefore

$$\|E\|=\frac{Kh^2}{|B_i|}\quad(24)$$

Where K is constant independent of h. Therefore $\|E\|=O\left(h^2\right)$ for the choice of parameters λ_1 and λ_2

whose sum is $\frac{1}{2}$. From the Eq.(25), it is observed that the proposed method is ε uniform convergent since the error is of order of the form $\|E\| = K^* h^2$, where K^* is independent of perturbation parameter ε.

4 Numerical Results

To demonstrate the applicability of the method, we consider two convective diffusion with dominated convection term exhibiting boundary layer at left end of the interval and one right end of the interval. The exact solution of the boundary value problem (1) with constant coefficients is given by

$$y(x) = \frac{\gamma - \phi e^{m_2}}{e^{m_1} - e^{m_2}} e^{m_1 x} + \frac{\phi e^{m_1} - \gamma}{e^{m_1} - e^{m_2}} e^{m_2 x}$$

where

$$m_1 = \frac{-(a - \delta b) + \sqrt{(a - \delta b)^2 - 4\varepsilon b}}{2\varepsilon}$$

$$m_2 = \frac{-(a - \delta b) - \sqrt{(a - \delta b)^2 - 4\varepsilon b}}{2\varepsilon}$$

Example 1.
Consider the following left end layer singularly perturbed of convection delayed dominated diffusion
$$\varepsilon y''(x) + 5y'(x) + y(x - \delta) = 0 \text{ on } \Omega = [0,1]$$
subject to the interval conditions
$$y(x) = 1 \text{ on } \delta \le x \le 0, y(1) = 0$$

In Table 1, represents the comparison between the present method and the method in [7] with fixed ε and δ. It is observed that the present method gives more accurate results than the method in [7]. Table 2 indicates the maximum absolute error for different N and ε, it is observed from the results, the method is ε-uniform convergent. Figure 1 displays the numerical solution for different ε values, the solution of the problem exhibits the boundary layer behavior to the left side while the perturbation parameter tends to zero.

Table 1. Numerical solution for Example 1. with $\varepsilon = 2^{-2}$ and δ=0.1, $\lambda_1 = \frac{1}{12}, \lambda_2 = \frac{5}{12}$

x	Exact	Proposed	Method in [7]
0.1	0.14379381821116	0.14380334377735	0.14085841827929
0.2	0.02067665891105	0.02067939843477	0.01984109173050
0.3	0.00297317208765	0.00297376299483	0.00279478220875
0.4	0.00042752012899	0.00042763342218	0.00039366596699
0.5	0.00006147117862	0.00006149154060	0.00005544852472
0.6	0.00000883567395	0.00000883918533	0.00000780774988
0.7	0.00000126708504	0.00000126767192	0.00000109714556
0.8	0.00000017883874	0.00000017893301	0.00000015190043
0.9	0.00000002242424	0.00000002243736	0.00000001875470

Table 2. Maximum absolute errors for the Example 1.
$\lambda_1 = \frac{1}{12}, \lambda_2 = \frac{5}{12}$

$\varepsilon \downarrow N \rightarrow$	128	256	512	1024
2^{-1}	3.90E-06	9.76E-07	2.44E-07	6.10E-08
2^{-2}	7.65E-06	1.91E-06	4.78E-07	1.19E-07
2^{-3}	1.52E-05	3.78E-06	9.46E-07	2.36E-07
2^{-4}	3.04E-05	7.55E-06	1.88E-06	4.70E-07
2^{-8}	1.33E-04	6.55E-05	2.77E-05	8.29E-06
2^{-12}	1.33E-04	6.65E-05	3.32E-05	1.66E-05
2^{-16}	1.33E-04	6.65E-05	3.32E-05	1.66E-05
2^{-32}	1.33E-04	6.65E-05	3.32E-05	1.66E-05

Example 2.
Consider the following variable coefficient of left end layer singularly perturbed problem
$$\varepsilon y''(x) + \left(1 - \frac{x}{2}\right) y'(x) - \frac{1}{2} y(x) = 0 \text{ on } \Omega = [0,1]$$
subject to the interval conditions
$$y(x) = 0 \text{ on } \delta \le x \le 0, y(1) = 1$$
A valid solution for the above problem is given by
$$u(x) = \frac{1}{2 - x} - e^{\frac{-\left(x - \frac{x^2}{4}\right)}{\varepsilon}}$$
Figure 2. indicates the numerical solution for different ε and δ values, it is observed that the boundary layer behavior is not only on perturbation parameter and also depends on the delay parameter.

4.1 Determination of fitting factor for right end problem

We assume that $(a(x) - \delta b(x)) \le M < 0$, M is a constant then, the Eq.(3) exhibits boundary layer at $x = 1$.

$$y(x) = y_0(x) + (\gamma - y_0(1)) e^{\frac{-(a(1) - \delta b(1))}{\varepsilon}(1 - x)} + o(\varepsilon) \quad (25)$$

As $h \rightarrow 0$, Eq.(26) becomes

$$\lim_{h \to 0} y(ih) = y_0(x) + (\gamma - y_0(1)) e^{-(a(1) - \delta b(1))\left(\frac{1}{\varepsilon} - i\rho\right)}$$

Introducing $\sigma(\rho)$ in Eq.(3) and by applying the same procedure as in Section(2) and simplifying, we get

$$\sigma_i(\rho) = \frac{\rho}{2}(\lambda_1 + \lambda_2)(a(x_i) - \delta b(x_i)) coth\left(\frac{(a(x_i) - \delta b(x_i))\rho}{2}\right)$$

$$\sigma(\rho) = \frac{\rho}{2}(\lambda_1 + \lambda_2)(a(1) - \delta b(1)) coth\left(\frac{(a(1) - \delta b(1))\rho}{2}\right)$$

is a constant fitting factor for right end boundary value problem.

Example 3.
Finally we consider the right end layer singularly perturbed of convection diffusion problem

$$\varepsilon y''(x)-5y'(x)+2y(x-\delta)=0 \text{ on } \Omega=[0,1]$$

subject to the interval conditions

$$y(x)=0 \text{ on } \delta\leq x\leq 0, y(1)=2$$

Table 3 gives the comparison between present method and the method in [7] for fixed ε and δ. It is observed that the method yields results better than the existing method in [7]. Figure 2 indicates the numerical solution for different ε and δ values. The perturbation parameter tends to zero the solution of the problem exhibits boundary layer behavior to the right side of the interval.

Acknowledgements

Authors would like to thank National Board for Higher Mathematics (NBHM), Government of India for providing financial support under the grant number 2/48(12)/2013/NBHM(R.P.)/R&D II/1084.

Table 3. Numerical solution for Example 2. with $\varepsilon=2^{-2}$ and δ=0.1 $\lambda_1 = 1/12, \lambda_2 = 5/12$

x	Exact	Proposed	Method in [7]
0.1	0.00000001824795	0.00000001827085	0.00000001297418
0.2	0.00000015942757	0.00000015960776	0.00000011682562
0.3	0.00000124681219	0.00000124804968	0.00000094810121
0.4	0.00000961694147	0.00000962512936	0.00000760202085
0.5	0.00007404065246	0.00007409319019	0.00006086311380
0.6	0.00056989545628	0.00057021895408	0.00048718987663
0.7	0.00438637260011	0.00438823990526	0.00389970919703
0.8	0.03376088858255	0.03377046939265	0.03121511401660
0.9	0.25984953872717	0.25988640673333	0.24986042277630

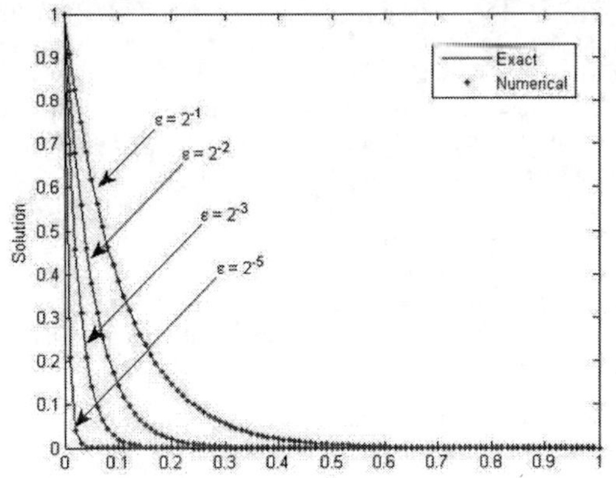

Figure 1. Numerical solution of the Example 1 for $\delta=0.1$

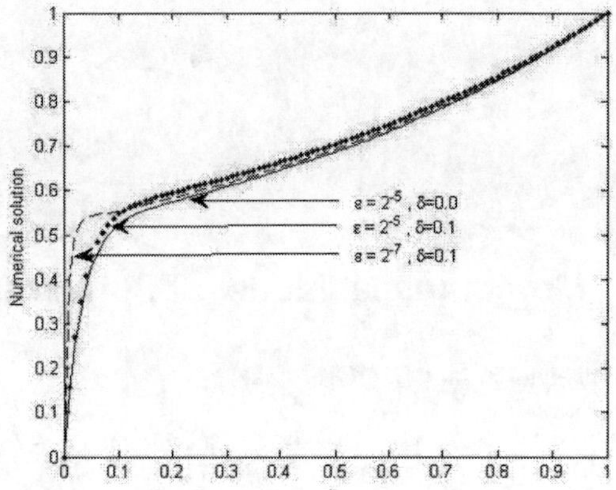

Figure 2. Numerical solution of the Example 2

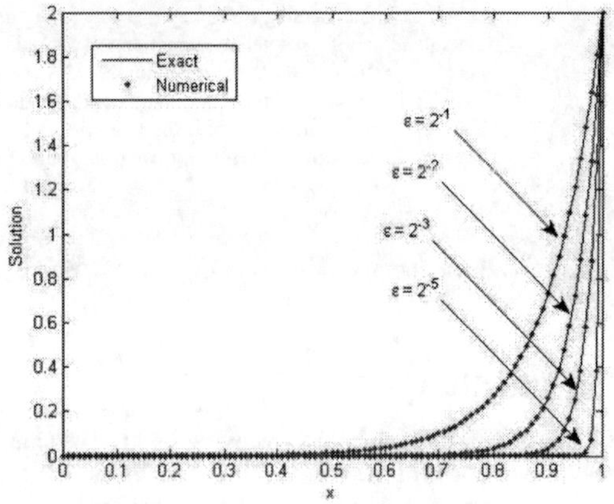

Figure3. Numerical solution of the Example 3 for $\delta=0.1$

References

1. C G. Lange , R M. Miura. SIAM Journal on Applied Mathematics **54(1)** 249-272 (1994)
2. M. K. Kadalbajoo, K. K. Sharma Applied Mathematics and Computation **157(1),** 11-28 (2004)
3. M. K .Kadalbajoo, K. K. Sharma. Applied Mathematics and Computation **197(2),** 692-707 (2008)
4. O'Malley Robert. *Academic Press.* (1974)
5. P.A. Farrell, A.F. Hegarty, J.J.H. Miller, E.O'Riordan, G.I. Shishkin, *CRC Press LLC* (2000)
6. T. Aziz, A. Khan, Journal of Computational and Applied Mathematics **147,** 445-452 (2002)
7. M. Sharma, A. Kaushik, C. Li. Journal of Mathematical Chemistry **52(9),** 2459-2474 (2014)

Mathematical Model of Pulp Washing Using Mathematica

Jitender Kumar[1,a], V.K. Kukreja[2]

[1]*Department of Mathematics, BGIET, Sangrur, 148001 (Punjab) INDIA*
[2]*Department of Mathematics, SLIET, Longowal, 148106 (Punjab) INDIA*

Abstract Good washing is the key to a successful closure of the water system in a pulp production line. An improved understanding of associated phenomena, particularly those related to the removal of the very last impurities, is therefore desirable. Therefore a mathematical model describing the desorption phenomenon. The system of governing partial differential equations is solved using Mathematica. Model is simulated using the industrial data. The output thus obtained is compared with the analytic and numerical results of earlier workers. Present technique is simple, elegant and convenient for solving two point boundary value problems with varying range of parameters.

Keywords: Peclet number, diffusion-dispersion, Mathematica

1 Introduction

An integral part of chemical recovery is the washing process. Methods of describing the efficiency of the various washing operations are necessary before an analysis of the overall consequences of recycling or mill closure is possible.Washing is simply bulk removal of the liquor surrounding the pulp fibers. In a laboratory washing cell, the mechanism of the displacement washing of unbleached Kraft pulp was investigated. The efficiency of displacement washing depends on the degree of mixing and also on the rate of desorption and diffusion of dissolved solids and chemicals from the pulp fibers. The separation of pulp and paper industry to produce its target production with high efficiency and less environmental load can only be met by initiating a meticulously planned research on mathematical methods. Modeling of pulp washing is done mainly using three approaches namely (a) Process modeling (b) Physical modeling (c) Statistical modeling. A complete review of the various process models used to describe pulp washing has been presented by Pekkanen and Norden [9]. Initially researchers Brenner [1] and Pellet [8] has introduced a mathematical model combining the effects of particle diffusion and axial dispersion. An extensive model related to mass transfer in fibrous particle was given by Grahs [2], it was also restricted for axial dispersion only. Moreover certain physical features of the fibers such as fiber porosity and fiber radius were ignored. Lapidus and Amundson [7] also followed two adsorption isotherms; one implies that equilibrium is established at each point in the bed while the other assumes the rate of adsorption is finite.

Washing models of pulp based on axial dispersion and particle diffusion along with adsorption isotherms for sodium and lignin are fairly established and also solved by earlier workers such as Arora et al. (2006, 2009), Babu et al. (2004), Brenner (1962), Kukreja (1996, 2009), Kumar (2002), Kumar et al. (2009), Potucek et al. (2006), Sotelo et al. (2004), Sridhar et al. (1999). Kukreja (1996) solved the washing model analytically by using Laplace transform technique and Kumar (2002) tried to solve numerically using finite difference technique.

The accuracy of the analytic solution undoubtedly exceeds the limit of applicability of the theory to real situations. Moreover, it is highly desirable to have a simple and consistent model of the transport phenomenon based on essential features of real situation. Keeping this modest goal in mind, the method of separation of variables is first applied on partial differential equation and then Laplace transform is taken of the equations.

2 Model based on axial dispersion

The displacement washing model based on the axial dispersion and particle diffusion describing the washing zone is given by

$$\frac{\partial c}{\partial t}+u\frac{\partial c}{\partial z}+\mu\frac{\partial n}{\partial t}=D_L\frac{\partial^2 c}{\partial z^2}, \qquad \dots (1)$$

[a] Corresponding author: jitenderrattan2005@gmail.com

with adsorption isotherm

$$n = kc. \quad \dots (2)$$

This equation represent the basis for the mathematical models of displacement washing, where t is the time from the commencement of the displacement, z is the distance from the point of introduction of the displacing fluid, $c = c(z,t)$ is the solute concentration, D_L is longitudinal dispersion coefficient, u is the average interstitial velocity of the fluid and L is thickness of the packed bed.

On account of unusual nature of displacement process, appropriate boundary conditions have been extensively discussed in the literature (Brenner, 1962; Shelly et al., 2006; Shirashi, 2001; Szukiewicz, 2000) like Danckwerts' conditions at the inlet and outlet of the bed:

$$c = c_s \quad \text{at } z = 0, \quad \dots (3)$$

$$\frac{\partial c}{\partial z} = 0 \quad \text{at } z = L. \quad \dots (4)$$

Accordingly, initial condition is given by

$$c(z,0) = n(z,0) = c_i \text{ for } 0 < t < \frac{L}{u}. \quad \dots (5)$$

2.1 Conversion of model in to dimensionless form:

Equations (1) to (5) can be put in dimensionless form using dimensionless variables:

$$C = \frac{c - c_s}{c_i - c_s};\ N = \frac{n - c_s}{c_i - c_s};$$

$$Z = \frac{z}{L};\ T = \frac{ut}{(1+\mu k)L}$$

The dimensionless time, T, corresponds physically to the number of pore displacements introduced into the medium since the starts of the experiment; that is, it is equal to the ratio of total fluid volume introduced to the free value of the bed.

By these means, equation (1) becomes

$$\frac{\partial C}{\partial T} + \frac{\partial C}{\partial Z} = \frac{1}{Pe}\frac{\partial^2 C}{\partial Z^2} \quad \dots (6)$$

where $Pe = \frac{uL}{D_L}$ is the Peclet number. The boundary conditions are now of the form

$$C(0,T) = 0 \text{ at } X = 0 \text{ for } T > 0 \quad \dots (7)$$

$$\frac{\partial C(1,T)}{\partial Z} = 0 \quad \text{at } X = 1 \text{ for } T > 0 \quad \dots (8)$$

While the initial condition is

$$C = 1 \text{ at } T = 0 \text{ for } 0 < Z \le 1 \quad \dots (9)$$

Now, our main aim is to estimate $C = C(Z,T)$ satisfying Eqs. (7)-(9), which will eventually lead to exit solute concentration $C_e = C_e(T) = C(1,T)$.

3 Solution of model

Method of separation of variables is used for solving Eq. (6). This method transforms the partial differential equation into a system of ordinary differential equation, each of which depends only on one of the functions and the solution is given as product of the functions.

Finally, the solute concentration at any location and time in the bed can be written as:

$$C(Z,T) = \sum_{n=1}^{\infty} \frac{\int_0^1 e^{-\frac{PeZ}{2}} sin(\alpha_n Z)dZ}{\int_0^1 sin^2(\alpha_n Z)dZ} \exp\left(\frac{PeZ}{2} - \frac{P_n^2 T}{Pe}\right) sin\alpha_n Z \quad \dots (10)$$

where the $\alpha_n\ (n = 1,2,3,\dots)$ are the positive roots, taken in order of increasing magnitude, of the transcendental equation

$$\tan\alpha = -\frac{2\alpha}{Pe} \text{ and } P_n = \sqrt{\frac{4\alpha_n^2 + Pe^2}{4}} \quad \dots (11)$$

The concentration of solute in the dimensionless form for any location and time can be found by Laplace transform as Kukreja [3].

$$\frac{c - C_s}{C_i - C_s} = \sum_{n=1}^{\infty} \frac{4\beta_n^2 e^{P_n T} e^{\frac{P_e Z}{2}} [P_e \sin\{(Z-1)\beta_n\} - 2\beta_n \cos\{(Z-1)\beta_n\}]}{P_e P_n \sin\beta_n (P_e^2 + 4\beta_n^2 + 2P_e)}$$

(9)

The concentration of solute in the dimensionless form for any location and time can be found by separation of variables as Grahs (1974).

$$C(x,T) = \exp\left\{\frac{B_0}{2}\left(x - \frac{T}{2}\right)\right\} \cdot \sum_{n=1}^{\infty} \frac{4w_n^2 \sin w_n x . \exp\left(-\frac{1}{B_0} w_n^2 T\right)}{(2w_n - \sin 2w_n)\left(\frac{B_0}{4} + w_n^2\right)}$$

Where w_n are the positive roots to the equation

$$\tan w_n = -\frac{4w_n}{B_0}$$

4 Results and Discussion

The dimensionless model equation (7) of longitudinal mixing is solved for Pe = 1.1 with Mathematica. The behavior of the dimensionless concentration with respect to the dimensionless time and dimensionless distance is shown in the Figure-1. The results at dimensionless distance at X = 1 are compared with work of previous researchers as shown in the figure 2. Brenner [1] The mathematical model developed through transport phenomena had solved the equation (7) (dimensionless form) analytically, while Grahs [2] had used orthogonal Comparison shows that Mathematica gives good accuracy and the magnitude of error is very low. It can be seen that at T = 0.0 the error is 0.001% which is minimum and oscillate up to 6.50% in such a way that at T = 0.1 error is 0.5%, at T = 0.2 error is 6.50%, at T = 0.3 error is 5.72%, at T = 0.4 error is 4.56%, at T = 0.5 error is 4.56%, at T = 0.6 error is 3.48%, at T = 0.7 error is 2.54%, at T = 0.8 error is 1.76%, at T = 0.9 error is .57%, at T = 1.0 error is 0.13%, at T = 1.1 error is 0.22%, at T = 1.2 error is 0.50%. Since the error is increasing up to time T = 0.2 and after that decrease at time T = 1.0 and for above at time T=1.0 again shows an increasing trend. Solutions from all the techniques analytic as well as numerical used by Brenner [1], Grahs [2] and "Mathematica" solver shows fluctuations in the results when time is increased above T = 1.2.

Hence comparison of the results for significant accuracy has been carried out up to time T = 1.2. However one very interesting observation is that the orthogonal collocation method used by Grahs [2] gives an increasing trend from T =0.5 to T =0.6 and thereafter amplitude starts decreasing consistently. This is contrary to analytical solution of Brenner [1] which shows a continuous decreasing trend from 1.0 to .0854. In the present investigation we also obtain the decreasing trend consistently as depicted in the analytical solution by Brenner [1].

The results at dimensionless distance X = 1, are obtained by using 22 meshes in both dimensionless time and dimensionless distance x are given in table-1. The results are also compared with results obtained by Kumar [4] using finite difference method. In this case, also results as obtained by Brenner [1] and numerical results by Grahs [2] and show better accuracy than Kumar [4].

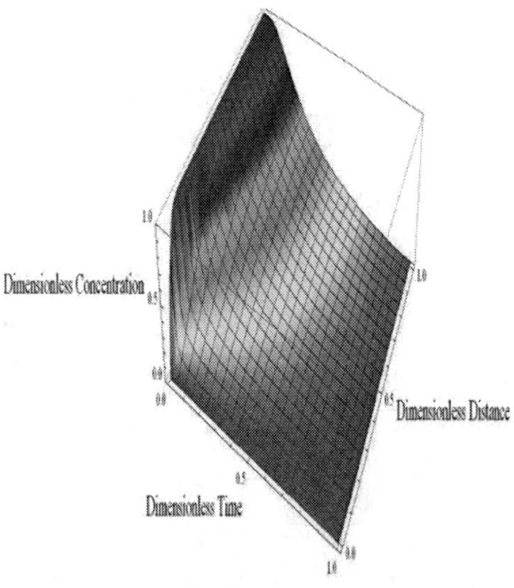

Figure 1. Solution of the Model equation (7) by Mathematica

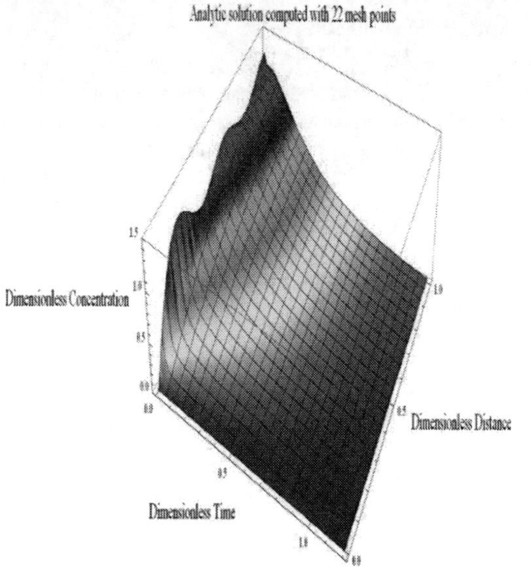

Figure 2. Solution of the Model equation (7) by Grahs (1974)

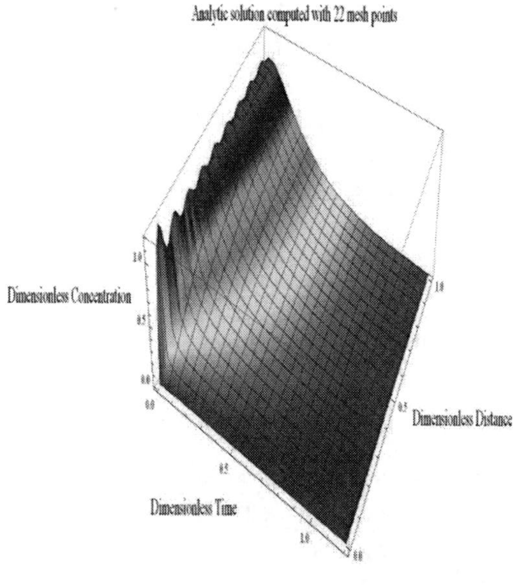

Figure 3. Solution of the Model equation (7) by Kukreja (1996)

DOI: 10.1051/matecconf/20165705008

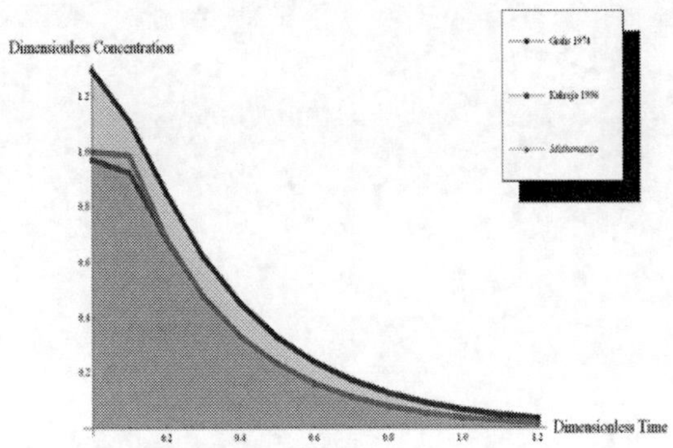

Figure 4. Comparison of Analytical solution of Brenner with Present investigation by Mathematica at Dimensionless distance X=1 at same $P_e = 1.1$

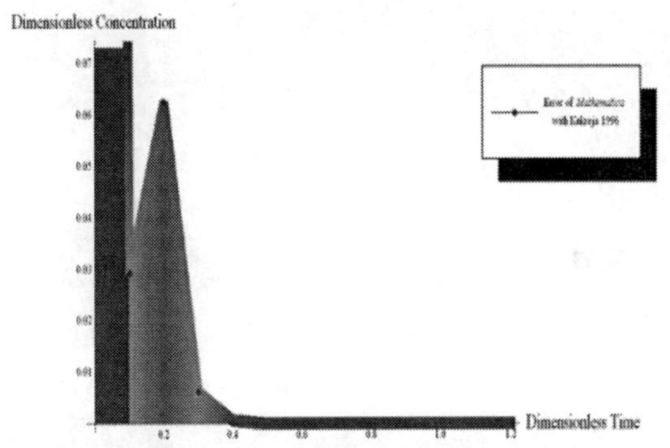

Figure 5. Error of Mathematica with Kukreja (1996)

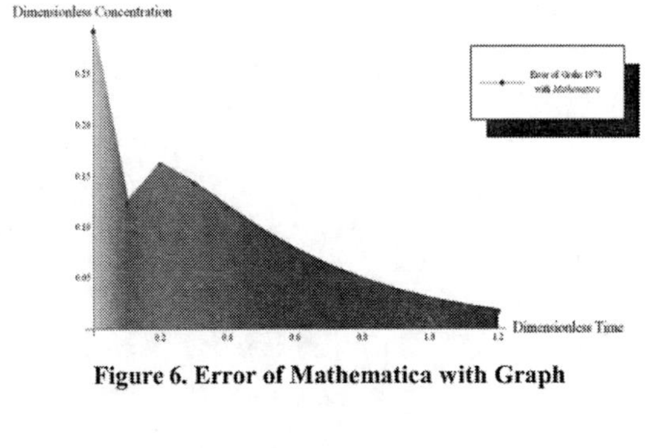

Figure 6. Error of Mathematica with Graph

Table1 Comparison of Analytical solution of Grahs (1974) and Kukreja (1996) with Present investigation by Mathematica at same $Pe = 1.1$

Time T	Grahs(1974) Analytical Solution $Pe = 1.1$	Kukreja (1996) Analytical Solution $Pe = 1.1$	Analytical Solution By Mathematica $Pe = 1.1$
0.00	1.2913	0.9708	1.0000
0.10	0.1087	0.9239	0.9864
0.20	0.8477	0.6805	0.6866
0.30	0.6203	0.4773	0.4779
0.40	0.4512	0.3326	0.3326
0.50	0.3279	0.2315	0.2315
0.60	0.2382	0.1611	0.1611
0.70	0.1731	0.1122	0.1122
0.80	0.1258	0.0781	0.0781
0.90	0.0914	0.0543	0.0543
1.00	0.0664	0.0378	0.0378
1.10	0.0482	0.0263	0.0263
1.20	0.0350	0.0183	0.0183

5 Conclusion

As the model so developed is mathematically complex to solve, so this is simplified for further investigation in order to compare some of the models proposed by Grahs [2] and Brenner [1] and Kumar [4] but with different numerical technique. The model with varying initial and boundary conditions is proposed to solve through Mathematica. The present technique is validated with the solutions of Grahs [2] based on numerical technique (Orthogonal Collocation) and Brenner [1] with analytical methods using Kumar [4]. Hence it may be concluded that the Mathematica used in the present investigation is simple, elegant and convenient for solving two point boundary value problems and average numerical errors. The algorithms in this solver are easy to set up, and so the method represents an advantage and good alternate to the available techniques for such type of problems.

6 Nomenclature

A_C: Area of cross section of washing zone, m^2

c: Concentration of the solute in the liquor, kg/m^3
C : Dimensionless solute concentration

D_L: Longitudinal dispersion coefficient, m^2/s
Pe : Peclet number (= uL/D_L) G,H : Constants
T : Dimensionless time
X: Dimensionless distance
u : Liquor speed in cake pores, m/s

References

1. Brenner H., "The diffusion model of longitudinal mixing in beds of finite length, Numerical values", Chemical Engineering Science, 17 , pp.229-243, (1962)
2. Grah L.E., "Displacement washing of packed beds of cellulose fibers, part-1: Mathematical model,
3. Kukreja V. K., "Modeling of washing of brown stock on rotary vacuum washer ", PhD. Thesis, University of Roorkee, Roorkee, India, (1996)
4. Kumar Mukesh, "Mathematical modeling of pulp washing systems and solutions", PhD. Thesis, Indian institute of Technology Roorkee, Roorkee, India, (2002)
5. Kumar Deepak, Kumar Vivek, Singh V.P, "Numerical Solution of Brown Stock Washer Problems in Paper Industry", 11th Punjab Science Congress, Thapar University Patiala, India, Feb. 7-9, (2008)
6. Lee P.F., "Optimizing the displacement washing of pads of wood pulp fibres", Tappi J., 62 (9), pp.75-78, (1979)
7. Lapidus L. and Amundson N.R., "Mathematics of Adsorption in beds, part-vi: The effect of longitudinal diffusion in ion exchange and chromatographic columns", J. of Physical Chemistry, 56, pp.984-988, (1952)
8. Pellett G. L., " Longitudinal dispersion, intra particle diffusion and liquid-phase mass transfer during flow through multi particle systems", Tappi J., 49(2), pp.75-82, (1966)
9. Pekkanen M. and Norden H. V., "Review of pulp washing models", Paperija Puu, 67(11), pp.689-696, (1985)

AUTHOR INDEX

AUTHOR INDEX